$$\int_0^\infty \frac{\sin ax}{x}\, dx = \begin{cases} \frac{1}{2}\pi; & a > 0 \\ 0; & a = 0 \\ -\frac{1}{2}\pi; & a < 0 \end{cases}$$

$$\int_0^\pi \sin^2 ax\, dx = \int_0^\pi \cos^2 ax\, dx = \frac{\pi}{2}$$

$$\int_0^\pi \sin mx \sin nx\, dx = \int_0^\pi \cos mx \cos nx\, dx = 0; \quad m \neq n,\ m \text{ and } n \text{ integers}$$

$$\int_0^\pi \sin mx \cos nx\, dx = \begin{cases} 0; & m + n \text{ even} \\ \dfrac{2m}{m^2 - n^2}; & m + n \text{ odd} \end{cases}$$

A short table of trigonometric identities

$$\sin(\alpha \pm \beta) = \sin \alpha \cos \beta \pm \cos \alpha \sin \beta$$

$$\cos(\alpha \pm \beta) = \cos \alpha \cos \beta \mp \sin \alpha \sin \beta$$

$$\cos(\alpha \pm 90°) = \mp \sin \alpha$$

$$\sin(\alpha \pm 90°) = \pm \cos \alpha$$

$$\cos \alpha \cos \beta = \tfrac{1}{2}\cos(\alpha + \beta) + \tfrac{1}{2}\cos(\alpha - \beta)$$

$$\sin \alpha \sin \beta = \tfrac{1}{2}\cos(\alpha - \beta) - \tfrac{1}{2}\cos(\alpha + \beta)$$

$$\sin \alpha \cos \beta = \tfrac{1}{2}\sin(\alpha + \beta) + \tfrac{1}{2}\sin(\alpha - \beta)$$

$$\sin 2\alpha = 2\sin \alpha \cos \alpha$$

$$\cos 2\alpha = 2\cos^2\alpha - 1 = 1 - 2\sin^2\alpha = \cos^2\alpha - \sin^2\alpha$$

$$\sin^2\alpha = \tfrac{1}{2}(1 - \cos 2\alpha)$$

$$\cos^2\alpha = \tfrac{1}{2}(1 + \cos 2\alpha)$$

$$\sin \alpha = \frac{e^{j\alpha} - e^{-j\alpha}}{j2}$$

$$\cos \alpha = \frac{e^{j\alpha} + e^{-j\alpha}}{2}$$

$$e^{\pm j\alpha} = \cos \alpha \pm j \sin \alpha$$

$$A \cos \alpha + B \sin \alpha = \sqrt{A^2 + B^2} \cos\left(\alpha + \tan^{-1}\frac{-B}{A}\right)$$

ENGINEERING CIRCUIT ANALYSIS

Circuits and Systems

Also available from McGraw-Hill

ENGINEERING CIRCUIT ANALYSIS

FIFTH EDITION

William H. Hayt, Jr.

Professor of Electrical Engineering, Emeritus
Purdue University

Jack E. Kemmerly

Professor of Engineering, Emeritus
California State University, Fullerton

McGRAW-HILL, INC.

New York   St. Louis   San Francisco   Auckland   Bogotá   Caracas
Lisbon   London   Madrid   Mexico   Milan   Montreal   New Delhi
Paris   San Juan   Singapore   Sydney   Tokyo   Toronto

ENGINEERING CIRCUIT ANALYSIS

1 2 3 4 5 6 7 8 9 0 DOW DOW 9 0 9 8 7 6 5 4 3

ISBN 0-07-027410-X

This book was set in Times Roman by Bi-Comp, Inc.
The editors were Anne T. Brown and Jack Maisel;
the designer was Joan Greenfield;
the production supervisor was Kathryn Porzio.
New drawings were done by Fine Line Illustrations, Inc.
R. R. Donnelley & Sons Company was printer and binder.

Library of Congress Cataloging-in-Publication Data

Hayt, William Hart, (date).
    Engineering circuit analysis/William H. Hayt, Jr., Jack E. Kemmerly.—5th ed.
    p.   cm.—(McGraw-Hill series in electrical and computer engineering. Circuits and systems)
    Includes index.
    ISBN 0-07-027410-X
    1. Electric circuit analysis.   2. Electric network analysis.
    I. Kemmerly, Jack E., (date).   II. Title.   III. Series.
    TK454.H4   1993
    621.319'2—dc20   92-37799

INTERNATIONAL EDITION

About the Authors

WILLIAM H. HAYT, JR. is Professor Emeritus of Electrical Engineering at Purdue University. He received his B.S. and M.S. at Purdue University and his Ph.D. from the University of Illinois. After spending four years in industry, Professor Hayt joined the faculty of Purdue University, where he served as Professor and Head of the School of Electrical Engineering. Besides *Engineering Circuit Analysis,* Professor Hayt has written three other texts, including *Engineering Electromagnetics,* Fifth Edition, 1989, McGraw-Hill. Professor Hayt is a member of many professional societies, including Eta Kappa Nu, Tau Beta Pi, Sigma Xi, Sigma Delta Chi, Fellow of IEEE, ASEE, and NAEB and is listed in *Who's Who in America*. While at Purdue, he received numerous teaching awards, including the university's Best Teacher Award.

JACK E. KEMMERLY is Professor Emeritus of Electrical Engineering at California State University, Fullerton. He received his B.S. magna cum laude from The Catholic University of America, M.S. from University of Denver, and Ph.D. from Purdue University. Professor Kemmerly first taught at Purdue University and later worked as principal engineer at the Aeronutronic Division of Ford Motor Company. He then joined California State University, Fullerton, where he was Professor, Chairman of the Faculty of Electrical Engineering, and Chairman of the Engineering Division. Professor Kemmerly is a member of Eta Kappa Nu, Tau Beta Pi, Sigma Xi, and ASEE and is a Senior Member of IEEE. He is also listed in *Who's Who in the West*. His pursuits outside of academe have included being an officer in the little league and a scoutmaster in the boy scouts.

Contents

Part One:
The Resistive
Circuit

16 State-Variable Analysis

**Part Six:
Signal Analysis**

17 Fourier Analysis

18 Fourier Transforms

19 Laplace Transform Techniques

Part Seven: Appendixes

Appendix 1/Determinants

Appendix 2/Matrices

Appendix 3/A Proof of Thévenin's Theorem

Appendix 4/Complex Numbers

Appendix 5/A SPICE Tutorial

Appendix 6/Answers to Odd-Numbered Problems

Preface

Reading this book is intended to be an enjoyable experience, even though the text is indeed scientifically rigorous and somewhat mathematical. Typical readers are in their early twenties, full of enthusiasm, and just beginning to study engineering—particularly *electrical* engineering.

We, the authors, are trying to sell the premise that circuit analysis can be fun. Not only is it useful and downright essential to the study of engineering, it is a marvelous education in logical thinking, good even for those who may never analyze another circuit in their professional lifetime. Students are truly impressed by all the excellent analytical tools that can be derived from only three simple scientific laws—Ohm's law and Kirchhoff's voltage and current laws.

In many colleges and universities, the introductory course in electrical engineering will be preceded or accompanied by an introductory physics course in which the basic concepts of electricity and magnetism are introduced, most often from the field aspect. Such a background is not a prerequisite, however. Instead, several of the requisite basic concepts of electricity and magnetism are discussed (or reviewed) in the first chapter. Only an introductory calculus course need be considered as a prerequisite—or possibly a co-requisite—to the reading of the book. Circuit elements are introduced and defined here in terms of their circuit equations; only incidental comments are offered about the pertinent field relationships. In the past, we have tried introducing the basic circuit analysis course with three or four weeks of electromagnetic field theory, so as to be able to define circuit elements more precisely in terms of Maxwell's equations. The results, especially in terms of students' acceptance, were not good.

We intend that this text be one from which students may teach the science of circuit analysis to themselves. It is written to the student, and not to the instructor, because the student is probably going to spend more time than the instructor in reading it. If at all possible, each new term is clearly defined when it is first introduced. The basic material appears toward the beginning of each chapter and is explained carefully and in detail; numerical examples are usually used to introduce and suggest general results. Drill problems appear at the ends of most sections; they are generally simple, and answers to the several parts are given in order. The more difficult problems appear at the ends of the chapters and follow the general order of presentation of the text material. These problems are occasionally used to introduce less important or more advanced topics through a guided step-by-step procedure, as well as to introduce topics which will appear in the following chapter. The introduction and resulting repetition are both important to the learning process. In all, there are 860 problems: 231 drill problems, each consisting of several parts, and 629 additional problems at the ends of the chapters. Most of these problems are new in this edition and, unfortunately for us authors, each problem had to be solved by both of us, independently. Then we argued over disagreements on answers until they were resolved.

The general order of the material has been selected so that the student may learn as many of the techniques of circuit analysis as possible in the simplest context, namely, the resistive circuit which is the subject of the first part of this text. Fundamental laws, a few theorems, and some elementary network topology enable most of the basic analytical techniques to be developed. Numerous examples and problems are possible since the solutions are not mathematically complicated. By means of many example problems, particular emphasis is placed on the use of Thévenin's theorem, a topic that is perennially troublesome to students. The extension of these techniques to more advanced circuits in subsequent parts of the text affords the opportunity for both review and generalization. Part 1 of the text may be covered in four to six weeks, depending on the students' background and ability, and on the course intensity.

Part 2 is devoted to the natural response and the complete response to dc excitation of the simpler RL, RC, and RLC circuits. Facility in differential and integral calculus is necessary, of course, but a background in differential equations is not really required. The unit-step function is introduced as an important singularity function in this part, but the introduction of the unit-impulse function is withheld until transform techniques are introduced in Chapter 18.

Part 3 of the text introduces the frequency domain and initiates operations with complex numbers by concentrating on sinusoidal analysis in the steady state. This part also includes a discussion of average power, rms values, and polyphase circuits, all of which are associated with the sinusoidal steady state.

In Part 4 the complex-frequency concept is introduced, and its use in relating the forced response and the natural response is emphasized. The determination of the complete response of sinusoidally excited circuits begins to tie together the material of the first three parts.

Part 5 begins with a consideration of magnetic coupling, which is basically a two-part phenomenon, and logically leads into a consideration of two-part network analysis and the linear modeling of various electronic devices, especially transistors.

Part 6 introduces more powerful techniques of network analysis. Some instructors are inclined to introduce these techniques, Laplace transforms in particular, much earlier in a basic circuit analysis course. However, we are very reluctant to do this, feeling as we do that this tends to discourage students from a deeper familiarity with the simpler circuits they encounter. The first of these analysis techniques is state-variable analysis. Following this is the Fourier series description of periodic waveforms. The treatment of Fourier series is then extended to nonperiodic forcing and response functions by use of the Fourier transform in Chapter 18. Numerous example problems illustrating the technique of convolving time functions—certainly not easy, bed-time reading for many students—are included in this chapter. The final chapter covers the more important Laplace transform techniques and their use in obtaining the complete response of more complicated circuits.

The material in this book is more than adequate for a two-semester course, but some selection may be made from the last four or five chapters. No material is included in the text which will not be of some value in the following term; thus, signal-flow graphs, the relationship of circuit theory to field theory, and advanced topological concepts are among those subjects which are relegated to subsequent courses.

A number of changes have been made in this edition: Chapters 1 and 2 of the previous edition have been combined into a new Chapter 1. We concluded that some of that old Chapter 1 was a bit too longwinded; so, with a sigh, we pared it down. The subject of state-variable analysis has been added, at the request of a considerable number of instructors and reviewers. Example problems—126 of them, mostly new—have been included in a different format. That is, they have been set off somewhat from the body of the text so that they can be identified or located quickly. Numerous other example problems are also included, but are woven into the text's narrative for illustrative purposes.

Computer-aided analysis is now included in this edition. Some instructors virtually *demand* that this topic be included; others, in almost equal numbers are adamantly opposed to its inclusion in an introductory text. So, we hope we have struck a proper compromise. Problems that are specifically intended for analysis using the popular SPICE (or PSpice®) programs are included at the *ends* of many of the chapter-end sets of problems; and they are identified by the designation (SPICE). In addition, Appendix 5 presents a simple tutorial on the use of SPICE, adequate for solving most of the analysis problems which are encountered here. For those who want a more detailed treatment, a new book is recommended as a companion text: Roger C. Conant's *Engineering Circuit Analysis with PSpice and PROBE* (available in PC and Macintosh versions), McGraw-Hill, Inc., 1993. The consequence of this method of handling computer-aided analysis is that those instructors who do not wish to use it at this level of instruction can easily omit it; whereas those who wish to use it can do so with alacrity.

A somewhat expanded treatment of the operational amplifier is introduced as one of today's most important devices, and it is used to provide examples of circuits containing dependent sources, such as the voltage follower, the integrator and differentiator, the multiplier, the inverting amplifier, and circuits which can simulate lossless *LC* circuits and voltage-gain transfer functions.

Added help is offered for both students and instructors: Appendix 6, which now includes answers to every odd-numbered problem; and a *Solutions Manual,* which includes detailed solutions of all drill problems and chapter-end problems, is available for the instructor. Moreover, a *Students' Solutions Manual* is available in which about 800 problems *similar* to the ones in the text, are stated and solved. So much for material that is new in this edition.

Throughout the book there is a logical trail leading from definition, through explanation, description, illustration, and numerical example, to problem-solving ability; and this new-found ability often tends to make students excited, and gets them to asking themselves, "Why does this happen? How is it related to last week's work? Where do we logically go next?" There is a tremendous amount of enthusiastic momentum in most beginning engineering students, and this may be preserved by providing frequent drill problems whose successful solution confirms progress in their minds by integrating the various sections into a coherent whole, by pointing out future applications and more advanced techniques, and by maintaining in them interested, inquisitive attitudes.

If the book occasionally appears to be informal, or even lighthearted, it is because we feel that it is not necessary to be dry or pompous to be educational. Amused smiles on the faces of our students are seldom obstacles to their absorbing information. If the writing of the text had its entertaining moments, then why not the reading too?

Much of the material in the text is based on courses taught at Purdue University; the California State University, Fullerton; and at Fort Lewis College in Durango.

We would like to express our thanks for the many useful comments and suggestions provided by colleagues who reviewed this text during the course of its development, especially to Roger H. Baumann, University of Massachusetts, Lowell; Richard B. Brown, Michigan Technological University; Roger C. Conant, University of Illinois at Chicago; James F. Delansky, The Pennsylvania State University; John A. Fleming, Texas A&M University; Yusuf Leblebici, University of Illinois at Urbana-Champaign; William Oliver, Boston University; Sheila Prasad, Northeastern University; and Rolf Schaumann, Portland State University.

We, the authors, have already thanked each other for our invaluable contributions. Modesty precludes further elaboration.

William H. Hayt, Jr.
Jack E. Kemmerly

ENGINEERING CIRCUIT ANALYSIS

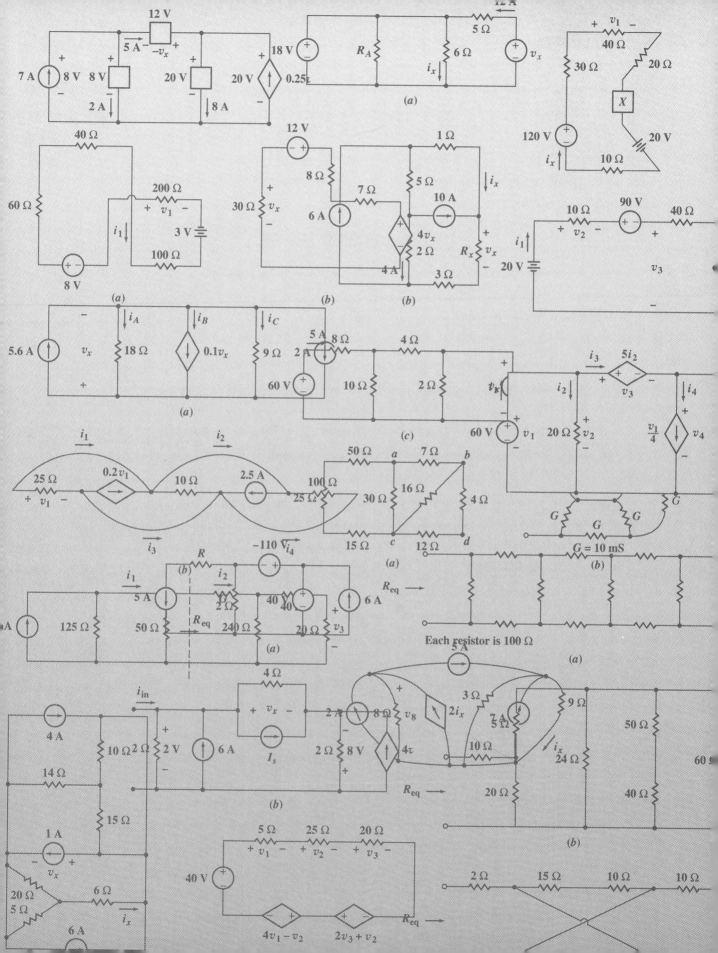

Part One:
The Resistive Circuit

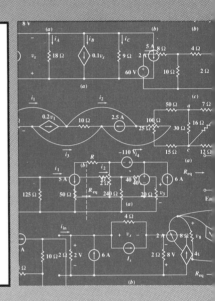

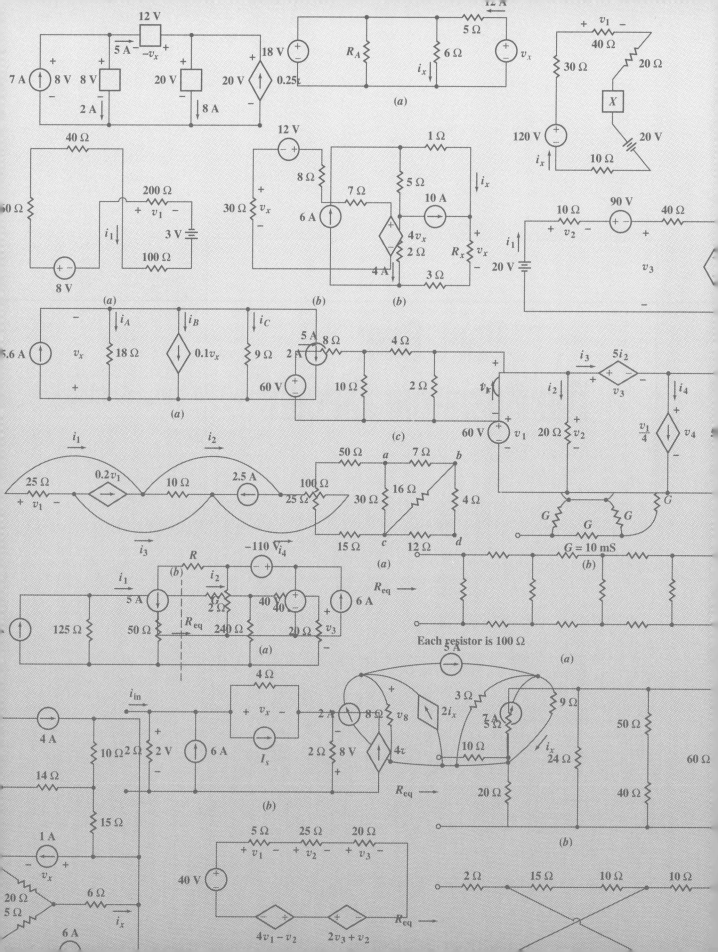

Units, Definitions, Experimental Laws, and Simple Circuits

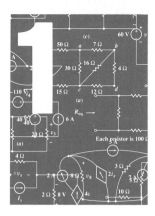

The purpose of this text, as its title implies, is to provide material that should lead to some proficiency in the subject of engineering circuit analysis. This topic is extremely useful for virtually every type of engineer as well as for many physicists and applied mathematicians; it is also stimulating, challenging, and quite enjoyable. The uninitiated may well ask immediately, "What *is* engineering circuit analysis?" The question is a fair one, and we may answer it by taking a quick look at a few dictionaries and professional journals, which provide the following revelations, including a rather awkward "official" definition of engineering:

engineering   a profession in which a knowledge of mathematical, natural, and social sciences gained by study, experience, and practice is applied with judgment to develop ways to utilize economically the materials and forces of nature for the benefit of society.

circuit   an interconnection of simple electrical devices in which there is at least one closed path in which current may flow.

analysis   a (mathematical) study of a complex entity and the interrelationship of its parts.

Thus we might be inclined to decide that "engineering circuit analysis" is a mathematical study of some useful interconnection of simple electrical devices in which there is at least one closed current path. This definition is essentially correct, although we cannot understand it fully until we clarify what is meant by "current" and "electrical devices," a task which we shall undertake shortly.

Not long ago,[1] a textbook of this kind would have been viewed as a strictly *electrical* engineering text. Now, however, it has become increasingly common for students of civil engineering, mechanical engineering, and the other engineering disciplines, as well as an occasional student of applied mathematics, computer sciences, biology, or physics, to study introductory circuit analysis with the electrical engineer; moreover, courses based largely on this subject are taken by engineering students even before they have become identified with a particular branch of engineering.

---

[1] At least it *seems* that way to the two jovial old authors.

If we have already entered or intend to enter an electrical engineering program, then circuit analysis may represent the introductory course in our chosen field. If we are associated with some other branch of engineering, then circuit analysis may represent a large fraction of our total study of electrical engineering; but it also enables us to continue our electrical work in electronics, instrumentation, and other areas. Most important, however, is the possibility given to us to broaden our educational base and become more informed members of a team which may be primarily concerned with the development of some electrical device or system. Effective communication within such a team can be achieved only if the language and definitions used are familiar to all.

Today's engineering graduates are not all employed solely to work on the technical design aspects of engineering problems. Their efforts now extend beyond the creation of better computers and radar systems to vigorous efforts to solve socioeconomic problems such as air and water pollution, urban planning, mass transportation, the discovery of new energy sources, and the conservation of existing natural resources, particularly oil and natural gas.

In order to contribute to the solution of these engineering problems, an engineer must acquire many skills, one of which is a knowledge of electric circuit analysis.

We shall begin this study by considering systems of units and several basic definitions and conventions. For those who have an elementary knowledge of basic electricity and magnetism, this early material should be rapid reading. After these introductory topics have been mastered, we can then turn our attention to a simple electric circuit.

1-2
Systems of units

We must first establish a common language. Engineers cannot communicate with one another in a meaningful way unless each term they use is clear and definite. It is also true that little learning can be achieved from a textbook which does not define carefully each new quantity as it is introduced. If we speak in the vague generalities of a television commercial—"gets clothes up to 40 percent whiter"—and do not bother to define whiteness or to provide units by which it may be measured, then we shall certainly be deterred from success in engineering, although we might sell a lot of detergent.

In order to state the value of some measurable quantity, we must give both a *number* and a *unit,* such as "3 inches." Fortunately we all use the same number system and know it well. This is not true for the units, and some time must be spent in becoming familiar with a suitable system of units. We must agree on a standard unit and be assured of its permanence and its general acceptability. The standard unit of length, for example, should not be defined in terms of the distance between two marks on a certain rubber band; this is not permanent, and furthermore everybody else is using another standard.

We shall also need to define each technical term at the time it is introduced, stating the definition in terms of previously defined units and quantities. Here the definition cannot always be as general as the more theoretically minded might wish. For instance, it will soon be necessary to define *voltage*. We must either accept a very complete and general definition, which we can neither appreciate nor understand now, or else adopt a less general but simpler definition which will satisfy our purposes for the present. By the time a more general definition is needed, our familiarity with the simpler concepts will help our understanding.

It will also become evident that many quantities are so closely related to each other that the first one defined needs a few subsequent definitions before it can be thoroughly understood. As an example, when the *circuit element* is defined, it is most convenient to define it in terms of a *current* and *voltage*; and when current and voltage are defined, it is helpful to do so with reference to a circuit element. None of these three definitions can be well understood until all have been stated. Therefore, our first definition of the circuit element may be somewhat inadequate, but then we shall define current and voltage in terms of a circuit element and, finally, go back and define a circuit element more carefully.

We have very little choice open to us with regard to a system of units. The one we shall use was adopted by the National Bureau of Standards in 1964; it is used by all the major professional engineering societies and is the language in which today's textbooks are written. This is the *International System of Units* (abbreviated SI in all languages), adopted by the General Conference on Weights and Measures in 1960. The SI is built upon six basic units: the meter, kilogram, second, ampere, kelvin, and candela. This, of course, is a "metric system," some form of which is now in common use in most of the technologically advanced countries of the world; its use is being strongly urged in the United States.

We shall look at the definitions of the meter, kilogram, second, and ampere in the following. Standard abbreviations for them and the other SI units that we shall be using are listed on the back inside cover and will be used throughout our discussions.

In the late 1700s the meter was defined to be exactly one ten-millionth of the distance from the earth's pole to its equator. The distance was marked off by two fine lines on a platinum-iridium bar which had been cooled to zero degrees Celsius (°C) (formerly centigrade). Although more accurate surveys have shown since that the marks on the bar do not represent this fraction of the earth's meridian exactly, the distance between the marks was nonetheless accepted internationally as the definition of the standard meter until 1960. In that year the General Conference based a more accurate definition of the meter (m) on a multiple of the wavelength of radiation of the orange line of krypton 86. Then in 1983 the meter was defined even more accurately as the distance that light travels through space in 1/299 792 458 second (to be defined in a second or two).

The basic unit of mass, the kilogram (kg), was defined in 1901 as the mass of a platinum block kept with the standard meter bar at the International Bureau of Weights and Measures in Sèvres, France. This definition was reaffirmed in 1960. The mass of this block is approximately 0.001 times the mass of 1 m^3 of pure water at 4°C.

The third basic unit, the second (s), was defined prior to 1956 as 1/86 400 of a mean solar day. At that time it was defined as 1/31 556 925.9747 of the tropical year 1900. Eight years later the second was defined more carefully as 9 192 631 770 periods of the transition frequency between the hyperfine levels $F = 4$, $m_F = 0$ and $F = 3$, $m_F = 0$ of the ground state $^2S_{1/2}$ of the atom of cesium 133, unperturbed by external fields. This latter definition is permanent and more reproducible than the former; it is also comprehensible only to atomic physicists. However, any of these definitions adequately describes the second with which we are all familiar.

The definition of the fourth basic unit, the ampere (A), will appear later in this chapter after we are more familiar with the basic properties of electricity. The remaining two basic units, the kelvin (K) and the candela (cd), are not of immediate concern to circuit analysts.[2]

The Sl incorporates the decimal system to relate larger and smaller units to the basic unit and uses standard prefixes to signify the various powers of 10. These are

atto- (a-, 10^{-18})	deci- (d-, 10^{-1})
femto- (f-, 10^{-15})	deka- (da-, 10^{1})
pico- (p-, 10^{-12})	hecto- (h-, 10^{2})
nano- (n-, 10^{-9})	kilo- (k-, 10^{3})
micro- (μ-, 10^{-6})	mega- (M-, 10^{6})
milli- (m-, 10^{-3})	giga- (G-, 10^{9})
centi- (c-, 10^{-2})	tera- (T-, 10^{12})

Those shown inside the block in the table are the ones used most frequently by students of electric circuit theory.

These prefixes are worth memorizing, for they will appear often, both in this text and in other scientific work. Thus, a millisecond (ms) is 0.001 second, and a kilometer (km) is 1000 m. It is apparent now that the gram (g) was originally established as the basic unit of mass and the kilogram then represented merely 1000 g. Now the kilogram is our basic unit, and we could describe the gram as a millikilogram if we wished to be confusing. Combinations of several prefixes, such as the millimicrosecond, are unacceptable; the term *nanosecond* should be used. Also officially frowned on is the use of *micron* for 10^{-6} m; the correct term is the *micrometer* (μm). The angstrom (Å), however, may be used for 10^{-10} m.

This power-of-10 relationship is not present in the so-called *British System of Units,* which unfortunately is in common use in this country. The fundamental British units are defined in terms of the SI units as follows: 1 inch (in) is exactly 0.0254 m, 1 pound-mass (lbm) is exactly 0.453 592 37 kg, and the second is common to both systems.

As a last item in our discussion of units, we consider the three derived units used to measure force, work or energy, and power. The newton (N) is the fundamental unit of force,[3] and it is the force required to accelerate a 1-kg mass by one meter per second per second (1 m/s²). A force of 1 N is equivalent to 0.224 81 pound of force (lbf), and the average nineteen-year-old male, having a mass of 68 kg, exerts a force of 667 N on the scales.

The fundamental unit of work or energy is the joule (J), defined as one newton-meter (N-m). The application of a constant 1-N force through a 1-m distance requires an energy expenditure of 1 J. The same amount of energy is required to lift this book, weighing about 10 N, a distance of approximately 10 cm. The joule is equivalent to 0.737 56 foot pound-force (ft-lbf). Other energy units include the calorie[4] (cal), equal to 4.1868 J; the British thermal unit (Btu), which is 1055.1 J; and the kilowatt-hour (kWh), equal to 3.6×10^6 J.

---

[2] An interesting discussion of the definitions of the basic units and measurement techniques is found in the article "Volts and amps are not what they used to be," by Paul Wallich, *IEEE Spectrum,* March 1987.

[3] It is worth noting that all units named after famous scientists have *abbreviations* beginning with capital letters.

[4] The calorie used with food, drink, and exercise is really a kilocalorie, 4186.8 J.

The last derived quantity with which we shall concern ourselves is power, the *rate* at which work is done or energy is expended. The fundamental unit of power is the watt (W), defined as 1 J/s. One watt is equivalent to 0.737 56 ft-lbf/s. It is also equivalent to 1/745.7 horsepower (hp), a unit which is now being phased out of engineering terminology.

> NOTE: Throughout the text, drill problems appear following sections in which a new principle is introduced, in order to allow students to test their understanding of the basic fact itself. The problems are useful in gaining familiarization with new terms and ideas and should all be worked. More general problems appear at the ends of the chapters. The answers to the drill problems are given in order. Answers are given to four significant figures if the first significant digit is 1, and to three significant figures if the first significant digit lies between 2 and 9, inclusive.

1-1. (*a*) What is the total surface area of a rectangular parallelepiped with dimensions of 0.6 mm $\times$ 200 μm $\times$ 10^5 nm? (*b*) A fast-food burger contains 562 kcal. How many Btu's are obtained if all the energy is converted to heat? (*c*) What acceleration in ft/s^2 is given to a 0.5-kg mass by a force of 1.5 N?

Ans: 0.4 mm^2; 2230 Btu; 9.84 ft/s^2

Drill Problem

Charge

1-3

Charge, current, voltage, and power

Our next task is to introduce and provide some preliminary definitions of the basic electrical quantities. We begin with *electric charge,* a concept often introduced by visualizing the following simple experiment.

Suppose that we take a small piece of some light material such as pith and suspend it by a fine thread. If we now rub a hard rubber comb with a woolen cloth and then touch the pith ball with the comb, we find that the pith ball tends to swing away from it; a force of repulsion exists between the comb and the pith ball. After laying down the comb and then approaching the pith ball with the woolen cloth, we can see that there is a force of attraction present between the pith ball and the woolen cloth.

We explain both of these forces on the pith ball by saying that they are *electrical* forces caused by the presence of *electric charges* on the pith ball, the comb, and the woolen cloth. Our experiment shows clearly that the electrical force may be one of either attraction or repulsion.

We explain the existence of electrical forces of both attraction and repulsion by the hypothesis that there are two kinds of charge, and that like charges repel and unlike charges attract. The two kinds of charge are called positive and negative, although we might have called them gold and black, or vitreous and resinous (as they were termed many years ago). Arbitrarily, the type of charge originally present on the comb was called negative by Benjamin Franklin, and that on the woolen cloth, positive.

We may now describe our experiment in these new terms. By rubbing the comb with the cloth, a negative charge is produced on the comb and a positive charge on the cloth. Touching the pith ball with the comb transfers some of its negative charge to the pith ball, and the force of repulsion between the like kinds of charge on the pith ball and comb causes the ball to move away. As we bring the positively charged woolen cloth near the negatively charged pith ball, a force of attraction between the two different kinds of charge is evident.

We also know now that all matter is made up of fundamental building blocks called atoms and that the atoms, in turn, are composed of different kinds of fundamental particles. The three most important particles are the electron, the proton, and the neutron. The electron possesses a negative charge, the proton possesses an equal-magnitude positive charge, and the neutron is neutral, or has no charge at all. As we rub the rubber comb with the woolen cloth, the comb acquires its negative charge because some of the electrons on the wool are rubbed off onto the comb; the cloth then has an insufficient number of electrons to maintain its electrical neutrality and thus behaves as a positive charge.

The mass of each of the three particles just named has been determined experimentally: $9.109\ 56 \times 10^{-31}$ kg for the electron and about 1840 times as large for the proton and the neutron.

Now we are ready to define the fundamental unit of charge, called the *coulomb* after Charles Coulomb, the first person to make careful quantitative measurements of the force between two charges. The coulomb can, of course, be defined in any way we wish as long as the definition is convenient, universally accepted, and permanent and does not contradict any previous definition. Again, this leaves us no freedom at all, because the definition which is already universally accepted is as follows: two small, identically charged particles which are separated by one meter in a vacuum and repel each other with a force of $10^{-7}c^2$ N possess an identical charge of either plus or minus one coulomb (C). The symbol c represents the velocity of light, $2.997\ 925 \times 10^8$ m/s. In terms of this unit, the charge of an electron is minus $1.602\ 18 \times 10^{-19}$ C, and 1 C (negative) therefore represents the combined charge of about 6.24×10^{18} electrons.

We shall symbolize charge by Q or q, the capital letter being reserved for a charge which does not change with time, or is a constant, and the lowercase letter representing the general case of a charge that may vary with time. We often call this latter case the *instantaneous* value of the charge and may emphasize its time dependence by writing it as $q(t)$. Note that $q(t)$ might represent a constant as a special case. This same use of capital and lowercase letters will be carried over to all other electrical quantities as well.

In their handwriting, many students[5] do not distinguish between uppercase and lowercase letters. This can have serious consequences, very few of them beneficial. For example, in electronics these four collector currents all mean different things: i_c, i_C, I_c, and I_C. Confusion is obviously ready to strike.

Current

The experiment described in the preceding paragraphs belongs to the field of electrostatics, which is concerned with the behavior of electric charges at rest. This is of interest to us only because it is a beginning and serves as a useful device to define charge.

One part of the experiment, however, departed from electrostatics, the process of transferring charge from the wool cloth to the comb or from the comb to the pith ball. This idea of "transfer of charge" or "charge in motion" is of vital importance to us in studying electric circuits, because, in moving a charge from

---

[5] Instructors are perfect.

place to place, we may also transfer energy from one point to another. The familiar cross-country power transmission line is a practical example of a device that transfers energy.

Of equal importance is the possibility of varying the *rate* at which the charge is transferred in order to communicate or transfer intelligence. This process is the basis of communication systems such as radio, television, and telemetry.

· Charge in motion represents a *current,* which we shall define more carefully in what follows. The current present in a discrete path, such as a metallic wire, has both a magnitude and a direction associated with it; it is a measure of the rate at which charge is moving past a given reference point in a specified direction.

Having specified the reference direction, we may now let $q(t)$ be the total charge which has passed the reference point since an arbitrary time $t = 0$, moving in a defined direction. A contribution to this total charge may be negative if negative charge is moving in the reference direction, or if positive charge is moving in the opposite direction.

A graph of the instantaneous value of the total charge might be similar to that shown in Fig. 1-1.

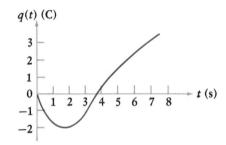

$q(t)$ (C)

Figure 1-1

A graph of the instantaneous value of the total charge $q(t)$ which has passed a given reference point since $t = 0$.

We are now ready to consider the *rate* at which charge is being transferred. In the time interval extending from t to $(t + \Delta t)$, the charge transferred past the reference point has increased from q to $(q + \Delta q)$. If the graph is decreasing in this interval, then Δq is a negative value. The rate at which charge is passing the reference point at time t is therefore very closely equal to $\Delta q/\Delta t$, and as the interval Δt decreases, the exact value of the rate is given by the derivative

$$\frac{dq}{dt} = \lim_{\Delta t \to 0} \frac{q(t + \Delta t) - q(t)}{\Delta t} = \lim_{\Delta t \to 0} \frac{\Delta q}{\Delta t}$$

We define the current at a specific point and flowing in a specified direction as the instantaneous rate at which net positive charge is moving past that point in the specified direction. Current is symbolized by I or i, and thus

$$i = \frac{dq}{dt} \tag{1}$$

The unit of current is the ampere (A), and 1 A corresponds to charge moving at the rate of 1 C/s.[6] The ampere was named after A. M. Ampère, a French

---

[6] The SI definition of one ampere is "that constant current which, if maintained in two straight parallel conductors of infinite length, of negligible circular cross section, and placed one meter apart in vacuum, would produce between these conductors a force equal to 2×10^{-7} newton per meter of length." Our definition is equivalent to this official one, and easier to understand from a circuit standpoint.

physicist of the early nineteenth century. It is often called an "amp," but this is informal and unofficial. The use of the lowercase letter i is again to be associated with an instantaneous value. Using the data of Fig. 1-1, the instantaneous current is given by the slope of the curve at every point. This current is plotted in Fig. 1-2.

Figure 1-2

The instantaneous current, $i = dq/dt$, where q is given in Fig. 1-1.

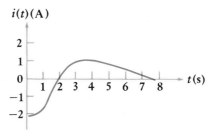

The charge transferred between time t_0 and t may be expressed as a definite integral,

$$q\Big|_{t_0}^{t} = \int_{t_0}^{t} i\, dt$$

The total charge transferred over all time is obtained by adding $q(t_0)$, the charge transferred up to the time t_0, to the preceding expression:

$$q = \int_{t_0}^{t} i\, dt + q(t_0) \tag{2}$$

Several different types of current are illustrated in Fig. 1-3. A current which is constant is termed a *direct current,* or simply dc, and is shown by Fig. 1-3a. We shall find many practical examples of currents which vary sinusoidally with time (Fig. 1-3b); currents of this form are present in normal household circuits. Such a current is often referred to as *alternating current,* or ac. Exponential currents and damped sinusoidal currents, sketched in Figs. 1-3c and d, will also be encountered later.

Figure 1-3

Several types of current: (a) Direct current, or dc. (b) Sinusoidal current, or ac. (c) Exponential current. (d) Damped sinusoidal current.

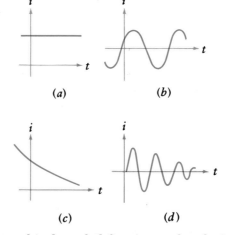

We establish a graphical symbol for current by placing an arrow next to the conductor. Thus, in Fig. 1-4a the direction of the arrow and the value "3 A" indicate *either* that a net positive charge of 3 C/s is moving to the right *or* that a net negative charge of -3 C/s is moving to the left each second. In Fig. 1-4b there are again two possibilities: either -3 A is flowing to the left or

Figure 1-4

Two methods of representation for the same current.

(a)   (b)

+3 A is flowing to the right. All four of these statements and both figures represent currents which are equivalent in their electrical effects, and we say that they are equal.

It is convenient to think of current as the motion of positive charge even though it is known that current flow in metallic conductors results from electron motion. In ionized gases, in electrolytic solutions, and in some semiconductor materials, positively charged elements in motion constitute part or all of the current. Thus, any definition of current can agree with the physical nature of conduction only part of the time. The definition and symbolism we have adopted are standard.

It is essential that we realize that the current arrow does not indicate the "actual" direction of current flow, but is simply part of a convention that allows us to talk about "the current in the wire" in an unambiguous manner. The arrow is a fundamental part of the *definition* of a current! Thus, to talk about the value of a current $i_1(t)$ without specifying the arrow is to discuss an unde-

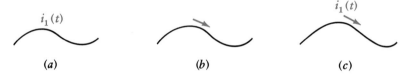

Figure 1-5

(a) (b) (c)

(*a*, *b*) Incomplete, improper, and incorrect definitions of a current.
(*c*) The correct definition of $i_1(t)$.

fined entity. Therefore, Figs. 1-5*a* and *b* are meaningless representations of $i_1(t)$, whereas Fig. 1-5*c* is the proper definitive symbology. Remember:

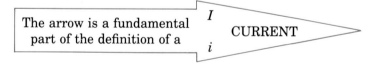

The arrow is a fundamental part of the definition of a I CURRENT i

Voltage

We must now begin to refer to a circuit element, and we shall define it in very general terms. Such electrical devices as fuses, light bulbs, resistors, batteries, capacitors, generators, and spark coils can be represented by combinations of simple circuit elements. We shall begin by showing a very general circuit element as a shapeless object possessing two terminals at which connections to other elements may be made (Fig. 1-6). This simple picture may serve as the

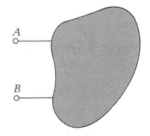

Figure 1-6

A general circuit element is characterized by a pair of terminals to which other general circuit elements may be connected.

definition of a general circuit element. There are two paths by which current may enter or leave the element. Later we shall define particular circuit elements by describing the electrical characteristics which may be observed at their pairs of terminals.

In Fig. 1-6, let us suppose that direct current is directed into terminal A, through the general element, and out of B. Let us also assume that the passage of charge through the element requires an expenditure of energy. We then say that an electrical *voltage* or a *potential difference* exists between the two terminals, or that there is a voltage or potential difference "across" the element. Thus, the voltage across a terminal pair is a measure of the work required to move charge through the element. Specifically, we shall define the voltage across the element as the work required to move a positive charge of 1 C from one terminal through the device to the other terminal.[7] The sign of the voltage will be discussed momentarily. The unit of voltage is the *volt* (V), and 1 V is the same as 1 J/C. Voltage is represented by V or v. We are indeed fortunate that the full name of the eighteenth-century Italian physicist, Alessandro Giuseppe Antonio Anastasio Volta, is not used for our unit of potential difference.

A voltage or potential difference can exist between a pair of electrical terminals whether a current is flowing or not. An automobile battery, for example, has a voltage of 12 V across its terminals even if nothing whatsoever is connected to the terminals.

According to the principle of conservation of energy, the energy which is expended in forcing the charges through the element must appear somewhere else. When we later meet specific circuit elements, we should note whether the energy is stored in some form which is readily available as electric energy or whether it changes irreversibly into heat, acoustic energy, or some other nonelectrical form.

We must now establish a convention by which we can distinguish between energy supplied to an element by some external source and energy which may be supplied by the element itself to some external device. We do this by our choice of a sign for the voltage of terminal A with respect to terminal B. If a positive current is entering terminal A of the element, and if an external source must expend energy to establish this current, then terminal A is positive with respect to terminal B. Alternatively, we may say also that terminal B is negative with respect to terminal A.

The sense of the voltage is indicated by a plus-minus pair of algebraic signs. In Fig. 1-7a, for example, the placement of the plus sign at terminal A indicates that terminal A is v volts positive with respect to terminal B. If we later find that v happens to have a numerical value of -5 V, then we may say either that

Figure 1-7

(a, b) Terminal B is 5 V positive with respect to terminal A; (c, d) terminal A is 5 V positive with respect to terminal B.

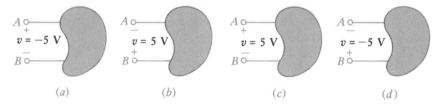

(a) (b) (c) (d)

---

[7] The SI definition of one volt is "the difference of electric potential between two points of conducting wire carrying a constant current of one ampere, when the power dissipated between these points is equal to one watt." Once again, our definition of the volt is equivalent to this official one, and easier to understand.

A is -5 V positive with respect to B or that B is 5 V positive with respect to terminal A. Other cases are shown and described in Figs. 1-7b, c, and d.

Just as we noted in our definition of current, it is essential to realize that the plus-minus pair of algebraic signs does not indicate the "actual" polarity of the voltage, but is simply part of a convention that enables us to talk unambiguously about "the voltage across the terminal pair." The definition of any voltage must include a plus-minus sign pair! Using a quantity $v_1(t)$ without specifying the location of the plus-minus sign pair is using an undefined term. Figures

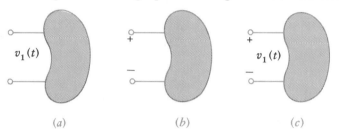

(a)   (b)   (c)

Figure 1-8

(a, b) These are inadequate definitions of a voltage. (c) A correct definition includes both a symbol for the variable and a plus-minus symbol pair.

1-8a and b do *not* serve as definitions of $v_1(t)$; Fig. 1-8c does. Remember:

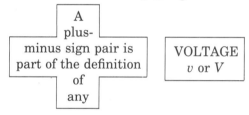

A plus-minus sign pair is part of the definition of any VOLTAGE v or V

Power

We now need to determine an expression for the power being absorbed by any circuit element in terms of the voltage across it and the current through it. Voltage has already been defined in terms of an energy expenditure, and power is the *rate* at which energy is expended. However, no statement can be made concerning energy transfer in any of the four cases shown in Fig. 1-7, for example, until the direction of the current is specified. Let us assume that a current arrow is placed alongside each upper lead, directed to the right, and labeled "+2 A"; then, since in both cases c and d terminal A is 5 V positive with respect to terminal B and since a positive current is entering terminal A, energy is being supplied to the element. In the remaining two cases, the element is delivering energy to some external device.

We have already defined power, and we shall represent it by P or p. If one joule of energy is expended in transferring one coulomb of charge through the device, then the rate of energy expenditure in transferring one coulomb of charge per second through the device is one watt. This absorbed power must be proportional both to the number of coulombs transferred per second, or current, and to the energy needed to transfer one coulomb through the element, or voltage. Thus,

$$p = vi \qquad (3)$$

Dimensionally, the right side of this equation is the product of joules per coulomb and coulombs per second, which produces the expected dimension of joules per second, or watts.

With a current arrow placed by each upper lead of Fig. 1-7, directed to the right and labeled "2 A," 10 W is absorbed by the element in c and d; but in a and b, −10 W is absorbed (or 10 W is generated).

The conventions for current, voltage, and power are summarized in Fig. 1-9. The sketch shows that if one terminal of the element is v volts positive

Figure 1-9

The power absorbed by the element is given by the product $p = vi$.

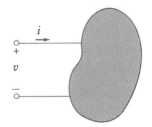

with respect to the other terminal, and if a current i is entering the element through the first terminal, then a power $p = vi$ is being *absorbed* by the element; it is also correct to say that a power $p = vi$ is being *delivered* to the element. When the current arrow is directed into the element at the plus-marked terminal, we satisfy the *passive sign convention*. This convention should be studied carefully, understood, and memorized. In other words, it says that if the current arrow and the voltage polarity signs are placed at the terminals of the element so that the current enters that end of the element marked with the positive sign, and if both the arrow and the sign pair are labeled with the appropriate algebraic quantities, then the power *absorbed by* the element can be expressed by the algebraic product of these two quantities. If the numerical value of the product is negative, then we say that the element is absorbing negative power, or that it is actually generating power and delivering it to some external element. For example, in Fig. 1-9 with $v = 5$ V and $i = −4$ A, then the element may be described as either absorbing −20 W or generating 20 W.

The three examples of Fig. 1-10 further illustrate these conventions.

Figure 1-10

(a) A power, $p = (2)(3) = 6$ W, is absorbed by the element. (b) A power, $p = (−2)(−3) = 6$ W, is absorbed by the element. (c) A power, $p = (4)(−5) = −20$ W, is absorbed by the element, or 20 W is delivered by the element.

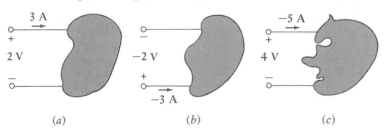

(a)   (b)   (c)

Drill Problems

1-2. The current $i_1(t)$ in Fig. 1-5c is given as $−2t$ A for $t \leq 0$, and $3t$ A for $t \geq 0$. Find (a) $i_1(−2.2)$; (b) $i_1(2.2)$; (c) the total charge that has passed along the conductor from left to right in the interval $−2 \leq t \leq 3$ s; (d) the average value of $i_1(t)$ in the interval $−2 \leq t \leq 3$ s.   *Ans:* 4.4 A; 6.6 A; 17.5 C; 3.5 A

1-3. In Fig. 1-8c, let $v_1(t) = 0.5 + \sin 400t$ V. Then determine (a) $v_1(1$ ms); (b) $v_1(10$ ms); (c) the energy required to move 3 C from the lower to the upper terminal at $t = 2$ ms.   *Ans:* 0.889 V; −0.257 V; −3.65 J

1-4. Find the power being: (a) delivered to the circuit element in Fig. 1-11a at $t = 5$ ms; (b) absorbed by the circuit element in Fig. 1-11b; (c) generated by the circuit element in Fig. 1-11c.   *Ans:* −15.53 W; 1.012 W; 6.65 W

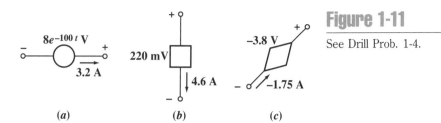

Figure 1-11

See Drill Prob. 1-4.

Using the concepts of current and voltage, it is now possible to be more specific in defining a circuit element.

It is important to differentiate between the physical device itself and the device's mathematical model which we shall use to analyze its behavior in a circuit. Let us agree that we will use the expression *circuit element* to refer to the mathematical model. The choice of a particular model for any real device must be made on the basis of experimental data or experience; we shall usually assume that this choice has already been made. We must first learn the methods of analysis of idealized circuits.

Now let us distinguish a *general circuit element* from a *simple circuit element* by the statement that a general circuit element may be composed of more than one simple circuit element, but that a simple circuit element cannot be further subdivided into other simple circuit elements. For brevity, we shall agree that the term *circuit element* generally refers to a simple circuit element. All the simple circuit elements that will be considered in the work that follows can be classified according to the relationship of the current through the element to the voltage across the element. For example, if the voltage across the element is directly proportional to the current through it, or $v = ki$, we shall call the element a *resistor*. Other types of simple circuit elements have terminal voltages which are proportional to the derivative of the current with respect to time, or to the integral of the current with respect to time. There are also elements in which the voltage is completely independent of the current, or the current is completely independent of the voltage; these are the *independent sources*. Furthermore, we shall need to define special kinds of sources in which the source voltage or current depends upon a current or voltage elsewhere in the circuit; such sources will be termed *dependent sources* or *controlled sources*.

By definition, a simple circuit element is the mathematical model of a two-terminal electrical device, and it can be completely characterized by its voltage-current relationship but cannot be subdivided into other two-terminal devices.

The first element which we shall need is an *independent voltage source*. It is characterized by a terminal voltage which is completely independent of the current through it. Thus, if we are given an independent voltage source and are notified that the terminal voltage is $50t^2$ V, we can be sure that at $t = 2$ s the voltage will be 200 V, regardless of the current that was flowing, is flowing, or is going to flow. The representation of an independent voltage source is shown in Fig. 1-12. The subscript s merely identifies the voltage as a "source" voltage.

A point worth repeating here is that the presence of the plus sign at the upper end of the symbol for the independent voltage source in Fig. 1-12 does not necessarily mean that the upper terminal is always positive with respect to the lower terminal. Instead, it means that the upper terminal is v_s volts

1-4

Types of circuits and circuit elements

Figure 1-12

The circuit symbol of an independent voltage source. (The circuit symbol of a dependent or controlled voltage source is shown in Figure 1-15*a*.)

positive with respect to the lower. If, at some instant, v_s happens to be negative, then the upper terminal is actually negative with respect to the lower at that instant.

If a current arrow labeled "*i*" is placed adjacent to the upper conductor of this source and directed to the left, then the current i is entering the terminal at which the positive sign is located, the passive sign convention is satisfied, and the source thus absorbs a power $p = v_s i$. More often than not, a source is expected to deliver power to a network and not to absorb it. Consequently, we might choose to direct the arrow to the right in order that $v_s i$ will represent the power delivered by the source. Either direction may be used.

The independent voltage source is an *ideal* source and does not represent exactly any real physical device, because the ideal source could theoretically deliver an infinite amount of energy from its terminals. Each coulomb passing through it receives an energy of v_s joules, and the number of coulombs per second is unlimited. This idealized voltage source does, however, furnish a reasonable approximation to several practical voltage sources. An automobile storage battery, for example, has a 12-V terminal voltage that remains essentially constant as long as the current through it does not exceed a few amperes. The small current may flow in either direction through the battery. If it is positive and flowing *out of* the positively marked terminal, then the battery is furnishing power to the headlights, say, as it discharges; but if the current is positive and flowing *into* the positive terminal, then the battery is charging or absorbing energy from the generator, or possibly a battery charger. An ordinary household electrical outlet also approximates an independent voltage source providing the voltage $v_s = 115\sqrt{2} \cos 2\pi 60t$ V; the representation is valid for currents less than perhaps 20 A.

An independent voltage source which has a constant terminal voltage is often termed an independent dc voltage source[8] and is represented by either symbol shown in Fig. 1-13. Note in Fig. 1-13*b* that, when the physical plate

Figure 1-13

Alternative representations of a constant, or dc, independent voltage source: (*a*) source is delivering 12 W; (*b*) battery is absorbing 12 W.

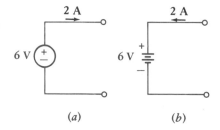

(*a*)   (*b*)

[8] Terms like *dc voltage source* and *dc current source* are commonly used. Literally, they mean "direct-current voltage source" and "direct-current current source," respectively. Although these terms may be redundant or misleading, the terminology is so widely used there's no point in fighting it.

structure of the battery is suggested, the longer plate is placed at the positive terminal; the plus and minus signs then represent redundant notation, but they are usually included anyway.

Another ideal source which we will need is the *independent current source.* Here, the current through the element is completely independent of the voltage across it. The symbol for an independent current source is shown in Fig. 1-14. If i_s is constant, we call the source an independent dc current source.

Figure 1-14

The circuit symbol of an independent current source. (The circuit symbol of a dependent or controlled current source is shown in Fig. 1-15b.)

Like the independent voltage source, the independent current source is at best a reasonable approximation for a physical element. In theory it can deliver infinite power from its terminals, because it produces the same finite current for any voltage across it, no matter how large that voltage may be. It is, however, a good approximation for many practical sources, particularly in electronic circuits. Also, the independent dc current source represents very closely the proton beam of a cyclotron which is operating at a constant beam current of perhaps 1 μA and will continue to deliver 1 μA to almost any device placed across its "terminals" (the beam and the earth).

The two types of ideal sources that we have discussed up to now are called *independent* sources because the value of the source quantity is not affected in any way by activities in the remainder of the circuit. This is in contrast with yet another kind of ideal source, the *dependent,* or *controlled, source,* in which the source quantity is determined by a voltage or current existing at some other location in the electrical system under examination. To distinguish between independent and dependent sources, we introduce the diamond symbols shown in Fig. 1-15. Sources such as these will appear in the equivalent electrical models for many electronic devices, such as transistors, operational amplifiers, and integrated circuits. We shall see all of these in the following chapters.

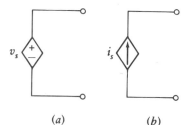

Figure 1-15

The diamond shape characterizes the circuit symbols for (*a*) the dependent voltage source and (*b*) the dependent current source.

(*a*)　　　　　(*b*)

Dependent and independent voltage and current sources are *active* elements; they are capable of delivering power to some external device. For the present we shall think of a *passive* element as one which is capable only of receiving power. However, we shall later see that several passive elements are able to store finite amounts of energy and then return that energy later to various external devices; and since we shall still wish to call such elements passive, it will be necessary to improve upon our two definitions then.

The interconnection of two or more simple circuit elements is called an electrical *network.* If the network contains at least one closed path, it is also an electric *circuit.* Every circuit is a network, but not all networks are circuits.

Figure 1-16

(a) An electrical network which is not a circuit. (b) A network which is a circuit.

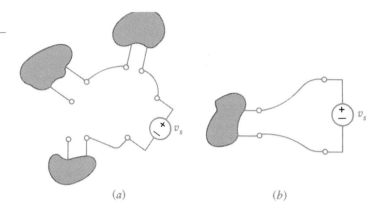

(a)   (b)

Figure 1-16a shows a network which is not a circuit, and Fig. 1-16b shows a network which is a circuit.

A network which contains at least one active element, such as an independent voltage or current source, is an *active network*. A network which does not contain any active elements is a *passive network*.

We have now defined what we mean by the term *circuit element,* and we have presented the definitions of several specific circuit elements, the independent and dependent voltage and current sources. Throughout the remainder of the book we will define only four additional circuit elements, all passive, called the *resistor,* the *inductor,* the *capacitor,* and a pair of *mutually coupled inductors.* These are all ideal elements. They are important because we may combine them into networks and circuits that represent real devices as accurately as we wish. Thus, the transistor whose physical construction is suggested by Fig. 1-17a, and whose electrical symbol is given in Fig. 1-17b, may be modeled by the single resistor and the single dependent current source of Fig. 1-17c if we need only approximate performance data at frequencies that are neither extremely high nor extremely low. Note that the dependent *current* source produces a current that depends on a *voltage* elsewhere in the circuit. Figure 1-17d illustrates a more accurate model for high-frequency applications that contains three resistors, two capacitors, and one dependent current source.[9]

Transistors such as these may constitute only one small part of an integrated circuit that is possibly 2 mm square and 0.2 mm thick and yet contains several thousand transistors plus thousands of resistors and capacitors. Thus we may have a physical device that is about the size of one letter on this page, but that might require a model composed of ten thousand ideal simple circuit elements.

Suitable models for the various physical devices that have wide practical importance are studied in courses in electronics, energy conversion, antennas, and other subjects appearing later in engineering curricula.

Drill Problem

1-5. Find the power absorbed by each element in the circuit of Fig. 1-18.

Ans: (left to right) −56 W; 16 W; −60 W; 160 W; −60 W

---

[9] We should note carefully that the transistor does not *actually* contain a dependent source. If we broke one open, we would search in vain for the source; but the transistor *acts as if* it contains a dependent source, thereby justifying our including it as a part of the model.

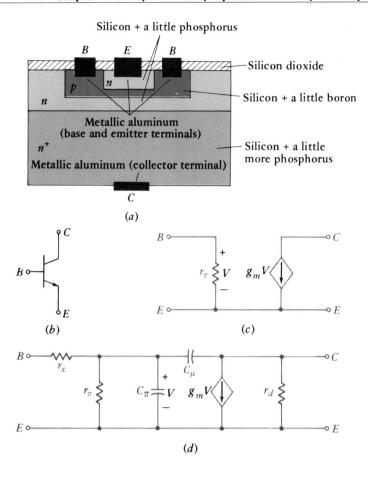

Silicon + a little phosphorus

Silicon dioxide

Silicon + a little boron

Metallic aluminum
(base and emitter terminals)

Silicon + a little
more phosphorus

Metallic aluminum (collector terminal)

(a)

(b)

(c)

(d)

Figure 1-17

(a) One possible physical configuration for an *npn* bipolar silicon transistor (not to scale). (b) The circuit symbol for the *npn* bipolar transistor. (c) A simple circuit model that is useful in the mid-frequency range. (d) A more accurate high-frequency model.

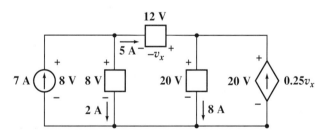

Figure 1-18

See Drill Prob. 1-5.

In the previous sections we became familiar with both independent and dependent voltage and current sources and were cautioned that they were idealized elements which could only be approximated in a real circuit. We are now ready to introduce another idealized element, the linear resistor. The resistor is the simplest passive element, and we begin our discussion by considering the work of an obscure German physicist, Georg Simon Ohm, who published a pamphlet in 1827 entitled "Die galvanische Kette mathematisch bearbeitet." [10] In it were

Ohm's law

---

[10] "The Galvanic Circuit Investigated Mathematically."

contained the results of one of the first efforts to measure currents and voltages and to describe and relate them mathematically. One result was a statement of the fundamental relationship we now call Ohm's law, even though it has since been shown that this result was discovered 46 years earlier in England by Henry Cavendish, a brilliant semirecluse. However, no one, including Ohm, we shall hope, knew of the work done by Cavendish, because it was not uncovered and published until long after both were dead.

Ohm's pamphlet received much undeserved criticism and ridicule for several years after its first publication, but it was later accepted and served to remove the obscurity associated with his name.

Ohm's law states that the voltage across many types of conducting materials is directly proportional to the current flowing through the material, or

$$v = Ri \tag{4}$$

where the constant of proportionality R is called the *resistance*. The unit of resistance is the *ohm*, which is 1 V/A and customarily abbreviated by a capital omega, Ω.

When this equation is plotted on v-versus-i axes, the graph is a straight line passing through the origin. The equation is a linear equation, and we shall consider it as the definition of a *linear resistor*. Hence, if the ratio of the current and voltage associated with any simple current element is a constant, then the element is a linear resistor and has a resistance equal to the voltage-to-current ratio. Resistance is normally considered to be a positive quantity, although negative resistances may be simulated with special circuitry.

Again, it must be emphasized that the linear resistor is an idealized circuit element; it is a mathematical model of a physical device. "Resistors" may be easily purchased or manufactured, but it is soon found that the voltage-current ratios of these physical devices are reasonably constant only within certain ranges of current, voltage, or power and depend also on temperature and other environmental factors. We shall usually refer to a linear resistor as simply a resistor, using the longer term only when the linear nature of the element needs emphasis. Any resistor which is nonlinear will always be described as such. Nonlinear resistors should not necessarily be considered as undesirable elements. Although it is true that their presence complicates an analysis, the performance of the device may depend on or be greatly improved by the nonlinearity. Zener diodes, tunnel diodes, and fuses are such elements.

Figure 1-19 shows the most common circuit symbol used for a resistor. In accordance with the voltage, current, and power conventions already adopted,

Figure 1-19

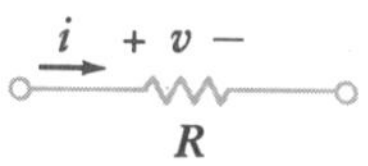

The circuit symbol for a resistor; $R = v/i$ and $p = vi = i^2R = v^2/R$.

the product of v and i gives the power absorbed by the resistor. That is, v and i are selected to satisfy the passive sign convention. The absorbed power appears physically as heat and is always positive; a (positive) resistor is a passive element that cannot deliver power or store energy. Alternative expressions for the absorbed power are

$$p = vi = i^2R = \frac{v^2}{R} \tag{5}$$

One of the authors (who prefers not to be identified further)[11] had the unfortunate experience of inadvertently connecting a 100-Ω, 2-W carbon resistor across a 110-V source. The ensuing flame, smoke, and fragmentation were rather disconcerting, demonstrating clearly that a practical resistor has definite limits to its ability to behave like the ideal linear model. In this case, the unfortunate resistor was called upon to absorb 121 W; since it was designed to handle only 2 W, its reaction was understandably violent.

The ratio of current to voltage is also a constant,

$$\frac{i}{v} = \frac{1}{R} = G \tag{6}$$

where G is called the *conductance*. The SI unit of conductance is the *siemens* (S), 1 A/V. An older, unofficial unit for conductance is the *mho,* abbreviated by an inverted capital omega, $\mho$. You will probably see its use on some circuit diagrams, in some catalogs, and in some texts. The same circuit symbol is used to represent both resistance and conductance. The absorbed power is again necessarily positive and may be expressed in terms of the conductance by

$$p = vi = v^2 G = \frac{i^2}{G} \tag{7}$$

Thus a 2-Ω resistor has a conductance of $\frac{1}{2}$ S, and if a current of 5 A is flowing through it, a voltage of 10 V is present across the terminals and a power of 50 W is being absorbed.

All the expressions given so far in this section were written in terms of instantaneous current, voltage, and power, such as $v = Ri$ and $p = vi$. Obviously, the current through and voltage across a resistor must both vary with time in the same manner. Thus, if $R = 10$ Ω and $v = 2 \sin 100t$ V, then $i = 0.2 \sin 100t$ A but the power is $0.4 \sin^2 100t$ W, and a simple sketch will illustrate the different nature of its variation with time. Although the current and voltage are each negative during certain time intervals, the absorbed power is *never* negative!

Resistance may be used as the basis for defining two commonly used terms, *short circuit* and *open circuit*. We define a short circuit as a resistance of zero ohms; then, since $v = Ri$, the voltage across a short circuit must be zero, although the current may have any value. In an analogous manner, we define an open circuit as an infinite resistance. It follows then that the current must be zero, regardless of the voltage across the open circuit.

Remember, when any formula is written containing v's and i's, all of these quantities must be defined on a circuit diagram, along with their reference polarities and directions, respectively. Otherwise, $v = -Ri$ would be just as applicable to some resistor as $v = Ri$.

1-6. With reference to the v and i defined in Fig. 1-19, find (*a*) R if $i = -1.6$ mA and $v = -6.3$ V; (*b*) the absorbed power if $v = -6.3$ V and $R = 21$ Ω; (*c*) i if $v = -8$ V and R is absorbing 0.24 W; (*d*) G if $v = -8$ V and R absorbs 3 mW.
Ans: 3.94 kΩ; 1.890 W; -30.0 mA; 46.9 μS

Drill Problem

---

[11] Name gladly furnished upon written request to W.H.H.

1-6

Kirchhoff's laws

We are now ready to consider the current and voltage relations in simple networks that result from the interconnection of two or more simple circuit elements. The elements will be connected together by electrical conductors, or leads, which have zero resistance, or are *perfectly conducting*. Since the network then appears as a number of simple elements and a set of connecting leads, it is called a *lumped-constant* network. A more difficult analysis problem arises when we are faced with a *distributed-constant* network, which essentially contains an infinite number of vanishingly small elements. Consideration of this latter type of network is mercifully relegated to later courses.

A point at which two or more elements have a common connection is called a *node*. Figure 1-20a shows a circuit containing three nodes. Sometimes networks are drawn so as to trap an unwary student into believing that there are more nodes present than is actually the case. This occurs when a node, such as node 1 in Fig. 1-20a, is shown as two separate junctions connected by a (zero-resistance) conductor, as in Fig. 1-20b. However, all that has been done is to spread the common point out into a common zero-resistance line. Thus, we must necessarily consider all of the perfectly conducting leads or portions of leads attached to the node as part of the node. Note also that every element has a node at each of its ends.

Figure 1-20

(*a*) A circuit containing three nodes and five branches. (*b*) Node 1 is redrawn to *look* like two nodes; it is still one node.

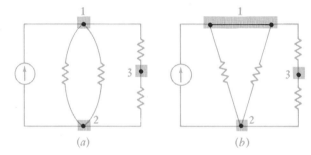

Suppose that we start at one node in a network and move through a simple element to the node at the other end, then continue from that node through a different element to the next node, and continue this movement until we have gone through as many elements as we wish. If no node was encountered more than once, then the set of nodes and elements that we have passed through is defined as a *path*. If the node at which we started is the same as the node on which we ended, then the path is, by definition, a *closed path* or a *loop*.

For example, in Fig. 1-20a, if we move from node 2 through the current source to node 1, and then through the upper right resistor to node 3, we have established a path; since we have not continued on to node 2 again, we do not have a closed path or a loop. If we proceeded from node 2 through the current source to node 1, down through the left resistor to node 2, and then up through the central resistor to node 1 again, we do not have a path, since a node (actually two nodes) was encountered more than once; we also do not have a loop, because a loop must be a path.

Another term whose use will prove convenient is *branch*. We define a branch as a single path in a network, composed of one simple element and the node at each end of that element. Thus, a path is a particular collection of branches. The circuit shown in Figs. 1-20a and b contains five branches.

We are now ready to consider the first of the two laws named for Gustav Robert Kirchhoff (two h's and two f's), a German university professor who was born about the time Ohm was doing his experimental work.[12] This axiomatic law is called *Kirchhoff's current law* (abbreviated KCL), and it states that

> the algebraic sum of the currents entering any node is zero.

We shall not offer any proof of the law in this text. However, it simply represents a mathematical statement of the fact that charge cannot accumulate at a node. That is, if there *were* a nonzero net current into a node, then the rate at which coulombs were accumulating at the node would not be zero. However, a node is not a circuit element, and it certainly cannot store, destroy, or generate charge. Hence, the currents must sum to zero.

Consider the node shown in Fig. 1-21. The algebraic sum of the four currents entering the node must be zero:

$$i_A + i_B - i_C - i_D = 0$$

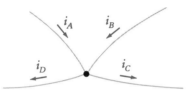

Figure 1-21

Kirchhoff's current law (KCL) enables us to write $i_A + i_B - i_C - i_D = 0$, $i_C + i_D - i_A - i_B = 0$, or $i_A + i_B = i_C + i_D$.

It is evident that the law could be equally well applied to the algebraic sum of the currents *leaving* any node:

$$-i_A - i_B + i_C + i_D = 0$$

We might also wish to equate the sum of the currents having reference arrows directed into the node to the sum of those directed out of the node:

$$i_A + i_B = i_C + i_D$$

A compact expression for Kirchhoff's current law is

$$\sum_{n=1}^{N} i_n = 0 \qquad (8)$$

and this is just a shorthand statement for

$$i_1 + i_2 + i_3 + \cdots + i_N = 0 \qquad (9)$$

Whether Eq. (8) or Eq. (9) is used, it is understood that the N current arrows either are all directed toward the node in question or are all directed away from it.

It is sometimes helpful to interpret Kirchhoff's current law in terms of a hydraulic analogy. Water, like charge, cannot be stored at a point, and thus if we identify a junction of several pipes as a node, it is evident that the number of gallons of water entering the node every second must equal the number of gallons leaving the node each second.

We now turn to *Kirchhoff's voltage law* (abbreviated KVL). This law states that

> the algebraic sum of the voltages around any closed path in a circuit is zero.

---

[12] Kirchhoff began his teaching career as a *Privat-Dozent,* or lecturer with official status but no salary, at the University of Berlin. This is not a popular option with today's young instructors.

Again, we must accept this law as an axiom, even though it is developed in introductory electromagnetic theory.

Current is a variable that is related to the charge flowing *through* a circuit element, whereas voltage is a measure of potential energy difference *across* the element. There is a single unique value for voltage in circuit theory. Thus, the energy required to move a unit charge from point A to point B in a circuit must have a value that is independent of the path taken from A to B.

In Fig. 1-22, if we carry a charge of 1 C from A to B through element 1, the reference polarity signs for v_1 show that we do v_1 joules of work. Now if, instead, we choose to proceed from A to B via C, then we expend $v_2 - v_3$ joules of energy.

Figure 1-22

The potential difference between points A and B is independent of the path selected, or $v_1 = v_2 - v_3$.

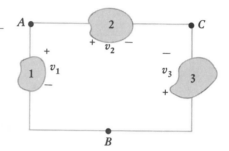

The work done, however, is independent of the path in a circuit, and these values must be equal. Any route leads to the same value for the voltage. Thus,

$$v_1 = v_2 - v_3 \qquad (10)$$

It follows that if we trace out a closed path, the algebraic sum of the voltages across the individual elements around it must be zero. Thus, we may write

$$\sum_{n=1}^{N} v_n = 0 \quad \text{or} \quad v_1 + v_2 + v_3 + \cdots + v_N = 0 \qquad (11)$$

Kirchhoff's voltage law is a consequence of the conservation of energy and the conservative property of the electric circuit. This law can also be interpreted in terms of a gravitational analogy. If a mass is moved around any *closed* path in a conservative gravitational field, the total work done on the mass is zero.

We may apply KVL to a circuit in several different ways. One method that leads to fewer equation-writing errors than others consists of moving mentally around the closed path in a clockwise direction and writing down directly the voltage of each element whose (+) terminal is entered, and writing down the negative of every voltage first met at the (−) sign. Applying this to the single loop of Fig. 1-22, we have

$$-v_1 + v_2 - v_3 = 0$$

which certainly agrees with our previous result, Eq. (10).

Example 1-1   As a last example of the use of KVL, let us attack the circuit of Fig. 1-23. There are eight circuit elements, and voltages with plus-minus pairs are shown across each element. Suppose that we want to find v_{R2}, the voltage across R_2.

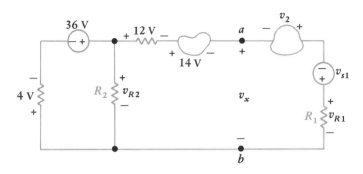

Figure 1-23

A circuit for which KVL shows that $v_x = 6$ V and $v_{R2} = 32$ V.

Solution: We may do this by writing one KVL equation around the loop on the left:

$$4 - 36 + v_{R2} = 0$$

finding that

$$v_{R2} = 32 \text{ V}$$

Finally, let us assume that we want to determine the value of v_x. We might think of this as the (algebraic) sum of the voltages across the three elements on the right, or we might consider it as the voltage across an ideal voltmeter connected between points a and b; but it is just as easy to let it be the voltage across the gap from a to b. We apply KVL, beginning in the lower left corner, moving up and across the top to a, through v_x to b, and through the conducting lead to the starting point:

$$4 - 36 + 12 + 14 + v_x = 0$$

so that

$$v_x = 6 \text{ V}$$

Knowing v_{R2}, we might have taken the shortcut through R_2,

$$-32 + 12 + 14 + v_x = 0$$

or $$v_x = 6 \text{ V}$$

once again. ■

1-7. Determine the number of branches and nodes in each of the circuits of Fig. 1-24. **Drill Problems**

Ans: 5, 3; 7, 5; 6, 4

1-8. Determine i_x in each of the circuits of Fig. 1-24. *Ans: 3 A; −8 A; 1 A*

1-9. Determine v_x in each of the circuits of Fig. 1-24. *Ans: 78 V; 80 V; 8 V*

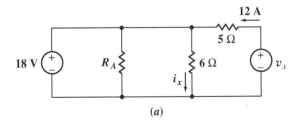

(a)

Figure 1-24

See Drill Probs. 1-7, 1-8, and 1-9. (*continues*)

Figure 1-24

(*continued*)

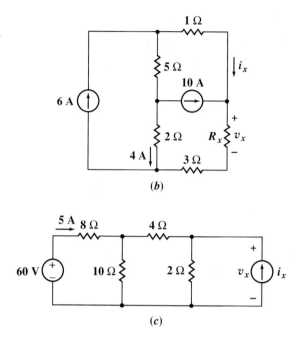

(*b*)

(*c*)

1-7

Analysis of a single-loop circuit

Having established Ohm's and Kirchhoff's laws, we may flex our analytical muscles by applying these tools to the analysis of a simple resistive circuit. Figure 1-25*a* illustrates how the series connection of two batteries and two resistors might look, as constructed by a youngster trying out a new soldering iron. Note that a connecting lead has been clamped onto the positive terminal of the left battery, and the other end of the wire has been soldered to one of the resistor leads. The terminal, the connecting leads, and the solder glob are all assumed to have zero resistance, and they constitute the node at the upper left-hand corner of the circuit diagram in Fig. 1-25*b*. Both batteries are replaced by ideal voltage sources; any resistances they may have are assumed to be small

Figure 1-25

(*a*) A sketch of a physical single-loop circuit containing four elements. Solder connections and connecting wires are indicated. (*b*) The circuit model with source voltages and resistance values given. (*c*) Current and voltage reference signs have been added to the circuit.

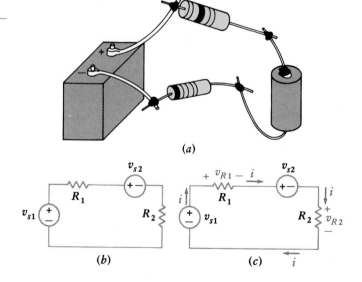

(*a*)

(*b*)   (*c*)

enough to neglect. If not, they may be incorporated into R_1 and R_2. The two resistors are assumed to be replaceable by ideal linear resistors.

We shall consider the resistance values and the source voltages of Fig. 1-25*b* to be known and attempt to determine the current through each element, the voltage across each element, and the power absorbed by each element. Our first step in the analysis is the assumption of reference directions for the unknown currents. Arbitrarily, let us select a clockwise current i which flows out of the upper terminal of the voltage source on the left. This choice is indicated by an arrow labeled i at that point in the circuit, as shown in Fig. 1-25*c*. A trivial application of Kirchhoff's current law assures us that this same current must also flow through every other element in the circuit. We may emphasize this fact this one time by placing several other current symbols about the circuit.

By definition, all the elements that carry the *same* current are said to be connected in *series*. Note that elements may carry *equal* currents and not be in series; two 100-W lamp bulbs in neighboring houses may very well carry equal currents, but they do not carry the *same* current and are not in series.

Our second step in the analysis is a choice of the voltage reference for each of the two resistors. We have already found that the application of Ohm's law without a minus sign, $v = Ri$, demands that the sense of the current and voltage be selected so that the current enters the terminal at which the positive voltage reference is located. This is the passive sign convention. If the choice of the current direction is arbitrary, then the sense of the voltage selection is fixed if we intend to use Ohm's law in the form $v = Ri$. The voltages v_{R1} and v_{R2} are shown in Fig. 1-25*c*.

The third step is the application of Kirchhoff's voltage law to the single closed path present. Let us decide to move around the circuit in the clockwise direction, beginning at the lower left corner, and to write down directly every voltage first met at its positive reference, and write down the negative of every voltage encountered at the negative terminal. Thus,

$$-v_{s1} + v_{R1} + v_{s2} + v_{R2} = 0$$

Finally, we apply Ohm's law to the resistive elements.

$$v_{R1} = R_1 i \quad \text{and} \quad v_{R2} = R_2 i$$

and obtain

$$-v_{s1} + R_1 i + v_{s2} + R_2 i = 0$$

This equation is solved for i, and thus

$$i = \frac{v_{s1} - v_{s2}}{R_1 + R_2}$$

where all the quantities on the right side are known and enable us to determine i. The voltage or power associated with any element may now be obtained in one step by applying $v = Ri$, $p = vi$, or $p = i^2 R$.

Example 1-2 Let us consider the numerical example illustrated in Fig. 1-26*a* and determine the power absorbed by each of the simple elements in the circuit.

Solution: Two batteries and two resistors are connected in a series circuit. The clockwise current and two resistor voltages are assigned to the circuit

Figure 1-26

(*a*) A given series circuit. (*b*) The circuit with current and voltage references assigned.

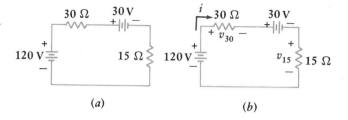

(*a*)   (*b*)

as indicated in Fig. 1-26*b*, and Kirchhoff's voltage law yields

$$-120 + v_{30} + 30 + v_{15} = 0$$

An application of Ohm's law to each resistor permits us to write

$$-120 + 30i + 30 + 15i = 0$$

from which

$$i = \frac{120 - 30}{30 + 15} = 2\ \text{A}$$

Thus, the voltages across the resistors are

$$v_{30} = 2(30) = 60\ \text{V} \qquad v_{15} = 2(15) = 30\ \text{V}$$

The power absorbed by each element is given by the product of the voltage across the element and the current flowing into the element terminal at which the positive voltage reference is located. For the 120-V battery, then, the power absorbed is

$$p_{120\text{V}} = 120(-2) = -240\ \text{W}$$

and thus 240 W is *delivered* to other elements in the circuit by this source. In a similar manner,

$$p_{30\text{V}} = 30(2) = 60\ \text{W}$$

and we find that this nominally active element is actually absorbing power delivered to it by the other battery (or being charged).

The power absorbed by each (positive) resistor is necessarily positive and may be calculated by

$$p_{30} = v_{30}i = 60(2) = 120\ \text{W}$$

or by

$$p_{30} = i^2 R = 2^2(30) = 120\ \text{W}$$

and

$$p_{15} = v_{15}i = i^2 R = 60\ \text{W}$$

The results check because the total power absorbed must be zero; or, in other words, the power delivered by the 120-V battery is exactly equal to the sum of the powers absorbed by the three other elements. A power balance is often a useful method of checking for careless mistakes.

Before leaving this example, it is important that we be convinced that our initial assumption of a current reference direction had nothing to do with the answers obtained. Let us assume a current i_x in a *counter*clockwise direction. Therefore, $i_x = -i$. Both resistor voltages must now be assigned opposite polarities, and we obtain

$$-120 - 30i_x + 30 - 15i_x = 0$$

and $i_x = -2$ A, $v_{x30} = -60$ V, and $v_{x15} = -30$ V. Since each voltage and current reference is reversed and each quantity is the negative of the previously obtained value, it is evident that the results are the same. Each absorbed power will be the same. ∎

Any random or convenient choice of current direction may be made, although clockwise currents are most often selected. Those who needlessly insist on positive answers may always go back and reverse the direction of the current arrow and rework the problem.

Example 1-3 Now let us complicate the analysis slightly by letting one of the voltage sources be a dependent source, as exemplified by Fig. 1-27. Once again we shall determine the power absorbed by each of the simple circuit elements.

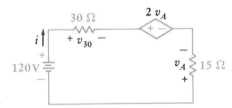

Figure 1-27

A current i and voltage v_{30} are assigned in a single-loop circuit containing a dependent source.

Solution: We again assign a reference direction for the current i and a reference polarity for the voltage v_{30}. There is no need to assign a voltage to the 15-Ω resistor, since the controlling voltage v_A for the dependent source is already available. It is worth noting, however, that the reference signs for v_A are reversed from those we would have assigned, and that Ohm's law for this element must thus be expressed as $v_A = -15i$. We apply Kirchhoff's voltage law around the loop,

$$-120 + v_{30} + 2v_A - v_A = 0$$

utilize Ohm's law twice,

$$v_{30} = 30i$$

$$v_A = -15i$$

and obtain

$$-120 + 30i - 30i + 15i = 0$$

and so

$$i = 8 \text{ A}$$

The power relationships show that the 120-V battery supplies 960 W, the dependent source supplies 1920 W, and the two resistors together dissipate 2880 W. ∎

More practical applications of the dependent source in equivalent circuits for the operational amplifier and the transistor will begin to appear in the following chapter.

Drill Problems

1-10. For the circuit shown in Fig. 1-28a, find (a) i_1; (b) v_1; (c) the power supplied by the 3-V battery.   *Ans:* 12.5 mA; −2.5 V; −37.5 mW

1-11. In the circuit of Fig. 1-28b, find the power absorbed by each of the five elements in the circuit.

Ans: (CW from left) 0.768 W; 1.920 W; 0.205 W; 0.1792 W; −3.07 W

Figure 1-28

See Drill Probs. 1-10 and 1-11.

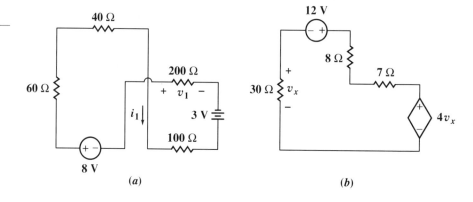

(a)   (b)

1-8

The single-node-pair circuit

The companion of the single-loop circuit discussed in the previous section is the single-node-pair circuit, in which any number of simple elements are connected between the same pair of nodes. An example of such a circuit is shown in Fig. 1-29a. The two current sources and the conductance values are known.

Example 1-4   Let us find the voltage, current, and power associated with each element in the circuit of Fig. 1-29a.

Solution: Our first step is now to assume a voltage across any element, assigning an arbitrary reference polarity. Then KVL forces us to recognize that the voltage across each branch is the same as across any other branch because a closed path proceeds through any branch from one node to the other and then is completed through any other branch. A total voltage of zero requires an identical voltage across every element. We shall say that elements having a common voltage across them are connected in *parallel*. Let us call this voltage v and arbitrarily select it as shown in Fig. 1-29b.

Two currents, flowing in the resistors, are then selected in conformance with the passive sign convention. These currents are also shown in Fig. 1-29b.

Figure 1-29

(a) A single-node-pair circuit. (b) A voltage and two currents are assigned.

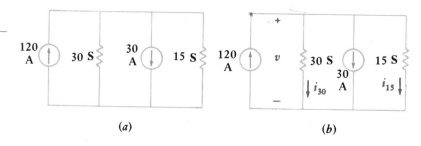

(a)   (b)

Our third step in the analysis of the single-node-pair circuit is the application of KCL to either of the two nodes in the circuit. It is usually clearer to apply it to the node at which the positive voltage reference is located, and thus we shall equate the algebraic sum of the currents leaving the upper node to zero:

$$-120 + i_{30} + 30 + i_{15} = 0$$

Finally, the current in each resistor is expressed in terms of v and the conductance of the resistor by Ohm's law,

$$i_{30} = 30v \quad \text{and} \quad i_{15} = 15v$$

and we obtain

$$-120 + 30v + 30 + 15v = 0$$

Thus,

$$v = 2\,\text{V}$$

and

$$i_{30} = 60\,\text{A} \quad \text{and} \quad i_{15} = 30\,\text{A}$$

The several values of absorbed power are now easily obtained. In the two resistors,

$$p_{30} = 30(2)^2 = 120\,\text{W} \qquad p_{15} = 15(2)^2 = 60\,\text{W}$$

and for the two sources,

$$p_{120A} = 120(-2) = -240\,\text{W} \qquad p_{30A} = 30(2) = 60\,\text{W}$$

Thus, the larger current source delivers 240 W to the other three elements in the circuit, and the conservation of energy is verified again.

∎

The similarity of this example to the one previously completed, illustrating the solution of the series circuit with independent sources only, should not have gone unnoticed. The numbers are all the same, but currents and voltages, resistances and conductances, and "series" and "parallel" are interchanged. This is an example of *duality,* and the two circuits are said to be *exact duals* of each other. If the element values or source values were changed in either circuit, without changing the configuration of the network, the two circuits would be duals, although not exact duals. We shall study and use duality later, and at this time should only suspect that any result we obtain in terms of current, voltage, and resistance in a series circuit will have its counterpart in terms of voltage, current, and conductance in a parallel circuit.

Now let us try our skills on a single-node-pair circuit containing a dependent source. In Fig. 1-30 the dependent current source is controlled by the current i_x in the 2-kΩ resistor.

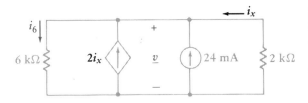

Figure 1-30

A voltage v and a current i_6 are assigned in a single-node-pair circuit containing a dependent source.

Example 1-5   Let us determine the values of v and the power absorbed by the independent current source in Fig. 1-30.

Solution: We have defined the voltage v, arbitrarily with the positive reference at the top, and a current i_6 in the 6-kΩ resistor. The sum of the currents leaving the upper node is zero, so that

$$i_6 - 2i_x - 0.024 - i_x = 0$$

We next apply Ohm's law to each resistor, noting that we are given the more common resistance values rather than conductances:

$$i_6 = \frac{v}{6000} \quad \text{and} \quad i_x = \frac{-v}{2000}$$

Therefore,

$$\frac{v}{6000} - 2\left(\frac{-v}{2000}\right) - 0.024 - \left(\frac{-v}{2000}\right) = 0$$

and
$$v = 600 \times 0.024 = 14.4 \text{ V}$$

Any other information we may want to find for this circuit is now easily obtained, usually in a single step. For example, the power supplied by the independent source is $p_{24} = 14.4(0.024) = 0.346$ W, and the current flowing to the right in the central conductor at the top is $i = -0.024 + (14.4/2000) = -0.0168$ A, or -16.8 mA. ∎

Drill Problems

1-12. For the single-node-pair circuit of Fig. 1-31a, find (a) i_A; (b) i_B; (c) i_C.

                                                 Ans: 3 A; -5.4 A; 6 A

1-13. For the single-node-pair circuit of Fig. 1-31b, find (a) i_1; (b) i_2; (c) i_3; (d) i_4.

                                         Ans: -2 A; 3 A; -8 A; -0.5 A

Figure 1-31

See Drill Probs. 1-12 and 1-13.

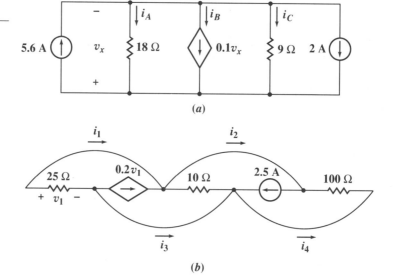

(a)

(b)

Some of the equation writing that we have been doing for the simple series and parallel circuits can be avoided. This is achieved by replacing relatively complicated resistor combinations by a single equivalent resistor whenever we are not specifically interested in the current, voltage, or power associated with any of the individual resistors in the combinations. All the current, voltage, and power relationships in the remainder of the circuit will be unchanged.

We first consider the series combination of N resistors shown schematically in Fig. 1-32a. The shading surrounding the resistors is intended to suggest that they are enclosed in a "black box," or perhaps in another room, and we wish to replace the N resistors by a single resistor with resistance R_{eq} so that the remainder of the circuit, in this case only the voltage source, does not realize that any change has been made. The source current, power, and, of course, voltage will be the same before and after the replacement.

1-9

Resistance and source combination

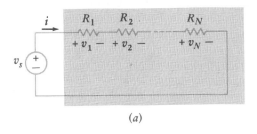

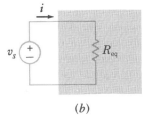

(a) (b)

Figure 1-32

(a) A circuit containing a series combination of N resistors. (b) A simpler equivalent circuit: $R_{eq} = R_1 + R_2 + \cdots + R_N$.

We apply Kirchhoff's voltage law

$$v_s = v_1 + v_2 + \cdots + v_N$$

and Ohm's law

$$v_s = R_1 i + R_2 i + \cdots + R_N i = (R_1 + R_2 + \cdots + R_N)i$$

and then compare this result with the simple equation applying to the equivalent circuit shown in Fig. 1-32b,

$$v_s = R_{eq} i$$

Thus, the value of the equivalent resistance for N series resistors is

$$R_{eq} = R_1 + R_2 + \cdots + R_N \qquad (12)$$

We are therefore able to replace a *two-terminal network* consisting of N series resistors with a single *two-terminal element* R_{eq} that has the same v-i relationship. No measurements that we can make outside the "black box" can tell which network is which.

It should be emphasized again that we might be particularly interested in the current, voltage, or power of one of the original elements, as would be the case when the voltage of a dependent voltage source depends upon, say, the voltage across R_3. Once R_3 is combined with several series resistors to form an equivalent resistance, then it is gone, and the voltage across it cannot be determined until R_3 is identified by removing it from the combination. It would have been better to look ahead and not make R_3 a part of the combination initially.

An inspection of the Kirchhoff voltage equation for a series circuit also shows two other possible simplifications. The order in which elements are placed in a series circuit makes no difference, and several voltage sources in series may be replaced by an equivalent voltage source having a voltage equal to the algebraic sum of the individual sources. There is usually little advantage in including a dependent voltage source in a series combination.

These simplifications may be illustrated by considering the circuit shown in Fig. 1-33.

Figure 1-33

(*a*) A given series circuit. (*b*) A simpler equivalent circuit.

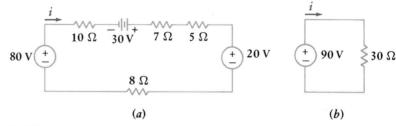

(*a*)   (*b*)

Example 1-6 Let us use resistance and source combinations to simplify the determination of the current *i* in Fig. 1-33*a*.

Solution: We first interchange the element positions in the circuit, being careful to preserve the proper sense of the sources, and then combine the three voltage sources into an equivalent 90-V source and the four resistors into an equivalent 30-Ω resistance, as shown in Fig. 1-33*b*. Thus, instead of writing

$$-80 + 10i - 30 + 7i + 5i + 20 + 8i = 0$$

we have simply

$$-90 + 30i = 0$$

and

$$i = 3 \text{ A}$$

In order to calculate the power delivered to the circuit by the 80-V source appearing in the given circuit, it is necessary to return to that circuit with the knowledge that the current is 3 A. The desired power is 240 W.

It is interesting to note that no element of the original circuit remains in the equivalent circuit, unless we are willing to count the interconnecting wires as elements. ■

Similar simplifications can be applied to parallel circuits.[13] A circuit containing *N* conductances in parallel, as in Fig. 1-34*a*, leads to the KCL equation

$$i_s = i_1 + i_2 + \cdots + i_N$$

or

$$i_s = G_1 v + G_2 v + \cdots + G_N v = (G_1 + G_2 + \cdots + G_N)v$$

whereas its equivalent in Fig. 1-34*b* gives

$$i_s = G_{eq} v$$

and thus

$$G_{eq} = G_1 + G_2 + \cdots + G_N$$

In terms of resistance instead of conductance,

$$\frac{1}{R_{eq}} = \frac{1}{R_1} + \frac{1}{R_2} + \cdots + \frac{1}{R_N}$$

or

$$R_{eq} = \frac{1}{1/R_1 + 1/R_2 + \cdots + 1/R_N} \tag{13}$$

[13] By the principle of duality.

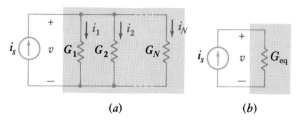

Figure 1-34

(a) A circuit containing N parallel resistors having conductances $G_1, G_2, \ldots, G_N$. (b) A simpler equivalent circuit: $G_{eq} = G_1 + G_2 + \cdots + G_N$.

(a) (b)

This last equation is probably the most often used means of combining parallel resistive elements. The parallel combination is often indicated by writing $R_{eq} = R_1 \| R_2 \| R_3$, for example.

The special case of only two parallel resistors

$$R_{eq} = R_1 \| R_2 = \frac{1}{1/R_1 + 1/R_2} = \frac{R_1 R_2}{R_1 + R_2} \tag{14}$$

is needed very often. The last form is worth memorizing.

Parallel current sources may also be combined by algebraically adding the individual currents, and the order of the parallel elements may be rearranged as desired.

The various combinations described in this section are used to simplify the circuit of Fig. 1-35a.

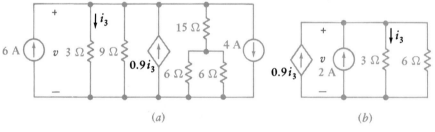

Figure 1-35

(a) A given circuit. (b) A simplified equivalent circuit.

(a) (b)

Example 1-7 Let us suppose that we wish to know the power and voltage of the dependent source in Fig. 1-35a.

Solution: We may just as well leave the dependent source alone and combine the remaining two sources into one 2-A source. The resistances are combined by beginning with the parallel combination of the two 6-Ω resistors into a 3-Ω resistor, followed by the series combination of 3 Ω and 15 Ω. The 18-Ω and 9-Ω elements combine in parallel to produce 6 Ω, and this is as far as we can proceed profitably. Certainly 6 Ω in parallel with 3 Ω is 2 Ω, but the current i_3 on which the source depends then disappears.[14]

From the equivalent circuit in Fig. 1-35b, we have

$$-0.9i_3 - 2 + i_3 + \frac{v}{6} = 0$$

and
$$v = 3i_3$$

yielding
$$i_3 = \tfrac{10}{3}\,\text{A}$$

and
$$v = 10\ \text{V}$$

---

[14] Of course, we could have preserved it by using the given circuit to write $i_3 = v/3$, thus expressing i_3 in terms of variables appearing in the final circuit.

Thus, the dependent source furnishes $v(0.9i_3) = 10(0.9 \times \frac{10}{3}) = 30$ W to the remainder of the circuit.

Now if we are belatedly asked for the power dissipated in the 15-Ω resistor, we must return to the original circuit. This resistor is in series with an equivalent 3-Ω resistor; a voltage of 10 V is across the 18-Ω total; thus, a current of $\frac{5}{9}$ A flows through the 15-Ω resistor and the power absorbed by this element is $(\frac{5}{9})^2(15)$, or 4.63 W. ∎

To conclude the discussion of parallel and series element combinations, we should consider the parallel combination of two voltage sources and the series combination of two current sources. For instance, what is the equivalent of a 5-V source in parallel with a 10-V source? By the definition of a voltage source, the voltage across the source cannot change; by Kirchhoff's voltage law, then, 5 equals 10 and we have hypothesized a physical impossibility. Thus, *ideal* voltage sources in parallel are permissible only when each has the same terminal voltage at every instant. Later, we shall see that *practical* voltage sources may be combined in parallel without any theoretical difficulty.

In a similar way, two current sources may not be placed in series unless each has the same current, including sign, for every instant of time.

A voltage source in parallel or series with a current source presents an interesting little intellectual diversion. The two possible cases are illustrated by Prob. 43 at the end of the chapter.

Three final comments on series and parallel combinations might be helpful. The first is illustrated by referring to Fig. 1-36a and asking, "Are v_s and R in series, or in parallel?" The answer is clearly, "Both." The two elements carry the same current and are therefore in series; they also enjoy the same voltage and consequently are in parallel. This simple circuit is the only one for which this is true.

The second comment is a word of caution. Circuits can be drawn by inexperi-

Figure 1-36

(a) These two circuit elements are both in series and in parallel.
(b) R_2 and R_3 are in parallel, and R_1 and R_8 are in series. (c) There are no circuit elements in series or in parallel.

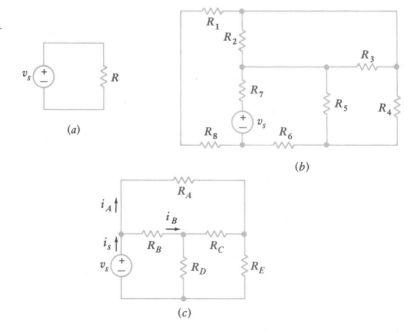

enced students or insidious instructors in such a way as to make series or parallel combinations difficult to spot. In Fig. 1-36b, for example, the only two resistors in parallel are R_2 and R_3, while the only two in series are R_1 and R_8. Of course, v_s and R_7 are also in series.

The final comment is simply that a simple circuit element need not be in series or parallel with any other simple circuit element in a circuit. For example, R_4 and R_5 in Fig. 1-36b are not in series or parallel with any other simple circuit element, and there are no simple circuit elements in Fig. 1-36c that are in series or parallel with any other simple circuit element.

1-14. An ohmmeter is an instrument that gives a value for the resistance seen between its two terminals. What is the correct reading if the instrument is attached to the network of Fig. 1-37a at points: (a) ac; (b) ab; (c) cd?

Drill Problems

Ans: 9 Ω; 5.69 Ω; 6.54 Ω

1-15. What resistance is measured at the terminals of the network in Fig. 1-37b if switch S is (a) open; (b) closed; (c) replaced by a 10-mS conductance?

Ans: 160 Ω; 60 Ω; 110 Ω

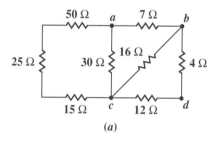

(a)

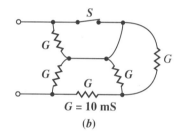

(b)

Figure 1-37

See Drill Probs. 1-14 and 1-15.

By combining resistances and sources, we have found one method of shortening the work of analyzing a circuit. Another useful shortcut is the application of the ideas of voltage and current division. Voltage division is used to express the voltage across one of several series resistors in terms of the voltage across the combination. In Fig. 1-38, the voltage across R_2 is obviously

1-10

Voltage and current division

$$v_2 = R_2 i = R_2 \frac{v}{R_1 + R_2}$$

or

$$\boxed{v_2 = \frac{R_2}{R_1 + R_2} v}$$

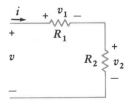

Figure 1-38

An illustration of voltage division, $v_2 = \dfrac{R_2}{R_1 + R_2} v$.

and the voltage across R_1 is, similarly,

$$v_1 = \frac{R_1}{R_1 + R_2} v$$

If the network of Fig. 1-38 is generalized by removing R_2 and replacing it with the series combination of R_2, R_3, . . . , R_N, then we have the general result for voltage division across a string of N series resistors,

$$v_1 = \frac{R_1}{R_1 + R_2 + \cdot \cdot \cdot + R_N} v \tag{15}$$

The voltage appearing across one of the series resistors is the total voltage times the ratio of its resistance to the total resistance. Voltage division and resistance combination may both be applied, as in the circuit shown in Fig. 1-39. We mentally combine the 3- and 6-Ω resistors, obtaining 2 Ω, and thus find that v_x is $\frac{2}{6}$ of 12 sin t, or 4 sin t V.

Figure 1-39

A numerical example illustrating resistance combination and voltage division. The wavy line within the source symbol indicates a sinusoidal variation with time.

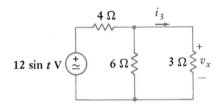

The dual of voltage division is current division. We now are given a total current supplied to several parallel conductances, as exemplified by the circuit of Fig. 1-40. The current flowing through G_2 is

$$i_2 = G_2 v = G_2 \frac{i}{G_1 + G_2}$$

or

$$i_2 = \frac{G_2}{G_1 + G_2} i$$

and, similarly,

$$i_1 = \frac{G_1}{G_1 + G_2} i$$

Figure 1-40

An illustration of current division, $i_2 = \dfrac{G_2}{G_1 + G_2} i = \dfrac{R_1}{R_1 + R_2} i.$

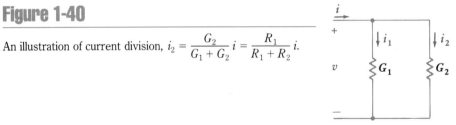

Thus the current flowing through either of two parallel conductances is the total current times the ratio of that conductance to the total conductance.

Since we are given the value of the resistance more often than the conductance, a more important form of the last result is obtained by replacing G_1 by

$1/R_1$ and G_2 by $1/R_2$,

$$i_2 = \frac{R_1}{R_1 + R_2} i \quad \text{and} \quad i_1 = \frac{R_2}{R_1 + R_2} i$$

Nature has not smiled on us here, for these last two equations have a factor which differs subtly from the factor used with voltage division, and some effort is going to be needed to avoid errors. Many students look on the expression for voltage division as "obvious" and that for current division as being "different." It also helps to realize that the larger of two parallel resistors always carries the smaller current.

We may also generalize these results by removing G_2 in Fig. 1-40 and replacing it with the parallel combination of $G_2, G_3, \ldots, G_N$. Thus, for N parallel conductances,

$$i_1 = \frac{G_1}{G_1 + G_2 + \cdots + G_N} i \tag{16}$$

In terms of resistance values, the result becomes

$$i_1 = \frac{1/R_1}{1/R_1 + 1/R_2 + \cdots + 1/R_N} i \tag{17}$$

Example 1-8 As an example of the use of both current division and resistance combination, let us return to the example of Fig. 1-39. We wish to write an expression for the current through the 3-Ω resistor.

Solution: The total current flowing into the 3- and 6-Ω combination is

$$i = \frac{12 \sin t}{4 + 6 \| 3}$$

and thus the desired current is

$$i_3 = \frac{12 \sin t}{4 + (6)(3)/(6 + 3)} \frac{6}{6 + 3} = \tfrac{4}{3} \sin t \qquad \blacksquare$$

Unfortunately, current division is sometimes applied when it is not applicable. As one example, let us consider again the circuit shown in Fig. 1-36c, a circuit that we have already agreed contains no circuit elements that are in series or in parallel. Without parallel resistors, there is no way that current division can be applied. Even so, there are too many students who take a quick look at resistors R_A and R_B and try to apply current division, writing an incorrect equation such as

$$i_A = i_s \frac{R_B}{R_A + R_B}$$

Remember, parallel resistors must be branches between the same pair of nodes.

1-16. In the circuit of Fig. 1-41: (a) use resistance combination methods to find R_{eq}; (b) use current division to find i_1; (c) find i_2; (d) find v_3.

Drill Problem

Ans: 50 Ω; 100 mA; 50 mA; 0.8 V

Figure 1-41

See Drill Prob. 1-16.

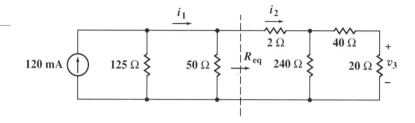

1-11

A practical example: the operational amplifier

We have now been introduced to enough basic laws and simple analytical techniques that we should be able to apply them successfully to some interesting practical circuits. In this section we shall begin to consider an electrical device called the *operational amplifier,* or *op-amp* for short.

The first operational amplifiers were built in the 1940s using vacuum tubes to perform the mathematical operations of addition, subtraction, multiplication, division, differentiation, and integration electrically, thus enabling the electrical solution of differential equations in the early analog computers.

Reduced to its essentials, the op-amp is just a voltage-controlled dependent voltage source. The dependent voltage source is drawn across the output terminals of the op-amp, and the voltage on which it depends is applied to the input terminals.

The symbol commonly used for an op-amp is shown in Fig. 1-42a. Two input terminals are shown on the left, and a single output terminal appears at the right. There is also a common terminal or node called *ground* that is not usually shown explicitly as a terminal of the op-amp itself. In practical circuits there are often numerous elements connected to the metallic case (or chassis) on which a circuit is built, and this chassis is then connected through a good conductor to the earth. Thus, the metallic case becomes the ground node. The symbol for the ground node appears several times at the bottom of Fig. 1-42b.

Figure 1-42

(a) The circuit symbol for an operational amplifier. (b) Input voltages v_1 and v_2, their difference v_i, and an output v_o are defined.

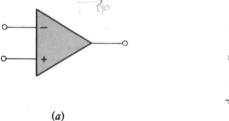

(a)

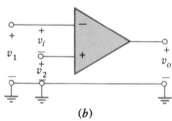

(b)

Although we may consider a single voltage signal or voltage source to be applied directly between the pair of input terminals, we can provide for a greater number of applications by establishing a voltage between each input terminal and ground. The input terminal identified by a minus sign is the *inverting input,* and a voltage v_1 is defined between the inverting input and ground, as shown in Fig. 1-42b. The voltage v_2 is defined between the *noninverting input* and ground. We shall see shortly that the voltage $(v_1 - v_2)$ appears greatly amplified and reversed in polarity between the output terminal and ground. If $v_2 = 0$, then v_1 appears amplified and inverted at the output; if $v_1 = 0$, then v_2 appears amplified, without a sign change, between the output terminal and ground. The amplification factor ranges from 10^4 to 10^7 for different op-amps; a typical value is 10^5. The difference between v_1 and v_2 is the input voltage $v_i = v_1 - v_2$.

An op-amp may cost as little as 20 cents, for which we obtain an integrated-circuit (IC) chip containing about 25 transistors and a dozen resistors, all packaged in a small can or ceramic wafer with 8 or 10 terminal pins for connection to the external circuit. In some cases, the IC chip may contain several op-amps. In addition to the output pins and the two inputs, other pins enable voltages to be supplied for the transistors and external adjustments to be made, to balance and compensate the op-amp. However, we are not concerned with the internal circuitry of the op-amp or the IC at this time, but only with the voltage and current relationships that exist at the external terminals. Thus, the connections shown in Fig. 1-42a are sufficient.

Figure 1-43 shows a useful model for an op-amp. The resistance between the input terminals is so large (10^5 to 10^{15} Ω) that we can safely represent it by an open circuit, as shown. In later examples we shall connect such an *input resistance* R_i between the two input terminals, but at this time we shall work with a model that is closer to the ideal. A voltage-controlled dependent voltage source provides an output voltage equal to A times the difference of the two input voltages. More exact models of an op-amp include an *output resistance* R_o, ranging from 1 to 1000 Ω, in series with the dependent source and the output terminal.

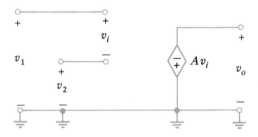

Figure 1-43

A simple model for an op-amp contains one dependent source and several terminals.

Let us suppose that $v_1 = 1$ μV, $v_2 = 0.6$ μV, and $A = 10^5$. Then, $v_i = 10^{-6} - 0.6 \times 10^{-6} = 0.4 \times 10^{-6}$ V, and $v_o = -10^5 \times 0.4 \times 10^{-6} = -0.04$ V. Note the polarity signs on the dependent source. If we now connect the noninverting input to ground, then $v_2 = 0$, $v_i = 10^{-6}$ V, and $v_o = -10^5 \times 10^{-6} = -0.1$ V; with the inverting input grounded, $v_1 = 0$, $v_i = -0.6 \times 10^{-6}$ V, and $v_o = -10^5(-0.6 \times 10^{-6}) = 0.06$ V.

The operational amplifier is seldom used in the bare form suggested by the preceding examples. There are usually several circuit elements connected in parallel or series with the input or the output, or between the output and the input. We shall use many of these practical circuits as examples as we study circuit analysis in the chapters that follow. Most of these practical op-amp circuits are given descriptive names. For convenience, all these named circuits are listed under "op-amp" in the index at the end of the text.

The first example is the *voltage follower* shown in Fig. 1-44a. Here an input signal v_s is connected between the noninverting input and ground so that $v_2 = v_s$. A short circuit is connected directly from the output to the inverting input, so that $v_1 = v_o$. Figure 1-44b shows the equivalent circuit, and it is seen to be a single-loop circuit even though there is an open circuit between the two input terminals. As a matter of fact, the loop current must be zero, and there is therefore no current anywhere in the circuit. It follows that KCL cannot give us any additional information, Ohm's law is worthless without any resistors, and we can only turn hopefully to KVL for some information about the relation-

Figure 1-44

(a) An op-amp is connected as a voltage follower. (b) The equivalent circuit is treated as a single-loop circuit with an open circuit and zero loop current.

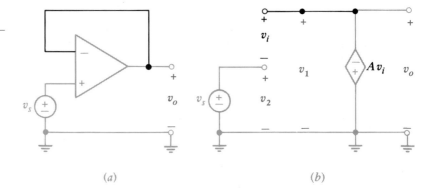

(a)   (b)

ship between the output voltage v_o and the input voltage v_s. Around the loop, we have

$$-v_s - v_i - Av_i = 0$$

Also,

$$v_o = -Av_i$$

or

$$v_i = -\frac{v_o}{A}$$

Therefore

$$-v_s + \frac{v_o}{A}(1 + A) = 0$$

and

$$v_o = \frac{A}{1 + A} v_s \qquad (18)$$

If $A = 10^5$, then $v_o = 0.999\,990 v_s$. For all practical purposes, therefore, $v_o = v_s$, and the output voltage "follows" the input voltage. The advantage of such a unity-gain amplifier is that the input draws negligible current and power from the source, while the output can supply reasonable currents (10 to 20 mA) and power (100 to 500 mW) to a load connected across the output. Thus, the load has little effect on the source, and for this reason the voltage follower is also known as a *buffer amplifier*.

Some specific numerical values for the voltages can be helpful. If $v_s = 1$ V and $A = 10^5$, then $v_o = 0.999\,990$ V, and the difference between the input voltages is found to be $v_i = -9.9999\ \mu$V.

The magnitude of v_i is quite small, and we often make an approximate analysis of an op-amp circuit by assuming that its value is zero, as well as the input current. If we do this for the voltage follower, then we can conclude immediately that $v_o = v_s$. This result is obtained more rigorously by letting the gain A approach infinity in Eq. (18).

Drill Problem

1-17. A voltage follower is operating with $v_s = 1.8$ mV. Find A if (a) $v_o = 1.7999$ mV; (b) $v_1 = 1.799\,926$ mV; (c) $v_i = -0.12\ \mu$V.   *Ans: 17 999; 24 323; 14 999*

1 A famous, mild-mannered reporter has a mass of 170 lbm, can leap a tall (400-ft) building at a single bound, and is as fast as a speeding (1200 ft/s) bullet. (*a*) What is his top speed in km/h? (*b*) What energy in joules must he impart to his leap to just clear the tall building? (*c*) For how many days would this energy power an electronic calculator requiring 80 mW? (*d*) Where was the reporter born?

Problems

2 The power supplied by a certain battery is a constant 6 W over the first 5 min, zero for the following 2 min, a value that increases linearly from zero to 10 W during the next 10 min, and a power that decreases linearly from 10 W to zero in the following 7 min. (*a*) What is the total energy in joules expended during this 24-min interval? (*b*) What is the average power in Btu/h during this time?

3 The total charge accumulated by a certain device is given as a function of time by $q = 18t^2 - 2t^4$ (in SI units). (*a*) What total charge is accumulated at $t = 2$ s? (*b*) What is the maximum charge accumulated in the interval $0 \le t \le 3$ s, and when does it occur? (*c*) At what rate is charge being accumulated at $t = 0.8$ s? (*d*) Sketch curves of q versus t and i versus t in the interval $0 \le t \le 3$ s.

4 The current $i_1(t)$ in Fig. 1-5c is given as $-2 + 3e^{-5t}$ A for $t < 0$, and $-2 + 3e^{3t}$ A for $t > 0$. Find (*a*) $i_1(-0.2)$; (*b*) $i_1(0.2)$; (*c*) those instants at which $i_1 = 0$; (*d*) the total charge that has passed from left to right along the conductor in the interval $-0.08 < t < 0.1$ s.

5 The waveform shown in Fig. 1-45 has a period of 10 s. (*a*) What is the average value of the current over one period? (*b*) How much charge is transferred in the interval $1 < t < 12$ s? (*c*) If $q(0) = 0$, sketch $q(t)$, $0 < t < 16$ s.

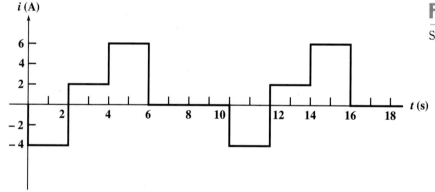

Figure 1-45

See Prob. 5.

6 Determine the power being absorbed by each of the circuit elements shown in Fig. 1-46.

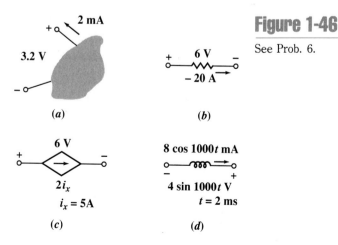

Figure 1-46

See Prob. 6.

7 Let $i = 3te^{-100t}$ mA and $v = (0.006 - 0.6t)e^{-100t}$ V for the circuit element of Fig. 1-9. (*a*) What power is being absorbed by the circuit element at $t = 5$ ms? (*b*) How much energy is delivered to the element in the interval $0 < t < \infty$?

8 Determine which of the five sources in Fig. 1-47 are being charged (absorbing positive power), and show that the algebraic sum of the five absorbed power values is zero.

Figure 1-47

See Prob. 8.

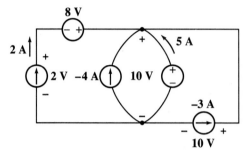

9 In Fig. 1-9, let $i = 3e^{-100t}$ A. Find the power being absorbed by the circuit element at $t = 8$ ms if v equals (*a*) $40i$; (*b*) $0.2\,di/dt$; (*c*) $30\int_0^t i\,dt + 20$ V.

10 Let $R = 1200\ \Omega$ for the resistor shown in Fig. 1-19. Find the power being absorbed by R at $t = 0.1$ s if (*a*) $i = 20e^{-12t}$ mA; (*b*) $v = 40\cos 20t$ V; (*c*) $vi = 8t^{1.5}$ VA.

11 A certain voltage is $+10$ V for 20 ms and -10 V for the succeeding 20 ms and continues oscillating back and forth between these two values at 20-ms intervals. The voltage is present across a 50-Ω resistor. Over any 40-ms interval find (*a*) the maximum value of the voltage; (*b*) the average value of the voltage; (*c*) the average value of the resistor current; (*d*) the maximum value of the absorbed power; (*e*) the average value of the absorbed power.

12 The resistance of a conductor having a length l and a uniform cross-sectional area A is given by $R = l/\sigma A$, where σ (sigma) is the electrical conductivity. If $\sigma = 5.8 \times 10^7$ S/m for copper: (*a*) what is the resistance of a #18 copper wire (diameter = 1.024 mm) that is 50 ft long? (*b*) If a circuit board has a copper-foil conducting ribbon 33 μm thick and 0.5 mm wide that can carry 3 A safely at 50°C, find the resistance of a 15-cm length of this ribbon and the power delivered to it by the 3-A current.

13 Find R and G in the circuit of Fig. 1-48*a* if the 5-A source is supplying 100 W and the 40-V source is supplying 500 W.

Figure 1-48

(*a*) See Prob. 13. (*b*) See Prob. 14.

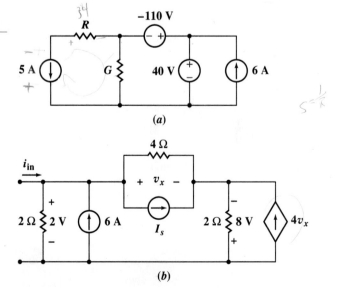

14 Use Ohm's and Kirchhoff's laws on the circuit of Fig. 1-48b to find (a) v_x; (b) i_{in}; (c) I_s; (d) the power provided by the dependent source.

15 (a) Use Kirchhoff's and Ohm's laws in a step-by-step procedure to evaluate all the currents and voltages in the circuit of Fig. 1-49. (b) Calculate the power absorbed by each of the five circuit elements and show that the sum is zero.

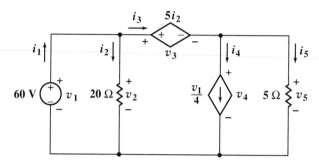

Figure 1-49

See Prob. 15.

16 With reference to the circuit shown in Fig. 1-50, find the power absorbed by each of the seven circuit elements.

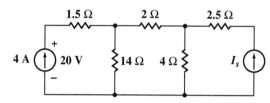

Figure 1-50

See Prob. 16.

17 A certain circuit contains six elements and four nodes, numbered 1, 2, 3, and 4. Each circuit element is connected between a different pair of nodes. The voltage v_{12} (+ reference at first-named node) is 12 V, and $v_{34} = -8$ V. Find v_{13}, v_{23}, and v_{24} if v_{14} equals (a) 0; (b) 6 V; (c) −6 V.

18 Find the power being absorbed by element X in Fig. 1-51 if it is a (a) 100-Ω resistor; (b) 40-V independent voltage source, + reference on top; (c) dependent voltage source labeled $25i_x$, + reference on top; (d) dependent voltage source labeled $0.8v_1$, + reference on top; (e) 2-A independent current source, arrow directed upward.

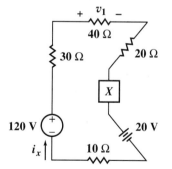

Figure 1-51

See Prob. 18.

19 Find i_1 in the circuit of Fig. 1-52 if the dependent voltage source is labeled: (a) $2v_2$; (b) $1.5v_3$; (c) $-15i_1$.

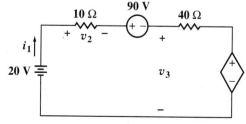

Figure 1-52

See Probs. 19 and 20.

20 Refer to the circuit of Fig. 1-52 and label the dependent source $1.8v_3$. Find v_3 if (a) the 90-V source generates 180 W; (b) the 90-V source absorbs 180 W; (c) the dependent source generates 100 W; (d) the dependent source absorbs 100 W.

21 For the battery charger modeled by the circuit of Fig. 1-53, find the value of the adjustable resistor R so that: (a) a charging current of 4 A flows; (b) a power of 25 W is delivered to the battery (0.035 Ω and 10.5 V); (c) a voltage of 11 V is present at the terminals of the battery (0.035 Ω and 10.5 V).

Figure 1-53

See Probs. 21 and 22.

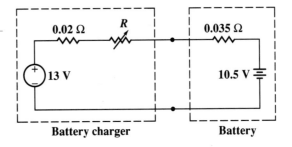

Battery charger   **Battery**

22 The circuit of Fig. 1-53 is modified by installing a dependent voltage source in series with the battery. Place the + reference at the bottom and let the control be $0.05i$, where i is the clockwise loop current. Find this current and the terminal voltage of the battery, including the dependent source, if $R = 0.5$ Ω.

23 Find the power absorbed by each of the six circuit elements in Fig. 1-54.

Figure 1-54

See Prob. 23.

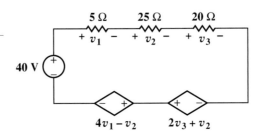

24 Find the power absorbed by each circuit element of Fig. 1-55 if the control for the dependent source is (a) $0.8i_x$; (b) $0.8i_y$.

Figure 1-55

See Prob. 24.

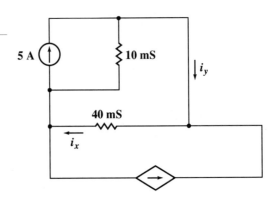

25 Find i_x in the circuit of Fig. 1-56.

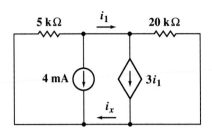

Figure 1-56

See Prob. 25.

26 Find the power absorbed by each element in the single-node-pair circuit of Fig. 1-57.

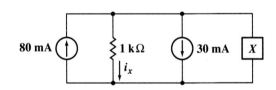

Figure 1-57

See Prob. 26.

27 Find the power absorbed by element X in the circuit of Fig. 1-58 if it is a (a) 4-kΩ resistor; (b) 20-mA independent current source, reference arrow downward; (c) dependent current source, reference arrow downward, labeled $2i_x$; (d) 60-V independent voltage source, + reference at top.

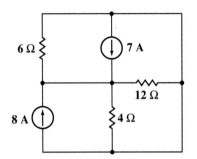

Figure 1-58

See Probs. 27 and 28.

28 (a) Let element X in Fig. 1-58 be an independent current source, arrow directed upward, labeled i_s. What is i_s if none of the four circuit elements absorbs any power? (b) Let element X be an independent voltage source, + reference on top, labeled v_s. What is v_s if the voltage source absorbs no power?

29 (a) Apply the techniques of single-node-pair analysis to the upper right node in Fig. 1-59 and find i_x. (b) Now work with the upper left node and find v_8. (c) How much power is the 5-A source generating?

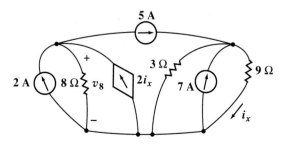

Figure 1-59

See Prob. 29.

30 Find R_{eq} for each of the resistive networks shown in Fig. 1-60.

Figure 1-60

See Prob. 30.

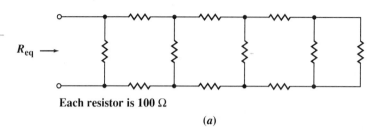

Each resistor is 100 Ω

(a)

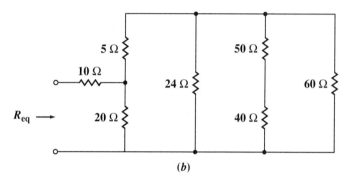

(b)

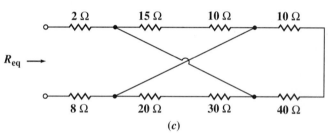

(c)

31 In the network shown in Fig. 1-61: (a) let $R = 80\ \Omega$ and find R_{eq}; (b) find R if $R_{eq} = 80\ \Omega$; (c) find R if $R = R_{eq}$.

Figure 1-61

See Prob. 31.

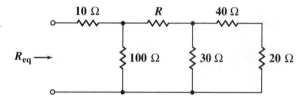

32 Show how to combine four 100-Ω resistors to obtain an equivalent resistance of (a) 25 Ω; (b) 60 Ω; (c) 40 Ω.

33 Find the power absorbed by each of the resistors in the circuit of Fig. 1-62.

Figure 1-62

See Prob. 33.

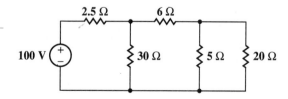

34 Use source- and resistor-combination techniques as a help in finding v_x and i_x in the circuit of Fig. 1-63.

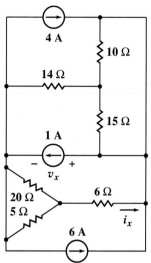

Figure 1-63

See Prob. 34.

35 Determine G_{in} for each network shown in Fig. 1-64. Values are all given in millisiemens.

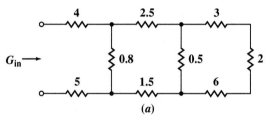

(a)

Figure 1-64

See Prob. 35.

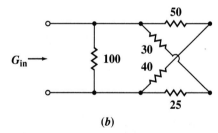

(b)

36 Use both resistance and source combinations, as well as current division, in the circuit of Fig. 1-65 to find the power absorbed by the 1-Ω, 10-Ω, and 13-Ω resistors.

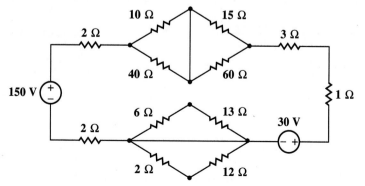

Figure 1-65

See Prob. 36.

37 Use current and voltage division on the circuit of Fig. 1-66 to find an expression for (a) v_2; (b) v_1; (c) i_4.

Figure 1-66

See Prob. 37.

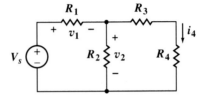

38 With reference to the circuit shown in Fig. 1-67: (a) let $v_s = 40$ V, $i_s = 0$, and find v_1; (b) let $v_s = 0$, $i_s = 3$ mA, and find i_2 and i_3.

Figure 1-67

See Prob. 38.

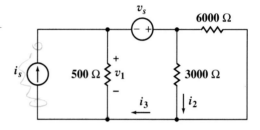

39 In Fig. 1-68: (a) let $v_x = 10$ V and find I_s; (b) let $I_s = 50$ A and find v_x; (c) calculate the ratio v_x/I_s.

Figure 1-68

See Prob. 39.

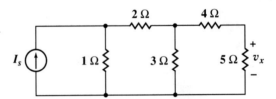

40 Determine how much power is absorbed by R_x in the circuit of Fig. 1-69.

Figure 1-69

See Prob. 40.

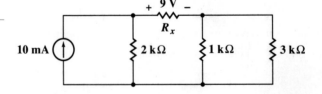

41 Use current and voltage division to help obtain an expression for v_5 in Fig. 1-70.

Figure 1-70

See Prob. 41.

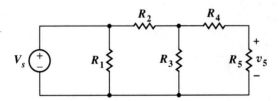

42 With reference to the circuit of Fig. 1-71, find (a) I_x if $I_1 = 12$ mA; (b) I_1 if $I_x = 12$ mA; (c) I_x if $I_2 = 15$ mA; (d) I_x if $I_s = 60$ mA.

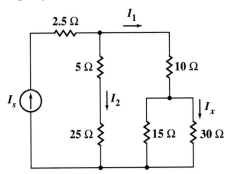

Figure 1-71

See Prob. 42.

43 The circuit shown in Fig. 1-72 contains several examples of independent current and voltage sources connected in series and in parallel. (a) Find the power absorbed by each source. (b) To what value should the 4-V source be changed to reduce the power supplied by the −5-A source to zero?

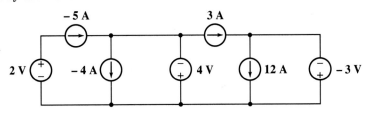

Figure 1-72

See Prob. 43.

44 Find v_o for the op-amp shown in Fig. 1-73 if $A = 10^5$ and (a) $R_i = \infty$ and $R_o = 20$ Ω; (b) $R_i = 1$ MΩ and $R_o = 0$.

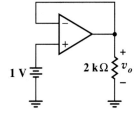

Figure 1-73

See Prob. 44.

45 To find the effect of reversing the input connections to a voltage follower, connect the noninverting input terminal directly to the output terminal, and install an independent voltage source V_s between the inverting input terminal and ground. Let the input resistance of the op-amp be infinite, the output resistance be zero, and A be 10^5. Find in terms of V_s: (a) v_o; (b) v_i; (c) v_2; (d) v_1.

46 Let $A = \infty$, $R_i = \infty$, and $R_o = 0$ for the op-amp of Fig. 1-74. (a) Find v_L. (b) Find v_L if the op-amp is removed and points a and b are connected directly.

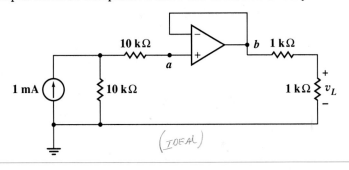

Figure 1-74

See Prob. 46.

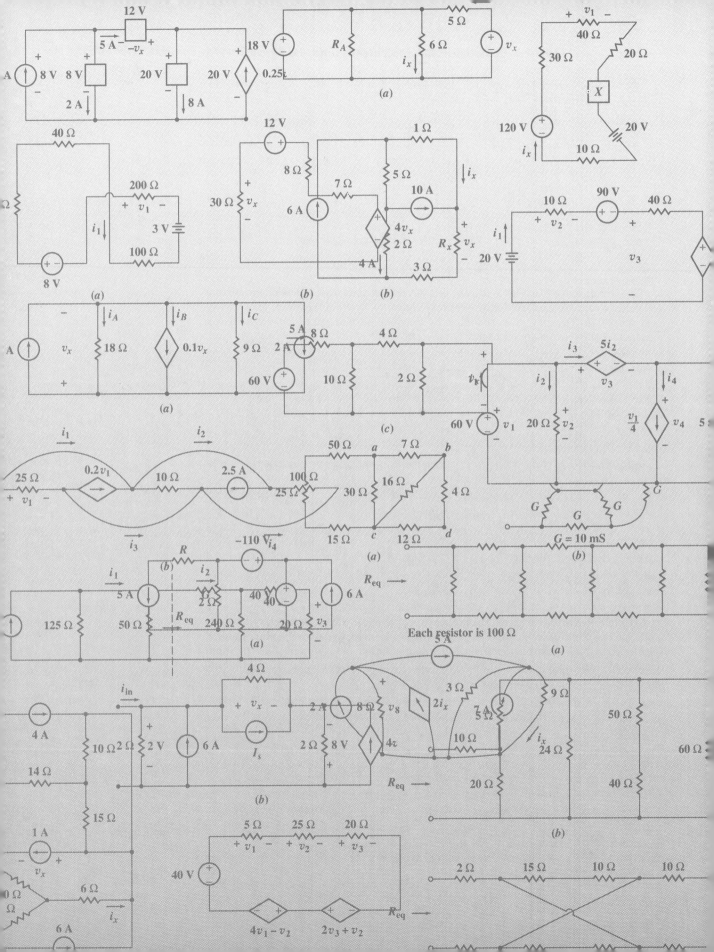

Some Useful Techniques of Circuit Analysis

We should now be familiar with Ohm's law and Kirchhoff's laws and their application in the analysis of simple series and parallel resistive circuits. When it will produce results more easily we should be able to combine resistors or sources in series or parallel and be able to use the principles of voltage and current division. Most of the circuits on which we have been practicing are simple and of questionable practical importance; they are useful in helping us learn to apply the fundamental laws. Now we must begin analyzing more complicated circuits.

These larger circuits may represent control circuits, communication systems, motors and generators, power distribution networks, or electronic systems containing commercially available integrated circuits. They also may be electric circuit models of nonelectrical systems.

It is evident that one of the primary goals of this chapter must be learning methods of simplifying the analysis of more complicated circuits. Among these methods will be superposition and loop, mesh, and nodal analysis. We shall also try to develop the ability to select the most *convenient* analysis method. Most often we are interested only in the detailed performance of an isolated portion of a complex circuit; a method of replacing the remainder of the circuit by a greatly simplified equivalent is then very desirable. The equivalent is often a single resistor in series or parallel with an ideal source. Thévenin's and Norton's theorems will enable us to effect this replacement.

We shall begin studying methods of simplifying circuit analysis by considering a powerful general method, that of nodal analysis.

In the previous chapter we considered the analysis of a simple circuit containing only two nodes. We found then that the major step of the analysis was taken as we obtained a single equation in terms of a single unknown quantity, the voltage between the pair of nodes. We shall now let the number of nodes increase, and correspondingly provide one additional unknown quantity and one additional equation for each added node. Thus, a three-node circuit should have two unknown voltages and two equations; a 10-node circuit will have nine unknown voltages and nine equations; and an N-node circuit will need $(N - 1)$ voltages and $(N - 1)$ equations.

We consider the mechanics of nodal analysis in this section, but the justification for our methods will not be developed until later in this chapter. As an example, let us consider the three-node circuit shown in Fig. 2-1a.

Figure 2-1

(a) A given three-node circuit.
(b) The circuit is redrawn to emphasize the three nodes, and each node is numbered. (c) A voltage, including polarity reference, is assigned between each node and the reference. (d) The voltage assignment is simplified by eliminating the polarity references; it is understood that each voltage is sensed positive relative to the reference node.

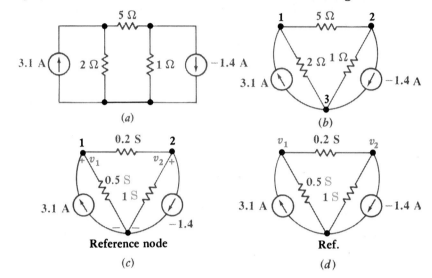

(a)

(b)

(c)

(d)

Example 2-1 We wish to obtain values for the unknown voltages across the various simple elements in Fig. 2-1a.

Solution: We may emphasize the locations of the three nodes by redrawing the circuit, as shown in Fig. 2-1b, where each node is identified by a number. We would now like to associate a voltage with each node, but we must remember that a voltage must be defined as existing between two nodes in a network. We thus select one node as a reference node, and then define a voltage between each remaining node and the reference node. Hence, we note again that there will be only $(N-1)$ voltages defined in an N-node circuit.

We choose node 3 as the reference node. Either of the other nodes *could* have been selected, but a little simplification in the resultant equations is obtained if the node to which the greatest number of branches is connected is identified as the reference node. If there is a ground node, it is usually most convenient to select it as the reference node. More often than not, the ground node appears as a common lead across the bottom of a circuit diagram.

The voltage of node 1 relative to the reference node 3 is defined as v_1, and v_2 is defined as the voltage of node 2 with respect to the reference node. These two voltages are sufficient, and the voltage between any other pair of nodes may be found in terms of them. For example, the voltage of node 1 with respect to node 2 is $(v_1 - v_2)$. The voltages v_1 and v_2 and their reference signs are shown in Fig. 2-1c. In this figure the resistance values have also been replaced with conductance values.

The circuit diagram is finally simplified in Fig. 2-1d by eliminating all voltage reference symbols. A reference node is plainly marked, and the voltage placed at each remaining node is understood to be the voltage of that node with respect to the reference node. This is the only situation for which we should use voltage symbols without the associated plus-minus sign pair, except for the battery symbol that was defined in Fig. 1-13b.

We must now apply Kirchhoff's current law to nodes 1 and 2. We do this by equating the total current leaving the node through the several conductances to the total source current entering the node. Thus,

$$0.5v_1 + 0.2(v_1 - v_2) = 3.1$$

or
$$0.7v_1 - 0.2v_2 = 3.1 \tag{1}$$

At node 2 we obtain

$$1v_2 + 0.2(v_2 - v_1) = 1.4$$

or
$$-0.2v_1 + 1.2v_2 = 1.4 \tag{2}$$

Equations (1) and (2) are the desired two equations in two unknowns, and they may be solved easily. The results are

$$v_1 = 5 \text{ V} \qquad v_2 = 2 \text{ V}$$

Also, the voltage of node 1 relative to node 2 is $(v_1 - v_2)$, or 3 V, and any current or power in the circuit may now be found in one step. For example, the current directed downward through the 0.5-S conductance is $0.5v_1$, or 2.5 A. ∎

Now let us increase the number of nodes by 1 and work a slightly more difficult problem.

Example 2-2   We shall find the three node voltages in the circuit of Fig. 2-2a, which is redrawn in Fig. 2-2b with the nodes identified, a convenient reference node chosen, and the node voltages specified.

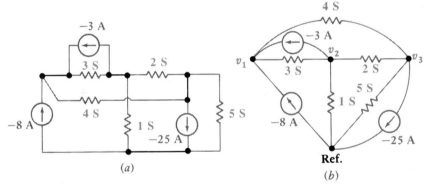

(a)

(b)

Figure 2-2

(*a*) A circuit containing four nodes and eight branches. (*b*) The same circuit redrawn with node voltages assigned.

Solution: We first sum the currents leaving node 1 in Fig. 2-2b:

$$3(v_1 - v_2) + 4(v_1 - v_3) - (-8) - (-3) = 0$$

or
$$7v_1 - 3v_2 - 4v_3 = -11 \tag{3}$$

At node 2:

$$3(v_2 - v_1) + 1v_2 + 2(v_2 - v_3) - 3 = 0$$

or
$$-3v_1 + 6v_2 - 2v_3 = 3 \tag{4}$$

and at node 3:

$$4(v_3 - v_1) + 2(v_3 - v_2) + 5v_3 - 25 = 0$$

or
$$-4v_1 - 2v_2 + 11v_3 = 25 \tag{5}$$

Equations (3) through (5) may be solved by a simple process of elimination of variables, or by Cramer's rule and determinants.[1] Using the latter method, we have

$$v_1 = \frac{\begin{vmatrix} -11 & -3 & -4 \\ 3 & 6 & -2 \\ 25 & -2 & 11 \end{vmatrix}}{\begin{vmatrix} 7 & -3 & -4 \\ -3 & 6 & -2 \\ -4 & -2 & 11 \end{vmatrix}}$$

Expanding the numerator and denominator determinants by minors along their first columns leads to

$$v_1 = \frac{-11\begin{vmatrix} 6 & -2 \\ -2 & 11 \end{vmatrix} - 3\begin{vmatrix} -3 & -4 \\ -2 & 11 \end{vmatrix} + 25\begin{vmatrix} -3 & -4 \\ 6 & -2 \end{vmatrix}}{7\begin{vmatrix} 6 & -2 \\ -2 & 11 \end{vmatrix} - (-3)\begin{vmatrix} -3 & -4 \\ -2 & 11 \end{vmatrix} + (-4)\begin{vmatrix} -3 & -4 \\ 6 & -2 \end{vmatrix}}$$

$$= \frac{-11(62) - 3(-41) + 25(30)}{7(62) + 3(-41) - 4(30)}$$

$$= \frac{-682 + 123 + 750}{434 - 123 - 120} = \frac{191}{191} = 1 \text{ V}$$

Similarly,

$$v_2 = \frac{\begin{vmatrix} 7 & -11 & -4 \\ -3 & 3 & -2 \\ -4 & 25 & 11 \end{vmatrix}}{191} = 2 \text{ V}$$

and

$$v_3 = \frac{\begin{vmatrix} 7 & -3 & -11 \\ -3 & 6 & 3 \\ -4 & -2 & 25 \end{vmatrix}}{191} = 3 \text{ V}$$

■

The denominator determinant is common to each of the preceding three evaluations. For circuits that do not contain either voltage sources or dependent sources (i.e., circuits containing only independent current sources), this denominator is the determinant of a matrix[2] that is defined as the *conductance matrix* of the circuit:

$$\mathbf{G} = \begin{bmatrix} 7 & -3 & -4 \\ -3 & 6 & -2 \\ -4 & -2 & 11 \end{bmatrix}$$

It should be noted that the nine elements of the matrix are the ordered array of the coefficients of Eqs. (3), (4), and (5), each of which is a conductance

---

[1] Appendix 1 provides a short review of determinants and the solution of a system of simultaneous linear equations by Cramer's rule.

[2] We shall not begin to manipulate matrices mathematically until Chap. 15; at that time a rudimentary knowledge of linear algebra is presumed.

value. The first row is composed of the coefficients of the KCL equation at the first node, the coefficients being given in the order of v_1, v_2, and v_3. The second row applies to the second node, and so on.

The conductance matrix is symmetrical about the major diagonal (upper left to lower right), and all elements not on this diagonal are negative, whereas all elements on it are positive. This is a general consequence of the systematic way in which we assigned variables, applied KCL, and ordered the equations, as well as of the reciprocity theorem, which we shall discuss in Chap. 15. For the present, we merely acknowledge the symmetry in those circuits that have only independent current sources and accept the check that it provides us in discovering errors we may have committed in writing circuit equations.

We still must see how voltage sources and dependent sources affect the strategy of nodal analysis. We now investigate the consequences of including a voltage source.

As a typical example, consider the circuit shown in Fig. 2-3. Our previous four-node circuit has been changed by replacing the 2-S conductance between nodes 2 and 3 by a 22-V voltage source. We still assign the same node-to-reference voltages v_1, v_2, and v_3. Previously, the next step was the application of KCL at each of the three nonreference nodes. If we try to do that once again, we see that we shall run into some difficulty at both nodes 2 and 3, for we do not know what the current is in the branch with the voltage source. There is no way by which we can express the current as a function of the voltage, for the definition of a voltage source is exactly that the voltage is *independent* of the current.

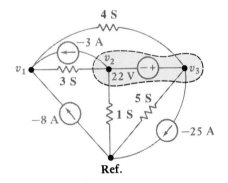

Figure 2-3

The 2-S conductance in the circuit of Fig. 2-2 is replaced by an independent voltage source. Kirchhoff's current law is used on the supernode enclosed by the broken line, and the source voltage is set equal to $v_3 - v_2$.

There are two ways out of these difficulties. The more difficult is to assign an unknown current to the branch which contains the voltage source, proceed to apply KCL three times, and then apply KVL once between nodes 2 and 3; the result is four equations in four unknowns for this example.

The easier method is to agree that we are primarily interested in the node voltages, so that we may avoid the current in the voltage-source branch that is causing our problems. We do this by treating node 2, node 3, and the voltage source together as a sort of supernode and applying KCL to both nodes at the same time. This is certainly possible, because, if the total current leaving node 2 is zero and the total current leaving node 3 is zero, then the total current leaving the totality of the two nodes is zero.

The supernode is indicated by the shaded region enclosed by the broken line in Fig. 2-3.

Example 2-3 Determine the value of the unknown node voltage v_1 in the circuit of Fig. 2-3.

Solution: We begin by setting the sum of the six currents leaving the supernode equal to zero. Beginning with the 3-S conductance branch and working clockwise, we have

$$3(v_2 - v_1) - 3 + 4(v_3 - v_1) - 25 + 5v_3 + 1v_2 = 0$$

or

$$-7v_1 + 4v_2 + 9v_3 = 28$$

The KCL equation at node 1 is unchanged from Eq. (3):

$$7v_1 - 3v_2 - 4v_3 = -11$$

We need one additional equation, since we have three unknowns, and it must utilize the fact that there is a 22-V voltage source between nodes 2 and 3,

$$v_3 - v_2 = 22$$

Rewriting these last three equations,

$$7v_1 - 3v_2 - 4v_3 = -11$$

$$-7v_1 + 4v_2 + 9v_3 = 28$$

$$-v_2 + v_3 = 22$$

the determinant solution for v_1 is

$$v_1 = \frac{\begin{vmatrix} -11 & -3 & -4 \\ 28 & 4 & 9 \\ 22 & -1 & 1 \end{vmatrix}}{\begin{vmatrix} 7 & -3 & -4 \\ -7 & 4 & 9 \\ 0 & -1 & 1 \end{vmatrix}} = \frac{-189}{42} = -4.5 \text{ V}$$

■

Note the lack of symmetry about the major diagonal in the denominator determinant as well as the fact that not all of the off-diagonal elements are negative. This is the result of the presence of the voltage source. Note also that it would not make much sense to call the denominator the determinant of the *conductance* matrix, because the bottom row comes from the equation $-v_2 + v_3 = 22$, and this equation does not depend on conductances in any way.

The presence of a voltage source thus reduces by 1 the number of nonreference nodes at which we must apply KCL, regardless of whether the voltage source extends between two nonreference nodes or is connected between a node and the reference.

Now let us consider a circuit containing a dependent source.

Example 2-4 Determine the value of the output voltage v_o in the circuit of Fig. 2-4*a*, in which an op-amp is connected as a voltage follower. We shall assume that the input voltage v_2 is 1 V.

Solution: This is the same circuit we investigated in the last section of Chap. 1, except that a finite load resistance, $R_L = 1$ kΩ, now appears between the output terminal and ground. We represent the op-amp by a

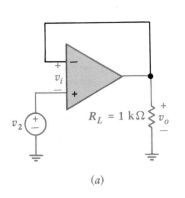

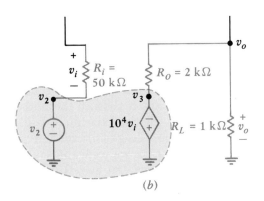

Figure 2-4

(*a*) A voltage follower feeds a finite load R_L. (*b*) The op-amp is replaced with an equivalent circuit that includes a noninfinite R_i and a nonzero R_o. Three node-to-reference voltages are assigned, and one supernode is indicated.

model that includes a noninfinite input resistance, $R_i = 50$ kΩ, and a nonzero output resistance, $R_o = 2$ kΩ, as shown in the circuit of Fig. 2-4*b*; we assume a typical value of $A = 10^4$.

Ground is selected as the reference node, and the three nonreference nodes are assigned the voltages v_2, v_o, and v_3. We note that the independent voltage source v_2 causes the v_2 node and the reference node to form a supernode; moreover, the dependent voltage source forces us to consider the v_3 node and the reference node as a supernode also. Thus, the v_2, v_3, and reference nodes form one large supernode, shown by the shaded, broken-line enclosure in Fig. 2-4*b*. Since the supernode includes the reference node, we will not write a KCL equation for it. The only KCL equation to be written is that at the v_o node. It is

$$\frac{v_o - v_2}{50\,000} + \frac{v_o - v_3}{2000} + \frac{v_o}{1000} = 0 \tag{6}$$

Since $v_2 = 1$ V, there are only two unknown node voltages, v_o and v_3, in Eq. (6), and no other (independent) KCL equations can be written. However, we must still express the voltage of each voltage source extending from node to node (and thus inside the broken-line enclosure) in terms of the node voltages, and we must also express the control of the dependent source (here, the voltage v_i) in terms of the node voltages.

First, we look at all the voltage sources inside the supernode. The source v_2 has been set equal to 1 V, and furthermore the node voltage itself was designated v_2. If we had been foolish enough to call it v_A, for example, then we should have to write the trivial equation $v_A = v_2$. Next is the source Av_i. Since it is connected between node 3 and ground, we have

$$v_3 = -10^4 v_i$$

Finally, we must relate the currents or voltages on which the controlled sources depend to the node voltages. Here, v_i is defined across R_i, and

$$v_i = v_o - v_2 = v_o - 1$$

To solve Eq. (6) for v_o, we let $v_3 = -10^4(v_o - 1)$, obtaining one equation in one unknown,

$$\frac{v_o - 1}{50\,000} + \frac{v_o + 10^4(v_o - 1)}{2000} + \frac{v_o}{1000} = 0$$

We find that $v_o = 0.999\,700$ V, and thus the output voltage closely equals the input, even for an op-amp that has relatively low gain, low input resistance, and high output resistance. ∎

In passing, we note that the negative of the first term in Eq. (6) is the current supplied by the 1-V source, here $(1 - v_o)/50\,000 = 6.00$ nA, an exceedingly small value that is not apt to disturb the most delicate of sources. In contrast, the output current is the third term in Eq. (6), $v_o/1000 = 1.000$ mA, over 10^5 times as large. Thus the voltage follower can deliver much more current and power to the load than it draws from the source. In doing so it is not violating the conservation of energy, but merely drawing power from the dc power supplies, which are usually not shown.

Let us summarize the method by which we may obtain a set of nodal equations for any resistive circuit:

1 Make a neat, simple circuit diagram. Indicate all element and source values. Each source should have its reference symbol.
2 Assuming that the circuit has N nodes, choose one of these nodes as a reference node. Then write the node voltages $v_1, v_2, \ldots, v_{N-1}$ at their respective nodes, remembering that each node voltage is understood to be measured with respect to the chosen reference.
3 If the circuit contains only current sources, apply Kirchhoff's current law at each nonreference node. To obtain the conductance matrix if a circuit has only independent current sources, equate the total current leaving each node through all conductances to the total source current entering that node, and order the terms from v_1 to v_{N-1}. For each dependent current source present, relate the source current and the controlling quantity to the variables $v_1, v_2, \ldots, v_{N-1}$, if they are not already in that form.
4 If the circuit contains voltage sources, form a supernode about each one by enclosing the source and its two terminals within a broken-line enclosure, thus reducing the number of nodes by 1 for each voltage source that is present. The assigned node voltages should not be changed. Using these assigned node-to-reference voltages, apply KCL at each of the nodes or supernodes (that do not contain the reference node) in this modified circuit. Relate each source voltage to the variables $v_1, v_2, \ldots, v_{N-1}$, if it is not already in that form.

With these suggestions in mind, let us consider the circuit displayed in Fig. 2-5, one which contains all four types of sources and has five nodes.

Example 2-5 Determine the values of the unknown node-to-reference voltages in the circuit of Fig. 2-5.

Solution: We select the central node as the reference, as shown, and assign v_1 to v_4 in a clockwise direction starting from the left node.

After establishing a supernode about each voltage source, we see that we need to write KCL equations only at node 2 and at the supernode containing both nodes 3 and 4 and the dependent voltage source. No extra equation need be written for the supernode which contains node 1 and the independent voltage source; it is obvious that $v_1 = -12$ V.

At node 2,

$$\frac{v_2 - v_1}{0.5} + \frac{v_2 - v_3}{2} = 14$$

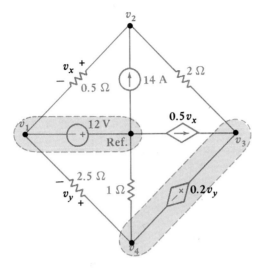

Figure 2-5

A five-node circuit containing all of the four different types of sources.

while at the 3-4 supernode,

$$\frac{v_3 - v_2}{2} - 0.5v_x + \frac{v_4}{1} + \frac{v_4 - v_1}{2.5} = 0$$

We next relate the source voltages to the node voltages:

$$v_1 = -12$$

$$v_3 - v_4 = 0.2v_y = 0.2(v_4 - v_1)$$

And finally we express the dependent current source in terms of the assigned variables:

$$0.5v_x = 0.5(v_2 - v_1)$$

Thus, we obtain a set of four equations in the four node voltages:

$$-2v_1 + 2.5v_2 - 0.5v_3 = 14$$

$$v_1 = -12$$

$$0.1v_1 - v_2 + 0.5v_3 + 1.4v_4 = 0$$

$$0.2v_1 + v_3 - 1.2v_4 = 0$$

to which the solutions are

$$v_1 = -12 \text{ V}$$

$$v_2 = -4 \text{ V}$$

$$v_3 = 0$$

$$v_4 = -2 \text{ V}$$ ■

2-1. Use nodal analysis to find v_x in the circuit of Fig. 2-6 if element A is (a) a 25-Ω resistor; (b) a 5-A current source, arrow pointing right; (c) a 10-V voltage source, + reference on the right; (d) a short circuit.

Drill Problems

Ans: 10.91 V; −10 V; 16.67 V; −13.33 V

Figure 2-6

See Drill Probs. 2-1 and 2-2.

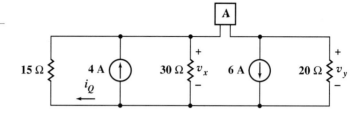

2-2. Use nodal analysis to find v_y in the circuit of Fig. 2-6 if element A is (a) a dependent current source, $1.5i_Q$, arrow pointing left; (b) a dependent voltage source, $0.5v_y$, positive reference on the right; (c) an open circuit; (d) a 10-Ω resistor in series with a 10-V voltage source, positive reference on the right.

Ans: -80 V; -20 V; -120 V; -35 V

2-3

Mesh analysis

The technique of nodal analysis described in the preceding section is completely general and can always be applied to any electrical network. This is not the only method for which a similar claim can be made, however. In particular, we shall meet a generalized nodal analysis method and a technique known as *loop analysis* in the concluding sections of this chapter.

First, however, let us consider a method known as *mesh analysis*. Even though this technique is not applicable to every network, it can be applied to most of the networks we shall need to analyze, and it is probably used more often than it should be; other methods are often simpler. Mesh analysis is applicable only to those networks which are planar, a term we hasten to define.

If it is possible to draw the diagram of a circuit on a plane surface in such a way that no branch passes over or under any other branch, then that circuit is said to be a *planar circuit*. Thus, Fig. 2-7a shows a planar network, Fig. 2-7b shows a nonplanar network, and Fig. 2-7c also shows a planar network, although it is drawn in such a way as to make it *appear* nonplanar at first glance.

Figure 2-7

(a) A planar network can be drawn on a plane surface without crossovers. (b) A nonplanar network cannot be drawn on a plane surface without at least one crossover. (c) A planar network can be drawn so that it may look nonplanar.

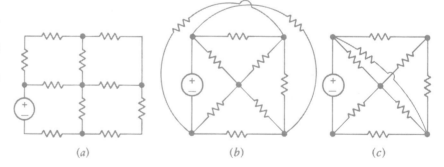

(a) (b) (c)

In the first chapter, the terms *path, closed path,* and *loop* were defined. Before we define a mesh, let us consider the sets of branches drawn with heavy lines in Fig. 2-8. The first set of branches is not a path, since four branches are connected to the center node, and it is of course also not a loop. The second set of branches does not constitute a path, since it is traversed only by passing through the central node twice. The remaining four paths are all loops. The circuit contains 11 branches.

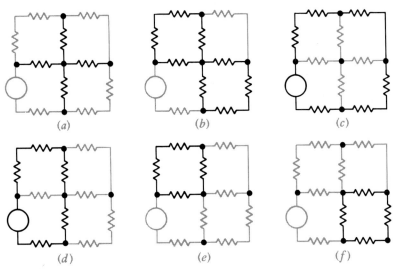

(a)   (b)   (c)

(d)   (e)   (f)

Figure 2-8

(a) The set of branches identified by the heavy lines is neither a path nor a loop. (b) The set of branches here is not a path, since it can be traversed only by passing through the central node twice. (c) This path is a loop but not a mesh, since it encloses other loops. (d) This path is also a loop but not a mesh. (e, f) Each of these paths is both a loop and a mesh.

The mesh is a property of a planar circuit and is not defined for a nonplanar circuit. We define a *mesh* as a loop which does not contain any other loops within it. Thus, the loops indicated in Figs. 2-8c and d are not meshes, whereas those of parts e and f are meshes. Once a circuit has been drawn neatly in planar form, it often has the appearance of a multipaned window; the boundary of each pane in the window may be considered to be a mesh.

If a network is planar, mesh analysis can be used to accomplish its analysis. This technique involves the concept of a *mesh current,* which we shall introduce by considering the analysis of the two-mesh circuit of Fig. 2-9.

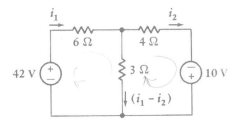

Figure 2-9

Two currents, i_1 and i_2, are assumed in a two-mesh circuit.

Example 2-6   Determine the values of the two unknown currents i_1 and i_2 in the circuit of Fig. 2-9.

Solution: As we did in the single-loop circuit, we shall begin by assuming a current through one of the branches. Let us call the current flowing to the right through the 6-Ω resistor i_1. We intend to apply Kirchhoff's voltage law around each of the two meshes, and the resulting two equations are sufficient to determine two unknown currents. Therefore we select a second current i_2 flowing to the right in the 4-Ω resistor. We might also choose to call the current flowing downward through the central branch i_3, but it is evident from Kirchhoff's current law that i_3 may be expressed in terms of the two previously assumed currents as $(i_1 - i_2)$. The assumed currents are shown in Fig. 2-9.

Following the method of solution for the single-loop circuit, we now apply Kirchhoff's voltage law to the left-hand mesh,

$$-42 + 6i_1 + 3(i_1 - i_2) = 0$$

or
$$9i_1 - 3i_2 = 42 \qquad (7)$$

and then to the right-hand mesh,

$$-3(i_1 - i_2) + 4i_2 - 10 = 0$$

or
$$-3i_1 + 7i_2 = 10 \qquad (8)$$

Equations (7) and (8) are independent equations; one cannot be derived from the other.[3] There are two equations and two unknowns, and the solution is easily obtained: i_1 is 6 A, i_2 is 4 A, and $(i_1 - i_2)$ is therefore 2 A. The voltage and power relationships may be quickly obtained if desired.

∎

If our circuit had contained M meshes, then we should have had to assume M branch currents and write M independent equations.[4] The solution in general may be systematically obtained through the use of determinants.

Now let us consider this same problem in a slightly different manner by using mesh currents. We define a *mesh current* as a current which flows only around the perimeter of a mesh.

Example 2-7   Repeat the problem of Example 2-6, but now use the technique of mesh analysis to determine the values of the two unknown currents i_1 and i_2 in the circuit of Fig. 2-10.

Figure 2-10

A clockwise mesh current is assigned to each mesh of a planar circuit.

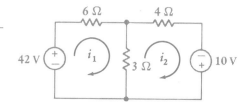

Solution: If we label the left-hand mesh of our problem as mesh 1, then we may establish a mesh current i_1 flowing in a clockwise direction about this mesh. A mesh current is indicated by a curved arrow that almost closes on itself and is drawn inside the appropriate mesh, as shown in Fig. 2-10. The mesh current i_2 is established in the remaining mesh, again in a clockwise direction. Although the directions are arbitrary, we shall always choose clockwise mesh currents because a certain error-minimizing symmetry then results in the equations.

We no longer have a current or current arrow shown directly on each branch in the circuit. The current through any branch must be determined by considering the mesh currents flowing in every mesh in which that branch appears. This is not difficult, because it is obvious that no branch can appear in more than two meshes. For example, the 3-Ω resistor appears in both meshes, and the current flowing downward through it is $(i_1 - i_2)$.

---

[3] It will be shown in Sec. 2-8 that mesh equations are always independent.
[4] The proof of this statement will be found in Sec. 2-8.

The 6-Ω resistor appears only in mesh 1, and the current flowing to the right in that branch is equal to the mesh current i_1. A mesh current may often be identified as a branch current, as i_1 and i_2 have been identified in this example. This is not always true, however, for consideration of a square nine-mesh network soon shows that the central mesh current cannot be identified as the current in *any* branch.

One of the greatest advantages in the use of mesh currents is the fact that Kirchhoff's current law is automatically satisfied. If a mesh current flows into a given node, it obviously flows out of it also.

We therefore may turn our attention to the application of KVL to each mesh. For the left-hand mesh,

$$-42 + 6i_1 + 3(i_1 - i_2) = 0$$

while for the right-hand mesh,

$$3(i_2 - i_1) + 4i_2 - 10 = 0$$

and these two equations are equivalent to Eqs. (7) and (8). ∎

Let us next consider the five-node, seven-branch, three-mesh circuit shown in Fig. 2-11. This is a slightly more complicated problem because of the additional mesh.

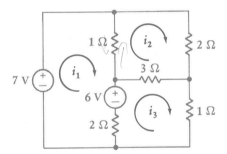

Figure 2-11

Mesh currents i_1, i_2, and i_3 are assumed in a five-node, seven-branch, three-mesh circuit.

Example 2-8   Use mesh analysis to determine the three unknown mesh currents in the circuit of Fig. 2-11.

Solution: The three required mesh currents are assigned as indicated in the figure, and we methodically apply KVL about each mesh:

$$-7 + 1(i_1 - i_2) + 6 + 2(i_1 - i_3) = 0$$
$$1(i_2 - i_1) + 2i_2 + 3(i_2 - i_3) = 0$$
$$2(i_3 - i_1) - 6 + 3(i_3 - i_2) + 1i_3 = 0$$

Simplifying,

$$3i_1 - i_2 - 2i_3 = 1$$
$$-i_1 + 6i_2 - 3i_3 = 0$$
$$-2i_1 - 3i_2 + 6i_3 = 6$$

and Cramer's rule leads to the formulation for i_3:

$$i_3 = \frac{\begin{vmatrix} 3 & -1 & 1 \\ -1 & 6 & 0 \\ -2 & -3 & 6 \end{vmatrix}}{\begin{vmatrix} 3 & -1 & -2 \\ -1 & 6 & -3 \\ -2 & -3 & 6 \end{vmatrix}} = \frac{117}{39} = 3 \text{ A}$$

The other mesh currents are $i_1 = 3$ A and $i_2 = 2$ A. ∎

Again we notice that we have a denominator determinant that is symmetrical about the major diagonal and has positive terms on the major diagonal and zero or negative terms off it. This occurs for circuits that contain only independent voltage sources when clockwise mesh currents are assigned; in such circuits the elements appearing in the first row of the determinant are the ordered coefficients of $i_1, i_2, \ldots, i_M$ in the KVL equation about the first mesh, the second row corresponds to the second mesh, and so on. This symmetrical array appearing in the denominator is the determinant of the *resistance matrix* of the network

$$\mathbf{R} = \begin{bmatrix} 3 & -1 & -2 \\ -1 & 6 & -3 \\ -2 & -3 & 6 \end{bmatrix}$$

How must we modify this straightforward procedure when a current source is present in the network? Taking our lead from nodal analysis (and duality), we should feel that there are two possible methods. First, we could assign an unknown voltage across the current source, apply KVL around each mesh as before, and then relate the source current to the assigned mesh currents. This is generally the more difficult approach.

A better technique is one that is quite similar to the supernode approach in nodal analysis. There we formed a supernode, completely enclosing the voltage source inside the supernode and reducing the number of nonreference nodes by 1 for each voltage source. Now we create a kind of "supermesh" from two meshes that have a current source as a common element; the current source is in the interior of the supermesh. We thus reduce the number of meshes by 1 for each current source present. If the current source lies on the perimeter of the circuit, then the single mesh in which it is found is ignored.[5] Kirchhoff's voltage law is applied only to those meshes or supermeshes in the modified network. Now let us consider an example using this procedure, applying it to the circuit of Fig. 2-12.

Example 2-9   Use the technique of mesh analysis to evaluate the three unknown mesh currents in Fig. 2-12.

Solution: Here we note that a 7-A independent current source is in the common boundary of two meshes. Mesh currents i_1, i_2, and i_3 are assigned, and ~~the current source causes us to create a supermesh whose interior is~~

---

[5] Such a current source is a common element with the mesh that "encloses" the outside of the entire circuit. Just as we do not write a nodal equation at the reference node, we do not write a KVL equation for this external mesh.

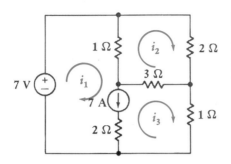

Figure 2-12

Mesh analysis is applied to this circuit containing a current source by writing the KVL equation about the loop: 7 V, 1 Ω, 3 Ω, 1 Ω.

and the current source causes us to create a supermesh whose interior is that of meshes 1 and 3. Applying KVL about this loop,

$$-7 + 1(i_1 - i_2) + 3(i_3 - i_2) + 1i_3 = 0$$

or

$$i_1 - 4i_2 + 4i_3 = 7 \qquad (9)$$

and around mesh 2,

$$1(i_2 - i_1) + 2i_2 + 3(i_2 - i_3) = 0$$

or

$$-i_1 + 6i_2 - 3i_3 = 0 \qquad (10)$$

Finally, the independent-source current is related to the assumed mesh currents,

$$i_1 - i_3 = 7 \qquad (11)$$

Solving Eqs. (9) through (11), we have

$$i_3 = \frac{\begin{vmatrix} -1 & 6 & 0 \\ 1 & -4 & 7 \\ 1 & 0 & 7 \end{vmatrix}}{\begin{vmatrix} -1 & 6 & -3 \\ 1 & -4 & 4 \\ 1 & 0 & -1 \end{vmatrix}} = \frac{28}{14} = 2 \text{ A}$$

We may also find that $i_1 = 9$ A and $i_2 = 2.5$ A. ∎

The presence of one or more dependent sources merely requires each of these source quantities and the variable on which it depends to be expressed in terms of the assigned mesh currents. In Fig. 2-13, for example, we note that both a dependent and an independent current source are included in the network. Let us see how their presence affects the analysis of the circuit and actually simplifies it.

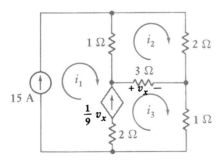

Figure 2-13

The presence of two current sources in this three-mesh circuit makes it necessary to apply KVL only once, around mesh 2.

Example 2-10 Use mesh analysis to evaluate the three unknown currents in the circuit of Fig. 2-13.

Solution: We first assign three mesh currents and then apply KVL around mesh 2:

$$1(i_2 - i_1) + 2i_2 + 3(i_2 - i_3) = 0$$

The current sources appear in meshes 1 and 3. Since the 15-A source is located on the perimeter of the circuit, we may eliminate mesh 1 from consideration. Then, the dependent current source is on the perimeter of the modified network, and thus we avoid any equation writing for mesh 3. Only mesh 2 remains, and we already have an equation for it. We therefore turn our attention to the source quantities, obtaining

$$i_1 = 15$$

and

$$\tfrac{1}{9} v_x = i_3 - i_1 = \tfrac{1}{9} [3(i_3 - i_2)]$$

Thus,

$$-i_1 + 6i_2 - 3i_3 = 0$$

$$i_1 = 15$$

$$-i_1 + \tfrac{1}{3} i_2 + \tfrac{2}{3} i_3 = 0$$

from which we have $i_1 = 15$, $i_2 = 11$, and $i_3 = 17$ A. We might note that we wasted a little time in assigning a mesh current i_1 to the left mesh; we should simply have indicated a mesh current and labeled it 15 A. ■

Let us summarize the method by which we may obtain a set of mesh equations for a resistive circuit:

1 Make certain that the network is a planar network. If it is nonplanar, mesh analysis is not applicable.
2 Make a neat, simple circuit diagram. Indicate all element and source values. Resistance values are preferable to conductance values. Each source should have its reference symbol.
3 Assuming that the circuit has M meshes, assign a clockwise mesh current in each mesh, $i_1, i_2, \ldots, i_M$.
4 If the circuit contains only voltage sources, apply Kirchhoff's voltage law around each mesh. To obtain the resistance matrix if a circuit has only independent voltage sources, equate the clockwise sum of all the resistor voltages to the counterclockwise sum of all the source voltages, and order the terms, from i_1 to i_M. For each dependent voltage source present, relate the source voltage and the controlling quantity to the variables $i_1, i_2, \ldots, i_M$, if they are not already in that form.
5 If the circuit contains current sources, create a supermesh for each current source that is common to two meshes by applying KVL around the larger loop formed by the branches that are not common to the two meshes; KVL need not be applied to a mesh containing a current source that lies on the perimeter of the entire circuit. The assigned mesh currents should not be changed. Relate each source current to the variables $i_1, i_2, \ldots, i_M$, if it is not already in that form.

2-3. Use mesh analysis to find i_1 in the circuit of Fig. 2-14 if element A is (a) an open circuit; (b) a 5-A independent current source, arrow directed to the right; (c) a 5-Ω resistor.   *Ans:* 3.00 A; 1.621 A; 3.76 A

2-4. Use mesh analysis to find v_3 in the circuit of Fig. 2-14 if element A is (a) a short circuit; (b) a 20-V independent voltage source, positive reference on the right; (c) a dependent voltage source, positive reference on the right, labeled $15i_1$.   *Ans:* 69.5 V; 73.7 V; 79.2 V

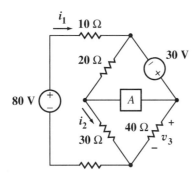

Figure 2-14

See Drill Probs. 2-3 and 2-4.

All the circuits which we have analyzed up to now (and which we shall analyze later) are linear circuits. At this time we must be more specific in defining a linear circuit. Having done this, we can then consider the most important consequence of linearity, the principle of superposition. This principle is very basic and will appear repeatedly in our study of linear circuit analysis. As a matter of fact, the nonapplicability of superposition to nonlinear circuits is the reason they are so difficult to analyze.

Linearity and super-position

The principle of superposition states that the response (a desired current or voltage) at any point in a linear circuit having more than one independent source can be obtained as the sum of the responses caused by the separate independent sources acting alone. In the following discussion, we shall investigate the meaning of "linear" and "acting alone." We shall also take note of a slightly broader form of the theorem.

Let us first define a *linear element* as a passive element that has a linear voltage-current relationship. By a "linear voltage-current relationship" we shall mean simply that multiplication of the time-varying current through the element by a constant K results in the multiplication of the time-varying voltage across the element by the same constant K. At this time, only one passive element has been defined, the resistor, and its voltage-current relationship

$$v(t) = Ri(t)$$

is obviously linear. As a matter of fact, if $v(t)$ is plotted as a function of $i(t)$, the result is a straight *line*. We shall see in Chap. 3 that the defining voltage-current equations for inductance and capacitance are also linear relationships, as is the defining equation for mutual inductance presented in Chap. 14.

We must also define a *linear dependent source* as a dependent current or voltage source whose output current or voltage is proportional only to the first power of some current or voltage variable in the circuit or to the sum of such quantities. That is, a dependent voltage source $v_s = 0.6i_1 - 14v_2$ is linear, but $v_s = 0.6i_1^2$ and $v_s = 0.6i_1v_2$ are not.

We may now define a *linear circuit* as a circuit composed entirely of independent sources, linear dependent sources, and linear elements. From this definition, it is possible to show[6] that "the response is proportional to the source," or that multiplication of all *independent* source voltages and currents by a constant K increases all the current and voltage responses by the same factor K (including the dependent source voltage or current outputs).

The most important consequence of linearity is superposition. Let us develop the superposition principle by considering first the circuit of Fig. 2-15,

Figure 2-15

A three-node circuit, containing two forcing functions, used to illustrate the superposition principle.

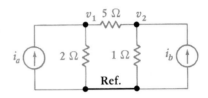

which contains two independent sources, the current generators which force the currents i_a and i_b into the circuit. Sources are often called *forcing functions* for this reason, and the voltages which they produce between node 1 or 2 and the reference node may be termed *response functions,* or simply *responses.* Both the forcing functions and the responses may be functions of time. The two nodal equations for this circuit are

$$0.7v_1 - 0.2v_2 = i_a \tag{12}$$

$$-0.2v_1 + 1.2v_2 = i_b \tag{13}$$

Now let us perform experiment x. We change the two forcing functions to i_{ax} and i_{bx}; the two unknown voltages will now be different, and we shall let them be v_{1x} and v_{2x}. Thus,

$$0.7v_{1x} - 0.2v_{2x} = i_{ax} \tag{14}$$

$$-0.2v_{1x} + 1.2v_{2x} = i_{bx} \tag{15}$$

We next perform experiment y by changing the source currents to i_{ay} and i_{by} and by letting the responses be v_{1y} and v_{2y}:

$$0.7v_{1y} - 0.2v_{2y} = i_{ay} \tag{16}$$

$$-0.2v_{1y} + 1.2v_{2y} = i_{by} \tag{17}$$

These three sets of equations describe the same circuit with different source currents. Let us *add* or *superpose* the last two sets of equations. Adding Eqs.

---

[6] The proof involves first showing that the use of nodal analysis on the linear circuit can produce only linear equations of the form:

$$a_1v_1 + a_2v_2 + \cdots + a_Nv_N = b$$

where the a_i are constants (combinations of resistance or conductance values, constants appearing in dependent source expressions, 0, or ± 1), the v_i are the unknown node voltages (responses), and b is an independent source value or a sum of independent source values. Given a set of such equations, if we multiply all the b's by K, then it is evident that the solution of this new set of equations will be the node voltages $Kv_1, Kv_2, \ldots, Kv_N$.

(14) and (16),

$$(0.7v_{1x} + 0.7v_{1y}) - (0.2v_{2x} + 0.2v_{2y}) = i_{ax} + i_{ay} \qquad (18)$$

$$0.7v_1 \quad - \quad 0.2v_2 \quad = \quad i_a \qquad (12)$$

and adding Eqs. (15) and (17),

$$-(0.2v_{1x} + 0.2v_{1y}) + (1.2v_{2x} + 1.2v_{2y}) = i_{bx} + i_{by} \qquad (19)$$

$$-0.2v_1 \quad + \quad 1.2v_2 \quad = \quad i_b \qquad (13)$$

where Eq. (12) has been written immediately below Eq. (18) and Eq. (13) below Eq. (19) for easy comparison.

The linearity of all these equations allows us to compare Eq. (18) with Eq. (12) and Eq. (19) with Eq. (13) and draw an interesting conclusion. If we select i_{ax} and i_{ay} such that their sum is i_a and select i_{bx} and i_{by} such that their sum is i_b, then the desired responses v_1 and v_2 may be found by *adding* v_{1x} to v_{1y} and v_{2x} to v_{2y}, respectively. In other words, we may perform experiment x and note the responses, perform experiment y and note the responses, and finally add the corresponding responses. These are the responses of the original circuit to independent sources which are the sums of the independent sources used in experiments x and y. This is the fundamental concept involved in the superposition principle.

It is evident that we may extend these results by breaking up either source current into as many pieces as we wish; there is no reason why we cannot perform experiments z and q also. It is necessary only that the algebraic sum of the pieces be equal to the original current.

The *superposition theorem* usually appears in a form similar to the following:

> In any linear resistive network containing several sources, the voltage across or the current through any resistor or source may be calculated by adding algebraically all the individual voltages or currents caused by the separate independent sources acting alone, with all other independent voltage sources replaced by short circuits and all other independent current sources replaced by open circuits.

Thus if there are N independent sources, we perform N experiments. Each independent source is active in only one experiment, and only one independent source is active in each experiment. An inactive independent voltage source is identical with a short circuit, and an inactive independent current source is an open circuit. Note that *dependent* sources are in general active in *every* experiment.

The circuit we have just used as an example, however, should indicate that a much stronger theorem might be written; a group of independent sources may be made active and inactive collectively, if we wish. For example, suppose there are three independent sources. The theorem states that we may find a given response by considering each of the three sources acting alone and adding the three results. Alternatively, we may find the response due to the first and second sources operating with the third inactive, and then add to this the response caused by the third source acting alone. This amounts to treating several sources collectively as a sort of supersource.

There is also no reason that an independent source must assume only its given value or a zero value in the several experiments; it is necessary only for the sum of the several values to be equal to the original value. An inactive source almost always leads to the simplest circuit, however.

Let us illustrate the application of the superposition principle by considering an example in which both types of independent source are present.

Example 2-11   For the circuit of Fig. 2-16, let us use superposition to write an expression for the unknown branch current i_x.

Figure 2-16

A circuit, containing both an independent current and an independent voltage source, which is easily analyzed by the superposition principle.

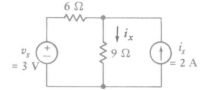

Solution: We may first set the current source equal to zero and obtain the portion of i_x due to the voltage source as 0.2 A. Then if we let the voltage source be zero and apply current division, the remaining portion of i_x is seen to be 0.8 A. We might write the answer in detail as

$$i_x = i_x \big|_{i_s=0} + i_x \big|_{v_s=0}$$

$$= \frac{3}{6+9} + 2\left(\frac{6}{6+9}\right) = 0.2 + 0.8 = 1.0 \text{ A} \qquad \blacksquare$$

As an example of the application of the superposition principle to a circuit containing a dependent source, consider Fig. 2-17.

Example 2-12   In the circuit of Fig. 2-17, use the superposition principle to determine the value of i_x.

Figure 2-17

Superposition may be used to analyze this circuit by first replacing the 3-A source by an open circuit and then replacing the 10-V source by a short circuit. The dependent voltage source is always active (unless $i_x = 0$).

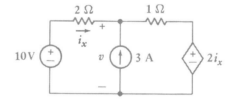

Solution: We first open-circuit the 3-A source. The single mesh equation is

$$-10 + 2i'_x + 1i'_x + 2i'_x = 0$$

so that

$$i'_x = 2 \text{ A}$$

Next, we short-circuit the 10-V source and write the single-node equation

$$\frac{v''}{2} + \frac{v'' - 2i''_x}{1} = 3$$

and relate the dependent-source-controlling quantity to v'':

$$v'' = -2i_x''$$

We find

$$i_x'' = -0.6\,\text{A}$$

and, thus,

$$i_x = i_x' + i_x'' = 2 - 0.6 = 1.4\,\text{A} \qquad \blacksquare$$

It usually turns out that *little if any time is saved in analyzing a circuit containing one or more dependent sources by use of the superposition principle,* for there must always be at least two sources in operation: one independent source and all the dependent sources.

We must constantly be aware of the limitations of superposition. It is applicable only to linear responses, and thus the most common nonlinear response—power—is not subject to superposition. For example, consider two 1-V batteries in series with a 1-Ω resistor. The power delivered to the resistor is obviously 4 W, but if we mistakenly try to apply superposition we might say that each battery alone furnished 1 W and thus the total power is 2 W. This is incorrect.

2-5. Use superposition to evaluate v_x in each of the circuits shown in Fig. 2-18.

Ans: 0 V; 96 V; −38.5 V

Drill Problem

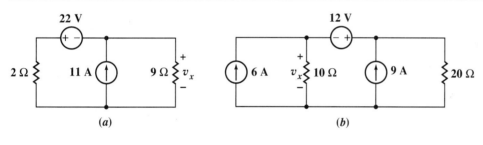

(a)　　　　　　　　　(b)

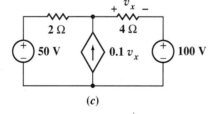

(c)

Figure 2-18

See Drill Prob. 2-5.

In all our previous work we have been making continual use of *ideal* voltage and current sources; it is now time to take a step closer to reality by considering *practical* sources. These sources will enable us to make more realistic representations of physical devices. Having defined the practical sources, we shall then study methods whereby practical current and voltage sources may be interchanged without affecting the remainder of the circuit. Such sources will be called *equivalent* sources. Our methods will be applicable for both independent and dependent sources, although we shall see later that they are not as useful for dependent sources.

2-5

▮▮▮▮▮▮▮▮▮▮

Source trans- formations

The ideal voltage source was defined as a device whose terminal voltage is independent of the current through it. A 1-V dc source produces a current of 1 A through a 1-Ω resistor and a current of 1 000 000 A through a 1-$\mu\Omega$ resistor; it may provide an unlimited amount of power. No such device exists practically, of course, and we agreed that a real physical voltage source might be represented by an ideal voltage source only as long as relatively small currents, or powers, were drawn from it. For example, an automobile storage battery may be approximated by an ideal dc voltage source if its current is limited to a few amperes. However, anyone who has ever tried to start an automobile with the headlights on must have observed that the lights dimmed perceptibly when the battery was asked to deliver the heavy starter current, 100 A or more, in addition to the headlight current. Under these conditions, an ideal voltage source may be a very poor representation of the storage battery.

To approximate a real device, the ideal voltage source must be modified to account for the lowering of its terminal voltage when large currents are drawn from it. Let us suppose that we observe experimentally that a storage battery has a terminal voltage of 12 V when no current is flowing through it and a reduced voltage of 11 V when 100 A is flowing. Thus, a more accurate model might be an ideal voltage source of 12 V in series with a resistor across which 1 V appears when 100 A flows through it. The resistor must be 0.01 Ω, and the ideal voltage source and this series resistor constitute a *practical voltage source*. Thus, we are using the series combination of two ideal circuit elements, an independent voltage source and a resistor, to model a real device.

We should not expect to find such an arrangement of ideal elements inside our storage battery, of course. Any real device is characterized by a certain current-voltage relationship at its terminals, and our problem is to develop some combination of ideal elements that can furnish a similar current-voltage characteristic, at least over some useful range of current, voltage, or power.

This particular practical voltage source is shown connected to a general load resistor in Fig. 2-19a. The terminal voltage of the practical source is the same as the voltage across R_L and is marked v_L.

Figure 2-19

(a) A practical source, which approximates the behavior of a certain 12-V storage battery, is shown connected to a load resistor R_L. (b) The relationship between i_L and v_L is linear.

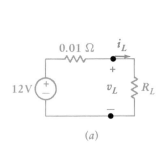

(a)

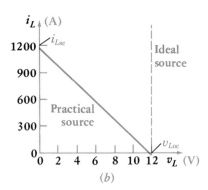

(b)

Figure 2-19b shows a plot of load current i_L as a function of the load voltage v_L for this practical source. The KVL equation for the circuit of Fig. 2-19a may be written in terms of i_L and v_L:

$$12 = 0.01i_L + v_L$$

and thus

$$i_L = 1200 - 100v_L$$

This is a linear equation in i_L and v_L, and the plot in Fig. 2-19b is a straight line. Each point on the line corresponds to a different value of R_L.

For example, the midpoint of the straight line is obtained when the load resistance is equal to the internal resistance of the practical source, or $R_L = 0.01\ \Omega$. Here, the load voltage is only one-half the ideal source voltage.

When $R_L = \infty$ and no current whatsoever is being drawn by the load, the practical source is open-circuited and the terminal voltage, or open-circuit voltage, is $v_{Loc} = 12$ V. If, on the other hand, $R_L = 0$, thereby short-circuiting the load terminals, then a load current or short-circuit current, $i_{Lsc} = 1200$ A, would flow. In practice, such an experiment would probably result in the destruction of the short circuit, the battery, and any measuring instruments incorporated in the circuit.

Since the plot of i_L versus v_L is a straight line for this practical voltage source, we should note that the values of v_{Loc} and i_{Lsc} uniquely determine the entire i_L-v_L curve.

The vertical broken line shows the i_L-v_L plot for an ideal voltage source; the terminal voltage remains constant for any value of load current. For the practical voltage source, the terminal voltage has a value near that of the ideal source only when the load current is relatively small.

Let us now consider a general practical voltage source, as shown in Fig. 2-20a. The voltage of the ideal source is v_s, and a resistance R_{sv}, called an *internal resistance* or *output resistance*, is placed in series with it. Again, we must note that the resistor is not one which is really present, or which we would wire or solder into the circuit, but merely serves to account for a terminal voltage that decreases as the load current increases. Its presence enables us to model the behavior of a physical voltage source more closely.

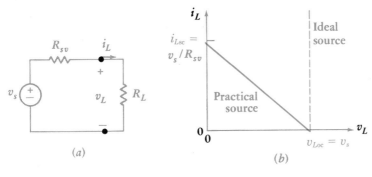

Figure 2-20

(a) A general practical voltage source connected to a load resistor R_L. (b) The terminal voltage decreases as i_L increases and $R_L = v_L/i_L$ decreases.

The linear relationship between v_L and i_L is

$$v_L = v_s - R_{sv} i_L \qquad (20)$$

and this is plotted in Fig. 2-20b. The open-circuit voltage and short-circuit current are

$$v_{Loc} = v_s \qquad (21)$$

$$i_{Lsc} = \frac{v_s}{R_{sv}} \qquad (22)$$

Once again, these values are the intercepts for the straight line in Fig. 2-20b, and they serve to define it completely.

An ideal current source is also nonexistent in the real world; there is no physical device which will deliver a constant current regardless of the load resistance to which it is connected or the voltage across its terminals. Certain transistor circuits will deliver a constant current to a wide range of load resistances, but the load resistance can always be made sufficiently large that the current through it becomes very small. Infinite power is simply never available.

A practical current source is defined as an ideal current source in parallel with an internal resistance R_{si}. Such a source is shown in Fig. 2-21a, and the current i_L and voltage v_L associated with a load resistance R_L are indicated. It is apparent that

$$i_L = i_s - \frac{v_L}{R_{si}} \tag{23}$$

which is again a linear relationship. The open-circuit voltage and the short-circuit current are

$$v_{Loc} = R_{si}\, i_s \tag{24}$$

$$i_{Lsc} = i_s \tag{25}$$

Figure 2-21

(a) A general practical current source connected to a load resistor R_L. (b) The load current provided by the practical current source is shown as a function of the load voltage.

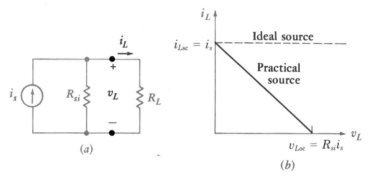

(a)

(b)

The variation of load current with changing load voltage may be investigated by changing the value of R_L as shown in Fig. 2-21b. The straight line is traversed from the short-circuit, or "northwest," end to the open-circuit termination at the "southeast" by increasing R_L from zero to infinite ohms. The midpoint occurs for $R_L = R_{si}$. It is evident that the load current i_L and the ideal source current i_s are approximately equal only for small values of load voltage, obtained with values of R_L that are small compared with R_{si}.

Having defined both practical sources, we are now ready to discuss their equivalence. We shall define two sources as being *equivalent* if they produce identical values of v_L and i_L when they are connected to identical values of R_L, no matter what the value of R_L may be. Since $R_L = 0$ and $R_L = \infty$ are two such values, equivalent sources provide the same open-circuit voltage and short-circuit current. In other words, if we are given two equivalent sources, one a practical voltage source and the other a practical current source, each enclosed in a black box with only a single pair of terminals on it, then there is no way in which we can differentiate between the boxes by measuring current or voltage in a resistive load.

The conditions for equivalence are now quickly established. Since the open-circuit voltages must be equal, from Eqs. (21) and (24) we have

$$v_{Loc} = v_s = R_{si}\, i_s \tag{26}$$

The short-circuit currents are also equal, and Eqs. (22) and (25) lead to

$$i_{Lsc} = \frac{v_s}{R_{sv}} = R_s$$

It follows that

$$R_{sv} = R_{si} = R_s \qquad (27)$$

and

$$v_s = R_s i_s \qquad (28)$$

where we now let R_s represent the internal resistance of either practical source.

To illustrate the use of these ideas, consider the practical current source shown in Fig. 2-22a. Since its internal resistance is 2 Ω, the internal resistance of the equivalent practical voltage source is also 2 Ω; the voltage of the ideal voltage source contained within the practical voltage source is (2)(3) = 6 V. The equivalent practical voltage source is shown in Fig. 2-22b.

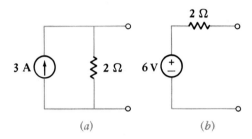

Figure 2-22

(a) A given practical current source. (b) The equivalent practical voltage source: $R_{sv} = R_{si} = R_s$; $v_s = R_s i_s$.

(a)   (b)

To check the equivalence, let us visualize a 4-Ω resistor connected to each source. In both cases a current of 1 A, a voltage of 4 V, and a power of 4 W are associated with the 4-Ω load. However, we should note very carefully that the ideal current source is delivering a total power of 12 W, while the ideal voltage source is delivering only 6 W. Furthermore, the internal resistance of the practical current source is absorbing 8 W, whereas the internal resistance of the practical voltage source is absorbing only 2 W. Thus we see that the two practical sources are equivalent only with respect to what transpires at the load terminals; they are *not* equivalent internally!

A very useful power theorem may be developed with reference to a practical voltage or current source. For the practical voltage source (Fig. 2-20a with $R_{sv} = R_s$), the power delivered to the load R_L is

$$p_L = i_L^2 R_L = \frac{v_s^2 R_L}{(R_s + R_L)^2}$$

To find the value of R_L that absorbs a maximum power from the given practical source, we differentiate with respect to R_L:

$$\frac{dp_L}{dR_L} = \frac{(R_s + R_L)^2 v_s^2 - v_s^2 R_L(2)(R_s + R_L)}{(R_s + R_L)^4}$$

and equate the derivative to zero, obtaining

$$2R_L(R_s + R_L) = (R_s + R_L)^2$$

or

$$R_s = R_L$$

Since the values $R_s = 0$ and $R_L = \infty$ both give a minimum ($p_L = 0$), and since we have already developed the equivalence between practical voltage and current sources, we have therefore proved the following *maximum power transfer theorem:*

An independent voltage source in series with a resistance R_s, or an independent current source in parallel with a resistance R_s, delivers a maximum power to that load resistance R_L for which $R_L = R_s$.

Thus, the maximum power transfer theorem tells us that a 2-Ω resistor draws the greatest power (4.5 W) from either practical source of Fig. 2-22, whereas a resistance of 0.01 Ω receives the maximum power (3.6 kW) in Fig. 2-19.

Drill Problems

2-6. (a) If $R_L = 2\ \Omega$ in Fig. 2-23a, find i_L and the power supplied by the 4-A source. (b) Transform the practical current source (8 Ω, 4 A) into a practical voltage source, and find i_L and the power supplied by the new ideal voltage source. (c) What value of R_L will absorb the maximum power, and what is the value of that power?   *Ans:* 2.6 A, 44.8 W; 2.6 A, 83.2 W; 8 Ω, 21.1 W

Figure 2-23

See Drill Probs. 2-6 and 2-7.

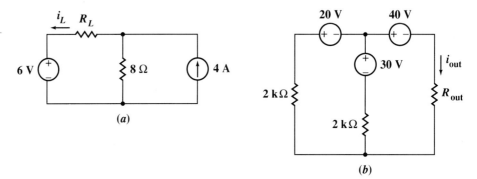

(a)

(b)

2-7. Transform the left two practical voltage sources in Fig. 2-23b into practical current sources, combine resistors and ideal current sources, then transform the resultant practical current sources into a practical voltage source, and combine the three ideal voltage sources. (a) If $R_{out} = 3$ kΩ, find the power delivered to it. (b) What is the maximum power that can be delivered to any R_{out}? (c) What two different values of R_{out} will have exactly 20 mW delivered to them?   *Ans:* 230 mW; 306 mW; 59.2 kΩ and 16.88 Ω

2-6

Thévenin's and Norton's theorems

Now that we have the superposition principle, it is possible to develop two more theorems which will greatly simplify the analysis of many linear circuits. The first of these theorems is named after M. L. Thévenin, a French engineer working in telegraphy, who published a statement of the theorem in 1883; the second may be considered a corollary of the first and is credited to E. L. Norton, a scientist formerly with the Bell Telephone Laboratories.[7]

Let us suppose that we need to make only a partial analysis of a circuit; perhaps we wish to determine the current, voltage, and power delivered to a single load resistor by the remainder of the circuit, which may consist of any number of sources and resistors; or perhaps we wish to find the response for

[7] Some questions have arisen regarding the true authorship of these theorems. It has been suggested that the theorem attributed to Thévenin was actually developed by Hermann von Helmholtz, a brilliant German physicist, in 1853, and that the theorem bearing Norton's name was originally developed by Hans Mayer, an electrical engineering professor at Cornell University, in 1926. However, the names "Thévenin's" theorem and "Norton's" theorem have been around a long time, and it's presumptuous for this one textbook to change them. So we will retain the old names, with apologies to Helmholtz and Mayer, if necessary.

different values of the load resistance. Thévenin's theorem then tells us that it is possible to replace everything except the load resistor by an equivalent circuit containing only an independent voltage source in series with a resistor; the response measured at the load resistor will be unchanged. Using Norton's theorem, we obtain an equivalent composed of an independent current source in parallel with a resistor.

It should thus be apparent that one of the main uses of Thévenin's and Norton's theorems is the replacement of a large part of a network, often a complicated and uninteresting part, by a very simple equivalent. The new, simpler circuit enables us to make rapid calculations of the voltage, current, and power which the original circuit is able to deliver to a load. It also helps us to choose the best value of this load resistance. In a transistor power amplifier, for example, the Thévenin or Norton equivalent enables us to determine the maximum power that can be taken from the amplifier and the type of load that is required to accomplish a maximum transfer of power or to obtain maximum practical voltage or current amplification.

Example 2-13 As an introductory example, consider the circuit shown in Fig. 2-24. We wish to determine the Thévenin and Norton equivalent circuits for that part of the circuit to the left of R_L.

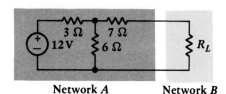

Figure 2-24

A simple resistive circuit is divided into network A, in which we have no detailed interest, and network B, a load resistor with which we are fascinated.

Network A Network B

Solution: The shaded regions separate the circuit into networks A and B; we shall assume that our main interest is in network B, which consists only of the load resistor R_L. Network A may be simplified by making repeated source transformations. We first treat the 12-V source and the 3-Ω resistor as a practical voltage source and replace it with a practical current source consisting of a 4-A source in parallel with 3 Ω. The parallel resistances are then combined into 2 Ω, and the practical current source which results is transformed back into a practical voltage source. The steps are indicated in Fig. 2-25, the final result appearing in Fig. 2-25d. From the viewpoint of the load resistor R_L, this circuit (the Thévenin equivalent) is equivalent to the original circuit; from our viewpoint, the circuit is much simpler and we can now easily compute the power delivered to the load. It is

$$p_L = \left(\frac{8}{9 + R_L}\right)^2 R_L$$

Furthermore we can see from the equivalent circuit that the maximum voltage which can be obtained across R_L is 8 V when $R_L = \infty$; a quick transformation of network A to a practical current source (the Norton equivalent) indicates that the maximum current which may be delivered to the load is $\frac{8}{9}$ A for $R_L = 0$; and the maximum power transfer theorem shows that a maximum power equal to $\frac{16}{9}$ W is delivered to R_L when R_L is 9 Ω. None of these facts is readily apparent from the original circuit. ∎

Figure 2-25

The source transformations and resistance combinations involved in simplifying network *A* are shown in order. The result, given in part *d*, is the Thévenin equivalent.

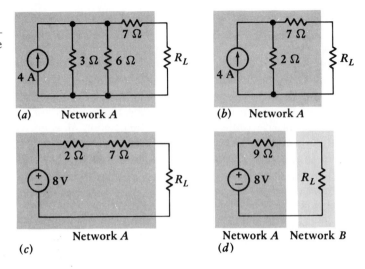

(a) Network *A*

(b) Network *A*

Network *A*
(c)

Network *A*   Network *B*
(d)

If network *A* had been more complicated, the number of source transformations and resistance combinations necessary to obtain the Thévenin or Norton equivalent could easily become prohibitive; also, with dependent sources present, the method of source transformation is usually inapplicable. Thévenin's and Norton's theorems allow us to find the equivalent circuit much more quickly and easily, even in more complicated circuits.

Let us now state *Thévenin's theorem* formally:

> Given any linear circuit, rearrange it in the form of two networks *A* and *B* that are connected together by two resistanceless conductors. If either network contains a dependent source, its control variable must be in the same network. Define a voltage v_{oc} as the open-circuit voltage which would appear across the terminals of *A* if *B* were disconnected so that no current is drawn from *A*. Then all the currents and voltages in *B* will remain unchanged if *A* is killed (all independent voltage and current sources in *A* replaced by short circuits and open circuits, respectively) and an independent voltage source v_{oc} is connected, with proper polarity, in series with the dead (inactive) *A* network.

The terms *killed* and *dead* are a little bloodthirsty, but they are descriptive and concise, and we shall use them in a friendly way. Moreover, it is possible that network *A* may only be sleeping, for it may still contain *dependent* sources which come to life whenever their controlling currents or voltages are nonzero.

Let us see if we can apply Thévenin's theorem successfully to the circuit we considered in Fig. 2-24. We have already found the Thévenin equivalent of the circuit to the left of R_L in Example 2-13, but now let us see if there is an easier way to obtain the same result.

Example 2-14   Use Thévenin's theorem in Fig. 2-24 to determine the Thévenin equivalent circuit for that part of the circuit to the left of R_L.

Solution: Disconnecting R_L, voltage division enables us to determine that v_{oc} is 8 V. Killing the *A* network, that is, replacing the 12-V source by a short circuit, we see looking back into the dead A network a 7-Ω resistor connected in series with the parallel combination of 6 Ω and 3 Ω. Thus the dead A network can be represented here by simply a 9-Ω resistor. This agrees with the previous result.   ∎

The equivalent circuit we have obtained is completely independent of the B network, because we have been instructed first to remove the B network and measure the open-circuit voltage produced by the A network, an operation which certainly does not depend on the B network in any way, and then to place the inactive A network in series with a voltage source v_{oc}. The B network is mentioned in the theorem and proof only to indicate that an equivalent for A may be obtained *no matter what arrangement of elements is connected to the A network*; the B network represents this general network.

A proof of Thévenin's theorem in the form in which we have stated it is rather lengthy, and therefore it has been placed in Appendix 3, where the curious or rigorous may peruse it.

There are several points about the theorem which deserve emphasis. First, the only restriction that we must impose on A or B, other than requiring that the original circuit composed of A and B be a linear circuit, is that all the dependent sources in A have their control variables in A, and similarly for B. No restrictions were imposed on the complexity of A or B; either one may contain any combination of independent voltage or current sources, linear dependent voltage or current sources, resistors, or any other circuit elements which are linear. The general nature of the theorem (and its proof) will enable it to be applied to networks containing inductors and capacitors, which are linear passive circuit elements to be defined in the following chapter. At this time, however, resistors are the only passive circuit elements which have been defined, and the application of Thévenin's theorem to resistive networks is a particularly simple special case. The dead A network can be represented by a single equivalent resistance, R_{th}. If A is an active *resistive* network, then it is obvious that the inactive A network may be replaced by a single equivalent resistance, which we shall also call the *Thévenin resistance,* since it is once again the resistance viewed at the terminals of the inactive A network.

Norton's theorem bears a close resemblance to Thévenin's theorem, another consequence of duality. As a matter of fact, the statements of the two theorems will be used as an example of dual language when the duality principle is discussed in the following chapter.

Norton's theorem may be stated as follows:

> Given any linear circuit, rearrange it in the form of two networks A and B that are connected together by two resistanceless conductors. If either network contains a dependent source, its control variable must be in that same network. Define a current i_{sc} as the short-circuit current which would appear at the terminals of A if B were short-circuited so that no voltage is provided by A. Then all the currents and voltages in B will remain unchanged if A is killed (all independent current and voltage sources in A replaced by open circuits and short circuits, respectively) and an independent current source i_{sc} is connected, with proper polarity, in parallel with the dead (inactive) A network.

The Norton equivalent of an active resistive network is the Norton current source i_{sc} in parallel with the Thévenin resistance R_{th}.

There is an important relationship between the Thévenin and Norton equivalents of an active resistive network. The relationship may be obtained by applying a source transformation to either equivalent network. For example, if we transform the Norton equivalent, we obtain a voltage source $R_{th}i_{sc}$ in series with the resistance R_{th}; this network is in the form of the Thévenin equivalent, and thus

$$v_{oc} = R_{th}i_{sc} \tag{29}$$

In resistive circuits containing *dependent* sources as well as independent sources, we shall often find it more convenient to determine either the Thévenin or Norton equivalent by finding both the open-circuit voltage and the short-circuit current and then determining the value of R_{th} as their quotient. It is therefore advisable to become adept at finding both open-circuit voltages and short-circuit currents, even in the simple problems which follow. If the Thévenin and Norton equivalents are determined independently, Eq. (29) may serve as a useful check.

Let us consider four examples of the determination of a Thévenin or Norton equivalent circuit.

Example 2-15   As a first example, let us find the Thévenin and Norton equivalent circuits for the network faced by the 1-kΩ resistor in Fig. 2-26a. That is, network B is this resistor, and network A is the remainder of the given circuit.

Figure 2-26

(a) A given circuit in which the 1-kΩ resistor is identified as network B. (b) The Thévenin equivalent is shown for network A. (c) The Norton equivalent is shown for network A.

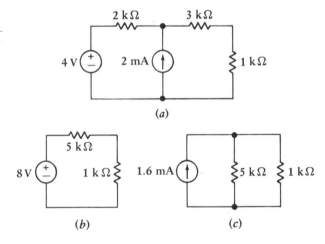

(a)

(b)   (c)

Solution: This circuit contains no dependent sources, and the easiest way to find the Thévenin equivalent is to determine R_{th} for the dead network directly, followed by a calculation of either v_{oc} or i_{sc}. We first kill both independent sources to determine the form of the dead A network. With the 4-V source short-circuited and the 2-mA source open-circuited, the result is the series combination of a 2-kΩ and a 3-kΩ resistor, or the equivalent, a 5-kΩ resistor. The open-circuit voltage is easily determined by superposition. With only the 4-V source operating, the open-circuit voltage is 4 V; when only the 2-mA source is on, the open-circuit voltage is also 4 V; with both independent sources operating, we see that $v_{oc} = 4 + 4 = 8$ V. This determines the Thévenin equivalent, shown in Fig. 2-26b, and from it the Norton equivalent of Fig. 2-26c can be drawn quickly. As a check, let us determine i_{sc} for the given circuit. We use superposition and a little current division:

$$i_{sc} = i_{sc}|_{4V} + i_{sc}|_{2mA} = \frac{4}{2+3} + (2)\frac{2}{2+3}$$

$$= 0.8 + 0.8 = 1.6\,\text{mA}$$

which completes the check.   ∎

As the second example, we consider the network A shown in Fig. 2-27a, which contains both independent and dependent sources.

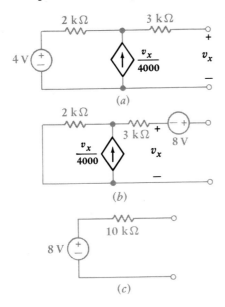

Figure 2-27

(a) A given network whose Thévenin equivalent is desired. (b) A possible, but rather useless, form of the Thévenin equivalent. (c) The best form of the Thévenin equivalent for this linear resistive network.

Example 2-16   Determine the Thévenin equivalent of the circuit shown in Fig. 2-27a.

Solution: The presence of the dependent source prevents us from determining R_{th} directly for the inactive network through resistance combination; instead, we find both v_{oc} and i_{sc}.

To find v_{oc} we note that $v_x = v_{oc}$ and that the dependent source current must pass through the 2-kΩ resistor, since there is an open circuit to the right. Summing voltages around the outer loop:

$$-4 + 2 \times 10^3 \left(\frac{-v_x}{4000}\right) + 3 \times 10^3(0) + v_x = 0$$

and $$v_x = 8 = v_{oc}$$

By Thévenin's theorem, then, the equivalent circuit could be formed with the dead A network in series with an 8-V source, as shown in Fig. 2-27b. This is correct, but is not very simple and not very helpful; in the case of linear resistive networks, we should certainly show a much simpler equivalent for the inactive A network, namely, R_{th}. We therefore seek i_{sc}. Upon short-circuiting the output terminals in Fig. 2-27a, it is apparent that

$v_x = 0$ and the dependent current source is zero. Hence, $i_{sc} = 4/(5 \times 10^3) = 0.8$ mA. Thus, $R_{th} = v_{oc}/i_{sc} = 8/(0.8 \times 10^{-3}) = 10$ kΩ, and the acceptable Thévenin equivalent of Fig. 2-27c is obtained. ∎

Another example for which we should find both v_{oc} and i_{sc} appears in Fig. 2-28. The circuit happens to be an op-amp connected as a voltage follower with $v_s = 5$ V, $R_i = \infty$, $A = 10^4$, and $R_o = 2$ kΩ.

Figure 2-28

The Thévenin equivalent of this voltage follower is desired for terminals a and b.

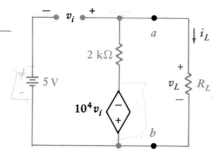

Example 2-17 Find the Thévenin equivalent of the voltage follower (the circuit to the left of the terminals marked a and b) in Fig. 2-28.

Solution: To find v_{Loc} we set $R_L = \infty$, or we simply consider it to be removed from the circuit. There now can be no current through the 2-kΩ resistor, and therefore

$$v_{Loc} = -10^4 v_i$$

where

$$v_i = v_{Loc} - 5$$

Therefore

$$v_{Loc} = \frac{10^4(5)}{10\ 001} = 5.00 \text{ V}$$

Next we need a value for i_{Lsc}, and so we replace R_L by a short circuit. Around the right mesh, KVL gives

$$10^4 v_i + 2000 i_{Lsc} = 0$$

while the application of KVL around the perimeter of the circuit gives

$$-5 - v_i = 0$$

Therefore

$$10^4(-5) + 2000 i_{Lsc} = 0$$

and

$$i_{Lsc} = 25.0 \text{ A}$$

We find R_{th} by the quotient

$$R_{th} = \frac{v_{Loc}}{i_{Lsc}} = \frac{5.00}{25.0} = 0.200 \text{ Ω}$$

Thus, the Thévenin equivalent presented to R_L at a and b by the voltage follower is 5.00 V in series with a very low value of resistance, 0.200 Ω. ∎

As our final example, let us consider a network having a dependent source but no independent source, such as that shown in Fig. 2-29a.

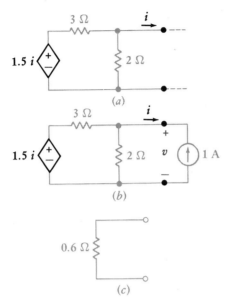

Figure 2-29

(a) A network, containing no independent sources, whose Thévenin equivalent is desired. (b) R_{th} is numerically equal to v. (c) The Thévenin equivalent of part a.

Example 2-18   Find the Thévenin equivalent of the circuit shown in Fig. 2-29a.

Solution: The network qualifies already as the dead A network, and $v_{\text{oc}} = 0$. We thus seek the value of R_{th} represented by this two-terminal network. However, we cannot find v_{oc} and i_{sc} and take their quotient, for there is no independent source in the network and both v_{oc} and i_{sc} are zero. Let us, therefore, be a little tricky. We apply a 1-A source externally, measure the resultant voltage, and then set $R_{\text{th}} = v/1$. Referring to Fig. 2-29b, we see that $i = -1$ and

$$\frac{v - 1.5(-1)}{3} + \frac{v}{2} = 1$$

so that

$$v = 0.6 \text{ V}$$

and

$$R_{\text{th}} = 0.6 \ \Omega$$

The Thévenin equivalent is shown in Fig. 2-29c.   ∎

We have now looked at four examples in which we determined a Thévenin or Norton equivalent circuit. The first example (Fig. 2-26) contained only independent sources and resistors, and we were able to use several different methods on it. One involved calculating R_{th} for the dead network and then v_{oc} for the live network. We could also have found R_{th} and i_{sc}, or v_{oc} and i_{sc}.

In the second and third examples (Figs. 2-27 and 2-28), both independent and dependent sources were present, and the only method we used required us to find v_{oc} and i_{sc}. We could not easily find R_{th} for the dead network, because the dependent source could not be made inactive.

The last example did not contain any independent sources, and we found R_{th} by applying 1 A and finding $v = 1 \times R_{\text{th}}$. We could also apply 1 V and determine $i = 1/R_{th}$. These Thévenin and Norton equivalents do not contain an independent source.

These important techniques and the types of circuits to which they may be applied most readily are indicated in Table 2-1.

Table 2-1

Recommended methods of finding Thévenin or Norton equivalents

	Circuit contains		
Methods			
R_{th} and v_{oc} or i_{sc} v_{oc} and i_{sc} $i = 1$ A or $v = 1$ V	✔ Possible —	— ✔ —	— — ✔

All possible methods do not appear in the table. We have used repeated source transformations on several networks when there were no dependent sources, and this is certainly an easy technique for networks that do not contain too many elements.

Two other methods have a certain appeal because they can be used for any of the three types of networks listed in the table. In the first, simply replace the B network by a voltage source v_s, define the current leaving its positive terminal as i, then analyze the A network to obtain i, and put the equation in the form $v_s = ai + b$. Evidently, $a = R_{th}$ and $b = v_{oc}$.

We could also apply a current source i_s, let its voltage be v, and then determine $v = ai_s + b$. Both of these last two procedures are universally applicable, but some other method can usually be found that is easier and more rapid.

Although we are devoting our attention almost entirely to the analysis of *linear* circuits, it is enlightening to know that Thévenin's and Norton's theorems are both valid if network B is nonlinear; only network A *must be* linear.

Drill Problems

2-8. Find the Thévenin equivalent for the network of (*a*) Fig. 2-30*a*; (*b*) Fig. 2-30*b*.   *Ans:* 130 V, 30 Ω; 125 V, 25 Ω

2-9. Find the Norton equivalent for the network of (*a*) Fig. 2-30*c*; (*b*) Fig. 2-30*d*.   *Ans:* doesn't exist (Thévenin is 100 V, 0 Ω); a0 V, 20 Ω

2-7

Trees and general nodal analysis[8]

In this section we shall generalize the method of nodal analysis that we have come to know and love. Since nodal analysis is applicable to any network, we cannot promise that we shall be able to solve a wider class of circuit problems. We can, however, look forward to being able to select a general nodal analysis method for any particular problem that may result in fewer equations and less work.

We must first extend our list of definitions relating to network topology. We begin by defining *topology* itself as a branch of geometry which is concerned

---

[8] This and the following section may be postponed if desired. They introduce analysis methods that are a little more general than those using node-to-reference voltages and mesh currents. Their use can lead to fewer equations or more useful current and voltage variables.

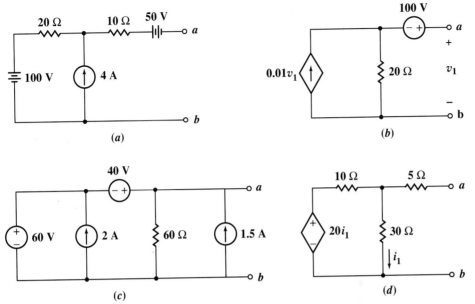

Figure 2-30

See Drill Probs. 2-8 and 2-9.

with those properties of a geometrical figure which are unchanged when the figure is twisted, bent, folded, stretched, squeezed, or tied in knots, with the provision that no parts of the figure are to be cut apart or to be joined together. A sphere and a tetrahedron are topologically identical, as are a square and a circle. In terms of electric circuits, then, we are not now concerned with the particular types of elements appearing in the circuit, but only with the way in which branches and nodes are arranged. As a matter of fact, we usually suppress the nature of the elements and simplify the drawing of the circuit by showing the elements as lines. The resultant drawing is called a *linear graph,* or simply a *graph.* A circuit and its graph are shown in Fig. 2-31. Note that all nodes are identified by heavy dots in the graph.

Since the topological properties of the circuit or its graph are unchanged when it is distorted, the three graphs shown in Fig. 2-32 are all topologically identical with the circuit and graph of Fig. 2-31.

Topological terms which we already know and have been using correctly are

node: a point at which two or more elements have a common connection.

path: a set of elements that may be traversed in order without passing through the same node twice.

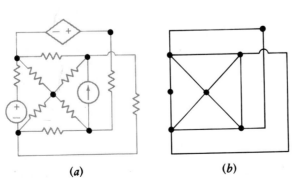

Figure 2-31

(*a*) A given circuit. (*b*) The linear graph of this circuit.

Figure 2-32

The three graphs shown are topologically identical to each other and to the graph of Fig. 2-31*b*, and each is a graph of the circuit shown in Fig. 2-31*a*.

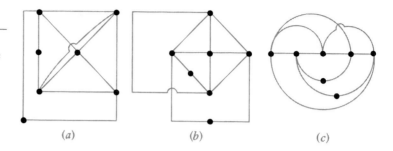

(*a*)   (*b*)   (*c*)

branch: a single path, containing one simple element, which connects one node to any other node.

loop: a closed path.

mesh: a loop which does not contain any other loops within it.

planar circuit: a circuit which may be drawn on a plane surface in such a way that no branch passes over or under any other branch.

nonplanar circuit: any circuit which is not planar.

The graphs of Fig. 2-32 each contain 12 branches and 7 nodes.

Three new properties of a linear graph must now be defined—a tree, a cotree, and a link. We define a *tree* as any set of branches which does not contain any loops and yet connects every node to every other node, not necessarily directly. There are usually a number of different trees which may be drawn for a network, and the number increases rapidly as the complexity of the network increases. The simple graph shown in Fig. 2-33*a* has eight possible trees, four of which are shown by heavy lines in Figs. 2-33*b*, *c*, *d*, and *e*.

Figure 2-33

(*a*) The linear graph of a three-node network. (*b, c, d, e*) Four of the eight different trees which may be drawn for this graph are shown by the heavy lines.

(*a*)   (*b*)   (*c*)   (*d*)   (*e*)

In Fig. 2-34*a* a more complex graph is shown. Figure 2-34*b* shows one possible tree, and Figs. 2-34*c* and *d* show sets of branches which are not trees because neither set satisfies the definition.

After a tree has been specified, those branches that are not part of the tree form the *cotree*, or complement of the tree. The lightly drawn branches in Figs. 2-33*b* to *e* show the cotrees that correspond to the heavier trees.

Figure 2-34

(*a*) A linear graph. (*b*) A possible tree for this graph. (*c, d*) These sets of branches do not satisfy the definition of a tree.

(*a*)   (*b*)   (*c*)   (*d*)

Once we understand the construction of a tree and its cotree, the concept of the link is very simple, for a *link* is any branch belonging to the cotree. It is evident that any particular branch may or may not be a link, depending on the particular tree which is selected.

The number of links in a graph may easily be related to the number of branches and nodes. If the graph has N nodes, then exactly $(N - 1)$ branches are required to construct a tree because the first branch chosen connects two nodes and each additional branch includes one more node. Thus, given B branches, the number of links L must be

$$L = B - (N - 1)$$

or
$$L = B - N + 1 \qquad\qquad (30)$$

There are L branches in the cotree and $(N - 1)$ branches in the tree.

In any of the graphs shown in Fig. 2-33, we note that $3 = 5 - 3 + 1$, and in the graph of Fig. 2-34b, $6 = 10 - 5 + 1$. A network may be in several disconnected parts, and Eq. (30) may be made more general by replacing $+1$ with $+S$, where S is the number of separate parts. However, it is also possible to connect two separate parts by a single conductor, thus causing two nodes to form one node; no current can flow through this single conductor. This process may be used to join any number of separate parts, and thus we shall not suffer any loss of generality if we restrict our attention to circuits for which $S = 1$.

We are now ready to discuss a method by which we may write a set of nodal equations that are independent and sufficient. The method will enable us to obtain many different sets of equations for the same network, and all the sets will be valid. However, the method does not provide us with every possible set of equations. Let us first describe the procedure, illustrate it by three examples, and then point out the reason that the equations are independent and sufficient.

Given a network, we should

1 Draw a graph and then identify a tree.
2 Place all voltage sources in the tree.
3 Place all current sources in the cotree.
4 Place all control-voltage branches for voltage-controlled dependent sources in the tree, if possible.
5 Place all control-current branches for current-controlled dependent sources in the cotree, if possible.

These last four steps effectively associate voltages with the tree and currents with the cotree.

We now assign a voltage variable (with its plus-minus pair) across each of the $(N - 1)$ branches in the tree. A branch containing a voltage source (dependent or independent) should be assigned that source voltage, and a branch containing a controlling voltage should be assigned that controlling voltage. The number of new variables that we have introduced is therefore equal to the number of branches in the tree $(N - 1)$, reduced by the number of voltage sources in the tree, and reduced also by the number of control voltages we were able to locate in the tree. In Example 2-21, we shall find that the number of new variables required *may* be zero.

Having a set of variables, we now need to write a set of equations that are sufficient to determine these variables. The equations are obtained through the application of KCL. Voltage sources are handled in the same way that they were in our earlier attack on nodal analysis; each voltage source and the two nodes at its terminals constitute a supernode or a part of a supernode. Kirchhoff's current law is then applied at all but one of the remaining nodes and supernodes. We set the sum of the currents leaving the node in all of the branches connected to it equal to zero. Each current is expressed in terms of the voltage variables we just assigned. One node may be ignored, just as was the case earlier for the reference node. Finally, in case there are current-controlled dependent sources, we must write an equation for each control current that relates it to the voltage variables; this also is no different from the procedure used before with nodal analysis.

Let us try out this process on the circuit shown in Fig. 2-35a. It contains four nodes and five branches, and its graph is shown in Fig. 2-35b.

Figure 2-35

(a) A circuit used as an example for general nodal analysis. (b) The graph of the given circuit. (c) The voltage source and the control voltage are placed in the tree, while the current source goes in the cotree. (d) The tree is completed and a voltage is assigned across each tree branch.

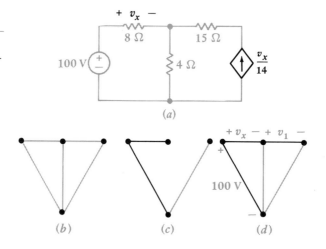

(a)

(b)   (c)   (d)

Example 2-19 Find the value of v_x in the circuit of Fig. 2-35a.

Solution: In accordance with steps 2 and 3 of the tree-drawing procedure, we place the voltage source in the tree and the current source in the cotree. Following step 4, we see that the v_x branch may also be placed in the tree, since it does not form any loop which would violate the definition of a tree. We have now arrived at the two tree branches and the single link shown in Fig. 2-35c, and we see that we do not yet have a tree, since the right node is not connected to the others by a path through tree branches. The only possible way to complete the tree is shown in Fig. 2-35d. The 100-V source voltage, the control voltage v_x, and a new voltage variable v_1 are next assigned to the three tree branches as shown.

We therefore have two unknowns, v_x and v_1, and it is obvious that we need to obtain two equations in terms of them. There are four nodes, but the presence of the voltage source causes two of them to form a single supernode. Kirchhoff's current law may be applied at any two of the three remaining nodes or supernodes. Let's attack the right node first. The current leaving to the left is $-v_1/15$, while that leaving downward is $-v_x/14$. Thus, our first equation is

$$-\frac{v_1}{15} - \frac{v_x}{14} = 0$$

The central node at the top looks easier than the supernode, and so we set the sum of the current to the left $(-v_x/8)$, the current to the right $(v_1/15)$, and the downward current through the 4-Ω resistor equal to zero. This latter current is given by the voltage across the resistor divided by 4 Ω, but there is no voltage labeled on that link. However, when a tree is constructed according to the definition, there is a path through it from any node to any other node. Then, since every branch in the tree is assigned a voltage, we may express the voltage across any link in terms of the tree-branch voltages. This downward current is therefore $(-v_x + 100)/4$, and we have the second equation,

$$-\frac{v_x}{8} + \frac{v_1}{15} + \frac{-v_x + 100}{4} = 0$$

The simultaneous solution of these two nodal equations gives

$$v_1 = -60 \text{ V} \qquad v_x = 56 \text{ V} \qquad \blacksquare$$

As a second example, let us reconsider a more complex circuit that we first analyzed by defining all node voltages with respect to a reference node. The circuit is that of Fig. 2-5, repeated as Fig. 2-36a.

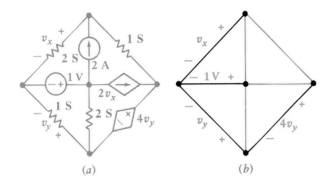

(a)   (b)

Figure 2-36

(a) The circuit of Fig. 2-5 is repeated. (b) A tree is chosen such that both voltage sources and both control voltages are tree branches.

Example 2-20   Find the values of v_x and v_y in the circuit of Fig. 2-36a.

Solution: We draw a tree so that both voltage sources and both control voltages appear as tree-branch voltages and, hence, as assigned variables. As it happens, these four branches constitute a tree, Fig. 2-36b, and tree-branch voltages v_x, 1, v_y, and $4v_y$ are chosen, as shown.

Both voltage sources define supernodes, and we apply KCL twice, once to the top node,

$$2v_x + 1(v_x - v_y - 4v_y) = 2$$

and once to the supernode consisting of the right node, the bottom node, and the dependent voltage source,

$$1v_y + 2(v_y - 1) + 1(4v_y + v_y - v_x) = 2v_x$$

Instead of the four equations we had previously, we have only two, and we find easily that $v_x = \frac{26}{9}$ V and $v_y = \frac{4}{3}$ V, both values agreeing with the earlier solution. $\blacksquare$

For a final example we consider the circuit of Fig. 2-37a.

Figure 2-37

(a) A circuit for which only one general nodal equation need be written.
(b) The tree and the tree-branch voltages used.

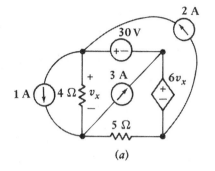
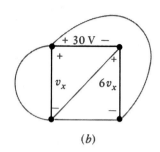

(a)   (b)

Example 2-21   Find the value of v_x in the circuit of Fig. 2-37a.

Solution: The two voltage sources and the control voltage establish the three-branch tree shown in Fig. 2-37b. Since the two upper nodes and the lower right node all join to form one supernode, we need write only one KCL equation. Selecting the lower left node, we have

$$-1 - \frac{v_x}{4} + 3 + \frac{-v_x + 30 + 6v_x}{5} = 0$$

and it follows that $v_x = -\frac{32}{3}$ V. In spite of the apparent complexity of this circuit, the use of general nodal analysis has led to an easy solution. Employing mesh currents or node-to-reference voltages would require more equations and more effort. ∎

We shall discuss the problem of finding the best analysis scheme in the following section.

If we should need to know some other voltage, current, or power in the previous example, one additional step would give the answer. For example, the power provided by the 3-A source is

$$3(-30 - \tfrac{32}{3}) = -122 \text{ W}$$

Let us conclude by discussing the sufficiency of the assumed set of tree-branch voltages and the independence of the nodal equations. If these tree-branch voltages are *sufficient,* then the voltage of every branch in either the tree or the cotree must be obtainable from a knowledge of the values of all the tree-branch voltages. This is certainly true for those branches in the tree. For the links we know that each link extends between two nodes, and, by definition, the tree must also connect those two nodes. Hence, every link voltage may also be established in terms of the tree-branch voltages.

Once the voltage across every branch in the circuit is known, then all the currents may be found by using either the given value of the current if the branch consists of a current source, by using Ohm's law if it is a resistive branch, or by using KCL and these current values if the branch happens to be a voltage source. Thus, all the voltages and currents are determined and sufficiency is demonstrated.

To demonstrate independency, let us satisfy ourselves by assuming the situation where the only sources in the network are independent current sources. As we have noticed earlier, independent voltage sources in the circuit result in fewer equations, while dependent sources usually necessitate a greater number of equations. For independent current sources only, there will then be precisely $(N - 1)$ nodal equations written in terms of $(N - 1)$ tree-branch voltages. To show that these $(N - 1)$ equations are *independent,* visualize the application of KCL to the $(N - 1)$ different nodes. Each time we write the KCL equation, there is a new tree branch involved—the one which connects that node to the remainder of the tree. Since that circuit element has not appeared in any previous equation, we must obtain an independent equation. This is true for each of the $(N - 1)$ nodes in turn, and hence we have $(N - 1)$ independent equations.

2-10. (*a*) How many trees may be constructed for the circuit of Fig. 2-38 that follow all five of the tree-drawing suggestions listed earlier? (*b*) Draw a suitable tree, write two equations in two unknowns, and find i_3. (*c*) What power is supplied by the dependent source?   *Ans:* 1; 7.2 A; 547 W

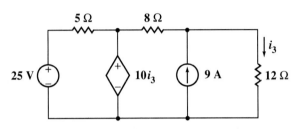

Figure 2-38

See Drill Prob. 2-10.

Now we shall consider the use of a tree to obtain a suitable set of loop equations. In some respects this is the dual of the method of writing nodal equations. Again it should be pointed out that, although we are able to guarantee that any set of equations we write will be both sufficient and independent, we should not expect that the method will lead directly to every possible set of equations.

We again begin by constructing a tree, and we use the same set of rules as we did for general nodal analysis. The objective for either nodal or loop analysis is to place voltages in the tree and currents in the cotree; this is a *mandatory* rule for sources and a desirable rule for controlling quantities.

Now, however, instead of assigning a voltage to each branch in the tree, we assign a current (including reference arrow, of course) to each element in the cotree or to each link. If there were 10 links, we would assign exactly 10 link currents. Any link that contains a current source is assigned that source current as the link current. Note that each link current may also be thought of as a loop current, for the link must extend between two specific nodes, and there must also be a path between those same two nodes through the tree. Thus, with each link there is associated a single specific loop that includes that one link and a unique path through the tree. It is evident that the assigned current may be thought of either as a loop current or as a link current. The link connotation is most helpful at the time the currents are being defined, for one must be established for each link; the loop interpretation is more convenient at equation-writing time, because we shall apply KVL around each loop.

Let us try out this process of defining link currents by reconsidering an earlier example we worked using mesh currents. The circuit is shown in Fig. 2-12 and redrawn as Fig. 2-39*a*. The tree selected is one of several that might be constructed for which the voltage source is in a tree branch and the current source is in a link. Let us first consider the link containing the current source. The loop associated with this link is the left-hand mesh, and so we show our

2-8

Links and loop analysis

Figure 2-39

(a) The circuit of Fig. 2-12 is shown again. (b) A tree is chosen such that the current source is in a link and the voltage source is in a tree branch.

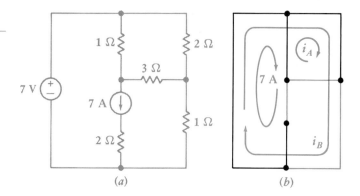

(a)   (b)

link current flowing about the perimeter of this mesh (Fig. 2-39b). An obvious choice for the symbol for this link current is "7 A." Remember that no other current can flow through this particular link, and thus its value must be exactly the strength of the current source.

We next turn our attention to the 3-Ω resistor link. The loop associated with it is the upper right-hand mesh, and this loop (or mesh) current is defined as i_A and also shown in Fig. 2-39b. The last link is the lower 1-Ω resistor, and the only path between its terminals through the tree is around the perimeter of the circuit. That link current is called i_B, and the arrow indicating its path and reference direction appears in Fig. 2-39b. It is not a mesh current.

Note that each link has only one current present in it, but a tree branch may have any number from 1 to the total number of link currents assigned. The use of long, almost closed, arrows to indicate the loops helps to indicate which loop currents flow through which tree branch and what their reference directions are.

A KVL equation must now be written around each of these loops. The variables used are the assigned link currents. Since the voltage across a current source cannot be expressed in terms of the source current, and since we have already used the value of the source current as the link current, we discard any loop containing a current source.

Example 2-22   For the example of Fig. 2-39, find the values of i_A and i_B.

Solution: We first traverse the i_A loop, proceeding clockwise from its lower left corner. The current going our way in the 1-Ω resistor is $(i_A - 7)$, in the 2-Ω element it is $(i_A + i_B)$, and in the link it is simply i_B. Thus

$$1(i_A - 7) + 2(i_A + i_B) + 3i_A = 0$$

For the i_B link, clockwise travel from the lower left corner leads to

$$-7 + 2(i_A + i_B) + 1i_B = 0$$

Traversal of the loop defined by the 7-A link is not required. Solving, we have $i_A = 0.5$ A, $i_B = 2$ A, once again. The solution has been achieved with one less equation than before!   ■

An example containing a dependent source appears in Fig. 2-40a.

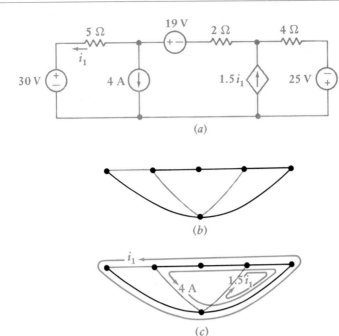

(a)

(b)

(c)

Figure 2-40

(a) A circuit for which i_1 may be found with one equation using general loop analysis. (b) The only tree that satisfies the rules outlined in Sec. 2-7. (c) The three link currents are shown with their loops.

Example 2-23 Evaluate i_1 in the circuit shown in Fig. 2-40a.

Solution: This circuit contains six nodes, and its tree therefore must have five branches. Since there are eight elements in the network, there are three links in the cotree. If we place the three voltage sources in the tree and the two current sources and the current control in the cotree, we are led to the tree shown in Fig. 2-40b. The source current of 4 A defines a loop as shown in Fig. 2-40c. The dependent source establishes the loop current $1.5i_1$ around the right mesh, and the control current i_1 gives us the remaining loop current about the perimeter of the circuit. Note that all three currents flow through the 4-Ω resistor.

We have only one unknown quantity, i_1, and after discarding the loops defined by the two current sources, we apply KVL around the outside of the circuit:

$$-30 + 5(-i_1) + 19 + 2(-i_1 - 4) + 4(-i_1 - 4 + 1.5i_1) - 25 = 0$$

Besides the three voltage sources, there are three resistors in this loop. The 5-Ω resistor has one loop current in it, since it is also a link; the 2-Ω resistor contains two loop currents; and the 4-Ω resistor has three. A carefully drawn set of loop currents is a necessity if errors in skipping currents, utilizing extra ones, or erring in choosing the correct direction are to be avoided. The foregoing equation is guaranteed, however, and it leads to $i_1 = -12$ A. ∎

How may we demonstrate sufficiency? Let us visualize a tree. It contains no loops and therefore contains at least two nodes to each of which only one tree branch is connected. The current in each of these two branches is easily found from the known link currents by applying KCL. If there are other nodes at which only one tree branch is connected, these tree-branch currents may also be immediately obtained. In the tree shown in Fig. 2-41, we thus have found the currents in branches a, b, c, and d. Now we move along the branches of the tree, finding the currents in tree branches e and f; the process may be continued until all the branch currents are determined. The link currents are therefore sufficient to determine all branch currents. It is helpful to look at the situation where an incorrect "tree" has been drawn which contains a loop. Even if all the link currents were zero, a current might still circulate about this "tree loop." Hence, the link currents could not determine this current, and they would not represent a sufficient set. Such a "tree" is by definition impossible.

Figure 2-41

A tree which is used as an example to illustrate the sufficiency of the link currents.

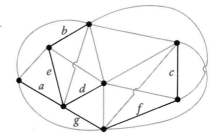

To demonstrate independence, let us satisfy ourselves by assuming the situation where the only sources in the network are independent voltage sources. As we have noticed earlier, independent current sources in the circuit result in fewer equations, while dependent sources usually necessitate a greater number of equations. If only independent voltage sources are present, there will then be precisely $(B - N + 1)$ loop equations written in terms of the $(B - N + 1)$ link currents. To show that these $(B - N + 1)$ loop equations are independent, it is necessary only to point out that each represents the application of KVL around a loop which contains one link not appearing in any other equation. We might visualize a different resistance $R_1, R_2, \ldots, R_{B-N+1}$ in each of these links, and it is then apparent that one equation can never be obtained from the others, since each contains one coefficient not appearing in any other equation.

Hence, the link currents are sufficient to enable a complete solution to be obtained, and the set of loop equations which we use to find the link currents is a set of independent equations.

Having looked at both general nodal analysis and loop analysis, we should now consider the advantages and disadvantages of each method so that an intelligent choice of a plan of attack can be made on a given analysis problem.

The nodal method in general requires $(N - 1)$ equations, but this number is reduced by 1 for each independent or dependent voltage source in a tree branch, and increased by 1 for each dependent source that is voltage-controlled by a link voltage, or current-controlled.

The loop method basically involves $(B - N + 1)$ equations. However, each independent or dependent current source in a link reduces this number by 1, while each dependent source that is current-controlled by a tree-branch current, or is voltage-controlled, increases the number by 1.

As a grand finale for this discussion, let us inspect the T-equivalent-circuit model for a transistor, shown in Fig. 2-42, to which are connected a sinusoidal source, 4 sin 1000t mV, and a 10-kΩ load.

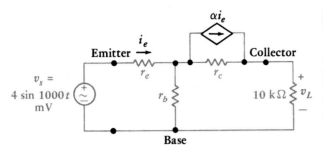

Figure 2-42

A sinusoidal voltage source and a 10-kΩ load are connected to the T-equivalent circuit of a transistor. The common connection between the input and output is at the base terminal of the transistor, and the arrangement is called the *common-base* configuration.

Example 2-24 Find the input (emitter) current i_e and the load voltage v_L in the circuit of Fig. 2-42, assuming typical values for the emitter resistance $r_e = 50$ Ω; the base resistance $r_b = 500$ Ω; the collector resistance $r_c = 20$ kΩ; and the common-base forward-current-transfer ratio $\alpha = 0.99$.

Solution: Although the details are requested in Drill Probs. 2-12 and 2-13, which follow, we should see readily that the analysis of this circuit might be accomplished by drawing trees requiring three general nodal equations $(N - 1 - 1 + 1)$ or two loop equations $(B - N + 1 - 1)$. We might also note that three equations are required in terms of node-to-reference voltages, as are three mesh equations.

No matter which method we choose, these results are obtained for this specific circuit:

$$i_e = 18.42 \sin 1000t \quad \mu A$$

$$v_L = 122.6 \sin 1000t \quad mV$$

and we therefore find that this transistor circuit provides a voltage gain (v_L/v_s) of 30.6, a current gain $(v_L/10\ 000i_e)$ of 0.666, and a power gain equal to the product, $30.6(0.666) = 20.4$. Higher gains could be secured by operating this transistor in a common-emitter configuration, as illustrated by the equivalent circuit of Prob. 43. ∎

2-11. Draw a suitable tree and use general loop analysis to find i_{10} in the circuit of (*a*) Fig. 2-43*a* by writing just one equation with i_{10} as the variable; (*b*) Fig. 2-43*b* by writing just two equations with i_{10} and i_3 as the variables.

Drill Problems

Ans: -4.00 mA; 4.69 A

Figure 2-43

See Drill Prob. 2-11.

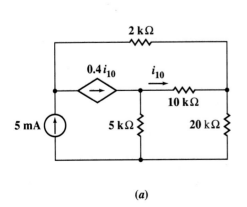

(*a*)

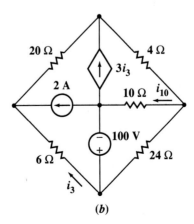

(*b*)

2-12. For the transistor amplifier equivalent circuit shown in Fig. 2-42, let $r_e = 50\ \Omega$, $r_b = 500\ \Omega$, $r_c = 20\ k\Omega$, and $\alpha = 0.99$, and find both i_e and v_L by drawing a suitable tree and using (a) two loop equations; (b) three nodal equations with a common reference node for the voltage; (c) three nodal equations without a common reference node.   *Ans: 18.42 sin 1000t μA; 122.6 sin 1000t mV*

2-13. Determine the Thévenin and Norton equivalent circuits presented to the 10-kΩ load in Fig. 2-42 by finding (a) the open-circuit value of v_L; (b) the (downward) short-circuit current; (c) the Thévenin equivalent resistance. All circuit values are given in Drill Prob. 2-12.

Ans: 147.6 sin 1000t mV; 72.2 sin 1000t μA; 2.05 kΩ

Problems

1 (a) Find v_2 if $0.1v_1 - 0.3v_2 - 0.4v_3 = 0$, $-0.5v_1 + 0.1v_2 = 4$, and $-0.2v_1 - 0.3v_2 + 0.4v_3 = 6$. ($b$) Evaluate the determinant:

$$\begin{vmatrix} 2 & 3 & 4 & 1 \\ 3 & 4 & 1 & 2 \\ 4 & 1 & 2 & 3 \\ 1 & -2 & 3 & 0 \end{vmatrix}$$

2 Use nodal analysis to find v_P in the circuit shown in Fig. 2-44.

Figure 2-44

See Prob. 2.

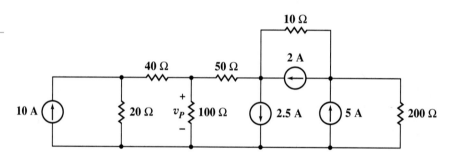

3 (a) Find v_A, v_B, and v_C if $v_A + v_B + v_C = 27$, $2v_B + 16 = v_A - 3v_C$, and $4v_C + 2v_A + 6 = 0$. (b) Evaluate the determinant:

$$\begin{vmatrix} 0 & 1 & 2 & 3 \\ 1 & 2 & 3 & 4 \\ 2 & 3 & 4 & 1 \\ 3 & 4 & 1 & 2 \end{vmatrix}$$

4 Use nodal analysis to find v_x in the circuit of Fig. 2-45.

Figure 2-45

See Prob. 4.

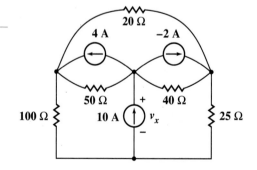

5 Use nodal analysis to find v_4 in the circuit shown in Fig. 2-46.

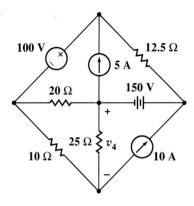

Figure 2-46

See Prob. 5.

6 With the help of nodal analysis on the circuit of Fig. 2-47, find (*a*) v_A; (*b*) the power dissipated in the 2.5-Ω resistor.

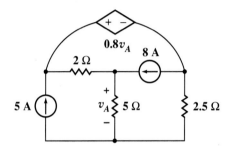

Figure 2-47

See Prob. 6.

7 Use nodal analysis to determine v_1 and the power being supplied by the dependent current source in the circuit shown in Fig. 2-48.

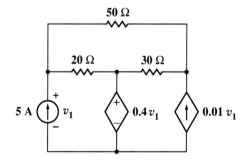

Figure 2-48

See Probs. 7 and 11.

8 In Fig. 2-49, use nodal analysis to find the value of k that will cause v_y to be zero.

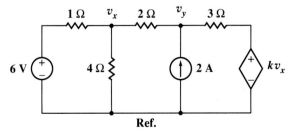

Figure 2-49

See Prob. 8.

9 Use mesh analysis to find i_x in the circuit shown in Fig. 2-50a.

Figure 2-50

See Probs. 9 and 10.

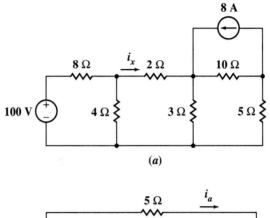

(a)

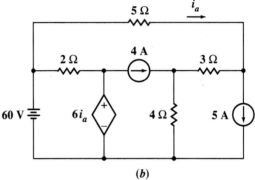

(b)

10 Calculate the power being dissipated in the 2-Ω resistor for the circuit of Fig. 2-50b.

11 Use mesh analysis on the circuit shown in Fig. 2-48 to find the power being supplied by the dependent voltage source.

12 Use mesh analysis to find i_x in the circuit shown in Fig. 2-51.

Figure 2-51

See Prob. 12.

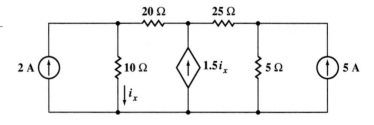

13 Use mesh analysis to help find the power generated by each of the five sources in Fig. 2-52.

Figure 2-52

See Prob. 13.

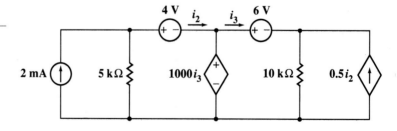

14 Find i_A in the circuit of Fig. 2-53.

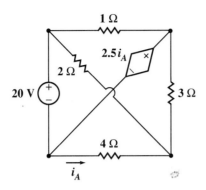

Figure 2-53

See Prob. 14.

15 With sources i_A and v_B on in the circuit of Fig. 2-54 and $v_C = 0$, $i_x = 20$ A; with i_A and v_C on and $v_B = 0$, $i_x = -5$ A; and finally, with all three sources on, $i_x = 12$ A. Find i_x if the only source operating is (a) i_A; (b) v_B; (c) v_C. (d) Find i_x if i_A and v_C are doubled in amplitude and v_B is reversed.

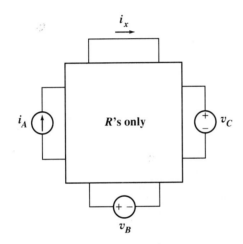

Figure 2-54

See Prob. 15.

16 Use superposition to find the value of v_x in the circuit of Fig. 2-55.

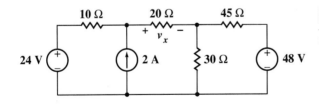

Figure 2-55

See Prob. 16.

17 Apply superposition to the circuit of Fig. 2-56 to find i_3.

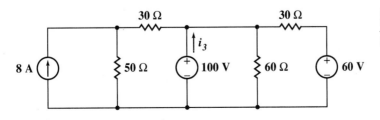

Figure 2-56

See Prob. 17.

18 Use superposition on the circuit shown in Fig. 2-57 to find the voltage V. Note that there is a dependent source present.

Figure 2-57

See Prob. 18.

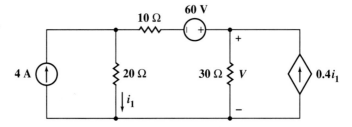

19 (a) Use the superposition theorem to find i_2 in the circuit shown in Fig. 2-58. (b) Calculate the power absorbed by each of the five circuit elements.

Figure 2-58

See Prob. 19.

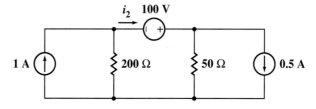

20 In the circuit shown in Fig. 2-59: (a) if $i_A = 10$ A and $i_B = 0$, then $v_3 = 80$ V; find v_3 if $i_A = 25$ A and $v_B = 0$. (b) If $i_A = 10$ A and $i_B = 25$ A, then $v_4 = 100$ V, while $v_4 = -50$ V if $i_A = 25$ A and $i_B = 10$ A; find v_4 if $i_A = 20$ A and $i_B = -10$ A.

Figure 2-59

See Prob. 20.

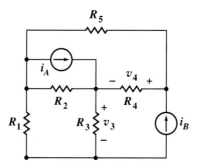

21 A certain practical dc voltage source can provide a current of 2.5 A when it is (momentarily) short-circuited, and can provide a power of 80 W to a 20-Ω load. Find (a) the open-circuit voltage and (b) the maximum power it could deliver to a well-chosen R_L. (c) What is the value of that R_L?

22 A practical current source provides 10 W to a 250-Ω load and 20 W to an 80-Ω load. A resistance R_L, with voltage v_L and current i_L, is connected to it. Find the values of R_L, v_L, and i_L if (a) $v_L i_L$ is a maximum; (b) v_L is a maximum; (c) i_L is a maximum.

23 Use source transformations and resistance combinations to simplify both networks of Fig. 2-60 until only two elements remain to the left of terminals a and b.

Figure 2-60

See Prob. 23.

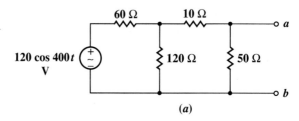

(a)

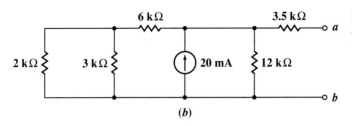

Figure 2-60

See Prob. 23.

(b)

24 If any value whatsoever may be selected for R_L in the circuit of Fig. 2-61, what is the maximum power that could be dissipated in R_L?

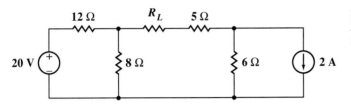

Figure 2-61

See Prob. 24.

25 (a) Find the Thévenin equivalent at terminals a and b for the network shown in Fig. 2-62. How much power would be delivered to a resistor connected to a and b if R_{ab} equals (b) 50 Ω; (c) 12.5 Ω?

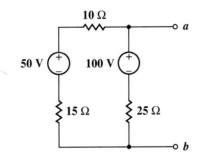

Figure 2-62

See Prob. 25.

26 For the network of Fig. 2-63: (a) remove terminal c and find the Norton equivalent seen at terminals a and b; (b) repeat for terminals b and c with a removed.

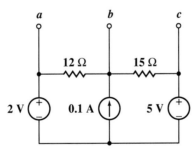

Figure 2-63

See Prob. 26.

27 Find the Thévenin equivalent of the network in Fig. 2-64 as viewed from terminals: (a) x and x'; (b) y and y'.

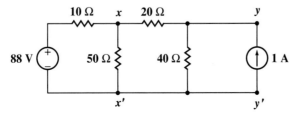

Figure 2-64

See Prob. 27.

28 (a) Find the Thévenin equivalent of the network shown in Fig. 2-65. (b) What power would be delivered to a load of 100 Ω at a and b?

Figure 2-65

See Prob. 28.

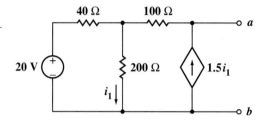

29 Find the Norton equivalent of the network shown in Fig. 2-66.

Figure 2-66

See Prob. 29.

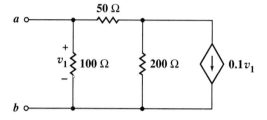

30 The voltage follower shown in Fig. 2-28 is modified by inserting a resistance R_i between the terminals between which v_i is defined. Find the new Thévenin equivalent in terms of R_i.

31 Find the Thévenin equivalent of the two-terminal network shown in Fig. 2-67.

Figure 2-67

See Prob. 31.

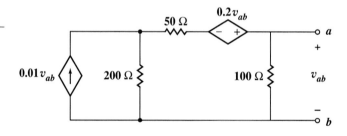

32 (a) Determine the Thévenin equivalent of the network shown in Fig. 2-68, and (b) find the power that can be drawn from it.

Figure 2-68

See Prob. 32.

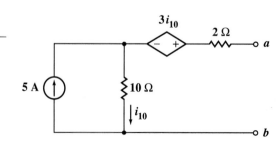

33 With reference to the circuit of Fig. 2-69: (a) determine that value of R_L to which a maximum power can be delivered, and (b) calculate the voltage across R_L then (+ reference at top).

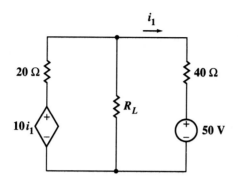

Figure 2-69

See Prob. 33.

34 (a) Find the Thévenin equivalent of the network shown in Fig. 2-70. (b) Repeat if the 10-A source is set equal to zero.

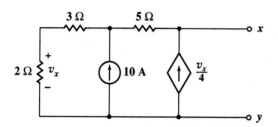

Figure 2-70

See Prob. 34.

35 (a) Construct all possible trees for the linear graph shown in Fig. 2-71. (b) If branches 1 and 2 are current sources and branch 3 is a voltage source, show all possible trees.

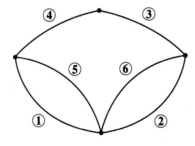

Figure 2-71

See Prob. 35.

36 Construct a tree for the circuit shown in Fig. 2-72 in which v_1 and v_2 are tree-branch voltages, write nodal equations, and solve for v_1.

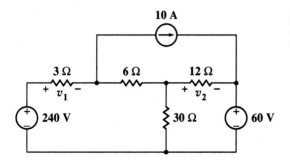

Figure 2-72

See Prob. 36.

37 Construct a suitable tree for the circuit of Fig. 2-73, assign tree-branch voltages, write KCL and control equations, and find i_2.

Figure 2-73

See Prob. 37.

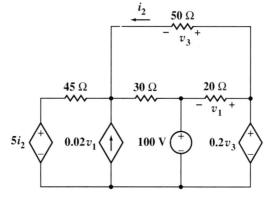

38 By constructing a suitable tree and using general nodal analysis on the circuit of Fig. 2-74, determine the value of V_2 that will result in $v_1 = 0$.

Figure 2-74

See Prob. 38.

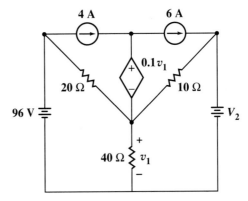

39 Construct a tree for the circuit shown in Fig. 2-75 in which i_1 and i_2 are link currents, write loop equations, and solve for i_1.

Figure 2-75

See Prob. 39.

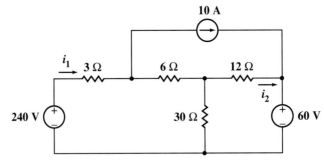

40 Select a tree for the circuit of Fig. 2-76 that leads to a single loop equation in i_1, and find i_1.

Figure 2-76

See Prob. 40.

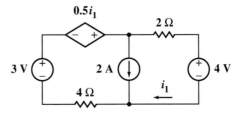

41 (*a*) Construct a tree for the circuit shown in Fig. 2-77 so that all the loop currents flow through the 7-Ω resistor. (*b*) Find i_5.

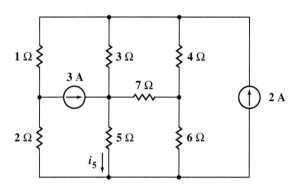

Figure 2-77

See Prob. 41.

42 Use general loop analysis on the circuit shown in Fig. 2-78 to find i_{40}.

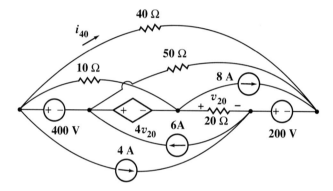

Figure 2-78

See Prob. 42.

43 Figure 2-79 shows one form of the equivalent circuit for a transistor amplifier. Determine the open-circuit value of v_2 and the output resistance (R_{th}) of the amplifier.

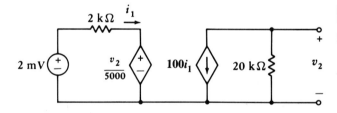

Figure 2-79

See Prob. 43.

44 (SPICE) Use SPICE to find v_3 in the circuit shown in Fig. 2-80.

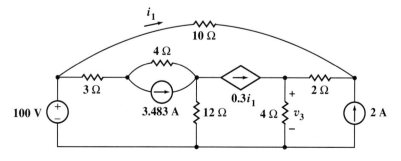

Figure 2-80

See Prob. 44.

45 (SPICE) Use SPICE to find i_5 in the circuit of Fig. 2-81.

Figure 2-81

See Prob. 45.

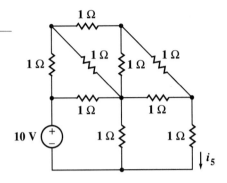

46 (SPICE) Use SPICE to find v_5 in the circuit shown in Fig. 2-82.

Figure 2-82

See Prob. 46.

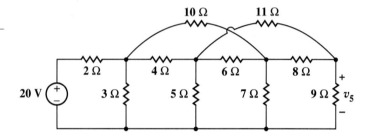

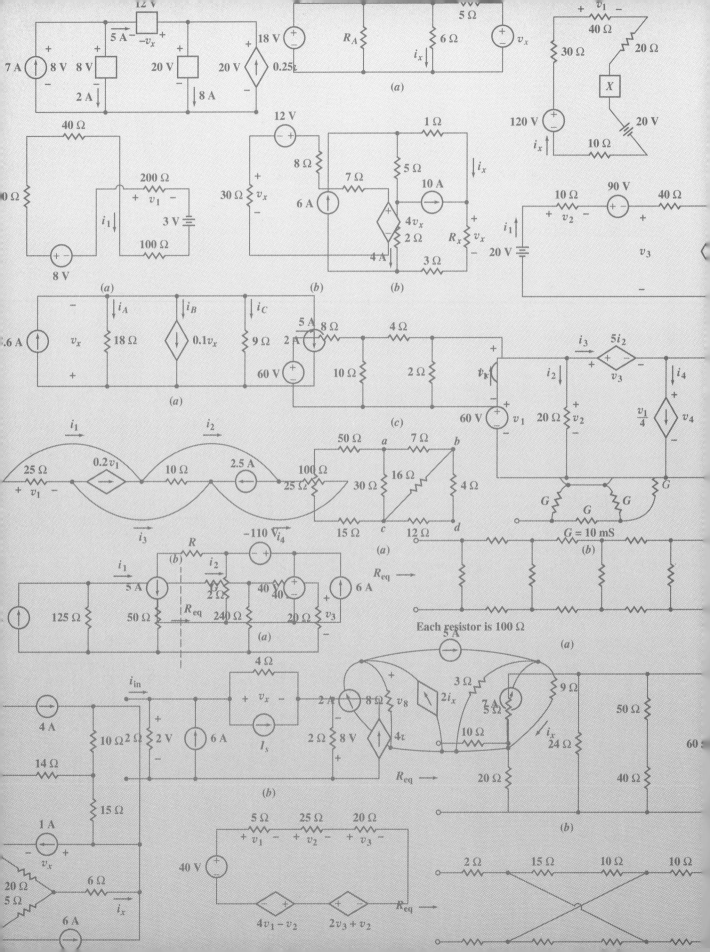

Part Two:
The Transient Circuit

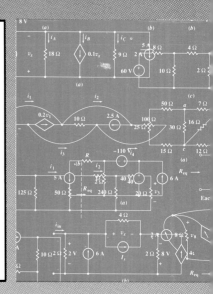

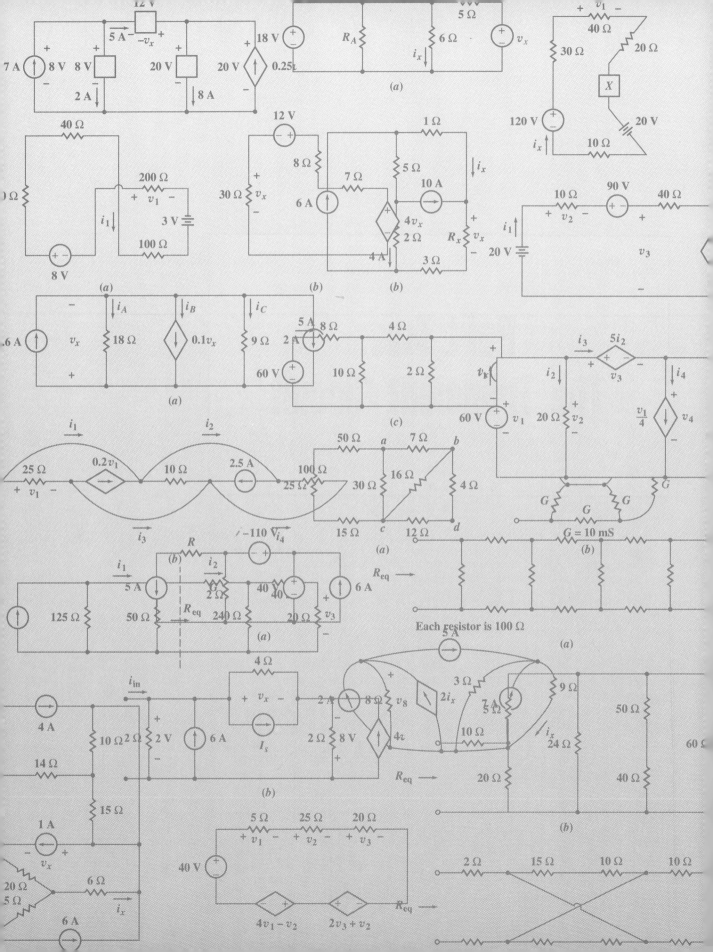

Inductance and Capacitance

We are now ready to begin the second major portion of our study of circuits. In this chapter we shall introduce two new simple circuit elements whose voltage-current relationships involve the rate of change of a voltage or current. Before we begin this new study, it will be worthwhile to pause for a moment and look back upon our study of the analysis of resistive circuits. A little bit of philosophical review might aid in our understanding of the coming work.

After setting up a satisfactory system of units, we began our discussion of electric circuits by defining current, voltage, and five simple circuit elements. The independent and dependent voltage and current sources were called *active elements* and the linear resistor was termed a *passive element,* although our definitions of *active* and *passive* are still slightly fuzzy and need to be brought into sharper focus. We now define an *active element* as an element which is capable of furnishing an average power greater than zero to some external device, where the average is taken over an infinite time interval. Ideal sources are active elements, and the operational amplifier is also an active device. A *passive element,* however, is defined as an element that cannot supply an average power that is greater than zero over an infinite time interval. The resistor falls into this category; the energy it receives is usually transformed into heat.

Each of these elements was defined in terms of the restrictions placed on its voltage-current relationship. In the case of the independent voltage source, for example, the terminal voltage must be completely independent of the current drawn from its terminals. We then considered circuits composed of the different building blocks. In general, we used only constant voltages and currents, but now that we have gained a familiarity with the basic analytical techniques by treating only the resistive circuit, we may begin to consider the much more interesting and practical circuits in which inductance and capacitance may be present and in which both the forcing functions and the responses usually vary with time.

Both the inductor, which is the subject of this and the following section, and the capacitor, which is discussed later in the chapter, are passive elements which are capable of storing and delivering finite amounts of energy. Unlike an ideal source, they cannot provide an unlimited amount of energy or a finite average power over an infinite time interval.

Although we shall define an inductor and inductance strictly from a circuit point of view, that is, by a voltage-current equation, a few comments about the historic development of the magnetic field may provide a better understanding of the definition. In the early 1800s the Danish scientist Oersted showed that a current-carrying conductor produced a magnetic field, or that compass needles were affected in the presence of a current-carrying conductor. In France, shortly thereafter, Ampère made some careful measurements which demonstrated that this magnetic field was linearly related to the current which produced it. The next step occurred some 20 years later when the English experimentalist Michael Faraday and the American inventor Joseph Henry discovered almost simultaneously[1] that a changing magnetic field could induce a voltage in a neighboring circuit. They showed that this voltage was proportional to the time rate of change of the current which produced the magnetic field. The constant of proportionality is what we now call the *inductance,* symbolized by L, and therefore

$$v = L \frac{di}{dt} \tag{1}$$

where we must realize that v and i are both functions of time. When we wish to emphasize this, we may do so by using the symbols $v(t)$ and $i(t)$.

The circuit symbol for the inductor is shown in Fig. 3-1, and it should be noted that the passive sign convention is used, just as it was with the resistor. The unit in which inductance is measured is the *henry* (H),[2] and the defining equation shows that the henry is just a shorter expression for a volt-second per ampere.

Figure 3-1

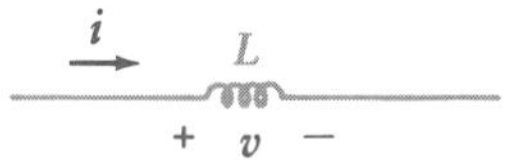

The reference signs for voltage and current are shown on the circuit symbol for an inductor: $v = L \, di/dt$.

The inductor whose inductance is defined by Eq. (1) is a mathematical model; it is an ideal element which we may use to approximate the behavior of a real device. A physical inductor may be constructed by winding a length of wire into a coil. This serves effectively to increase the current which is causing the magnetic field and also to increase the "number" of neighboring circuits into which Faraday's voltage may be induced. The result of this twofold effect is that the inductance of a coil is approximately proportional to the square of the number of complete turns made by the conductor out of which it is formed. For example, an inductor, or "coil," that has the form of a long helix of very small pitch is found to have an inductance of $\mu N^2 A/s$, where A is the cross-sectional area, s is the axial length of the helix, N is the number of complete turns of wire, and μ (mu) is a constant of the material inside the helix, called the *permeability.* For free space (and very closely for air), $\mu = \mu_0 = 4\pi \times 10^{-7}$ H/m.

Physical inductors should be available for you to view in your laboratory course. Topics such as the magnetic flux and the permeability, and the methods of using the characteristics of the physical coil to calculate a suitable inductance for the mathematical model, are treated in both physics courses and courses in electromagnetic field theory.

[1] Faraday won.
[2] An empty victory.

It is also possible to assemble electronic networks that do not contain any inductors and yet can provide the *v-i* relationship of Eq. (1) at their input terminals. We shall look at one example of this in Sec. 6-8.

Let us now scrutinize Eq. (1) to determine some of the electrical characteristics of this mathematical model. This equation shows that the voltage across an inductor is proportional to the time rate of change of the current through it. In particular, it shows that there is no voltage across an inductor carrying a constant current, regardless of the magnitude of this current. Accordingly, we may view an inductor as a "short circuit to dc."

Another fact which is evidenced by this equation is related to an infinite rate of change of the inductor current, such as that caused by an instantaneous change in current from one finite value to some other finite value. This sudden or discontinuous change in the current must be associated with an infinite voltage across the inductor. In other words, if we wish to produce an abrupt change in an inductor current, we must apply an infinite voltage. Although an infinite-voltage forcing function might be acceptable theoretically, it can never be a part of the phenomena displayed by a real physical device. As we shall see shortly, an abrupt change in the inductor current also requires an abrupt change in the energy stored in the inductor, and this sudden change in energy requires infinite power at that instant; infinite power is again not a part of the real physical world. In order to avoid infinite voltage and infinite power, an inductor current must not be allowed to jump instantaneously from one value to another.

If an attempt is made to open-circuit a physical inductor through which a finite current is flowing, an arc may appear across the switch. The stored energy is dissipated in ionizing the air in the path of the arc. This is useful in the ignition system of some automobiles, where the current through the spark coil is interrupted by the distributor and the arc appears across the spark plug.

We shall not consider any circuits at the present time in which an inductor is suddenly open-circuited. It should be pointed out, however, that we shall remove this restriction later when we hypothesize the existence of a voltage forcing function or response which does become infinite instantaneously.

Equation (1) may also be interpreted (and solved, if necessary) by graphical methods. Let us base an example of this technique on Fig. 3-2a.

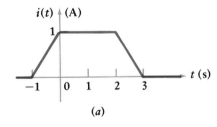

(a)

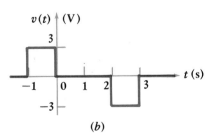

(b)

Figure 3-2

(a) The current waveform in a 3-H inductor. (b) The corresponding voltage waveform, $v = 3\,di/dt$.

Example 3-1   Given the waveform of the current in a 3-H inductor, as shown in Fig. 3-2a, determine the inductor voltage and sketch it.

Solution: The current is zero prior to $t = -1$ s, increases linearly to 1 A in the next second, remains at 1 A for 2 s, and then decreases to zero in the next second, remaining zero thereafter. If this current is present in a 3-H inductor, and if the voltage and current are defined so as to satisfy the passive sign convention, then we may use Eq. (1) to obtain the voltage waveform. Since the current is zero and constant for $t < -1$, the voltage is zero in this interval. The current then begins to increase at the linear rate of 1 A/s, and thus a constant voltage of 3 V is produced. During the following 2-s interval, the current is constant and the voltage is therefore zero. The final decrease of the current causes a negative 3 V and no response thereafter. The voltage waveform is sketched in Fig. 3-2b on the same time scale.   ■

Let us now investigate the effect of a more rapid rise and decay of the current between the zero and l-A values.

Example 3-2   Find the inductor voltage that results from applying the current waveform shown in Fig. 3-3a.

Figure 3-3

(a) The time required for the current of Fig. 3-2a to change from 0 to 1 and from 1 to 0 is decreased by a factor of 10. (b) The resultant voltage waveform. Note that the pulse widths are exaggerated slightly for clarity.

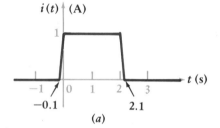

(a)

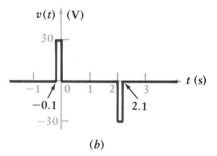

(b)

Solution: Note that the intervals required for the rise and fall have decreased to 0.1 s. Thus, the derivative must be 10 times as great in magnitude. This condition is shown in the current and voltage sketches of Figs. 3-3a and b. In the voltage waveforms of Figs. 3-2b and 3-3b, it is interesting to note that the area under each voltage pulse is 3 V-s.   ■

A further decrease in the length of these two intervals will produce a proportionally larger voltage magnitude, but only within the interval in which the current is increasing or decreasing. An abrupt change in the current will cause the infinite voltage "spikes" (each having an area of 3 V-s) that are suggested by the waveforms of Figs. 3-4a and b; or, from the equally valid but

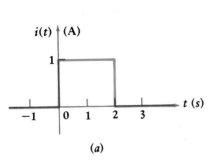

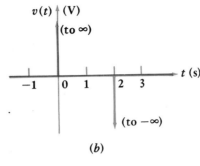

(a)　　　　(b)

Figure 3-4

(a) The time required for the current of Fig. 3-2a to change from 0 to 1 and from 1 to 0 is decreased to zero; the rise and fall are abrupt. (b) The associated voltage across the 3-H inductor consists of a positive and a negative infinite spike.

opposite point of view, these infinite voltage spikes are required to produce the abrupt changes in the current. It will be convenient later to provide such infinite voltages (and currents), and we shall then call them *impulses*; for the present, however, we shall stay closer to physical reality by not permitting infinite voltage, current, or power. An abrupt change in the inductor current is therefore temporarily outlawed.

Drill Problems

3-1. For the circuit of Fig. 3-5a, find (a) i_1; (b) i_2; (c) i_3.　　*Ans:* 5 A; −2 A; 2.1 A

3-2. The current through a 0.2-H inductor is shown as a function of time in Fig. 3-5b. Assume the passive sign convention, and find v_L at t equal to: (a) 0; (b) 2 ms; (c) 6 ms.　　*Ans:* 0.4 V; 0.2 V; −0.267 V

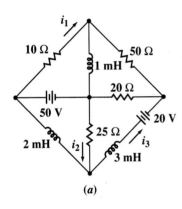

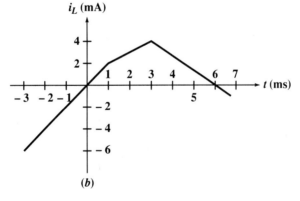

(a)　　　　(b)

Figure 3-5

See Drill Probs. 3-1 and 3-2.

3-3

Integral relationships for the inductor

We have defined inductance by a simple differential equation,

$$v = L \frac{di}{dt} \qquad (2)$$

and we have been able to draw several conclusions about the characteristics of an inductor from this relationship. For example, we have found that we may consider an inductor to be a short circuit to direct current, and we have agreed that we cannot permit an inductor current to change abruptly from one value to another, because this would require that an infinite voltage and power be associated with the inductor. The defining equation for inductance still contains more information, however. Rewritten in a slightly different form,

$$di = \frac{1}{L} v \, dt$$

it invites integration. Let us first consider the limits to be placed on the two integrals. We desire the current i at time t, and this pair of quantities therefore provides the upper limits on the integrals appearing on the left and right side of the equation, respectively; the lower limits may also be kept general by merely assuming that the current is $i(t_0)$ at time t_0. Thus

$$\int_{i(t_0)}^{i(t)} di = \frac{1}{L} \int_{t_0}^{t} v \, dt$$

or

$$i(t) - i(t_0) = \frac{1}{L} \int_{t_0}^{t} v \, dt$$

and

$$i(t) = \frac{1}{L} \int_{t_0}^{t} v \, dt + i(t_0) \tag{3}$$

Equation (2) expresses the inductor voltage in terms of the current, whereas Eq. (3) gives the current in terms of the voltage. Other forms are also possible for this latter equation. We may write the integral as an indefinite integral and include a constant of integration k:

$$i(t) = \frac{1}{L} \int v \, dt + k \tag{4}$$

We also may assume that we are solving a realistic problem in which the selection of t_0 as $-\infty$ ensures no current or energy in the inductor. Thus, if $i(t_0) = i(-\infty) = 0$, then

$$i(t) = \frac{1}{L} \int_{-\infty}^{t} v \, dt \tag{5}$$

Let us consider the use of these several integrals by working a simple example.

Example 3-3   Suppose that the voltage across a 2-H inductor is known to be $6 \cos 5t$ V. What information is then available about the inductor current?

Solution: From Eq. (3),

$$i(t) = \frac{1}{2} \int_{t_0}^{t} 6 \cos 5t \, dt + i(t_0)$$

or

$$i(t) = \tfrac{1}{2} \left(\tfrac{6}{5}\right) \sin 5t - \tfrac{1}{2} \left(\tfrac{6}{5}\right) \sin 5t_0 + i(t_0)$$

$$= 0.6 \sin 5t - 0.6 \sin 5t_0 + i(t_0)$$

The first term indicates that the inductor current varies sinusoidally; the second and third terms together merely represent a constant which becomes known when the current is numerically specified at some instant of time. Let us assume that the statement of our example problem also shows us that the current is 1 A at $t = -\pi/2$ s. We thus identify t_0 as $-\pi/2$ with $i(t_0) = 1$, and find that

$$i(t) = 0.6 \sin 5t - 0.6 \sin (-2.5\pi) + 1$$

or $\qquad\qquad\qquad\qquad i(t) = 0.6 \sin 5t + 1.6$

We may obtain the same result from Eq. (4). We have

$$i(t) = 0.6 \sin 5t + k$$

and we establish the numerical value of k by forcing the current to be 1 A at $t = -\pi/2$:

$$1 = 0.6 \sin (-2.5\pi) + k$$

or $\qquad\qquad\qquad\qquad k = 1 + 0.6 = 1.6$

and $\qquad\qquad\qquad\qquad i(t) = 0.6 \sin 5t + 1.6$

once more.

Equation (5) is going to cause trouble with this particular voltage. We based the equation on the assumption that the current was zero when $t = -\infty$. To be sure, this must be true in the real, physical world, but we are working in the land of the mathematical model; our elements and forcing functions are all idealized. The difficulty arises after we integrate, obtaining

$$i(t) = 0.6 \sin 5t \Big|_{-\infty}^{t}$$

and attempt to evaluate the integral at the lower limit:

$$i(t) = 0.6 \sin 5t - 0.6 \sin (-\infty)$$

The sine of $\pm\infty$ is indeterminate; we might just as well represent it by an unknown constant:

$$i(t) = 0.6 \sin 5t + k$$

We see that this result is identical with that which we obtained when we assumed an arbitrary constant of integration in Eq. (4). ∎

We should not make any snap judgments, based on this example, as to which single form we are going to use forever after; each has its advantages, depending on the problem and the application. Equation (3) represents a long, general method, but it shows clearly that the constant of integration is a current. Equation (4) is a somewhat more concise expression of Eq. (3), but the nature of the integration constant is suppressed. Finally, Eq. (5) is an excellent expression, since no constant is necessary; however, it applies only when the current is zero at $t = -\infty$ and when the analytical expression for the current is not indeterminate there.

Let us now turn our attention to power and energy. The absorbed power is given by the current-voltage product

$$p = vi = Li\frac{di}{dt} \qquad \text{W}$$

The energy w_L accepted by the inductor is stored in the magnetic field around the coil and is expressed by the integral of the power over the desired time interval,

$$\int_{t_0}^{t} p \, dt = L\int_{t_0}^{t} i\frac{di}{dt} \, dt = L\int_{i(t_0)}^{i(t)} i \, di = \frac{1}{2}L\{[i(t)]^2 - [i(t_0)]^2\}$$

and thus

$$w_L(t) - w_L(t_0) = \tfrac{1}{2}L\,\{[i(t)]^2 - [i(t_0)]^2\} \qquad \text{J} \tag{6}$$

where we have again assumed that the current is $i(t_0)$ at time t_0. In using the energy expression, it is customary to assume that a value of t_0 is selected at which the current is zero; it is also customary to assume that the energy is zero at this time. We then have simply

$$w_L(t) = \tfrac{1}{2}Li^2 \tag{7}$$

where we now understand that our reference for zero energy is any time at which the inductor current is zero. At any subsequent time at which the current is zero, we also find no energy stored in the coil. Whenever the current is not zero, and regardless of its direction or sign, energy is stored in the inductor. It follows, therefore, that power must be delivered to the inductor for a part of the time and recovered from the inductor later. All the stored energy may be recovered from an ideal inductor; there are no storage charges or agent's commissions in the mathematical model. A physical coil, however, must be constructed out of real wire and thus will always have a resistance associated with it. Energy can no longer be stored and recovered without loss.

These ideas may be illustrated by a simple example. In Fig. 3-6, a 3-H inductor is shown in series with a 0.1-Ω resistor and a sinusoidal current source, $i_s = 12 \sin (\pi t/6)$ A. The resistor may be interpreted, if we wish, as the resistance of the wire which must be associated with the physical coil.

Figure 3-6

A sinusoidal current is applied as a forcing function to a series RL circuit.

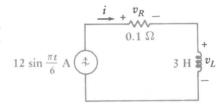

Example 3-4   Let us first find the maximum energy stored in the inductor of Fig. 3-6, and then calculate how much energy is dissipated in the resistor in the time during which the energy is being stored and recovered in the inductor.

Solution: The voltage across the resistor is given by

$$v_R = Ri = 1.2 \sin \frac{\pi}{6}t$$

and the voltage across the inductor is found by applying the defining equation for inductance,

$$v_L = L\frac{di}{dt} = 3\frac{d}{dt}\left(12 \sin \frac{\pi}{6} t\right) = 6\pi \cos \frac{\pi}{6} t$$

The energy stored in the inductor is

$$w_L = \frac{1}{2}Li^2 = 216 \sin^2 \frac{\pi}{6} t \qquad \text{J}$$

and it is apparent that this energy increases from zero at $t = 0$ to 216 J at $t = 3$ s. Thus, the maximum energy stored in the inductor is 216 J.

During the next 3 s, the energy leaves the inductor completely. Let us see what price we have paid in this coil for the privilege of storing and removing 216 J in these 6 s. The power dissipated in the resistor is easily found as

$$p_R = i^2R = 14.4 \sin^2 \frac{\pi}{6}t$$

and the energy converted into heat in the resistor within this 6-s interval is therefore

$$w_R = \int_0^6 p_R \, dt = \int_0^6 14.4 \sin^2 \frac{\pi}{6}t \, dt$$

or

$$w_R = \int_0^6 14.4 \left(\frac{1}{2}\right)\left(1 - \cos\frac{\pi}{3}t\right) dt = 43.2 \text{ J}$$

Thus, we have expended 43.2 J in the process of storing and recovering 216 J in a 6-s interval. This represents 20 percent of the maximum stored energy, but it is a reasonable value for many coils having this large an inductance. For coils having an inductance of about 100 μH, we should expect a figure closer to 2 or 3 percent. In Chap. 13 we shall formalize this concept by defining a quality factor Q that is proportional to the ratio of the maximum energy stored to the energy lost per period. ∎

Let us now recapitulate by listing several characteristics of an inductor which result from its defining equation:

1 There is no voltage across an inductor if the current through it is not changing with time. An inductor is therefore a short circuit to dc.

2 A finite amount of energy can be stored in an inductor even if the voltage across the inductor is zero, such as when the current through it is constant.

3 It is impossible to change the current through an inductor by a finite amount in zero time, for this requires an infinite voltage across the inductor. It will be advantageous later to *hypothesize* that such a voltage may be generated and applied to an inductor, but for the present we shall avoid such a forcing function or response. An inductor resists an abrupt change in the current through it in a manner analogous to the way a mass resists an abrupt change in its velocity.

4 The inductor never dissipates energy, but only stores it. Although this is true for the mathematical model, it is not true for a physical inductor.

Drill Problem

3-3. Let $L = 25$ mH for the inductor of Fig. 3-1. (a) Find v at $t = 12$ ms if $i = 10te^{-100t}$ A. (b) Find i at $t = 0.1$ s if $v = 6e^{-12t}$ V and $i(0) = 10$ A. If $i = 8(1 - e^{-40t})$ mA, find (c) the power being delivered to the inductor at $t = 50$ ms, and (d) the energy stored in the inductor at $t = 40$ ms.

Ans: -15.06 mV; 24.0 A; 7.49 μW; 0.510 μJ

3-4

The capacitor

Our next passive circuit element is the capacitor. We define *capacitance C* by the voltage-current relationship

$$i = C\frac{dv}{dt} \tag{8}$$

where v and i satisfy the conventions for a passive element, as shown in Fig. 3-7. From Eq. (8), we may determine the unit of capacitance as an ampere-second per volt, or coulomb per volt, but we shall now define the *farad* (F) as one coulomb per volt.

Figure 3-7

The current and voltage reference marks are shown on the circuit symbol for a capacitor so that $i = C \, dv/dt$.

The capacitor whose capacitance is defined by Eq. (8) is again a mathematical model of a real device. The construction of the physical device is suggested by the circuit symbol shown in Fig. 3-7, in much the same way that the helical symbol used for the inductor represents the coiled wire in that physical element. A capacitor, physically, consists of two conducting surfaces on which charge may be stored, separated by a thin insulating layer which has a very large resistance. If we assume that this resistance is sufficiently large that it may be considered infinite, then equal and opposite charges placed on the capacitor "plates" can never recombine, at least by any path *within* the element. Let us visualize some external device, such as a current source, connected to this capacitor and causing a positive current to flow into one plate of the capacitor and out of the other plate. Equal currents are entering and leaving the two terminals of the element, and this is no more than we expect for any circuit element. Now let us examine the interior of the capacitor. The positive current entering one plate represents positive charge moving toward that plate through its terminal lead; this charge cannot pass through the interior of the capacitor, and it therefore accumulates on the plate. As a matter of fact, the current and the increasing charge are related by the familiar equation

$$i = \frac{dq}{dt}$$

Now let us pose ourselves a troublesome problem by considering this plate as an overgrown node and applying Kirchhoff's current law. It apparently does not hold; current is approaching the plate from the external circuit, but it is not flowing out of the plate into the "internal circuit." This dilemma bothered a famous Scottish scientist, James Clerk Maxwell, more than a century ago, and the unified electromagnetic theory which he then developed hypothesizes a displacement current which is present wherever an electric field or a voltage is varying with time. The displacement current flowing internally beween the capacitor plates is exactly equal to the conduction current flowing in the capacitor leads; Kirchhoff's current law is therefore satisfied if we include both conduction and displacement currents. However, circuit analysis is not concerned with this internal displacement current, and since it is fortunately equal to the conduction current, we may consider Maxwell's hypothesis as relating the conduction current to the changing voltage across the capacitor.

The relationship is linear, and the constant of proportionality is obviously the capacitance C:

$$i_{\text{disp}} = i = C\frac{dv}{dt}$$

A capacitor constructed of two parallel conducting plates of area A, separated by a distance d, has a capacitance $C = \varepsilon A/d$, where ε is the *permittivity,* a constant of the insulating material between the plates, and where the linear dimensions of the conducting plates are all very much greater than d. For air or vacuum, $\varepsilon = \varepsilon_0 = 8.854$ pF/m.

The concepts of the electric field and displacement current, and the generalized form of Kirchhoff's current law, are more appropriate subjects for courses in physics and electromagnetic field theory, as is the determination of a suitable mathematical model to represent a specific physical capacitor.

Several important characteristics of our new mathematical model can be discovered from the defining equation, Eq. (8). A constant voltage across a capacitor requires zero current passing through it; a capacitor is thus an "open circuit to dc." This fact is certainly represented by the capacitor symbol. It is also apparent that a sudden jump in the voltage requires an infinite current. Just as we outlawed abrupt changes in inductor currents and the associated infinite voltages on physical grounds, we shall not permit abrupt changes in capacitor voltage; the infinite current (and infinite power) which results is nonphysical. (We shall remove this restriction at the time we assume the existence of the current impulse.)

The capacitor voltage may be expressed in terms of the current by integrating Eq. (8). We first obtain

$$dv = \frac{1}{C}i\,dt$$

and then integrate between the times t_0 and t and between the corresponding voltages $v(t_0)$ and $v(t)$:

$$v(t) = \frac{1}{C}\int_{t_0}^{t} i\,dt + v(t_0) \tag{9}$$

Equation (9) may also be written as an indefinite integral plus a constant of integration:

$$v(t) = \frac{1}{C}\int i\,dt + k \tag{10}$$

Finally, in many real problems, t_0 may be selected as $-\infty$ and $v(-\infty)$ as zero:

$$v(t) = \frac{1}{C}\int_{-\infty}^{t} i\,dt \tag{11}$$

Since the integral of the current over any time interval is the charge accumulated in that period on the capacitor plate into which the current is flowing, it is apparent that capacitance might have been defined by

$$q = Cv$$

The similarity between the several integral equations introduced in this section and those appearing in our discussion of inductance is striking and suggests that the duality we observed between mesh and nodal equations in resistive networks may be extended to include inductance and capacitance as well. The principle of duality will be presented and discussed later in this chapter.

Example 3-5   In order to illustrate the use of the several integral equations displayed in the preceding discussion, let us find the capacitor voltage which is associated with the current shown graphically in Fig. 3-8a. We shall assume that the single 20-mA rectangular pulse of 2-ms duration is applied to a 5-μF capacitor.

Figure 3-8

(a) The current waveform applied to a 5-μF capacitor. (b) The resultant voltage waveform, easily obtained by integrating graphically.

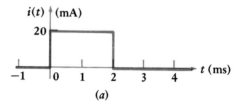

(a)

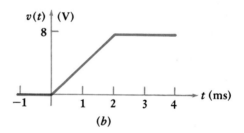

(b)

Solution: Interpreting Eq. (9) graphically, we know that the difference between the values of the voltage at t and t_0 is proportional to the area under the current curve between these same two values of time. The proportionality constant is $1/C$. The area can be obtained from Fig. 3-8a by inspection for desired values of t_0 and t. Thus, if $t_0 = -0.5$ and $t = 0.5$ (in ms),

$$v(0.5) = 2 + v(-0.5)$$

or, if $t_0 = 0$ and $t = 3$,

$$v(3) = 8 + v(0)$$

We may express our results in more general terms by dividing the interesting range of time into several intervals. Let us select our starting point t_0 prior to zero time. Then the first interval of t is selected between t_0 and zero:

$$v(t) = 0 + v(t_0) \qquad t_0 \le t \le 0$$

and since our waveform implies that no current has ever been applied to this capacitor since the Creation,

$$v(t_0) = 0$$

and, thus,

$$v(t) = 0 \qquad t \le 0$$

If we now consider the time interval represented by the rectangular pulse, we obtain

$$v(t) = 4000t \qquad 0 \leq t \leq 2 \text{ ms}$$

For the semi-infinite interval following the pulse, we have

$$v(t) = 8 \qquad t \geq 2 \text{ ms}$$

The results for these three intervals therefore provide us with analytical expressions for the capacitor voltage at any time after $t = t_0$; the time t_0, however, may be selected as early as we wish. The results are expressed much more simply in a sketch than by these analytical expressions, as shown in Fig. 3-8b. ∎

The power delivered to a capacitor is

$$p = vi = Cv\frac{dv}{dt}$$

and the energy stored in its electric field is therefore

$$\int_{t_0}^{t} p \, dt = C\int_{t_0}^{t} v\frac{dv}{dt}\, dt = C\int_{v(t_0)}^{v(t)} v \, dv = \frac{1}{2}C\{[v(t)]^2 - [v(t_0)]^2\}$$

and thus

$$w_C(t) - w_C(t_0) = \tfrac{1}{2}C\{[v(t)]^2 - [v(t_0)]^2\} \qquad \text{J} \qquad (12)$$

where the stored energy is $w_C(t_0)$ and the voltage is $v(t_0)$ at t_0. If we select a zero-energy reference at t_0, implying that the capacitor voltage is also zero at that instant, then

$$w_C(t) = \tfrac{1}{2}Cv^2 \qquad (13)$$

Let us consider a simple numerical example. As sketched in Fig. 3-9, we shall assume a sinusoidal voltage source in parallel with a 1-MΩ resistor and a 20-μF capacitor. The parallel resistor may be assumed to represent the resistance of the insulator or dielectric between the plates of the physical capacitor.

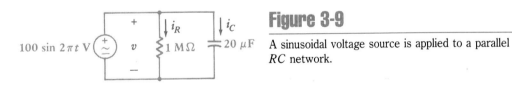

Figure 3-9

A sinusoidal voltage source is applied to a parallel RC network.

Example 3-6 Determine the maximum energy stored in the capacitor of Fig. 3-9 and the energy dissipated in the resistor over the interval $0 \leq t \leq 0.5$ s.

Solution: The current through the resistor is

$$i_R = \frac{v}{R} = 10^{-4} \sin 2\pi t \qquad \text{A}$$

while the current through the capacitor is

$$i_C = C\frac{dv}{dt} = 20 \times 10^{-6}\frac{d}{dt}(100 \sin 2\pi t)$$

$$= 4\pi \times 10^{-3} \cos 2\pi t \qquad \text{A}$$

We next obtain the energy stored in the capacitor,

$$w_C(t) = \tfrac{1}{2}Cv^2 = 0.1 \sin^2 2\pi t \qquad \text{J}$$

and see that the energy increases from zero at $t = 0$ to a maximum of 0.1 J at $t = \tfrac{1}{4}$ s and then decreases to zero in another $\tfrac{1}{4}$ s. During this $\tfrac{1}{2}$-s interval, the energy dissipated in the resistor is

$$w_R = \int_0^{0.5} p_R \, dt = \int_0^{0.5} 10^{-2} \sin^2 2\pi t \, dt = 2.5 \text{ mJ}$$

We conclude that $w_{C,\text{max}} = 0.1$ J and $w_R = 2.5$ mJ. Thus, an energy equal to 2.5 percent of the maximum stored energy is lost in the process of storing and removing the energy in the ideal capacitor. Much smaller values are possible in "low-loss" capacitors, but these smaller percentages are customarily associated with much smaller capacitors. ■

Some of the important characteristics of a capacitor are now apparent:

> 1 There is no current through a capacitor if the voltage across it is not changing with time. A capacitor is therefore an open circuit to dc.
> 2 A finite amount of energy can be stored in a capacitor even if the current through the capacitor is zero, such as when the voltage across it is constant.
> 3 It is impossible to change the voltage across a capacitor by a finite amount in zero time, for this requires an infinite current through the capacitor. It will be advantageous later to *hypothesize* that such a current may be generated and applied to a capacitor, but for the present we shall avoid such a forcing function or response. A capacitor resists an abrupt change in the voltage across it in a manner analogous to the way a spring resists an abrupt change in its displacement.
> 4 The capacitor never dissipates energy, but only stores it. Although this is true for the mathematical model, it is not true for a physical capacitor.

It is interesting to anticipate our discussion of duality by rereading the previous four statements with certain words replaced by their "duals." If capacitor and inductor, capacitance and inductance, voltage and current, across and through, open circuit and short circuit, spring and mass, and displacement and velocity are interchanged (in either direction), the four statements previously given for inductors are obtained.

We conclude our introduction to the ideal capacitor by seeing how it might be used in conjunction with an ideal op-amp. Up to now we have discussed the op-amp only as it is used as a voltage follower, a device having a very low output resistance and an output voltage that is virtually *equal* to the input voltage. We will now show that an ideal capacitor and an ideal resistor can be combined with an op-amp to form a device having an output voltage that is proportional to the *time integral* of the input.

To create this integrator, we ground the noninverting input of an op-amp that has infinite input resistance and zero output resistance, install an ideal capacitor as a feedback element from the output back to the inverting input,

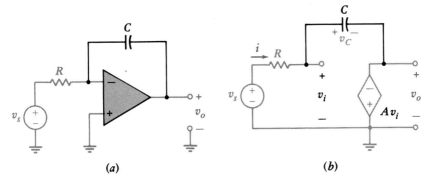

Figure 3-10

(*a*) An op-amp is connected as an integrator. (*b*) The equivalent circuit, assuming $R_i = \infty$ and $R_o = 0$.

and connect a signal source v_s to the inverting input through an ideal resistor, as shown in Fig. 3-10*a*. With $R_i = \infty$ and $R_o = 0$, we obtain the equivalent circuit shown in Fig. 3-10*b*.

The output voltage is related to the voltage v_i by

$$v_o = -Av_i$$

so that

$$v_i = \frac{-v_o}{A} \tag{14}$$

Let us relate the output v_o to the signal voltage v_s by first assuming that A is infinite. It follows from Eq. (14) that $v_i = 0$. Therefore, we see from Fig. 3-10*b* that

$$i = \frac{v_s}{R}$$

Also, with $v_i = 0$, the capacitor voltage v_C is equal to $(-v_o)$, and

$$v_C = -v_o = \frac{1}{C}\int_0^t i \, dt + v_C(0) = \frac{1}{C}\int_0^t \frac{v_s}{R} dt + v_C(0)$$

or

$$v_o = -\frac{1}{RC}\int_0^t v_s \, dt - v_C(0) \tag{15}$$

We therefore have combined a resistor, a capacitor, and an op-amp to form an *integrator*. Note that the first term of the output is $1/RC$ times the *negative* of the integral of the input from $t = 0$ to t, and the second term is the negative of the initial value of v_C. The value of $(RC)^{-1}$ can be made equal to unity if we wish by choosing $R = 1$ MΩ and $C = 1$ μF, for example; or other selections may be made that will increase or decrease the output voltage. The sign change is often convenient when the integrator is used for the simulation of engineering systems. However, the sign of the output may also be changed by using the inverting amplifier whose circuit appears in Chap. 6.

We should note in passing that, when the value of A is assumed to increase without limit and become infinite, the voltage of the inverting terminal, relative to ground, approaches zero. That is, v_i becomes virtually zero, as we concluded from Eq. (14). It is often said that the inverting terminal of the op-amp is then a *virtual ground*.

The initial voltage $v_C(0)$ appearing in Eq. (15) may be included in the integrator circuit by adding a battery and a normally closed switch, as indicated in Fig. 3-11. In practical circuits, both the switch and the initial voltage are usually electronic devices, such as transistors or other op-amps.

Figure 3-11

The inclusion of a normally closed switch and an ideal voltage source enables the value of $v_o(0^+)$ to be set at $-v_C(0^-)$.

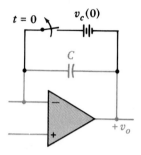

If we do not make the assumption that A is infinite, then we may write one KVL equation around the perimeter of the circuit in Fig. 3-10b,

$$-v_s + Ri + \frac{1}{C}\int_0^t i\,dt + v_C(0) + v_o = 0$$

use $i = (v_s - v_i)/R$ to eliminate i, and then use Eq. (14) to eliminate v_i. The result is

$$\left(1 + \frac{1}{A}\right)v_o = -\frac{1}{RC}\int_0^t \left(v_s + \frac{v_o}{A}\right)dt - v_C(0) \tag{16}$$

As $A \to \infty$, this results in an equation identical to Eq. (15).

Before we leave the integrator circuit, we might anticipate a question from an inquisitive reader, "Could we use an inductor in place of the capacitor and obtain a differentiator?" Indeed we could, but circuit designers usually avoid the use of inductors whenever possible because of their size, weight, cost, and associated resistance and capacitance. Instead, it is possible to interchange the positions of the resistor and capacitor in Fig. 3-10a and obtain a differentiator. The analysis of this circuit is the subject of Prob. 10.

Drill Problems

3-4. Assume that v and i satisfy the passive sign convention for a 2.5-μF capacitor. At $t = 2$ ms, find (a) i if $v = 20e^{-500t}$ V; (b) v if $i = 20e^{-500t}$ mA and $v(0) = 4$ V; and (c) i if the stored energy is $w = 20e^{-500t}$ μJ for $t > 0$.

Ans: -9.17 mA; 14.11 V; ± 1.516 mA

3-5. In the circuit of Fig. 3-10a, let v_s be a single triangular pulse, rising linearly from 0 at $t = 0$ to 12 mV at $t = 1$ s, and then decreasing linearly to 0 at $t = 2$ s. Let $C = 0.5$ μF and $R = 100$ kΩ, and assume that $v_C(0) = 0$. Find the value of v_o at t equal to (a) 0.5 s; (b) 1 s; (c) 1.5 s; (d) 2 s; (e) 2.5 s.

Ans: -30 mV; -120 mV; -210 mV; -240 mV; -240 mV

Now that we have added the inductor and capacitor to our list of passive circuit elements, we need to decide whether or not the methods we have developed for resistive circuit analysis are still valid. It will also be convenient to learn how to replace series and parallel combinations of either of these elements with simpler equivalents, just as we did with resistors in Chap. 1.

We look first at Kirchhoff's two laws, both of which are axiomatic. However, when we hypothesized these two laws, we did so with no restrictions as to the types of elements constituting the network. Both, therefore, remain valid.

Now we may extend the procedures we have derived for reducing various combinations of resistors into one equivalent resistor to the analogous cases of inductors and capacitors. We shall first consider an ideal voltage source applied to the series combination of N inductors, as shown in Fig. 3-12a. We desire a

3-5

Inductance and capacitance combinations

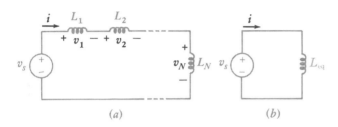

(a)　(b)

Figure 3-12

(a) A circuit containing N inductors in series. (b) The desired equivalent circuit, in which $L_{eq} = L_1 + L_2 + \cdots + L_N$.

single equivalent inductor, with inductance L_{eq}, which may replace the series combination so that the source current $i(t)$ is unchanged. The equivalent circuit is sketched in Fig. 3-12b. For the original circuit,

$$v_s = v_1 + v_2 + \cdots + v_N$$

$$= L_1 \frac{di}{dt} + L_2 \frac{di}{dt} + \cdots + L_N \frac{di}{dt}$$

$$= (L_1 + L_2 + \cdots + L_N) \frac{di}{dt}$$

or, written more concisely,

$$v_s = \sum_{n=1}^{N} v_n = \sum_{n=1}^{N} L_n \frac{di}{dt} = \frac{di}{dt} \sum_{n=1}^{N} L_n$$

But for the equivalent circuit we have

$$v_s = L_{eq} \frac{di}{dt}$$

and thus the equivalent inductance is

$$L_{eq} = (L_1 + L_2 + \cdots + L_N)$$

or

$$L_{eq} = \sum_{n=1}^{N} L_n$$

The inductor which is equivalent to several inductors connected in series is one whose inductance is the sum of the inductances in the original circuit. This is exactly the same result we obtained for resistors in series.

Figure 3-13

(a) The parallel combination of N inductors. (b) The equivalent circuit, where $L_{eq} = 1/[(1/L_1) + (1/L_2) + \cdots + (1/L_N)]$.

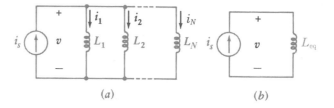

(a)   (b)

The combination of a number of parallel inductors is accomplished by writing the single nodal equation for the original circuit, shown in Fig. 3-13a,

$$i_s = \sum_{n=1}^{N} i_n = \sum_{n=1}^{N} \left[\frac{1}{L_n} \int_{t_0}^{t} v\, dt + i_n(t_0) \right]$$

$$= \left(\sum_{n=1}^{N} \frac{1}{L_n} \right) \int_{t_0}^{t} v\, dt + \sum_{n=1}^{N} i_n(t_0)$$

and comparing it with the result for the equivalent circuit of Fig. 3-13b,

$$i_s = \frac{1}{L_{eq}} \int_{t_0}^{t} v\, dt + i_s(t_0)$$

Since Kirchhoff's current law demands that $i_s(t_0)$ be equal to the sum of the branch currents at t_0, the two integral terms must also be equal; hence,

$$L_{eq} = \frac{1}{1/L_1 + 1/L_2 + \cdots + 1/L_N}$$

For the special case of two inductors in parallel,

$$L_{eq} = \frac{L_1 L_2}{L_1 + L_2}$$

and we note that inductors in parallel combine exactly as do resistors in parallel.

In order to find a capacitor which is equivalent to N capacitors in series, we use the circuit of Fig. 3-14a and its equivalent in Fig. 3-14b to write

$$v_s = \sum_{n=1}^{N} v_n = \sum_{n=1}^{N} \left[\frac{1}{C_n} \int_{t_0}^{t} i\, dt + v_n(t_0) \right]$$

$$= \left(\sum_{n=1}^{N} \frac{1}{C_n} \right) \int_{t_0}^{t} i\, dt + \sum_{n=1}^{N} v_n(t_0)$$

and
$$v_s = \frac{1}{C_{eq}} \int_{t_0}^{t} i\, dt + v_s(t_0)$$

Figure 3-14

(a) A circuit containing N capacitors in series. (b) The desired equivalent, $C_{eq} = 1/[(1/C_1) + (1/C_2) + \cdots + (1/C_N)]$.

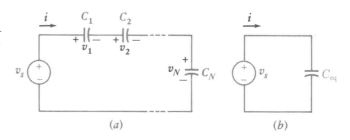

(a)   (b)

However, Kirchhoff's voltage law establishes the equality of $v_s(t_0)$ and the sum of the capacitor voltages at t_0; thus

$$C_{eq} = \frac{1}{1/C_1 + 1/C_2 + \cdot \cdot \cdot + 1/C_N}$$

and capacitors in series combine as do conductances in series, or resistors in *parallel*. The special case of two capacitors in *series*, of course, yields

$$C_{eq} = \frac{C_1 C_2}{C_1 + C_2}$$

Finally, the circuits of Fig. 3-15 enable us to establish the value of the capacitance of the capacitor which is equivalent to N parallel capacitors as

$$C_{eq} = C_1 + C_2 + \cdot \cdot \cdot + C_N$$

and it is no great source of amazement to note that capacitors in parallel combine in the same manner in which we combine resistors in series, that is, by simply adding all the individual capacitances.

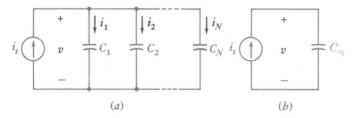

(a) (b)

Figure 3-15

(a) The parallel combination of N capacitors. (b) The equivalent circuit, where $C_{eq} = C_1 + C_2 + \cdot \cdot \cdot + C_N$.

Example 3-7 As an example in which some simplification may be achieved by combining like elements, simplify the network of Fig. 3-16a.

Solution: The 6- and 3-μF series capacitors are first combined into a 2-μF equivalent, and this capacitor is then combined with the 1-μF element with which it is in parallel to yield an equivalent capacitance of 3 μF. In addition, the 3- and 2-H inductors are replaced by an equivalent 1.2-H inductor, which is then added to the 0.8-H element to give a total equivalent inductance of 2 H. The much simpler (and probably less expensive) equivalent network is shown in Fig. 3-16b. ∎

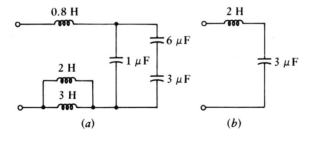

(a) (b)

Figure 3-16

(a) A given LC network. (b) A simpler equivalent circuit.

The network shown in Fig. 3-17 contains three inductors and three capacitors, but no series or parallel combinations of either the inductors or the capacitors can be achieved. Simplification of this network cannot be accomplished at this time.

Next let us turn to mesh, loop, and nodal analysis. Since we already know that we may safely apply Kirchhoff's laws, we should have little difficulty in

Figure 3-17

An *LC* network in which no series or parallel combinations of either the inductors or the capacitors can be made.

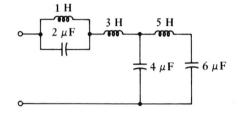

writing a set of equations that are both sufficient and independent. They will be constant-coefficient linear integrodifferential equations, however, which are hard enough to pronounce, let alone solve. Consequently, we shall write them now to gain familiarity with the use of Kirchhoff's laws in *RLC* circuits and discuss the solution of the simpler cases in the following chapters.

Example 3-8   Let us now attempt to write nodal equations for the circuit of Fig. 3-18.

Figure 3-18

A four-node *RLC* circuit with node voltages assigned.

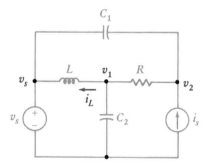

Solution: Node voltages are chosen as indicated, and we sum currents leaving the central node:

$$\frac{1}{L}\int_{t_0}^{t} (v_1 - v_s)\,dt + i_L(t_0) + \frac{v_1 - v_2}{R} + C_2\frac{dv_1}{dt} = 0$$

where $i_L(t_0)$ is the value of the inductor current at the time the integration begins, or the initial value. At the right-hand node,

$$C_1\frac{d(v_2 - v_s)}{dt} + \frac{v_2 - v_1}{R} - i_s = 0$$

Rewriting these two equations, we have

$$\frac{v_1}{R} + C_2\frac{dv_1}{dt} + \frac{1}{L}\int_{t_0}^{t} v_1\,dt - \frac{v_2}{R} = \frac{1}{L}\int_{t_0}^{t} v_s\,dt - i_L(t_0)$$

$$-\frac{v_1}{R} + \frac{v_2}{R} + C_1\frac{dv_2}{dt} = C_1\frac{dv_s}{dt} + i_s$$

These are the promised integrodifferential equations, and we may note several interesting points about them. First, the source voltage v_s happens to enter the equations as an integral and as a derivative, but not simply as v_s. Since both sources are specified for all time, we should be able to evaluate the derivative or integral. Second, the initial value of the inductor current, $i_L(t_0)$, acts as a (constant) source current at the center node. ∎

We shall not attempt the solution of equations of this type at this time. It is worthwhile pointing out, however, that when the two voltage forcing functions are sinusoidal functions of time, it will be possible to define a voltage-current ratio (called *impedance*) or a current-voltage ratio (called *admittance*) for each of the three passive elements. The factors operating on the two node voltages in the preceding equations will then become simple multiplying factors, and the equations will be linear *algebraic* equations once again. These we may solve by determinants or a simple elimination of variables as before.

Drill Problem

3-6. (a) Find L_{eq} for the network of Fig. 3-19a. (b) Find C_{eq} for the network of Fig. 3-19b. (c) If $v_C(t) = 4 \cos 10^5 t$ V in the circuit of Fig. 3-19c, find $v_s(t)$.

Ans: 4 μH; 3.18 μF; $-2.4 \cos 10^5 t$ V

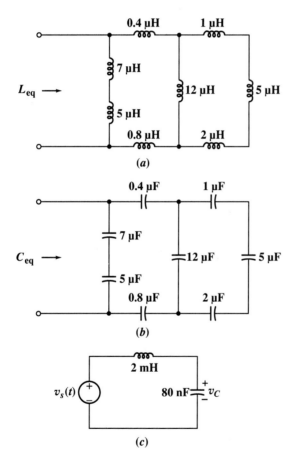

Figure 3-19

See Drill Prob. 3-6.

3-6

Duality

Duality was mentioned earlier in connection with resistive circuits, and more recently in the discussion of inductance and capacitance; the comments made were introductory and, like the man who tried to pet the alligator, a little offhand. Now we may make an exact definition and then use the definition to recognize or construct dual circuits and thus avoid the labor of analyzing both a circuit and its dual.

We shall define duality in terms of the circuit equations. Two circuits are *duals* if the mesh equations that characterize one of them have the same mathematical form as the nodal equations that characterize the other. They are said to be *exact duals* if each mesh equation of the one circuit is numerically identical with the corresponding nodal equation of the other; the current and voltage variables themselves cannot be identical, of course. *Duality* itself merely refers to any of the properties exhibited by dual circuits.

Let us interpret the definition and use it to construct an exact dual circuit by writing the two mesh equations for the circuit shown in Fig. 3-20. Two mesh currents i_1 and i_2 are assigned, and the mesh equations are

$$3i_1 + 4\frac{di_1}{dt} - 4\frac{di_2}{dt} = 2\cos 6t \tag{17}$$

$$-4\frac{di_1}{dt} + 4\frac{di_2}{dt} + \frac{1}{8}\int_0^t i_2\,dt + 5i_2 = -10 \tag{18}$$

Note that the capacitor voltage v_C is assumed to be 10 V at $t = 0$.

Figure 3-20

A given circuit to which the definition of duality may be applied to determine the dual circuit.

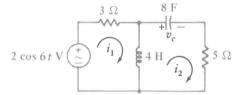

We may now construct the two equations which describe the exact dual of the given circuit. We wish these to be nodal equations, and we thus begin by replacing the mesh currents i_1 and i_2 in Eqs. (17) and (18) by two node-to-reference voltages v_1 and v_2. We obtain

$$3v_1 + 4\frac{dv_1}{dt} - 4\frac{dv_2}{dt} = 2\cos 6t \tag{19}$$

$$-4\frac{dv_1}{dt} + 4\frac{dv_2}{dt} + \frac{1}{8}\int_0^t v_2\,dt + 5v_1 = -10 \tag{20}$$

and we now seek the circuit represented by these two nodal equations.

Let us first draw a line to represent the reference node, and then we may establish two nodes at which the positive references for v_1 and v_2 are located. Equation (19) indicates that a current source of 2 cos 6t A is connected between node 1 and the reference node, oriented to provide a current entering node 1. This equation also shows that a 3-S conductance appears between node 1 and the reference node. Turning to Eq. (20), we first consider the nonmutual terms, or those terms which do not appear in Eq. (19), and they instruct us to connect an 8-H inductor and a 5-S conductance (in parallel) between node 2 and the reference. The two similar terms in Eqs. (19) and (20) represent a 4-F capacitor present mutually at nodes 1 and 2; the circuit is completed by connecting this capacitor between the two nodes. The constant term on the right side of Eq. (20) is the value of the inductor current at $t = 0$; thus, $i_L(0) = 10$ A. The dual circuit is shown in Fig. 3-21; since the two sets of equations are numerically identical, the circuits are exact duals.

Dual circuits may be obtained more readily than by this method, for the equations need not be written. In order to construct the dual of a given circuit, we think of the circuit in terms of its mesh equations. With each mesh we must associate a nonreference node, and, in addition, we must supply the reference

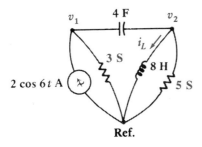

Figure 3-21

The exact dual of the circuit of Fig. 3-20.

node. On a diagram of the given circuit we therefore place a node in the center of each mesh and supply the reference node as a line near the diagram or a loop enclosing the diagram. Each element which appears jointly in two meshes is a *mutual* element and gives rise to identical terms, except for sign, in the two corresponding mesh equations. It must be replaced by an element which supplies the dual term in the two corresponding nodal equations. This dual element must therefore be connected directly between the two nonreference nodes which are within the meshes in which the given mutual element appears. The nature of the dual element itself is easily determined; the mathematical form of the equations will be the same only if inductance is replaced by capacitance, capacitance by inductance, conductance by resistance, and resistance by conductance. Thus, the 4-H inductor which is common to meshes 1 and 2 in the circuit of Fig. 3-20 appears as a 4-F capacitor connected directly between nodes 1 and 2 in the dual circuit.

Elements which appear only in one mesh must have duals which appear between the corresponding node and the reference node. Referring again to Fig. 3-20, the voltage source $2 \cos 6t$ V appears only in mesh 1; its dual is a current source $2 \cos 6t$ A, which is connected only to node 1 and the reference node. Since the voltage source is clockwise-sensed, the current source must be into-the-nonreference-node-sensed. Finally, provision must be made for the dual of the initial voltage present across the 8-F capacitor in the given circuit. The equations have shown us that the dual of this initial voltage across the capacitor is an initial current through the inductor in the dual circuit; the numerical values are the same, and the correct sign of the initial current may be determined most readily by considering both the initial voltage in the given circuit and the initial current in the dual circuit as sources. Thus, if v_C in the given circuit is treated as a source, it would appear as $-v_C$ on the right side of the mesh equation; in the dual circuit, treating the current i_L as a source would yield a term $-i_L$ on the right side of the nodal equation. Since each has the same sign when treated as a source, then, if $v_C(0) = 10$ V, $i_L(0)$ must be 10 A.

The circuit of Fig. 3-20 is repeated in Fig. 3-22, and its exact dual is constructed on the circuit diagram itself by merely drawing the dual of each

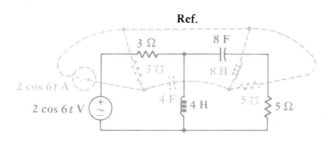

Figure 3-22

The dual of the circuit of Fig. 3-20 is constructed directly from the circuit diagram.

given element between the two nodes which are inside the two meshes that are common to the given element. A reference node which surrounds the given circuit may be helpful. After the dual circuit is redrawn in more standard form, it appears as shown in Fig. 3-21.

An additional example of the construction of a dual circuit is shown in Figs. 3-23a and b. Since no particular element values are specified, these two circuits are duals, but not necessarily exact duals. The original circuit may be recovered from the dual by placing a node in the center of each of the five meshes of Fig. 3-23b and proceeding as before.

Figure 3-23

(a) The dual (in color) of a given circuit (in black) is constructed on the given circuit. (b) The dual circuit is drawn in more conventional form.

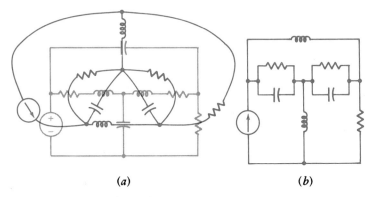

       (a)   (b)

The concept of duality may also be carried over into the language by which we describe circuit analysis or operation. One example of this was discussed previously in Sec. 3-4, and the duals of several words appeared there. Most of these pairs are obvious; whenever there is any question as to the dual of a word or phrase, the dual circuit may always be drawn or visualized and then described in similar language. For example, if we are given a voltage source in series with a capacitor, we might wish to make the important statement, "The voltage source causes a current to flow through the capacitor"; the dual statement is, "The current source causes a voltage to exist across the inductor." The dual of a less carefully worded statement, such as "The current goes round and round the series circuit," often requires a little inventiveness.[3]

Practice in using dual language can be obtained by reading Thévenin's theorem in this sense; Norton's theorem should result.

We have spoken of dual elements, dual language, and dual circuits. What about a dual *network*? Consider a resistor R and an inductor L in series. The dual of this two-terminal network exists and is most readily obtained by connecting some ideal source to the given network. The dual circuit is then obtained as the dual source in parallel with a conductance G, $G = R$, and a capacitance C, $C = L$. We consider the dual network as the two-terminal network that is connected to the dual source; it is thus a pair of terminals between which G and C are connected in parallel.

Before leaving the definition of duality, it should be pointed out that duality is defined on the basis of mesh and nodal equations. Since nonplanar circuits cannot be described by a system of mesh equations, a circuit which cannot be drawn in planar form does not possess a dual.

We shall use duality principally to reduce the work which we must do to analyze the simple standard circuits. After we have analyzed the series RL

---

[3] Someone has suggested, "The voltage is across all over the parallel circuit."

circuit, then the parallel RC circuit requires less attention, not because it is less important, but because the analysis of the dual network is already known. Since the analysis of some complicated circuit is not apt to be well known, duality will usually not provide us with any quick solution.

Drill Problem

3-7. Write the single nodal equation for the circuit of Fig. 3-24a, and show, by direct substitution, that $v = -80e^{-10^6t}$ mV is a solution. Knowing this, refer to Fig. 3-24b and find (a) v_1; (b) v_2; and (c) i.

Ans: $-8e^{-10^6t}$ mV; $16 \times 10^{-3}e^{-10^6t}$ V; $-80e^{-10^6t}$ mA

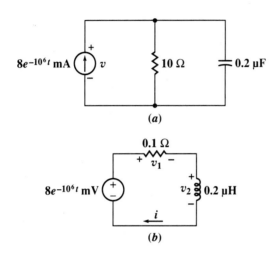

Figure 3-24

See Drill Prob. 3-7.

3-7

Linearity and its consequences again

In the previous chapter we learned that the principle of superposition is a necessary consequence of the linear nature of the resistive circuits which we were analyzing. The resistive circuits are linear because the voltage-current relationship for the resistor is linear and Kirchhoff's laws are linear.

We now wish to show that the benefits of linearity apply to RLC circuits as well. In accordance with our previous definition of a linear circuit, these circuits are also linear because the voltage-current relationships for the inductor and capacitor are linear relationships. For the inductor, we have

$$v = L\frac{di}{dt}$$

and multiplication of the current by some constant K leads to a voltage which is also greater by a factor K. In the integral formulation,

$$i = \frac{1}{L}\int_{t_0}^{t} v\,dt + i_L(t_0)$$

it can be seen that, if each term is to increase by a factor of K, then the initial value of the current must also increase by this same factor. That is, the factor K applies not only to the current and voltage at time t but also to their past values.

A corresponding investigation of the capacitor shows that it, too, is linear. Thus, a circuit composed of independent sources, linear dependent sources, and linear resistors, inductors, and capacitors is a linear circuit.

In this linear circuit the response is again proportional to the forcing function. The proof of this statement is accomplished by first writing a general system of integrodifferential equations, say, in terms of loop currents. Let us place all the terms having the form of Ri, $L\,di/dt$, and $(1/C)\int i\,dt$ on the left side of each equation and keep the independent source voltages on the right side. As a simple example, one of the equations might have the form

$$Ri + L\frac{di}{dt} + \frac{1}{C}\int_{t_0}^{t} i\,dt + v_C(t_0) = v_s$$

If every independent source is now increased by a factor K, then the right side of each equation is greater by the factor K. Now each term on the left side is either a linear term involving some loop current or an initial capacitor voltage. In order to cause all the responses (loop currents) to increase by a factor K, it is apparent that we must also increase the initial capacitor voltages by a factor K. That is, we must treat the *initial capacitor voltage as an independent source voltage* and increase it also by a factor K. In a similar manner, initial inductor currents must be treated as independent source currents in nodal analysis.

The principle of proportionality between source and response is thus extensible to the general RLC circuit, and it follows that the principle of superposition is also applicable. It should be emphasized that initial inductor currents and capacitor voltages must be treated as independent sources in applying the superposition principle; each initial value must take its turn in being rendered inactive.

Before we can apply the superposition principle to RLC circuits, however, it is first necessary to develop methods of solving the equations describing these circuits when only one independent source is present. At this time we should feel convinced that a linear circuit will possess a response whose amplitude is proportional to the amplitude of the source. We should be prepared to apply superposition later, considering an inductor current or capacitor voltage specified at $t = t_0$ as a source which must be killed when its turn comes.

Thévenin's and Norton's theorems are based on the linearity of the initial circuit, the applicability of Kirchhoff's laws, and the superposition principle. The general RLC circuit conforms perfectly to these requirements, and it follows, therefore, that all linear circuits which contain any combinations of independent voltage and current sources, linear dependent voltage and current sources, and linear resistors, inductors, and capacitors may be analyzed with the use of these two theorems, if we wish. It is not necessary to repeat the theorems here, for they were previously stated in a manner that is equally applicable to the general RLC circuit.

Problems

1 With reference to Fig. 3-25: (*a*) sketch v_L as a function of time, $0 < t < 60$ ms; (*b*) find the value of time at which the inductor is absorbing a maximum power; (*c*) find the value of time at which it is supplying a maximum power; and (*d*) find the energy stored in the inductor at $t = 40$ ms.

2 In Fig. 3-1, let $L = 50$ mH, with $i = 0$ for $t < 0$ and $80te^{-100t}$ mA for $t > 0$. Find the maximum values of $|i|$ and $|v|$, and the time at which each maximum occurs.

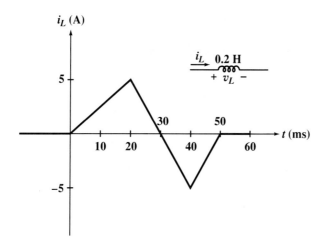

Figure 3-25

See Prob. 1.

3 (a) If $i_s = 0.4t^2$ A for $t > 0$ in the circuit of Fig. 3-26a, find and sketch $v_{in}(t)$ for $t > 0$.
(b) If $v_s = 40t$ V for $t > 0$ and $i_L(0) = 5$ A, find and sketch $i_{in}(t)$ for $t > 0$ in the circuit
of Fig. 3-26b.

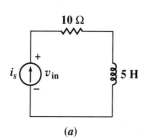
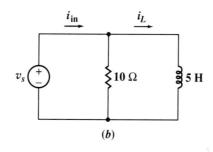

(a)   (b)

Figure 3-26

See Prob. 3.

4 The voltage $20 \cos 1000t$ V is applied to a 25-mH inductor. If the inductor current is
zero at $t = 0$, find and sketch ($0 \le t \le 2\pi$ ms): (a) the power being absorbed by the
inductor; (b) the energy stored in the inductor.

5 The voltage v_L across a 0.2-H inductor is 100 V for $0 < t \le 10$ ms; decreases linearly
to zero in the interval $10 < t < 20$ ms; is 0 for $20 \le t < 30$ ms; is 100 V for $30 < t < 40$
ms; and is zero thereafter. Assume the passive sign convention for v_L and i_L. (a) Calculate
i_L at $t = 8$ ms if $i_L(0) = -2$ A. (b) Determine the stored energy at $t = 22$ ms if $i_L(0) = 0$.
(c) If the circuit shown in Fig. 3-27 has been connected for a very long time, find i_x.

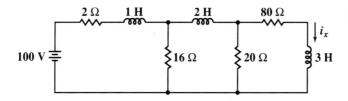

Figure 3-27

See Prob. 5.

6 The voltage across a 5-H inductor is $v_L = 10(e^{-t} - e^{-2t})$ V. If $i_L(0) = 80$ mA and v_L and
i_L satisfy the passive sign convention, find (a) $v_L(1\text{ s})$; (b) $i_L(1\text{ s})$; and (c) $i_L(\infty)$.

7 (a) If the capacitor shown in Fig. 3-7 has a capacitance of 0.2 μF, let $v_C = 5 +
3 \cos^2 200t$ V, and find $i_C(t)$. (b) What is the maximum energy stored in the capacitor?
(c) If $i_C = 0$ for $t < 0$ and $i_C = 8e^{-100t}$ mA for $t > 0$, find $v_C(t)$ for $t > 0$. (d) If
$i_C = 8e^{-100t}$ mA for $t > 0$ and $v_C(0) = 100$ V, find $v_C(t)$ for $t > 0$.

8 The current waveform shown for $t > 0$ in Fig. 3-28 is applied to a 2-mF capacitor. Given that $v_C(0) = 250$ V, and assuming the passive sign convention, during what time interval is the value of v_C between 2000 and 2100 V?

Figure 3-28

See Prob. 8.

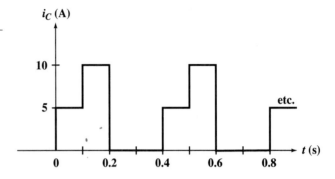

9 A resistance R is connected in parallel with a 1-μF capacitor. For any $t \geq 0$, the energy stored in the capacitor is $20e^{-1000t}$ mJ. (a) Find R. (b) By integration, show that the energy dissipated in R over the interval $0 \leq t < \infty$ is 0.02 J.

10 Interchange the location of R and C in the circuit of Fig. 3-10a, and assume that $R_i = \infty$, $R_o = 0$, and $A = \infty$ for the op-amp. (a) Find $v_o(t)$ as a function of $v_s(t)$. (b) Obtain an equation relating $v_o(t)$ and $v_s(t)$ if A is not assumed to be infinite.

11 In the circuit of Fig. 3-10a, let $R = 0.5$ MΩ, $C = 2$ μF, $R_i = \infty$, and $R_o = 0$. Suppose that we wish the output to be $v_o = \cos 10t - 1$ V. Differentiate Eq. (16) to obtain the necessary $v_s(t)$ if (a) $A = 2000$ and (b) A is infinite.

12 A long time after all connections have been made in the circuit shown in Fig. 3-29, find v_x if (a) a capacitor is present between x and y and (b) an inductor is present between x and y.

Figure 3-29

See Probs. 12 and 22.

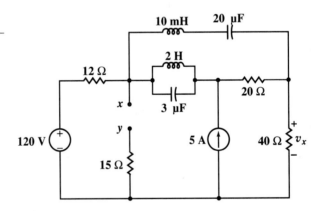

13 Refer to the network shown in Fig. 3-30 and find (a) R_{eq} if each element is a 10-Ω resistor; (b) L_{eq} if each element is a 10-H inductor; and (c) C_{eq} if each element is a 10-F capacitor.

14 In Fig. 3-31, let elements A, B, C, and D be (a) 1-H, 2-H, 3-H, and 4-H inductors, respectively, and find the input inductance with x-x' first open-circuited and then short-circuited; (b) 1-F, 2-F, 3-F, and 4-F capacitors, respectively, and find the input capacitance with x-x' first open-circuited and then short-circuited.

15 Given a boxful of 1-nF capacitors, and using as few capacitors as possible, show how it is possible to obtain an equivalent capacitance of (a) 2.25 nF; (b) 0.75 nF; (c) 0.45 nF.

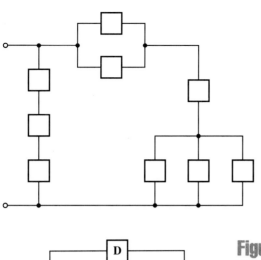

Figure 3-30

See Prob. 13.

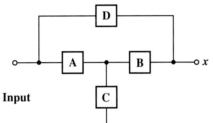

Figure 3-31

See Prob. 14.

16 With reference to the circuit shown in Fig. 3-32, find (a) w_L; (b) w_C; (c) the voltage across each circuit element; (d) the current in each circuit element.

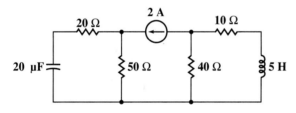

Figure 3-32

See Probs. 16 and 27.

17 Let $v_s = 400t^2$ V for $t > 0$ and $i_L(0) = 0.5$ A in the circuit of Fig. 3-33. At $t = 0.4$ s, find the values of energy: (a) stored in the capacitor; (b) stored in the inductor; and (c) dissipated by the resistor since $t = 0$.

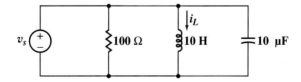

Figure 3-33

See Prob. 17.

18 In the circuit shown in Fig. 3-34, let $i_s = 60e^{-200t}$ mA with $i_1(0) = 20$ mA. (a) Find $v(t)$ for all t. (b) Find $i_1(t)$ for $t \geq 0$. (c) Find $i_2(t)$ for $t \geq 0$.

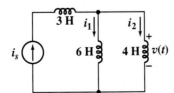

Figure 3-34

See Prob. 18.

19 Let $v_s = 100e^{-80t}$ V and $v_1(0) = 20$ V in the circuit of Fig. 3-35. (a) Find $i(t)$ for all t. (b) Find $v_1(t)$ for $t \geq 0$. (c) Find $v_2(t)$ for $t \geq 0$.

Figure 3-35

See Probs. 19 and 24.

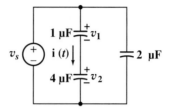

20 (a) Write nodal equations for the circuit of Fig. 3-36. (b) Write mesh equations for the same circuit.

Figure 3-36

See Probs. 20 and 21.

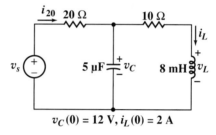

$v_C(0) = 12$ V, $i_L(0) = 2$ A

21 (a) Draw the exact dual of the circuit shown in Fig. 3-36. Specify the dual variables and the dual initial conditions. (b) Write nodal equations for the dual circuit. (c) Write mesh equations for the dual circuit.

22 Draw the exact dual of the circuit shown in Fig. 3-29. Draw the circuit in a neat, clean form with square corners, a recognizable reference node, and no crossovers.

23 Draw the exact dual of the circuit shown in Fig. 3-37. Keep it neat!

Figure 3-37

See Prob. 23.

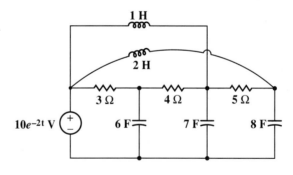

24 (a) Draw the exact dual of the circuit given for Prob. 19, including variables. (b) Write the dual of the problem statement for Prob. 19. (c) Solve your new Prob. 19.

25 Construct a tree for the circuit shown in Fig. 3-38 that not only satisfies the criteria listed in Secs. 2-7 and 2-8, but also places all capacitors in the tree and all inductors in the cotree. (a) Assign tree-branch voltages, and write a set of nodal equations. Assume there is no energy storage at $t = 0$. (b) Assign link currents, and write a set of loop equations, again assuming no energy storage at $t = 0$.

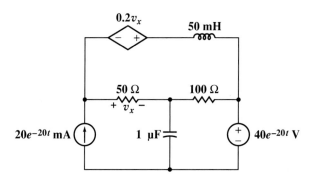

Figure 3-38

See Prob. 25.

26 If it is assumed that all the sources in the circuit of Fig. 3-39 have been connected and operating for a very long time, use the superposition principle to find $v_C(t)$ and $v_L(t)$.

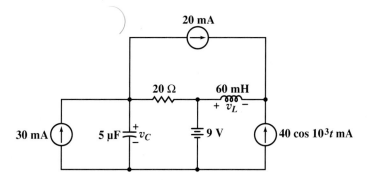

Figure 3-39

See Prob. 26.

27 (SPICE) Repeat Prob. 16a and b if a 30-Ω resistor is connected between the upper terminals of the capacitor and inductor in the circuit of Fig. 3-32. Use SPICE to find the capacitor voltage and inductor current.

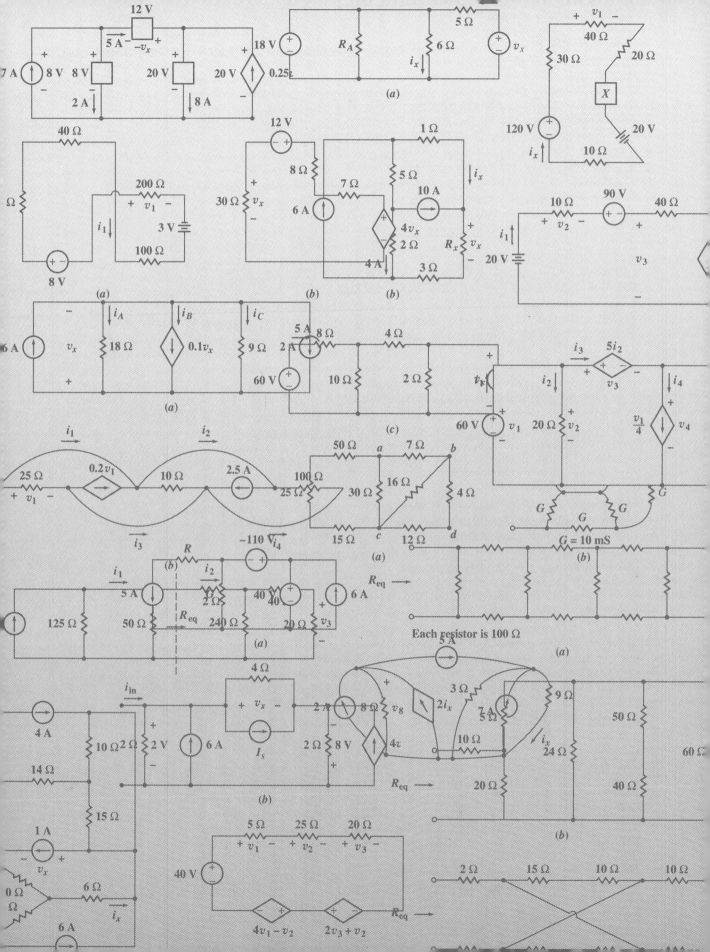

Source-Free *RL* and *RC* Circuits

In the previous chapter we wrote equations governing the response of several circuits containing both inductance and capacitance, but we did not solve any of them. At this time we are ready to proceed with the solution for the simpler circuits. We shall restrict our attention to certain circuits which contain only resistors and inductors or only resistors and capacitors, and which contain no sources. We shall, however, allow stored energy to be present in the inductors or capacitors, for without such energy every response would be zero.

Although the circuits which we are about to consider have a very elementary appearance, they are also of practical importance. Networks of this form find use as coupling networks in electronic amplifiers, as compensating networks in automatic control systems and operational amplifiers, as equalizing networks in communications channels, and in many other ways. A familiarity with these simple circuits will enable us to predict the accuracy with which the output of an amplifier can follow an input which is changing rapidly with time or to predict how quickly the speed of a motor will change in response to a change in its field current. Our knowledge of the performance of the simple *RL* and *RC* circuits will also enable us to suggest modifications to the amplifier or motor in order to obtain a more desirable response.

The analysis of such circuits is dependent upon the formulation and solution of the integrodifferential equations which characterize the circuits. We shall call the special type of equation we obtain a *homogeneous linear differential equation*, which is simply a differential equation in which every term is of the first degree in the dependent variable or one of its derivatives. A *solution* is obtained when we have found an expression for the dependent variable, as a function of time, which satisfies the differential equation and also satisfies the prescribed energy distribution in the inductors or capacitors at a prescribed instant of time, usually $t = 0$.

The solution of the differential equation represents a response of the circuit, and it is known by many names. Since this response depends upon the general "nature" of the circuit (the types of elements, their sizes, the interconnection of the elements), it is often called a *natural response*. It is also obvious that any real circuit we construct cannot store energy forever; the resistances necessarily associated with inductors and capacitors will eventually convert all stored energy into heat. The response must eventually die out, and it is therefore referred to as the *transient response*. Finally, we must also be familiar with the mathematicians' contribution to the nomenclature; they call the solution of a homogeneous linear differential equation a *complementary function.* When we consider independent sources acting on a circuit, part of the response will partake of the nature of the particular source (or forcing function) used; this part of the response, called the *particular solution* or the *forced response,* will be "complemented" by the complementary response produced in the source-free circuit, and the sum of the complementary function and the particular solution will be the complete response. That is, the complete response is the sum of the natural response, studied in this chapter, and the forced response, which we shall examine in Chap. 5. The source-free response may be called the *natural response,* the *transient response,* the *free response,* or the *complementary function,* but because of its more descriptive nature, we shall most often call it the *natural response.*

    We shall consider several different methods of solving these differential equations. This mathematics, however, is not circuit analysis. Our greatest interest lies in the solutions themselves, their meaning, and their interpretation, and we shall try to become sufficiently familiar with the form of the response that we are able to write down answers for new circuits by just plain thinking. Although complicated analytical methods are needed when simpler methods fail, an engineer must always remember that these complex techniques are only tools with which meaningful, informative answers may be obtained; they do not constitute engineering in themselves.

4-2

The simple *RL* circuit

We shall begin our study of transient analysis by considering the simple series *RL* circuit shown in Fig. 4-1. Let us designate the time-varying current as $i(t)$; we shall let the value of $i(t)$ at $t = 0$ be prescribed as I_0. That is, $i(0) = I_0$. We therefore have

$$v_R + v_L = Ri + L\frac{di}{dt} = 0$$

or

$$\frac{di}{dt} + \frac{R}{L}i = 0 \tag{1}$$

and we must determine an expression for $i(t)$ which satisfies this equation and also has the value I_0 at $t = 0$. The solution may be obtained by several different methods.

Figure 4-1

A series *RL* circuit for which $i(t)$ is to be determined, subject to the initial condition that $i(0) = I_0$.

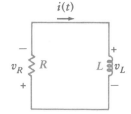

    One very direct method of solving a differential equation consists of writing the equation in such a way that the variables are separated, and then integrating each side of the equation. The variables in Eq. (1) are i and t, and it is apparent that the equation may be multiplied by dt, divided by i, and arranged with the variables separated:

$$\frac{di}{i} = -\frac{R}{L}dt \qquad (2)$$

Since the current is I_0 at $t = 0$ and $i(t)$ at time t, we may equate the two definite integrals which are obtained by integrating each side between the corresponding limits:

$$\int_{I_0}^{i(t)} \frac{di}{i} = \int_0^t -\frac{R}{L}dt$$

and therefore,

$$\ln i \Big|_{I_0}^{i} = -\frac{R}{L}t \Big|_0^t$$

or

$$\ln i - \ln I_0 = -\frac{R}{L}(t - 0)$$

Therefore,

$$i(t) = I_0 e^{-Rt/L} \qquad (3)$$

We check our solution by first showing that substitution of Eq. (3) in Eq. (1) yields the identity $0 \equiv 0$ and then showing that substitution of $t = 0$ in Eq. (3) produces $i(0) = I_0$. Both steps are necessary; the solution must satisfy the differential equation which characterizes the circuit, and it must also satisfy the initial condition.

The solution may also be obtained by a slight variation of the method just described. After separating the variables, we may obtain the *indefinite* integral of each side of Eq. (2) if we also include a constant of integration. Thus,

$$\int \frac{di}{i} = -\int \frac{R}{L}dt + K$$

and integration gives us

$$\ln i = -\frac{R}{L}t + K \qquad (4)$$

The constant K cannot be evaluated by substitution of Eq. (4) in the original differential equation (1); the identity $0 \equiv 0$ will result, because Eq. (4) is a solution of Eq. (1) for *any* value of K. The constant of integration must be selected to satisfy the initial condition $i(0) = I_0$. Thus, at $t = 0$, Eq. (4) becomes

$$\ln I_0 = K$$

and we use this value for K in Eq. (4) to obtain the desired response

$$\ln i = -\frac{R}{L}t + \ln I_0$$

or

$$i(t) = I_0 e^{-Rt/L}$$

as before.

Either of these methods can be used when the variables can be separated, but this is only occasionally possible. In the remaining cases we shall rely on a very powerful method, the success of which will depend upon our intuition or experience. We simply guess or assume a form for the solution and then test our assumptions, first by substitution in the differential equation, and then by applying the given initial conditions. Since we cannot be expected to guess the exact numerical expression for the solution, we shall assume a solution containing several unknown constants and select the values for these constants in order to satisfy the differential equation and the initial conditions. Many of the differential equations encountered in circuit analysis have a solution which may be represented by the exponential function or by the sum of several exponential functions. Let us assume a solution of Eq. (1) in exponential form,

$$i(t) = Ae^{s_1 t} \tag{5}$$

where A and s_1 are constants to be determined. After substituting this assumed solution in Eq. (1), we have

$$As_1 e^{s_1 t} + \frac{R}{L} e^{s_1 t} = 0$$

or

$$\left(s_1 + \frac{R}{L}\right) Ae^{s_1 t} = 0$$

In order to satisfy this equation for all values of time, it is necessary that either $A = 0$, or $s_1 = -\infty$, or $s_1 = -R/L$. But if $A = 0$ or $s_1 = -\infty$, then every response is zero; neither can be a solution to our problem. Therefore, we must choose

$$s_1 = -\frac{R}{L}$$

and our assumed solution takes on the form

$$i(t) = Ae^{-Rt/L}$$

The remaining constant must be evaluated by applying the initial condition $i = I_0$ at $t = 0$. Thus,

$$I_0 = A$$

and the final form of the assumed solution is

$$i(t) = I_0 e^{-Rt/L}$$

once again.

We shall not consider any other methods for solving Eq. (1), although a number of other techniques may be used. Watch for them in a study of differential equations.

Before we turn our attention to the interpretation of the response, let us check the power and energy relationships in this circuit. The power being dissipated in the resistor is

$$p_R = i^2 R = I_0^2 R e^{-2Rt/L}$$

and the total energy turned into heat in the resistor is found by integrating the instantaneous power from zero time to infinite time:

$$W_R = \int_0^\infty p_R dt = I_0^2 R \int_0^\infty e^{-2Rt/L} dt$$

$$= I_0^2 R \left(\frac{-L}{2R}\right) e^{-2Rt/L} \Big|_0^\infty = \frac{1}{2} L I_0^2$$

This is the result we expect, because the total energy stored initially in the inductor is $\frac{1}{2}LI_0^2$ and there is no energy stored in the inductor at infinite time. All the initial energy is accounted for by dissipation in the resistor.

Drill Problems

4-1. The circuits shown in Fig. 4-2 have each been in the form shown for a long time. The switches in parts *a* and *b* of the figure are opened at $t = 0$. The switch in part *c* is a single-pole double-throw switch that is drawn to indicate that it closes one circuit before opening the other. It is called a *make-before-break* switch. Review the characteristics of an inductor at the end of Sec. 3-3, and determine i_L in each circuit at the instant just *before* the switch changes.

Ans: 2 A; 2.4 A; 4 A

4-2. Determine the value of i_L in each circuit of Fig. 4-2 at the instant just *after* the switch changes.

Ans: 2 A; 2.4 A; 4 A

4-3. Find v in each circuit of Fig. 4-2 at the instant just *after* the switch changes.

Ans: 40 V; −96 V; −48 V

4-4. Let $R = 40\ \Omega$, $L = 20$ mH, and $I_0 = 30$ mA in the circuit shown in Fig. 4-1. Find (*a*) $i(1$ ms); (*b*) $v_L(0.8$ ms); (*c*) $w_L(0.4$ ms).

Ans: 4.06 mA; −0.242 V; 1.817 μJ

Figure 4-2

See Drill Probs. 4-1, 4-2, 4-3, and 4-6.

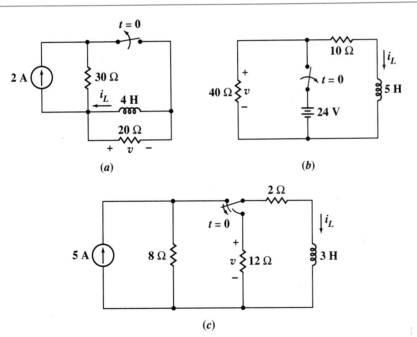

(a)

(b)

(c)

4-3

Properties of the exponential response

Let us now consider the nature of the response in the series *RL* circuit. We found that the current is represented by

$$i(t) = I_0 e^{-Rt/L} \tag{6}$$

At zero time, the current is the assumed value I_0, and as time increases, the current decreases and approaches zero. The shape of this decaying exponential is seen by a plot of $i(t)/I_0$ versus t, as shown in Fig. 4-3. Since the function we are plotting is $e^{-Rt/L}$, the curve will not change if R/L does not change. Thus, the same curve must be obtained for every series *RL* circuit having the same R/L or L/R ratio. Let us see how this ratio affects the shape of the curve.

Figure 4-3

A plot of $e^{-Rt/L}$ versus t.

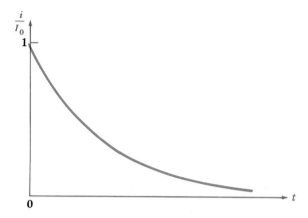

If we double the ratio of L to R, then the exponent will be unchanged if t is also doubled. In other words, the original response will occur at a later time, and the new curve is obtained by moving each point on the original curve twice as far to the right. With this larger L/R ratio, the current takes longer to decay to any given fraction of its original value. We might have a tendency to say that the "width" of the curve is doubled, or that the width is proportional to L/R. However, we should have to define our term *width*, because each curve extends from $t = 0$ to ∞. Instead, let us consider the time that would be required for the current to drop to zero *if it continued to drop at its initial rate.*

The initial rate of decay is found by evaluating the derivative at zero time:

$$\frac{d}{dt}\frac{i}{I_0}\bigg|_{t=0} = -\frac{R}{L}e^{-Rt/L}\bigg|_{t=0} = -\frac{R}{L}$$

We designate the value of time it takes for i/I_0 to drop from unity to zero, assuming a constant rate of decay, by the Greek letter τ (*tau*). Thus,

$$\frac{R}{L}\tau = 1$$

or
$$\tau = \frac{L}{R} \qquad\qquad (7)$$

The ratio L/R has the units of seconds, since the exponent $-Rt/L$ must be dimensionless. This value of time τ is called the *time constant*; it is shown in Fig. 4-4. It is apparent that the time constant of a series RL circuit may easily be found graphically from the response curve; it is necessary only to draw the tangent to the curve at $t = 0$ and determine the intercept of this tangent line with the time axis. This is often a convenient way of approximating the time constant from the display on an oscilloscope.

An equally important interpretation of the time constant τ is obtained by determining the value of $i(t)/I_0$ at $t = \tau$. We have

$$\frac{i(\tau)}{I_0} = e^{-1} = 0.368 \qquad\text{or}\qquad i(\tau) = 0.368 I_0$$

Thus, in one time constant the response has dropped to 36.8 percent of its initial value; the value of τ may also be determined graphically from this fact, as indicated by Fig. 4-5. It is convenient to measure the decay of the current at

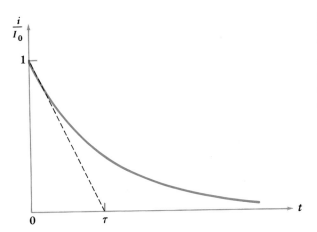

Figure 4-4

The time constant τ is L/R for a series *RL* circuit. It is the time required for the response curve to drop to zero if it decays at a constant rate which is equal to its initial rate of decay.

intervals of one time constant, and recourse to a hand calculator or a table of negative exponentials shows that $i(t)/I_0$ is 0.368 at $t = \tau$, 0.135 at $t = 2\tau$, 0.0498 at $t = 3\tau$, 0.0183 at $t = 4\tau$, and 0.0067 at $t = 5\tau$. At some point three to five time constants after zero time, most of us would agree that the current is a negligible fraction of its former self. Thus, if we are asked, "How long does it take for the current to decay to zero?" our answer might be, "About five time constants." At the end of that time interval the current is less than 1 percent of its original value.

Why does a larger value of the time constant L/R produce a response curve which decays more slowly? Let us consider the effect of each element. An increase in L allows a greater energy storage for the same initial current, and this larger energy requires a longer time to be dissipated in the resistor. We may also increase L/R by reducing R. In this case, the power flowing into the resistor is less for the same initial current; again, a greater time is required to dissipate the stored energy.

In terms of the time constant τ, the response of the series *RL* circuit may be written simply as

$$i(t) = I_0 e^{-t/\tau}$$

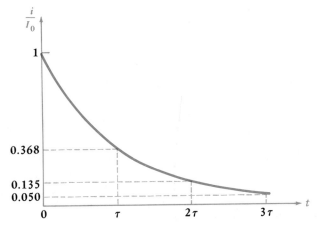

Figure 4-5

The current in a series *RL* circuit is 36.8, 13.5, and 5 percent of its initial value at τ, 2τ, and 3τ, respectively.

Drill Problem

4-5. In a source-free series RL circuit, find the numerical value of the ratio: (a) $i(2\tau)/i(\tau)$, (b) $i(0.5\tau)/i(0)$, and (c) t/τ if $i(t)/i(0) = 0.2$; (d) t/τ if $i(0) - i(t) = i(0) \ln 2$.

Ans: 0.368; 0.607; 1.609; 1.181

4-4

A more general *RL* circuit

It is not difficult to extend the results obtained for the series RL circuit to a circuit containing any number of resistors and one inductor. We fix our attention on the two terminals of the inductor and determine the equivalent resistance across these terminals. The circuit is thus reduced to the simple series case. As an example, consider the circuit shown in Fig. 4-6. The equivalent resistance which the inductor faces is

$$R_{eq} = R_3 + R_4 + \frac{R_1 R_2}{R_1 + R_2}$$

and the time constant is therefore

$$\tau = \frac{L}{R_{eq}}$$

The inductor current i_L is

$$i_L = i_L(0)e^{-t/\tau} \tag{8}$$

Figure 4-6

A source-free circuit containing one inductor and several resistors is analyzed by determining the time constant $\tau = L/R_{eq}$.

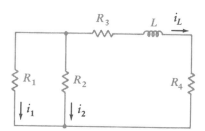

and Eq. (8) represents what we might call the *basic solution* to the problem. It is quite possible that some current or voltage other than i_L is needed, such as the current i_2 in R_2. We can always apply Kirchhoff's laws and Ohm's law to the resistive portion of the circuit without any difficulty, but current division provides the quickest answer in this circuit:

$$i_2 = -\frac{R_1}{R_1 + R_2} i_L(0)e^{-t/\tau}$$

It may also happen that we know the initial value of some current other than the inductor current. Since the current in a resistor may change instantaneously, we shall indicate the instant *after* any change that might have occurred at $t = 0$ by the use of the symbol 0^+; in more mathematical language, $i(0^+)$ is the limit from the right of $i_1(t)$ as t approaches zero. Thus, if we are given the initial value of i_1 as $i_1(0^+)$, then it is apparent that the initial value of i_2 is

$$i_2(0^+) = i_1(0^+)\frac{R_1}{R_2}$$

From these values, we obtain the necessary initial value of $i_L(0)$ [or $i_L(0^-)$ or $i_L(0^+)$]:

$$i_L(0^+) = -[i_1(0^+) + i_2(0^+)] = -\frac{R_1 + R_2}{R_2}i_1(0^+)$$

and the expression for i_2 becomes

$$i_2 = i_1(0^+)\frac{R_1}{R_2}e^{-t/\tau}$$

Let us see if we can obtain this last expression more directly. Since the inductor current decays exponentially as $e^{-t/\tau}$, every current throughout the circuit must follow the same functional behavior. This is made clear by considering the inductor current as a source current which is being applied to a resistive network. Every current and voltage in the resistive network must have the same time dependence. Using these ideas, we therefore express i_2 as

$$i_2 = Ae^{-t/\tau}$$

where

$$\tau = \frac{L}{R_{eq}}$$

and A must be determined from a knowledge of the initial value of i_2. Since $i_1(0^+)$ is known, the voltage across R_1 and R_2 is known, and

$$i_2(0^+) = i_1(0^+)\frac{R_1}{R_2}$$

Therefore,

$$i_2 = i_1(0^+)\frac{R_1}{R_2}e^{-t/\tau}$$

A similar sequence of steps will provide a rapid solution to a large number of problems. We first recognize the time dependence of the response as an exponential decay, determine the appropriate time constant by combining resistances, write the solution with an unknown amplitude, and then determine the amplitude from a given initial condition.

This same technique is also applicable to a circuit which contains one resistor and any number of inductors, as well as to those special circuits containing two or more inductors and also two or more resistors that may be simplified by resistance or inductance combination until the simplified circuits have only one inductor or one resistor.

Example 4-1　As an example of such a circuit, let us determine both i_1 and i_2 in the circuit shown in Fig. 4-7.

Solution: After $t = 0$, when the voltage source is disconnected, we easily calculate an equivalent inductance,

$$L_{eq} = \frac{2 \times 3}{2 + 3} + 1 = 2.2 \text{ mH}$$

an equivalent resistance,

$$R_{eq} = \frac{90(60 + 120)}{90 + 180} + 50 = 110 \ \Omega$$

Figure 4-7

After $t = 0$, this circuit simplifies to an equivalent resistance of 110 Ω in series with $L_{eq} = 2.2$ mH.

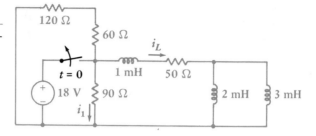

and the time constant,

$$\tau = \frac{L_{eq}}{R_{eq}} = \frac{2.2 \times 10^{-3}}{110} = 20 \ \mu s$$

Thus, the form of the natural response is $Ae^{-50\,000t}$. With the independent source connected ($t < 0$), i_L is $\frac{18}{50}$, or 0.36 A, while i_1 is $\frac{18}{90}$, or 0.2 A. At $t = 0^+$, i_L must still be 0.36 A, but i_1 will jump to a new value determined by $i_L(0^+)$. Thus,

$$i_1(0^+) = -i_L(0^+)\tfrac{180}{270} = -0.24 \text{ A}$$

Hence,

$$i_L = \begin{cases} 0.36 \text{ A} & (t < 0) \\ 0.36e^{-50\,000t} \text{ A} & (t > 0) \end{cases}$$

and

$$i_1 = \begin{cases} 0.2 \text{ A} & (t < 0) \\ -0.24e^{-50\,000t} \text{ A} & (t > 0) \end{cases}$$

$\blacksquare$

In idealized circuits in which a pure inductance loop is present, such as that through the 2- and 3-mH inductors of Fig. 4-7, a constant current may continue to circulate as $t \rightarrow \infty$. The current through either one of these inductors is not necessarily of the form $Ae^{-t/\tau}$, but takes the more general form $A_1 + A_2e^{-t/\tau}$. This unimportant special case is illustrated by Prob. 14 at the end of this chapter.

We have now considered the task of finding the natural response of any circuit which can be represented by an equivalent inductor in series with an equivalent resistor. A circuit containing several resistors and several inductors does not in general possess a form which allows either the resistors or the inductors to be combined into single equivalent elements. There is no single negative exponential term or single time constant associated with the circuit. Rather, there will, in general, be several negative exponential terms, the number of terms being equal to the number of inductors that remain after all possible inductor combinations have been made. The natural response of these more complex circuits is obtained by using techniques that we shall study later. One of the methods will appear toward the end of Chap. 12; it is based on the concept of complex frequency. The most powerful methods rely on the use of Fourier or Laplace transforms and will arise in Chaps. 18 and 19.

4-6. After $t = 0$, each of the portions of the circuits in Fig. 4-2 that contain the **Drill Problems** inductor is source-free. Find values for i_L and v at $t = 0.2$ s in (*a*) Fig. 4-2*a*; (*b*) Fig. 4-2*b*; (*c*) Fig. 4-2*c*.

Ans: 0.736 A, 14.72 V; 0.325 A, −12.99 V; 1.573 A, −18.88 V

4-7. At $t = 0.15$ s in the circuit of Fig. 4-8, find the value of (*a*) i_L; (*b*) i_1; (*c*) i_2.

Ans: 0.756 A; 0; 1.244 A

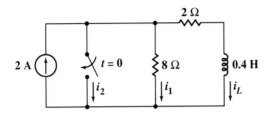

Figure 4-8

See Drill Prob. 4-7.

The series combination of a resistor and a capacitor has a greater practical importance than does the combination of a resistor and an inductor. When engineers have any freedom of choice between using a capacitor and using an inductor in the coupling network of an electronic amplifier, in the compensation networks of an automatic control system, or in the synthesis of an equalizing network, for example, they choose the *RC* network over the *RL* network whenever possible. The reasons for this choice are the smaller losses present in a physical capacitor, the lower cost, the better approximation which the mathematical model makes to the physical element it is intended to represent, and the smaller size and lighter weight as exemplified by capacitors in hybrid and integrated circuits.

4-5

The simple *RC* circuit

Let us see how closely the analysis of the parallel (or is it series?) *RC* circuit corresponds to that of the *RL* circuit. The *RC* circuit is shown in Fig. 4-9. We shall assume an initial stored energy in the capacitor by selecting

$$v(0) = V_0$$

The total current leaving the node at the top of the circuit diagram must be zero, and, therefore,

$$C\frac{dv}{dt} + \frac{v}{R} = 0$$

Division by C gives us

$$\frac{dv}{dt} + \frac{v}{RC} = 0 \tag{9}$$

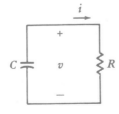

Figure 4-9

A parallel *RC* circuit for which $v(t)$ is to be determined, subject to the initial condition that $v(0) = V_0$.

Equation (9) has a familiar form; comparison with Eq. (1),

$$\frac{di}{dt} + \frac{R}{L}i = 0 \tag{1}$$

shows that the replacement of i by v and L/R by RC produces the identical equation we considered previously. It should, for the RC circuit we are now analyzing is the dual of the RL circuit we considered first. This duality forces $v(t)$ for the RC circuit and $i(t)$ for the RL circuit to have identical expressions if the resistance of one circuit is equal to the reciprocal of the resistance of the other circuit and if L is numerically equal to C. Thus, the response of the RL circuit,

$$i(t) = i(0)e^{-Rt/L} = I_0 e^{-Rt/L}$$

enables us to write immediately

$$v(t) = v(0)e^{-t/RC} = V_0 e^{-t/RC} \tag{10}$$

for the RC circuit.

Now let us suppose that we had selected the current i as our variable in the RC circuit, rather than the voltage v. Applying Kirchhoff's voltage law,

$$\frac{1}{C}\int_{t_0}^{t} i\,dt - v(t_0) + Ri = 0$$

we obtain an integral equation and not a differential equation. However, if we take the time derivative of both sides of this equation,

$$\frac{i}{C} + R\frac{di}{dt} = 0 \tag{11}$$

and replace i by v/R, we obtain Eq. (9) again:

$$\frac{v}{RC} + \frac{dv}{dt} = 0$$

Equation (11) could have been used as our starting point, but duality would not have appeared as naturally.

Let us discuss the physical nature of the voltage response of the RC circuit as expressed by Eq. (10). At $t = 0$ we obtain the correct initial condition, and as t becomes infinite the voltage approaches zero. This latter result agrees with our thinking that if there were any voltage remaining across the capacitor, then energy would continue to flow into the resistor and be dissipated as heat. Thus, a final voltage of zero is necessary. The time constant of the RC circuit may be found by using the duality relationships on the expression for the time constant of the RL circuit, or it may be found by simply noting the time at which the response has dropped to 36.8 percent of its initial value:

$$\frac{\tau}{RC} = 1$$

and
$$\tau = RC \tag{12}$$

Our familiarity with the negative exponential and the significance of the time constant τ enables us to sketch the response curve readily (Fig. 4-10).

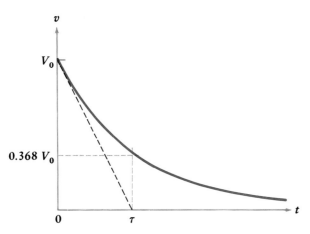

Figure 4-10

The capacitor voltage $v(t)$ in the parallel RC circuit is plotted as a function of time. The initial value of $v(t)$ is assumed to be V_0.

Larger values of R or C provide larger time constants and slower dissipation of the stored energy. A larger resistance will dissipate a smaller power[1] with a given voltage across it, thus requiring a greater time to convert the stored energy into heat; a larger capacitance stores a larger energy with a given voltage across it, again requiring a greater time to lose this initial energy.

4-8. Determine $v(0^+)$ for each circuit shown in Fig. 4-11.

Drill Problems

Ans: 40 V; 50 V; 20 V

4-9. Find $i(0^+)$ and $v(2\ \text{ms})$ for each circuit in Fig. 4-11.

Ans: 16 mA, 17.97 V; 62.5 mA, 14.33 V; 0, 2.71 V

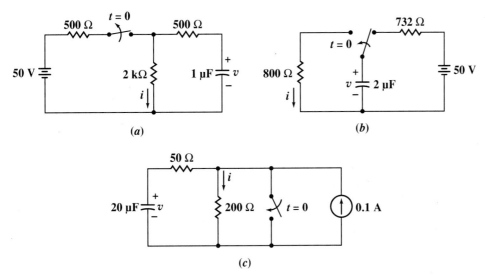

Figure 4-11

See Drill Probs. 4-8 and 4-9.

(a)

(b)

(c)

[1] "Greater resistance leads to less dissipation" might be the scholar's motto.

4-6

A more general *RC* circuit

Many of the *RC* circuits for which we would like to find the natural response contain more than a single resistor and capacitor. Just as we did for the *RL* circuits, we first consider those cases in which the given circuit may be reduced to an equivalent circuit consisting of only one resistor and one capacitor.

Let us suppose first that we are faced with a circuit containing only one capacitor, but any number of resistors. It is possible to replace the two-terminal resistive network which is across the capacitor terminals by an equivalent resistor, and we may then write down the expression for the capacitor voltage immediately.

Example 4-2 Let us find $v(0^+)$ and $i_1(0^+)$ for the circuit shown in Fig. 4-12*a*.

Figure 4-12

(*a*) A given circuit containing one capacitor and several resistors. (*b*) The resistors have been replaced by a single equivalent resistor; the time constant is now obvious.

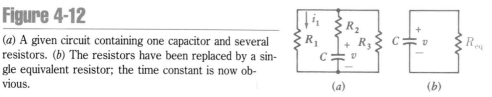

(*a*)   (*b*)

Solution: We first simplify the circuit of Fig. 4-12*a* to that of Fig. 4-12*b*, enabling us to write

$$v = V_0 e^{-t/R_{\text{eq}}C}$$

where

$$v(0) = V_0 \qquad \text{and} \qquad R_{\text{eq}} = R_2 + \frac{R_1 R_3}{R_1 + R_3}$$

Every current and voltage in the resistive portion of the network must have the form $Ae^{-t/R_{\text{eq}}C}$, where A is the initial value of that current or voltage. Thus, the current in R_1, for example, may be expressed as

$$i_1 = i_1(0^+)e^{-t/\tau}$$

where

$$\tau = \left(R_2 + \frac{R_1 R_3}{R_1 + R_3} \right) C$$

and $i_1(0^+)$ remains to be determined from some initial condition. Suppose that $v(0)$ is given. Since v cannot change instantaneously, we may think of the capacitor as being replaced by an independent dc source, $v(0)$. Thus,

$$i_1(0^+) = \frac{v(0)}{R_2 + R_1 R_3/(R_1 + R_3)} \frac{R_3}{R_1 + R_3}$$

The solution is obtained by collecting these results. ∎

Another special case includes those circuits containing one resistor and any number of capacitors. The resistor voltage is easily obtained by establishing the value of the equivalent capacitance and determining the time constant. Once again our mathematically perfect elements may lead to phenomena which would not exist in a physical circuit. Here, two capacitors in series may have equal and opposite voltages across each element and yet have zero voltage across the combination. Thus the general form of the voltage across either is $A_1 + A_2 e^{-t/\tau}$, while the voltage across the series combination continues to be

$Ae^{-t/\tau}$. An example of such a situation is provided by Prob. 27 at the end of the chapter.

Some circuits containing a number of both resistors and capacitors may be replaced by an equivalent circuit containing only one resistor and one capacitor; it is necessary that the original circuit be one which can be broken into two parts, one containing all resistors and the other containing all capacitors, such that the two parts are connected by only two ideal conductors. This is not possible in general.

More complicated circuits that cannot be reduced to simple series *RC* circuits will be considered later, in Chaps. 12, 18, and 19.

4-10. Find values of v_C and v_o in the circuit of Fig. 4-13 at t equal to (*a*) 0⁻; (*b*) 0⁺; (*c*) 1.3 ms.　　　*Ans:* 100 V, 38.4 V; 100 V, 25.6 V; 59.5 V, 15.22 V

Drill Problem

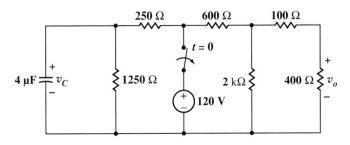

Figure 4-13

See Drill Prob. 4-10.

---

1 After having been closed for a long time, the switch in the circuit of Fig. 4-14*a* is opened at $t = 0$. (*a*) Find $i_L(t)$ for $t > 0$. (*b*) Evaluate $i_L(10\text{ ms})$. (*c*) Find t_1 if $i_L(t_1) = 0.5i_L(0)$.

Problems

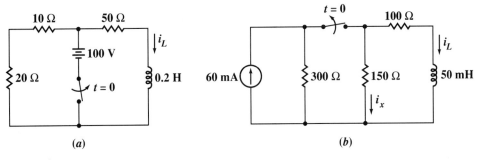

(*a*)　　　　　　　　　　　　　(*b*)

Figure 4-14

(*a*) See Prob. 1. (*b*) See Prob. 2.

2 The switch in Fig. 4-14*b* opens at $t = 0$ after having been closed for time immemorial. Find i_L and i_x at (*a*) $t = 0^-$; (*b*) $t = 0^+$; (*c*) $t = 0.3$ ms.

3 After being in the configuration shown for hours, the switch in the circuit of Fig. 4-15 is closed at $t = 0$. At $t = 5\ \mu$s, calculate: (*a*) i_L; (*b*) i_{SW}.

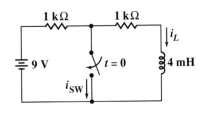

Figure 4-15

See Prob. 3.

4 The switch of Fig. 4-16 has been open for a long time before it closes at $t = 0$. For the time interval $-5 < t < 5$ μs, sketch: (a) $i_L(t)$; (b) $i_x(t)$.

Figure 4-16

See Prob. 4.

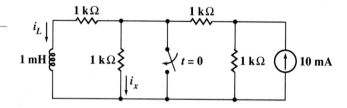

5 A 0.2-H inductor is in parallel with a 100-Ω resistor. The inductor current is 4 A at $t = 0$. (a) Find $i_L(t)$ at $t = 0.8$ ms. (b) If another 100-Ω resistor is connected in parallel with the inductor at $t = 1$ ms, calculate i_L at $t = 2$ ms.

6 A 20-mH inductor is in parallel with a 1-kΩ resistor. Let the value of the loop current be 40 mA at $t = 0$. (a) At what time will the current be 10 mA? (b) What series resistance should be switched into the circuit at $t = 10$ μs so that the current is 10 mA at $t = 15$ μs?

7 Figure 4-3 shows a plot of i/I_0 as a function of t. (a) Determine the values of t/τ at which i/I_0 is 0.1, 0.01, and 0.001. (b) If a tangent to the curve is drawn at the point where $t/\tau = 1$, where will it intersect the t/τ axis?

8 In the network of Fig. 4-17, initial values are $i_1(0) = 20$ mA and $i_2(0) = 15$ mA. (a) Determine $v(0)$. (b) Find $v(15$ μs$)$. (c) At what time is $v(t) = 0.1v(0)$?

Figure 4-17

See Prob. 8.

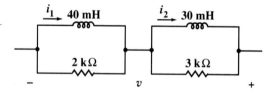

9 Select values for R_1 and R_2 in the circuit of Fig. 4-18 so that $v_R(0^+) = 10$ V and $v_R(1$ ms$) = 5$ V.

Figure 4-18

See Prob. 9.

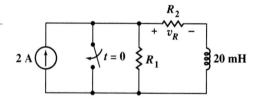

10 The switch in the circuit shown in Fig. 4-19 has been open for a long time before it closes at $t = 0$. (a) Find $i_L(t)$ for $t > 0$. (b) Sketch $v_x(t)$ for $-4 < t < 4$ ms.

Figure 4-19

See Prob. 10.

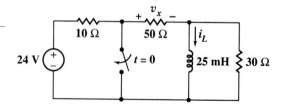

11 If $i_L(0) = 10$ A in the circuit of Fig. 4-20, find $i_L(t)$ for $t > 0$.

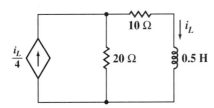

Figure 4-20

See Prob. 11.

12 Refer to the circuit of Fig. 4-21 and determine i_1 at $t = -0.1, 0.03,$ and 0.1 s. Prepare a sketch of i_1 versus t, $-0.1 < t < 1$ s.

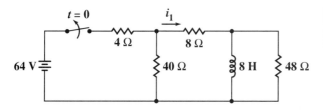

Figure 4-21

See Prob. 12.

13 A circuit consists of a 0.5-H inductor, a 10-Ω resistor, and a 40-Ω resistor in series. The inductor current is 4 A at $t = 0$. (*a*) Calculate $i_L(15$ ms). (*b*) The 40-Ω resistor is short-circuited at $t = 15$ ms. Calculate $i_L(30$ ms).

14 The circuit shown in Fig. 4-22 contains two inductors in parallel, thus providing the opportunity of a trapped current circulating around the inductive loop. Let $i_1(0^-) = 10$ A and $i_2(0^-) = 20$ A. (*a*) Find $i_1(0^+)$, $i_2(0^+)$, and $i(0^+)$. (*b*) Determine the time constant τ for $i(t)$. (*c*) Find $i(t)$, $t > 0$. (*d*) Find $v(t)$. (*e*) Find $i_1(t)$ and $i_2(t)$ from $v(t)$ and the initial values. (*f*) Show that the stored energy at $t = 0$ is equal to the sum of the energy dissipated in the resistive network between $t = 0$ and $t = \infty$, plus the energy stored in the inductors at $t = \infty$.

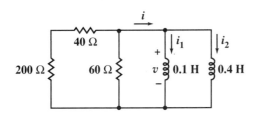

Figure 4-22

See Prob. 14.

15 (*a*) Find $v_C(t)$ for all time in the circuit of Fig. 4-23. (*b*) At what time is $v_C = 0.1v_C(0)$?

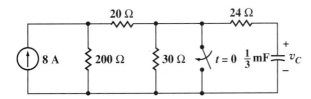

Figure 4-23

See Prob. 15.

16 The circuit of Fig. 4-24 has been in the form shown since noon yesterday. The switch is opened at exactly 10:00 a.m. Find i_1 and v_C at (*a*) 9:59 a.m.; (*b*) 10:05 a.m.

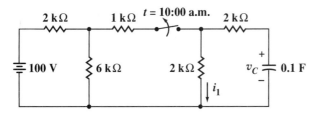

Figure 4-24

See Prob. 16.

17 A 4-A current source, a 20-Ω resistor, and a 5-μF capacitor are all in parallel. The amplitude of the current source drops suddenly to zero (becoming a 0-A current source) at $t = 0$. At what time has (a) the capacitor voltage dropped to one-half of its initial value, and (b) the energy stored in the capacitor dropped to one-half of its initial value?

18 Determine $v_C(t)$ and $i_C(t)$ for the circuit of Fig. 4-25 and sketch both curves on the same time axis, $-0.1 < t < 0.1$ s.

Figure 4-25

See Prob. 18.

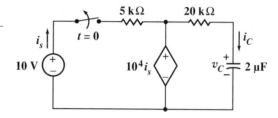

19 After being in the configuration shown for a long time, the switch in Fig. 4-26 is opened at $t = 0$. Determine values for (a) $i_s(0^-)$; (b) $i_x(0^-)$; (c) $i_x(0^+)$; (d) $i_s(0^+)$; (e) $i_x(0.4 \text{ s})$.

Figure 4-26

See Prob. 19.

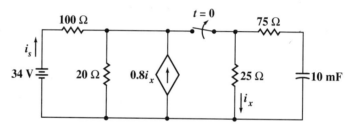

20 After being closed for a long time, the switch in the circuit of Fig. 4-27 is opened at $t = 0$. (a) Find $v_C(t)$ for $t > 0$. (b) Calculate values for $i_A(-100 \text{ μs})$ and $i_A(100 \text{ μs})$.

Figure 4-27

See Prob. 20.

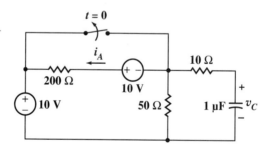

21 Many moons after the circuit of Fig. 4-28 was first assembled, its switch is closed at $t = 0$. (a) Find $i_1(t)$ for $t < 0$. (b) Find $i_1(t)$ for $t > 0$.

Figure 4-28

See Prob. 21.

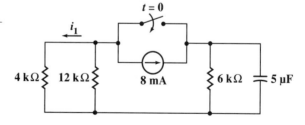

22 A long time after the circuit of Fig. 4-29 was assembled, both switches are opened simultaneously at $t = 0$, as indicated. (a) Obtain an expression for v_{out} for $t > 0$. (b) Obtain values for v_{out} at $t = 0^+$, 1 μs, and 5 μs.

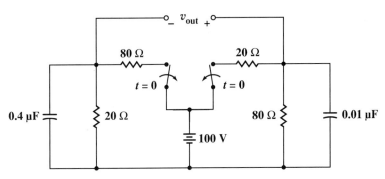

Figure 4-29

See Prob. 22.

23 Assume that the circuit shown in Fig. 4-30 has been in the form shown for a very long time. Find $v_C(t)$ for all t after the switch opens.

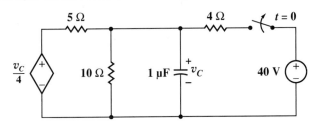

Figure 4-30

See Prob. 23.

24 Determine values for R_0 and R_1 in the circuit of Fig. 4-31 so that $v_C = 50$ V at $t = 0.5$ ms and $v_C = 25$ V at $t = 2$ ms.

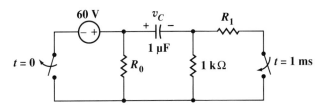

Figure 4-31

See Prob. 24.

25 For the circuit shown in Fig. 4-32, determine $v_C(t)$ for (a) $t < 0$; (b) $t > 0$.

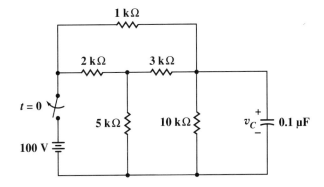

Figure 4-32

See Prob. 25.

26 Find $i_1(t)$ for $t < 0$ and $t > 0$ in the circuit of Fig. 4-33.

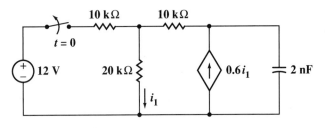

Figure 4-33

See Prob. 26.

27 The switch in Fig. 4-34 is moved from A to B at $t = 0$ after being at A for a long time. This places the two capacitors in series, thus allowing equal and opposite dc voltages to be trapped on the capacitors. (a) Determine $v_1(0^-)$, $v_2(0^-)$, and $v_R(0^-)$. (b) Find $v_1(0^+)$, $v_2(0^+)$, and $v_R(0^+)$. (c) Determine the time constant of $v_R(t)$. (d) Find $v_R(t)$, $t > 0$. (e) Find $i(t)$. (f) Find $v_1(t)$ and $v_2(t)$ from $i(t)$ and the initial values. (g) Show that the stored energy at $t = \infty$ plus the total energy dissipated in the 20-kΩ resistor is equal to the energy stored in the capacitors at $t = 0$.

Figure 4-34

See Prob. 27.

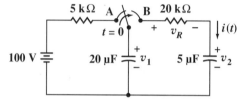

28 The value of i_s in the circuit of Fig. 4-35 is 1 mA for $t < 0$, and zero for $t > 0$. Find $v_x(t)$ for (a) $t < 0$; (b) $t > 0$.

Figure 4-35

See Prob. 28.

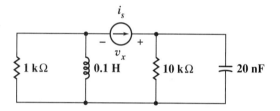

29 The value of v_s in the circuit of Fig. 4-36 is 20 V for $t < 0$, and zero for $t > 0$. Find $i_x(t)$ for (a) $t < 0$; (b) $t > 0$.

Figure 4-36

See Prob. 29.

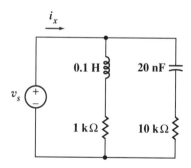

30 (SPICE) Let $i_L(0) = 20$ A in the circuit shown in Fig. 4-37. Use TSTEP = 1 ms in a SPICE program to find i_L at $t = 20$ ms.

Figure 4-37

See Prob. 30.

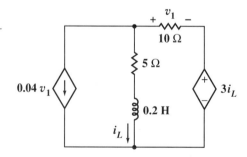

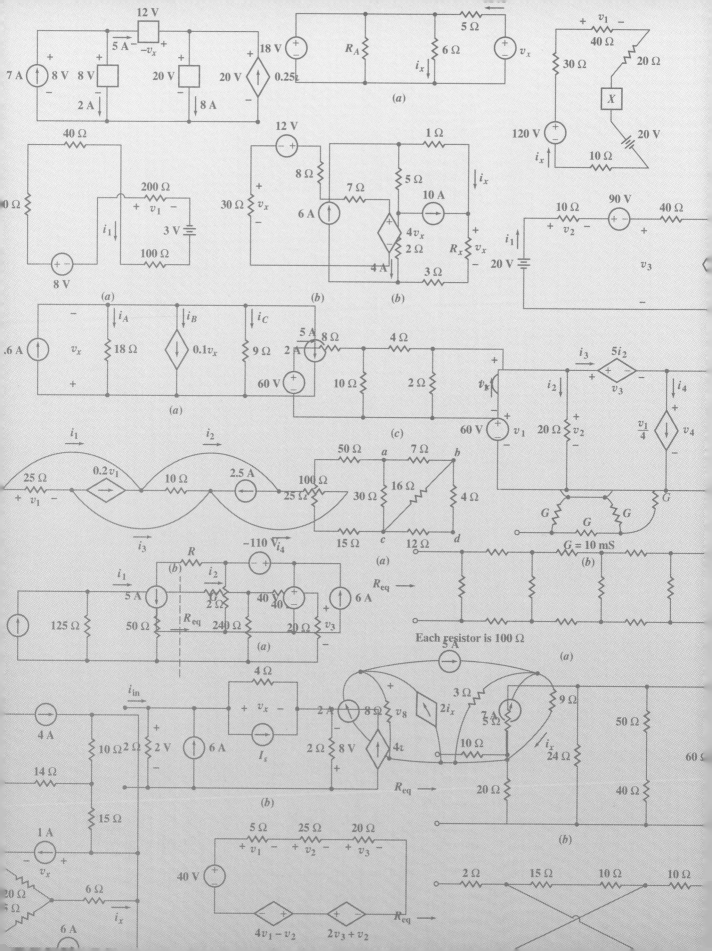

The Application of the Unit-Step Forcing Function

Introduction

We have just spent a chapter's worth of our time studying the response of *RL* and *RC* circuits when no sources or forcing functions were present. We termed this response the *natural response,* because its form depends only on the nature of the circuit. The reason that any response at all is obtained arises from the presence of initial energy storage within the inductive or capacitive elements in the circuit. In many of the examples and problems, we were confronted with circuits containing sources and switches; we were informed that certain switching operations were performed at $t = 0$ in order to remove all the sources from the circuit, while leaving known amounts of energy stored here and there. In other words, we have been solving problems in which energy sources are suddenly *removed* from the circuit; now we must consider that type of response which results when energy sources are suddenly *applied* to a circuit.

We shall devote this chapter to a study of the response which occurs when the energy sources which are suddenly applied are dc sources. After we have studied sinusoidal and exponential sources, we may then consider the general problem of the sudden application of a more general source. Since every electrical device is intended to be energized at least once, and since most devices are turned on and off many times in the course of their lifetimes, it should be evident that our study will be applicable to many practical cases. Even though we are now restricting ourselves to dc sources, there are still innumerable cases in which these simpler examples correspond to the operation of physical devices. For example, the first circuit we shall analyze may be considered to represent the buildup of the field current when a dc motor is started. The generation and use of the rectangular voltage pulses needed to represent a number or a command in a digital computer provide many examples in the field of electronic or transistor circuitry. Similar circuits are found in the synchronization and sweep circuits of television receivers, in communication systems using pulse modulation, and in radar systems, to name but a few examples. Furthermore, an important part of the analysis of most servomechanisms is the determination of their responses to suddenly applied constant inputs.

The unit-step forcing function

We have been speaking of the "sudden application" of an energy source, and by this phrase we imply its application in zero time. The operation of a switch in series with a battery is thus equivalent to a forcing function which is zero up to the instant that the switch is closed and is equal to the battery voltage thereafter. The forcing function has a break, or discontinuity, at the instant the switch is closed. Certain special forcing functions which are discontinuous or have discontinuous derivatives are called *singularity functions,* the two most important of these singularity functions being the unit-step function and the unit-impulse function. The unit-step function is the subject of this section; the unit impulse is discussed in Chaps. 18 and 19.

We define the *unit-step forcing function* as a function of time which is zero for all values of its argument less than zero and which is unity for all positive values of its argument. If we let $(t - t_0)$ be the argument and represent the unit-step function by u, then $u(t - t_0)$ must be zero for all values of t less than t_0, and it must be unity for all values of t greater than t_0. At $t = t_0$, $u(t - t_0)$ changes abruptly from 0 to 1. Its value at $t = t_0$ is not defined, but its value is known for all instants of time that are arbitrarily close to $t = t_0$. We often indicate this by writing $u(t_0^-) = 0$ and $u(t_0^+) = 1$. The concise mathematical definition of the unit-step forcing function is

$$u(t - t_0) = \begin{cases} 0 & t < t_0 \\ 1 & t > t_0 \end{cases}$$

and the function is shown graphically in Fig. 5-1. Note that a vertical line of

Figure 5-1

The unit-step forcing function, $u(t - t_0)$.

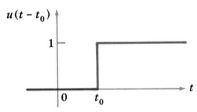

unit length is shown at $t = t_0$. Although this "riser" is not strictly a part of the definition of the unit step, it is usually shown in each drawing.

We also note that the unit step need not be a *time* function, although our attention in this chapter will be directed only to time functions. For example, $u(x - x_0)$ might be used to denote a *unit-step function* which is not a unit-step *forcing* function because it is not a function of time t. Rather, it is a function of x, where x might be a distance in meters, for example, or a frequency, as we shall see in Chap. 18.

Very often in circuit analysis a discontinuity or a switching action takes place at an instant that is defined as $t = 0$. In that case $t_0 = 0$, and we then represent the corresponding unit-step forcing function by $u(t - 0)$, or more

Figure 5-2

The unit-step forcing function $u(t)$ is shown as a function of t.

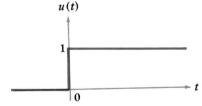

simply $u(t)$. This is shown in Fig. 5-2. Thus

$$u(t) = \begin{cases} 0 & t < 0 \\ 1 & t > 0 \end{cases}$$

The unit-step forcing function is in itself dimensionless. If we wish it to represent a voltage, it is necessary to multiply $u(t - t_0)$ by some constant voltage, such as V_0. Thus, $v(t) = V_0 u(t - t_0)$ is an ideal voltage source which is zero before $t = t_0$ and a constant V_0 after $t = t_0$. This forcing function is shown connected to a general network in Fig. 5-3a.

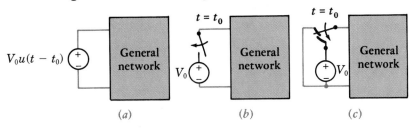

$V_0 u(t - t_0)$

$t = t_0$

General network

V_0

$t = t_0$

General network

General network

V_0

(a)　　　　　(b)　　　　　(c)

Figure 5-3

(a) A voltage-step forcing function is shown as the source driving a general network. (b) A simple circuit which, although not the exact equivalent of part a, may be used as its equivalent in many cases. (c) An exact equivalent of part a.

We should now logically ask what physical source is the equivalent of this discontinuous forcing function. By *equivalent,* we mean simply that the voltage-current characteristics of the two networks are identical. For the step-voltage source of Fig. 5-3a, the voltage-current characteristic is quite simple: the voltage is zero prior to $t = t_0$, it is V_0 after $t = t_0$, and the current may be any (finite) value in either time interval. Our first thoughts might produce the attempt at an equivalent shown in Fig. 5-3b, a dc source V_0 in series with a switch which closes at $t = t_0$. This network is not equivalent for $t < t_0$, however, because the voltage across the battery and switch is completely unspecified in this time interval. The "equivalent" source is an open circuit, and the voltage across it may be anything. After $t = t_0$, the networks are equivalent, and if this is the only time interval in which we are interested, and if the initial currents which flow from the two networks are identical at $t = t_0$, then Fig. 5-3b becomes a useful equivalent of Fig. 5-3a.

In order to obtain an exact equivalent for the voltage-step forcing function, we may provide a single-pole double-throw switch. Before $t = t_0$, the switch serves to ensure zero voltage across the input terminals of the general network. After $t = t_0$, the switch is thrown to provide a constant input voltage V_0. At $t = t_0$, the voltage is indeterminate (as is the step forcing function), and the battery is momentarily short-circuited (it is fortunate that we are dealing with mathematical models!). This exact equivalent of Fig. 5-3a is shown in Fig. 5-3c.

Before concluding our discussion of equivalence, it is enlightening to consider the exact equivalent of a battery and a switch. What is the voltage-step forcing function which is equivalent to Fig. 5-3b? We are searching for some arrangement which changes suddenly from an open circuit to a constant voltage; a change in resistance is involved, and this is the crux of our difficulty. The step function enables us to change a voltage (or a current) discontinuously, but here we need a changing resistance as well. The equivalent therefore must contain a resistance or conductance step function, a passive element which is time-varying. Although we might construct such an element with the unit-step function, it should be apparent that the end product is a switch; a switch is merely a resistance which changes instantaneously from zero to infinite ohms, or vice versa. Thus, we conclude that the exact equivalent of a battery and switch in series must be a battery in series with some representation of a time-varying resistance; no arrangement of voltage- and current-step forcing functions is able to provide us with the exact equivalent.[1]

---

[1] An equivalent may always be determined *if some information about the general network is available* (the voltage across the switch for $t < t_0$); we assume no a priori knowledge about the general network.

Figure 5-4

(*a*) A current-step forcing function is applied to a general network. (*b*) A simple circuit which, although not the exact equivalent of part *a*, may be used as its equivalent in many cases.

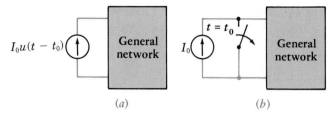

(*a*)   (*b*)

Figure 5-4*a* shows a current-step forcing function driving a general network. If we attempt to replace this circuit by a dc source in parallel with a switch (which *opens* at $t = t_0$), we must realize that the circuits are equivalent after $t = t_0$ but that the responses are alike after $t = t_0$ only if the initial conditions are the same. Judiciously, then, we may often use the circuits of Figs. 5-4*a* and *b* interchangeably. The exact equivalent of Fig. 5-4*a* is the dual of the circuit of Fig. 5-3*c*; the exact equivalent of Fig. 5-4*b* cannot be constructed with current- and voltage-step forcing functions alone.[2]

Some very useful forcing functions may be obtained by manipulating the unit-step forcing function. Let us define a rectangular voltage pulse by the following conditions:

$$v(t) = \begin{cases} 0 & t < t_0 \\ V_0 & t_0 < t < t_1 \\ 0 & t_1 < t \end{cases}$$

The pulse is drawn in Fig. 5-5. Can this pulse be represented in terms of the

Figure 5-5

A useful forcing function, the rectangular voltage pulse.

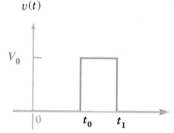

unit-step forcing function? Let us consider the difference of the two unit steps, $u(t - t_0) - u(t - t_1)$. The two step functions are shown in Fig. 5-6*a*, and their

Figure 5-6

(*a*) The unit steps $u(t - t_0)$ and $-u(t - t_1)$. (*b*) A source which yields the rectangular voltage pulse of Fig. 5-5.

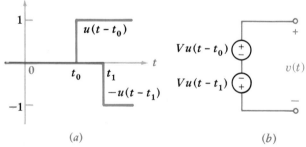

(*a*)   (*b*)

difference is obviously a rectangular pulse. The source $V_0 u(t - t_0) - V_0 u(t - t_1)$ which will provide us with the desired voltage is indicated in Fig. 5-6*b*.

If we have a sinusoidal voltage source $V_m \sin \omega t$ which is suddenly connected to a network at $t = t_0$, then an appropriate voltage forcing function would be $u(t) = V_m u(t - t_0) \sin \omega t$. If we wish to represent one burst of energy from a

---

[2] The equivalent can be drawn if the current through the switch prior to $t = t_0$ is known.

radar transmitter, we may turn the sinusoidal source off $\frac{1}{10}$ μs later by a second unit-step forcing function. The voltage pulse is thus

$$u(t) = V_m[u(t - t_0) - u(t - t_0 - 10^{-7})] \sin \omega t$$

This forcing function is sketched in Fig. 5-7.

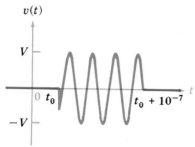

Figure 5-7

A radio-frequency pulse, described by $v(t) = V_m[u(t - t_0) - u(t - t_0 - 10^{-7})] \sin \omega t$. The sinusoidal frequency within the pulse shown is about 36 MHz, a value which is too low for radar, but about right for constructing legible drawings.

As a final introductory remark, we should note that the unit-step forcing function must be considered only as the mathematical model of an actual switching operation. No physical resistor, inductor, or capacitor behaves entirely like its idealized circuit element; we also cannot perform a switching operation in zero time. However, switching times less than 1 ns are common in many circuits, and this time is often sufficiently short compared with the time constants in the rest of the circuit that it may be ignored.

Drill Problems

5-1. Evaluate each of the following at $t = 0.8$: (a) $3u(t) - 2u(-t) + 0.8u(1 - t)$; (b) $[4u(t)]^{u(-t)}$; (c) $2u(t) \sin \pi t$.
$\qquad$ *Ans: 3.8; 1; 1.176*

5-2. For the circuit of Fig. 5-8, find i_1 at t equal to (a) -2 s; (b) -0.5 s; (c) 0.5 s; (d) 2 s.
$\qquad$ *Ans: 2 A; 5 A; 4.33 A; 3 A*

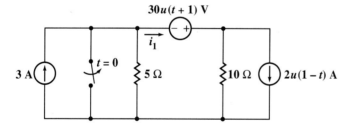

Figure 5-8

See Drill Prob. 5-2.

5-3

A first look at the driven RL circuit

We are now ready to subject a simple network to the sudden application of a dc source. The circuit consists of a battery whose voltage is V_0 in series with a switch, a resistor R, and an inductor L. The switch is closed at $t = 0$, as indicated on the circuit diagram of Fig. 5-9a. It is evident that the current $i(t)$ is zero before $t = 0$, and we are therefore able to replace the battery and switch by a voltage-step forcing function $V_0u(t)$, which also produces no response prior to $t = 0$. After $t = 0$, the two circuits are obviously identical. Hence, we seek the current $i(t)$ either in the given circuit of Fig. 5-9a or in the equivalent circuit of Fig. 5-9b.

We shall find $i(t)$ at this time by writing the appropriate circuit equation and then solving it by separation of the variables and integration. After we obtain the answer and investigate the two parts of which it is composed, we shall next spend some time (the following section) in learning the general

Figure 5-9

(a) The given circuit. (b) An equivalent circuit, possessing the same response $i(t)$ for all time.

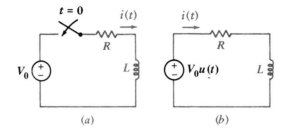

(a)   (b)

significance of these two terms. We can then construct the solution to this problem very easily; moreover, we shall be able to apply the general principles behind this simpler method to produce more rapid and more meaningful solutions to every problem involving the sudden application of any source. Let us now proceed with the more formal method of solution.

Applying Kirchhoff's voltage law to the circuit of Fig. 5-9b, we have

$$Ri + L\frac{di}{dt} = V_0 u(t)$$

Since the unit-step forcing function is discontinuous at $t = 0$, we shall first consider the solution for $t < 0$ and then for $t > 0$. It is evident that the application of zero voltage since $t = -\infty$ has not produced any response, and, therefore,

$$i(t) = 0 \qquad t < 0$$

For positive time, however, $u(t)$ is unity and we must solve the equation

$$Ri + L\frac{di}{dt} = V_0 \qquad t > 0$$

The variables may be separated in several simple algebraic steps, yielding

$$\frac{L\,di}{V_0 - Ri} = dt$$

and each side may be integrated directly:

$$-\frac{L}{R}\ln(V_0 - Ri) = t + k$$

In order to evaluate k, an initial condition must be invoked. Prior to $t = 0$, $i(t)$ is zero, and thus $i(0^-) = 0$; since the current in an inductor cannot change by a finite amount in zero time without being associated with an infinite voltage, we thus have $i(0^+) = 0$. Setting $i = 0$ at $t = 0$, we obtain

$$-\frac{L}{R}\ln V_0 = k$$

and, hence,

$$-\frac{L}{R}[\ln(V_0 - Ri) - \ln V_0] = t$$

Rearranging,

$$\frac{V_0 - Ri}{V_0} = e^{-Rt/L}$$

or

$$i = \frac{V_0}{R} - \frac{V_0}{R} e^{-Rt/L} \qquad t > 0$$

Thus, an expression for the response valid for all t would be

$$i = \left(\frac{V_0}{R} - \frac{V_0}{R} e^{-Rt/L} \right) u(t) \qquad\qquad (1)$$

This is the desired solution, but it has not been obtained in the simplest manner. In order to establish a more direct procedure, let us try to interpret the two terms appearing in Eq. (1). The exponential term has the functional form of the natural response of the RL circuit; it is a negative exponential, it approaches zero as time increases, and it is characterized by the time constant L/R. The *functional form* of this part of the response is thus identical with that which is obtained in the source-free circuit. However, the amplitude of this exponential term depends on V_0. We might generalize, then, that the response will be the sum of two terms, where one term has a functional form which is identical with that of the source-free response, but has an amplitude which depends on the forcing function. Now let us consider the nature of the second term of the response.

Equation (1) also contains a constant term, V_0/R. Why is it present? The answer is simple: the natural response approaches zero as the energy is gradually dissipated, but the total response must not approach zero. Eventually the circuit behaves as a resistor and an inductor in series with a battery, and a direct current V_0/R flows. This current is a part of the response which is directly attributable to the forcing function, and we call it the *forced response*. It is the response which is present a long time after the switch is closed.

The complete response is composed of two parts, the natural response and the forced response. The natural response is a characteristic of the circuit and not of the sources. Its form may be found by considering the source-free circuit, and it has an amplitude which depends on the initial amplitude of the source and the initial energy storage. The forced response has the characteristics of the forcing function; it is found by pretending that all switches have been thrown a long time ago. Since we are presently concerned only with switches and dc sources, the forced response is merely the solution of a simple dc circuit problem.

The reason for the two responses, forced and natural, may also be seen from physical arguments. We know that our circuit will eventually assume the forced response. However, at the instant the switches are thrown, the initial inductor currents (or, in other circuits, the voltages across the capacitors) will have values which depend only on the energy stored in these elements. These currents or voltages cannot be expected to be the same as the currents and voltages demanded by the forced response. Hence, there must be a transient period during which the currents and voltages change from their given initial values to their required final values. The portion of the response which provides the transition from initial to final values is the natural response (often called the *transient* response, as we found earlier). If we describe the response of the *source-free* simple RL circuit in these terms, then we should say that the forced response is zero and that the natural response serves to connect the initial response produced by stored energy with the zero value of the forced response. This description is appropriate only for those circuits in which the natural response eventually dies out. This always occurs in physical circuits where some resistance is associated with every element, but there are a number of "pathologic" circuits in which the natural response is nonvanishing as time becomes infinite. Those circuits in which trapped currents circulate around inductive loops, or voltages are trapped in series strings of capacitors, are examples.

Now let us search out the mathematical basis for dividing the response into a natural response and a forced response.

Drill Problem

5-3. The voltage source $60 - 40u(t)$ V is in series with a 10-Ω resistor and a 50-mH inductor. Let superposition help in finding the magnitudes of the inductor current and voltage at t equal to (a) 0^-; (b) 0^+; (c) ∞; (d) 3 ms.

Ans: 6 A, 0 V; 6 A, 40 V; 2 A, 0 V; 4.20 A, 22.0 V

5-4

The natural response and the forced response

There is also an excellent mathematical reason for considering the complete response to be composed of two parts—the forced response and the natural response. The reason is based on the fact that the solution of any linear differential equation may be expressed as the sum of two parts: the complementary solution (natural response) and the particular solution (forced response). Without delving into the general theory of differential equations, let us consider a general equation of the type met in the previous section:

$$\frac{di}{dt} + Pi = Q$$

or

$$di + Pi\,dt = Q\,dt \qquad (2)$$

We may identify Q as a forcing function and express it as $Q(t)$ to emphasize its general time dependence. In all our circuits, P will be a positive constant, but the remarks that follow about the solution of Eq. (2) are equally valid for the cases in which P is a general function of time. Let us simplify the discussion by assuming that P is a positive constant. Later, we shall also assume that Q is constant, thus restricting ourselves to dc forcing functions.

In any standard text on elementary differential equations, it is shown that if both sides of Eq. (2) are multiplied by a so-called integrating factor, then each side becomes an exact differential which can be integrated directly to obtain the solution. We are not separating the variables, but merely arranging them in such a way that integration is possible. For this equation, the integrating factor is $e^{\int P\,dt}$, or e^{Pt}, since P is a constant. We multiply each side of the equation by this integrating factor and obtain

$$e^{Pt}\,di + iPe^{Pt}\,dt = Qe^{Pt}\,dt$$

The form of the left side may now be improved when it is recognized as the exact differential of ie^{Pt}:

$$d(ie^{Pt}) = e^{Pt}\,di + iPe^{Pt}\,dt$$

and, thus,

$$d(ie^{Pt}) = Qe^{Pt}\,dt$$

We may now integrate each side, finding

$$ie^{Pt} = \int Qe^{Pt}\,dt + A$$

where A is a constant of integration. Since this constant is explicitly shown, we should remember that no integration constant needs to be added to the remaining integral at the time when we evaluate it. Multiplication by e^{-Pt} produces the solution for $i(t)$,

$$i = e^{-Pt} \int Q e^{Pt} dt + A e^{-Pt} \qquad (3)$$

If $Q(t)$, the forcing function, is known, then it remains only to evaluate the integral to obtain the exact functional form for $i(t)$. We shall not evaluate such an integral for each problem, however; instead, we are interested in using Eq. (3) as an exemplary solution from which we shall draw several very general conclusions.

We should note first that, for a source-free circuit, Q must be zero, and the solution is the natural response

$$i_n = A e^{-Pt} \qquad (4)$$

We shall find that the constant P is never negative; its value depends only on the passive circuit elements[3] and their interconnection in the circuit. The natural response therefore approaches zero as time increases without limit. It must do so, of course, in the simple RL series circuit, because the initial energy is gradually dissipated in the resistor. There are also idealized, nonphysical circuits in which P is zero; in these circuits the natural response does not die out, but approaches a constant value, as exemplified by trapped currents or voltages.

We therefore find that one of the two terms making up the complete response has the form of the natural response; it has an amplitude which will depend on (but not usually be equal to) the initial value of the complete response and thus on the initial value of the forcing function also.

We next observe that the first term of Eq. (3) depends on the functional form of $Q(t)$, the forcing function. Whenever we have a circuit in which the natural response dies out as t becomes infinite, this first term must completely describe the form of the response after the natural response has disappeared. This term we shall call the *forced response;* it is also called the *steady-state response,* the *particular solution,* or the *particular integral.*

For the present, we have elected to consider only those problems involving the sudden application of dc sources, and $Q(t)$ will therefore be a constant for all values of time after the switch in Fig. 5-9a has been closed. If we wish, we can now evaluate the integral in Eq. (3), obtaining the forced response

$$i_f = \frac{Q}{P}$$

or the complete response

$$i(t) = \frac{Q}{P} + A e^{-Pt}$$

For the RL series circuit, Q/P is the constant current V_0/R and $1/P$ is the time constant τ. We should see that the forced response might have been obtained without evaluating the integral, because it must be the complete response at infinite time; it is merely the source voltage divided by the series resistance. The forced response is thus obtained by inspection.

---

[3] If the circuit contains a dependent source or a negative resistance, P may be negative.

In the following section we shall consider several examples in which we find the complete response of an RL circuit by obtaining the natural and forced responses and then adding them.

Drill Problem

5-4. A voltage source, $v_s = 20e^{-100t}u(t)$ V, is in series with a 200-Ω resistor and a 4-H inductor. After substituting the correct quantities in Eq. (3), find the magnitude of the inductor current at t equal to (a) 0^-; (b) 0^+; (c) 8 ms; (d) 15 ms.   *Ans:* 0; 0; 22.1 mA; 24.9 mA

5-5

RL circuits

Let us use the simple RL series circuit to illustrate how to determine the complete response by the addition of the natural and forced responses. This circuit, shown in Fig. 5-10, has been analyzed earlier, but by a longer method. The desired response is the current $i(t)$, and we first express this current as the sum of the natural and the forced current,

$$i = i_n + i_f$$

The functional form of the natural response must be the same as that obtained without any sources. We therefore replace the step-voltage source by a short circuit and recognize the old RL series loop. Thus,

$$i_n = Ae^{-Rt/L}$$

where the amplitude A is yet to be determined.

We next consider the forced response, that part of the response which depends upon the nature of the forcing function itself. In this particular problem the forced response must be constant, because the source is a constant V_0 for all positive values of time. After the natural response has died out, therefore, there can be no voltage across the inductor; hence, a voltage V_0 appears across R, and the forced response is simply

$$i_f = \frac{V_0}{R}$$

Note that the forced response is determined completely; there is no unknown amplitude. We next combine the two responses to get

$$i = Ae^{-Rt/L} + \frac{V_0}{R}$$

and apply the initial condition to evaluate A. The current is zero prior to $t = 0$, and it cannot change value instantaneously, since it is the current flowing through an inductor. Thus, the current is zero immediately after $t = 0$, and

$$0 = A + \frac{V_0}{R}$$

Figure 5-10

A series RL circuit that is used to illustrate the method by which the complete response is obtained as the sum of the natural and forced responses.

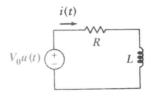

and
$$i = \frac{V_0}{R}(1 - e^{-Rt/L}) \qquad (5)$$

Note carefully that A is not the initial value of i, since $A = -V_0/R$, while $i(0) = 0$. In Chap. 4, where the circuits were source-free, A was indeed the initial value of the response. When forcing functions are present, however, we must first find the initial value of the response and then substitute this in the equation for the complete response to find A.

This response is plotted in Fig. 5-11, and we can see the manner in which

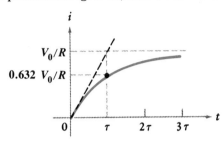

Figure 5-11

The current expressed by Eq. (5) is shown graphically. A line extending the initial slope meets the constant forced response at $t = \tau$.

the current builds up from its initial value of zero to its final value of V_0/R. The transition is effectively accomplished in a time 3τ. If our circuit represents the field coil of a large dc motor, we might have $L = 10$ H and $R = 20$ Ω, obtaining $\tau = 0.5$ s. The field current is thus established in about 1.5 s. In one time constant, the current has attained 63.2 percent of its final value.

Now let us apply this method to a more complicated circuit.

Example 5-1   Determine $i(t)$ for all values of time in the circuit of Fig. 5-12.

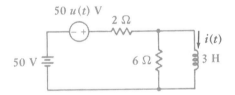

Figure 5-12

The circuit of Example 5-1.

Solution: The circuit contains a dc voltage source as well as a step-voltage source. We might choose to replace everything to the left of the inductor by the Thévenin equivalent, but instead let us merely recognize the form of that equivalent as a resistor in series with some voltage source. The circuit contains only one energy-storage element, the inductor. We first note that

$$\tau = \frac{L}{R_{eq}} = \frac{3}{1.5} = 2 \text{ s}$$

and recall that

$$i = i_f + i_n$$

The natural response is therefore a negative exponential as before. That is

$$i_n = Ae^{-t/2} \qquad t > 0$$

The forced response must be that produced by a constant voltage of 100 V. The forced response is constant, and no voltage is present across the inductor; it behaves as a short circuit, and, therefore,

$$i_f = \tfrac{100}{2} = 50$$

Thus,

$$i = 50 + Ae^{-0.5t} \qquad t > 0$$

In order to evaluate A, we must establish the initial value of the inductor current. Prior to $t = 0$, this current is 25 A, and it cannot change instantaneously. Thus,

$$25 = 50 + A \qquad \text{or} \qquad A = -25$$

Hence,

$$i = 50 - 25e^{-0.5t} \qquad t > 0$$

We complete the solution by also stating

$$i = 25 \qquad t < 0$$

or by writing a single expression valid for all t,

$$i = 25 + 25(1 - e^{-0.5t})u(t) \qquad \text{A}$$

The complete response is sketched in Fig. 5-13. Note how the natural

Figure 5-13

The response $i(t)$ of the circuit shown in Fig. 5-12 is sketched for values of time less and greater than zero.

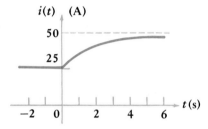

$i(t)$ (A)

response serves to connect the response for $t < 0$ with the constant forced response.

∎

As a final example of this method by which the complete response of any circuit subjected to a transient may be written down almost by inspection, let us again consider the simple RL series circuit.

Example 5-2 We want to find the current response in a simple series RL circuit when the forcing function is a rectangular voltage pulse of amplitude V_0 and duration t_0.

Solution: We represent the forcing function as the sum of two step-voltage sources $V_0 u(t)$ and $-V_0 u(t - t_0)$, as indicated in Figs. 5-14a and b, and we

Figure 5-14

(a) A rectangular voltage pulse which is to be used as the forcing function in a simple series RL circuit. (b) The series RL circuit, showing the representation of the forcing function by the series combination of two independent voltage-step sources. The current $i(t)$ is desired.

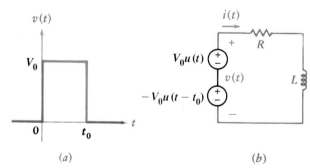

(a)

(b)

plan to obtain the response by using the superposition principle. Suppose we designate that part of $i(t)$ which is due to the upper source $V_0 u(t)$ acting alone by the symbol $i_1(t)$ and then let $i_2(t)$ represent that part due to $-V_0 u(t - t_0)$ acting alone. Then,

$$i(t) = i_1(t) + i_2(t)$$

Our object is now to write each of the partial responses i_1 and i_2 as the sum of a natural and a forced response. The response $i_1(t)$ is familiar; this problem was solved in Eq. (5):

$$i_1(t) = \frac{V_0}{R}(1 - e^{-Rt/L}) \qquad t > 0$$

Note that the range of t in which this solution is valid, $t > 0$, is indicated.

We now turn our attention to the lower source and its response $i_2(t)$. Only the polarity of the source and the time of its application are different. There is thus no need to determine the form of the natural response and the forced response; the solution for $i_1(t)$ enables us to write

$$i_2(t) = -\frac{V_0}{R}[1 - e^{-R(t-t_0)/L}] \qquad t > t_0$$

where the applicable range of t, $t > t_0$, must again be indicated.

We now add the two solutions, but do so carefully, since each is valid over a different interval of time. Thus,

$$i(t) = \frac{V_0}{R}(1 - e^{-Rt/L}) \qquad 0 < t < t_0$$

$$i(t) = \frac{V_0}{R}(1 - e^{-Rt/L}) - \frac{V_0}{R}(1 - e^{-R(t-t_0)/L}) \qquad t > t_0$$

or

$$i(t) = \frac{V_0}{R}e^{-Rt/L}(e^{Rt_0/L} - 1) \qquad t > t_0$$

The solution is completed by stating that $i(t)$ is zero for negative t and sketching the response as a function of time. The type of curve obtained depends upon the relative values of t_0 and the time constant τ. Two possible curves are shown in Fig. 5-15. The left curve is drawn for the case where

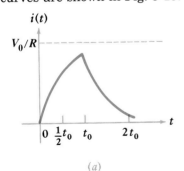

(a)

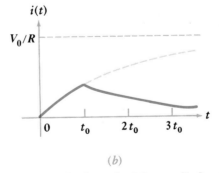

(b)

Figure 5-15

Two possible response curves are shown for the circuit of Fig. 5-14b. (a) τ is selected as $t_0/2$. (b) τ is selected as $2t_0$.

the time constant is only one-half as large as the length of the applied pulse; the rising portion of the exponential has therefore almost reached V_0/R before the decaying exponential begins. The opposite situation is shown to the right; there, the time constant is twice t_0 and the response never has a chance to reach the larger amplitudes. ∎

The procedure we have been using to find the response of an RL circuit after dc sources have been switched on or off, or in or out of the circuit, at some instant of time (say, $t = 0$) is summarized in the following. We assume that the circuit is reducible to a single equivalent resistance R_{eq} in series with a single equivalent inductance L_{eq} when all independent sources are set equal to zero. The response we seek is represented by $f(t)$.

---

1 With all independent sources killed, simplify the circuit to determine R_{eq}, L_{eq}, and the time constant $\tau = L_{eq}/R_{eq}$.

2 Viewing L_{eq} as a short circuit, use dc-analysis methods to find $i_L(0^-)$, the inductor current just prior to the discontinuity.

3 Again viewing L_{eq} as a short circuit, use dc-analysis methods to find the forced response. This is the value approached by $f(t)$ as $t \to \infty$; we represent it by $f(\infty)$.

4 Write the total response as the sum of the forced and natural responses: $f(t) = f(\infty) + Ae^{-t/\tau}$.

5 Find $f(0^+)$ by using the condition that $i_L(0^+) = i_L(0^-)$. If desired, L_{eq} may be replaced by a current source $i_L(0^+)$ [an open circuit if $i_L(0^+) = 0$] for this calculation. With the exception of inductor currents (and capacitor voltages), other currents and voltages in the circuit may change abruptly.

6 Then $f(0^+) = f(\infty) + A$, and $f(t) = f(\infty) + [f(0^+) - f(\infty)]e^{-t/\tau}$, or total response = final value + (initial value − final value)$e^{-t/\tau}$.

Drill Problems

5-5. For the circuit shown in Fig. 5-16a, find i_1, v_1, and i_2 at t equal to (a) 0^-; (b) 0^+; (c) ∞; (d) 50 ms.

           Ans: 0, 0, 0; 0, 288 V, 0; 7.2 A, 0, 5.76 A; 4.55 A, 105.9 V, 3.64 A

5-6. The circuit shown in Fig. 5-16b has been in the form shown for a very long time. The switch opens at $t = 0$. Find i_R at t equal to (a) 0^-; (b) 0^+; (c) ∞; (d) 1.5 ms.   *Ans:* 0; 10 mA; 4 mA; 5.34 mA

Figure 5-16

(a) See Drill Prob. 5-5. (b) See Drill Prob. 5-6.

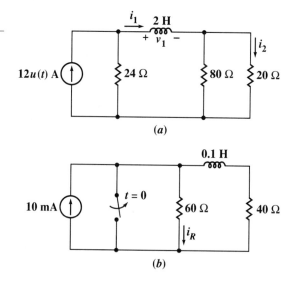

(a)

(b)

The complete response of any RC circuit may also be obtained as the sum of the natural and the forced response. We illustrate this by working an example completely.

5-6

RC circuits

Example 5-3   Figure 5-17 shows a friendly little circuit containing two batteries,

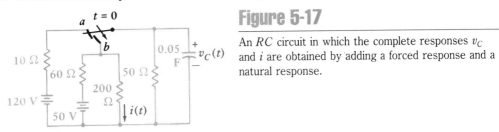

Figure 5-17

An RC circuit in which the complete responses v_C and i are obtained by adding a forced response and a natural response.

four resistors, one capacitor, and a switch that is assumed to have been in position a for a long time. Our task is to find the capacitor voltage $v_C(t)$ and the current $i(t)$ in the 200-Ω resistor for all time.

Solution: First, we decide that any transient response that resulted from the original movement of the switch to a has disappeared, leaving only a forced response caused by the 120-V source. We are asked for $v_C(t)$, and we thus begin by finding this forced response prior to $t = 0$ with the switch at position a. The voltages throughout the circuit are all constant, and there is thus no current through the capacitor. Simple voltage division gives us the initial voltage,

$$v_C(0) = \frac{50}{50 + 10}(120) = 100 \text{ V}$$

Since the capacitor voltage cannot change instantaneously, this voltage is equally valid at $t = 0^-$ and $t = 0^+$.

The switch is now thrown to b, and the complete response is

$$v_C = v_{Cf} + v_{Cn}$$

The form of the natural response is obtained by replacing the 50-V source by a short circuit and evaluating the equivalent resistance to find the time constant:

$$v_{Cn} = Ae^{-t/R_{eq}C}$$

where

$$R_{eq} = \frac{1}{\frac{1}{50} + \frac{1}{200} + \frac{1}{60}} = 24 \text{ }\Omega$$

or

$$v_{Cn} = Ae^{-t/1.2}$$

In order to evaluate the forced response with the switch at b, we wait until all the voltages and currents have stopped changing, thus treating the capacitor as an open circuit, and use voltage division once more:

$$v_{Cf} = \frac{(50)(200)/(50 + 200)}{60 + (50)(200)/(50 + 200)}(50) = 20$$

Thus,

$$v_C = 20 + Ae^{-t/1.2}$$

and from the initial condition already obtained,

$$100 = 20 + A$$

or $$\qquad v_C = 20 + 80e^{-t/1.2} \qquad t > 0$$

This response is sketched in Fig. 5-18a; again the natural response is seen to form a transition from the initial to the final response.

Figure 5-18

The responses (a) v_C and (b) i are plotted as functions of time for the circuit of Fig. 5-17.

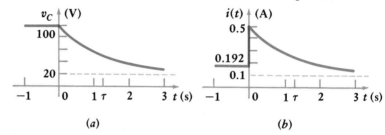

(a)   (b)

Next we attack $i(t)$. This response need not remain constant during the instant of switching. With the contact at a, it is evident that $i = \frac{50}{260} = 0.1923$ A. When the switch moves to position b, the forced response for this current becomes

$$i_f = \frac{50}{60 + (50)(200)/(50 + 200)} \left(\frac{50}{50 + 200} \right) = 0.1 \, \text{A}$$

The form of the natural response is the same as that which we already determined for the capacitor voltage:

$$i_n = Ae^{-t/1.2}$$

Combining the forced and natural responses, we obtain

$$i = 0.1 + Ae^{-t/1.2}$$

To evaluate A, we need to know $i(0^+)$. This is found by fixing our attention on the energy-storage element, here, the capacitor; for the fact that v_C must remain 100 V during the switching interval is the governing condition establishing the other currents and voltages at $t = 0^+$. Since $v_C(0^+) = 100$ V, and since the capacitor is in parallel with the 200-Ω resistor, we find $i(0^+) = 0.5$ A, $A = 0.4$, and thus

$$i(t) = 0.1923 \qquad t < 0$$

$$i(t) = 0.1 + 0.4e^{-t/1.2} \qquad t > 0$$

or $$\qquad i(t) = 0.1923 + (-0.0923 + 0.4e^{-t/1.2})u(t) \qquad \text{A}$$

where the last expression is correct for all t.

The complete response for all t may also be written concisely by using $u(-t)$, which is unity for $t < 0$ and 0 for $t > 0$. Thus,

$$i(t) = 0.1923u(-t) + (0.1 + 0.4e^{-t/1.2})u(t) \qquad \text{A}$$

This response is sketched in Fig. 5-18b. Note that only four numbers are needed to write the functional form of the response for this single-energy-storage-element circuit, or to prepare the sketch: the constant value prior to switching (0.1923 A), the instantaneous value just after switching (0.5 A), the constant forced response (0.1 A), and the time constant (1.2 s). The appropriate negative exponential function is then easily written or drawn.   ■

We conclude by listing the duals of the statements given at the end of Sec. 5-5.

The procedure we have been using to find the response of an RC circuit after dc sources have been switched on or off, or in or out of the circuit, at some instant of time—say, $t = 0$—is summarized in the following. We assume that the circuit is reducible to a single equivalent resistance R_{eq} in parallel with a single equivalent capacitance C_{eq} when all independent sources are set equal to zero. The response we seek is represented by $f(t)$.

---

1 With all independent sources killed, simplify the circuit to determine R_{eq}, C_{eq}, and the time constant $\tau = R_{eq}C_{eq}$.
2 Viewing C_{eq} as an open circuit, use dc-analysis methods to find $v_C(0^-)$, the capacitor voltage just prior to the discontinuity.
3 Again viewing C_{eq} as an open circuit, use dc-analysis methods to find the forced response. This is the value approached by $f(t)$ as $t \rightarrow \infty$; we represent it by $f(\infty)$.
4 Write the total response as the sum of the forced and natural responses: $f(t) = f(\infty) + Ae^{-t/\tau}$.
5 Find $f(0^+)$ by using the condition that $v_C(0^+) = v_C(0^-)$. If desired, C_{eq} may be replaced by a voltage source $v_C(0^+)$ [a short circuit if $v_C(0^+) = 0$] for this calculation. With the exception of capacitor voltages (and inductor currents), other voltages and currents in the circuit may change abruptly.
6 Then $f(0^+) = f(\infty) + A$, and $f(t) = f(\infty) + [f(0^+) - f(\infty)]e^{-t/\tau}$, or total response = final value + (initial value − final value)$e^{-t/\tau}$.

5-7. For the circuit of Fig. 5-19, find $v_C(t)$ at t equal to (a) 0^-; (b) 0^+; (c) ∞; **Drill Problems** (d) 0.08 s.   *Ans:* 20 V; 20 V; 28 V; 24.4 V

5-8. For the circuit of Fig. 5-19, find $i_R(t)$ at t equal to (a) 0^-; (b) 0^+; (c) ∞; (d) 0.08 s.   *Ans:* −0.8 mA; −0.4 mA; −0.72 mA; −0.576 mA

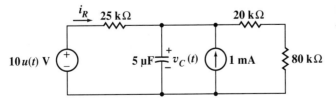

Figure 5-19

See Drill Probs. 5-7 and 5-8.

Problems

1 The source values in the circuit of Fig. 5-20 are $v_A = 300u(t - 1)$ V, $v_B = -120u(t + 1)$ V, and $i_C = 3u(-t)$ A. Find i_1 at $t = -1.5, -0.5, 0.5,$ and 1.5 s.

Figure 5-20

See Probs. 1 and 2.

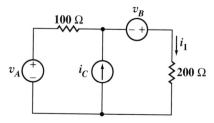

2 Source values for Fig. 5-20 are $v_A = 600tu(t + 1)$ V, $v_B = 600(t + 1)u(t)$ V, and $i_C = 6(t - 1)u(t - 1)$ A. (a) Find i_1 at $t = -1.5, -0.5, 0.5,$ and 1.5 s. (b) Sketch i_1 versus t, $-2.5 < t < 2.5$ s.

3 At $t = 2$, find the value of (a) $2u(1 - t) - 3u(t - 1) + 4u(t + 1)$; (b) $[5 - u(t)][2 + u(3 - t)][1 - u(1 - t)]$; (c) $4e^{-u(3-t)}u(3 - t)$.

4 Find i_x for $t < 0$ and for $t > 0$ in the circuit of Fig. 5-21 if the unknown branch contains: (a) a normally open switch that closes at $t = 0$, in series with a 60-V battery, + reference at top; (b) a voltage source, $60u(t)$ V, + reference at top.

Figure 5-21

See Prob. 4.

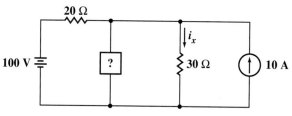

5 Find i_x in the circuit of Fig. 5-22 at 1-s intervals from $t = -0.5$ s to $t = 3.5$ s.

Figure 5-22

See Prob. 5.

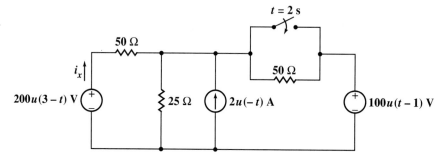

6 The switch in Fig. 5-23 is at position A for $t < 0$. At $t = 0$ it moves to B, and then it moves to C at $t = 4$ s and on to D at $t = 6$ s, where it remains. Sketch $v(t)$ as a function of time and express it as a sum of step forcing functions.

Figure 5-23

See Prob. 6.

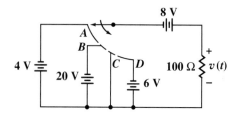

7 Refer to the circuit shown in Fig. 5-24 and (a) find $i_L(t)$; (b) use the expression for $i_L(t)$ to find $v_L(t)$.

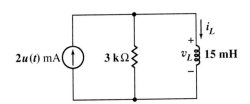

Figure 5-24

See Prob. 7.

8 Find i_L in the circuit of Fig. 5-25 at t equal to (a) -0.5 s; (b) 0.5 s; (c) 1.5 s.

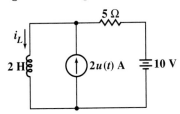

Figure 5-25

See Prob. 8.

9 The switch shown in Fig. 5-26 has been closed for a long time. (a) Find i_L for $t < 0$. (b) Find $i_L(t)$ for all t after the switch opens at $t = 0$.

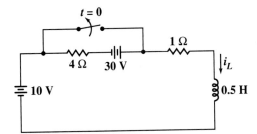

Figure 5-26

See Prob. 9.

10 The switch in Fig. 5-27 has been open for a long time. (a) Find i_L for $t < 0$. (b) Find $i_L(t)$ for all t after the switch closes at $t = 0$.

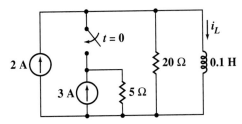

Figure 5-27

See Prob. 10.

11 Equation (3) in Sec. 5-4 represents the general solution of the driven RL series circuit, where Q is a function of time in general and A and P are constants. Let $R = 125\ \Omega$ and $L = 5$ H, and find $i(t)$ for $t > 0$ if the voltage forcing function $LQ(t)$ is (a) 10 V; (b) $10u(t)$ V; (c) $10 + 10u(t)$ V; (d) $10u(t) \cos 50t$ V.

12 With reference to the circuit shown in Fig. 5-28, obtain an algebraic expression for and also sketch: (a) $i_L(t)$; (b) $v_1(t)$.

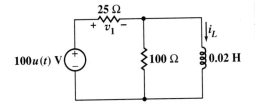

Figure 5-28

See Prob. 12.

13 For the circuit shown in Fig. 5-29, find values for i_L and v_1 at t equal to (a) 0^-; (b) 0^+; (c) ∞; (d) 0.2 ms.

Figure 5-29

See Prob. 13.

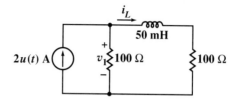

14 The switch shown in Fig. 5-30 has been closed for a very long time. (*a*) Find i_L for $t < 0$. (*b*) Just after the switch is opened, find $i_L(0^+)$. (*c*) Find $i_L(\infty)$. (*d*) Derive an expression for $i_L(t)$ for $t > 0$.

Figure 5-30

See Prob. 14.

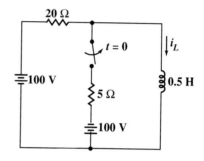

15 Find i_L for all t in the circuit of Fig. 5-31.

Figure 5-31

See Prob. 15.

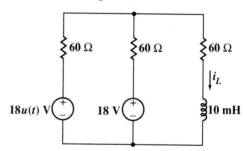

16 Assume that the switch in Fig. 5-32 has been closed for a long time and then opens at $t = 0$. Find i_x at t equal to (*a*) 0^-; (*b*) 0^+; (*c*) 40 ms.

Figure 5-32

See Probs. 16 and 17.

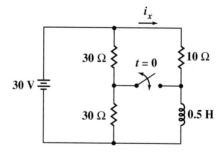

17 Assume that the switch in Fig. 5-32 has been open for a long time and then closes at $t = 0$. Find i_x at t equal to (*a*) 0^-; (*b*) 0^+; (*c*) 40 ms.

18 Find $v_x(t)$ for all t in the circuit of Fig. 5-33.

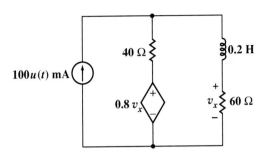

Figure 5-33

See Prob. 18.

19 With reference to the circuit shown in Fig. 5-34, find (a) $i_L(t)$; (b) $i_1(t)$.

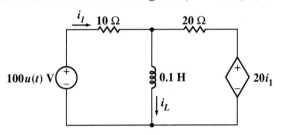

Figure 5-34

See Prob. 19.

20 Find v_C in the circuit shown in Fig. 5-35 at $t = -2$ μs and $t = +2$ μs.

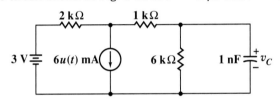

Figure 5-35

See Prob. 20.

21 After being open for a long time, the switch shown in Fig. 5-36 closes at $t = 0$. Find i_A for all time.

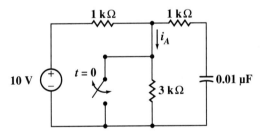

Figure 5-36

See Probs. 21 and 22.

22 After being closed for a long time, the switch shown in Fig. 5-36 opens at $t = 0$. Find i_A for all time.

23 Let $v_s = -12u(-t) + 24u(t)$ V in the circuit of Fig. 5-37. Over the time interval -5 ms $< t < 5$ ms, find an algebraic expression for and sketch (a) $v_C(t)$; (b) $i_{in}(t)$.

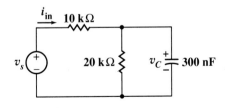

Figure 5-37

See Prob. 23.

24 The switch in the circuit of Fig. 5-38 has been open for a long time. It closes suddenly at $t = 0$. Find i_{in} at t equal to (a) -1.5 s; (b) 1.5 s.

Figure 5-38

See Prob. 24.

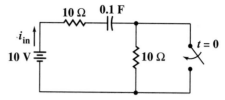

25 Find the value of $v_C(t)$ at $t = 0.4$ and 0.8 s in the circuit of Fig. 5-39.

Figure 5-39

See Prob. 25.

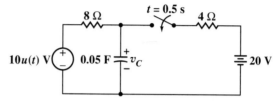

26 Find v_C for $t > 0$ in the circuit of Fig. 5-40.

Figure 5-40

See Prob. 26.

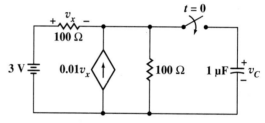

27 In the circuit of Fig. 5-41, find $v_R(t)$ for (a) $t < 0$; (b) $t > 0$. Now assume that the switch has been *closed* for a very long time and opens at $t = 0$. Find $v_R(t)$ for (c) $t < 0$; (d) $t > 0$.

Figure 5-41

See Prob. 27.

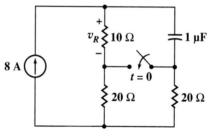

28 In the circuit of Fig. 5-42: (a) find $v_C(t)$ for all time, and (b) sketch $v_C(t)$ for $-1 < t < 2$ s.

Figure 5-42

See Prob. 28.

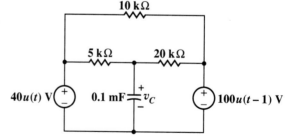

29 Find the first instant of time after $t = 0$ at which $v_x = 0$ in the circuit of Fig. 5-43.

Figure 5-43

See Prob. 29.

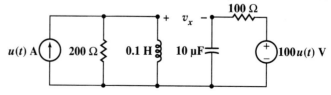

30 The switch in Fig. 5-44 has been at A for a long time. It is moved to B at $t = 0$, and back to A at $t = 1$ ms. Find R_1 and R_2 so that $v_C(1\text{ ms}) = 8$ V and $v_C(2\text{ ms}) = 1$ V.

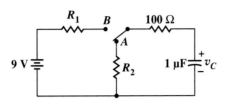

Figure 5-44

See Prob. 30.

31 Assume that the op-amp shown in Fig. 5-45 is ideal, and find $v_o(t)$ for all t.

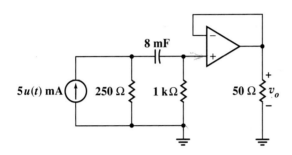

Figure 5-45

See Prob. 31.

32 Assume that the op-amp shown in Fig. 5-46 is ideal, and find $v_x(t)$ for all t.

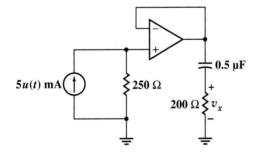

Figure 5-46

See Prob. 32.

33 Assume that the op-amp shown in Fig. 5-47 is ideal, and find $v_o(t)$ for all t.

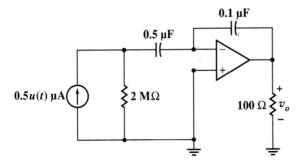

Figure 5-47

See Prob. 33.

34 Assume that the op-amp shown in Fig. 5-48 is ideal, and that $v_C(0) = 0$. Find $v_o(t)$ for all t.

Figure 5-48

See Prob. 34.

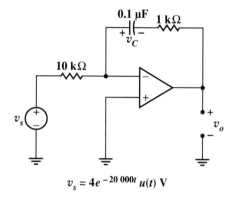

$$v_s = 4e^{-20\,000t}\,u(t)\text{ V}$$

35 (SPICE) (a) Find $i_L(0)$ for the RL circuit of Fig. 5-49. (b) Using SPICE and the initial value found in part a, determine i_L at $t = 50$ ms. Let TSTEP = 2 ms.

Figure 5-49

See Prob. 35.

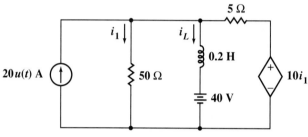

36 (SPICE) (a) Find $v_C(0)$ for the RC circuit of Fig. 5-50. (b) Using SPICE and the initial value found in part a, determine v_C at $t = 50$ ms. Let TSTEP = 2 ms.

Figure 5-50

See Prob. 36.

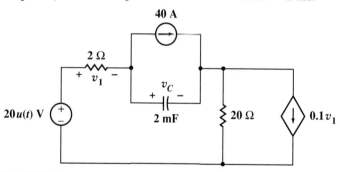

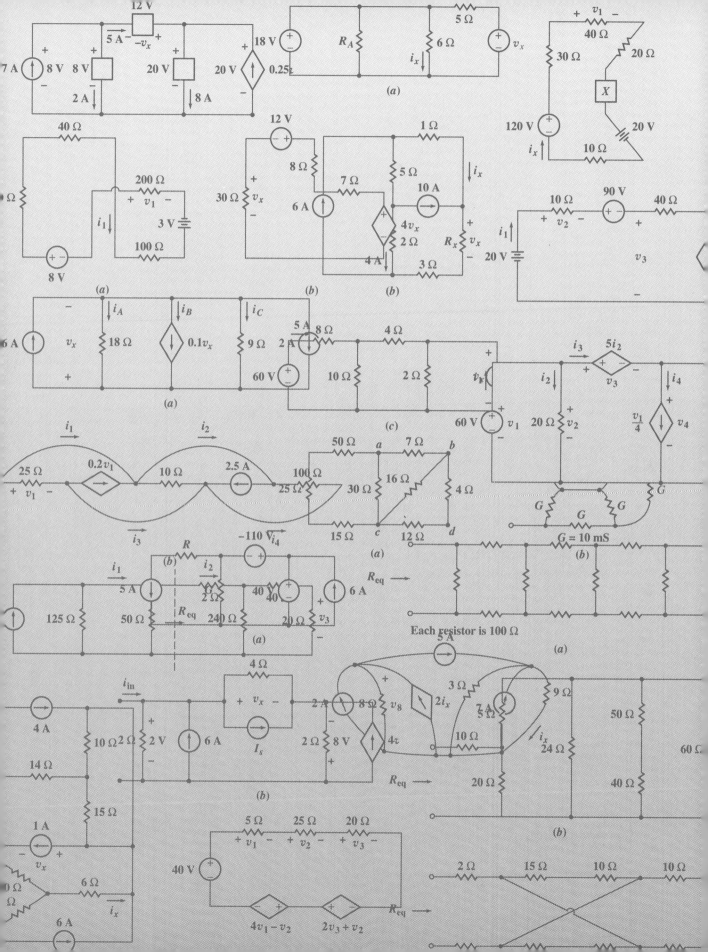

The *RLC* Circuit

It would be very pleasant to learn that the detailed study we have just completed for the *RL* and *RC* circuits will make the analysis of the *RLC* circuit a simple task; unfortunately, however, the analysis remains difficult. The presence of inductance and capacitance in the same circuit produces at least a *second-order system,* a system characterized by one linear differential equation that includes a second-order derivative, or by two simultaneous linear first-order differential equations. This increase in order will make it necessary to evaluate two arbitrary constants. Furthermore, it becomes necessary to determine initial conditions for derivatives. And finally, we shall see that the presence of inductance and capacitance in the same circuit leads to a response which takes on different functional forms for circuits which have the same configuration but different element values. With this cheerful news, let us quickly review the methods and results we found useful for first-order systems, in order that we may extend this information as intelligently as possible to the second-order system.

We first considered the source-free first-order system. The response was termed the *natural response,* and it was determined completely by the types of passive elements in the network, by the manner in which they were interconnected, and by the initial conditions which were established by the stored energy. The natural response was invariably an exponentially decreasing function of time, and this response approached a constant value as time became infinite. The constant was usually zero, except in those circuits where paralleled inductors or series-connected capacitors allowed trapped currents or voltages to appear.

The addition of sources to the first-order system resulted in a two-part response, the familiar natural response and an additional term we called the *forced response.* This latter term was intimately related to the forcing function; its functional form was that of the forcing function itself, plus the integral and first derivative of the forcing function.[1] Since we treated only a constant forcing function, we have not needed to devote much attention to the proper form of the forced response; this problem will not arise until sinusoidal forcing functions

---

[1] Higher-order derivatives will appear in higher-order systems, and, strictly speaking, we should say that all derivatives are present, although possibly with zero amplitude. Forcing functions which do not possess a finite number of different derivatives are exceptions which we shall not consider; the singularity functions are exceptions to the exceptions.

are encountered in the following chapter. To the known forced response, we added the correct expression for the natural response, complete except for a multiplicative constant. This constant was evaluated to make the total response fit the prescribed initial conditions.

We now turn to circuits which are characterized by linear second-order differential equations. Our first task is the determination of the natural response. This is most conveniently done by considering first the source-free circuit. We may then include dc sources, switches, or step sources in the circuit, representing the total response once again as the sum of the natural response and the (usually constant) forced response. The second-order system that we are about to analyze is fundamentally the same as any lumped-constant mechanical second-order system. Our results, for example, will be of direct use to a mechanical engineer who is interested in the displacement of a spring-supported mass subjected to viscous damping, a system which can approximate the vertical motion of an automobile, with shock absorbers providing the damping. Also the results may be interesting to someone who is excited by the motion of a simple pendulum or a torsional pendulum. Our results are still applicable, although less directly, to any distributed-parameter second-order system, such as a short-circuited transmission line, a diving board, a flute, or the ecology of the lemming.

6-2

The source-free parallel circuit

Our first goal is the determination of the natural response of a simple circuit formed by connecting R, L, and C in parallel; this modest goal will be reached after completing this and the next three sections. This particular combination of ideal elements is a suitable model for portions of many communications networks. It represents, for example, an important part of some of the electronic amplifiers found in every radio receiver, and it enables the amplifiers to produce a large voltage amplification over a narrow band of signal frequencies with nearly zero amplification outside this band. Frequency selectivity of this kind enables us to listen to the transmission of one station while rejecting the transmission of any other station. Other applications include the use of parallel RLC circuits in multiplexing filters, harmonic-suppression filters, and so forth. But even a simple discussion of these principles requires an understanding of such terms as *resonance, frequency response,* and *impedance,* which we have not yet discussed. Let it suffice to say, therefore, that an understanding of the natural behavior of the parallel RLC circuit is fundamentally important to future studies of communications networks and filter design.

When a physical inductor is connected in parallel with a capacitor and the inductor has associated with it a nonzero ohmic resistance, the resulting network can be shown to have an equivalent circuit model like that shown in Fig. 6-1. Energy losses in the physical inductor are taken into account by the

Figure 6-1

The source-free parallel RLC circuit.

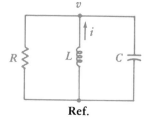

presence of the ideal resistor, whose resistance R is dependent upon (but not equal to) the ohmic resistance of the inductor.

In the following analysis we shall assume that energy may be stored initially in both the inductor and the capacitor, and thus that nonzero initial values of both inductor current and capacitor voltage may be present. With reference to the circuit of Fig. 6-1, we may then write the single nodal equation

$$\frac{v}{R} + \frac{1}{L}\int_{t_0}^{t} v \, dt - i(t_0) + C\frac{dv}{dt} = 0 \tag{1}$$

Note that the minus sign is a consequence of the assumed direction for i. We must solve Eq. (1) subject to the initial conditions

$$i(0^+) = I_0 \tag{2}$$

$$v(0^+) = V_0 \tag{3}$$

When both sides of Eq. (1) are differentiated once with respect to time, the result is the linear second-order homogeneous differential equation

$$C\frac{d^2v}{dt^2} + \frac{1}{R}\frac{dv}{dt} + \frac{1}{L}v = 0 \tag{4}$$

whose solution $v(t)$ is the desired natural response.

There are a number of interesting ways to solve Eq. (4). Most of these methods we shall leave to a course in differential equations, selecting only the quickest and simplest method to use now. We shall assume a solution, relying upon our intuition and modest experience to select one of the several possible forms which are suitable. Our experience with the first-order equation should suggest that we at least try the exponential form once more. Moreover, the form of Eq. (4) indicates that this may work, because we must add three terms—the second derivative, the first derivative, and the function itself, each multiplied by a constant factor—and achieve a sum of zero. A function whose derivatives have the same form as the function itself is obviously a sensible choice. With every hope for success, then, we assume

$$v = Ae^{st} \tag{5}$$

where we shall be as general as possible by allowing A and s to be complex[2] numbers if necessary. Substituting Eq. (5) in Eq. (4), we obtain

$$CAs^2e^{st} + \frac{1}{R}Ase^{st} + \frac{1}{L}Ae^{st} = 0$$

or
$$Ae^{st}\left(Cs^2 + \frac{1}{R}s + \frac{1}{L}\right) = 0$$

In order for this equation to be satisfied for all time, at least one of the three factors must be zero. If either of the first two factors is put equal to zero, then $v(t) = 0$. This is a trivial solution of the differential equation which cannot satisfy our given initial conditions. We therefore equate the remaining factor

---

[2] There is no cause for panic. Complex numbers will appear in this chapter only in an introductory way in the derivations. Their use as a tool will be necessary in Chap. 8.

to zero:

$$Cs^2 + \frac{1}{R}s + \frac{1}{L} = 0 \tag{6}$$

This equation is usually called the *auxiliary equation* or the *characteristic equation* by mathematicians. If it can be satisfied, then our assumed solution is correct. Since Eq. (6) is a quadratic equation, there are two solutions, identified as s_1 and s_2:

$$s_1 = -\frac{1}{2RC} + \sqrt{\left(\frac{1}{2RC}\right)^2 - \frac{1}{LC}} \tag{7}$$

and

$$s_2 = -\frac{1}{2RC} - \sqrt{\left(\frac{1}{2RC}\right)^2 - \frac{1}{LC}} \tag{8}$$

If either of these two values is used for s in the assumed solution, then that solution satisfies the given differential equation; it thus becomes a valid solution of the differential equation.

Let us assume that we replace s by s_1 in Eq. (5), obtaining

$$v_1 = A_1 e^{s_1 t}$$

and, similarly,

$$v_2 = A_2 e^{s_2 t}$$

The former satisfies the differential equation

$$C\frac{d^2v_1}{dt^2} + \frac{1}{R}\frac{dv_1}{dt} + \frac{1}{L}v_1 = 0$$

and the latter satisfies

$$C\frac{d^2v_2}{dt^2} + \frac{1}{R}\frac{dv_2}{dt} + \frac{1}{L}v_2 = 0$$

Adding these two differential equations and combining similar terms, we have

$$C\frac{d^2(v_1 + v_2)}{dt^2} + \frac{1}{R}\frac{d(v_1 + v_2)}{dt} + \frac{1}{L}(v_1 + v_2) = 0$$

Linearity triumphs, and it is seen that the sum of the two solutions is also a solution. We thus have the form of the natural response

$$v = A_1 e^{s_1 t} + A_2 e^{s_2 t} \tag{9}$$

where s_1 and s_2 are given by Eqs. (7) and (8); A_1 and A_2 are two arbitrary constants which are to be selected to satisfy the two specified initial conditions.

The form of the natural response as given in Eq. (9) can hardly be expected to bring forth any expressions of interested amazement, for, in its present form, it offers little insight into the nature of the curve we might obtain if $v(t)$ were plotted as a function of time. The relative amplitudes of A_1 and A_2, for example, will certainly be important in determining the shape of the response curve. Furthermore, the constants s_1 and s_2 can be real numbers or conjugate complex numbers, depending upon the values of R, L, and C in the given network. These two cases will produce fundamentally different response forms. Therefore it will be helpful to make some simplifying substitutions in Eq. (9) for the sake of conceptual clarity.

Since the exponents s_1t and s_2t must be dimensionless, s_1 and s_2 must have the unit of some dimensionless quantity "per second." From Eqs. (7) and (8) it is apparent that the units of $1/2RC$ and $1/\sqrt{LC}$ must also be s^{-1}. Units of this type are called *frequencies*. Although we shall expand this concept in much more detail in Chap. 12, we shall introduce several of the terms now.

Let us represent $1/\sqrt{LC}$ by ω_0 (omega):

$$\omega_0 = \frac{1}{\sqrt{LC}} \tag{10}$$

and reserve the term *resonant frequency*[3] for it. On the other hand, we shall call $1/2RC$ the *neper frequency,* or the *exponential damping coefficient,* and represent it by the symbol α (alpha):

$$\alpha = \frac{1}{2RC} \tag{11}$$

This latter descriptive expression is used because α is a measure of how rapidly the natural response decays or damps out to its steady, final value (usually zero).[4] Finally, s, s_1, and s_2, which are quantities that will form the basis for some of our later work, will be called *complex frequencies*.

We should note that s_1, s_2, α, and ω_0 are merely symbols used to simplify the discussion of *RLC* circuits; they are not mysterious new properties of any kind. It is easier, for example, to say "alpha" than it is to say "the reciprocal of $2RC$."

Let us collect these results. The natural response of the parallel *RLC* circuit is

$$v = A_1 e^{s_1 t} + A_2 e^{s_2 t} \tag{9}$$

where

$$s_1 = -\alpha + \sqrt{\alpha^2 - \omega_0^2} \tag{12}$$

$$s_2 = -\alpha - \sqrt{\alpha^2 - \omega_0^2} \tag{13}$$

$$\alpha = \frac{1}{2RC} \tag{11}$$

$$\omega_0 = \frac{1}{\sqrt{LC}} \tag{10}$$

and A_1 and A_2 must be found by applying the given initial conditions.

The response described by the preceding equations applies not only to the voltage $v(t)$, but also to the current flowing in each of the three circuit elements. The values of the constants, A_1 and A_2 for $v(t)$, will, of course, be different from those for the currents.

It is now apparent that the nature of the response depends upon the relative magnitudes of α and ω_0. The radical appearing in the expressions for s_1 and s_2

---

[3] More accurately, the resonant *radian* frequency.
[4] The ratio of α to ω_0 is called the *damping ratio* by control system engineers and is designated by ζ (zeta). We shall encounter it again in Chap. 13.

will be real when α is greater than ω_0, imaginary when α is less than ω_0, and zero when α and ω_0 are equal. Each of these cases will be considered separately in the following three sections.

6-1. A parallel RLC circuit contains a 100-Ω resistor and has the parameter values $\alpha = 1000$ s^{-1}, $\omega_0 = 800$ rad/s. Find (*a*) C; (*b*) L; (*c*) s_1; (*d*) s_2.

Ans: 5 μF; 0.3125 H; -400 s^{-1}; -1600 s^{-1}

6-3

The overdamped parallel *RLC* circuit

A comparison of Eqs. (10) and (11) in Sec. 6-2 shows that α will be greater than ω_0 if $LC > 4R^2C^2$. In this case the radical used in calculating s_1 and s_2 will be real, and both s_1 and s_2 will be real. Moreover, the following inequalities,

$$\sqrt{\alpha^2 - \omega_0^2} < \alpha$$

$$(-\alpha - \sqrt{\alpha^2 - \omega_0^2}) < (-\alpha + \sqrt{\alpha^2 - \omega_0^2}) < 0$$

may be applied to Eqs. (12) and (13) to show that both s_1 and s_2 are *negative* real numbers. Thus, the response $v(t)$ can be expressed as the (algebraic) sum of two decreasing exponential terms, both of which approach zero as time increases without limit. In fact, since the absolute value of s_2 is larger than that of s_1, the term containing s_2 has the more rapid rate of decrease, and, for large values of time, we may write the limiting expression

$$v(t) \rightarrow A_1 e^{s_1 t} \rightarrow 0 \qquad t \rightarrow \infty$$

In order to discuss the method by which the arbitrary constants A_1 and A_2 are selected to conform with the initial conditions, and in order to provide a typical example of a response curve, let us talk our way through a numerical example. We shall select a parallel RLC circuit for which $R = 6$ Ω, $L = 7$ H, and, for ease of computation, $C =$ the impractically large value $\frac{1}{42}$ F. The initial energy storage is specified by choosing an initial voltage across the circuit $v(0) = 0$ and an initial inductor current $i(0) = 10$ A, where v and i are defined in Fig. 6-2.

Figure 6-2

A parallel *RLC* circuit used as a numerical example. The circuit is overdamped.

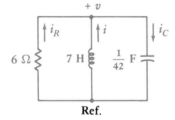

We may easily determine the values of the several parameters

$$\alpha = 3.5 \qquad \omega_0 = \sqrt{6}$$
$$\qquad\qquad\qquad (\text{all s}^{-1})$$
$$s_1 = -1 \qquad s_2 = -6$$

and immediately write the general form of the natural response

$$v(t) = A_1 e^{-t} + A_2 e^{-6t} \tag{14}$$

Only the evaluation of the two constants A_1 and A_2 remains. If we knew the response $v(t)$ at two different values of time, these two pairs of values could be substituted in Eq. (14) and A_1 and A_2 easily found. However, we know only one instantaneous value of $v(t)$,

$$v(0) = 0$$

and, therefore,

$$0 = A_1 + A_2 \tag{15}$$

We shall obtain a second equation relating A_1 and A_2 by taking the derivative of $v(t)$ with respect to time in Eq. (14), determining the initial value of this derivative through the use of the remaining initial condition $i(0) = 10$, and equating the results. Taking the derivative of both sides of Eq. (14),

$$\frac{dv}{dt} = -A_1 e^{-t} - 6A_2 e^{-6t}$$

and evaluating the derivative at $t = 0$,

$$\frac{dv}{dt}\bigg|_{t=0} = -A_1 - 6A_2$$

we next pause to consider how the initial value of the derivative can be found numerically. This next step is always suggested by the derivative itself; dv/dt suggests capacitor current, for

$$i_C = C\frac{dv}{dt}$$

Thus,

$$\frac{dv}{dt}\bigg|_{t=0} = \frac{i_C(0)}{C} = \frac{i(0) + i_R(0)}{C} = \frac{i(0)}{C} = 420 \text{ V/s}$$

since zero initial voltage across the resistor requires zero initial current through it. We thus have our second equation,

$$420 = -A_1 - 6A_2 \tag{16}$$

and simultaneous solution of Eqs. (15) and (16) provides the two amplitudes $A_1 = 84$ and $A_2 = -84$. Therefore, the final numerical solution for the natural response is

$$v(t) = 84(e^{-t} - e^{-6t}) \tag{17}$$

We shall spend a little more time with this result shortly, but first let us consider the evaluation of A_1 and A_2 for other conditions of initial energy storage, including initial energy storage in the capacitor.

Example 6-1　Find $v_C(t)$ after $t = 0$ in the circuit of Fig. 6-3a.

Solution: Let us first find the parameter values, α, ω_0, s_1, and s_2. The element values are identified by considering the source-free circuit present after $t = 0$. We have $L = 5$ mH, $R = 200$ Ω, and $C = 20$ nF. Therefore, $\alpha = 1/2RC = 125\ 000$ s^{-1}, $\omega_0 = 1/\sqrt{LC} = 100\ 000$ rad/s, and $s_{1,2} = -\alpha \pm \sqrt{\alpha^2 - \omega_0^2}$, yielding $s_1 = -50\ 000$ s^{-1} and $s_2 = -200\ 000$ s^{-1}. The form

Figure 6-3

(a) The *RLC* circuit analyzed in Example 6-1. (b) The equivalent circuit at $t = 0^-$ enables us to find $i_L(0^-) = -0.3$ A and $v_C(0^-) = 60$ V. (c) The equivalent circuit at $t = 0^+$.

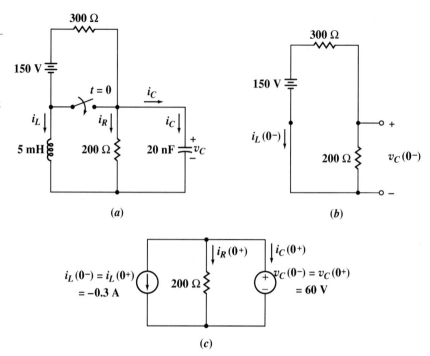

(a)

(b)

(c)

of the solution is therefore,

$$v_C(t) = A_1 e^{-50\,000t} + A_2 e^{-200\,000t}$$

Next we need two initial conditions, $v_C(0^+)$ and $dv_C/dt\big|_{t=0^+}$. Our attack on the initial values always begins with the energy-storage elements at the instant just before the discontinuity occurs—here, the capacitor and the inductor at $t = 0^-$. The inductor current and the capacitor voltage remain the same during the switching interval, $0^- < t < 0^+$. For clarity, we show two equivalent circuits, the first at $t = 0^-$ in Fig. 6-3*b*, and the second at $t = 0^+$ in Fig. 6-3*c*. The first is used to find the values of $i_L(0^-)$ and $v_C(0^-)$, and the second shows these two values represented as independent sources.

From Fig. 6-3*b*, a nice little single-loop resistive circuit, we see easily that $i_L(0^-) = -150/(200 + 300) = -0.3$ A, while $v_C(0^-) = 200(0.3) = 60$ V. These two values then appear as the two source values in Fig. 6-3*c*. Note that the switch has closed in the interval from $t = 0^-$ to $t = 0^+$, removing the 300-Ω resistor from consideration. This circuit also identifies the current values, $i_R(0^+)$ and $i_C(0^+)$.

Thus, we already have $v_C(0^+) = 60$ V, and we need only the value of $dv_C/dt\big|_{t=0^+}$. This is easily related to the capacitor current, so that

$$\frac{dv_C}{dt}\bigg|_{t=0^+} = \frac{1}{C} i_C(0^+) = \frac{1}{C}[-i_L(0^+) - i_R(0^+)]$$

or

$$\frac{dv_C}{dt}\bigg|_{t=0^+} = \frac{10^9}{20}\left(0.3 - \frac{60}{200}\right) = 0$$

We now use these initial values of i_C and dv_C/dt in our equation for $v_C(t)$:

$$v_C(0^+) = 60 = A_1 + A_2$$

and

$$dv_C/dt|_{t=0^+} = 0 = -50\,000A_1 - 200\,000A_2$$

Therefore, $A_1 = -4A_2$, so that we obtain $A_1 = 80$ and $A_2 = -20$. The solution to our problem is

$$v_C(t) = 80e^{-50\,000t} - 20e^{-200\,000t} \text{ V} \qquad t > 0 \qquad \blacksquare$$

Now let us return to Eq. (17) and see what additional information we can glean from this equation without calculating unduly. We note that $v(t)$ is zero at $t = 0$, a comforting check on our original assumption. We may also interpret the first exponential term as having a time constant of 1 s and the other exponential, a time constant of $\frac{1}{6}$ s. Each starts with unity amplitude, but the latter decays more rapidly; $v(t)$ is thus never negative. As time becomes infinite, each term approaches zero, and the response itself dies out as it should. We therefore have a response curve which is zero at $t = 0$, is zero at $t = \infty$, and is never negative; since it is not everywhere zero, it must possess at least one maximum, and this is not a difficult point to determine exactly. We differentiate the response

$$\frac{dv}{dt} = 84(-e^{-t} + 6e^{-6t})$$

set the derivative equal to zero to determine the time t_m at which the voltage becomes maximum,

$$0 = -e^{-t_m} + 6e^{-6t_m}$$

manipulate once,

$$e^{5t_m} = 6$$

and obtain

$$t_m = 0.358 \text{ s}$$

and

$$v(t_m) = 48.9 \text{ V}$$

A reasonable sketch of the response may be made by plotting the two exponential terms $84e^{-t}$ and $84e^{-6t}$ and then taking their difference. The usefulness of this technique is indicated by the curves of Fig. 6-4; the two exponentials are shown lightly, and their difference, the total response $v(t)$, is drawn as a colored line. The curves also verify our previous prediction that the functional behavior of $v(t)$ for very large t is $84e^{-t}$, the exponential term containing the smaller magnitude of s_1 and s_2.

Another question that frequently arises in the consideration of network response is concerned with the length of time it takes for the transient part of the response to disappear (or damp out). In practice, it is often desirable to have this transient response approach zero as rapidly as possible, that is, to minimize the *settling time* t_s. Theoretically, of course, t_s is infinite, because $v(t)$ never settles to zero in a finite time. However, a negligible response is present after the magnitude of $v(t)$ has settled to values that remain less than 1 percent

Figure 6-4

The response $v(t) = 84(e^{-t} - e^{-6t})$ of the network shown in Fig. 6-2.

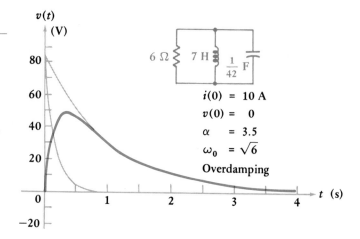

$i(0) = 10$ A
$v(0) = 0$
$\alpha = 3.5$
$\omega_0 = \sqrt{6}$
Overdamping

of its maximum absolute value $|v_m|$. The time which is required for this to occur we define as the settling time.[5] Since $|v_m| = v_m = 48.9$ V for our example, the settling time is the time required for the response to drop to 0.489 V. Substituting this value for $v(t)$ in Eq. (17) and neglecting the second exponential term, known to be negligible here, the settling time is found to be 5.15 s.

In comparison with the responses which we shall obtain in the following two sections, this is a comparatively large settling time; the damping time is overly long, and the response is called *overdamped*. We shall refer to the case for which α is greater than ω_0 as the *overdamped* case. Now let us see what happens as α is decreased.

Drill Problems

6-2. Refer to the circuit shown in Fig. 6-2 and let $v(0) = 50$ V and $i(0) = -10$ A. In the expression for $v(t)$ find (*a*) A_1; (*b*) A_2; (*c*) $v(0.2 \text{ s})$.

Ans: -94 V; 144 V; -33.6 V

6-3. After being open for a long time, the switch in Fig. 6-5 closes at $t = 0$. Find (*a*) $i_L(0^-)$; (*b*) $v_C(0^-)$; (*c*) $i_R(0^+)$; (*d*) $i_C(0^+)$; (*e*) $v_C(0.2)$.

Ans: 1 A; 48 V; 2 A; -3 A; -17.54 V

Figure 6-5

See Drill Prob. 6-3.

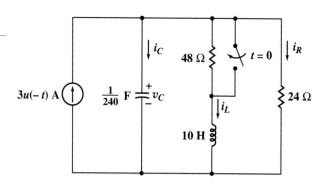

[5] The 1 percent level is somewhat arbitrary. Some people prefer 2 percent or 5 percent.

The overdamped case is characterized by

$$\alpha > \omega_0$$

or

$$LC > 4R^2C^2$$

and leads to negative real values for s_1 and s_2 and to a response expressed as the algebraic sum of two negative exponentials. Typical forms of the response $v(t)$ are obtained through the numerical examples in the last section and in the drill problems following it.

Now let us adjust the element values until α and ω_0 are equal. This is a very special case which is termed *critical damping*. If we were to attempt to construct a parallel *RLC* circuit that is critically damped, we would be attempting an essentially impossible task, for we could never make α exactly equal to ω_0. The result of such an attempt would result in an overdamped circuit, or an underdamped circuit, which we shall discuss in the next section. For completeness, however, we shall discuss the critically damped circuit here, because it shows an interesting transition between overdamping and underdamping.

Critical damping is achieved when

$$\left.\begin{array}{l} \alpha = \omega_0 \\ LC = 4R^2C^2 \\ L = 4R^2C \end{array}\right\} \text{ critical damping}$$

or
or

It is obvious that we may produce critical damping by changing the value of any of the three elements in the numerical example discussed at the beginning of Sec. 6-3. We shall select R, increasing its value until critical damping is obtained, and thus leave ω_0 unchanged. The necessary value of R is $7\sqrt{6}/2\ \Omega$; L is still 7 H, and C remains $\frac{1}{42}$ F. We thus find

$$\alpha = \omega_0 = \sqrt{6}$$

$$s_1 = s_2 = -\sqrt{6}$$

and we blithely construct the response as the sum of two exponentials,

$$v(t) \overset{?}{=} A_1 e^{-\sqrt{6}t} + A_2 e^{-\sqrt{6}t}$$

which may be written as

$$v(t) \overset{?}{=} A_3 e^{-\sqrt{6}t}$$

At this point, some of us should feel we have lost our way. We have a response which contains only one arbitrary constant, but there are two initial conditions, $v(0) = 0$ and $i(0) = 10$, which must be satisfied by this single constant. This is in general impossible. In our case, for example, the first initial condition requires A_3 to be zero, and it is then impossible to satisfy the second initial condition.

Our mathematics and our electricity have been unimpeachable; therefore, if a mistake has not led to our difficulties, we must have begun with an incorrect assumption, and only one assumption has been made. We originally hypothesized that the differential equation could be solved by assuming an exponential solution, and this turns out to be incorrect for this single special case of critical damping. When $\alpha = \omega_0$, the differential equation, Eq. (4), becomes

$$\frac{d^2v}{dt^2} + 2\alpha\frac{dv}{dt} + \alpha^2 v = 0$$

The solution of this equation is not a tremendously difficult process, but we shall avoid developing it here, since the equation is a standard type found in the usual differential-equation texts. The solution is

$$v = e^{-\alpha t}(A_1 t + A_2) \tag{18}$$

It should be noted that the solution may be expressed as the sum of two terms, where one term is the familiar negative exponential but the second is t times a negative exponential. We should also note that the solution contains the *two* expected arbitrary constants.

Let us now complete our numerical example. After we substitute the known value of α in Eq. (18), obtaining

$$v = A_1 t e^{-\sqrt{6}t} + A_2 e^{-\sqrt{6}t}$$

we establish the values of A_1 and A_2 by first imposing the initial condition on $v(t)$ itself, $v(0) = 0$. Thus, $A_2 = 0$. This simple result occurs because the initial value of the response $v(t)$ was selected as zero; the more general case, which leads to an equation determining A_2, may be expected to arise in the drill problems. The second initial condition must be applied to the derivative dv/dt just as in the overdamped case. We therefore differentiate, remembering that $A_2 = 0$:

$$\frac{dv}{dt} = A_1 t(-\sqrt{6})e^{-\sqrt{6}t} + A_1 e^{-\sqrt{6}t}$$

evaluate at $t = 0$:

$$\left.\frac{dv}{dt}\right|_{t=0} = A_1$$

and express the derivative in terms of the initial capacitor current:

$$\left.\frac{dv}{dt}\right|_{t=0} = \frac{i_C(0)}{C} = \frac{i_R(0)}{C} + \frac{i(0)}{C}$$

where reference directions for i_C, i_R, and i are defined in Fig. 6-2. Thus,

$$A_1 = 420$$

The response is, therefore,

$$v(t) = 420te^{-2.45t} \tag{19}$$

Before plotting this response in detail, let us again try to anticipate its form by qualitative reasoning. The specified initial value is zero, and Eq. (19) concurs. It is not immediately apparent that the response also approaches zero as t becomes infinitely large, because $te^{-2.45t}$ is an indeterminate form. However, this minor obstacle is easily overcome by use of L'Hôpital's rule. Thus,

$$\lim_{t\to\infty} v(t) = 420\lim_{t\to\infty}\frac{t}{e^{2.45t}} = 420\lim_{t\to\infty}\frac{1}{2.45e^{2.45t}} = 0$$

and once again we have a response which begins and ends at zero and has positive values at all other times. A maximum value v_m again occurs at time t_m; for our example,

$$t_m = 0.408 \text{ s} \quad \text{and} \quad v_m = 63.1 \text{ V}$$

This maximum is larger than that obtained in the overdamped case and is a result of the smaller losses that occur in the larger resistor; the time of the maximum response is slightly later than it was with overdamping. The settling time may also be determined by solving

$$\frac{v_m}{100} = 420 t_s e^{-2.45 t_s}$$

for t_s (by trial-and-error methods or a calculator's SOLVE routine):

$$t_s = 3.12 \text{ s}$$

which is a considerably smaller value than that which arose in the overdamped case (5.15 s). As a matter of fact, it can be shown that, for given values of L and C, the selection of that value of R which provides critical damping will always give a shorter settling time than any choice of R which produces an overdamped response. However, a slight improvement (reduction) in settling time may be obtained by a further slight increase in resistance; a slightly underdamped response which will undershoot the zero axis before it dies out will yield the shortest settling time.

The response curve for critical damping is drawn in Fig. 6-6; it may be

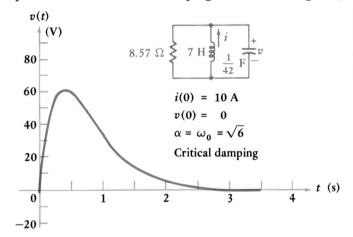

Figure 6-6

The response $v(t) = 420 t e^{-2.45t}$ of the network shown in Fig. 6-2 with R changed to provide critical damping.

compared with the overdamped (and underdamped) case by reference to Fig. 6-9.

6-4. (*a*) Choose R_1 in the circuit of Fig. 6-7 so that the response after $t = 0$ will **Drill Problem** be critically damped. (*b*) Now select R_2 to obtain $v(0) = 100$ V. (*c*) Find $v(t)$ at $t = 1$ ms.
                                            Ans: 1 kΩ; 250 Ω; −212 V

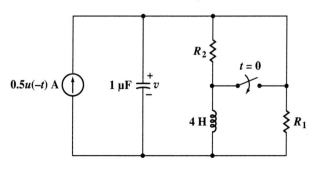

Figure 6-7

See Drill Prob. 6-4.

6-5

The underdamped parallel *RLC* circuit

Let us continue the process begun in the last section by increasing R once more. Thus, the damping coefficient α decreases while ω_0 remains constant, α^2 becomes smaller than ω_0^2, and the radicand appearing in the expressions for s_1 and s_2 becomes negative. This causes the response to take on quite a different character, but it is fortunately not necessary to return to the basic differential equation again. By using complex numbers, the exponential response turns into a damped sinusoidal response; this response is composed entirely of real quantities, the complex quantities being necessary only for the derivation.[6]

We therefore begin with the exponential form

$$v(t) = A_1 e^{s_1 t} + A_2 e^{s_2 t}$$

where

$$s_{1,2} = -\alpha \pm \sqrt{\alpha^2 - \omega_0^2}$$

and then let

$$\sqrt{\alpha^2 - \omega_0^2} = \sqrt{-1}\sqrt{\omega_0^2 - \alpha^2} = j\sqrt{\omega_0^2 - \alpha^2}$$

where

$$j = \sqrt{-1}$$

We now take the new radical, which is real for the underdamped case, and call it ω_d, the *natural resonant frequency*:

$$\omega_d = \sqrt{\omega_0^2 - \alpha^2}$$

Collecting, the response may now be written as

$$v(t) = e^{-\alpha t}(A_1 e^{j\omega_d t} + A_2 e^{-j\omega_d t})$$

or, in the longer but equivalent form,

$$v(t) = e^{-\alpha t}\left\{(A_1 + A_2)\left[\frac{e^{j\omega_d t} + e^{-j\omega_d t}}{2}\right] + j(A_1 - A_2)\left[\frac{e^{j\omega_d t} - e^{-j\omega_d t}}{j2}\right]\right\}$$

Two of the most important identities in the field of complex numbers, identities which are later proved in Appendix 4, may now be readily applied. The first square bracket in the preceding equation is identically equal to $\cos \omega_d t$, and the second is identically $\sin \omega_d t$. Hence,

$$v(t) = e^{-\alpha t}[(A_1 + A_2)\cos \omega_d t + j(A_1 - A_2)\sin \omega_d t]$$

and the multiplying factors may be assigned new symbols:

$$v(t) = e^{-\alpha t}(B_1 \cos \omega_d t + B_2 \sin \omega_d t) \tag{20}$$

If we are dealing with the underdamped case, we have now left complex numbers behind. This is true since α, ω_d, and t are real quantities, $v(t)$ itself must be a real quantity (which might be presented on an oscilloscope, a voltmeter, or a sheet of graph paper), and thus B_1 and B_2 are real quantities. Equation (20) is the desired functional form for the underdamped response, and its

---

[6] An introduction to the use of complex numbers appears in Chap. 8 and Appendix 4. At that time we shall emphasize the more general nature of complex quantities by identifying them with boldface type; no special symbolism need be adopted in these few pages.

validity may be checked by direct substitution in the original differential equation; this exercise is left to the doubters. The two real constants B_1 and B_2 are again selected to fit the given initial conditions.

Let us increase the resistance in our example from $7\sqrt{6}/2$, or 8.57 Ω, to 10.5 Ω; L and C are unchanged. Thus,

$$\alpha = \frac{1}{2RC} = 2$$

$$\omega_0 = \frac{1}{\sqrt{LC}} = \sqrt{6}$$

and
$$\omega_d = \sqrt{\omega_0^2 - \alpha^2} = \sqrt{2} \text{ rad/s}$$

Except for the evaluation of the arbitrary constants, the response is now known:

$$v(t) = e^{-2t}(B_1 \cos \sqrt{2}t + B_2 \sin \sqrt{2}t)$$

The determination of the two constants proceeds as before. If we again assume that $v(0) = 0$ and $i(0) = 10$, then B_1 must be zero. Hence

$$v(t) = B_2 e^{-2t} \sin \sqrt{2}t$$

The derivative is

$$\frac{dv}{dt} = \sqrt{2}B_2 e^{-2t} \cos \sqrt{2}t - 2B_2 e^{-2t} \sin \sqrt{2}t$$

and at $t = 0$ it becomes

$$\left.\frac{dv}{dt}\right|_{t=0} = \sqrt{2}B_2 = \frac{i_C(0)}{C} = 420$$

where i_C is defined in Fig. 6-2. Therefore,

$$v(t) = 210\sqrt{2}e^{-2t} \sin \sqrt{2}t$$

Notice that, as before, this response function has an initial value of zero, because of the initial voltage condition we imposed, and a final value of zero, because the exponential term vanishes for large values of t. As t increases from zero through small positive values, $v(t)$ increases as $210\sqrt{2} \sin \sqrt{2}t$, because the exponential term remains essentially equal to unity. But at a time t_m, the exponential function begins to decrease more rapidly than $\sin \sqrt{2}t$ is increasing; thus $v(t)$ reaches a maximum v_m and begins to decrease. We should note that t_m is not the value of t for which $\sin \sqrt{2}t$ is a maximum, but must occur somewhat before $\sin \sqrt{2}t$ reaches its maximum.

When $t = \pi/\sqrt{2}$, $v(t)$ is zero. Thus, in the interval $\pi/\sqrt{2} < t < \sqrt{2}\pi$, the response is negative, becoming zero again at $t = \sqrt{2}\pi$. Hence, $v(t)$ is an *oscillatory* function of time and crosses the time axis an infinite number of times at $t = n\pi/\sqrt{2}$, where n is any positive integer. In our example, however, the response is only slightly underdamped, and the exponential term causes the function to die out so rapidly that most of the zero crossings will not be evident in a sketch.

The oscillatory nature of the response becomes more noticeable as α decreases. If α is zero, which corresponds to an infinitely large resistance, then $v(t)$ is an undamped sinusoid which oscillates with constant amplitude. There is never a time at which $v(t)$ drops and stays below 1 percent of its maximum value; the settling time is therefore infinite. This is not perpetual motion; we have merely assumed an initial energy in the circuit and have not provided any means to dissipate this energy. It is transferred from its initial location in the inductor to the capacitor, then returns to the inductor, and so on, forever.

A noninfinite R in the parallel RLC circuit acts as a kind of electrical transfer agent. Every time energy is transferred from L to C or from C to L, the agent exacts its commission. Before long, the agent has taken all the energy, wantonly dissipating every last joule. The L and C are left without a joule of their own, *sans* voltage and *sans* current.

Actual parallel RLC circuits can be made to have effective values of R so large that a natural undamped sinusoidal response can be maintained for years without supplying any additional energy. We can also build active networks which introduce a sufficient amount of energy during each oscillation of $v(t)$ so that a sinusoidal response which is nearly perfect can be maintained for as long as we wish. This circuit is a sinusoidal oscillator, or signal generator, which is an important laboratory instrument. Section 6-8 develops an operational-amplifier version of a sinusoidal oscillator.

Returning to our specific numerical problem, differentiation locates the first maximum of $v(t)$,

$$v_{m1} = 71.8 \text{ V} \qquad \text{at} \qquad t_{m1} = 0.435 \text{ s}$$

the succeeding minimum,

$$v_{m2} = -0.845 \text{ V} \qquad \text{at} \qquad t_{m2} = 2.657 \text{ s}$$

and so on. The response curve is shown in Fig. 6-8.

Figure 6-8

The response $v(t) = 210\sqrt{2}e^{-2t} \sin \sqrt{2}t$ of the network shown in Fig. 6-2 with R increased to produce an underdamped response.

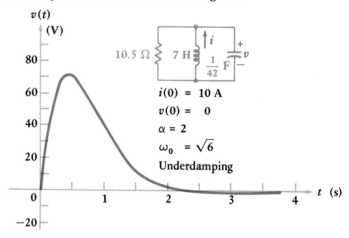

$i(0) = 10 \text{ A}$

$v(0) = 0$

$\alpha = 2$

$\omega_0 = \sqrt{6}$

Underdamping

The settling time may be obtained by a trial-and-error solution, and it turns out to be 2.92 s, somewhat smaller than for critical damping. Note that t_s is *greater* than t_{m2} because the magnitude of v_{m2} is greater than 1 percent of the magnitude of v_{m1}. This suggests that a slight decrease in R would reduce the magnitude of the undershoot and permit t_s to be less than t_{m2}. Problem 31 at the end of this chapter is provided so that each student may satisfy the inevitable curiosity prompted by these remarks and determine the numerical value of the minimum possible settling time for this circuit, as well as the value of R that produces it.

The overdamped, critically damped, and underdamped responses for this

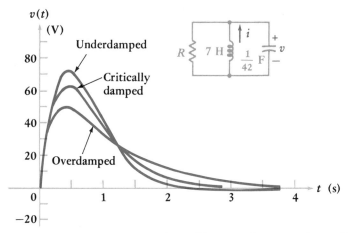

Figure 6-9

Three response curves for a parallel *RLC* circuit for which $\omega_0 = \sqrt{6}$, $v(0) = 0$, $i(0) = 10$ A, and α is 3.5 (overdamped), 2.45 (critically damped), and 2 (underdamped).

network are shown on the same graph in Fig. 6-9. A comparison of the three curves makes the following general conclusions plausible:

1 When the damping is changed by adjusting the size of the parallel resistance, the maximum magnitude of the response is greater with smaller damping.
2 The response becomes oscillatory when underdamping is present, and the minimum settling time is obtained for slight underdamping.

6-5. The switch in Fig. 6-10 has been in the left position for a long time; it is moved to the right at $t = 0$. Find (*a*) dv/dt at $t = 0^+$; (*b*) v at $t = 1$ ms; (*c*) t_0, the first value of t greater than zero at which $v = 0$.

Drill Problem

Ans: -1400 V/s; 0.695 V; 1.609 ms

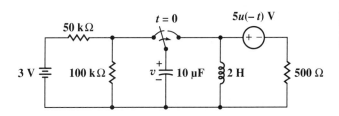

Figure 6-10

See Drill Prob. 6-5.

We now wish to determine the natural response of a circuit model composed of an ideal resistor, an ideal inductor, and an ideal capacitor connected in series. The ideal resistor may represent a physical resistor connected into a series *LC* or *RLC* circuit; it may represent the ohmic losses and the losses in the ferromagnetic core of the inductor; or it may be used to represent all these and other energy-absorbing devices. In a special case, the resistance of the ideal resistor may even be exactly equal to the measured resistance of the wire out of which the physical inductor is made.

6-6

The source-free series *RLC* circuit

The series *RLC* circuit is the dual of the parallel *RLC* circuit, and this single fact is sufficient to make its analysis a trivial affair. Figure 6-11*a* shows the series circuit. The fundamental integrodifferential equation is

$$L\frac{di}{dt} + Ri + \frac{1}{C}\int_{t_0}^{t} i\,dt - v_C(t_0) = 0$$

and should be compared with the analogous equation for the parallel *RLC*

Figure 6-11

(a) The series *RLC* circuit which is the dual of (b) a parallel *RLC* circuit. Element values are, of course, not identical in the two circuits.

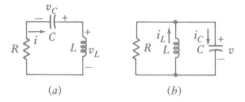

(a)　　　　(b)

circuit, drawn again in Fig. 6-11b,

$$C\frac{dv}{dt} + \frac{1}{R}v + \frac{1}{L}\int_{t_0}^{t} v\,dt - i_L(t_0) = 0$$

The respective second-order equations obtained by differentiating these two equations with respect to time are also duals:

$$L\frac{d^2 i}{dt^2} + R\frac{di}{dt} + \frac{1}{C}i = 0 \tag{21}$$

$$C\frac{d^2 v}{dt^2} + \frac{1}{R}\frac{dv}{dt} + \frac{1}{L}v = 0 \tag{22}$$

It is apparent that our complete discussion of the parallel *RLC* circuit is directly applicable to the series *RLC* circuit; the initial conditions on capacitor voltage and inductor current are equivalent to the initial conditions on inductor current and capacitor voltage; the voltage response becomes a current response. It is quite possible to reread the previous four sections (including the drill problems) using dual language and thereby obtain a complete description of the series *RLC* circuit.[7] This process, however, is apt to induce a mild neurosis after the first few paragraphs and does not really seem to be necessary.

A brief résumé of the series circuit response is easily collected. In terms of the circuit shown in Fig. 6-11a, the overdamped response is

$$i(t) = A_1 e^{s_1 t} + A_2 e^{s_2 t}$$

where

$$s_{1,2} = -\frac{R}{2L} \pm \sqrt{\left(\frac{R}{2L}\right)^2 - \frac{1}{LC}}$$

$$= -\alpha \pm \sqrt{\alpha^2 - \omega_0^2}$$

and thus

$$\alpha = \frac{R}{2L}$$

$$\omega_0 = \frac{1}{\sqrt{LC}}$$

[7] In fact, during the writing of the first edition of this text, the authors wrote these first sections to describe the series *RLC* circuit. But, after deciding that it would be better to present the analysis of the more practical parallel *RLC* circuit first, it was easy to go back to the original writing and replace it with its dual. The numerical values of several of the elements were also scaled, a process described later in Chap. 13.

The form of the critically damped response is

$$i(t) = e^{-\alpha t}(A_1 t + A_2)$$

and the underdamped case may be written

$$i(t) = e^{-\alpha t}(B_1 \cos \omega_d t + B_2 \sin \omega_d t)$$

where

$$\omega_d = \sqrt{\omega_0^2 - \alpha^2}$$

It is evident that if we work in terms of the parameters α, ω_0, and ω_d, the mathematical forms of the responses for the dual situations are identical. An increase in α in either the series or parallel circuit, while keeping ω_0 constant, tends toward an overdamped response. The only caution that we need exert is in the computation of α, which is $1/2RC$ for the parallel circuit and $R/2L$ for the series circuit; thus, α is increased by increasing the series resistance or decreasing the parallel resistance. Lest we forget,

$$\alpha = \frac{1}{2RC} \qquad \alpha = \frac{R}{2L}$$

$$\text{(parallel)} \quad \text{(series)}$$

Now let us consider a numerical example.

Example 6-2 Given a series *RLC* circuit in which $L = 1$ H, $R = 2$ kΩ, $C = \frac{1}{401}$ μF, $i(0) = 2$ mA, and $v_C(0) = 2$ V, find and sketch $i(t)$.

Solution: We find that $\alpha = R/2L = 1000$ s^{-1} and $\omega_0 = 1/\sqrt{LC} = 20\,025$ s^{-1}. Thus an underdamped response is indicated; we therefore calculate the value of ω_d and obtain 20 000. Except for the evaluation of the two arbitrary constants, the response is now known:

$$i(t) = e^{-1000t}(B_1 \cos 20\,000t + B_2 \sin 20\,000t)$$

By applying the given initial value of the current, we find

$$B_1 = 0.002$$

and thus

$$i(t) = e^{-1000t}(0.002 \cos 20\,000t + B_2 \sin 20\,000t)$$

The remaining initial condition must be applied to the derivative; thus,

$$\frac{di}{dt} = e^{-1000t}(-40 \sin 20\,000t + 20\,000B_2 \cos 20\,000t$$

$$- 2 \cos 20\,000t - 1000B_2 \sin 20\,000t)$$

and

$$\left.\frac{di}{dt}\right|_{t=0} = 20\,000B_2 - 2 = \frac{v_L(0)}{L} = \frac{v_C(0) - Ri(0)}{L}$$

$$= \frac{2 - 2000(0.002)}{1} = -2 \text{ A/s}$$

or

$$B_2 = 0$$

The desired response is, therefore,

$$i(t) = 2e^{-1000t} \cos 20\ 000t \qquad \text{mA} \qquad \blacksquare$$

This response is more oscillatory, or shows less damping, than any we have considered up to this time, and the direct calculation of enough points to graph a smooth response curve is a tedious undertaking. A good sketch may be made by first drawing in the two portions of the exponential *envelope*, $0.002e^{-1000t}$ and $-0.002e^{-1000t}$, as shown by the broken lines in Fig. 6-12. The location of the

Figure 6-12

The current response in an under-damped series RLC circuit for which $\alpha = 1000$ s^{-1}, $\omega_0 = 20\ 000$ s^{-1}, $i(0) = 2$ mA, and $v_C(0) = 2$ V. The graphical construction is simplified by drawing in the envelope, shown as a pair of broken lines.

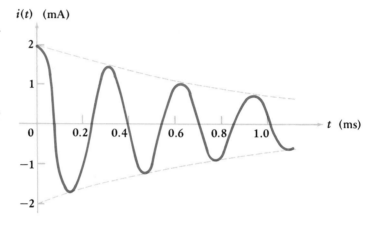

quarter-cycle points of the sinusoidal wave at $20\ 000t = 0$, $\pi/2$, π, etc., or $t = 0.078\ 54k$ ms, $k = 0, 1, 2, \ldots$, by light marks on the time axis then permits the oscillatory curve to be sketched in quickly.

The settling time can be determined easily here by using the upper portion of the envelope. That is, we set $0.002e^{-1000t_s}$ equal to 1 percent of its maximum value, 0.002. Thus, $e^{-1000t_s} = 0.01$, and $t_s = 461$ μs is the approximate value that is usually used.

Drill Problems

6-6. With reference to the circuit shown in Fig. 6-13, find (*a*) α; (*b*) ω_0; (*c*) $i(0^+)$; (*d*) $di/dt|_{t=0^+}$; (*e*) $i(12$ ms).   *Ans:* 100 s^{-1}; 224 rad/s; 1 A; 0; -0.1204 A

6-7. Change the value of the capacitor in Fig. 6-13 to $\frac{1}{3200}$ F, and repeat Drill Prob. 6-6.   *Ans:* 100 s^{-1}; 80 rad/s; 1 A; 0; 0.776 A

Figure 6-13

See Drill Probs. 6-6 and 6-7.

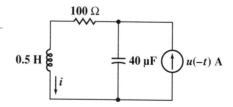

We must now consider those _RLC_ circuits in which dc sources are switched into the network and produce forced responses that do not necessarily vanish as time becomes infinite. The general solution is obtained by the same procedure that was followed in _RL_ and _RC_ circuits: the forced response is determined completely; the natural response is obtained as a suitable functional form containing the appropriate number of arbitrary constants; the complete response is written as the sum of the forced and the natural responses; and the initial conditions are then determined and applied to the complete response to find the values of the constants. It is this last step which is quite frequently the most troublesome to students. Consequently, although the determination of the initial conditions is basically no different for a circuit containing dc sources from what it is for the source-free circuits which we have already covered in some detail, this topic will receive particular emphasis in the examples that follow.

Most of the confusion in determining and applying the initial conditions arises for the simple reason that we do not have a rigorous set of rules laid down for us to follow. At some point in each analysis, a situation usually arises in which some thinking is involved that is more or less unique to that particular problem. This originality and flexibility of thought, as simple as it is to achieve after several problems' worth of practice, is the source of the difficulty.

The complete response (arbitrarily assumed to be a voltage response) of a second-order system consists of a forced response,

$$v_f(t) = V_f$$

which is a constant for dc excitation, and a natural response,

$$v_n(t) = Ae^{s_1 t} + Be^{s_2 t}$$

Thus,

$$v(t) = V_f + Ae^{s_1 t} + Be^{s_2 t}$$

We shall now assume that s_1, s_2, and V_f have already been determined from the circuit and the given forcing functions; A and B remain to be found. The last equation shows the functional interdependence of A, B, v, and t, and substitution of the known value of v at $t = 0^+$ thus provides us with a single equation relating A and B, $v(0^+) = V_f + A + B$. This is the easy part. Another relationship between A and B is necessary, unfortunately, and this is normally obtained by taking the derivative of the response,

$$\frac{dv}{dt} = 0 + s_1 Ae^{s_1 t} + s_2 Be^{s_2 t}$$

and inserting in it the known value of dv/dt at $t = 0^+$. There is no reason that this process cannot be continued; a second derivative might be taken, and a third relationship between A and B will then result if the value of d^2v/dt^2 at $t = 0^+$ is used. This value is _not_ usually known, however, in a second-order system; as a matter of fact, we are much more likely to use this method to _find_ the initial value of the second derivative if we should need it. We thus have only two equations relating A and B, and these may be solved simultaneously to evaluate the two constants.

The only remaining problem is that of determining the values of v and dv/dt at $t = 0^+$. Let us suppose that v is a capacitor voltage, v_C. Since $i_C =$

$C\ dv_C/dt$, we should recognize the relationship between the initial value of dv/dt and the initial value of some capacitor current. If we can establish a value for this initial capacitor current, then we shall automatically establish the value of dv/dt. Students are usually able to get $v(0^+)$ very easily, but are inclined to stumble a bit in finding the initial value of dv/dt. If we had selected an inductor current i_L as our response, then the initial value of di_L/dt would be intimately related to the initial value of some inductor voltage. Variables other than capacitor voltages and inductor currents are determined by expressing their initial values and the initial values of their derivatives in terms of the corresponding values for v_C and i_L.

We shall illustrate the procedure and find all these values by the careful analysis of the circuit shown in Fig. 6-14. To simplify the analysis, an unrealistically large capacitance is used again.

Figure 6-14

An *RLC* circuit which is used to illustrate several procedures by which the initial conditions may be obtained. The desired response is nominally taken to be $v_C(t)$.

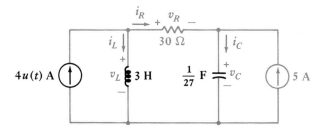

Example 6-3   There are three passive elements in the circuit shown in Fig. 6-14, and a voltage and a current are defined for each. Find the values of these six quantities at both $t = 0^-$ and $t = 0^+$. We shall solve this problem by two different methods.

First Solution: Our object is to find the value of each current and voltage at both $t = 0^-$ and $t = 0^+$. Once these quantities are known, the initial values of the derivatives may be found easily. We shall employ a logical step-by-step method first.

At $t = 0^-$, only the right-hand current source is active. The circuit is assumed to have been in this state forever, and all currents and voltages are constant. Thus, a constant current through the inductor requires zero voltage across it:

$$v_L(0^-) = 0$$

and a constant voltage across the capacitor requires zero current through it:

$$i_C(0^-) = 0$$

We next apply Kirchhoff's current law to the right-hand node to obtain

$$i_R(0^-) = -5\ \text{A}$$

which also yields

$$v_R(0^-) = -150\ \text{V}$$

We may now use Kirchhoff's voltage law around the central mesh, finding

$$v_C(0^-) = 150\ \text{V}$$

while KCL enables us to find the inductor current,

$$i_L(0^-) = 5 \text{ A}$$

Although the derivatives at $t = 0^-$ are of little interest to us, it is evident that they are all zero.

Now let time increase an incremental amount. During the interval from $t = 0^-$ to $t = 0^+$, the left-hand current source becomes active and many of the voltage and current values at $t = 0^-$ will change abruptly. However, we should *begin by focusing our attention on those quantities which cannot change, inductor current and capacitor voltage*. Both of these must remain constant during the switching interval. Thus,

$$i_L(0^+) = 5 \text{ A} \quad \text{and} \quad v_C(0^+) = 150 \text{ V}$$

Since two currents are now known at the left node, we next obtain[8]

$$i_R(0^+) = -1 \text{ A} \quad \text{and} \quad v_R(0^+) = -30 \text{ V}$$

Thus,

$$i_C(0^+) = 4 \text{ A} \quad \text{and} \quad v_L(0^+) = 120 \text{ V}$$

and we have our six initial values at $t = 0^-$ and six more at $t = 0^+$. Among these last six values, only the capacitor voltage and the inductor current are unchanged from the $t = 0^-$ values.

Second Solution: Now let us turn to a slightly different method by which all these currents and voltages may be evaluated at $t = 0^-$ and $t = 0^+$. As we did in solving Example 6-1 for the overdamped circuit, we construct two equivalent circuits, one of which is valid for the steady-state condition reached at $t = 0^-$, and a second of which is valid during the switching interval. The discussion which follows relies on some of the reasoning we just completed with the first solution and, for that reason, appears shorter than it would be if it were presented first. Fair is fair.

Prior to the switching operation, only direct currents and voltages exist in the circuit, and the inductor may therefore be replaced by a short circuit, its dc equivalent, while the capacitor is replaced by an open circuit. Redrawn in this manner, the circuit of Fig. 6-14 appears as shown in Fig. 6-15a. Only the current source at the right is active, and its 5 A flows through the resistor and the inductor. We therefore have $i_R(0^-) = -5 \text{ A}$ and $v_R(0^-) = -150 \text{ V}$, $i_L(0^-) = 5 \text{ A}$ and $v_L(0^-) = 0$, and $i_C(0^-) = 0$ and $v_C(0^-) = 150 \text{ V}$, as before.

We now turn to the problem of drawing an equivalent circuit which will assist us in determining the several voltages and currents at $t = 0^+$. Each capacitor voltage and each inductor current must remain constant during the switching interval. These conditions are ensured by replacing the inductor by a current source and the capacitor by a voltage source. Each source serves to maintain a constant response during the discontinuity. The equivalent circuit of Fig. 6-15b results; it should be noted that the left-hand current source is now 4 A.

---

[8] This current is the only one of the four remaining quantities which can be obtained in one step. In more complicated circuits, it is quite possible that none of the remaining initial values can be obtained with a single step; either circuit equations must then be written or a simpler equivalent resistive circuit must be drawn which can be analyzed by writing simultaneous equations. This latter method will be described shortly.

Figure 6-15

(*a*) A simple circuit which is the equivalent of the circuit of Fig. 6-14 for $t = 0^-$. (*b*) Another equivalent of the circuit of Fig. 6-14, valid during the switching interval, $t = 0^-$ to $t = 0^+$.

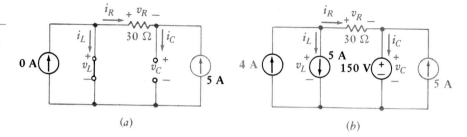

(*a*)   (*b*)

The voltages and currents at $t = 0^+$ are obtained by analyzing this dc circuit. The solution is not difficult, but the relatively large number of sources present in the network does produce a somewhat strange sight. However, problems of this type were solved in Chap. 2, and nothing new is involved. Attacking the currents first, we begin at the upper left node and see that $i_R(0^+) = 4 - 5 = -1$ A. Moving to the upper right node, we find that $i_C(0^+) = -1 + 5 = 4$ A. And, of course, $i_L(0^+) = 5$ A.

Next we consider the voltages. Using Ohm's law, we see that $v_R(0^+) = 30(-1) = -30$ V. For the inductor, KVL gives us $v_L(0^+) = -30 + 150 = 120$ V. Finally, including $v_C(0^+) = 150$ V, we have all the values at $t = 0^+$. ■

Example 6-4   Complete the determination of the initial conditions in the circuit of Fig. 6-14 by finding values at $t = 0^+$ for the first derivatives of the three voltage and three current variables defined on the circuit diagram.

Solution: We begin with the two energy-storage elements. For the inductor,

$$v_L = L \frac{di_L}{dt}$$

and, specifically,

$$v_L(0^+) = L \frac{di_L}{dt}\Big|_{t=0^+}$$

Thus,

$$\frac{di_L}{dt}\Big|_{t=0^+} = \frac{v_L(0^+)}{L} = \frac{120}{3} = 40 \text{ A/s}$$

Similarly,

$$\frac{dv_C}{dt}\Big|_{t=0^+} = \frac{i_C(0^+)}{C} = \frac{4}{\frac{1}{27}} = 108 \text{ V/s}$$

The other four derivatives may be determined by realizing that KCL and KVL are both satisfied by the derivatives also. For example, at the left-hand node in Fig. 6-14,

$$4 - i_L - i_R = 0 \qquad t > 0$$

and thus,

$$0 - \frac{di_L}{dt} - \frac{di_R}{dt} = 0 \qquad t > 0$$

and therefore,

$$\left.\frac{di_R}{dt}\right|_{t=0^+} = -40 \text{ A/s}$$

The three remaining initial values of the derivatives are found to be

$$\left.\frac{dv_R}{dt}\right|_{t=0^+} = -1200 \text{ V/s}$$

$$\left.\frac{dv_L}{dt}\right|_{t=0^+} = -1092 \text{ V/s}$$

and

$$\left.\frac{di_C}{dt}\right|_{t=0^+} = -40 \text{ A/s} \quad\blacksquare$$

Before leaving this problem of the determination of the necessary initial values, it should be pointed out that at least one other powerful method of determining them has been omitted: we could have written general nodal or loop equations for the original circuit. Then, the substitution of the known zero values of inductor voltage and capacitor current at $t = 0^-$ would uncover several other response values at $t = 0^-$ and enable the remainder to be found easily. A similar analysis at $t = 0^+$ must then be made. This is an important method, and it becomes a necessary one in more complicated circuits which cannot be analyzed by the simpler step-by-step procedures we have followed. However, we must leave a few topics to be covered at the time operational methods of circuit analysis are introduced later.

Now let us briefly complete the determination of the response $v_C(t)$ for the original circuit of Fig. 6-14. With both sources dead, the circuit appears as a series *RLC* circuit and s_1 and s_2 are easily found to be -1 and -9, respectively. The forced response may be found by inspection or, if necessary, by drawing the dc equivalent, which is similar to Fig. 6-15a, with the addition of a 4-A current source. The forced response is 150 V. Thus,

$$v_C(t) = 150 + Ae^{-t} + Be^{-9t}$$

and

$$v_C(0^+) = 150 = 150 + A + B$$

or

$$A + B = 0$$

Then,

$$\frac{dv_C}{dt} = -Ae^{-t} - 9Be^{-9t}$$

and

$$\left.\frac{dv_C}{dt}\right|_{t=0^+} = 108 = -A - 9B$$

Finally,

$$A = 13.5 \qquad B = -13.5$$

and

$$v_C(t) = 150 + 13.5(e^{-t} - e^{-9t})$$

In summary then, whenever we wish to determine the transient behavior of a simple three-element *RLC* circuit, we must first decide whether we are confronted with a series or a parallel circuit, so that we may use the correct relationship for α. The two equations are

$$\alpha = \frac{1}{2RC} \qquad \text{(parallel } RLC\text{)}$$

$$\alpha = \frac{R}{2L} \qquad \text{(series } RLC\text{)}$$

Our second decision is made after comparing α with ω_0, which is given for either circuit by

$$\omega_0 = \frac{1}{\sqrt{LC}}$$

If $\alpha > \omega_0$, the circuit is overdamped, and the natural response has the form:

$$f_n(t) = A_1 e^{s_1 t} + A_2 e^{s_2 t}$$

where

$$s_{1,2} = -\alpha \pm \sqrt{\alpha^2 - \omega_0^2}$$

If $\alpha = \omega_0$, then the circuit is critically damped and

$$f_n(t) = e^{-\alpha t}(A_1 t + A_2)$$

And finally, if $\alpha < \omega_0$, then we are faced with the underdamped response,

$$f_n(t) = e^{-\alpha t}(A_1 \cos \omega_d t + A_2 \sin \omega_d t)$$

where

$$\omega_d = \sqrt{\omega_0^2 - \alpha^2}$$

Our last decision depends on the independent sources. If there are none acting in the circuit after the switching or discontinuity is completed, then the circuit is source-free and the natural response accounts for the complete response. If independent sources are still present, then the circuit is driven and a forced response must be determined. The complete response is then the sum

$$f(t) = f_f(t) + f_n(t)$$

This is applicable to any current or voltage in the circuit.

Drill Problems

6-8. Let $i_s = 10u(-t) - 20u(t)$ A in Fig. 6-16a. Find (a) $i_L(0^-)$; (b) $v_C(0^+)$; (c) $v_R(0^+)$; (d) $i_{L,f}$; (e) $i_L(0.1$ ms). *Ans:* 10 A; 200 V; 200 V; −20 A; 2.07 A

6-9. Let $v_s = 10 + 20u(t)$ V in the circuit of Fig. 6-16b. Find (a) $i_L(0)$; (b) $v_C(0)$; (c) $i_{L,f}$; (d) $i_L(0.1$ s). *Ans:* 0.2 A; 10 V; 0.6 A; 0.319 A

Figure 6-16

See Drill Probs. 6-8 and 6-9.

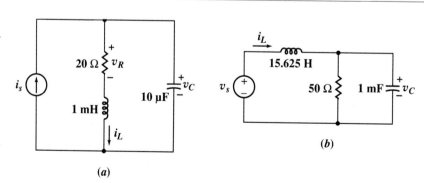

(a)   (b)

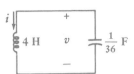

Figure 6-17

This circuit is lossless, and it provides the undamped response $v = 2 \sin 3t$ V, if $v(0) = 0$ and $i(0) = -\frac{1}{6}$ A.

If the value of the resistance in a parallel *RLC* circuit becomes infinite, or that in a series *RLC* circuit becomes zero, we have a simple *LC* loop in which an oscillatory response can be maintained forever. Let us look briefly at an example of such a circuit, and then discuss another means of obtaining an identical response without the need of supplying any inductance.

Consider the circuit of Fig. 6-17, in which the large values $L = 4$ H and $C = \frac{1}{36}$ F are used so that the mathematics will be simple. We shall let $i(0) = -\frac{1}{6}$ A and $v(0) = 0$. The circuit is source-free, $\alpha = 0$, and $\omega_0^2 = 9$. Thus, $\omega_d = \omega_0 = 3$, and the voltage is

$$v = A \cos 3t + B \sin 3t$$

Since $v(0) = 0$, we see that $A = 0$. Next,

$$\left. \frac{dv}{dt} \right|_{t=0} = 3B = -\frac{i(0)}{C}$$

But $i(0) = -\frac{1}{6}$ A, $C = \frac{1}{36}$ F, and therefore $dv/dt = 6$ V/s at $t = 0$. We must have $B = 2$ V and

$$v = 2 \sin 3t \qquad \text{V}$$

which is an undamped sinusoidal response.

Now let us see how we might obtain this voltage without using an *LC* circuit. Our intentions are to write the differential equation that v satisfies and then to develop a configuration of op-amps that will yield the solution of the equation. Although we are working with a specific example, the technique is a general one that can be used to solve any linear homogeneous differential equation.

For the *LC* circuit of Fig. 6-17, we select v as our variable and set the sum of the downward inductor and capacitor currents equal to zero:

$$\frac{1}{4} \int_{t_0}^{t} v \, dt - \frac{1}{6} + \frac{1}{36} \frac{dv}{dt} = 0$$

Differentiating once, we have

$$\frac{1}{4} v + \frac{1}{36} \frac{d^2v}{dt^2} = 0$$

or

$$\frac{d^2v}{dt^2} = -9v$$

In order to solve this equation, let us make use of the operational amplifier as an integrator twice. We assume that the highest-order derivative appearing in the differential equation—here, d^2v/dt^2—is available at some point in our configuration of op-amps, say, at an arbitrary point A. We now make use of the integrator, with $RC = 1$, as discussed at the end of Sec. 3-4. The input is d^2v/dt^2, and the output must be $-dv/dt$, where we assume a sign change in the integrator. The initial value of dv/dt is 6 V/s, as we showed when we first analyzed the circuit, and thus an initial value of -6 V must be set in the integrator. The negative of the first derivative now forms the input to a second integrator. Its output is therefore $v(t)$, and the initial value is $v(0) = 0$. Now it only remains to multiply v by -9 to obtain the second derivative we assumed at point A. This is amplification by 9 with a sign change, and it is easily accomplished by using the op-amp as an inverting amplifier.

Figure 6-18 shows the circuit of an inverting amplifier. For an ideal op-amp, both the input current and the input voltage are zero. Thus, the current

Figure 6-18

The inverting operational amplifier provides a gain $v_o/v_s = -R_f/R_1$, with an ideal op-amp.

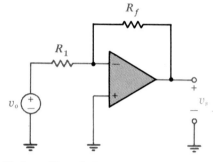

going "east" through R_1 is v_s/R_1, while that traveling west through R_f is v_o/R_f. Since their sum is zero, we have

$$\frac{v_o}{v_s} = -\frac{R_f}{R_1}$$

Thus, we can achieve a gain of -9 by setting $R_f = 90$ kΩ and $R_1 = 10$ kΩ, for example.

If we let R be 1 MΩ and C be 1 μF in each of the integrators, then

$$v_o = -\int_0^t v_s\, dt + v_o(0)$$

in each case. The output of the inverting amplifier now forms the assumed input at point A, leading to the configuration of op-amps shown in Fig. 6-19. If

Figure 6-19

Two integrators and an inverting amplifier are connected to provide the solution of the differential equation $d^2v/dt^2 = -9v$.

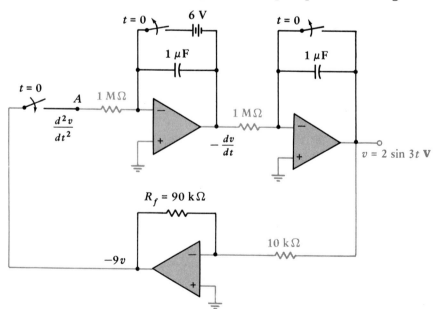

the left switch is closed at $t = 0$ while the two initial-condition switches are opened at the same time, the output of the second integrator will be the undamped sine wave $v = 2 \sin 3t$ V.

Note that both the *LC* circuit of Fig. 6-17 and the op-amp circuit of Fig. 6-19 have the same output, but the op-amp circuit does not contain a single inductor. It simply *acts* as though it contained an inductor, providing the appropriate sinusoidal voltage between its output terminal and ground. This can be a considerable practical or economic advantage in circuit design.

6-10. Give new values for R_f and the two initial voltages in the circuit of Fig. 6-19 if the output represents the voltage $v(t)$ in the circuit of Fig. 6-20. **Drill Problem**

Ans: 250 kΩ; 400 V; 10 V

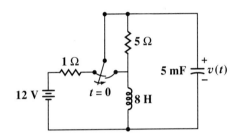

Figure 6-20

See Drill Prob. 6-10.

1 A source-free parallel *RLC* circuit contains an inductor for which the product $\omega_0 L$ is 10 Ω. If $s_1 = -6$ s^{-1} and $s_2 = -8$ s^{-1}, find R, L, and C. **Problems**

2 The capacitor current in the circuit of Fig. 6-21 is $i_C = 40e^{-100t} - 30e^{-200t}$ mA. If $C = 1$ mF and $v(0) = -0.5$ V, find (*a*) $v(t)$; (*b*) $i_R(t)$; (*c*) $i(t)$.

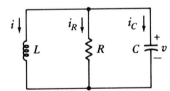

Figure 6-21

See Probs. 2, 3, 4, and 10.

3 In the circuit of Fig. 6-21, let $L = 5$ H, $R = 8$ Ω, $C = 12.5$ mF, and $v(0^+) = 40$ V. Find (*a*) $v(t)$ if $i(0^+) = 8$ A; (*b*) $i(t)$ if $i_C(0^+) = 8$ A.

4 With reference to the circuit of Fig. 6-21, let $i(0) = 40$ A and $v(0) = 40$ V. If $L = 12.5$ mH, $R = 0.1$ Ω, and $C = 0.2$ F: (*a*) find $v(t)$, and (*b*) sketch it for $0 < t < 0.3$ s.

5 Obtain an expression for $i_L(t)$ in the circuit of Fig. 6-22 that is valid for all t.

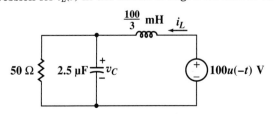

Figure 6-22

See Probs. 5, 8 and 19.

6 Find $i_L(t)$ for $t \geq 0$ in the circuit shown in Fig. 6-23.

Figure 6-23

See Probs. 6 and 9.

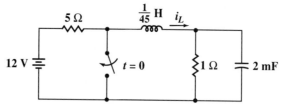

7 The circuit of Fig. 6-24 has been in the condition shown for a long time. After the switch closes at $t = 0$, find (a) $v(t)$; (b) $i(t)$; (c) the settling time for $v(t)$.

Figure 6-24

See Prob. 7.

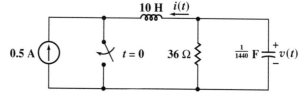

8 (a) What new value of resistance should be used in the circuit of Fig. 6-22 to achieve critical damping? (b) Using this value of resistance, find $v_C(t)$ for $t > 0$.

9 Change the inductance value in the circuit of Fig. 6-23 until the circuit is critically damped. (a) What is the new inductance? (b) Find i_L at $t = 5$ ms. (c) Find the settling time.

10 In the circuit of Fig. 6-21, let $v(0) = -400$ V and $i(0) = 0.1$ A. If $L = 5$ mH, $C = 10$ nF, and the circuit is critically damped: (a) find R; (b) find $|i|_{max}$; (c) find i_{max}.

11 Find $i_C(t)$ for $t > 0$ in the circuit shown in Fig. 6-25.

Figure 6-25

See Probs. 11 and 25.

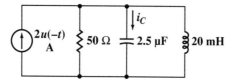

12 For the circuit shown in Fig. 6-26, find (a) $i_L(0^+)$; (b) $v_C(0^+)$; (c) $di_L/dt|_{t=0^+}$; (d) $dv_C/dt|_{t=0^+}$; (e) $v_C(t)$. (f) Sketch $v_C(t)$, $-0.1 < t < 2$ s.

Figure 6-26

See Prob. 12.

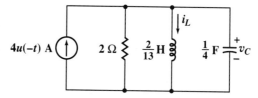

13 After being open for a long time, the switch in the circuit of Fig. 6-27 is closed at $t = 0$. For $t > 0$, find (a) $v_C(t)$; (b) $i_{SW}(t)$.

Figure 6-27

See Prob. 13.

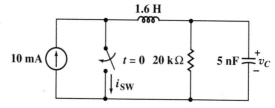

14 Let $\omega_d = 6$ rad/s in the circuit of Fig. 6-28. (a) Find L. (b) Obtain an expression for $i_L(t)$ valid for all t. (c) Sketch $i_L(t)$, $-0.1 < t < 0.6$ s.

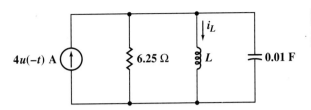

Figure 6-28

See Prob. 14.

15 Find $i_1(t)$ for $t > 0$ in the circuit of Fig. 6-29.

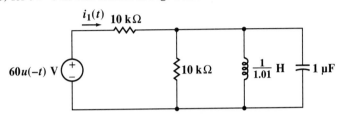

Figure 6-29

See Prob. 15.

16 (_a_) Find $v(t)$ for $t > 0$ for the circuit shown in Fig. 6-30. (_b_) Make a quick sketch of $v(t)$ over the time interval $0 < t < 0.1$ s.

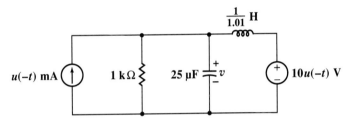

Figure 6-30

See Prob. 16.

17 Find $i_L(t)$ for $t > 0$ in the circuit of Fig. 6-31.

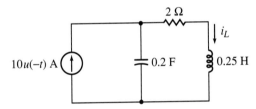

Figure 6-31

See Probs. 17 and 23.

18 Find v_C, v_R, and v_L at $t = 40$ ms in the circuit shown in Fig. 6-32.

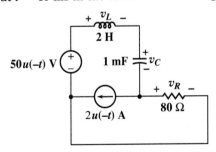

Figure 6-32

See Prob. 18.

19 Write the dual of Prob. 5, including the dual of the circuit shown in Fig. 6-22. Solve your dual problem.

20 In the circuit shown in Fig. 6-11a, let $R = 200$ Ω and $C = 1$ μF with the circuit critically damped. If $v_C(0) = -10$ V and $i_L(0) = -150$ mA, find (_a_) $v_C(t)$; (_b_) $|v_C|_{\max}$; (_c_) $v_{C,\max}$.

21 For the circuit of Fig. 6-33, $t > 0$, find (_a_) $i_L(t)$; (_b_) $v_C(t)$.

Figure 6-33

See Probs. 21 and 26.

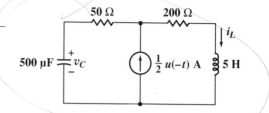

22 (a) Find $i_L(t)$ for $t > 0$ in the circuit shown in Fig. 6-34. (b) Find $|i_L|_{max}$ and $i_{L,max}$.

Figure 6-34

See Prob. 22.

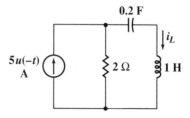

23 The source in the circuit shown in Fig. 6-31 is changed to $10u(t)$ A. Find $i_L(t)$.

24 (a) Find $i_L(t)$ for all t in the circuit of Fig. 6-35. (b) At what instant of time after $t = 0$ is $i_L(t) = 0$?

Figure 6-35

See Prob. 24.

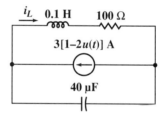

25 Replace the source shown in Fig. 6-25 with $i_s = 2[1 + u(t)]$ A and find $i_C(t)$ for $t > 0$.

26 Replace the source in the circuit of Fig. 6-33 with $i_s = 0.5[1 - 2u(t)]$ A and find $i_L(t)$.

27 The switch in the circuit of Fig. 6-36 has been closed for a very long time. It opens at $t = 0$. Find $v_C(t)$ for $t > 0$.

Figure 6-36

See Prob. 27.

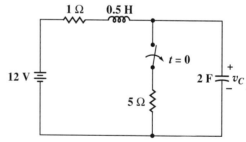

28 (a) Find $v_C(t)$ for $t > 0$ in the circuit shown in Fig. 6-37. (b) Sketch $v_C(t)$ versus t, $-0.1 < t < 2$ ms.

Figure 6-37

See Prob. 28.

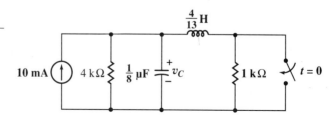

29 Find $i_s(t)$ for $t > 0$ in the circuit of Fig. 6-38 if $v_s(t)$ equals (*a*) $10u(-t)$ V; (*b*) $10u(t)$ V.

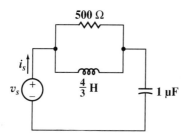

Figure 6-38

See Prob. 29.

30 Find $i_R(t)$ for $t > 0$ in the circuit of Fig. 6-39 if $v_s(t)$ equals (*a*) $10u(-t)$ V; (*b*) $10u(t)$ V.

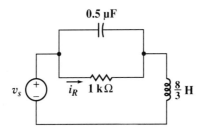

Figure 6-39

See Prob. 30.

31 Determine the value of R for the underdamped circuit of Sec. 6-5 ($L = 7$ H, $C = \frac{1}{42}$ F, $i(0) = 10$ A, $v(0) = 0$) that will lead to a minimum value of the settling time t_s. What is the value of t_s?

32 A source-free *RL* circuit contains a 20-Ω resistor and a 5-H inductor. If the initial value of the inductor current is 2 A: (*a*) write the differential equation for i for $t > 0$, and (*b*) design an op-amp integrator to provide $i(t)$ as the output, using $R_1 = 1$ MΩ and $C_f = 1$ μF.

33 Refer to Fig. 6-40, and design an op-amp circuit whose output will be $i(t)$ for $t > 0$.

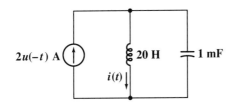

Figure 6-40

See Prob. 33.

34 (SPICE) Given the initial conditions $i(0) = 10$ A and $v(0) = 0$, the solution for $v(t)$ in the circuit of Fig. 6-2 was found to be $v(t) = 84(e^{-t} - e^{-6t})$ V for $t > 0$. Use SPICE to find $v(t)$ at $t = 0.5$ and 2 s if TSTEP is selected as 0.01 s.

35 (SPICE) Modify Prob. 34 by installing a 0.1-Ω resistor between the bottom of the inductor and the reference node. Let $v(0) = 0$ and $i(0) = 10$ A. (*a*) Use SPICE with TSTEP = 0.01 s to find values for $v(t)$ at $t = 0.5$ and 2 s. (*b*) Compare these results with those obtained from Eq. (17), obtained when the 0.1-Ω resistor is not present.

36 (SPICE) Use SPICE analysis methods to find reasonable values for v_C in the circuit of Fig. 6-41 at $t = 0.1$, 0.4, and 1 s.

Figure 6-41

See Prob. 36.

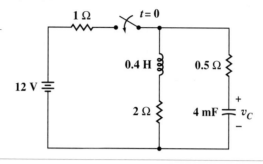

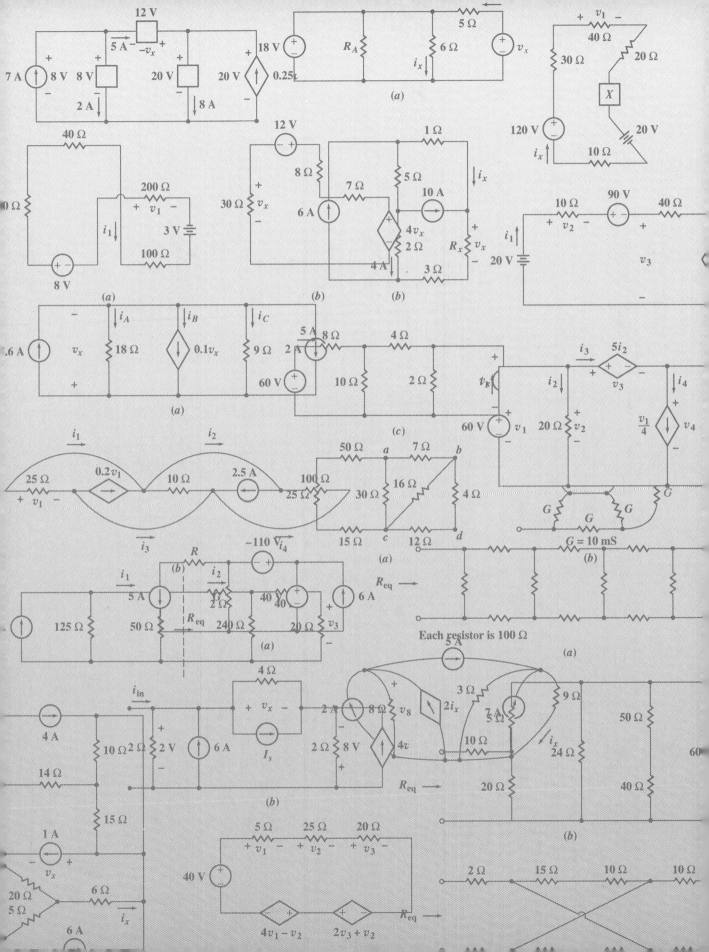

Part Three:
Sinusoidal Analysis

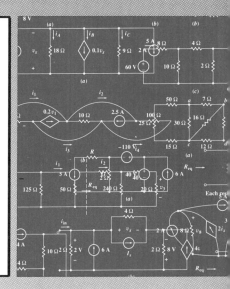

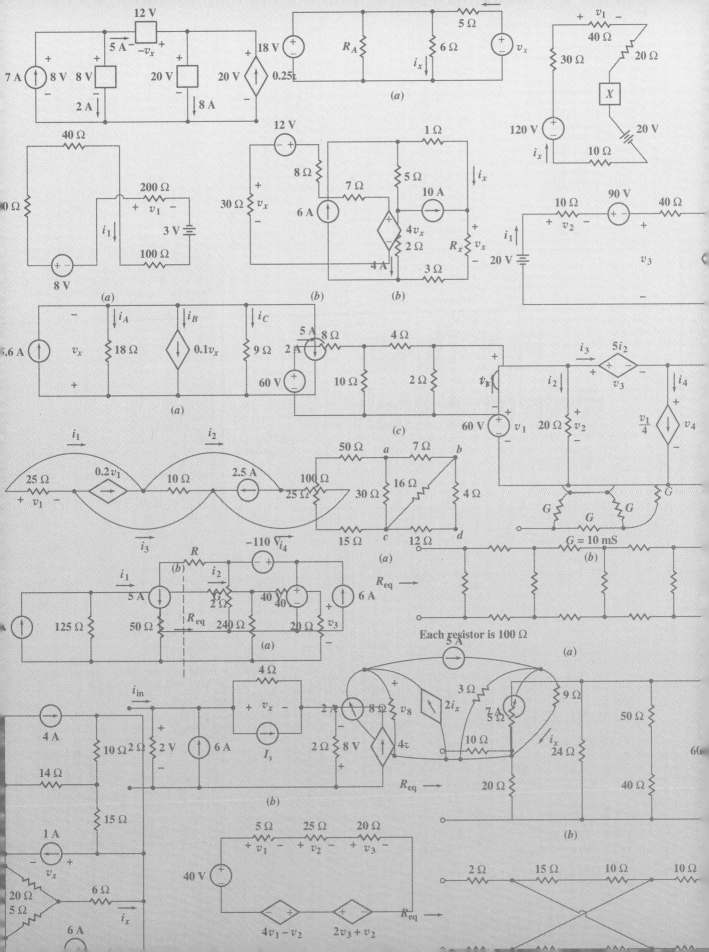

The Sinusoidal Forcing Function

Introduction

The complete response of a linear electric circuit is composed of two parts, the natural response and the forced response. The first part of our study was devoted to the resistive circuit, in which only the forced response was required or present. For simplicity, we usually restricted our forcing functions to dc sources, and we therefore became exceedingly familiar with the various techniques useful in finding the dc forced response. We then passed on to the next part and considered the natural response of a number of different circuits containing one or two energy-storage elements. Without undue strain we were then able to determine the complete response of these circuits by adding the natural response—which is characteristic of the circuit and not of the forcing function—to the forced response produced by dc forcing functions, the only forced response with which we are familiar. We are therefore now in a position where our mastery of the natural response is greater than our knowledge of the forced response.

In this third part of our study we shall extend our knowledge of the forced response by considering the sinusoidal forcing function.

Why should we select the sinusoidal forcing function as the second functional form to study? Why not the linear function, the exponential function, or a modified Bessel function of the second kind? There are many reasons for the choice of the sinusoid, and any one of them would probably be sufficient to lead us in this direction.

One of these reasons is apparent from the results of the preceding chapter: the natural response of an underdamped second-order system is a damped sinusoid, and if no losses are present it is a pure sinusoid. The sinusoid thus appears naturally (as does the negative exponential). Indeed, Nature in general seems to have a decidedly sinusoidal character; the motion of a pendulum, the bouncing of a ball, the vibration of a guitar string, the back-and-forth political trends in any country, and the ripples on the surface of a stein of chocolate milk will always display a reasonably sinusoidal character.

Perhaps it was observations of these natural phenomena that led the great French mathematician Fourier to his discovery of the important analytical method embodied in the Fourier theorem. In Chap. 17 we shall see that this theorem enables us to represent most of the useful mathematical functions of time which repeat themselves f_0 times per second by the sum of an infinite number of sinusoidal time functions with frequencies that are integral multiples of f_0; the given periodic function $f(t)$ can also be approximated as closely as we wish by the sum of a finite number of such terms, even though the graph of $f(t)$ might look very nonsinusoidal. The accuracy of such an approximation is illustrated by Drill Prob. 7-1.

This decomposition of a periodic forcing function into a number of appropriately chosen sinusoidal forcing functions is a very powerful analytical method, for it enables us to superimpose the partial responses produced in any linear circuit by the respective sinusoidal components and thereby obtain the desired response caused by the given periodic forcing function. Thus, another reason for studying the response to a sinusoidal forcing function is found in the dependence of other forcing functions on sinusoidal analysis.

A third reason is found in an important mathematical property of the sinusoidal function. Its derivatives and integrals are also all sinusoids.[1] Since the forced response takes on the form of the forcing function, its integral, and its derivatives, the sinusoidal forcing function will produce a sinusoidal forced response throughout a linear circuit. The sinusoidal forcing function thus allows a much easier mathematical analysis than does almost every other forcing function.

Finally, the sinusoidal forcing function has important practical applications. It is an easy function to generate and is the waveform used predominantly throughout the electric power industry; and every electrical laboratory contains a number of sinusoidal generators which operate throughout a tremendous range of useful frequencies.

Drill Problem

7-1. The Fourier theorem, which we shall study in Chap. 17, shows that the periodic triangular waveform, shown in Fig. 7-1, and the infinite sum of sine terms,

$$v_1(t) = \frac{8}{\pi^2}\left(\sin \pi t - \frac{1}{3^2}\sin 3\pi t + \frac{1}{5^2}\sin 5\pi t - \frac{1}{7^2}\sin 7\pi t + \cdots\right)$$

are equal. At $t = 0.4$ s: (a) calculate the exact value of $v_1(t)$. Calculate the approximate value obtained from the infinite sum of sine terms if the only terms used are (b) the first three; (c) the first four; (d) the first five.

Ans: 0.800; 0.824; 0.814; 0.805

Figure 7-1

See Drill Prob. 7-1.

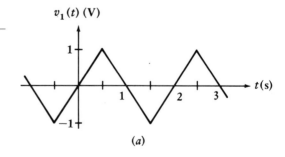

(a)

---

[1] We are using the term *sinusoid* collectively here to include cosinusoidal functions of time also. After all, a cosine function can be written as a sine function if the angle is increased by 90°.

In this section we shall define the trigonometric nomenclature which is used to describe sinusoidal (or cosinusoidal) functions. The definitions should be familiar to most of us, and if we remember a little trigonometry, the section can be read very rapidly.

Let us consider a sinusoidally varying voltage

$$v(t) = V_m \sin \omega t$$

shown graphically in Figs. 7-2a and b. The *amplitude* of the sine wave is V_m, and the *argument* is ωt. The *radian frequency,* or *angular frequency,* is ω. In Fig. 7-2a, $V_m \sin \omega t$ is plotted as a function of the argument ωt, and the periodic nature of the sine wave is evident. The function repeats itself every 2π radians, and its *period* is therefore 2π radians. In Fig. 7-2b, $V_m \sin \omega t$ is plotted as a function of t and the *period* is now T. The period may also be expressed in degrees, or occasionally in other units such as centimeters or inches. A sine wave having a period T must execute $1/T$ periods each second; its *frequency f* is $1/T$ hertz, abbreviated Hz. Thus, one hertz is identical to one cycle per second, a term whose use is now discouraged because so many people incorrectly used "cycle" for "cycle per second." Thus,

$$f = \frac{1}{T}$$

and since

$$\omega T = 2\pi$$

we obtain the common relationship between frequency and radian frequency,

$$\omega = 2\pi f$$

A more general form of the sinusoid,

$$v(t) = V_m \sin (\omega t + \theta) \tag{1}$$

7-2

Characteristics of sinusoids

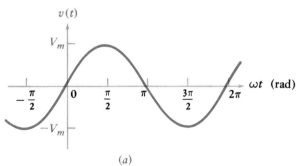

(a)

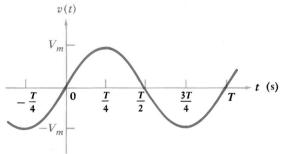

(b)

Figure 7-2

The sinusoidal function $v(t) = V_m \sin \omega t$ is plotted (a) versus ωt and (b) versus t.

includes a *phase angle* θ in its argument $(\omega t + \theta)$. Equation (1) is plotted in Fig. 7-3 as a function of ωt, and the phase angle appears as the number of radians by which the original sine wave, shown as a broken line in the sketch, is shifted to the left, or earlier in time. Since corresponding points on the sinusoid $V_m \sin (\omega t + \theta)$ occur θ rad, or θ/ω seconds, earlier, we say that $V_m \sin (\omega t + \theta)$ *leads* $V_m \sin \omega t$ by θ rad. Conversely, it is correct to describe $\sin \omega t$ as *lagging* $\sin (\omega t + \theta)$ by θ rad, as leading $\sin (\omega t + \theta)$ by $-\theta$ rad, or as leading $\sin (\omega t - \theta)$ by θ rad.

Figure 7-3

The sine wave $V_m \sin (\omega t + \theta)$ leads $V_m \sin \omega t$ by θ rad.

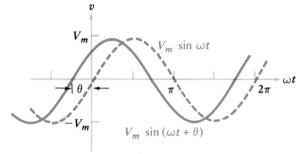

In either case, leading or lagging, we say that the sinusoids are *out of phase*. If the phase angles are equal, they are said to be *in phase*.

In electrical engineering, the phase angle is commonly given in degrees, rather than radians, and no confusion will arise if the degree symbol is always used. Thus, instead of writing

$$v = 100 \sin \left(2\pi 1000t - \frac{\pi}{6} \right)$$

we customarily use

$$v = 100 \sin (2\pi 1000t - 30°)$$

In evaluating this expression at a specific instant of time, say, $t = 10^{-4}$ s, $2\pi 1000t$ becomes 0.2π *radians,* and this should be expressed as $36°$ before $30°$ is subtracted from it. Don't confuse your apples with your oranges.

Two sinusoidal waves that are to be compared in phase must both be written as sine waves, or both as cosine waves; both waves must be written with positive amplitudes; and each must have the same frequency. It is also evident that multiples of $360°$ may be added to or subtracted from the argument of any sinusoidal function without changing the value of the function. Hence, we may say that

$$v_1 = V_{m1} \sin (5t - 30°)$$

lags

$$v_2 = V_{m2} \cos (5t + 10°)$$
$$= V_{m2} \sin (5t + 90° + 10°)$$
$$= V_{m2} \sin (5t + 100°)$$

by $130°$, and it is also correct to say that v_1 leads v_2 by $230°$, since v_2 may be written as

$$v_2 = V_{m2} \sin (5t - 260°)$$

We assume that V_{m1} and V_{m2} are both positive quantities. Normally, the difference in phase between two sinusoids is expressed by that angle which is less than or equal to 180° in magnitude.

The concept of a leading or lagging relationship between two sinusoids will be used extensively, and the relationship should be recognizable both mathematically and graphically.

Drill Problems

7-2. Find the angle by which i_1 lags v_1 if $v_1 = 120 \cos (120\pi t - 40°)$ V and i_1 equals (a) $2.5 \cos (120\pi t + 20°)$ A; (b) $1.4 \sin (120\pi t - 70°)$ A; (c) $-0.8 \cos (120\pi t - 110°)$ A.   *Ans:* $-60°$; $120°$; $-110°$

7-3. Find A, B, C, and ϕ if $40 \cos (100t - 40°) - 20 \sin (100t + 170°) = A \cos 100t + B \sin 100t = C \cos (100t + \phi)$.   *Ans:* 27.2; 45.4; 52.9; $-59.1°$

Now that we are familiar with the mathematical characteristics of sinusoids and can describe and compare them intelligently, we are ready to apply a sinusoidal forcing function to a simple circuit and to obtain the forced response. We shall first write the differential equation which applies to the given circuit. The complete solution of this equation is composed of two parts, the complementary solution (which we call the *natural response*) and the particular integral (or *forced response*). The natural response is independent of the mathematical form of the forcing function and depends only upon the type of circuit, the element values, and the initial conditions. We find it by setting all the forcing functions equal to zero, thus reducing the equation to the simpler linear *homogeneous* differential equation. We have already determined the natural response of many RL, RC, and RLC circuits.

The forced response has the mathematical form of the forcing function, plus all its derivatives and its first integral. From this knowledge, it is apparent that one of the methods by which the forced response may be found is to assume a solution composed of a sum of such functions, where each function has an unknown amplitude to be determined by direct substitution in the differential equation. This is a lengthy method, but it is the one which we shall use in this chapter to introduce sinusoidal analysis because it involves a minimum of new concepts. However, if the simpler method to be described in the following chapters were not available, circuit analysis would be an impractical, useless art.

The term *steady-state response* is used synonymously with *forced response,* and the circuits we are about to analyze are commonly said to be in the "sinusoidal steady state." Unfortunately, *steady state* carries the connotation of "not changing with time" in the minds of many students. This is true for dc forcing functions, but the sinusoidal steady-state response is definitely changing with time. The steady state simply refers to the condition which is reached after the transient or natural response has died out.

Now let us consider the series RL circuit shown in Fig. 7-4. The sinusoidal source voltage $v_s = V_m \cos \omega t$ has been switched into the circuit at some remote time in the past, and the natural response has died out completely. We seek the forced response, or steady-state response, and it must satisfy the differential equation

$$L \frac{di}{dt} + Ri = V_m \cos \omega t$$

7-3

Forced response to sinusoidal forcing functions

Figure 7-4

A series RL circuit for which the forced response is desired.

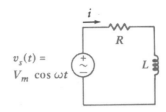

$$v_s(t) = V_m \cos \omega t$$

The functional form of the forced response is next obtained by integration and repeated differentiation of the forcing function. Only two different forms are obtained, $\sin \omega t$ and $\cos \omega t$. The forced response must therefore have the general form

$$i(t) = I_1 \cos \omega t + I_2 \sin \omega t$$

where I_1 and I_2 are real constants whose values depend upon V_m, R, L, and ω. No constant or exponential function can be present. Substituting the assumed form for the solution in the differential equation yields

$$L(-I_1\omega \sin \omega t + I_2\omega \cos \omega t) + R(I_1 \cos \omega t + I_2 \sin \omega t) = V_m \cos \omega t$$

If we collect the cosine and sine terms, we obtain

$$(-LI_1\omega + RI_2) \sin \omega t + (LI_2\omega + RI_1 - V_m) \cos \omega t = 0$$

This equation must be true for all values of t, and this can be achieved only if the factors multiplying $\cos \omega t$ and $\sin \omega t$ are each zero. Thus,

$$-\omega LI_1 + RI_2 = 0 \qquad \text{and} \qquad \omega LI_2 + RI_1 - V_m = 0$$

and simultaneous solution for I_1 and I_2 leads to

$$I_1 = \frac{RV_m}{R^2 + \omega^2 L^2} \qquad I_2 = \frac{\omega L V_m}{R^2 + \omega^2 L^2}$$

Thus, the forced response is obtained:

$$i(t) = \frac{RV_m}{R^2 + \omega^2 L^2} \cos \omega t + \frac{\omega L V_m}{R^2 + \omega^2 L^2} \sin \omega t \tag{2}$$

This expression is slightly cumbersome, however, and a clearer picture of the response can be obtained by expressing the response as a single sinusoid or cosinusoid with a phase angle. Let us select the cosine function in anticipation of the method presented in the following chapter:

$$i(t) = A \cos (\omega t - \theta) \tag{3}$$

At least two methods of obtaining the values of A and θ should suggest themselves. We might substitute Eq. (3) directly in the original differential equation, or we could simply equate the two solutions, Eqs. (2) and (3). Let us select the latter method—since the former makes an excellent problem for the end of the chapter—and equate Eqs. (2) and (3) after expanding the function $\cos (\omega t - \theta)$:

$$A \cos \theta \cos \omega t + A \sin \theta \sin \omega t = \frac{RV_m}{R^2 + \omega^2 L^2} \cos \omega t + \frac{\omega L V_m}{R^2 + \omega^2 L^2} \sin \omega t$$

Thus, again collecting the coefficients of cos ωt and sin ωt and setting them equal to zero, we find

$$A\cos\theta = \frac{RV_m}{R^2 + \omega^2 L^2} \qquad A\sin\theta = \frac{\omega L V_m}{R^2 + \omega^2 L^2}$$

To find A and θ, we divide one equation by the other:

$$\frac{A\sin\theta}{A\cos\theta} = \tan\theta = \frac{\omega L}{R}$$

and also square both equations and add the results:

$$A^2\cos^2\theta + A^2\sin^2\theta = A^2 = \frac{R^2 V_m^2}{(R^2+\omega^2 L^2)^2} + \frac{\omega^2 L^2 V_m^2}{(R^2+\omega^2 L^2)^2} = \frac{V_m^2}{R^2+\omega^2 L^2}$$

Hence,

$$\theta = \tan^{-1}\frac{\omega L}{R}$$

and

$$A = \frac{V_m}{\sqrt{R^2 + \omega^2 L^2}}$$

The alternative form of the forced response therefore becomes

$$i(t) = \frac{V_m}{\sqrt{R^2 + \omega^2 L^2}}\cos\left(\omega t - \tan^{-1}\frac{\omega L}{R}\right) \qquad (4)$$

The electrical characteristics of the response $i(t)$ should now be considered. The amplitude of the response is proportional to the amplitude of the forcing function; if it were not, the linearity concept would have to be discarded. The amplitude of the response also decreases as R, L, or ω is increased, but not proportionately. This is confirmed by the differential equation, for an increase in R, L, or di/dt requires a decrease in current amplitude if the source-voltage amplitude is not changed. The current is seen to lag the applied voltage by $\tan^{-1}(\omega L/R)$, an angle between 0 and 90°. When $\omega = 0$ or $L = 0$, the current must be in phase with the voltage; since the former situation is direct current and the latter provides a resistive circuit, the result is expected. If $R = 0$, the current lags the voltage by 90°; then $v_s = L(di/dt)$, and the derivative-integral relationship between the sine and cosine indicates the validity of the 90° phase difference. In an inductor alone, then, the current lags the voltage by exactly 90°, if the passive sign convention is satisfied. In a similar manner we can show that the current through a capacitor *leads* the voltage across it by 90°.

The applied voltage and the resultant current are both plotted on the same ωt axis in Fig. 7-5, but arbitrary current and voltage ordinates are assumed. The fact that the current lags the voltage in this simple RL circuit is now visually apparent. We shall later be able to show easily that this phase relationship is found at the input of every circuit composed only of inductors and resistors.[2]

[2] Once upon a time the symbol E (for electromotive force) was used to designate a voltage. Then every student learned the phrase "ELI the ICE man" as a reminder that voltage leads current in an inductive circuit while current leads voltage in a capacitive circuit, and they all lived happily ever after.

Figure 7-5

The applied sinusoidal forcing function (color) and the resultant sinusoidal current response (black) of the series RL circuit shown in Fig. 7-4.

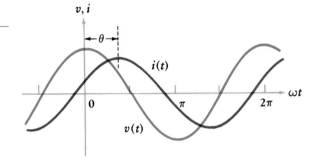

The phase difference between the current and voltage depends upon the ratio of the quantity ωL to R. We call ωL the *inductive reactance* of the inductor; it is measured in ohms, and it is a measure of the opposition which is offered by the inductor to the passage of a sinusoidal current. Much more will be said about reactance in the following chapter.

Let us see how we can apply the results of this general analysis to a specific circuit that is not just a simple series loop.

Example 7-1   Find the current i_L in the circuit shown in Fig. 7-6a.

Figure 7-6

(*a*) The circuit for Example 7-1. The current i_L is desired. (*b*) The Thévenin equivalent is desired at terminals a and b.

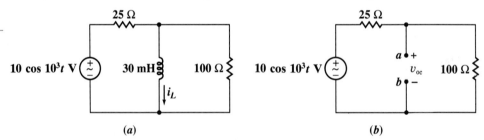

         (*a*)   (*b*)

Solution: Although this circuit has a sinusoidal source and a single inductor, it contains two resistors and is not a single loop. In order to apply the results of the preceding analysis, we first need to look at the remainder of the circuit from the terminals of the inductor. We therefore seek the Thévenin equivalent as viewed from terminals a and b in Fig. 7-6b.

The open-circuit voltage v_{oc} is

$$v_{oc} = (10 \cos 10^3 t) \frac{100}{100 + 25} = 8 \cos 10^3 t \quad \text{V}$$

Since there are no dependent sources in sight, we find R_{th} by killing the independent source and calculating the resistance of the passive network, $R_{th} = (25 \times 100)/(25 + 100) = 20 \ \Omega$.

Now we do have a series RL circuit, with $L = 30$ mH, $R_{th} = 20 \ \Omega$, and a source voltage of 8 cos $10^3 t$ V. Thus,

$$i_L = \frac{8}{\sqrt{20^2 + (10^3 \times 30 \times 10^{-3})^2}} \cos \left(10^3 t - \tan^{-1} \frac{30}{20} \right)$$

$$= 0.222 \cos (10^3 t - 56.3°) \quad \text{A} \quad \blacksquare$$

The method by which we found the sinusoidal steady-state response for the general series RL circuit was not a trivial problem. We might think of the analytical complications as arising through the presence of the inductor; if both the passive elements had been resistors, the analysis would have been ridiculously easy, even with the sinusoidal forcing function present. The reason the analysis would be so easy results from the simple voltage-current relationship specified by Ohm's law. The voltage-current relationship for an inductor is not as simple, however; instead of solving an algebraic equation, we were faced with a nonhomogeneous differential equation. It would be quite impractical to analyze every circuit by the method described in the example, and in the following chapter we shall therefore take steps to simplify the analysis. Our result will be an algebraic relationship between sinusoidal current and sinusoidal voltage for inductors and capacitors as well as resistors, and we shall be able to produce a set of *algebraic* equations for a circuit of any complexity. The constants and the variables in the equations will be complex numbers rather than real numbers, but the analysis of any circuit in the sinusoidal steady state becomes almost as easy as the analysis of a similar resistive circuit.

7-4. Element values in the circuit of Fig. 7-4 are $R = 30\ \Omega$ and $L = 0.5$ H. If the source voltage is $v_s = 100 \cos 50t$ V, find (a) $i(t)$; (b) $v_L(t)$, the voltage across L, + reference at the top; (c) $v_R(t)$, the voltage across R, + reference at the left; (d) the power being supplied by the source at $t = 0.5$ s.
Ans: $2.56 \cos (50t - 39.8°)$ A; $64.0 \cos (50t + 50.2°)$ V; $76.8 \cos (50t - 39.8°)$ V; 171.8 W

Drill Problems

7-5. Let $v_s = 40 \cos 8000t$ V in the circuit of Fig. 7-7. Use Thévenin's theorem where it will do the most good, and find the value at $t = 0$ for (a) i_L; (b) v_L; (c) i_R; (d) i_s.　　　*Ans:* 18.71 mA; 15.97 V; 5.32 mA; 24.0 mA

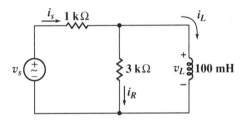

Figure 7-7

See Drill Prob. 7-5.

1 A sine wave, $f(t)$, is zero and increasing at $t = 2.1$ ms, and the succeeding positive maximum of 8.5 occurs at $t = 7.5$ ms. Express the wave in the form $f(t)$ equals (a) $C_1 \sin (\omega t + \phi)$, where ϕ is positive, as small as possible, and in degrees; (b) $C_2 \cos (\omega t + \beta)$, where β has the smallest possible magnitude and is in degrees; (c) $C_3 \cos \omega t + C_4 \sin \omega t$.

Problems

2 (a) If $-10 \cos \omega t + 4 \sin \omega t = A \cos (\omega t + \phi)$, where $A > 0$ and $-180° < \phi \le 180°$, find A and ϕ. (b) If $200 \cos (5t + 130°) = F \cos 5t + G \sin 5t$, find F and G. (c) Find three values of t, $0 \le t \le 1$ s, at which $i(t) = 5 \cos 10t - 3 \sin 10t = 0$. (d) In what time interval between $t = 0$ and $t = 10$ ms is $10 \cos 100\pi t \ge 12 \sin 100\pi t$?

3 Given the two sinusoidal waveforms, $f(t) = -50 \cos \omega t - 30 \sin \omega t$ and $g(t) = 55 \cos \omega t - 15 \sin \omega t$, find (a) the amplitude of each, and (b) the phase angle by which $f(t)$ leads $g(t)$.

4 Carry out the exercise threatened in the text by substituting the assumed current response of Eq. (3), $i(t) = A \cos (\omega t - \theta)$, directly in the differential equation,

$L(di/dt) + Ri = V_m \cos \omega t$, to show that values for A and θ are obtained which agree with Eq. (4).

5 Let $v_s = 20 \cos 500t$ V in the circuit of Fig. 7-8. After simplifying the circuit a little, find $i_L(t)$.

Figure 7-8

See Prob. 5.

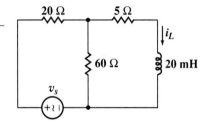

6 If $i_s = 0.4 \cos 500t$ A in the circuit shown in Fig. 7-9, simplify the circuit until it is in the form of Fig. 7-4 and then find (a) $i_L(t)$; (b) $i_x(t)$.

Figure 7-9

See Prob. 6.

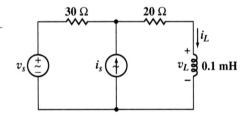

7 A sinuoidal voltage source $v_s = 100 \cos 10^5 t$ V, a 500-Ω resistor, and an 8-mH inductor are in series. Determine those instants of time, $0 \le t < \frac{1}{2}T$, at which zero power is being: (a) delivered to the resistor; (b) delivered to the inductor; (c) generated by the source.

8 In the circuit of Fig. 7-10, let $v_s = 3 \cos 10^5 t$ V and $i_s = 0.1 \cos 10^5 t$ A. After making use of superposition and Thévenin's theorem, find the instantaneous values of i_L and v_L at $t = 10$ μs.

Figure 7-10

See Prob. 8.

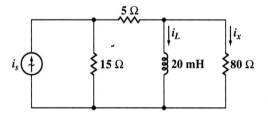

9 Find $i_L(t)$ in the circuit shown in Fig. 7-11.

Figure 7-11

See Prob. 9.

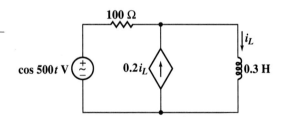

10 Both voltage sources in the circuit of Fig. 7-12 are given by $120 \cos 120 \pi t$ V. (a) Find an expression for the instantaneous energy stored in the inductor, and (b) use it to find the average value of the stored energy.

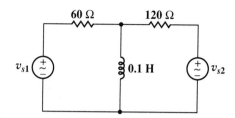

Figure 7-12

See Probs. 10 and 11.

11 In the circuit of Fig. 7-12, the voltage sources are v_{s1} = 120 cos 400t V and v_{s2} = 180 cos 200t V. Find the downward inductor current.

12 Assume that the op-amp in Fig. 7-13 is ideal (R_i = ∞, R_o = 0, and A = ∞). Note also that the integrator input has two signals applied to it, $-V_m$ cos ωt and v_{out}. If the product R_1C_1 is set equal to the ratio L/R in the circuit of Fig. 7-4, show that v_{out} equals the voltage across R (+ reference on the left) in Fig. 7-4.

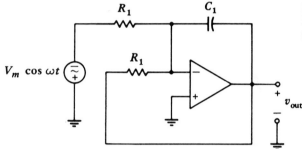

Figure 7-13

See Prob. 12.

13 A voltage source V_m cos ωt, a resistor R, and a capacitor C are all in series. (a) Write an integrodifferential equation in terms of the loop current i and then differentiate it to obtain the differential equation for the circuit. (b) Assume a suitable general form for the forced response $i(t)$, substitute it in the differential equation, and determine the exact form of the forced response.

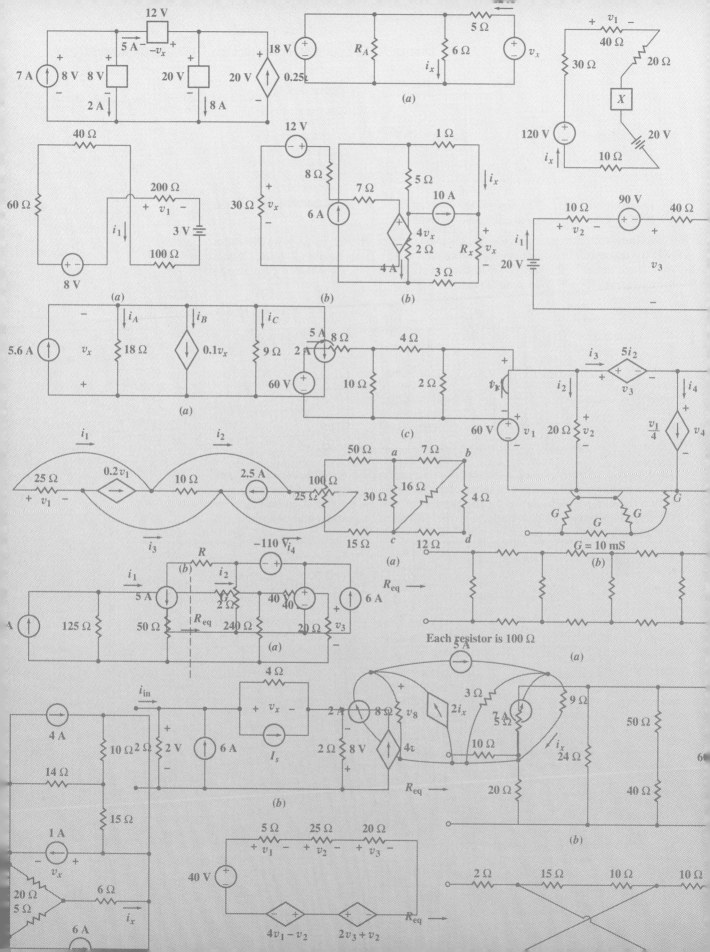

The Phasor Concept

Throughout the earlier portions of our study of circuit analysis, we devoted our entire attention to the resistive circuit. However, we might remember that we were often promised that those methods which we were applying to resistive circuits would later prove applicable to circuits containing inductors and capacitors as well. In this chapter we shall lay the descriptive groundwork which will make this prediction come true. We shall develop a method for representing a sinusoidal forcing function or a sinusoidal response by a complex-number symbolism called a *phasor transform,* or simply a *phasor.* This is nothing more than a number which, by specifying both the amplitude and the phase angle of a sinusoid, characterizes that sinusoid just as completely as if it were expressed as an analytical function of time. By working with phasors, rather than with derivatives and integrals of sinusoids as we did in the preceding chapter, we shall effect a truly remarkable simplification in the steady-state sinusoidal analysis of general *RLC* circuits. The simplification should become apparent toward the end of this chapter.

The use of a mathematical transformation to simplify a problem should not be a new idea to us. For example, we are all familiar with the procedure of using logarithms to simplify calculator-less multiplication and division. In order to multiply several numbers together, we first determined the logarithm of each of the numbers, or "transformed" the numbers into an alternative mathematical description. We might now describe that operation as obtaining the "logarithmic transform." We then added all the logarithms to obtain the logarithm of the desired product. Then, finally, we found the antilogarithm, a process which might be termed an "inverse transformation"; the antilogarithm was our desired answer. Our solution carried us from the domain of everyday numbers to the logarithmic domain, and back again.

Other familiar examples of transform operations may be found in the alternative representations of a circle as a mathematical equation, as a geometric figure on a rectangular-coordinate plane, or merely as a set of three numbers, where it is understood that the first is the *x*-coordinate value of the center, the second is the *y*-coordinate value, and the third is the magnitude of the radius. Each of the three representations contains exactly the same information, and once the rules of the transformations are laid down in analytic geometry, we find no difficulty in passing from the algebraic domain to the geometric domain or to the "domain of the ordered triplet."

Few other transforms with which we are familiar provide the simplification that can be achieved with the phasor concept.

8-2

The complex forcing function

We are now ready to think about applying a complex forcing function (that is, one that has both a real and an imaginary part) to an electrical network.[1] It may seem strange, but we shall find that the use of complex quantities in sinusoidal steady-state analysis leads to methods which are much simpler than those involving purely real quantities. We should expect a complex forcing function to produce a complex response; we might even suspect, and suspect correctly, that the real part of the forcing function will produce the real part of the response, while the imaginary portion of the forcing function will result in the imaginary portion of the response. Our goal in this section is to prove, or at least demonstrate, that these suspicions are correct.

Let us first discuss the problem in rather general terms, thus indicating the method by which we might prove our assertions if we were to construct a general network and analyze it by means of a set of simultaneous equations. In Fig. 8-1, a sinusoidal source

$$V_m \cos (\omega t + \theta) \tag{1}$$

is connected to a general network, which we shall arbitrarily assume to be passive in order to avoid complicating our use of the superposition principle later. A current response in some other branch of the network is to be determined. The parameters appearing in (1) are all real quantities.

Figure 8-1

The sinusoidal forcing function $V_m \cos (\omega t + \theta)$ produces the steady-state sinusoidal response $I_m \cos (\omega t + \phi)$.

The discussion in Chap. 7 of the method whereby the response to a sinusoidal forcing function may be determined, through the assumption of a sinusoidal form with arbitrary amplitude and arbitrary phase angle, shows that the response may be represented by

$$I_m \cos (\omega t + \phi) \tag{2}$$

A sinusoidal forcing function always produces a sinusoidal forced response of the same frequency in a linear circuit.

Now let us change our time reference by shifting the phase of the forcing function by 90°, or changing the instant that we call $t = 0$. Thus, the forcing function

$$V_m \cos (\omega t + \theta - 90°) = V_m \sin (\omega t + \theta) \tag{3}$$

when applied to the same network will produce a corresponding response

$$I_m \cos (\omega t + \phi - 90°) = I_m \sin (\omega t + \phi) \tag{4}$$

We must next depart from physical reality by applying an imaginary forcing function, one which cannot be applied in the laboratory but can be applied mathematically.

---

[1] Appendix 4 defines the complex number and related terms, describes the arithmetic operations on complex numbers, and develops Euler's identity and the exponential and polar forms.

We construct an imaginary source very simply; it is only necessary to multiply the source, as it appears in Eq. (3), by j, the imaginary operator. We thus apply

$$jV_m \sin (\omega t + \theta) \tag{5}$$

What is the response? If we had doubled the source, then the principle of linearity would require that we double the response; multiplication of the forcing function by a constant k would result in the multiplication of the response by the same constant k. The fact that this constant is the imaginary operator j does not destroy this relationship, even though our earlier definition and discussion of linearity did not specifically include complex constants. It is now more realistic to conclude that it did not specifically *exclude* them, for the entire discussion is equally applicable if all the constants in the equations are complex. The response to the imaginary source of (5) is thus

$$jI_m \sin (\omega t + \phi) \tag{6}$$

The imaginary source and response are indicated in Fig. 8-2.

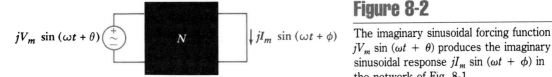

$$jV_m \sin (\omega t + \theta) \qquad N \qquad \downarrow jI_m \sin (\omega t + \phi)$$

Figure 8-2

The imaginary sinusoidal forcing function $jV_m \sin (\omega t + \theta)$ produces the imaginary sinusoidal response $jI_m \sin (\omega t + \phi)$ in the network of Fig. 8-1.

We have applied a real source and obtained a real response; we have also applied an imaginary source and obtained an imaginary response. Now we may use the superposition theorem to find the response to that complex forcing function which is the sum of the real and imaginary forcing functions. The applicability of superposition, of course, is guaranteed by the linearity of the circuit and does not depend on the form of the forcing functions. Thus, the sum of the forcing functions of (1) and (5),

$$V_m \cos (\omega t + \theta) + jV_m \sin (\omega t + \theta) \tag{7}$$

must therefore produce a response which is the sum of (2) and (6),

$$I_m \sin (\omega t + \phi) + jI_m \sin (\omega t + \phi) \tag{8}$$

The complex source and response may be represented more simply by applying Euler's identity. The source of (7) thus becomes

$$V_m e^{j(\omega t + \theta)} \tag{9}$$

and the response of (8) is

$$I_m e^{j(\omega t + \phi)} \tag{10}$$

The complex source and response are illustrated in Fig. 8-3.

There are several important conclusions to be drawn from this general example. A real, an imaginary, or a complex forcing function will produce a

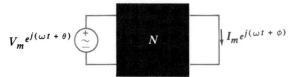

$$V_m e^{j(\omega t + \theta)} \qquad N \qquad \downarrow I_m e^{j(\omega t + \phi)}$$

Figure 8-3

The complex forcing function $V_m e^{j(\omega t + \theta)}$ produces the complex response $I_m e^{j(\omega t + \phi)}$ in the network of Fig. 8-1.

real, an imaginary, or a complex response, respectively. Moreover, a complex forcing function may be considered, by the use of Euler's identity and the superposition theorem, as the sum of a real and an imaginary forcing function; thus the real part of the complex response is produced by the real part of the complex forcing function, while the imaginary part of the response is caused by the imaginary part of the complex forcing function.

Instead of applying a real forcing function to obtain the desired real response, we apply a complex forcing function whose real part is the given real forcing function; we obtain a complex response whose real part is the desired real response. Through this procedure, the integrodifferential equations describing the steady-state response of a circuit will become simple algebraic equations.

Let us try out this idea on the simple RL series circuit shown in Fig. 8-4. The real source $V_m \cos \omega t$ is applied; the real response $i(t)$ is desired.

Figure 8-4

A simple circuit in the sinusoidal steady state is to be analyzed by the application of a complex forcing function.

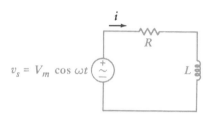

We first construct the complex forcing function which, upon the application of Euler's identity, yields the given real forcing function. Since

$$\cos \omega t = \text{Re } e^{j\omega t}$$

the necessary complex source is

$$V_m e^{j\omega t}$$

The complex response which results is expressed in terms of an unknown amplitude I_m and an unknown phase angle ϕ:

$$I_m e^{j(\omega t + \phi)}$$

Writing the differential equation for this particular circuit,

$$Ri + L\frac{di}{dt} = v_s$$

we insert our complex expressions for v_s and i:

$$RI_m e^{j(\omega t + \phi)} + L\frac{d}{dt}(I_m e^{j(\omega t + \phi)}) = V_m e^{j\omega t}$$

take the indicated derivative:

$$RI_m e^{j(\omega t + \phi)} + j\omega L I_m e^{j(\omega t + \phi)} = V_m e^{j\omega t}$$

and obtain a complex *algebraic* equation. In order to determine the value of I_m and ϕ, we divide throughout by the common factor $e^{j\omega t}$:

$$RI_m e^{j\phi} + j\omega L I_m e^{j\phi} = V_m \qquad (11)$$

factor the left side:

$$I_m e^{j\phi}(R + j\omega L) = V_m$$

rearrange:

$$I_m e^{j\phi} = \frac{V_m}{R + j\omega L}$$

and identify I_m and ϕ by expressing the right side of the equation in exponential or polar form:

$$I_m e^{j\phi} = \frac{V_m}{\sqrt{R^2 + \omega^2 L^2}} e^{j[-\tan^{-1}(\omega L/R)]} \qquad (12)$$

Thus,

$$I_m = \frac{V_m}{\sqrt{R^2 + \omega^2 L^2}}$$

and

$$\phi = -\tan^{-1}\frac{\omega L}{R}$$

The complex response is given by Eq. (12). Since I_m and ϕ are readily identified, we can write the expression for $i(t)$ immediately. However, if we feel like using a more rigorous approach, we may obtain the real response $i(t)$ by reinserting the $e^{j\omega t}$ factor on both sides of Eq. (12) and taking the real part, obtained by applying Euler's omnipotent formula. Thus,

$$i(t) = I_m \cos(\omega t + \phi)$$

$$= \frac{V_m}{\sqrt{R^2 + \omega^2 L^2}} \cos\left(\omega t - \tan^{-1}\frac{\omega L}{R}\right)$$

which agrees with the response obtained for this same circuit in Eq. (4) in the previous chapter.

Now let us try out these ideas on a numerical example.

Example 8-1   Find the complex voltage across the series combination of a 500-Ω resistor and a 95-mH inductor if the complex current $8e^{j3000t}$ mA flows through the two elements in series.

Solution: We could determine the result by simply plugging the given data into the results of the preceding paragraphs, but instead let us work through the example using the specified numerical values. If we let the unknown complex voltage have an unknown amplitude V_m and an unknown phase angle ϕ, then, at the given frequency of 3000 rad/s, we may express this voltage as

$$V_m e^{j(3000t+\phi)}$$

and equate it to the sum of the resistor and inductor voltages:

$$V_m e^{j(3000t+\phi)} = (500)0.008e^{j3000t} + (0.095)\frac{d(0.008e^{j3000t})}{dt}$$

Taking the derivative, we have

$$V_m e^{j(3000t+\phi)} = 4e^{j3000t} + j2.28e^{j3000t}$$

When we collect these expressions and factor out the exponential term, we have

$$V_m e^{j\phi} = 4 + j2.28$$

But, since

$$4 + j2.28 = 4.60e^{j29.7°}$$

we see that $V_m = 4.60$ V and $\phi = 29.7°$, so that the desired voltage is

$$4.60e^{j(3000t+29.7°)} \quad \text{V}$$

If we are asked for the real response, we need only take the real part of the complex response:

$$\text{Re} \, (4.60e^{j(3000t+29.7°)}) = 4.60 \cos{(3000t + 29.7°)} \quad \text{V} \qquad ▪$$

Although we have successfully worked a sinusoidal steady-state problem by applying a complex forcing function and obtaining a complex response, we have not taken advantage of the full power of the complex representation. In order to do so, we must carry the concept of the complex source or response one additional step and define the quantity called a *phasor*. This we hasten to do in the following section.

Drill Problems

8-1. (If you have trouble working this drill problem, you should study Appendix 4.) Evaluate and express the result in rectangular form: (a) $[(2/\underline{30°})(5/\underline{-110°})](1 + j2)$; (b) $(5/\underline{-200°}) + 4/\underline{20°}$. Evaluate and express the result in polar form: (c) $(2 - j7)/(3 - j1)$; (d) $8 - j4 + [(5/\underline{80°})/(2/\underline{20°})]$.
Ans: $21.4 - j6.38$; $-0.940 + j3.08$; $2.30/\underline{-55.6°}$; $9.43/\underline{-11.22°}$

8-2. If the use of the passive sign convention is specified, find the (a) complex voltage that results when the complex current $4e^{j800t}$ A is applied to the series combination of a 1-mF capacitor and a 2-Ω resistor; (b) complex current that results when the complex voltage $100e^{j2000t}$ V is applied to the parallel combination of a 10-mH inductor and a 50-Ω resistor.
Ans: $9.43e^{j(800t-32.0°)}$ V; $5.39e^{j(2000t-68.2°)}$ A

8-3

The phasor

A sinusoidal current or voltage *at a given frequency* is characterized by only two parameters, an amplitude and a phase angle. The complex representation of the voltage or current is also characterized by these same two parameters. For example, the assumed sinusoidal form of the current response in Example 8-1 was

$$I_m \cos{(\omega t + \phi)}$$

and the corresponding representation of this current in complex form is

$$I_m e^{j(\omega t + \phi)}$$

Once I_m and ϕ are specified, the current is exactly defined. Throughout any linear circuit operating in the sinusoidal steady state at a single frequency ω, every current or voltage may be characterized completely by a knowledge of its amplitude and phase angle. Moreover, the complex representation of every voltage and current will contain the same factor $e^{j\omega t}$. The factor is actually superfluous. Since it is the same for every quantity, it contains no useful information. Of course, the value of the frequency may be recognized by inspecting one of these factors, but it is a lot simpler to write down the value of the frequency near the circuit diagram once and for all and avoid carrying redundant information throughout the solution. Thus, we could simplify the voltage source and the current response of the example by representing them concisely as

$$V_m \quad \text{or} \quad V_m e^{j0°}$$

and

$$I_m e^{j\phi}$$

These complex quantities are usually written in polar form rather than exponential form in order to achieve a slight additional saving of time and effort. Thus, the source voltage

$$v(t) = V_m \cos \omega t$$

we now express in complex form as

$$V_m \underline{/0°}$$

and the current response

$$i(t) = I_m \cos (\omega t + \phi)$$

becomes

$$I_m \underline{/\phi}$$

This abbreviated complex representation is called a *phasor*.[2] Let us review the steps by which a real sinusoidal voltage or current is transformed into a phasor, and then we shall be able to define a phasor more meaningfully and to assign a symbol to represent it.

A real sinusoidal current

$$i(t) = I_m \cos (\omega t + \phi)$$

is expressed as the real part of a complex quantity by an equivalent form of Euler's identity

$$i(t) = \text{Re} \left(I_m e^{j(\omega t + \phi)} \right)$$

We then represent the current as a complex quantity by dropping the instruction Re, thus adding an imaginary component to the current without affecting the real component; further simplification is achieved by suppressing the factor $e^{j\omega t}$:

$$\mathbf{I} = I_m e^{j\phi}$$

and writing the result in polar form:

$$\mathbf{I} = I_m \underline{/\phi}$$

This abbreviated complex representation is the *phasor representation;* phasors are complex quantities and hence are printed in boldface type. Capital letters are used for the phasor representation of an electrical quantity because the phasor is not an instantaneous function of time; it contains only amplitude and phase information. We recognize this difference in viewpoint by referring to $i(t)$ as a *time-domain representation* and terming the phasor $\mathbf{I}$ a *frequency-domain representation*. It should be noted that the frequency-domain expression of a current or voltage does not explicitly include the frequency; however, we might

---

[2] Contrary to popular opinion, the phasor was not invented by *Star Trek*'s Captain Kirk.

think of the frequency as being so fundamental in the frequency domain that it is emphasized by its omission.[3]

The process by which we change $i(t)$ into $\mathbf{I}$ is called a *phasor transformation* from the time domain to the frequency domain. The mathematical steps of the time-domain to frequency-domain transformation are as follows:

1 Given the sinusoidal function $i(t)$ in the time domain, write $i(t)$ as a cosine wave with a phase angle. For example, $\sin \omega t$ should be written as $\cos (\omega t - 90°)$.
2 Express the cosine wave as the real part of a complex quantity by using Euler's identity.
3 Drop Re.
4 Suppress $e^{j\omega t}$.

In practice, we find it very easy to jump directly from the first step to the last by extracting the amplitude and phase angle of the cosine wave from its time-domain expression. The four steps are listed for technical fulfillment only.

Let us consider a simple example.

Example 8-2 Let us transform the time-domain voltage $v(t) = 100 \cos (400t - 30°)$ into the frequency domain.

Solution: The time-domain expression is already in the form of a cosine wave with a phase angle, and the rigorous time-domain to frequency-domain transformation is accomplished by taking the real part of the complex representation,

$$v(t) = \text{Re } (100e^{j(400t-30°)})$$

and dropping Re and suppressing $e^{j\omega t} = e^{j400t}$:

$$\mathbf{V} = 100 \, \underline{/-30°}$$

However, it is much simpler to identify the 100 and $-30°$ in the time-domain cosine representation, and to write $\mathbf{V} = 100 \, \underline{/-30°}$ directly.

In a similar fashion, the time-domain current

$$i(t) = 5 \sin (377t + 150°)$$

transforms into the phasor

$$\mathbf{I} = 5 \, \underline{/60°}$$

after it is written as a cosine by subtracting 90° from the argument. ∎

Before we consider the analysis of circuits in the sinusoidal steady state through the use of phasors, it is necessary to learn how to shift our transformation smoothly into reverse to return to the time domain from the frequency domain. The process is exactly the reverse of the previous sequence. Thus, the detailed steps in the frequency-domain to time-domain transformation are as follows:

1 Given the phasor current $\mathbf{I}$ in polar form in the frequency domain, write the complex expression in exponential form.

---

[3] Very little local mail in this country includes "U.S.A." in its address.

2 Reinsert (multiply by) the factor $e^{j\omega t}$.

3 Replace the real-part operator Re.

4 Obtain the time-domain representation by applying Euler's identity. The resultant cosine-wave expression may be changed to a sine wave, if desired, by increasing the argument by 90°.

Once again, we should be able to bypass the mathematics and write the desired time-domain expression by utilizing the amplitude and phase angle of the polar form. Thus, given the phasor voltage

$$\mathbf{V} = 115\ \underline{/-45°}$$

we would write the time-domain equivalent directly:

$$v(t) = 115\ \cos{(\omega t - 45°)}$$

As a sine wave, $v(t)$ could be written

$$v(t) = 115\ \sin{(\omega t + 45°)}$$

In anticipation of the methods of applying phasors in the analysis of the sinusoidal steady-state circuit, we may treat ourselves to a quick preview by returning to the example of the *RL* series circuit. A number of steps after writing the applicable differential equation, we arrived at Eq. (11), rewritten here:

$$RI_m e^{j\phi} + j\omega L I_m e^{j\phi} = V_m$$

If we substitute phasors for the current:

$$\mathbf{I} = I_m\ \underline{/\phi}$$

and the voltage:

$$\mathbf{V} = V_m\ \underline{/0°}$$

we obtain

$$R\mathbf{I} + j\omega L\mathbf{I} = \mathbf{V}$$

or

$$(R + j\omega L)\mathbf{I} = \mathbf{V} \qquad (13)$$

a complex algebraic equation in which the current and voltage are expressed in phasor form. This equation is only slightly more complicated than Ohm's law for a single resistor. The next time we analyze this circuit, we shall begin with Eq. (13).

Drill Problems

8-3. Transform each of the following functions of time into phasor form: (*a*) $-5\sin{(580t - 110°)}$; (*b*) $3\cos{600t} - 5\sin{(600t + 110°)}$; (*c*) $8\cos{(4t - 30°)} + 4\sin{(4t - 100°)}$.   *Ans:* $5\underline{/-20°}$; $2.41\underline{/-134.8°}$; $4.46\underline{/-47.9°}$

8-4. Let $\omega = 2000$ rad/s and $t = 1$ ms. Find the instantaneous value of each of the currents given here in phasor form: (*a*) $j10$ A; (*b*) $20 + j10$ A; (*c*) $20 + j(10\underline{/20°})$ A.   *Ans:* -9.09 A; -17.42 A; -15.44 A

8-4

Phasor relationships for R, L, and C

Now that we are able to transform into and out of the frequency domain, we can proceed to our simplification of sinusoidal steady-state analysis by establishing the relationship between the phasor voltage and phasor current for each of the three passive elements. We shall begin with the defining equation for each of the elements—a time-domain relationship—and then let both the current and the voltage become complex quantities. After suppressing $e^{j\omega t}$ throughout the equation, the desired relationship between the phasor voltage and phasor current will become apparent.

The resistor provides the simplest case. In the time domain, as indicated by Fig. 8-5a, the defining equation is

$$v(t) = Ri(t) \tag{14}$$

Figure 8-5

A resistor and its associated voltage and current in (a) the time domain, $v = Ri$; and (b) the frequency domain, $\mathbf{V} = R\mathbf{I}$.

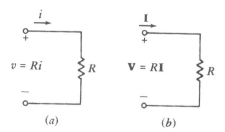

(a)   (b)

Now let us apply the complex voltage

$$V_m e^{j(\omega t + \theta)} = V_m \cos(\omega t + \theta) + jV_m \sin(\omega t + \theta) \tag{15}$$

and assume the complex current response

$$I_m e^{j(\omega t + \phi)} = I_m \cos(\omega t + \phi) + jI_m \sin(\omega t + \phi) \tag{16}$$

and obtain

$$V_m e^{j(\omega t + \theta)} = RI_m e^{j(\omega t + \phi)}$$

By dividing throughout by $e^{j\omega t}$ (or suppressing $e^{j\omega t}$ on both sides of the equation), we find

$$V_m e^{j\theta} = RI_m e^{j\phi}$$

or, in polar form,

$$V_m \underline{/\theta} = RI_m \underline{/\phi}$$

But $V_m \underline{/\theta}$ and $I_m \underline{/\phi}$ merely represent the general voltage and current phasors $\mathbf{V}$ and $\mathbf{I}$. Thus,

$$\mathbf{V} = R\mathbf{I} \tag{17}$$

The voltage-current relationship in phasor form for a resistor has the same form as the relationship between the time-domain voltage and current. The defining equation in phasor form is illustrated in Fig. 8-5b. The equality of the angles θ and ϕ is apparent, and the current and voltage are thus in phase.

As an example of the use of both the time-domain and frequency-domain relationships, let us assume that a voltage of $8 \cos(100t - 50°)$ V is across a 4-Ω resistor. Working in the time domain, we find that the current must be

$$i(t) = \frac{v(t)}{R} = 2 \cos(100t - 50°) \quad \text{A}$$

The phasor form of the same voltage is $8\underline{/-50°}$ V, and therefore

$$\mathbf{I} = \frac{\mathbf{V}}{R} = 2\underline{/-50°} \text{ A}$$

If we transform this answer back to the time domain, it is evident that the same expression for the current is obtained.

It is apparent that there is no saving in time or effort when a resistive circuit is analyzed in the frequency domain. As a matter of fact, if it is necessary to transform a given time-domain source to the frequency domain and then translate the desired response back to the time domain, we should be much better off working completely in the time domain. This definitely does not apply to any circuit containing resistance and either inductance or capacitance, unless the complexity of the problem calls forth the digital computer, a machine that is quite indifferent to tedious calculations.

Let us now turn to the inductor. The time-domain network is shown in Fig. 8-6a, and the defining equation, a time-domain expression, is

$$v(t) = L\frac{di(t)}{dt} \tag{18}$$

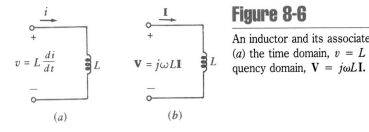

(a)   (b)

Figure 8-6

An inductor and its associated voltage and current in (a) the time domain, $v = L\, di/dt$; and (b) the frequency domain, $\mathbf{V} = j\omega L\mathbf{I}$.

After substituting the complex voltage equation (15) and complex current equation (16) in Eq. (18), we have

$$V_m e^{j(\omega t+\theta)} = L\frac{d}{dt}I_m e^{j(\omega t+\phi)}$$

Taking the indicated derivative:

$$V_m e^{j(\omega t+\theta)} = j\omega L I_m e^{j(\omega t+\phi)}$$

and suppressing $e^{j\omega t}$:

$$V_m e^{j\theta} = j\omega L I_m e^{j\phi}$$

we obtain the desired phasor relationship

$$\mathbf{V} = j\omega L\mathbf{I} \tag{19}$$

The time-domain differential equation (18) has become the algebraic equation (19) in the frequency domain. The phasor relationship is indicated in Fig. 8-6b. Note that the angle of the factor $j\omega L$ is exactly $+90°$ and that $\mathbf{I}$ must therefore lag $\mathbf{V}$ by 90° in an inductor.

Example 8-3   To illustrate our mastery of the phasor relationship, let us apply the voltage $8\underline{/-50°}$ V at a frequency $\omega = 100$ rad/s to a 4-H inductor and determine the phasor current.

Solution: We make use of the expression we just obtained for the inductor,

$$\mathbf{I} = \frac{\mathbf{V}}{j\omega L} = \frac{8\underline{/-50°}}{j100(4)} = -j0.02\;\underline{/-50°}$$

or

$$\mathbf{I} = 0.02\underline{/-140°}\quad\text{A}$$

If we express this current in the time domain, it becomes

$$i(t) = 0.02\cos(100t - 140°)\quad\text{A}$$

This response is also easily obtained by working entirely in the time domain; it is not so easily obtained if resistance or capacitance is combined with the inductance.   ∎

The final element we must consider is the capacitor. The definition of capacitance, a familiar time-domain expression, is

$$i(t) = C\frac{dv(t)}{dt} \tag{20}$$

The equivalent expression in the frequency domain is obtained once more by letting $v(t)$ and $i(t)$ be the complex quantities of Eqs. (15) and (16), taking the indicated derivative, suppressing $e^{j\omega t}$, and recognizing the phasors $\mathbf{V}$ and $\mathbf{I}$. It is

$$\mathbf{I} = j\omega C\mathbf{V} \tag{21}$$

Thus, $\mathbf{I}$ leads $\mathbf{V}$ by 90° in a capacitor. This, of course, does not mean that a current response is present one-quarter of a period earlier than the voltage that caused it! We are studying steady-state response, and we find that the current maximum is caused by the increasing voltage that occurs 90° earlier than the voltage maximum.

If the phasor voltage $8\underline{/-50°}$ V is applied to a 4-F capacitor at $\omega = 100$ rad/s, the phasor current is

$$\mathbf{I} = j100(4)(8\underline{/-50°}) = 3200\underline{/40°}\quad\text{A}$$

The amplitude of the current is tremendous, but the assumed size of the capacitor is also unrealistic. If a 4-F capacitor were constructed of two flat plates separated 1 mm in air, each plate would have the area of about 85 000 football fields.[4]

The time-domain and frequency-domain representations are compared in Figs. 8-7a and b. We have now obtained the $\mathbf{V}$-$\mathbf{I}$ relationships for the three passive elements. These results are summarized in Table 8-1, where the time-domain v-i expressions and the frequency-domain $\mathbf{V}$-$\mathbf{I}$ relationships are shown in adjacent columns for the three circuit elements. All the phasor equations are algebraic. Each is also linear, and the equations relating to inductance and capacitance bear a great similarity to Ohm's law. In fact, we shall indeed *use* them as we use Ohm's law.

Before we do so, we must show that phasors satisfy Kirchhoff's two laws. Kirchhoff's voltage law in the time domain is

$$v_1(t) + v_2(t) + \cdots + v_N(t) = 0$$

[4] Including both end zones.

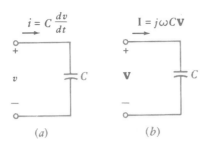

Figure 8-7

(a) The time-domain and (b) the frequency-domain relationships between capacitor current and voltage.

Time domain		Frequency domain	
	$v = Ri$	$\mathbf{V} = \mathbf{RI}$	
	$v = L\dfrac{di}{dt}$	$\mathbf{V} = j\omega L\mathbf{I}$	
	$v = \dfrac{1}{C}\int i\,dt$	$\mathbf{V} = \dfrac{1}{j\omega C}\mathbf{I}$	

Table 8-1

A comparison and summary of the relationships between v and i in the time domain and $\mathbf{V}$ and $\mathbf{I}$ in the frequency domain for R, L, and C.

We now use Euler's identity to replace each real voltage by the complex voltage having the same real part, suppress $e^{j\omega t}$ throughout, and obtain

$$\mathbf{V}_1 + \mathbf{V}_2 + \cdots + \mathbf{V}_N = 0$$

Kirchhoff's current law is shown to hold for phasor currents by a similar argument.

Now let us look briefly at the series RL circuit that we have considered several times before. The circuit is shown in Fig. 8-8, and a phasor current and several phasor voltages are indicated. We may obtain the desired response, a time-domain current, by first finding the phasor current. The method is similar to that used in analyzing our first single-loop resistive circuit. From Kirchhoff's voltage law,

$$\mathbf{V}_R + \mathbf{V}_L = \mathbf{V}_s$$

and the recently obtained **V-I** relationships for the elements, we have

$$R\mathbf{I} + j\omega L\mathbf{I} = \mathbf{V}_s$$

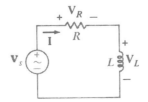

Figure 8-8

The series RL circuit with a phasor voltage applied.

The phasor current is then found in terms of the source voltage $\mathbf{V}_s$:

$$\mathbf{I} = \frac{\mathbf{V}_s}{R + j\omega L}$$

Let us select a source-voltage amplitude of V_m and phase angle of $0°$; the latter merely represents the simplest possible choice of a reference. Thus,

$$\mathbf{I} = \frac{V_m \underline{/0°}}{R + j\omega L}$$

The current may be transformed to the time domain by first writing it in polar form:

$$\mathbf{I} = \frac{V_m}{\sqrt{R^2 + \omega^2 L^2}} \underline{/-\tan^{-1}(\omega L/R)}$$

and then following the familiar sequence of steps to obtain in a very simple manner the same result we obtained the "hard way" as Eq. (4) of Sec. 7-3.

Drill Problem

8-5. In Fig. 8-9, let $\omega = 1200$ rad/s, $\mathbf{I}_C = 1.2\underline{/28°}$ A, and $\mathbf{I}_L = 3\underline{/53°}$ A. Find (a) $\mathbf{I}_s$; (b) $\mathbf{V}_s$; (c) $i_R(t)$.

Ans: $2.33\underline{/-31.0°}$ A; $34.9\underline{/74.5°}$ V; $3.99 \cos(1200t + 17.42°)$ A

Figure 8-9

See Drill Prob. 8-5.

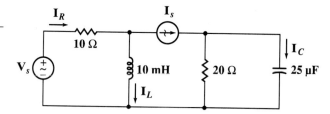

8-5

Impedance

The current-voltage relationships for the three passive elements in the frequency domain are (assuming that the passive sign convention is satisfied)

$$\mathbf{V} = R\mathbf{I} \qquad \mathbf{V} = j\omega L\mathbf{I} \qquad \mathbf{V} = \frac{\mathbf{I}}{j\omega C}$$

If these equations are written as phasor-voltage–phasor-current ratios

$$\frac{\mathbf{V}}{\mathbf{I}} = R \qquad \frac{\mathbf{V}}{\mathbf{I}} = j\omega L \qquad \frac{\mathbf{V}}{\mathbf{I}} = \frac{1}{j\omega C}$$

we find that these ratios are simple functions of the element values—and frequency also, in the case of inductance and capacitance. We treat these ratios in the same manner that we treat resistances, with the exception that they are complex quantities and all algebraic manipulations must be those appropriate for complex numbers.

Let us define the ratio of the phasor voltage to the phasor current as *impedance*, symbolized by the letter $\mathbf{Z}$. The impedance is a complex quantity having the dimensions of ohms. Impedance is not a phasor and cannot be transformed to the time domain by multiplying by $e^{j\omega t}$ and taking the real part. Instead, we think of an inductor as being represented in the time domain by its inductance L and in the frequency domain by its impedance $j\omega L$. A capacitor in the time domain has a capacitance C; in the frequency domain, it has an impedance $1/j\omega C$. Impedance is a part of the frequency domain and not a concept which is a part of the time domain.

The validity of Kirchhoff's two laws in the frequency domain enables it to be easily demonstrated that impedances may be combined in series and parallel by the same rules we have already established for resistances. For example, at $\omega = 10^4$ rad/s, a 5-mH inductor in series with a 100-μF capacitor may be replaced by the single impedance which is the sum of the individual impedances. The impedance of the inductor is

$$\mathbf{Z}_L = j\omega L = j50 \ \Omega$$

the impedance of the capacitor is

$$\mathbf{Z}_C = \frac{1}{j\omega C} = -j1 \ \Omega$$

and the impedance of the series combination is therefore

$$\mathbf{Z}_{eq} = j50 - j1 = j49 \ \Omega$$

The impedance of inductors and capacitors is a function of frequency, and this equivalent impedance is thus applicable only at the single frequency at which it was calculated, $\omega = 10\,000$. At $\omega = 5000$, $\mathbf{Z}_{eq} = j23 \ \Omega$.

The *parallel* combination of these same two elements at $\omega = 10^4$ yields an impedance which is the product divided by the sum,

$$\mathbf{Z}_{eq} = \frac{(j50)(-j1)}{j50 - j1} = \frac{50}{j49} = -j1.020 \ \Omega$$

At $\omega = 5000$, the parallel equivalent is $-j2.17 \ \Omega$.

The complex number or quantity representing impedance may be expressed in either polar or rectangular form. In polar form, an impedance, such as $\mathbf{Z} = 100\underline{/-60°}$, is described as having an impedance magnitude of 100 Ω and a phase angle of $-60°$. The same impedance in rectangular form, $50 - j86.6$, is said to have a *resistive component,* or *resistance,* of 50 Ω and a *reactive component,* or *reactance,* of -86.6 Ω. The resistive component is the real part of the impedance, and the reactive component is the imaginary component of the impedance, including sign, but of course excluding the imaginary operator.

It is important to note that the resistive component of the impedance is not necessarily equal to the resistance of the resistor which is present in the network. For example, a 10-Ω resistor and a 5-H inductor in series at $\omega = 4$ have an equivalent impedance $\mathbf{Z} = 10 + j20$ Ω, or, in polar form, $22.4\underline{/63.4°}$ Ω. In this case, the resistive component of the impedance is equal to the resistance of the series resistor because the network is a simple series network. However, if these same two elements are placed in parallel, the equivalent impedance is $10(j20)/(10 + j20)$, or $8 + j4$ Ω. The resistive component of the impedance is now 8 Ω.

No special symbol is assigned for impedance magnitude or phase angle. A general form for an impedance in polar form might be

$$\mathbf{Z} = |\mathbf{Z}|\underline{/\theta}$$

In rectangular form, the resistive component is represented by R and the reactive component by X. Thus,

$$\mathbf{Z} = R + jX$$

Let us now use the impedance concept to analyze an *RLC* circuit.

Example 8-4 Find the current $i(t)$ in the circuit shown in Fig. 8-10a.

Solution: The circuit is shown in the time domain, and the time-domain current response $i(t)$ is required. However, the analysis should be carried out in the frequency domain. We therefore begin by drawing a frequency-domain circuit. The source is transformed to the frequency domain, becoming $40\underline{/-90°}$ V; the response is transformed to the frequency domain, being represented as $\mathbf{I}$; and the impedances of the inductor and capacitor, determined at $\omega = 3000$, are $j1$ and $-j2$ kΩ, respectively. The frequency-domain circuit is shown in Fig. 8-10b.

Figure 8-10

(a) An *RLC* circuit for which the sinusoidal forced response $i(t)$ is desired. (b) The frequency-domain equivalent of the given circuit at $\omega = 3000$ rad/s.

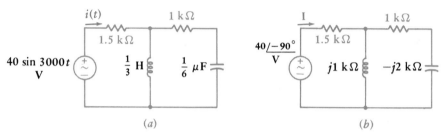

(a) (b)

The equivalent impedance offered to the source is now calculated:

$$\mathbf{Z}_{eq} = 1.5 + \frac{(j1)(1 - j2)}{j1 + 1 - j2} = 1.5 + \frac{2 + j1}{1 - j1}$$

$$= 1.5 + \frac{2 + j1}{1 - j1}\frac{1 + j1}{1 + j1} = 1.5 + \frac{1 + j3}{2}$$

$$= 2 + j1.5 = 2.5\underline{/36.9°} \text{ k}\Omega$$

The phasor current is thus

$$\mathbf{I} = \frac{\mathbf{V}_s}{\mathbf{Z}_{eq}} = \frac{40\underline{/-90°}}{2.5\underline{/36.9°}} = 16\underline{/-126.9°} \text{ mA}$$

Upon transforming the current to the time domain, the desired response is obtained:

$$i(t) = 16 \cos{(3000t - 126.9°)} \qquad \text{mA}$$

If the capacitor current is desired, current division should be applied in the frequency domain. ∎

Before we begin to write great numbers of equations in the time domain or in the frequency domain, it is most important that we shun the construction of equations which are partly in the time domain, partly in the frequency domain, and wholly incorrect. One clue that a faux pas of this type has been committed is the sight of both a complex number and a t in the same equation, except when the factor $e^{j\omega t}$ is also present. And, since $e^{j\omega t}$ plays a much bigger role in derivations than it is in applications, it is pretty safe to say that students who find they have just created an equation containing j and t, or $\underline{/\quad}$ and t, have created a monster that they and the world would be better off without.

For example, two equations back, we saw

$$\mathbf{I} = \frac{\mathbf{V}_s}{\mathbf{Z}_{eq}} = \frac{40\underline{/-90°}}{2.5\underline{/36.9°}} = 16\underline{/-126.9°} \text{ mA}$$

Please do not try anything like the following:

$$i(t) = \frac{40 \sin 3000t}{2.5 / 36.9°} \qquad \text{(No! No! No!)}$$

or

$$i(t) = \frac{40 \sin 3000t}{2 + j1.5} \qquad \text{(also, No!)}$$

8-6. With reference to the network shown in Fig. 8-11*a* , find the input imped-   **Drill Problems**
ance $\mathbf{Z}_{\text{in}}$ that would be measured between terminals: (*a*) *a* and *g*; (*b*) *b* and *g*;
(*c*) *a* and *b*.　　　　　　　*Ans:* 2.81 + *j*4.49 Ω; 1.798 − *j*1.124 Ω; 0.1124 − *j*3.82 Ω

8-7. In the frequency-domain circuit of Fig. 8-11*ḃ*, find (*a*) $\mathbf{I}_1$; (*b*) $\mathbf{I}_2$; (*c*) $\mathbf{I}_3$.
　　　　　　　　　　　　　　　　　Ans: 28.3 / 45° A; 20 / 90° A; 20 / 0° A

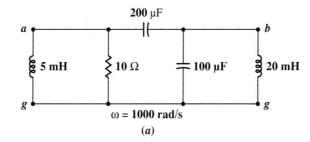

ω = 1000 rad/s

(*a*)

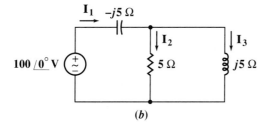

(*b*)

Figure 8-11

(*a*) See Drill Prob. 8-6. (*b*) See Drill Prob. 8-7.

Just as conductance, the reciprocal of resistance, proved to be a useful quantity   **8-6**
in the analysis of resistive circuits, so does the reciprocal of impedance offer
some convenience in the sinusoidal steady-state analysis of a general *RLC*
circuit. We define the *admittance* **Y** of a circuit element as the ratio of phasor   **Admittance**
current to phasor voltage (assuming that the passive sign convention is sat-
isfied):

$$\mathbf{Y} = \frac{\mathbf{I}}{\mathbf{V}}$$

and thus

$$\mathbf{Y} = \frac{1}{\mathbf{Z}}$$

The real part of the admittance is the *conductance G,* and the imaginary part
of the admittance is the *susceptance B.* Thus,

$$\mathbf{Y} = G + jB = \frac{1}{\mathbf{Z}} = \frac{1}{R + jX} \qquad (22)$$

Equation (22) should be scrutinized carefully; it does *not* state that the real part of the admittance is equal to the reciprocal of the real part of the impedance or that the imaginary part of the admittance is equal to the reciprocal of the imaginary part of the impedance.

Admittance, conductance, and susceptance are all measured in siemens. An impedance

$$\mathbf{Z} = 1 - j2 \ \Omega$$

which might be represented, for example, by a 1-Ω resistor in series with a 0.1-μF capacitor at $\omega = 5$ Mrad/s, possesses an admittance

$$\mathbf{Y} = \frac{1}{\mathbf{Z}} = \frac{1}{1 - j2} = \frac{1}{1 - j2} \frac{1 + j2}{1 + j2} = 0.2 + j0.4 \ \text{S}$$

Without stopping to inspect a formal proof, it should be apparent that the equivalent admittance of a network consisting of a number of parallel branches is the sum of the admittances of the individual branches. Thus, the numerical value of the admittance just shown might be obtained from a conductance of 0.2 S in parallel with a positive susceptance of 0.4 S. The former could be represented by a 5-Ω resistor and the latter by a 0.08-μF capacitor at $\omega = 5$ Mrad/s, since the admittance of a capacitor is evidently $j\omega C$.

As a check on our analysis, let us compute the impedance of this latest network, a 5-Ω resistor in parallel with a 0.08-μF capacitor at $\omega = 5$ Mrad/s. The equivalent impedance is

$$\mathbf{Z} = \frac{5(1/j\omega C)}{5 + 1/j\omega C} = \frac{5(-j2.5)}{5 - j2.5} = 1 - j2 \ \Omega$$

as before. These two networks represent only two of an infinite number of different networks which possess this same impedance and admittance at this frequency. They do, however, represent the only two 2-element networks, and thus might be considered to be the two simplest networks having an impedance of $1 - j2 \ \Omega$ and an admittance of $0.2 + j0.4$ S at $\omega = 5 \times 10^6$ rad/s.

The term *immittance*, a combination of the words *impedance* and *admittance*, is sometimes used as a general term for both impedance and admittance. For example, it is evident that a knowledge of the phasor voltage across a known immittance enables the current through the immittance to be calculated.

Drill Problems

8-8. Determine the admittance (in rectangular form) of (*a*) an impedance $\mathbf{Z} = 1000 + j400 \ \Omega$; (*b*) a network consisting of the parallel combination of an 800-Ω resistor, a 1-mH inductor, and a 2-nF capacitor, if $\omega = 1$ Mrad/s; (*c*) a network consisting of the series combination of an 800-Ω resistor, a 1-mH inductor, and a 2-nF capacitor, if $\omega = 1$ Mrad/s.

Ans: $0.862 - j0.345$ mS; $1.125 + j1$ mS; $0.899 - j0.562$ mS

8-9. A 20-μF capacitor is in series with the parallel combination of a resistor R and a 15-mH inductor at $\omega = 1$ krad/s. (*a*) Find the admittance of the network if $R = 80 \ \Omega$. (*b*) If $\mathbf{Y} = G + j25$ mS, find R.   *Ans:* $2.14 + j28.0$ mS; $25.4 \ \Omega$

1 Convert these complex numbers to rectangular form: (a) $5\underline{/-110°}$; (b) $6e^{j160°}$; (c) $(3 + j6)(2\underline{/50°})$. Convert to polar form: (d) $-100 - j40$; (e) $2\underline{/50°} + 3\underline{/-120°}$.

2 Carry out the indicated calculations, and express the result in polar form: (a) $40\underline{/-50°} - 18\underline{/25°}$; ($b$) $3 + \dfrac{2}{j} + \dfrac{2 - j5}{1 + j2}$. Express in rectangular form: (c) $(2.1\underline{/25°})^3$; (d) $0.7e^{j0.3}$.

3 In the circuit of Fig. 8-12a, let i_C be expressed as the complex response $20e^{j(40t+30°)}$ A, and express v_s as a complex forcing function.

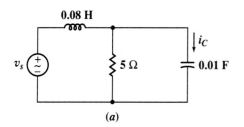

0.08 H

v_s $5\ \Omega$ i_C 0.01 F

(a)

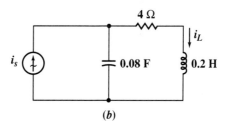

$4\ \Omega$

i_s 0.08 F i_L 0.2 H

(b)

Figure 8-12

(a) See Prob. 3. (b) See Prob. 4.

4 In the circuit of Fig. 8-12b, let i_L be expressed as the complex response $20e^{j(10t+25°)}$ A, and express the source current $i_s(t)$ as a complex forcing function.

5 In a linear network, such as that shown in Fig. 8-1, a sinusoidal source voltage, $v_s = 80 \cos(500t - 20°)$ V, produces the output current $i_{out} = 5 \cos(500t + 12°)$ A. Find i_{out} if v_s equals (a) $40 \cos(500t + 10°)$ V; (b) $40 \sin(500t + 10°)$ V; (c) $40e^{j(500t+10°)}$ V; (d) $(50 + j20)e^{j500t}$ V.

6 Express each of the following currents as a phasor: (a) $12 \sin(400t + 110°)$ A; (b) $-7 \sin 800t - 3 \cos 800t$ A; (c) $4 \cos(200t - 30°) - 5 \cos(200t + 20°)$ A. If $\omega = 600$ rad/s, find the instantaneous value of each of these voltages at $t = 5$ ms: (d) $70\underline{/30°}$ V; (e) $-60 + j40$ V.

7 Let $\omega = 4$ krad/s, and determine the instantaneous value of i_x at $t = 1$ ms if $\mathbf{I}_x$ equals (a) $5\underline{/-80°}$ A; (b) $-4 + j1.5$ A. Express in polar form the phasor voltage $\mathbf{V}_x$ if $v_x(t)$ equals (c) $50 \sin(250t - 40°)$ V; (d) $20 \cos 108t - 30 \sin 108t$ V; (e) $33 \cos(80t - 50°) + 41 \cos(80t - 75°)$ V.

8 The phasor voltages $\mathbf{V}_1 = 10\underline{/90°}$ mV at $\omega = 500$ rad/s and $\mathbf{V}_2 = 8\underline{/90°}$ mV at $\omega = 1200$ rad/s are added together in an op-amp. If the op-amp multiplies this input by a factor of -5, find the output at $t = 0.5$ ms.

9 If $\omega = 500$ rad/s and $\mathbf{I}_L = 2.5\underline{/40°}$ A in the circuit of Fig. 8-13, find $v_s(t)$.

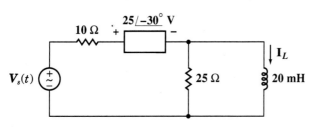

$25\underline{/-30°}$ V

$10\ \Omega$

$\mathbf{V}_s(t)$ $25\ \Omega$ $\mathbf{I}_L$ 20 mH

Figure 8-13

See Prob. 9.

10 Let $\omega = 5$ krad/s in the circuit of Fig. 8-14. Find (a) $v_1(t)$; (b) $v_2(t)$; (c) $v_3(t)$.

Figure 8-14

See Prob. 10.

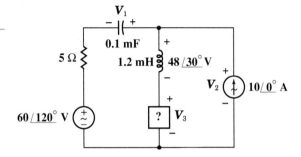

11 A phasor current of $1\underline{/0°}$ A is flowing through the series combination of 1 Ω, 1 H, and 1 F. At what frequency is the amplitude of the voltage across the network twice the amplitude of the voltage across the resistor?

12 Find v_x in the circuit shown in Fig. 8-15.

Figure 8-15

See Probs. 12 and 20.

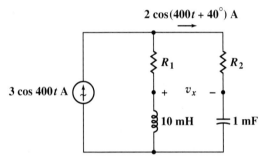

13 A black box with yellow stripes contains two current sources, $\mathbf{I}_{s1}$ and $\mathbf{I}_{s2}$. The output voltage is identified as $\mathbf{V}_{\text{out}}$. If $\mathbf{I}_{s1} = 2\underline{/20°}$ A and $\mathbf{I}_{s2} = 3\underline{/-30°}$ A, then $\mathbf{V}_{\text{out}} = 80\underline{/10°}$ V. However, if $\mathbf{I}_{s1} = \mathbf{I}_{s2} = 4\underline{/40°}$ A, then $\mathbf{V}_{\text{out}} = 90 - j30$ V. Find $\mathbf{V}_{\text{out}}$ if $\mathbf{I}_{s1} = 2.5\underline{/-60°}$ A and $\mathbf{I}_{s2} = 2.5\underline{/60°}$ A.

14 Find $\mathbf{Z}_{\text{in}}$ at terminals a and b in Fig. 8-16 if ω equals (a) 800 rad/s; (b) 1600 rad/s.

Figure 8-16

See Probs. 14 and 16.

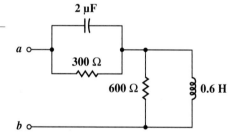

15 Let $\omega = 100$ rad/s in the circuit of Fig. 8-17. Find (a) $\mathbf{Z}_{\text{in}}$; (b) $\mathbf{Z}_{\text{in}}$ if a short circuit is connected from x to y.

Figure 8-17

See Prob. 15.

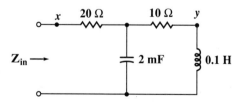

16 If a voltage source $v_s = 120 \cos 800t$ V is connected to terminals a and b in Fig. 8-16 (+ reference at the top), what current flows to the right in the 300-Ω resistance?

17 Find **V** in Fig. 8-18 if the box contains (a) 3 Ω in series with 2 mH; (b) 3 Ω in series with 125 μF; (c) 3 Ω, 2 mH, and 125 μF in series; (d) 3 Ω, 2 mH, and 125 μF in series, but $\omega = 4$ krad/s.

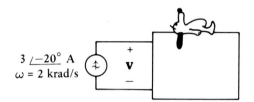

Figure 8-18

See Prob. 17.

$3 \underline{/-20°}$ A
$\omega = 2$ krad/s

V

18 A 10-H inductor, a 200-Ω resistor, and a capacitor C are in parallel. (a) Find the impedance of the parallel combination at $\omega = 100$ rad/s if $C = 20$ μF. (b) If the magnitude of the impedance is 125 Ω at $\omega = 100$ rad/s, find C. (c) At what two values of ω is the magnitude of the impedance equal to 100 Ω if $C = 20$ μF?

19 A 20-mH inductor and a 30-Ω resistor are in parallel. Find the frequency ω at which: (a) $|\mathbf{Z}_{\text{in}}| = 25$ Ω; (b) angle ($\mathbf{Z}_{\text{in}}$) = 25°; (c) Re ($\mathbf{Z}_{\text{in}}$) = 25 Ω; (d) Im ($\mathbf{Z}_{\text{in}}$) = 10 Ω.

20 Find R_1 and R_2 in the circuit of Fig. 8-15.

21 A two-element network has an input impedance of $200 + j80$ Ω at $\omega = 1200$ rad/s. What capacitance C should be placed in parallel with the network to provide an input impedance with (a) zero reactance? (b) a magnitude of 100 Ω?

22 For the network of Fig. 8-19, find $\mathbf{Z}_{\text{in}}$ at $\omega = 4$ rad/s if terminals a and b are (a) open-circuited; (b) short-circuited.

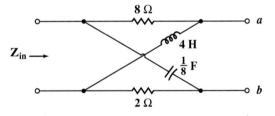

Figure 8-19

See Prob. 22.

8 Ω

a

4 H

$\dfrac{1}{8}$ F

$\mathbf{Z}_{\text{in}} \longrightarrow$

2 Ω

b

23 Find the input admittance $\mathbf{Y}_{ab}$ of the network shown in Fig. 8-20 and draw it as the parallel combination of a resistance R and an inductance L, giving values for R and L if $\omega = 1$ rad/s.

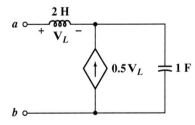

Figure 8-20

See Prob. 23.

2 H

a

+ $\mathbf{V}_L$ −

$0.5\mathbf{V}_L$ 1 F

b

24 A 5-Ω resistance, a 20-mH inductance, and a 2-mF capacitance form a series network having terminals a and b. (a) Work with admittances to determine what size capacitance should be connected between a and b so that $\mathbf{Z}_{\text{in},ab} = R_{\text{in},ab} + j0$ at $\omega = 500$ rad/s. (b) What is $R_{\text{in},ab}$? (c) With your C in place, what is $\mathbf{Y}_{\text{in},ab}$ at $\omega = 1000$ rad/s?

25 In the network shown in Fig. 8-21, find the frequency at which (a) R_{in} = 550 Ω; (b) X_{in} = 50 Ω; (c) G_{in} = 1.8 mS; (d) B_{in} = −150 μS.

Figure 8-21

See Prob. 25.

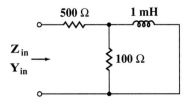

26 Two admittances, Y_1 = 3 + $j4$ mS and Y_2 = 5 + $j2$ mS, are in parallel, and a third admittance, Y_3 = 2 − $j4$ mS, is in series with the parallel combination. If a current I_1 = 0.1$\underline{/30°}$ A is flowing through Y_1, find the magnitude of the voltage across (a) Y_1; (b) Y_2; (c) Y_3; (d) the entire network.

27 The admittance of the parallel combination of a 10-Ω resistance and a 50-μF capacitance at $ω$ = 1 krad/s is the same as the admittance of R_1 and C_1 in series at that frequency. (a) Find R_1 and C_1. (b) Repeat for $ω$ = 2 krad/s.

28 A cartesian coordinate plane contains a horizontal axis on which G_{in} is given in siemens, and a vertical axis along which B_{in} is measured, also in S. Let Y_{in} represent the series combination of a 1-Ω resistor and a 0.1-F capacitor. (a) Find Y_{in}, G_{in}, and B_{in} as functions of $ω$. (b) Locate the coordinate pairs (G_{in}, B_{in}) on the plane at the frequency values $ω$ = 0, 1, 2, 5, 10, 20, and 10^6 rad/s.

29 (SPICE) Let V_s = 100$\underline{/0°}$ V at f = 50 Hz in the circuit shown in Fig. 8-22. Use a SPICE analysis to find the phasor voltage V_x if element X is a (a) 30-Ω resistor; (b) 0.5-H inductor; (c) 50-μF capacitor.

Figure 8-22

See Prob. 29.

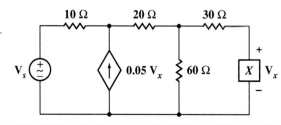

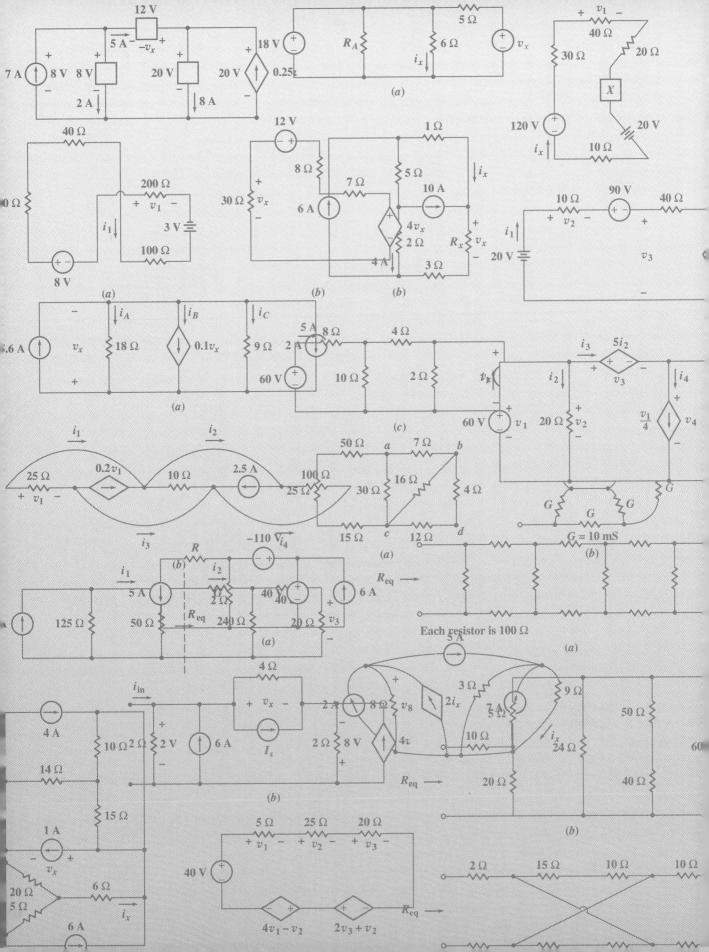

The Sinusoidal Steady-State Response

In Chap. 1 and, particularly, Chap. 2 we learned several methods which are useful in analyzing resistive circuits. No matter how complex the resistive circuit, we are able to determine any desired response by using nodal, mesh, or loop analysis, superposition, source transformations, or Thévenin's or Norton's theorem. Sometimes one method is sufficient, but more often we find it convenient to combine several methods to obtain the response in the most direct manner. We now wish to extend these techniques to the analysis of circuits in the sinusoidal steady state, and we have already seen that impedances combine in the same manner as do resistances. The extension of the techniques of resistive circuit analysis has been promised several times, and we must now find out why the extension is justified and practice its use.

Let us first review the arguments by which we accepted nodal analysis for a purely resistive N-node circuit. After designating a reference node and assigning voltage variables between each of the $N - 1$ remaining nodes and the reference, we applied Kirchhoff's current law to each of these $N - 1$ nodes. The application of Ohm's law to all the resistors then led to $N - 1$ equations in $N - 1$ unknowns if no voltage sources or dependent sources were present; if they were, additional equations were written in accordance with the definitions of the types of sources involved.

Now, is a similar procedure valid in terms of phasors and impedances for the sinusoidal steady state? We already know that both of Kirchhoff's laws are valid for phasors; also, we have an Ohm-like law for the passive elements, $\mathbf{V} = \mathbf{ZI}$. In other words, the laws upon which nodal analysis rests are true for phasors, and we may proceed, therefore, to analyze circuits by nodal techniques in the sinusoidal steady state. It is also evident that mesh- and loop-analysis methods are valid as well.

Let us consider an example of nodal analysis.

Example 9-1 We desire the time-domain node voltages $v_1(t)$ and $v_2(t)$ in the circuit shown in Fig. 9-1.

Solution: In this circuit, each passive element is specified by its impedance, although the analysis might be simplified slightly by using admittance

Figure 9-1

A frequency-domain circuit for which node voltages V_1 and V_2 are identified.

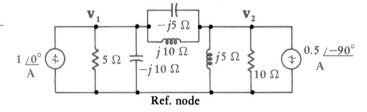

values. Two current sources are given as phasors, and phasor node voltages V_1 and V_2 are indicated. At the left node we apply KCL and $I = V/Z$:

$$\frac{V_1}{5} + \frac{V_1}{-j10} + \frac{V_1 - V_2}{-j5} + \frac{V_1 - V_2}{j10} = 1 + j0$$

At the right node,

$$\frac{V_2 - V_1}{-j5} + \frac{V_2 - V_1}{j10} + \frac{V_2}{j5} + \frac{V_2}{10} = -(-j0.5)$$

Combining terms, we have

$$(0.2 + j0.2)V_1 - j0.1V_2 = 1 \qquad (1)$$

and

$$-j0.1V_1 + (0.1 - j0.1)V_2 = j0.5 \qquad (2)$$

Using determinants to solve Eqs. (1) and (2), we obtain

$$V_1 = \frac{\begin{vmatrix} 1 & -j0.1 \\ j0.5 & (0.1 - j0.1) \end{vmatrix}}{\begin{vmatrix} (0.2 + j0.2) & -j0.1 \\ -j0.1 & (0.1 - j0.1) \end{vmatrix}}$$

$$= \frac{0.1 - j0.1 - 0.05}{0.02 - j0.02 + j0.02 + 0.02 + 0.01}$$

$$= \frac{0.05 - j0.1}{0.05} = 1 - j2 \text{ V}$$

and

$$V_2 = \frac{\begin{vmatrix} 0.2 + j0.2 & 1 \\ -j0.1 & j0.5 \end{vmatrix}}{0.05} = \frac{-0.1 + j0.1 + j0.1}{0.05}$$

$$= -2 + j4 \text{ V}$$

The time-domain solutions are therefore obtained by expressing V_1 and V_2 in polar form:

$$V_1 = 2.24\underline{/-63.4°} \qquad V_2 = 4.47\underline{/116.6°}$$

and passing to the time domain:

$$v_1(t) = 2.24 \cos(\omega t - 63.4°) \qquad \text{V}$$

$$v_2(t) = 4.47 \cos(\omega t + 116.6°) \qquad \text{V}$$

Note that the value of ω would have to be known in order to compute the impedance values given on the circuit diagram. Also, both sources are assumed to operate at the same frequency. ∎

Now let us look at an example of loop or mesh analysis.

With only the right source active, current division helps us to obtain

$$\mathbf{V}_{1R} = (-0.5\underline{/-90°}) \left(\frac{2 + j4}{4 - j2 - j10 + 2 + j4} \right) (4 - j2) = -1$$

Summing, then

$$\mathbf{V}_1 = 2 - j2 - 1 = 1 - j2 \text{ V}$$

which agrees with our previous result.   ∎

We might also see whether or not Thévenin's theorem can help the analysis of this circuit (Fig. 9-5).

Example 9-4   Let us determine the Thévenin equivalent faced by the $-j10\text{-}\Omega$ impedance.

Solution: The open-circuit voltage (+ reference to left) is

$$\mathbf{V}_{oc} = (1\underline{/0°})(4 - j2) + (0.5\underline{/-90°})(2 + j4)$$

$$= 4 - j2 + 2 - j1 = 6 - j3 \text{ V}$$

The impedance of the inactive circuit, as viewed from the load terminals, is simply the sum of the two remaining impedances. Hence,

$$\mathbf{Z}_{th} = 6 + j2 \ \Omega$$

Thus, when we reconnect the circuit, the current directed from node 1 toward node 2 through the $-j10\text{-}\Omega$ load is

$$\mathbf{I}_{12} = \frac{6 - j3}{6 + j2 - j10} = 0.6 + j0.3 \text{ A}$$

Subtracting this from the left source current, the downward current through the $(4 - j2)\text{-}\Omega$ branch is found:

$$\mathbf{I}_1 = 1 - 0.6 - j0.3 = 0.4 - j0.3 \text{ A}$$

and, thus,

$$\mathbf{V}_1 = (0.4 - j0.3)(4 - j2) = 1 - j2 \text{ V}$$ ∎

We might have been cleverer and used Norton's theorem on the three elements on the right, assuming that our chief interest is in $\mathbf{V}_1$. Source transformations can also be used repeatedly to simplify the circuit. Thus, all the shortcuts and tricks that arose in Chaps. 1 and 2 are available for circuit analysis in the frequency domain. The slight additional complexity that is apparent now arises from the necessity of using complex numbers and not from any more involved theoretical considerations.

Finally, we should also be pleased to hear that these same techniques will be applicable to the forced response of circuits driven by damped sinusoidal forcing functions, exponential forcing functions, and forcing functions having a *complex frequency* in general. Thus, we shall meet these same techniques again in Chap. 12.

Drill Problems

9-3. If superposition is used on the circuit of Fig. 9-3, find $\mathbf{V}_1$ with (*a*) only the $20\underline{/0°}$-mA source operating; (*b*) only the $50\underline{/-90°}$-mA source operating.

Ans: $0.1951 - j0.556$ V; $0.780 + j0.976$ V

9-4. For the circuit shown in Fig. 9-6, find the (*a*) open-circuit voltage $\mathbf{V}_{ab}$; (*b*) downward current in a short circuit between *a* and *b*; (*c*) Thévenin-equivalent impedance $\mathbf{Z}_{ab}$.   *Ans:* $16.77\underline{/-33.4°}$ V; $2.60 + j1.500$ A; $2.5 - j5$ Ω

Figure 9-6

See Drill Prob. 9-4.

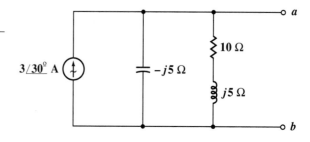

9-4

Phasor diagrams

The *phasor diagram* is a name given to a sketch in the complex plane showing the relationships of the phasor voltages and phasor currents throughout a specific circuit. It also provides a graphical method for solving certain problems which may be used to check more exact analytical methods. It proves to be of considerable help in simplifying the analytical work in certain symmetrical polyphase problems (Chap. 11) by enabling us to recognize the symmetry and to apply it in a helpful way. In the following chapter we shall encounter similar diagrams which display the complex power relationships in the sinusoidal steady state. The use of other complex planes will appear in connection with complex frequency in Chap. 12.

We are already familiar with the use of the complex plane in the graphical identification of complex numbers and in their addition and subtraction. Since phasor voltages and currents are complex numbers, they may also be identified as points in a complex plane. For example, the phasor voltage $\mathbf{V}_1 = 6 + j8 = 10\underline{/53.1°}$ V is identified on the complex voltage plane shown in Fig. 9-7. The

Figure 9-7

A simple phasor diagram shows the single voltage phasor $\mathbf{V}_1 = 6 + j8 = 10\underline{/53.1°}$ V.

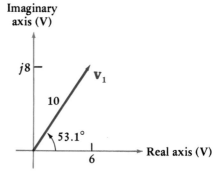

axes are the real voltage axis and the imaginary voltage axis; the voltage $\mathbf{V}_1$ is located by an arrow drawn from the origin. Since addition and subtraction are particularly easy to perform and display on a complex plane, it is apparent that phasors may be easily added and subtracted in a phasor diagram. Multiplication and division result in the addition and subtraction of angles and a change of amplitude; the latter is less clearly shown, since the amplitude change depends on the amplitude of each phasor and on the scale of the diagram. Figure 9-8a shows the sum of $\mathbf{V}_1$ and a second phasor voltage $\mathbf{V}_2 = 3 - j4 = 5\underline{/-53.1°}$ V, and Fig. 9-8b shows the current $\mathbf{I}_1$, which is the product of $\mathbf{V}_1$ and the admittance $\mathbf{Y} = 1 + j1$ S.

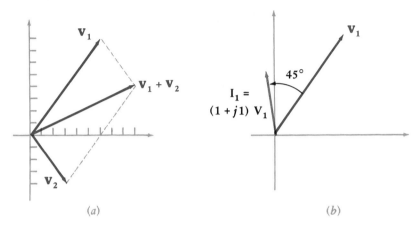

(a)   (b)

Figure 9-8

(a) A phasor diagram showing the sum of $\mathbf{V}_1 = 6 + j8$ V and $\mathbf{V}_2 = 3 - j4$ V, $\mathbf{V}_1 + \mathbf{V}_2 = 9 + j4 = 9.85\underline{/24.0°}$ V. (b) The phasor diagram shows $\mathbf{V}_1$ and $\mathbf{I}_1$, where $\mathbf{I}_1 = \mathbf{Y}\mathbf{V}_1$ and $\mathbf{Y} = 1 + j1$ S.

This last phasor diagram shows both current and voltage phasors on the same complex plane; it is understood that each will have its own amplitude scale, but a common angle scale. For example, a phasor voltage 1 cm long might represent 100 V, while a phasor current 1 cm long could indicate 3 mA.

The phasor diagram also offers an interesting interpretation of the time-domain to frequency-domain transformation, since the diagram may be interpreted from either the time- or the frequency-domain viewpoint. Up to this time, it is obvious that we have been using the frequency-domain interpretation, because we have been showing phasors directly on the phasor diagram. However, let us proceed to a time-domain viewpoint by first showing the phasor voltage $\mathbf{V} = V_m\underline{/\alpha}$ as sketched in Fig. 9-9a. In order to transform $\mathbf{V}$ to the time

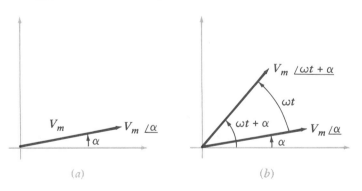

(a)   (b)

Figure 9-9

(a) The phasor voltage $V_m\underline{/\alpha}$. (b) The complex voltage $V_m\underline{/\omega t + \alpha}$ is shown as a phasor at a particular instant of time. This phasor leads $V_m\underline{/\alpha}$ by ωt rad.

domain, the next necessary step is the multiplication of the phasor by $e^{j\omega t}$; thus we now have the complex voltage $V_m e^{j\alpha} e^{j\omega t} = V_m\underline{/\omega t + \alpha}$. This voltage may also be interpreted as a phasor, one which possesses a phase angle that increases linearly with time. On a phasor diagram it therefore represents a rotating line segment, the instantaneous position being ωt rad ahead (counterclockwise) of $V_m\underline{/\alpha}$. Both $V_m\underline{/\alpha}$ and $V_m\underline{/\omega t + \alpha}$ are shown on the phasor diagram of Fig. 9-9b.

The passage to the time domain is now completed by taking the real part of $V_m\underline{/\omega t + \alpha}$. The real part of this complex quantity, however, is merely the projection of $V_m\underline{/\omega t + \alpha}$ on the real axis.

In summary, then, the frequency-domain phasor appears on the phasor diagram, and the transformation to the time domain is accomplished by allowing the phasor to rotate in a counterclockwise direction at an angular velocity of ω rad/s and then visualizing the projection on the real axis. It is helpful to think of the arrow representing the phasor **V** on the phasor diagram as the photographic snapshot, taken at $\omega t = 0$, of a rotating arrow whose projection on the real axis is the instantaneous voltage $v(t)$.

Let us now construct the phasor diagrams for several simple circuits. The series RLC circuit shown in Fig. 9-10a has several different voltages associated with it, but only a single current. The phasor diagram is constructed most easily by employing the single current as the reference phasor. Let us arbitrarily select $\mathbf{I} = I_m\underline{/0°}$ and place it along the real axis of the phasor diagram, Fig. 9-10b. The resistor, capacitor, and inductor voltages may next be calculated and placed on the diagram, where the 90° phase relationships stand out clearly. The sum of these three voltages is the source voltage, and for this circuit, which is in the resonant condition[1] where $\mathbf{Z}_C = -\mathbf{Z}_L$, the source voltage and resistor voltage are equal. The total voltage across the resistor and inductor or resistor and capacitor is easily obtained from the phasor diagram.

Figure 9-10

(a) A series RLC circuit shown in the frequency domain. (b) The associated phasor diagram, drawn with the single mesh current as the reference phasor.

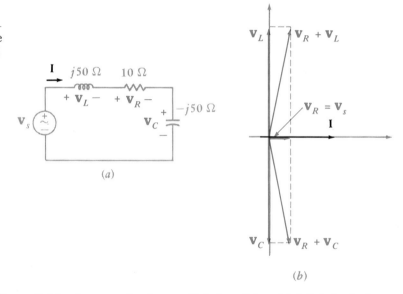

(a)

(b)

Figure 9-11a shows a simple parallel circuit in which it is logical to use the single voltage between the two nodes as a reference phasor. Suppose that $\mathbf{V} = 1\underline{/0°}$ V. The resistor current, $\mathbf{I}_R = 0.2\underline{/0°}$ A, is in phase with this voltage, and the capacitor current, $\mathbf{I}_C = j0.1$ A, leads the reference voltage by 90°. After these two currents are added to the phasor diagram, shown as Fig. 9-11b, they may be summed to obtain the source current. The result is $\mathbf{I}_s = 0.2 + j0.1$ A.

If the source current were specified initially as, for example, $1\underline{/0°}$ A and the node voltage is not initially known, it is still convenient to begin construction of the phasor diagram by assuming a node voltage—say, $\mathbf{V} = 1\underline{/0°}$ V once again—and using it as the reference phasor. The diagram is then completed as before, and the source current which flows as a result of the assumed node voltage is again found to be $0.2 + j0.1$ A. The true source current is $1\underline{/0°}$ A,

---

[1] Resonance will be defined in Chap. 13.

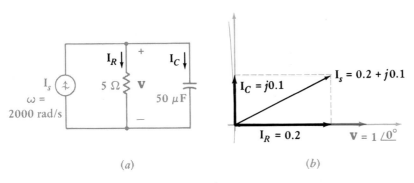

Figure 9-11

(a) A parallel RC circuit. (b) The phasor diagram for this circuit; the node voltage $\mathbf{V}$ is used as a convenient reference phasor.

however, and thus the true node voltage is obtained by multiplying the assumed node voltage by $1\underline{/0°}/(0.2 + j0.1)$; the true node voltage is therefore $4 - j2$ V $= \sqrt{20}\ \underline{/-26.6°}$. The assumed voltage leads to a phasor diagram which differs from the true phasor diagram by a change of scale (the assumed diagram is smaller by a factor of $1/\sqrt{20}$) and an angular rotation (the assumed diagram is rotated counterclockwise through 26.6°).

Phasor diagrams are usually very simple to construct, and most sinusoidal steady-state analyses will be more meaningful if such diagrams are included. Additional examples of the use of phasor diagrams will appear frequently throughout the remainder of our study.

Drill Problems

9-5. Select some convenient value for $\mathbf{V}$, such as $10\underline{/0°}$ V, in the circuit of Fig. 9-12a, and construct a phasor diagram showing $\mathbf{I}_R$, $\mathbf{I}_L$, and $\mathbf{I}_C$. By combining these currents, determine the angle by which $\mathbf{I}_s$ leads (a) $\mathbf{I}_R$; (b) $\mathbf{I}_C$; (c) $\mathbf{I}_x$.
Ans: 45°; −45°; 71.6°

9-6. Select some convenient reference value for $\mathbf{I}_C$ in Fig. 9-12b, draw a phasor diagram showing $\mathbf{V}_R$, $\mathbf{V}_2$, $\mathbf{V}_1$, and $\mathbf{V}_s$, and measure the ratio of the lengths of (a) $\mathbf{V}_s$ to $\mathbf{V}_1$; (b) $\mathbf{V}_1$ to $\mathbf{V}_2$; (c) $\mathbf{V}_s$ to $\mathbf{V}_R$.
Ans: 1.90; 1.00; 2.12

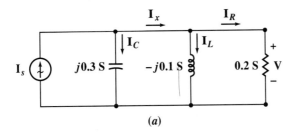

(a)

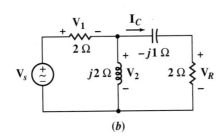

(b)

Figure 9-12

(a) See Drill Prob. 9-5. (b) See Drill Prob. 9-6.

9-5

Response as a function of ω

We will now consider methods of obtaining and presenting the response of a circuit with sinusoidal excitation as a function of the radian frequency ω. With the possible exception of the 60-Hz power area in which frequency is a constant and the load is the variable, sinusoidal frequency response is extremely important in almost every branch of electrical engineering as well as in related areas, such as the theory of mechanical vibrations or automatic control.

Let us suppose that we have a circuit which is excited by a single source $\mathbf{V}_s = V_s\underline{/\theta}$. This phasor voltage may also be transformed into the time-domain source voltage $V_s \cos(\omega t + \theta)$. Somewhere in the circuit exists the desired response, say, a current $\mathbf{I}$. As we know, this phasor response is a complex number, and its value cannot be specified in general without the use of two quantities: either a real part and an imaginary part, or an amplitude and a phase angle. The latter pair of quantities is more useful and more easily determined experimentally, and is the information which we shall obtain analytically as a function of frequency. The data may be presented as two curves, the magnitude of the response as a function of ω and the phase angle of the response as a function of ω. We often normalize the curves by plotting the magnitude of the current-voltage ratio and the phase angle of the current-voltage ratio versus ω. It is evident that an alternative description of the resultant curves is the magnitude and phase angle of an admittance as a function of frequency. The admittance might be an input admittance or, if the current and voltage are measured at different locations in a circuit, a *transfer* admittance. A normalized voltage response to a current source may be similarly presented as the magnitude and phase angle of an input or transfer impedance versus ω. Other possibilities are voltage-voltage ratios (voltage gains) or current-current ratios (current gains). Let us consider the details of this process by thoroughly discussing several examples.

For the first example, we select the series RL circuit. The phasor voltage $\mathbf{V}_s$ is therefore applied to this simple circuit, and the phasor current $\mathbf{I}$ (leaving the positively marked end of the voltage source) is selected as the desired response. We are dealing with the forced response only, and the familiar phasor methods enable the current to be obtained:

$$\mathbf{I} = \frac{\mathbf{V}_s}{R + j\omega L}$$

Let us immediately express this result in normalized form as a ratio of current to voltage, that is, as an input admittance:

$$\mathbf{Y} = \frac{\mathbf{I}}{\mathbf{V}_s}$$

or

$$\mathbf{Y} = \frac{1}{R + j\omega L} \tag{3}$$

If we like, we may consider the admittance as the current produced by a source voltage $1\underline{/0°}$ V. The magnitude of the response is

$$|\mathbf{Y}| = \frac{1}{\sqrt{R^2 + \omega^2 L^2}} \tag{4}$$

while the angle of the response is found to be

$$\text{ang } \mathbf{Y} = -\tan^{-1}\frac{\omega L}{R} \tag{5}$$

Equations (4) and (5) are the analytical expressions for the magnitude and phase angle of the response as functions of ω; we now desire to present this same information graphically.

First consider the magnitude curve. It is important to note that we are plotting the absolute value of some quantity versus ω, and the entire curve must therefore lie *above* the ω axis. The response curve is constructed by noting that the value of the response at zero frequency is $1/R$, the initial slope is zero, and the response approaches zero as frequency approaches infinity; the graph of the magnitude of the response as a function of ω is shown in Fig. 9-13a.

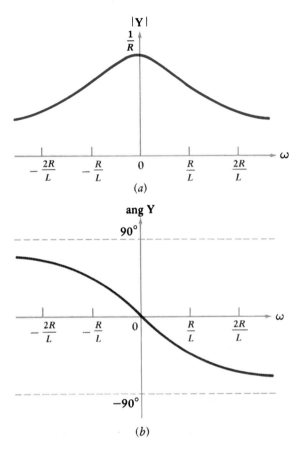

(a)

(b)

Figure 9-13

(a) The magnitude of $\mathbf{Y} = \mathbf{I}/\mathbf{V}_s$ and (b) the angle of $\mathbf{Y}$ are plotted as functions of ω for a series RL circuit with sinusoidal excitation.

For the sake of generality and completeness, the response is shown for both positive and negative values of frequency; the symmetry results from the fact that Eq. (4) indicates that $|\mathbf{Y}|$ is unchanged when ω is replaced by $(-\omega)$. The physical interpretation of a negative radian frequency, such as $\omega = -100$ rad/s, depends on the time-domain function, and it may always be obtained by inspection of the time-domain expression. Suppose, for example, that we consider the voltage $v(t) = 50 \cos (\omega t + 30°)$. If $\omega = 100$, the voltage is $v(t) = 50 \cos (100t + 30°)$, but if $\omega = -100$, $v(t) = 50 \cos (-100t + 30°)$ or $50 \cos (100t - 30°)$. These voltages have different values at $t = 1$ ms, for example. Any sinusoidal response may be treated in a similar manner.

The second part of the response, the phase angle of $\mathbf{Y}$ versus ω, is an inverse tangent function. The tangent function itself is quite familiar, and we should have no difficulty turning that curve on its side. Drawing in asymptotes of $+90°$ and $-90°$ is helpful. The response curve is shown in Fig. 9-13b. The points at which $\omega = \pm R/L$ are marked on both the magnitude and phase curves. At these frequencies the magnitude is 0.707 times the maximum magnitude at zero frequency and the phase angle has a magnitude of 45°. At the frequency at which the admittance magnitude is 0.707 times its maximum value, the current magnitude is 0.707 times its maximum value, and the average power supplied by the source is 0.707^2, or 0.5, times its maximum value. It is not very strange that $\omega = R/L$ is identified as a *half-power frequency*.

As a second example, let us select a parallel LC circuit driven by a sinusoidal current source, as illustrated in Fig. 9-14a. The voltage response $\mathbf{V}$ is easily obtained:

$$\mathbf{V} = \mathbf{I}_s \frac{(j\omega L)(1/j\omega C)}{j\omega L - j(1/\omega C)}$$

and it may be expressed as an input impedance

$$\mathbf{Z} = \frac{\mathbf{V}}{\mathbf{I}_s} = \frac{L/C}{j(\omega L - 1/\omega C)}$$

or

$$\mathbf{Z} = -j\frac{1}{C}\frac{\omega}{\omega^2 - 1/LC} \tag{6}$$

Figure 9-14

(*a*) A sinusoidally excited parallel LC circuit. (*b*) The magnitude of the input impedance, $\mathbf{Z} = \mathbf{V}/\mathbf{I}_s$, and (*c*) the angle of the input impedance are plotted as functions of ω.

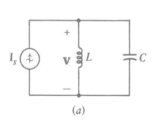

(*a*)

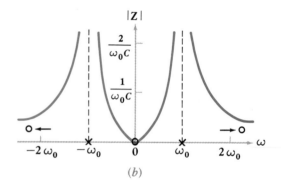

(*b*)

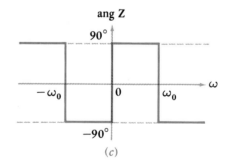

(*c*)

By letting

$$\omega_0 = \frac{1}{\sqrt{LC}}$$

and factoring the expression for the input impedance, the magnitude of the impedance may be written in a form which enables those frequencies to be identified at which the response is zero or infinite:

$$|\mathbf{Z}| = \frac{1}{C}\frac{|\omega|}{|(\omega - \omega_0)(\omega + \omega_0)|} \tag{7}$$

Such frequencies are termed *critical frequencies,* and their early identification simplifies the construction of the response curve. We note first that the response has zero amplitude at $\omega = 0$; when this happens, we say that the response has a *zero* at $\omega = 0$, and we also describe the frequency at which it occurs as a zero. Response of infinite amplitude is noted at $\omega = \omega_0$ and $-\omega_0$; these frequencies are called *poles,* and the response is said to have a pole at each of these frequencies. Finally, we note that the response approaches zero as $\omega \to \infty$, and thus $\omega = \pm\infty$ is also a zero.[2]

The locations of the critical frequencies should be marked on the ω axis, by using small circles for the zeros and crosses for the poles. Poles or zeros at infinite frequency should be indicated by an arrow near the axis, as shown in Fig. 9-14*b*. The actual drawing of the graph is made easier by adding broken vertical lines as asymptotes at each pole location. The completed graph of magnitude versus ω is shown in Fig. 9-14*b*; the slope at the origin is *not* zero.

An inspection of Eq. (6) shows that the phase angle of the input impedance must be either $+90°$ or $-90°$; no other values are possible, as must apparently be the case for any circuit composed entirely of inductors and capacitors. An analytical expression for ang **Z** would therefore consist of a series of statements that the angle is $+90°$ or $-90°$ in certain frequency ranges. It is simpler to present the information graphically, as shown in Fig. 9-14*c*. Although this curve is only a collection of horizontal straight line segments, errors are often made in its construction, and it is a good idea to make certain that it can be drawn directly from an inspection of Eq. (6).

Example 9-5   Construct plots of the amplitude and phase of $\mathbf{Z}_{in}$ versus ω for the *LC* circuit of Fig. 9-15.

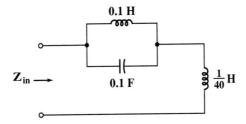

$\mathbf{Z}_{in} \longrightarrow$

0.1 H

0.1 F

$\frac{1}{40}$ H

Figure 9-15

An *LC* circuit whose frequency response is determined in Example 9-5.

Solution: We begin by finding the analytical expression for $\mathbf{Z}_{in}$,

$$\mathbf{Z}_{in} = \frac{(j\omega/10)(10/j\omega)}{j\omega/10 + 10/j\omega} + \frac{j\omega}{40} = \frac{1}{(100 - \omega^2)/j10\omega} + \frac{j\omega}{40}$$

$$= \frac{j10\omega}{100 - \omega^2} + \frac{j\omega}{40} = j\omega\left[\frac{100 - \omega^2 + 400}{40(100 - \omega^2)}\right]$$

$$= \frac{j\omega}{40}\left[\frac{\omega^2 - 500}{\omega^2 - 100}\right]$$

Therefore,

$$\mathbf{Z}_{in} = j\frac{\omega(\omega - 22.4)(\omega + 22.4)}{40(\omega - 10)(\omega + 10)} \tag{8}$$

[2] It is customary to consider plus infinity and minus infinity as being the same point. The phase angle of the response at very large positive and negative values of ω need not be the same, however.

We now may write an expression for the magnitude of $\mathbf{Z}_{in}$,

$$|\mathbf{Z}_{in}| = \left| \frac{\omega(\omega - 22.4)(\omega + 22.4)}{40(\omega - 10)(\omega + 10)} \right|$$

We note the presence of zeros at $\omega = 0$, -22.4, and 22.4 rad/s; poles are found at $\omega = -10$ and 10 rad/s. Since $|\mathbf{Z}_{in}|$ approaches infinity as ω approaches infinity, there is also a pole at $\omega = \pm\infty$. These six critical frequencies are indicated on the sketch of Fig. 9-16a, and the response curve is roughed in.

Figure 9-16

For the *LC* network of Fig. 9-15, sketches are shown of (*a*) $|\mathbf{Z}_{in}|$ versus ω; (*b*) ang $\mathbf{Z}_{in}$ versus ω.

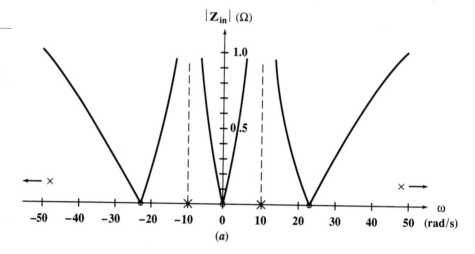

(*a*)

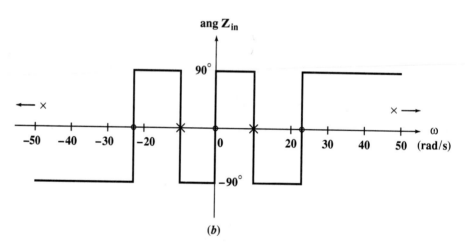

(*b*)

From Eq. (8), we can see that a large (in magnitude) negative value of ω, such as $\omega = -100$, produces a negative imaginary $\mathbf{Z}_{in}$, here $\mathbf{Z}_{in} = -j2.40\ \Omega$; hence the phase angle is $-90°$ for $-\infty < \omega < -22.4$. The angle alternates between $-90°$ and $+90°$ every time the frequency passes a pole or a zero. The resultant phase plot is shown as Fig. 9-16b. ∎

Drill Problems

9-7. For the circuit shown in Fig. 9-17, sketch, as a function of ω: (*a*) $|\mathbf{V}_1|$; (*b*) ang $\mathbf{V}_1$; (*c*) $|\mathbf{I}_2|$.   *Ans:* $|\mathbf{V}_1(j2)| = 33.8$ V; ang $\mathbf{V}_1(j2) = -5.44°$; $|\mathbf{I}_2(j2)| = 0.358$ A

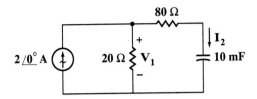

Figure 9-17

See Drill Probs. 9-7 and 9-8.

9-8. Replace the two resistors in the circuit of Fig. 9-17 with two 10-mH inductors and let V_1 be the desired response. Sketch $|V_1|$ and ang V_1 versus ω. Identify all the critical frequencies.

$$Ans: \mathbf{V}_1(j80) = 2.06\underline{/-90°} \text{ V}; 0, \pm70.7 \text{ rad/s}; \pm100 \text{ rad/s}$$

Problems

1 Use phasors and nodal analysis on the circuit of Fig. 9-18 to find $\mathbf{V}_2$.

Figure 9-18

See Probs. 1 and 2.

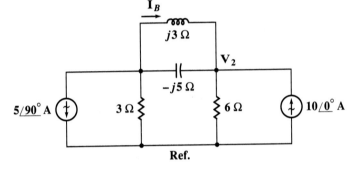

2 Use phasors and mesh analysis on the circuit of Fig. 9-18 to find $\mathbf{I}_B$.

3 Find $v_x(t)$ in the circuit of Fig. 9-19 if $v_{s1} = 20 \cos 1000t$ V and $v_{s2} = 20 \sin 1000t$ V.

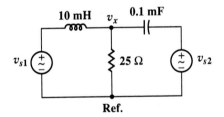

Figure 9-19

See Prob. 3.

4 (a) Find $\mathbf{V}_3$ in the circuit shown in Fig. 9-20. (b) To what identical values should the three capacitive impedances be changed so that $\mathbf{V}_3$ is 180° out of phase with the source voltage?

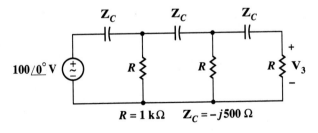

Figure 9-20

See Prob. 4.

5 Use mesh analysis to find $i_x(t)$ in the circuit shown in Fig. 9-21.

Figure 9-21

See Probs. 5 and 6.

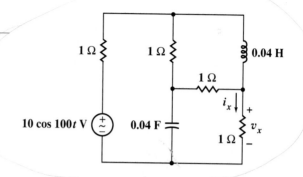

6 Find $v_x(t)$ for the circuit of Fig. 9-21 using phasors and nodal analysis.

7 (*a*) Construct a tree for the circuit of Fig. 9-22 so that i_1 is a link current, assign a complete set of link currents, and find $i_1(t)$. (*b*) Construct another tree in which v_1 is a tree-branch voltage, assign a complete set of tree-branch voltages, and find $v_1(t)$.

Figure 9-22

See Probs. 7 and 11.

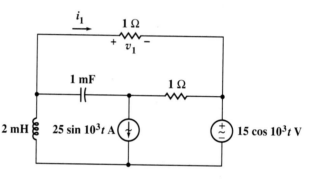

8 The op-amp shown in Fig. 9-23 has an infinite input impedance, zero output imped-ance, and a large but finite (positive, real) gain, $\mathbf{A} = -\mathbf{V}_o/\mathbf{V}_i$. (*a*) Construct a basic differentiator by letting $\mathbf{Z}_f = R_f$, find $\mathbf{V}_o/\mathbf{V}_s$, and then show that $\mathbf{V}_o/\mathbf{V}_s \to -j\omega C_1 R_f$ as $\mathbf{A} \to \infty$. (*b*) Let $\mathbf{Z}_f$ represent C_f and R_f in parallel, find $\mathbf{V}_o/\mathbf{V}_s$, and then show that $\mathbf{V}_o/\mathbf{V}_s \to -j\omega C_1 R_f/(1 + j\omega C_f R_f)$ as $\mathbf{A} \to \infty$.

Figure 9-23

See Prob. 8.

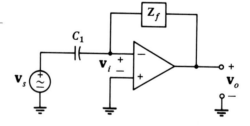

9 Find the frequency-domain Thévenin equivalent of the network shown in Fig. 9-24*a*. Show the result as $\mathbf{V}_{th}$ in series with $\mathbf{Z}_{th}$.

Figure 9-24

See Probs. 9 and 10.

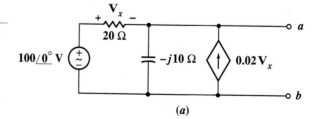

(*a*)

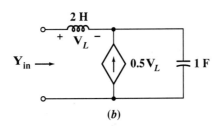

Figure 9-24

(*continued*)

(b)

10 Find the input admittance of the circuit shown in Fig. 9-24b, and represent it as the parallel combination of a resistance R and an inductance L, giving values for R and L if $\omega = 1$ rad/s.

11 With reference to the circuit of Fig. 9-22, think superposition and find that part of $v_1(t)$ due to (a) the voltage source acting alone; (b) the current source acting alone.

12 Use $\omega = 1$ rad/s, and find the Norton equivalent of the network shown in Fig. 9-25. Construct the Norton equivalent as a current source $\mathbf{I}_N$ in parallel with a resistance R_N and either an inductance L_N or a capacitance C_N.

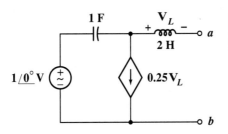

Figure 9-25

See Prob. 12.

13 In the circuit of Fig. 9-26, let $i_{s1} = 2\cos 200t$ A, $i_{s2} = 1\cos 100t$ A, and $v_{s3} = 2\sin 200t$ V. Find $v_L(t)$.

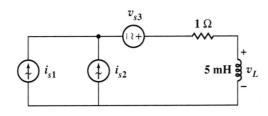

Figure 9-26

See Prob. 13.

14 Find the Thévenin-equivalent circuit for Fig. 9-27.

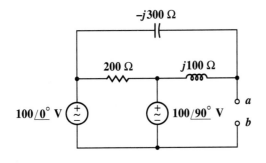

Figure 9-27

See Prob. 14.

15 (a) Calculate values for $\mathbf{I}_L$, $\mathbf{I}_R$, $\mathbf{I}_C$, $\mathbf{V}_L$, $\mathbf{V}_R$, and $\mathbf{V}_C$ (plus $\mathbf{V}_s$) for the circuit shown in Fig. 9-28. (b) Using scales of 50 V to 1 in and 25 A to 1 in, show all seven quantities on a phasor diagram, and indicate that $\mathbf{I}_L = \mathbf{I}_R + \mathbf{I}_C$ and $\mathbf{V}_s = \mathbf{V}_L + \mathbf{V}_R$.

Figure 9-28

See Prob. 15.

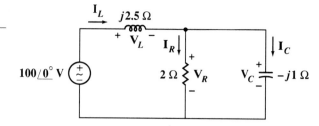

16 In the circuit of Fig. 9-29, find values for (a) $\mathbf{I}_1$, $\mathbf{I}_2$, and $\mathbf{I}_3$. (b) Show $\mathbf{V}_s$, $\mathbf{I}_1$, $\mathbf{I}_2$, and $\mathbf{I}_3$ on a phasor diagram (scales of 50 V/in and 2 A/in work fine). (c) Find $\mathbf{I}_s$ graphically and give its amplitude and phase angle.

Figure 9-29

See Prob. 16.

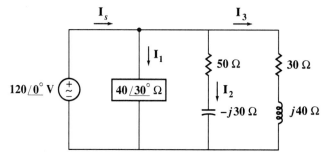

17 In the circuit sketched in Fig. 9-30, it is known that $|\mathbf{I}_1| = 5$ A and $|\mathbf{I}_2| = 7$ A. Find $\mathbf{I}_1$ and $\mathbf{I}_2$ with the help of compass, ruler, straightedge, protractor, and all that fun stuff.

Figure 9-30

See Prob. 17.

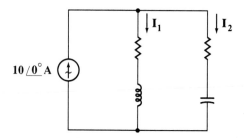

18 Let $\mathbf{V}_1 = 100\underline{/0°}$ V, $|\mathbf{V}_2| = 140$ V, and $|\mathbf{V}_1 + \mathbf{V}_2| = 120$ V. Use graphical methods to find two possible values for the angle of $\mathbf{V}_2$.

19 Plot $|\mathbf{V}_{\text{out}}|$ versus ω for the circuit shown in Fig. 9-31. Cover the frequency range from $\omega = 800$ to $\omega = 1200$ rad/s.

Figure 9-31

See Prob. 19.

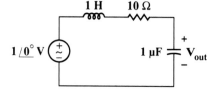

20 Make sketches of $|\mathbf{Z}_{\text{in}}|$, R_{in}, and X_{in} for the two-terminal network of Fig. 9-32.

Figure 9-32

See Probs. 20 and 21.

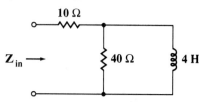

21 Replace the 4-H inductor in the circuit shown in Fig. 9-32 with a 2.5-μF capacitor and then rework Prob. 20.

22 (a) Plot a curve of $|\mathbf{Z}_{in}|$ versus ω for the network shown in Fig. 9-33. Cover the frequency range $0.2 \leq \omega \leq 5$ rad/s. (b) What value does $|\mathbf{Z}_{in}|$ approach as ω approaches infinity?

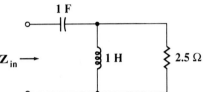

Figure 9-33

See Prob. 22.

23 Determine the critical frequencies of the input impedance illustrated in Fig. 9-34 and make a rough sketch of $|\mathbf{Z}_{in}|$ versus ω, $-3000 < \omega < 3000$ rad/s.

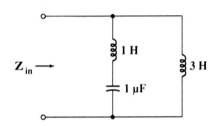

Figure 9-34

See Prob. 23.

24 For the network shown in Fig. 9-35, sketch a curve of $|X_{in}|$ versus ω, $-5 < \omega < 5$ rad/s.

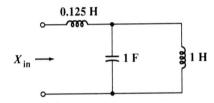

Figure 9-35

See Prob. 24.

25 Construct sketches of both the magnitude and phase angle versus ω for the input impedance of a network consisting of the parallel combination of two branches. One branch is the series combination of a 2-Ω resistor and a 1-mH inductor, and the other is the series combination of a 2-Ω resistor and a 0.25-mF capacitor.

26 (SPICE) Modify the network of Fig. 9-15 by installing a 0.1-Ω resistor in series with the 0.1-H inductor. One procedure for using SPICE to obtain values for $|\mathbf{Z}_{in}|$ is to connect a $1\underline{/0°}$-A source to the input terminals and find $|\mathbf{V}_{in}|$. (a) Find $|\mathbf{Z}_{in}|$ at $\omega = 5$ rad/s. (b) Compare this value with that obtained without the added resistance.

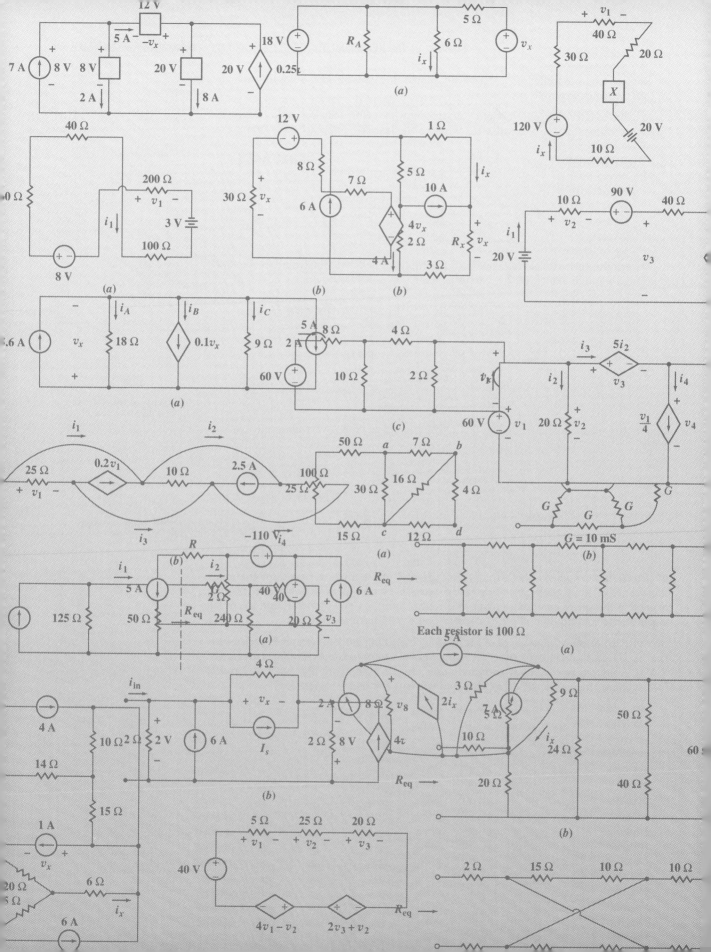

Average Power and RMS Values

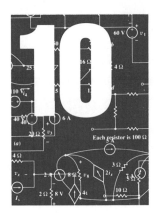

Nearly all problems in circuit analysis are concerned with applying one or more sources of electric energy to a circuit and then quantitatively determining one or more responses throughout the circuit. A response may be a current or a voltage, but we are also interested in the amount of energy supplied from the sources, in the amount of energy dissipated or stored within the circuit, and in the manner in which energy is delivered to the points at which the responses are determined. Primarily, however, we are concerned with the *rate* at which energy is being generated and absorbed; our attention must now be directed to *power*.

We shall begin by considering instantaneous power, the product of the time-domain voltage and time-domain current associated with the element or network of interest. The instantaneous power is sometimes quite useful in its own right, because its maximum value might have to be limited in order to avoid exceeding the safe or useful operating range of a physical device. For example, transistor and vacuum-tube power amplifiers both produce a distorted output, and speakers give a distorted sound, when the peak power exceeds a certain limiting value. However, we are mainly interested in instantaneous power for the simple reason that it provides us with the means to calculate a more important quantity, the average power. In a similar way, the progress of a cross-country automobile trip is best described by the average velocity; our interest in the instantaneous velocity is limited to the avoidance of maximum velocities which will endanger our safety or arouse the highway patrol.

In practical problems we shall deal with values of average power which range from the small fraction of a picowatt available in a telemetry signal from outer space, to the few watts of audio power supplied to the speakers in a high-fidelity stereo system, the several hundred watts required to invigorate the morning coffeepot, or the 10 billion watts generated at the Grand Coulee Dam.

Our discussion will not be concerned entirely with the average power delivered by a sinusoidal current or voltage; we shall therefore define a quantity called the *effective value,* a mathematical measure of the effectiveness of other waveforms in delivering power. Our study of power will be completed by considering the descriptive quantities power factor and complex power, two concepts which introduce the practical and economic aspects associated with the distribution of electric power.

10-2

Instantaneous power

The power, as a function of time, delivered to any device is given by the product of the instantaneous voltage across the device and the instantaneous current through it, as we well know; the passive sign convention is assumed. Thus,

$$p = vi \qquad (1)$$

A knowledge of both the current and the voltage is presumed. If the device in question is a resistor of resistance R, then the power may be expressed solely in terms of either the current or the voltage:

$$p = vi = i^2R = \frac{v^2}{R} \qquad (2)$$

If the voltage and current are associated with a device which is entirely inductive, then

$$p = vi = Li\frac{di}{dt} = \frac{1}{L}v\int_{-\infty}^{t} v\,dt \qquad (3)$$

where we have arbitrarily assumed that the voltage is zero at $t = -\infty$. In the case of a capacitor,

$$p = vi = Cv\frac{dv}{dt} = \frac{1}{C}i\int_{-\infty}^{t} i\,dt \qquad (4)$$

where a like assumption about the current is made. This listing of equations for power in terms of only a current or a voltage soon becomes unwieldy, however, as we begin to consider more general networks. The listing is also quite unnecessary, for we need only find *both* the current and voltage at the network terminals. As an example, we may consider the series RL circuit, as shown in Fig. 10-1, excited by a step-voltage source. The familiar current response is

$$i(t) = \frac{V_0}{R}(1 - e^{-Rt/L})u(t)$$

Figure 10-1

The instantaneous power that is delivered to R is $p_R = i^2R = (V_0^2/R)(1 - e^{-Rt/L})^2u(t)$.

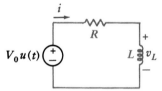

and thus the total power delivered by the source or absorbed by the passive network is

$$p = vi = \frac{V_0^2}{R}(1 - e^{-Rt/L})u(t)$$

since the square of the unit-step function is obviously the unit-step function itself.

The power delivered to the resistor is

$$p_R = i^2R = \frac{V_0^2}{R}(1 - e^{-Rt/L})^2u(t)$$

In order to determine the power absorbed by the inductor, we first obtain the inductor voltage:

$$v_L = L \frac{di}{dt}$$

$$= V_0 e^{-Rt/L} u(t) + \frac{LV_0}{R}(1 - e^{-Rt/L}) \frac{du(t)}{dt}$$

$$= V_0 e^{-Rt/L} u(t)$$

since $du(t)/dt$ is zero for $t > 0$ and $(1 - e^{-Rt/L})$ is zero at $t = 0$. The power absorbed by the inductor is thus

$$p_L = v_L i = \frac{V_0^2}{R} e^{-Rt/L}(1 - e^{-Rt/L})u(t)$$

Only a few algebraic manipulations are required to show that

$$p = p_R + p_L$$

which serves to check the accuracy of our work.

The majority of the problems which involve power calculations are perhaps those which deal with circuits excited by sinusoidal forcing functions in the steady state. As we have been told previously, even when periodic nonsinusoidal forcing functions are employed, it is possible to resolve the problem into a number of subproblems in which the forcing functions *are* sinusoidal. The special case of the sinusoid therefore deserves special attention.

Let us change the voltage source in the circuit of Fig. 10-1 to the sinusoidal source $V_m \cos \omega t$. The familiar time-domain response is

$$i(t) = I_m \cos (\omega t + \phi)$$

where

$$I_m = \frac{V_m}{\sqrt{R^2 + \omega^2 L^2}} \qquad \text{and} \qquad \phi = -\tan^{-1} \frac{\omega L}{R}$$

The instantaneous power delivered to the entire circuit in the sinusoidal steady state is, therefore,

$$p = vi = V_m I_m \cos (\omega t + \phi) \cos \omega t$$

which we shall find convenient to rewrite in a form obtained by using the trigonometric identity for the product of two cosine functions. Thus,

$$p = \frac{V_m I_m}{2} [\cos (2\omega t + \phi) + \cos \phi]$$

$$= \frac{V_m I_m}{2} \cos \phi + \frac{V_m I_m}{2} \cos (2\omega t + \phi)$$

The last equation possesses several characteristics which are true in general for circuits in the sinusoidal steady state. One term, the first, is not a function of time; and a second term is included which has a cyclic variation at *twice* the applied frequency. Since this term is a cosine wave, and since sine waves and cosine waves have average values which are zero (when averaged over an integral number of periods), this introductory example may serve to indicate that the *average* power is $\frac{1}{2}V_mI_m\cos\phi$. This is true, and we shall now establish this relationship in more general terms.

Drill Problems

10-1. A voltage source, $40 + 60u(t)$ V, a 5-μF capacitor, and a 200-Ω resistor are in series. At $t = 1.2$ ms, find the power being absorbed by the (*a*) capacitor; (*b*) resistor; (*c*) $60u(t)$-V portion of the source.   *Ans:* 7.40 W; 1.633 W; -5.42 W

10-2. A current source of $12\cos 2000t$ A, a 200-Ω resistor, and a 0.2-H inductor are in parallel. Assume steady-state conditions exist. At $t = 1$ ms, find the power being absorbed by the (*a*) resistor; (*b*) inductor; (*c*) sinusoidal source.
                       Ans: 13.98 kW; -5.63 kW; -8.35 kW

10-3

Average power

When we speak of an average value for the instantaneous power, the time interval over which the averaging process takes place must be clearly defined. Let us first select a general interval of time from t_1 to t_2. We may then obtain the average value by integrating $p(t)$ from t_1 to t_2 and dividing the result by the time interval $t_2 - t_1$. Thus,

$$P = \frac{1}{t_2 - t_1}\int_{t_1}^{t_2} p(t)\,dt \tag{5}$$

The average value is denoted by the capital letter P, since it is not a function of time, and it usually appears without any specific subscripts that identify it as an average value. Although P is not a function of time, it *is* a function of t_1 and t_2, the two instants of time which define the interval of integration. This dependence of P on a specific time interval may be expressed in a simpler manner if $p(t)$ is a periodic function. We shall consider this important case first.

Let us assume that our forcing function and the circuit responses are all periodic; a steady-state condition has been reached, although not necessarily the sinusoidal steady state. We may define a *periodic* function $f(t)$ mathematically by requiring that

$$f(t) = f(t + T) \tag{6}$$

where T is the period. We now show that the average value of the instantaneous power as expressed by Eq. (5) may be computed over an interval of one period having an arbitrary beginning.

A general periodic waveform is shown in Fig. 10-2 and identified as $p(t)$. We first compute the average power by integrating from t_1 to a time t_2 which is one period later, $t_2 = t_1 + T$:

Figure 10-2

The average value P of a periodic power function $p(t)$ is the same over any period T.

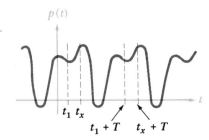

and then by integrating from some other time t_x to $t_x + T$:

$$P_x = \frac{1}{T} \int_{t_x}^{t_x+T} p(t)\, dt$$

The equality of P_1 and P_x should be evident from the graphical interpretation of the integrals; the area which represents the integral to be evaluated in determining P_x is smaller by the area from t_1 to t_x, but greater by the area from $t_1 + T$ to $t_x + T$. The periodic nature of the curve requires these two areas to be equal. Thus, the average power may be computed by integrating the instantaneous power over any interval which is one period in length and then dividing by the period:

$$P = \frac{1}{T} \int_{t_x}^{t_x+T} p\, dt \tag{7}$$

It is important to note that we might also integrate over any integral number of periods, provided that we divide by this same integral number of periods. Thus,

$$P = \frac{1}{nT} \int_{t_x}^{t_x+nT} p\, dt \qquad n = 1, 2, 3, \ldots \tag{8}$$

If we carry this concept to the extreme by integrating over all time, another useful result is obtained. We first provide ourselves with symmetrical limits on the integral

$$P = \frac{1}{nT} \int_{-nT/2}^{nT/2} p\, dt$$

and then take the limit as n becomes infinite,

$$P = \lim_{n \to \infty} \frac{1}{nT} \int_{-nT/2}^{nT/2} p\, dt$$

If $p(t)$ is a mathematically well-behaved function, as all *physical* forcing functions and responses are, it is apparent that if a large integer n is replaced by a slightly larger number which is not an integer, then the value of the integral and of P is changed by a negligible amount; moreover, the error decreases as n increases. Without justifying this step rigorously, we therefore replace the discrete variable nT by the continuous variable τ:

$$P = \lim_{\tau \to \infty} \frac{1}{\tau} \int_{-\tau/2}^{\tau/2} p\, dt \tag{9}$$

We shall find it convenient on several occasions to integrate periodic functions over this "infinite period." Examples of the use of Eqs. (7), (8), and (9) follow.

Let us illustrate the calculation of the average power of a periodic wave by finding the average power delivered to a resistor R by the (periodic) sawtooth current waveform shown in Fig. 10-3a. We have

$$i(t) = \frac{I_m}{T} t \qquad 0 < t \leq T$$

$$i(t) = \frac{I_m}{T} (t - T) \qquad T < t \leq 2T$$

Figure 10-3

(a) A sawtooth current waveform and (b) the instantaneous power waveform it produces in a resistor R.

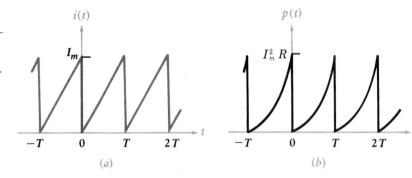

and so on; and

$$p(t) = \frac{1}{T^2} I_m^2 R t^2 \qquad 0 < t \le T$$

$$p(t) = \frac{1}{T^2} I_m^2 R (t - T)^2 \qquad T < t \le 2T$$

and so on, as sketched in Fig. 10-3b. Integrating over the simplest range of one period, from $t = 0$ to $t = T$, we have

$$P = \frac{1}{T} \int_0^T \frac{I_m^2 R}{T^2} t^2 \, dt = \frac{1}{3} I_m^2 R$$

The selection of other ranges of one period, such as from $t = 0.1T$ to $t = 1.1T$, would produce the same answer. Integration from 0 to $2T$ and division by $2T$—that is, the application of Eq. (8) with $n = 2$ and $t_x = 0$—would also provide the same answer.

Now let us obtain the general result for the sinusoidal steady state. We shall assume the general sinusoidal voltage

$$v(t) = V_m \cos (\omega t + \theta)$$

and current

$$i(t) = I_m \cos (\omega t + \phi)$$

associated with the device in question. The instantaneous power is

$$p(t) = V_m I_m \cos (\omega t + \theta) \cos (\omega t + \phi)$$

Again expressing the product of two cosine functions as one-half the sum of the cosine of the difference angle and the cosine of the sum angle,

$$p(t) = \tfrac{1}{2} V_m I_m \cos(\theta - \phi) + \tfrac{1}{2} V_m I_m \cos (2\omega t + \theta + \phi) \tag{10}$$

we may save ourselves some integration by an inspection of the result. The first term is a constant, independent of t. The remaining term is a cosine function; $p(t)$ is therefore periodic, and its period is $\tfrac{1}{2}T$. Note that the period T is associated with the given current and voltage, and not with the power; the power function has a period $\tfrac{1}{2}T$. However, we may integrate over an interval of T to determine the average value if we wish; it is necessary only that we also divide by T. Our familiarity with cosine and sine waves, however, shows that the average value of either over a period is zero. There is thus no need to integrate Eq. (10) formally; by inspection, the average value of the second term is zero over a period T (or $\frac{1}{2}T$) and the average value of the first term, a constant, must be that constant itself. Thus,

$$P = \tfrac{1}{2}V_m I_m \cos(\theta - \phi) \qquad (11)$$

This important result, introduced in the previous section for a specific circuit, is therefore quite general for the sinusoidal steady state. The average power is one-half the product of the crest amplitude of the voltage, the crest amplitude of the current, and the cosine of the phase-angle difference between the current and the voltage; the sense of the difference is immaterial.

Let us try our hands on a numerical example.

Example 10-1 We are given the time-domain voltage $v = 4\cos(\pi t/6)$ V, and we wish to find the power relationships that result when the corresponding phasor voltage $\mathbf{V} = 4\underline{/0°}$ V is applied across an impedance $\mathbf{Z} = 2\underline{/60°}\ \Omega$.

Solution: The phasor current is $\mathbf{V/Z} = 2\underline{/-60°}$ A, and the average power is

$$P = \tfrac{1}{2}(4)(2)\cos 60° = 2\ \text{W}$$

The time-domain voltage,

$$v(t) = 4\cos\frac{\pi t}{6}$$

time-domain current,

$$i(t) = 2\cos\left(\frac{\pi t}{6} - 60°\right)$$

and instantaneous power,

$$p(t) = 8\cos\frac{\pi t}{6}\cos\left(\frac{\pi t}{6} - 60°\right) = 2 + 4\cos\left(\frac{\pi t}{3} - 60°\right)$$

are all sketched on the same time axis in Fig. 10-4. Both the 2-W average value of the power and its period of 6 s, one-half the period of either the current or the voltage, are evident. The zero value of the instantaneous power at each instant when either the voltage or current is zero is also apparent. ∎

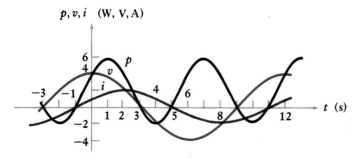

p, v, i (W, V, A)

Figure 10-4

Curves of $v(t)$, $i(t)$, and $p(t)$ are plotted as functions of time for a simple circuit in which the phasor voltage $\mathbf{V} = 4\underline{/0°}$ V is applied to the impedance $\mathbf{Z} = 2\underline{/60°}\ \Omega$ at $\omega = \pi/6$ rad/s.

Two special cases are worth isolating for consideration, the average power delivered to an ideal resistor, and that to an ideal reactor (any combination of only capacitors and inductors).

The phase-angle difference between the current through and the voltage across a pure resistor is zero, and therefore

$$P_R = \tfrac{1}{2} V_m I_m$$

or
$$P_R = \tfrac{1}{2} I_m^2 R \tag{12}$$

or
$$P_R = \frac{V_m^2}{2R} \tag{13}$$

The last two formulas, enabling us to determine the average power delivered to a pure resistance from a knowledge of either the sinusoidal current or voltage, are simple and important. They are often misused. The most common error is made in trying to apply them in cases where, say, the voltage included in Eq. (13) is *not the voltage across the resistor*. If care is taken to use the current through the resistor in Eq. (12) and the voltage across the resistor in Eq. (13), satisfactory operation is guaranteed. Also, do not forget the factor of $\tfrac{1}{2}$!

The average power delivered to any device which is purely reactive must be zero. This is evident from the 90° phase difference which must exist between current and voltage; hence, $\cos(\theta - \phi) = 0$ and

$$P_X = 0$$

The *average* power delivered to any network composed entirely of ideal inductors and capacitors is zero; the *instantaneous* power is zero only at specific instants. Thus, power flows into the network for a part of the cycle and out of the network during another portion of the cycle, with *no* power lost.

Example 10-2 Find the average power being delivered to an impedance $\mathbf{Z}_L = 8 - j11\ \Omega$ by a current $\mathbf{I} = 5\underline{/20°}$ A.

Solution: We may find the solution quite rapidly by using Eq. (12). Only the 8-Ω resistor enters the average-power calculation, and

$$P = \tfrac{1}{2}(5^2)\,8 = 100\ \text{W}$$

since no average power can be absorbed by the $-j11\ \Omega$. Note also that if a current is given in rectangular form, say, $\mathbf{I} = 2 + j5$ A, then the magnitude squared is $2^2 + 5^2$, and the average power delivered to $\mathbf{Z}_L = 8 - j11\ \Omega$ would be

$$P = \tfrac{1}{2}(2^2 + 5^2)8 = 116\ \text{W} \qquad\blacksquare$$

Example 10-3 As a futher example illustrating these power relationships, let us consider the circuit shown in Fig. 10-5. We need the average power absorbed by each of the three passive elements and the average power supplied by each source.

Solution: The values of $\mathbf{I}_1$ and $\mathbf{I}_2$ are found by any of several methods, such as mesh analysis, nodal analysis, or superposition. They are

$$\mathbf{I}_1 = 5 - j10 = 11.18\underline{/-63.4°}$$

$$\mathbf{I}_2 = 5 - j5 = 7.07\underline{/-45°}$$

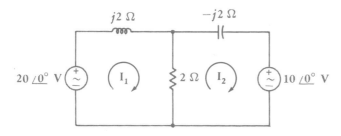

Figure 10-5

The average power delivered to each reactive element is zero in the sinusoidal steady state.

The downward current through the 2-Ω resistor is

$$\mathbf{I}_1 - \mathbf{I}_2 = -j5 = 5\underline{/-90°}$$

so that $I_m = 5$ A, and the average power absorbed by the resistor is found most easily by Eq. (12):

$$P_R = \tfrac{1}{2}I_m^2 R = \tfrac{1}{2}(5^2)2 = 25 \text{ W}$$

This result may be checked by using Eq. (11) or Eq. (13). The average power absorbed by each reactive element is zero.

We next turn to the left source. The voltage $20\underline{/0°}$ and current $11.18\underline{/-63.4°}$ satisfy the *active* sign convention, and thus the power *delivered by* this source is

$$P_{\text{left}} = \tfrac{1}{2}(20)(11.18)\cos(0° + 63.4°) = 50 \text{ W}$$

In a similar manner, we find the power *absorbed* by the right source,

$$P_{\text{right}} = \tfrac{1}{2}(10)(7.07)\cos(0° + 45°) = 25 \text{ W}$$

Since 50 = 25 + 25, the power relations check. ∎

In Sec. 2-5 of Chap. 2, we considered the maximum power transfer theorem as it applied to resistive loads and resistive source impedances. For a Thévenin source $\mathbf{V}_s$ and impedance $\mathbf{Z}_{\text{th}} = R_{\text{th}} + jX_{\text{th}}$ connected to a load $\mathbf{Z}_L = R_L + jX_L$, it may be shown readily (see Prob. 10 in this chapter) that the average power delivered to the load is a maximum when $R_L = R_{\text{th}}$ and $X_L = -X_{\text{th}}$, that is, when $\mathbf{Z}_L = \mathbf{Z}_{\text{th}}^*$. This result is often dignified by calling it the *maximum power transfer theorem for the sinusoidal steady state*:

> An independent voltage source in series with an impedance $\mathbf{Z}_{\text{th}}$ or an independent current source in parallel with an impedance $\mathbf{Z}_{\text{th}}$ delivers a maximum average power to that load impedance $\mathbf{Z}_L$ which is the conjugate of $\mathbf{Z}_{\text{th}}$, or $\mathbf{Z}_L = \mathbf{Z}_{\text{th}}^*$.

It is apparent that the resistive condition considered in Chap. 2 is merely a special case.

We must now pay some attention to *nonperiodic* functions. One practical example of a nonperiodic power function for which an average power value is desired is the power output of a radio telescope which is directed toward a "radio star." Another is the sum of a number of periodic functions, each function having a different period, such that no greater common period can be found for the combination. For example, the current

$$i(t) = \sin t + \sin \pi t \tag{14}$$

is nonperiodic because the ratio of the periods of the two sine waves is an irrational number. At $t = 0$, both terms are zero and increasing. But the first term is zero and increasing only when $t = 2\pi n$, where n is an integer, and thus periodicity demands that πt or $\pi(2\pi n)$ must equal $2\pi m$, where m is also an integer. No solution (integral values for both m and n) for this equation is possible. It may be illuminating to compare the nonperiodic expression in Eq. (14) with the *periodic* function

$$i(t) = \sin t + \sin 3.14t \qquad (15)$$

where 3.14 is an exact decimal expression and is *not* to be interpreted as 3.141 592 .... With a little effort, it can be shown that the period of this current wave is 100π s.[1]

The average value of the power delivered to a 1-Ω resistor by either a periodic current such as Eq. (15) or a nonperiodic current such as Eq. (14) may be found by integrating over an infinite interval. Much of the actual integration can be avoided because of our thorough knowledge of the average values of simple functions. We therefore obtain the instantaneous power delivered by the current in Eq. (14) by applying Eq. (9):

$$P = \lim_{\tau \to \infty} \frac{1}{\tau} \int_{-\tau/2}^{\tau/2} (\sin^2 t + \sin^2 \pi t + 2 \sin t \sin \pi t) \, dt$$

We now consider P as the sum of three average values. The average value of $\sin^2 t$ over an infinite interval is found by replacing $\sin^2 t$ by $(\frac{1}{2} - \frac{1}{2} \cos 2t)$; the average is obviously $\frac{1}{2}$. Similarly, the average value of $\sin^2 \pi t$ is also $\frac{1}{2}$. And the last term can be expressed as the sum of two cosine functions, each of which must certainly have an average value of zero. Thus,

$$P = \tfrac{1}{2} + \tfrac{1}{2} = 1 \text{ W}$$

An identical result is obtained for the periodic current, Eq. (15).

Applying this same method to a current function which is the sum of several sinusoids of *different periods* and arbitrary amplitudes,

$$i(t) = I_{m1} \cos \omega_1 t + I_{m2} \cos \omega_2 t + \cdots + I_{mN} \cos \omega_N t \qquad (16)$$

we find the average power delivered to a resistance R,

$$P = \tfrac{1}{2}(I_{m1}^2 + I_{m2}^2 + \cdots + I_{mN}^2)R \qquad (17)$$

The result is unchanged if an arbitrary phase angle is assigned to each component of the current. This important result is surprisingly simple when we think of the steps required for its derivation: squaring the current function, integrating, and taking the limit. The result is also just plain surprising, because it shows that, *in this special case of a current such as Eq. (16), superposition is applicable to power.* Superposition is *not* applicable for a current which is the sum of two direct currents, nor is it applicable for a current which is the sum of two sinusoids of the same frequency.

Example 10-4 Find the average power delivered to a 4-Ω resistor by the current $i_1 = 2 \cos 10t - 3 \cos 20t$ A.

---

[1] $T_1 = 2\pi$ and $T_2 = 2\pi/3.14$. Therefore we seek integral values of m and n such that $2\pi n = 2\pi m/3.14$, or $3.14n = m$, or $(\frac{314}{100})n = m$, or $157n = 50m$. Thus, the smallest integral values for n and m are $n = 50$ and $m = 157$. The period is therefore $T = 2\pi n = 100\pi$, or $T = 2\pi(157/3.14) = 100\pi$ s.

Solution: Since the two cosine terms are at different frequencies, the two average-power values may be calculated separately and added. Thus, this current delivers $\frac{1}{2}(2^2)4 + \frac{1}{2}(3^2)4 = 8 + 18 = 26$ W to a 4-Ω resistor. ∎

Example 10-5 Find the average power delivered to a 4-Ω resistor by the current $i_2 = 2 \cos 10t - 3 \cos 10t$ A.

Solution: Here, the two components of the current are at the same frequency, and they must therefore be combined into a single sinusoid at that frequency. Thus, $i_2 = 2 \cos 10t - 3 \cos 10t = -\cos 10t$ delivers only $\frac{1}{2}(1^2)4 = 2$ W of average power to that same resistor. ∎

10-3. Determine the average power delivered to each of the boxed networks in the circuit of Fig. 10-6.　　Ans: 208 W; 61.5 W; 123.1 W

Drill Problems

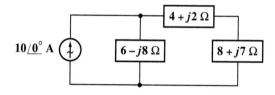

Figure 10-6

See Drill Prob. 3.

10-4. Find the average power delivered to a 5-Ω resistor by each of the periodic waveforms shown in Fig. 10-7.　　Ans: 208 W; 333 W; 160.0 W

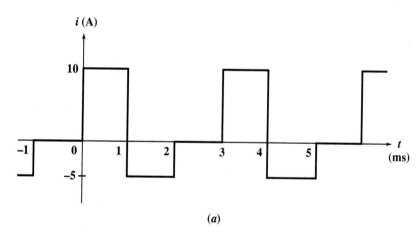

(a)

Figure 10-7

See Drill Probs. 4 and 6, and Prob. 16.

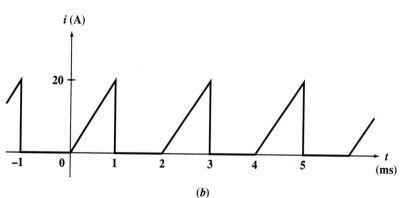

(b)

Figure 10-7 (*continued*)

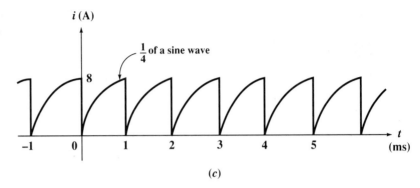

(*c*)

10-5. A voltage source v_s is connected across a 4-Ω resistor. Find the average power absorbed by the resistor if v_s equals (*a*) 8 sin 200*t* V; (*b*) 8 sin 200*t* − 6 cos (200*t* − 45°) V; (*c*) 8 sin 200*t* − 4 sin 100*t* V; (*d*) 8 sin 200*t* − 6 cos (200*t* − 45°) − 5 sin 100*t* + 4 V.   *Ans:* 8.00 W; 4.01 W; 10.00 W; 11.14 W

10-4

Effective values of current and voltage

Most of us are aware that the voltage available at the power outlets in our homes is a sinusoidal voltage having a frequency of 60 Hz and a "voltage" of 115 V. But what is meant by "115 volts"? This is certainly not the instantaneous value of the voltage, for the voltage is not a constant. The value of 115 V is also not the amplitude which we have been symbolizing as V_m; if we displayed the voltage waveform on a calibrated cathode-ray oscilloscope, we should find that the amplitude of this voltage at one of our ac outlets is $115\sqrt{2}$, or 162.6, V. We also cannot fit the concept of an average value to the 115 V, because the average value of the sine wave is zero. We might come a little closer by trying the magnitude of the average over a positive or negative half cycle; by using a rectifier-type voltmeter at the outlet, we should measure 103.5 V. As it turns out, however, the 115 V is the *effective value* of this sinusoidal voltage. This value is a measure of the effectiveness of a voltage source in delivering power to a resistive load.

Let us now proceed to define the effective value of any periodic waveform representing either a current or a voltage. We shall consider the sinusoidal waveform as only a special, albeit practically important, case. Let us arbitrarily define effective value in terms of a current waveform, although a voltage could equally well be selected.

The *effective value* of any periodic current is equal to the value of the direct current which, flowing through an R-ohm resistor, delivers the same average power to the resistor as does the periodic current.

In other words, we allow the given periodic current to flow through the resistor, determine the instantaneous power i^2R, and then find the average value of i^2R over a period; this is the average power. We then cause a direct current to flow through this same resistor and adjust the value of the direct current until the same value of average power is obtained. The resulting magnitude of the direct current is equal to the effective value of the given periodic current. These ideas are illustrated in Fig. 10-8.

The general mathematical expression for the effective value of $i(t)$ is now

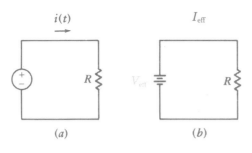

Figure 10-8

If the resistor receives the same average power in parts a and b, then the effective value of $i(t)$ is equal to I_{eff}, and the effective value of $v(t)$ is equal to V_{eff}.

easily obtained. The average power delivered to the resistor by the periodic current $i(t)$ is

$$P = \frac{1}{T} \int_0^T i^2 R \, dt = \frac{R}{T} \int_0^T i^2 \, dt$$

where the period of $i(t)$ is T. The power delivered by the direct current is

$$P = I_{\text{eff}}^2 R$$

Equating the power expressions and solving for I_{eff}, we get

$$I_{\text{eff}} = \sqrt{\frac{1}{T} \int_0^T i^2 dt} \tag{18}$$

The result is independent of the resistance R, as it must be to provide us with a worthwhile concept. A similar expression is obtained for the effective value of a periodic voltage by replacing i and I_{eff} by v and V_{eff}, respectively.

Notice that the effective value is obtained by first squaring the time function, then taking the average value of the squared function over a period, and finally taking the square root of the average of the squared function. In abbreviated language, the operation involved in finding an effective value is the (square) *root* of the *mean* of the *square*; for this reason, the effective value is often called the *root-mean-square* value, or simply the rms value.

The most important special case is that of the sinusoidal waveform. Let us select the sinusoidal current

$$i(t) = I_m \cos(\omega t + \phi)$$

which has a period

$$T = \frac{2\pi}{\omega}$$

and substitute in Eq. (18) to obtain the effective value

$$I_{\text{eff}} = \sqrt{\frac{1}{T} \int_0^T I_m^2 \cos^2(\omega t + \phi) \, dt}$$

$$= I_m \sqrt{\frac{\omega}{2\pi} \int_0^{2\pi/\omega} \left[\frac{1}{2} + \frac{1}{2} \cos(2\omega t + 2\phi)\right] dt}$$

$$= I_m \sqrt{\frac{\omega}{4\pi} [t]_0^{2\pi/\omega}} = \frac{I_m}{\sqrt{2}}$$

Thus the effective value of a sinusoidal current is a real quantity which is independent of the phase angle and numerically equal to $1/\sqrt{2} = 0.707$ times the amplitude of the current. A current $\sqrt{2} \cos(\omega t + \phi)$ A, therefore, has an effective value of 1 A and will deliver the same power to any resistor as will a direct current of 1 A.

It should be noted carefully that the $\sqrt{2}$ factor that we obtained as the ratio of the amplitude of the periodic current to the effective value is applicable only when the periodic function is sinusoidal. For the sawtooth waveform of Fig. 10-3, for example, the effective value is equal to the maximum value divided by $\sqrt{3}$. The factor by which the maximum value must be divided to obtain the effective value depends on the mathematical form of the given periodic function; it may be either rational or irrational, depending on the nature of the function.

The use of the effective value also simplifies slightly the expression for the average power delivered by a sinusoidal current or voltage by avoiding use of the factor $\frac{1}{2}$. For example, the average power delivered to an R-ohm resistor by a sinusoidal current is

$$P = \tfrac{1}{2} I_m^2 R$$

Since $I_{\text{eff}} = I_m/\sqrt{2}$, the average power may be written as

$$P = I_{\text{eff}}^2 R \tag{19}$$

The other power expressions may also be written in terms of effective values:

$$P = V_{\text{eff}} I_{\text{eff}} \cos(\theta - \phi) \tag{20}$$

$$P = \frac{V_{\text{eff}}^2}{R} \tag{21}$$

The fact that the effective value is defined in terms of an equivalent dc quantity provides us with average power formulas for resistive circuits which are identical with those used in dc analysis.

Although we have succeeded in eliminating the factor $\frac{1}{2}$ from our average-power relationships, we must now take care to determine whether a sinusoidal quantity is expressed in terms of its amplitude or its effective value. In practice, the effective value is usually used in the fields of power transmission or distribution and of rotating machinery; in the areas of electronics and communications, the amplitude is more often used. We shall assume that the amplitude is specified unless the term rms is explicitly used.

In the sinusoidal steady state, phasor voltages and currents may be given either as effective values or as amplitudes; the two expressions differ only by a factor $\sqrt{2}$. The voltage $50\underline{/30°}$ V is expressed in terms of an amplitude; as an rms voltage, we should describe the same voltage as $35.4\underline{/30°}$ V rms.

In order to determine the effective value of a periodic or nonperiodic waveform which is composed of the sum of a number of sinusoids of different frequencies, we may use the appropriate average-power relationship of Eq. (17), developed in the previous section, rewritten in terms of the effective values of the several components:

$$P = (I_{1\,\text{eff}}^2 + I_{2\,\text{eff}}^2 + \cdots + I_{N\,\text{eff}}^2)R \tag{22}$$

These results indicate that if a sinusoidal current of 5 A rms at 60 Hz flows through a 2-Ω resistor, an average power of $5^2(2) = 50$ W is absorbed by the resistor; if a second current—say, 3 A rms at 120 Hz—is also present, the absorbed power is $3^2(2) + 50 = 68$ W; however, if the second current is also at 60 Hz, then the absorbed power may have any value between 8 W and 128 W, depending on the relative phase of the two current components.

We therefore have found the effective value of a current which is composed of any number of sinusoidal currents of *different* frequencies,

$$I_{\text{eff}} = \sqrt{I_{1\,\text{eff}}^2 + I_{2\,\text{eff}}^2 + \cdots + I_{N\,\text{eff}}^2} \tag{23}$$

The total current may or may not be periodic; the result is the same. The effective value of the sum of the 60- and 120-Hz currents in the preceding example is 5.83 A; the effective value of the sum of the two 60-Hz currents may have any value between 2 and 8 A.

Drill Problems

10-6. Use Eq. (18) to calculate the effective values of the three periodic waveforms shown in Fig. 10-7.   *Ans:* 6.45 A; 8.16 A; 5.66 A

10-7. Calculate the effective value of each of the periodic voltages: (*a*) 6 cos 25*t*; (*b*) 6 cos 25*t* + 4 sin (25*t* + 30°); (*c*) 6 cos 25*t* + 5 cos² 25*t*; (*d*) 6 cos 25*t* + 5 sin 30*t* + 4 V.   *Ans:* 4.24 V; 6.16 V; 5.23 V; 6.82 V

10-5

Apparent power and power factor

Historically, the introduction of the concepts of apparent power and power factor can be traced to the electric power industry, where large amounts of electric energy must be transferred from one point to another; the efficiency with which this transfer is effected is related directly to the cost of the electric energy, which is eventually paid by the consumer. Customers who provide loads which result in a relatively poor transmission efficiency must pay a greater price for each kilowatthour (kWh) of electric energy they actually receive and use. In a similar way, customers who require a costlier investment in transmission and distribution equipment by the power company will also pay more for each kilowatthour unless the company is benevolent and enjoys losing money.

Let us first define *apparent power* and *power factor* and then show briefly how these terms are related to the aforementioned economic situations. We shall assume that the sinusoidal voltage

$$v = V_m \cos(\omega t + \theta)$$

is applied to a network and the resultant sinusoidal current is

$$i = I_m \cos(\omega t + \phi)$$

The phase angle by which the voltage leads the current is therefore $(\theta - \phi)$. The average power delivered to the network, assuming a passive sign convention at its input terminals, may be expressed either in terms of the maximum values:

$$P = \tfrac{1}{2}V_m I_m \cos(\theta - \phi)$$

or in terms of the effective values:

$$P = V_{\text{eff}} I_{\text{eff}} \cos(\theta - \phi)$$

If our applied voltage and current responses had been dc quantities, the average power delivered to the network would have been given simply by the product of the voltage and the current. Applying this dc technique to the sinusoidal problem, we should obtain a value for the absorbed power which is "apparently" given by the familiar product $V_{\text{eff}}I_{\text{eff}}$. However, this product of the effective values of the voltage and current is not the average power; we define it as the *apparent power*. Dimensionally, apparent power must be measured in the same units as real power, since $\cos(\theta - \phi)$ is dimensionless; but in order to avoid confusion, the term *voltamperes*, or VA, is applied to the apparent power. Since $\cos(\theta - \phi)$ cannot have a magnitude greater than unity, it is evident that the magnitude of the real power can never be greater than the magnitude of the apparent power.

Apparent power is not a concept which is limited to sinusoidal forcing functions and responses. It may be determined for any current and voltage waveshapes by simply taking the product of the effective values of the current and voltage.

The ratio of the real or average power to the apparent power is called the *power factor*, symbolized by PF. Hence,

$$\text{PF} = \frac{\text{average power}}{\text{apparent power}} = \frac{P}{V_{\text{eff}}I_{\text{eff}}}$$

In the sinusoidal case, the power factor is simply $\cos(\theta - \phi)$, where $(\theta - \phi)$ is the angle by which the voltage leads the current. This relationship is the reason why the angle $(\theta - \phi)$ is often referred to as the *PF angle*.

For a purely resistive load, the voltage and current are in phase, $(\theta - \phi)$ is zero, and the PF is unity. The apparent power and the average power are equal. Unity PF, however, may also be achieved for loads which contain both inductance and capacitance if the element values and the operating frequency are selected to provide an input impedance having a zero phase angle.

A purely reactive load, that is, one containing no resistance, will cause a phase difference between the voltage and current of either plus or minus 90°, and the PF is therefore zero.

Between these two extreme cases there are the general networks for which the PF can range from zero to unity. A PF of 0.5, for example, indicates a load having an input impedance with a phase angle of either 60° or −60°; the former describes an inductive load, since the voltage leads the current by 60°, while the latter refers to a capacitive load. The ambiguity in the exact nature of the load is resolved by referring to a leading PF or a lagging PF, the terms *leading* or *lagging* referring to the *phase of the current with respect to the voltage*. Thus, an inductive load will have a lagging PF and a capacitive load a leading PF.

Before considering the practical consequences of these ideas, we can practice the necessary techniques by analyzing the circuit illustrated in Fig. 10-9.

Example 10-6 Calculate values for the average power delivered to each of the two loads shown in Fig. 10-9, the apparent power supplied by the source, and the power factor of the combined loads.

Figure 10-9

A circuit in which we seek the average power delivered to each element, the apparent power supplied by the source, and the power factor of the combined load.

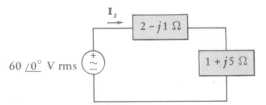

Solution: The source current is

$$\mathbf{I}_s = \frac{60\underline{/0^\circ}}{2 - j1 + 1 + j5} = 12\underline{/-53.1^\circ} \text{ A rms}$$

Therefore, the source supplies an average power of

$$P_s = (60)(12) \cos [0^\circ - (-53.1^\circ)] = 432 \text{ W}$$

The upper load receives an average power

$$P_{\text{upper}} = 12^2(2) = 288 \text{ W}$$

For the right-hand load, we find an average power of

$$P_{\text{right}} = 12^2(1) = 144 \text{ W}$$

Thus, the source provides 432 W, of which 288 W is dissipated in the upper load and 144 W in the load to the right. The power balance is correct.

The source supplies an apparent power of 60(12) = 720 VA.

Finally, the power factor of the combined loads is found by considering the voltage and current associated with the combined loads. This power factor is, of course, identical to the power factor for the source. Thus,

$$\text{PF} = \frac{P}{V_{\text{eff}}I_{\text{eff}}} = \frac{432}{60(12)} = 0.6$$

We might also combine the two loads in series, obtaining $(3 + j4)\ \Omega$ or $5\underline{/53.1^\circ}\ \Omega$. We therefore identify 53.1° as the PF angle, and thus have a PF of $\cos 53.1° = 0.6$. We also note that the combined load is inductive, and the PF is therefore 0.6 lagging.   ■

The practical importance of these new terms is shown by several different situations we shall see in the following. Let us first assume that we have a sinusoidal ac generator, which is a rotating machine driven by some other device whose output is a mechanical torque, such as a steam turbine, an electric motor, or an internal-combustion engine. We shall let our generator produce an output voltage of 200 V rms at 60 Hz. Suppose now that an additional rating of the generator is stated as a maximum power output of 1 kW. The generator would therefore be capable of delivering an rms current of 5 A to a resistive load. If, however, a load requiring 1 kW at a lagging power factor of 0.5 is connected to the generator, then an rms current of 10 A is necessary. As the PF decreases, greater and greater currents must be delivered to the load if operation at 200 V and 1 kW is to be maintained. If our generator were correctly and economically designed to furnish safely a maximum current of 5 A, then these greater currents would cause unsatisfactory operation, such as causing the insulation to overheat and begin smoking, which could be injurious to its health.

The rating of the generator is more informatively given in terms of apparent power in voltamperes. Thus a 1000-VA rating at 200 V indicates that the generator can deliver a maximum current of 5 A at rated voltage; the power it delivers depends on the load, and in an extreme case might be zero. An apparent power rating is equivalent to a current rating when operation is at a constant voltage.

When electric power is being supplied to large industrial consumers by a power company, the company will frequently include a PF clause in its rate schedules. Under this clause, an additional charge is made to the consumer whenever the PF drops below a certain specified value, usually about 0.85 lagging. Very little industrial power is consumed at leading PFs, because of the nature of the typical industrial loads. There are several reasons that force the power company to make this additional charge for low PFs. In the first place, it is evident that larger current-carrying capacity must be built into its generators in order to provide the larger currents that go with lower-PF operation at constant power and constant voltage. Another reason is found in the increased losses in its transmission and distribution system.

As an example, let us suppose that a certain consumer is using an average power of 11 kW at unity PF and 220 V rms. We also assume a total resistance of 0.2 Ω in the transmission lines through which the power is delivered to the consumer. An rms current of 50 A therefore flows in the load and in the lines, producing a line loss of 500 W. In order to supply 11 kW to the consumer, the power company must generate 11.5 kW (at the higher voltage, 230 V). Since the energy is necessarily metered at the location of each consumer, this consumer would be billed for 95.6 percent of the energy which the power company actually produced.

Now let us hypothesize another consumer, also requiring 11 kW, but at a PF angle of 60° lagging. This consumer forces the power company to push 100 A through the load and (of particular interest to the company) through the line resistance. The line losses are now found to be 2 kW, and the customer's meter indicates only 84.6 percent of the actual energy generated. This figure departs from 100 percent by more than the power company will tolerate; this costs it money. Of course, the transmission losses might be reduced by using heavier transmission lines, which have lower resistance, but this costs more money too. The power company's solution to this problem is to encourage operation at PFs which exceed 0.9 lagging, by offering slightly reduced rates, and to discourage operation at PFs which are less than 0.85 lagging, by invoking increased rates.

The power drawn by most homes is used at reasonably high PFs (and reasonably small power levels); no charge is usually made for low-PF operation.

Besides paying for the actual energy consumed and for operation at excessively low PFs, industrial consumers are also billed for inordinate *demand*.[2] An energy of 100 kWh is delivered much more economically as 5 kW for 20 h than it is as 20 kW for 5 h, particularly if everyone else demands large amounts of power at the same time.

Drill Problem

10-8. A 440-V rms source supplies power to a load $\mathbf{Z}_L = 10 + j2$ Ω through a transmission line having a total resistance of 1.5 Ω. Find (a) the average and apparent power supplied to the load; (b) the average and apparent power lost in the transmission line; (c) the average and apparent power supplied by the source; (d) the power factor at which the source operates. (e) Does the average power supplied by the source equal the sum of the average powers lost in the transmission line and delivered to the load? (f) Does the apparent power supplied by the source equal the sum of the apparent powers lost in the transmission line and delivered to the load?
Ans: 14 209 W, 14 491 VA; 2131 W, 2131 VA; 16 341 W, 16 586 VA; 0.985 lag; yes; no

[2] Several electrical utilities have made experimental installations of demand kilowatthour meters on residential customer premises.

Some simplification in power calculations is achieved if power is considered to be a complex quantity. The magnitude of the complex power will be found to be the apparent power, and the real part of the complex power will be shown to be the (real) average power. The new quantity, the imaginary part of the complex power, we shall call *reactive power*.

We define complex power with reference to a general sinusoidal voltage $\mathbf{V}_{eff} = V_{eff}\underline{/\theta}$ across a pair of terminals and a general sinusoidal current $\mathbf{I}_{eff} = I_{eff}\underline{/\phi}$ flowing into one of the terminals in such a way as to satisfy the passive sign convention. The average power P absorbed by the two-terminal network is thus

$$P = V_{eff}I_{eff}\cos(\theta - \phi)$$

Complex nomenclature is next introduced by making use of Euler's formula in the same way as we did in introducing phasors. We express P as

$$P = V_{eff}I_{eff}\operatorname{Re}[e^{j(\theta-\phi)}]$$

or

$$P = \operatorname{Re}[V_{eff}e^{j\theta}I_{eff}e^{-j\phi}]$$

The phasor voltage may now be recognized as the first two factors within the brackets in the preceding equation, but the second two factors do not quite correspond to the phasor current, because the angle includes a minus sign, which is not present in the expression for the phasor current. That is, the phasor current is

$$\mathbf{I}_{eff} = I_{eff}e^{j\phi}$$

and we therefore must make use of conjugate notation:

$$\mathbf{I}_{eff}^* = I_{eff}e^{-j\phi}$$

Hence

$$P = \operatorname{Re}[\mathbf{V}_{eff}\mathbf{I}_{eff}^*]$$

and we may now let power become complex by defining the complex power $\mathbf{S}$ as

$$\mathbf{S} = \mathbf{V}_{eff}\mathbf{I}_{eff}^* \qquad (24)$$

If we first inspect the polar or exponential form of the complex power,

$$\mathbf{S} = V_{eff}I_{eff}e^{j(\theta-\phi)}$$

it is evident that the magnitude of $\mathbf{S}$ is the apparent power and the angle of $\mathbf{S}$ is the PF angle, that is, the angle by which the voltage leads the current.

In rectangular form, we have

$$\mathbf{S} = P + jQ \qquad (25)$$

where P is the real average power, as before. The imaginary part of the complex power is symbolized as Q and is termed the *reactive power*. The dimensions of Q are obviously the same as those of the real power P, the complex power $\mathbf{S}$, and the apparent power $|\mathbf{S}|$. In order to avoid confusion with these other quantities, the unit of Q is defined as the *var* (abbreviated VAR), standing for volt-ampere reactive. From Eq. (24), it is seen that

$$Q = V_{eff}I_{eff}\sin(\theta - \phi)$$

Another interpretation of the reactive power may be seen by constructing a phasor diagram containing $\mathbf{V}_{\text{eff}}$ and $\mathbf{I}_{\text{eff}}$ as shown in Fig. 10-10. If the phasor current is resolved into two components, one in phase with the voltage, having a magnitude $I_{\text{eff}} \cos(\theta - \phi)$, and one 90° out of phase with the voltage, with magnitude equal to $I_{\text{eff}} \sin|\theta - \phi|$, then it is clear that the real power is given by the product of the magnitude of the voltage phasor and the component of the phasor current which is in phase with the voltage. Moreover, the product of the magnitude of the voltage phasor and the component of the phasor current which is 90° out of phase with the voltage is the reactive power Q. It is common to speak of the component of a phasor which is 90° out of phase with some other phasor as a *quadrature component*. Thus Q is simply V_{eff} times the quadrature component of $\mathbf{I}_{\text{eff}}$. Q is also known as the *quadrature power*.

Figure 10-10

The current phasor $\mathbf{I}_{\text{eff}}$ is resolved into two components, one in phase with the voltage phasor $\mathbf{V}_{\text{eff}}$ and the other 90° out of phase with the voltage phasor. This latter component is called a *quadrature component*.

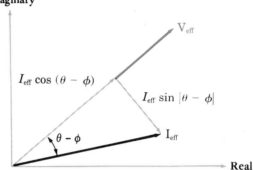

The sign of the reactive power characterizes the nature of a passive load at which $\mathbf{V}_{\text{eff}}$ and $\mathbf{I}_{\text{eff}}$ are specified. If the load is inductive, then $(\theta - \phi)$ is an angle between 0 and 90°, the sine of this angle is positive, and the reactive power is positive. A capacitive load results in a negative reactive power.

Just as a wattmeter reads the average real power drawn by a load, a varmeter will read the average reactive power Q drawn by the load. Both quantities may be metered simultaneously. In addition, watthourmeters and varhourmeters may be used simultaneously to record real and reactive energy used by any consumer during any desired time interval. From these records the average PF may be determined and the consumer's bill may be adjusted accordingly.

It is easy to show that the complex power delivered to several interconnected loads is the sum of the complex powers delivered to each of the individual loads, no matter how the loads are interconnected. For example, consider the two loads shown connected in parallel in Fig. 10-11. If rms values are assumed, the complex power drawn by the combined load is

$$\mathbf{S} = \mathbf{V}\mathbf{I}^* = \mathbf{V}(\mathbf{I}_1 + \mathbf{I}_2)^* = \mathbf{V}(\mathbf{I}_1^* + \mathbf{I}_2^*)$$

Figure 10-11

A circuit used to show that the complex power drawn by two parallel loads is the sum of the complex powers drawn by the individual loads.

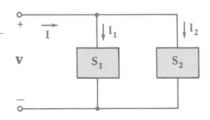

and thus

$$\mathbf{S} = \mathbf{VI}_1^* + \mathbf{VI}_2^*$$

as stated.

These new ideas can be clarified by a practical numerical example.

Example 10-7 Let us suppose that an industrial consumer is operating a 50-kW (67.1-hp) induction motor at a lagging PF of 0.8. The source voltage is 230 V rms. In order to obtain lower electrical rates, the customer wishes to raise the PF to 0.95 lagging. Specify an arrangement by which this may be done.

Solution: Although the PF might be raised by increasing the real power and maintaining the reactive power constant, this would not result in a lower bill and is not a cure which interests the consumer. A purely reactive load must be added to the system, and it is clear that it must be added in parallel, since the supply voltage to the induction motor must not change. The circuit of Fig. 10-11 is thus applicable if we interpret $\mathbf{S}_1$ as the induction motor's complex power and $\mathbf{S}_2$ as the complex power drawn by the corrective device.

The complex power supplied to the induction motor must have a real part of 50 kW and an angle of $\cos^{-1}(0.8)$, or 36.9°. Hence,

$$\mathbf{S}_1 = \frac{50\,\underline{/36.9^\circ}}{0.8} = 50 + j37.5 \text{ kVA}$$

In order to achieve a PF of 0.95, the total complex power must become

$$\mathbf{S} = \frac{50}{0.95}\,\underline{/\cos^{-1}(0.95)} = 50 + j16.43 \text{ kVA}$$

Thus, the complex power drawn by the corrective load is

$$\mathbf{S}_2 = -j21.07 \text{ kVA}$$

The necessary load impedance $\mathbf{Z}_2$ may be found in several simple steps. We select a phase angle of 0° for the voltage source, and therefore the current drawn by $\mathbf{Z}_2$ is

$$\mathbf{I}_2^* = \frac{\mathbf{S}_2}{\mathbf{V}} = \frac{-j21\,070}{230} = -j91.6 \text{ A}$$

or

$$\mathbf{I}_2 = j91.6 \text{ A}$$

Therefore,

$$\mathbf{Z}_2 = \frac{\mathbf{V}_2}{\mathbf{I}_2} = \frac{230}{j91.6} = -j2.51 \text{ }\Omega$$

If the operating frequency is 60 Hz, this load can be provided by a 1056-μF capacitor connected in parallel with the motor. ∎

A capacitive load may also be simulated by a so-called synchronous capacitor, a type of rotating machine. This is usually the more economical procedure for small capacitive reactances (large capacitances). Whatever device is selected, however, its initial cost, maintenance, and depreciation must be covered by the reduction in the electric bill.

Drill Problem

10-9. For the circuit shown in Fig. 10-12, find the complex power absorbed by the (*a*) 1-Ω resistor; (*b*) −*j*10-Ω capacitor; (*c*) 5 + *j*10-Ω impedance; (*d*) source.

Ans: 26.6 + *j*0 VA; 0 − *j*1331 VA; 532 + *j*1065 VA; −559 + *j*266 VA

Figure 10-12

See Drill Prob. 9.

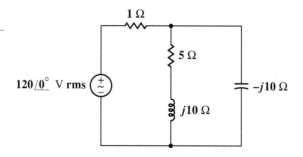

Problems

1 A current source, $i_s(t) = 2 \cos 500t$ A, a 50-Ω resistor, and a 25-μF capacitor are in parallel. Find the power being supplied by the source, being absorbed by the resistor, and being absorbed by the capacitor, all at $t = \pi/2$ ms.

2 The current $i = 2t^2 − 1$ A, $1 \leq t \leq 3$ s, is flowing through a certain circuit element. (*a*) If the element is a 4-H inductor, what energy is delivered to it in the given time interval? (*b*) If the element is a 0.2-F capacitor with $v(1) = 2$ V, what power is being delivered to it at $t = 2$ s?

3 If $v_C(0) = −2$ V and $i(0) = 4$ A in the circuit of Fig. 10-13, find the power being absorbed by the capacitor at t equal to (*a*) 0^+; (*b*) 0.2 s; (*c*) 0.4 s.

Figure 10-13

See Prob. 3.

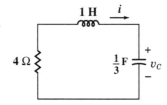

4 The circuit shown in Fig. 10-14 has reached steady-state conditions. Find the power being absorbed by each of the four circuit elements at $t = 0.1$ s.

Figure 10-14

See Prob. 4.

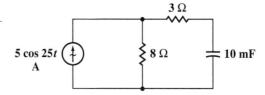

5 Find the average power being absorbed by each of the five circuit elements shown in Fig. 10-15.

Figure 10-15

See Prob. 5.

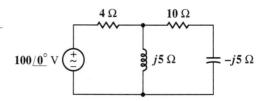

6 Calculate the average power generated by each source and the average power delivered to each impedance in the circuit of Fig. 10-16.

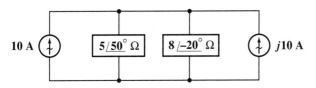

Figure 10-16

See Prob. 6.

7 In the circuit shown in Fig. 10-17, find the average power being (*a*) dissipated in the 3-Ω resistor; (*b*) generated by the source.

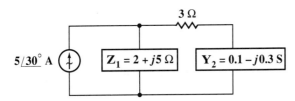

Figure 10-17

See Prob. 7.

8 Find the average power absorbed by each of the five circuit elements shown in Fig. 10-18.

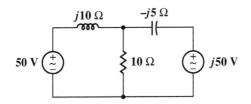

Figure 10-18

See Prob. 8.

9 Determine the average power supplied by the dependent source in the circuit of Fig. 10-19.

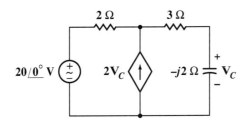

Figure 10-19

See Prob. 9.

10 A frequency-domain Thévenin equivalent circuit consists of a sinusoidal source $\mathbf{V}_{th}$ in series with an impedance $\mathbf{Z}_{th} = R_{th} + jX_{th}$. Specify the conditions on a load $\mathbf{Z}_L = R_L + jX_L$ if it is to receive a maximum average power subject to the constraint that (*a*) $X_{th} = 0$; (*b*) R_L and X_L may be selected independently; (*c*) R_L is fixed (not equal to R_{th}); (*d*) X_L is fixed (independent of X_{th}); (*e*) $X_L = 0$.

11 For the circuit of Fig. 10-20: (*a*) what value of $\mathbf{Z}_L$ will absorb a maximum average power? (*b*) What is the value of this maximum power?

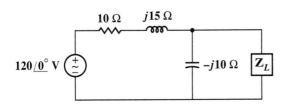

Figure 10-20

See Probs. 11 and 12.

12 For the circuit of Fig. 10-20, it is required that the load be a pure resistance R_L. What value of R_L will absorb a maximum average power, and what is the value of this power?

13 Find the average power supplied by the dependent source of Fig. 10-21.

Figure 10-21

See Prob. 13.

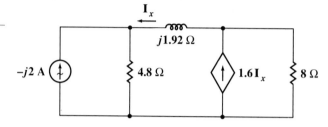

14 For the network of Fig. 10-22: (a) what impedance $\mathbf{Z}_L$ should be connected between a and b so that a maximum average power will be absorbed by it? (b) What is this maximum average power?

Figure 10-22

See Prob. 14.

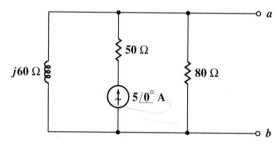

15 Find the value of R_L in the circuit of Fig. 10-23 that will absorb a maximum power, and specify the value of that power.

Figure 10-23

See Prob. 15.

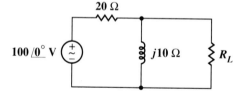

16 (a) Calculate the average value of each of the waveforms shown in Fig. 10-7. (b) If each of these waveforms is now squared, find the average value of each of the new periodic waveforms (in A²).

17 Find the effective value of (a) $v(t) = 10 + 9 \cos 100t + 6 \sin 100t$; (b) the waveform appearing as Fig. 10-24. (c) Also find the average value of this waveform.

Figure 10-24

See Prob. 17.

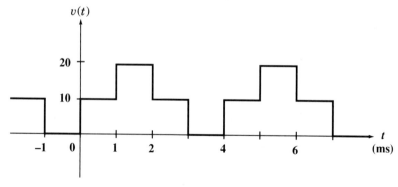

18 Find the effective value of (a) $g(t) = 2 + 3 \cos 100t + 4 \cos (100t - 120°)$; (b) $h(t) = 2 + 3 \cos 100t + 4 \cos (101t - 120°)$; (c) the waveform of Fig. 10-25.

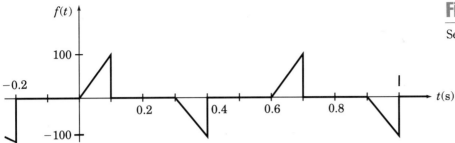

Figure 10-25

See Prob. 18.

19 Given the periodic waveform $f(t) = (2 - 3 \cos 100t)^2$, find (a) its average value; (b) its rms value.

20 A voltage waveform has a period of 5 s, and it is expressed as $v(t) = 10t[u(t) - u(t - 2)] + 16e^{-0.5(t-3)}[u(t - 3) - u(t - 5)]$ V in the interval $0 < t < 5$ s. Find the effective value of the waveform.

21 Four ideal independent voltage sources, $A \cos 10t$, $B \sin (10t + 45°)$, $C \cos 40t$, and the constant D, are connected in series with a 4-Ω resistor. Find the average power dissipated in the resistor if (a) $A = B = 10$ V, $C = D = 0$; (b) $A = C = 10$ V, $B = D = 0$; (c) $A = 10$ V, $B = -10$ V, $C = D = 0$; (d) $A = B = C = 10$ V, $D = 0$; (e) $A = B = C = D = 10$ V.

22 Each of the waveforms shown in Fig. 10-26 has a period of 3 s. They are also somewhat similar. (a) Calculate the average value of each one. (b) Determine the two effective values.

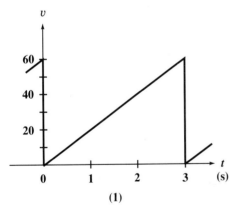

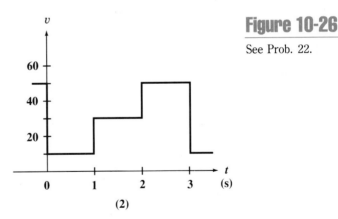

Figure 10-26

See Prob. 22.

(1)

(2)

23 In Fig. 10-27, let $\mathbf{I} = 4\underline{/35°}$ A rms, and find the average power being supplied: (a) by the source; (b) to the 20-Ω resistor; (c) to the load. Find the apparent power being supplied: (d) by the source; (e) to the 20-Ω resistor; (f) to the load. (g) What is the load PF?

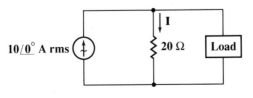

$10\underline{/0°}$ A rms

I

20 Ω

Load

Figure 10-27

See Prob. 23.

24 (a) Find the power factor at which the source in the circuit of Fig. 10-28 is operating. (b) Find the average power being supplied by the source. (c) What size capacitor should be placed in parallel with the source to cause its power factor to be unity?

Figure 10-28

See Prob. 24.

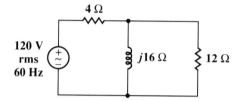

25 In the circuit shown in Fig. 10-29, let $\mathbf{Z}_A = 5 + j2\ \Omega$, $\mathbf{Z}_B = 20 - j10\ \Omega$, $\mathbf{Z}_C = 10\underline{/30°}$ Ω, and $\mathbf{Z}_D = 10\underline{/-60°}\ \Omega$. Find the apparent power delivered to each load and the apparent power generated by the source.

Figure 10-29

See Prob. 25.

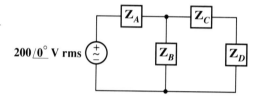

26 The load in Fig. 10-30 draws 10 kVA at PF = 0.8 lagging. If $|\mathbf{I}_L| = 40$ A rms, what must be the value of C to cause the source to operate at PF = 0.9 lagging?

Figure 10-30

See Prob. 26.

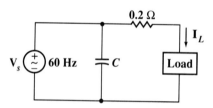

27 Let us visualize a network operating at $f = 50$ Hz that utilizes loads connected in series and carrying a common current of $10\underline{/0°}$ A rms. Such a system is the dual of one operating with parallel loads and a common voltage. In the series system, a load would be turned off by short-circuiting it; open circuits would cause all kinds of fireworks. Two loads are on this particular system: $\mathbf{Z}_1 = 30\underline{/15°}\ \Omega$ and $\mathbf{Z}_2 = 40\underline{/40°}\ \Omega$. (a) At what PF is the source operating? (b) What size capacitor should be installed in the series circuit to cause a lagging power factor of 0.9?

28 A 250-V rms system is supplying three parallel loads. One draws 20 kW at unity power factor, a second uses 25 kVA at PF = 0.8 lagging, and the third requires a power of 30 kW at a lagging PF of 0.75. (a) Find the total power supplied by the source. (b) Find the total apparent power supplied by the source. (c) At what PF does the source operate?

29 Analyze the circuit of Fig. 10-31 to find the complex power absorbed by each of the five circuit elements.

Figure 10-31

See Prob. 29.

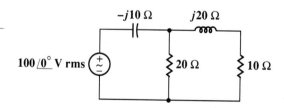

30 Both sources shown in Fig. 10-32 are operating at the same frequency. Find the complex power generated by each source and the complex power absorbed by each passive circuit element.

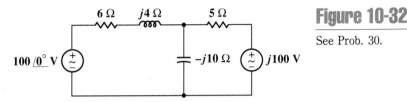

Figure 10-32

See Prob. 30.

31 Find the complex power being delivered to a load that (*a*) draws 500 VA at a leading PF of 0.75; (*b*) draws 500 W at a leading PF of 0.75; (*c*) draws −500 VAR at a PF of 0.75.

32 A capacitive impedance, $\mathbf{Z}_C = -j120 \ \Omega$, is in parallel with a load $\mathbf{Z}_L$. The parallel combination is supplied by a source, $\mathbf{V}_s = 400\underline{/0°}$ V rms, that generates a complex power of $1.6 + j0.5$ kVA. Find the (*a*) complex power delivered to $\mathbf{Z}_L$; (*b*) PF of $\mathbf{Z}_L$; (*c*) PF of the source.

33 A source of 230 V rms is supplying three loads in parallel: 1.2 kVA at a lagging PF of 0.8, 1.6 kVA at a lagging PF of 0.9, and 900 W at unity PF. Find (*a*) the amplitude of the source current; (*b*) the PF at which the source is operating; (*c*) the complex power being furnished by the source.

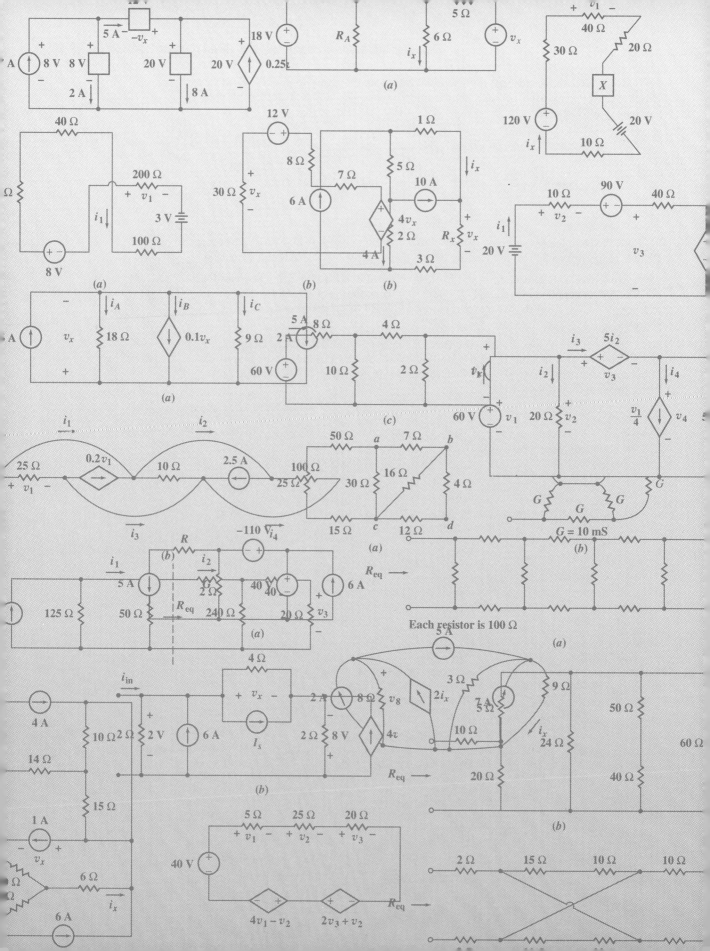

Polyphase Circuits

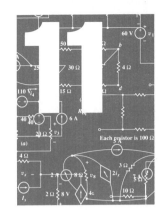

One of the reasons for studying the sinusoidal steady state is that most household and industrial electric power is utilized as alternating current. The sinusoidal waveform may characterize a special mathematical function, but it represents a very common and very useful forcing function. In the same way, a *polyphase* source is even more specialized, but we consider it because almost the entire output of the electric power industry in this country is generated and distributed as polyphase power at a 60-Hz frequency. Before defining our terms carefully, let us look briefly at the most common polyphase system, a balanced three-phase system. The source has perhaps three terminals, and voltmeter measurements will show that sinusoidal voltages of equal amplitude are present between any two terminals. However, these voltages are not in phase; it will be easily shown later that each of the three voltages is 120° out of phase with each of the other two, the sign of the phase angle depending on the sense of the voltages. A balanced load draws power equally from the three phases, but when one of the voltages is instantaneously zero, the phase relationship shows that the other two must each be at half amplitude. At no instant does the instantaneous power drawn by the total load reach zero; in fact, this total instantaneous power is constant. This is an advantage in rotating machinery, for it keeps the torque on the rotor much more constant than it would be if a single-phase source were used. There is less vibration.

There are also advantages in using rotating machinery to generate three-phase power rather than single-phase power, and there are economical advantages in favor of the transmission of power in a three-phase system.

The use of a higher number of phases, such as 6- and 12-phase systems, is limited almost entirely to the supply of power to large rectifiers. Here, the rectifiers convert the alternating current to direct current, which is required for certain processes such as electrolysis. The rectifier output is a direct current plus a smaller pulsating component, or ripple, which decreases as the number of phases increases.

Almost without exception, polyphase systems in practice contain sources which may be closely approximated by ideal voltage sources or by ideal voltage sources in series with small internal impedances. Three-phase current sources are extremely rare.

It is convenient to describe polyphase voltages and currents using a double-subscript notation. With this notation, a voltage or current, such as $\mathbf{V}_{ab}$ or $\mathbf{I}_{aA}$, has more meaning than if it were indicated simply as $\mathbf{V}_3$ or $\mathbf{I}_x$. By definition, let the voltage of point a with respect to point b be $\mathbf{V}_{ab}$. Thus, the plus sign is located at a, as indicated in Fig. 11-1a. We therefore consider the double subscripts to be *equivalent to* a plus-minus sign pair. The use of both would be redundant. With reference to Fig. 11-1b, it is now obvious that $\mathbf{V}_{ad} = \mathbf{V}_{ab} + \mathbf{V}_{cd}$. The advantage of the double-subscript notation lies in the fact that Kirchhoff's voltage law requires the voltage between two points to be the same, regardless of the path chosen between the points; thus $\mathbf{V}_{ad} = \mathbf{V}_{ab} + \mathbf{V}_{bd} = \mathbf{V}_{ac} + \mathbf{V}_{cd} = \mathbf{V}_{ab} + \mathbf{V}_{bc} + \mathbf{V}_{cd}$, and so forth. It is apparent that KVL may be satisfied without reference to the circuit diagram; correct equations may be written even though a point, or subscript letter, is included which is not marked on the diagram. For example, we might have written $\mathbf{V}_{ad} = \mathbf{V}_{ax} + \mathbf{V}_{xd}$, where x identifies the location of any interesting point of our choice.

Figure 11-1

(a) The definition of the voltage $\mathbf{V}_{ab}$. (b) $\mathbf{V}_{ad} = \mathbf{V}_{ab} + \mathbf{V}_{bc} + \mathbf{V}_{cd} = \mathbf{V}_{ab} + \mathbf{V}_{cd}$.

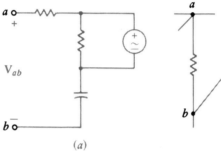 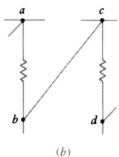

(a)   (b)

One possible representation of a three-phase system of voltages[1] will be found to be that of Fig. 11-2. Let us assume that the voltages $\mathbf{V}_{an}$, $\mathbf{V}_{bn}$, and $\mathbf{V}_{cn}$ are known:

$$\mathbf{V}_{an} = 100\,\underline{/0°}\text{ V rms}$$

$$\mathbf{V}_{bn} = 100\,\underline{/-120°}$$

$$\mathbf{V}_{cn} = 100\,\underline{/-240°}$$

Figure 11-2

A network used as a numerical example of double-subscript voltage notation.

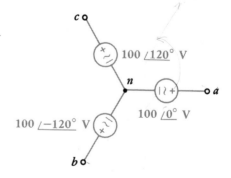

---

[1] Rms values of currents and voltages will be used throughout this chapter.

and thus the voltage $\mathbf{V}_{ab}$ may be found, with an eye on the subscripts:

$$\mathbf{V}_{ab} = \mathbf{V}_{an} + \mathbf{V}_{nb} = \mathbf{V}_{an} - \mathbf{V}_{bn}$$
$$= 100\,\underline{/0°} - 100\,\underline{/-120°}$$
$$= 100 - (-50 - j86.6)$$
$$= 173.2\,\underline{/30°}$$

The three given voltages and the construction of the phasor $\mathbf{V}_{ab}$ are shown on the phasor diagram of Fig. 11-3.

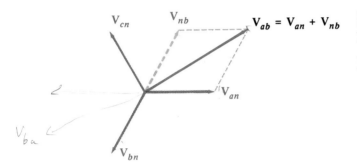

Figure 11-3

This phasor diagram illustrates the graphical use of the double-subscript voltage convention to obtain $\mathbf{V}_{ab}$ for the network of Fig. 11-2.

A double-subscript notation may also be applied to currents. We define the current $\mathbf{I}_{ab}$ as the current flowing from a to b *by the direct path*. In every complete circuit we consider, there must of course be at least two possible paths between the points a and b, and we agree that we shall not use double-subscript notation unless it is obvious that one path is much shorter, or much more direct. Usually this path is through a single element. Thus, the current $\mathbf{I}_{ab}$ is correctly indicated in Fig. 11-4. In fact, we do not even need the direction arrow when talking about this current; the subscripts *tell* us the direction. However, the mere identification of a current as $\mathbf{I}_{cd}$ would cause confusion.

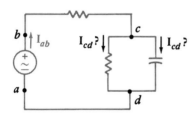

Figure 11-4

An illustration of the use and misuse of the double-subscript convention for current notation.

Before considering polyphase systems, we shall make use of double-subscript notation to help with the analysis of a special single-phase system.

Drill Problems

11-1. Let $\mathbf{V}_{ab} = 100\,\underline{/0°}$ V, $\mathbf{V}_{bd} = 40\,\underline{/80°}$ V, and $\mathbf{V}_{ca} = 70\,\underline{/200°}$ V. Find (a) $\mathbf{V}_{ad}$; (b)$\mathbf{V}_{bc}$; (c) $\mathbf{V}_{cd}$.
Ans: 114.0$\underline{/20.2°}$ V; 41.8$\underline{/145.0°}$ V ; 44.0$\underline{/20.6°}$ V

11-2. Refer to the circuit shown in Fig. 11-5 and let $I_{fj} = 3$ A, $I_{de} = 2$ A, and $I_{hd} = -6$ A. Find (a) I_{cd}; (b) I_{ef}; (c) I_{ij}.
Ans: −3 A; 7 A; 7 A

Figure 11-5

See Drill Prob. 11-2.

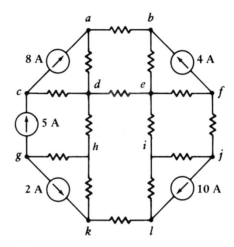

11-2

Single-phase three-wire systems

A single-phase three-wire source is defined as a source having three output terminals, such as a, n, and b in Fig. 11-6a, at which the phasor voltages $\mathbf{V}_{an}$ and $\mathbf{V}_{nb}$ are equal. The source may therefore be represented by the combination of two identical voltage sources; in Fig. 11-6b, $\mathbf{V}_{an} = \mathbf{V}_{nb} = \mathbf{V}_1$. It is apparent that $\mathbf{V}_{ab} = 2\mathbf{V}_{an} = 2\mathbf{V}_{nb}$, and we therefore have a source to which loads operating at either of two voltages may be connected. The normal household system is single-phase three-wire, permitting the operation of both 115-V and 230-V appliances. The higher-voltage appliances are normally those drawing larger amounts of power, and thus they cause currents in the lines which are only half those which operation at the same power and half the voltage would produce. Smaller-diameter wire may consequently be used safely in the appliance, the household distribution system, and the distribution system of the utility company.

Figure 11-6

(a) A single-phase three-wire source. (b) The representation of a single-phase three-wire source by two identical voltage sources.

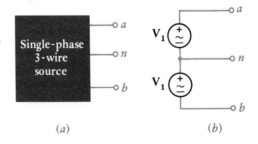

The name *single-phase* arises because the voltages $\mathbf{V}_{an}$ and $\mathbf{V}_{nb}$, being equal, must have the same phase angle. From another viewpoint, however, the voltages between the outer wires and the central wire, which is usually referred to as the *neutral*, are exactly 180° out of phase. That is, $\mathbf{V}_{an} = -\mathbf{V}_{bn}$, and $\mathbf{V}_{an} + \mathbf{V}_{bn} = 0$. Later, we shall see that balanced polyphase systems are characterized by a set of voltages of equal *amplitude* whose (phasor) sum is zero. From this viewpoint, then, the single-phase three-wire system is really a balanced two-phase system. *Two-phase,* however, is a term that is traditionally reserved for a relatively unimportant unbalanced system utilizing two voltage sources 90° out of phase; we shall not discuss it further.

Let us now consider a single-phase three-wire system which contains identical loads $\mathbf{Z}_p$ between each outer wire and the neutral (Fig. 11-7). We shall first assume that the wires connecting the source to the load are perfect conductors.

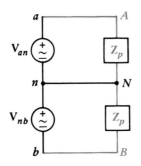

Figure 11-7

A simple single-phase three-wire system. The two loads are identical, and the neutral current is zero.

Since

$$\mathbf{V}_{an} = \mathbf{V}_{nb}$$

then,

$$\mathbf{I}_{aA} = \frac{\mathbf{V}_{an}}{\mathbf{Z}_p} = \mathbf{I}_{Bb} = \frac{\mathbf{V}_{nb}}{\mathbf{Z}_p}$$

and therefore

$$\mathbf{I}_{nN} = \mathbf{I}_{Bb} + \mathbf{I}_{Aa} = \mathbf{I}_{Bb} - \mathbf{I}_{aA} = 0$$

Thus there is no current in the neutral wire, and it could be removed without changing any current or voltage in the system. This result is achieved through the equality of the two loads and of the two sources.

We next consider the effect of a finite impedance in each of the wires. If lines aA and bB each have the same impedance, this impedance may be added to $\mathbf{Z}_p$, resulting in two equal loads once more, and zero neutral current. Now let us allow the neutral wire to possess some impedance $\mathbf{Z}_n$. Without carrying out any detailed analysis, superposition should show us that the symmetry of the circuit will still cause zero neutral current. Moreover, the addition of any impedance connected directly from one of the outer lines to the other outer line also yields a symmetrical circuit and zero neutral current. Thus, zero neutral current is a consequence of a balanced, or symmetrical, load; any impedance in the neutral wire does not destroy the symmetry.

The most general single-phase three-wire system will contain unequal loads between each outside line and the neutral and another load directly between the two outer lines; the impedances of the two outer lines may be expected to be approximately equal, but the neutral impedance is often slightly larger. Let us consider an example of such a system.

Example 11-1 We are asked to analyze the system shown in Fig. 11-8 in order to determine the power delivered to each of the three loads and the power lost in the neutral wire and each of the two lines.

Figure 11-8

A typical single-phase three-wire system.

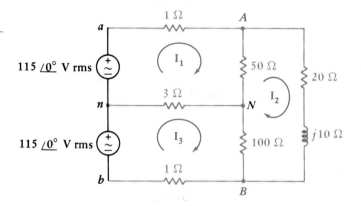

Solution: The analysis of the circuit may be achieved by assigning mesh currents and writing the appropriate equations. The three mesh currents are

$$\mathbf{I}_1 = \frac{\begin{vmatrix} 115 & -50 & -3 \\ 0 & 170+j10 & -100 \\ 115 & -100 & 104 \end{vmatrix}}{\begin{vmatrix} 54 & -50 & -3 \\ -50 & 170+j10 & -100 \\ -3 & -100 & 104 \end{vmatrix}} = 11.24\underline{/-19.83°}\text{ A rms}$$

$$\mathbf{I}_2 = \frac{\begin{vmatrix} 54 & 115 & -3 \\ -50 & 0 & -100 \\ -3 & 115 & 104 \end{vmatrix}}{\text{denom}} = 9.39\underline{/-24.47°}$$

$$\mathbf{I}_3 = \frac{\begin{vmatrix} 54 & -50 & 115 \\ -50 & 170+j10 & 0 \\ -3 & -100 & 115 \end{vmatrix}}{\text{denom}} = 10.37\underline{/-21.80°}$$

The currents in the outer lines are thus

$$\mathbf{I}_{aA} = \mathbf{I}_1 = 11.24\underline{/-19.83°}\text{ A rms}$$

$$\mathbf{I}_{bB} = -\mathbf{I}_3 = 10.37\underline{/158.20°}$$

and the smaller neutral current is

$$\mathbf{I}_{nN} = \mathbf{I}_3 - \mathbf{I}_1 = 0.946\underline{/-177.7°}\text{ A rms}$$

The power drawn by each load may be determined:

$$P_{50} = |\mathbf{I}_1 - \mathbf{I}_2|^2(50) = 206\text{ W}$$

which could represent two 100-W lamps in parallel. Also,

$$P_{100} = |\mathbf{I}_3 - \mathbf{I}_2|^2(100) = 117\text{ W}$$

which might represent one 100-W lamp. Finally,

$$P_{20+j10} = |\mathbf{I}_2|^2(20) = 1763 \text{ W}$$

which we may think of as a 2-hp induction motor. The total load power is 2086 W. The loss in each of the wires is next found:

$$P_{aA} = |\mathbf{I}_1|^2(1) = 126 \text{ W}$$

$$P_{bB} = |\mathbf{I}_3|^2(1) = 108 \text{ W}$$

$$P_{nN} = |\mathbf{I}_{nN}|^2(3) = 3 \text{ W}$$

giving a total line loss of 237 W. The wires are evidently quite long; otherwise, the relatively high power loss in the two outer lines would cause a dangerous temperature rise. The total generated power must therefore be 206 + 117 + 1763 + 237, or 2323 W, and this may be checked by finding the power delivered by each voltage source:

$$P_{an} = 115(11.24) \cos 19.83° = 1216 \text{ W}$$

$$P_{bn} = 115(10.37) \cos 21.80° = 1107 \text{ W}$$

or a total of 2323 W. The transmission efficiency for this system is

$$\text{Eff.} = \frac{2086}{2086 + 237} = 89.8\%$$

This value would be unbelievable for a steam engine or an internal-combustion engine, but it is too low for a well-designed distribution system. Larger-diameter wires should be used if the source and the load cannot be placed closer to each other.

A phasor diagram showing the two source voltages, the currents in the outer lines, and the current in the neutral is constructed in Fig. 11-9. The fact that $\mathbf{I}_{aA} + \mathbf{I}_{bB} + \mathbf{I}_{nN} = 0$ is indicated on the diagram.

■

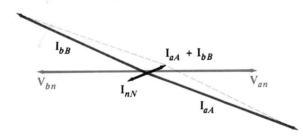

Figure 11-9

The source voltages and three of the currents in the circuit of Fig. 11-8 are shown on a phasor diagram. Note that $\mathbf{I}_{aA} + \mathbf{I}_{bB} + \mathbf{I}_{nN} = 0$.

Drill Problem

11-3. In Fig. 11-8, add 1.5-Ω resistance to each of the two outer lines, and 2.5-Ω resistance to the neutral wire. Find the average power delivered to each of the three loads.   *Ans:* 153.1 W; 95.8 W; 1374 W

11-3

Three-phase Y-Y connection

Three-phase sources have three terminals, called the *line* terminals, and they may or may not have a fourth terminal, the *neutral* connection. We shall begin by discussing a three-phase source which does have a neutral connection. It may be represented by three ideal voltage sources connected in a Y, as shown in Fig. 11-10; terminals a, b, c, and n are available. We shall consider only balanced three-phase sources, which may be defined as having

$$|\mathbf{V}_{an}| = |\mathbf{V}_{bn}| = |\mathbf{V}_{cn}|$$

and

$$\mathbf{V}_{an} + \mathbf{V}_{bn} + \mathbf{V}_{cn} = 0$$

These three voltages, each existing between one line and the neutral, are called *phase voltages*. If we arbitrarily choose $\mathbf{V}_{an}$ as the reference, or define

$$\mathbf{V}_{an} = V_p\underline{/0°}$$

where we shall consistently use V_p to represent the rms *amplitude* of any of the phase voltages, then the definition of the three-phase source indicates that either

$$\mathbf{V}_{bn} = V_p\underline{/-120°} \qquad \mathbf{V}_{cn} = V_p\underline{/-240°}$$

or

$$\mathbf{V}_{bn} = V_p\underline{/120°} \qquad \mathbf{V}_{cn} = V_p\underline{/240°}$$

Figure 11-10

A Y-connected three-phase four-wire source.

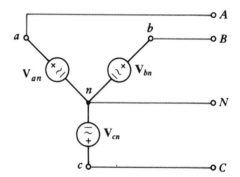

The former is called *positive phase sequence*, or *abc* phase sequence, and is shown in Fig. 11-11*a*; the latter is termed *negative phase sequence*, or *cba* phase sequence, and is indicated by the phasor diagram of Fig. 11-11*b*. It is apparent that the phase sequence of a physical three-phase source depends on the arbitrary choice of the three terminals to be lettered a, b, and c. They may always

Figure 11-11

(*a*) Positive, or *abc*, phase sequence.
(*b*) Negative, or *cba*, phase sequence.

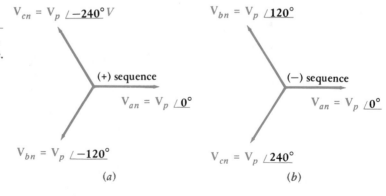

$$V_{cn} = V_p\underline{/-240°}\,V \qquad\qquad V_{bn} = V_p\underline{/120°}$$

(+) sequence   (−) sequence

$$V_{an} = V_p\underline{/0°} \qquad\qquad V_{an} = V_p\underline{/0°}$$

$$V_{bn} = V_p\underline{/-120°} \qquad\qquad V_{cn} = V_p\underline{/240°}$$

(*a*)   (*b*)

be chosen to provide positive phase sequence, and we shall assume that this has been done in most of the systems we consider.

Let us next find the line-to-line voltages (or simply "line" voltages) which are present when the phase voltages are those of Fig. 11-11a. It is easiest to do this with the help of a phasor diagram, since the angles are all multiples of 30°. The necessary construction is shown in Fig. 11-12; the results are

$$\mathbf{V}_{ab} = \sqrt{3}V_p\underline{/30°}$$

$$\mathbf{V}_{bc} = \sqrt{3}V_p\underline{/-90°}$$

$$\mathbf{V}_{ca} = \sqrt{3}V_p\underline{/-210°}.$$

Kirchhoff's voltage law requires the sum of these three voltages to be zero, and it is zero.

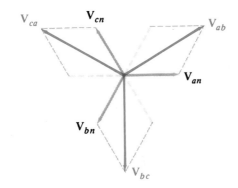

Figure 11-12

A phasor diagram which is used to determine the line voltages from the given phase voltages.

If the rms amplitude of any of the line voltages is denoted by V_L, then one of the important characteristics of the Y-connected three-phase source may be expressed as

$$V_L = \sqrt{3}\,V_p$$

Note that with positive phase sequence, $\mathbf{V}_{an}$ leads $\mathbf{V}_{bn}$ and $\mathbf{V}_{bn}$ leads $\mathbf{V}_{cn}$, in each case by 120°, and also that $\mathbf{V}_{ab}$ leads $\mathbf{V}_{bc}$ and $\mathbf{V}_{bc}$ leads $\mathbf{V}_{ca}$, again by 120°. The statement is true for negative sequence if "lags" is substituted for "leads."

Now let us connect a balanced Y-connected three-phase load to our source, using three lines and a neutral, as drawn in Fig. 11-13. The load is represented by an impedance $\mathbf{Z}_p$ between each line and the neutral. The three line currents

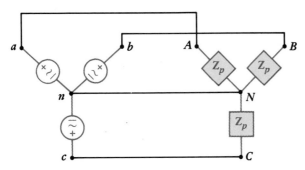

Figure 11-13

A balanced three-phase system, connected Y-Y and including a neutral.

326   **Part Three: Sinusoidal Analysis**

are found very easily, since we really have three single-phase circuits which possess one common lead:

$$\mathbf{I}_{aA} = \frac{\mathbf{V}_{an}}{\mathbf{Z}_p}$$

$$\mathbf{I}_{bB} = \frac{\mathbf{V}_{bn}}{\mathbf{Z}_p} = \frac{\mathbf{V}_{an}\,\underline{/-120°}}{\mathbf{Z}_p} = \mathbf{I}_{aA}\,\underline{/-120°}$$

$$\mathbf{I}_{cC} = \mathbf{I}_{aA}\,\underline{/-240°}$$

and therefore

$$\mathbf{I}_{Nn} = \mathbf{I}_{aA} + \mathbf{I}_{bB} + \mathbf{I}_{cC} = 0$$

Thus, the neutral carries no current if the source and load are both balanced and if the four wires have zero impedance. How will this change if an impedance $\mathbf{Z}_L$ is inserted in series with each of the three lines and an impedance $\mathbf{Z}_n$ is inserted in the neutral? Evidently, the line impedances may be combined with the three load impedances; this effective load is still balanced, and a perfectly conducting neutral wire could be removed. Thus, if no change is produced in the system with a short circuit or an open circuit between n and N, any impedance may be inserted in the neutral and the neutral current will remain zero.

It follows that, if we have balanced sources, balanced loads, and balanced line impedances, a neutral wire of any impedance may be replaced by any other impedance, including a short circuit and an open circuit; the replacement will not affect the system's voltages or currents. It is often helpful to *visualize* a short circuit between the two neutral points, whether a neutral wire is actually present or not; the problem is then reduced to three single-phase problems, all identical except for the consistent difference in phase angle. We say that we thus work the problem on a "per-phase" basis.

Let us work several examples involving a balanced three-phase system having a Y-Y connection.

Example 11-2   A straightforward problem is suggested by the circuit of Fig. 11-14; we are asked to find the several currents and voltages throughout the circuit and to find the total power.

Solution: Since one of the source phase voltages is given, and since positive phase sequence is assumed, the three phase voltages are

$$\mathbf{V}_{an} = 200\underline{/0°} \qquad \mathbf{V}_{bn} = 200\underline{/-120°} \qquad \mathbf{V}_{cn} = 200\underline{/-240°}$$

Figure 11-14

A balanced three-phase three-wire Y-Y connected system.

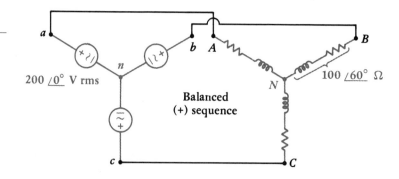

The line voltage is $200\sqrt{3}$, or 346 V rms; the phase angle of each line voltage can be determined by constructing a phasor diagram, as before. As a matter of fact, the phasor diagram of Fig. 11-12 is applicable, and $\mathbf{V}_{ab}$ is $346/\underline{30°}$ V.

Let us work with phase A. The line current is

$$\mathbf{I}_{aA} = \frac{\mathbf{V}_{an}}{\mathbf{Z}_p} = \frac{200/\underline{0°}}{100/\underline{60°}} = 2/\underline{-60°} \text{ A rms}$$

and the power absorbed by this phase is, therefore,

$$P_{AN} = 200(2) \cos (0° + 60°) = 200 \text{ W}$$

Thus, the total power drawn by the three-phase load is 600 W. The problem is completed by drawing a phasor diagram and reading from it the appropriate phase angles which apply to the other line voltages and currents. The completed diagram is shown in Fig. 11-15.　∎

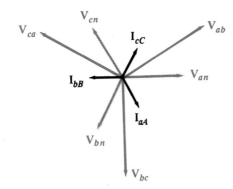

Figure 11-15

The phasor diagram that is applicable to the circuit of Fig. 11-14.

We may also use per-phase methods to work problems in what might be called the backward direction.

Example 11-3　Suppose that we have a balanced three-phase system with a line voltage of 300 V rms, and we know that it is supplying a balanced Y-connected load with 1200 W at a leading PF of 0.8. What are the line current and the per-phase load impedance?

Solution: It is evident that the phase voltage is $300/\sqrt{3}$ V rms and the per-phase power is 400 W. Thus the line current may be found from the power relationship

$$400 = \frac{300}{\sqrt{3}} (I_L)(0.8)$$

and the line current is therefore 2.89 A rms. The phase impedance is given by

$$|\mathbf{Z}_p| = \frac{V_p}{I_L} = \frac{300/\sqrt{3}}{2.89} = 60 \ \Omega$$

Since the PF is 0.8, leading, the impedance phase angle is $-36.9°$, and $\mathbf{Z}_p = 60/\underline{-36.9°} \ \Omega$.　∎

More complicated loads can be handled easily, since the problems reduce to simpler single-phase problems.

Example 11-4   Suppose that a balanced 600-W lighting load is added (in parallel) to the system of Example 11-3. Let us again determine the line current.

Solution: A suitable per-phase circuit is first sketched, as shown in Fig. 11-16.

Figure 11-16

The per-phase circuit that is used to analyze a balanced three-phase example.

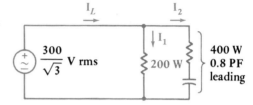

The amplitude of the lighting current is determined by

$$200 = \frac{300}{\sqrt{3}} |\mathbf{I}_1| \cos 0°$$

and

$$|\mathbf{I}_1| = 1.155$$

In a similar way, the amplitude of the capacitive load current is found to be unchanged from its previous value,

$$|\mathbf{I}_2| = 2.89$$

If we assume that the phase with which we are working has a phase voltage with an angle of 0°, then

$$\mathbf{I}_1 = 1.155\underline{/0°} \qquad \mathbf{I}_2 = 2.89\underline{/+36.9°}$$

and the line current is

$$\mathbf{I}_L = \mathbf{I}_1 + \mathbf{I}_2 = 3.87\underline{/+26.6°} \text{ A rms}$$

Furthermore, the power generated by this phase of the source is

$$P_p = \frac{300}{\sqrt{3}} 3.87 \cos(+26.6°) = 600 \text{ W}$$

which checks with the original hypothesis.   ∎

If an *unbalanced* Y-connected load is present in an otherwise balanced three-phase system, the circuit may still be analyzed on a per-phase basis *if* the neutral wire is present and *if* it has zero impedance. If either of these conditions is not met, other methods must be used, such as mesh or nodal analysis. However, engineers who spend most of their time with unbalanced three-phase systems will find the use of *symmetrical components* a great time-saver. We shall not discuss this method here.

Drill Problems

11-4. A balanced three-phase three-wire system has a Y-connected load. Each phase contains three loads in parallel: $-j100$ Ω, 100 Ω, and $50 + j50$ Ω. Assume positive phase sequence with $\mathbf{V}_{ab} = 400\underline{/0°}$ V. Find (*a*) $\mathbf{V}_{an}$; (*b*) $\mathbf{I}_{aA}$; (*c*) the power factor of the balanced load; (*d*) the total power drawn by the load.

Ans: $231\underline{/-30°}$ V; $4.62\underline{/-30°}$ A; 1.000; 3200 W

11-5. A balanced three-phase three-wire system has a line voltage of 500 V. Two balanced Y-connected loads are present. One is a capacitive load with $7 - j2$ Ω per phase, and the other is an inductive load with $4 + j2$ Ω per phase. Find (a) the phase voltage; (b) the line current; (c) the total power drawn by the load; (d) the power factor at which the source is operating. ·

Ans: 289 V; 97.5 A; 83.0 kW; 0.983 lag

11-6. Three balanced Y-connected loads are installed on a balanced three-phase four-wire system. Load 1 draws a total power of 6 kW at unity PF, load 2 requires 10 kVA at PF = 0.96 lagging, and load 3 needs 7 kW at 0.85 lagging. If the phase voltage at the loads is 135 V, if each line has a resistance of 0.1 Ω, and if the neutral has a resistance of 1 Ω, find (a) the total power drawn by the loads; (b) the combined PF of the loads; (c) the total power lost in the three lines; (d) the phase voltage at the source; (e) the power factor at which the source is operating.   *Ans:* 22.6 kW; 0.954 lag; 1027 W; 140.6 V; 0.957 lag

---

A three-phase load is more apt to be found Δ-connected than Y-connected. One reason for this, at least for the case of an unbalanced load, is the flexibility with which loads may be added or removed on a single phase. This is difficult (or impossible) to do with a Y-connected three-wire load.

Let us consider a balanced Δ-connected load which consists of an impedance $\mathbf{Z}_p$ inserted between each pair of lines. We shall assume a three-wire system for obvious reasons. With reference to Fig. 11-17, let us assume known line voltages

11-4

The delta (Δ) connection

$$V_L = |\mathbf{V}_{ab}| = |\mathbf{V}_{bc}| = |\mathbf{V}_{ca}|$$

or known phase voltages

$$V_p = |\mathbf{V}_{an}| = |\mathbf{V}_{bn}| = |\mathbf{V}_{cn}|$$

where

$$V_L = \sqrt{3}V_p \quad \text{and} \quad \mathbf{V}_{ab} = \sqrt{3}V_p \underline{/30°}$$

and so forth, as before. Since the voltage across each branch of the Δ is known, the *phase currents* are found:

$$\mathbf{I}_{AB} = \frac{\mathbf{V}_{ab}}{\mathbf{Z}_p} \qquad \mathbf{I}_{BC} = \frac{\mathbf{V}_{bc}}{\mathbf{Z}_p} \qquad \mathbf{I}_{CA} = \frac{\mathbf{V}_{ca}}{\mathbf{Z}_p}$$

and their differences provide us with the line currents, such as

$$\mathbf{I}_{aA} = \mathbf{I}_{AB} - \mathbf{I}_{CA}$$

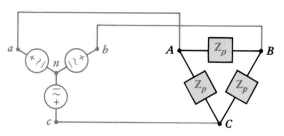

Figure 11-17

A balanced Δ-connected load is present on a three-wire three-phase system. The source happens to be Y-connected.

The three phase currents are of equal amplitude:

$$I_p = |\mathbf{I}_{AB}| = |\mathbf{I}_{BC}| = |\mathbf{I}_{CA}|$$

The line currents are also equal in amplitude, because the phase currents are equal in amplitude and 120° out of phase. The symmetry is apparent from the phasor diagram of Fig. 11-18. We thus have

$$I_L = |\mathbf{I}_{aA}| = |\mathbf{I}_{bB}| = |\mathbf{I}_{cC}|$$

and $$I_L = \sqrt{3}I_p$$

Figure 11-18

A phasor diagram which could apply to the circuit of Fig. 11-17 if $\mathbf{Z}_p$ were an inductive impedance.

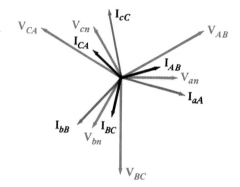

Let us disregard the source for the moment and consider only the balanced load. If the load is Δ-connected, then the phase voltage and the line voltage are indistinguishable, but the line current is larger than the phase current by a factor of $\sqrt{3}$; with a Y-connected load, however, the phase current and the line current refer to the same current, and the line voltage is greater than the phase voltage by a factor of $\sqrt{3}$.

The solution of three-phase problems will be speedily accomplished if the $\sqrt{3}$'s are used properly. Let us consider a typical numerical example.

Example 11-5 We are asked to determine the amplitude of the line current in a 300-V rms three-phase system which supplies 1200 W to a Δ-connected load at a lagging PF of 0.8.

Solution: Let us again consider a single phase. It draws 400 W, 0.8 lagging PF, at a 300-V rms line voltage. Thus,

$$400 = 300(I_p)(0.8)$$

and $$I_p = 1.667 \text{ A rms}$$

and the relationship between phase currents and line currents yields

$$I_L = \sqrt{3}(1.667) = 2.89 \text{ A rms}$$

Furthermore, the phase angle of the load is $\cos^{-1}(0.8) = 36.9°$, and therefore the impedance in each phase must be

$$\mathbf{Z}_p = \frac{300}{1.667} \underline{/36.9°} = 180 \underline{/36.9°} \ \Omega$$

∎

Now let us change the statement of the problem slightly.

Example 11-6 The load in Example 11-5 is Y-connected instead of Δ-connected. Again we are asked for the phase impedances.

Solution: On a per-phase basis, we now have a phase voltage of $300/\sqrt{3}$ V rms, a power of 400 W, and a lagging PF of 0.8. Thus,

$$400 = \frac{300}{\sqrt{3}}(I_p)(0.8)$$

and $\qquad I_p = 2.89 \qquad$ or $\qquad I_L = 2.89$ A rms

The phase angle of the load is again 36.9°, and thus the impedance in each phase of the Y is

$$\mathbf{Z}_p = \frac{300/\sqrt{3}}{2.89} \underline{/36.9°} = 60 \underline{/36.9°} \ \Omega \qquad \blacksquare$$

The $\sqrt{3}$ factor not only relates phase and line quantities but also appears in a useful expression for the total power drawn by any balanced three-phase load. If we assume a Y-connected load with a power-factor angle θ, then the power taken by any phase is

$$P_p = V_p I_p \cos\theta = V_p I_L \cos\theta = \frac{V_L}{\sqrt{3}} I_L \cos\theta$$

and the total power is

$$P = 3P_p = \sqrt{3} V_L I_L \cos\theta$$

In a similar way, the power delivered to each phase of a Δ-connected load is

$$P_p = V_p I_p \cos\theta = V_L I_p \ \cos\theta = V_L \frac{I_L}{\sqrt{3}} \ \cos\theta$$

giving a total power

$$P = 3P_p$$

$$P = \sqrt{3} V_L I_L \ \cos\theta \qquad (1)$$

Thus Eq. (1) enables us to calculate the total power delivered to a balanced load from a knowledge of the magnitude of the line voltage, of the line current, and of the phase angle of the load impedance (or admittance), regardless of whether the load is Y-connected or Δ-connected. The line current in Examples 11-5 and 11-6 can now be obtained in two simple steps:

$$1200 = \sqrt{3}(300)(I_L)(0.8)$$

Therefore

$$I_L = \frac{5}{\sqrt{3}} = 2.89 \text{ A rms}$$

The source may also be connected in a Δ configuration. This is not typical, however, for a slight unbalance in the source phases can lead to large currents circulating in the Δ loop. For example, let us call the three single-phase sources $\mathbf{V}_{ab}$, $\mathbf{V}_{bc}$, and $\mathbf{V}_{cd}$. Before closing the Δ by connecting d to a, let us determine the unbalance by measuring the sum $\mathbf{V}_{ab} + \mathbf{V}_{bc} + \mathbf{V}_{ca}$. Suppose that the amplitude of the resultant is only 1 percent of the line voltage. The circulating current is thus approximately $\frac{1}{3}$ percent of the line voltage divided by the internal impedance of any source. How large is this impedance apt to be? It must depend on the current that the source is expected to deliver with a negligible drop in terminal voltage. If we assume that this maximum current causes a 1 percent drop in the terminal voltage, then it is seen that the circulating current is one-third of the maximum current. This reduces the useful current capacity of the source and also increases the losses in the system.

We should also note that balanced three-phase sources may be transformed from Y to Δ, or vice versa, without affecting the load currents or voltages. The necessary relationships between the line and phase voltages are shown in Fig. 11-12 for the case where $\mathbf{V}_{an}$ has a reference phase angle of $0°$. This transformation enables us to use whichever source connection we prefer, and all the load relationships will be correct. Of course, we cannot specify any currents or voltages within the source until we know how it is actually connected.

Drill Problems

11-7. Each phase of a balanced three-phase Δ-connected load consists of a 0.2-H inductor in series with the parallel combination of a 5-μF capacitor and a 200-Ω resistance. Assume zero line resistance and a phase voltage of 200 V at $\omega = 400$ rad/s. Find (a) the phase current; (b) the line current; (c) the total power absorbed by the load. *Ans:* 1.158 A; 2.01 A; 693 W

11-8. A balanced three-phase three-wire system is terminated with two Δ-connected loads in parallel. Load 1 draws 40 kVA at a lagging PF of 0.8, while load 2 absorbs 24 kW at a leading PF of 0.9. Assume no line resistance, and let $\mathbf{V}_{ab} = 440\underline{/30°}$ V. Find (a) the total power drawn by the loads; (b) the phase current $\mathbf{I}_{AB1}$ for the lagging load; (c) $\mathbf{I}_{AB2}$; (d) $\mathbf{I}_{aA}$. *Ans:* 56.0 kW; 30.3$\underline{/-6.87°}$ A; 20.2$\underline{/55.8°}$ A; 75.3$\underline{/-12.46°}$ A

Problems

1 In the circuit graph shown in Fig. 11-19, some currents are $\mathbf{I}_{14} = 5\underline{/0°}$, $\mathbf{I}_{15} = 2\underline{/90°}$, $\mathbf{I}_{13} = 4\underline{/120°}$, $\mathbf{I}_{25} = 3\underline{/30°}$, $\mathbf{I}_{23} = -j4$, and $\mathbf{I}_{34} = 1 - j1$, all in amperes. Find (a) $\mathbf{I}_{35}$; (b) $\mathbf{I}_{24}$.

Figure 11-19

See Probs. 1 and 2.

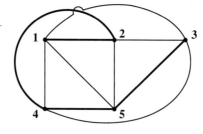

2 In the circuit graph shown in Fig. 11-19, some voltages are $\mathbf{V}_{12} = 100\underline{/0°}$, $\mathbf{V}_{45} = 60\underline{/75°}$, $\mathbf{V}_{42} = 80\underline{/120°}$, and $\mathbf{V}_{35} = -j120$, all in volts. Find (a) $\mathbf{V}_{25}$; (b) $\mathbf{V}_{13}$.

3 The 230/460-V rms 60-Hz three-wire system shown in Fig. 11-20 supplies power to three loads: load AN draws a complex power of $10\underline{/40°}$ kVA, load NB uses $8\underline{/10°}$ kVA, and load AB requires $4\underline{/-80°}$ kVA. Find the two line currents and the neutral current.

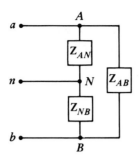

Figure 11-20

See Prob. 3.

4 An inefficient three-wire single-phase system has source voltages of $\mathbf{V}_{an} = \mathbf{V}_{nb} = 720\underline{/0°}$ V rms, line resistances $R_{aA} = R_{bB} = 1\ \Omega$ with $R_{nN} = 10\ \Omega$, and loads $\mathbf{Z}_{AN} = 10 + j3\ \Omega$, $\mathbf{Z}_{NB} = 8 + j2\ \Omega$, and $\mathbf{Z}_{AB} = 18 + j0\ \Omega$. Find (a) $\mathbf{I}_{aA}$; (b) $\mathbf{I}_{nN}$; (c) $P_{\text{wiring,total}}$; (d) $P_{\text{gen,total}}$.

5 A balanced three-wire single-phase system has loads $\mathbf{Z}_{AN} = \mathbf{Z}_{NB} = 10\ \Omega$, and a load $\mathbf{Z}_{AB} = 16 + j12\ \Omega$. The three lines may be assumed to be resistanceless. Let $\mathbf{V}_{an} = \mathbf{V}_{nb} = 120\underline{/0°}$ V rms. (a) Find I_{aA} and I_{nN}. (b) The system is unbalanced by connecting another 10-Ω resistance in parallel with $\mathbf{Z}_{AN}$. Find I_{aA}, I_{bB}, and I_{nN}.

6 A balanced three-wire single-phase system has source voltages $\mathbf{V}_{an} = \mathbf{V}_{nb} = 200\underline{/0°}$ V rms, zero line and neutral resistance, and loads $\mathbf{Z}_{AN} = \mathbf{Z}_{NB} = 12 + j3\ \Omega$. Find $\mathbf{Z}_{AB}$ so that: (a) $X_{AB} = 0$ and $I_{aA} = 30$ A rms; (b) $R_{AB} = 0$ and ang $\mathbf{I}_{aA} = 0°$.

7 In the balanced three-wire single-phase system of Fig. 11-21, let $V_{AN} = 220$ V rms at 60 Hz. (a) What size should C be to provide a unity–power-factor load? (b) How many kVA does C handle?

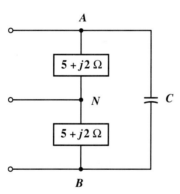

Figure 11-21

See Prob. 7.

8 Figure 11-22 shows a balanced three-phase three-wire system with positive phase sequence. Let $\mathbf{V}_{BC} = 120\underline{/60°}$ V rms and $R_w = 0.6\ \Omega$. If the total load (including wire resistance) draws 5 kVA at PF = 0.8 lagging, find (a) the total power lost in the line resistance, and (b) $\mathbf{V}_{an}$.

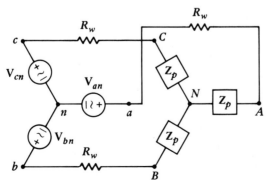

Figure 11-22

See Probs. 8 to 13.

9 Let $\mathbf{V}_{an} = 2300\underline{/0°}$ V rms in the balanced system shown in Fig. 11-22, and set $R_w = 2$ Ω. Assume positive phase sequence with the source supplying a total complex power of $\mathbf{S} = 100 + j30$ kVA. Find (a) $\mathbf{I}_{aA}$; (b) $\mathbf{V}_{AN}$; (c) $\mathbf{Z}_p$; (d) the transmission efficiency.

10 In the balanced three-phase system of Fig. 11-22, let $\mathbf{Z}_p = 12 + j5$ Ω and $\mathbf{I}_{bB} = 20\underline{/0°}$ A rms with (+) phase sequence. If the source is operating with a power factor of 0.935, find (a) R_w; (b) $\mathbf{V}_{bn}$; (c) $\mathbf{V}_{AB}$; (d) the complex power supplied by the source.

11 The phase impedance $\mathbf{Z}_p$ in the system shown in Fig. 11-22 consists of an impedance of $75\underline{/25°}$ Ω in parallel with a 25-μF capacitance. Let $\mathbf{V}_{an} = 240\underline{/0°}$ V rms at 60 Hz, and $R_w = 2$ Ω. Find (a) $\mathbf{I}_{aA}$; (b) P_{wires}; (c) P_{load}; (d) the source power factor.

12 A lossless neutral conductor is installed between nodes n and N in the three-phase system shown in Fig. 11-22. Assume a balanced system with (+) phase sequence, but connect unbalanced loads: $\mathbf{Z}_{AN} = 8 + j6$ Ω, $\mathbf{Z}_{BN} = 12 - j16$ Ω, and $\mathbf{Z}_{CN} = 5$ Ω. If $\mathbf{V}_{an} = 120\underline{/0°}$ V rms and $R_w = 0.5$ Ω, find $\mathbf{I}_{nN}$.

13 The balanced three-phase system shown in Fig. 11-22 has $R_w = 0$ and $\mathbf{Z}_p = 10 + j5$ Ω per phase. (a) At what power factor is the source operating? (b) Assuming $f = 60$ Hz, what size capacitor must be placed in parallel with each phase impedance to raise the PF to 0.93 lagging? (c) How much reactive power is drawn by each capacitor if the line voltage at the load is 440 V rms?

14 Figure 11-23 shows a balanced three-wire three-phase circuit. Let $R_w = 0$ and $\mathbf{V}_{an} = 200\underline{/60°}$ V rms. Each phase of the load absorbs a complex power, $\mathbf{S}_p = 2 - j1$ kVA. If (+) phase sequence is assumed, find: (a) $\mathbf{V}_{bc}$; (b) $\mathbf{Z}_p$; (c) $\mathbf{I}_{aA}$.

Figure 11-23

See Probs. 14 to 17.

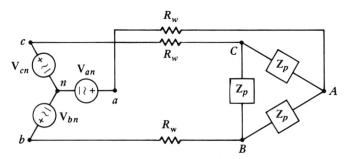

15 The balanced Δ load of Fig. 11-23 requires 15 kVA at a lagging PF of 0.8. Assume (+) phase sequence with $\mathbf{V}_{BC} = 180\underline{/30°}$ V rms. If $R_w = 0.75$ Ω, find (a) $\mathbf{V}_{bc}$; (b) the total complex power generated by the source.

16 The load in the balanced system of Fig. 11-23 draws a total complex power of $3 + j1.8$ kVA, while the source generates $3.45 + j1.8$ kVA. If $R_w = 5$ Ω, find (a) I_{aA}; (b) I_{AB}; (c) V_{an}.

17 The balanced three-phase Y-connected source in Fig. 11-23 has $\mathbf{V}_{an} = 140\underline{/0°}$ V rms with (+) phase sequence. Let $R_w = 0$. The balanced three-phase load draws 15 kW and +9 kVAR. Find (a) $\mathbf{V}_{AB}$; (b) $\mathbf{I}_{AB}$; (c) $\mathbf{I}_{aA}$.

18 The source in Fig. 11-24 is balanced and exhibits (+) phase sequence. Find (a) $\mathbf{I}_{aA}$; (b) $\mathbf{I}_{bB}$; (c) $\mathbf{I}_{cC}$; (d) the total complex power supplied by the source.

Figure 11-24

See Probs. 18 and 19.

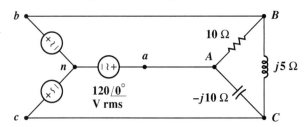

19 Add 1 Ω of resistance in each of the lines of Fig. 11-24 and again work Prob. 18.

20 (SPICE) Let $f = 60$ Hz in the circuit shown in Fig. 11-25. Use SPICE to determine the (rms) phasor value of the current: (a) $\mathbf{I}_{aA}$; (b) $\mathbf{I}_{bB}$; (c) $\mathbf{I}_{nN}$.

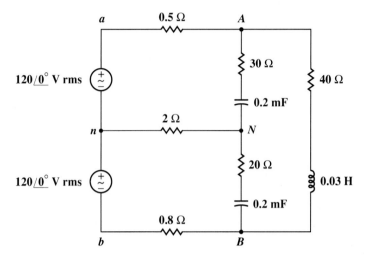

Figure 11-25

See Prob. 20.

21 (SPICE) In the three-phase system of Fig. 11-26, assume a balanced source with a (+) phase sequence. Let $f = 60$ Hz. Find the magnitude of (a) $\mathbf{V}_{AN}$; (b) $\mathbf{V}_{BN}$; (c) $\mathbf{V}_{CN}$.

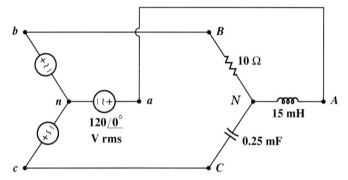

Figure 11-26

See Prob. 21.

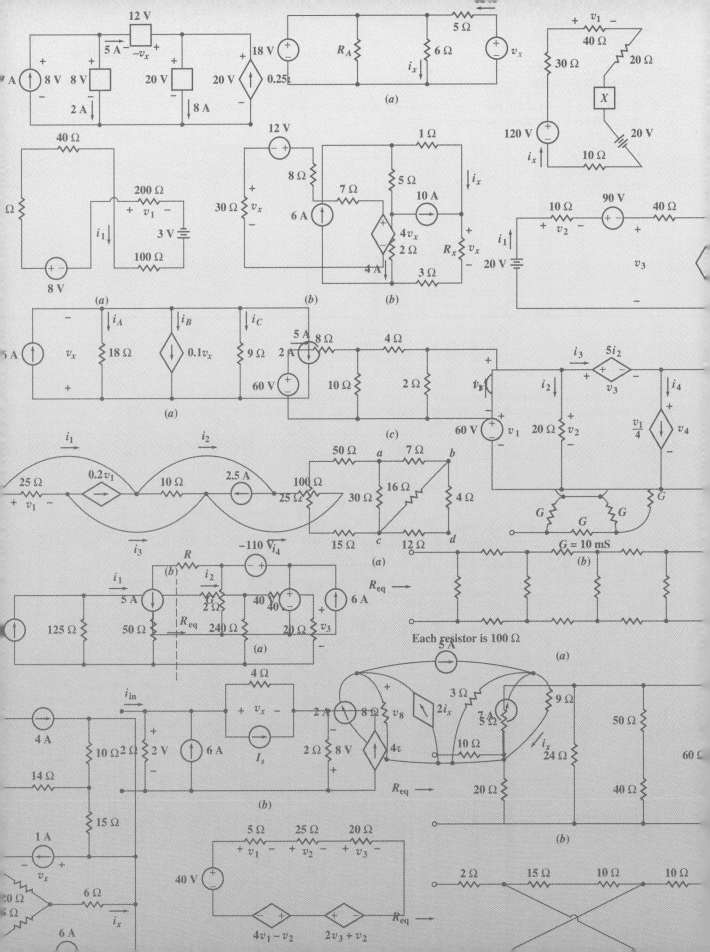

Part Four:
Complex Frequency

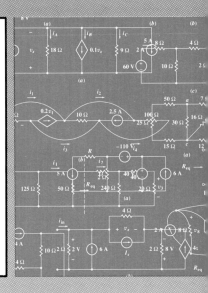

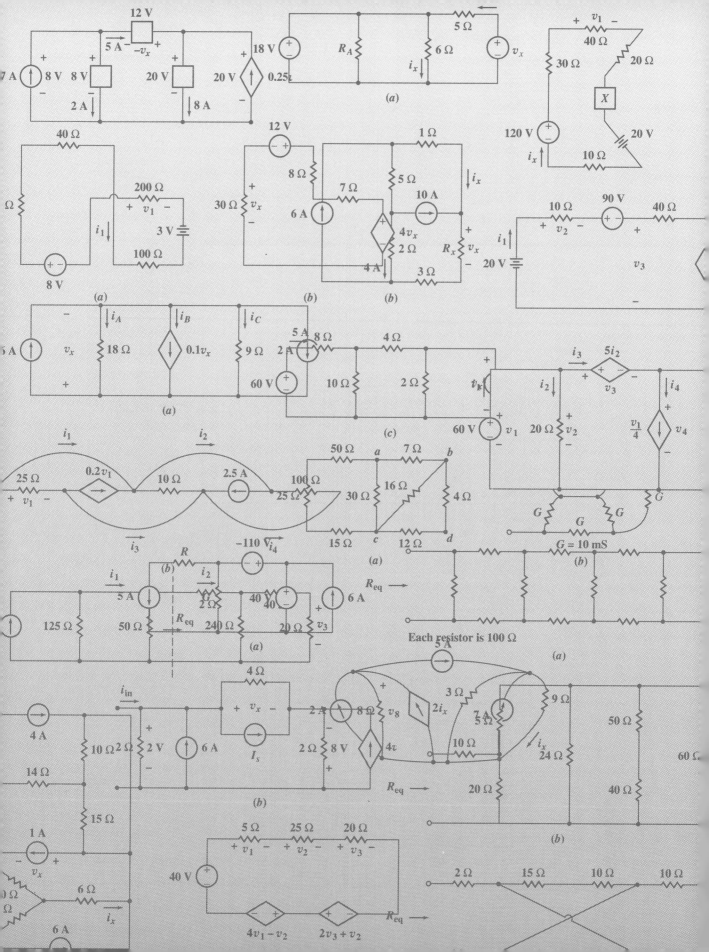

Complex Frequency

We are now about to begin the fourth major portion of our study of circuit analysis, a discussion of the concept of complex frequency. This, we shall see, is a remarkably unifying concept which will enable us to tie together all our previously developed analytical techniques into one neat package. Resistive circuit analysis, steady-state sinusoidal analysis, transient analysis, the forced response, the complete response, and the analysis of circuits excited by exponential forcing functions and exponentially damped sinusoidal forcing functions will all become special cases of the general techniques of circuit analysis which are associated with the complex-frequency concept.

We introduce complex frequency by considering an exponentially damped sinusoidal function, such as the voltage

$$v(t) = V_m e^{\sigma t} \cos(\omega t + \theta) \qquad (1)$$

where σ (sigma) is a real quantity and is usually negative. Although we refer to this function as being "damped," it is possible that the sinusoidal amplitude may increase if σ is positive; the more practical case is that of the damped function. Our work with the natural response of the RLC circuit also indicates that σ is the negative of the exponential damping coefficient.

We may first construct a constant voltage from Eq. (1) by letting $\sigma = \omega = 0$:

$$v(t) = V_m \cos\theta = V_0 \qquad (2)$$

If we set only σ equal to zero, then we obtain a general sinusoidal voltage

$$v(t) = V_m \cos(\omega t + \theta) \qquad (3)$$

And if $\omega = 0$, we have the exponential voltage

$$v(t) = V_m (\cos\theta) e^{\sigma t} = V_0 e^{\sigma t} \qquad (4)$$

Thus, the damped sinusoid of Eq. (1) includes as special cases the dc Eq. (2), sinusoidal Eq. (3), and exponential Eq. (4) functions.

Some additional insight into the significance of σ can be obtained by comparing the exponential function of Eq. (4) with the complex representation of a sinusoidal function with a zero-degree phase angle,

$$v(t) = V_0 e^{j\omega t} \qquad (5)$$

339

It is apparent that the two functions, Eqs. (4) and (5), have much in common. The only difference is that the exponent in Eq. (4) is real and the one in Eq. (5) is imaginary. The similarity between the two functions is emphasized by describing σ as a "frequency." This choice of terminology will be discussed in detail in the following sections, but for now we need merely note that σ is specifically termed the *real part* of the complex frequency. It should not be called the "real frequency," however, for this is a term which is more suitable for f (or, loosely, for ω). We shall also refer to σ as the *neper frequency*, the name arising from the dimensionless unit of the exponent of e. Thus, given e^{7t}, the dimensions of $7t$ are nepers (Np), and 7 is the neper frequency in nepers per second. The neper itself was named after Napier and his napierian logarithm system; the spelling of his name is historically uncertain.

Let us now think in terms of a forcing function or source that has a damped sinusoidal or exponential form. When $\sigma < 0$, the amplitude of the forcing function can have extremely large values at times in the distant past. That is, we have considered the forced response as that produced by a forcing function applied since $t = -\infty$; its application at some finite time gives rise to a transient response in addition to the forced response. Since the infinite amplitude of the forcing function at $t = -\infty$ may make us somewhat uncomfortable, we might note that initial conditions can be established in any circuit so that the application of a specified forcing function at a specified instant of time produces thereafter a response identical to the forced response without any transient response. Examples of this will appear later. From a practical point of view, we know that, although it may be impossible in the laboratory to generate damped sinusoidal or exponential forcing functions which are accurate for all time, we can produce approximations that are satisfactory for circuits whose transient responses do not last very long.

The forced response of a network to a general forcing function of the form of Eq. (1) is found very simply by using a method almost identical with that used for the sinusoidal forcing function; we shall discuss the method in Sec. 12-3. When we are able to find the forced response to this damped sinusoid, we should realize that we shall also have found the forced response to a dc voltage, an exponential voltage, and a sinusoidal voltage. Now let us see how we may consider σ and ω as the real and imaginary parts of a complex frequency.

12-2

Complex frequency

Let us first provide ourselves with a purely mathematical definition of complex frequency and then gradually develop a physical interpretation as the chapter progresses. We say that any function which may be written in the form

$$\mathbf{f}(t) = \mathbf{K}e^{\mathbf{s}t} \tag{6}$$

where $\mathbf{K}$ and $\mathbf{s}$ are complex constants (independent of time) is characterized by the *complex frequency* $\mathbf{s}$. The complex frequency $\mathbf{s}$ is therefore simply the factor which multiplies t in this complex exponential representation. Until we are able to determine the complex frequency of a given function by inspection, it is necessary to write the function in the form of Eq. (6).

We may apply this definition first to the more familiar forcing functions. For example, a constant voltage

$$v(t) = V_0$$

may be written in the form

$$v(t) = V_0 e^{(0)t}$$

The complex frequency of a dc voltage or current is thus zero; $\mathbf{s} = 0$.

The next simple case is the exponential function

$$v(t) = V_0 e^{\sigma t}$$

which is already in the required form. The complex frequency of this voltage is therefore σ; $\mathbf{s} = \sigma + j0$.

Now let us consider a sinusoidal voltage, one which may provide a slight surprise. Given

$$v(t) = V_m \cos(\omega t + \theta)$$

it is necessary to find an equivalent expression in terms of the complex exponential. From our past experience, we therefore use the formula we derived from Euler's identity,

$$\cos(\omega t + \theta) = \tfrac{1}{2}\left(e^{j(\omega t + \theta)} + e^{-j(\omega t + \theta)}\right)$$

and obtain

$$v(t) = \tfrac{1}{2}V_m\left(e^{j(\omega t + \theta)} + e^{-j(\omega t + \theta)}\right)$$

$$= \left(\tfrac{1}{2}V_m e^{j\theta}\right)e^{j\omega t} + \left(\tfrac{1}{2}V_m e^{-j\theta}\right)e^{-j\omega t}$$

or

$$v(t) = \mathbf{K}_1 e^{\mathbf{s}_1 t} + \mathbf{K}_2 e^{\mathbf{s}_2 t}$$

We have the *sum* of two complex exponentials, and *two* complex frequencies are therefore present, one for each term. The complex frequency of the first term is $\mathbf{s} = \mathbf{s}_1 = j\omega$, and that of the second term is $\mathbf{s} = \mathbf{s}_2 = -j\omega$. These two values of $\mathbf{s}$ are conjugates, or $\mathbf{s}_2 = \mathbf{s}_1^*$; and the two values of $\mathbf{K}$ are also conjugates: $\mathbf{K}_1 = \tfrac{1}{2}V_m e^{j\theta}$ and $\mathbf{K}_2 = \mathbf{K}_1^* = \tfrac{1}{2}V_m e^{-j\theta}$. The entire first term and the entire second term are therefore conjugates, which we should expect inasmuch as their sum must be a real quantity.

Finally, let us determine the complex frequency or frequencies associated with the exponentially damped sinusoidal function, Eq. (1). We again use Euler's formula to obtain a complex exponential representation:

$$v(t) = V_m e^{\sigma t}\cos(\omega t + \theta)$$

$$= \tfrac{1}{2}V_m e^{\sigma t}\left(e^{j(\omega t + \theta)} + e^{-j(\omega t + \theta)}\right)$$

and thus

$$v(t) = \tfrac{1}{2}V_m e^{j\theta}e^{j(\sigma + j\omega)t} + \tfrac{1}{2}V_m e^{-j\theta}e^{j(\sigma - j\omega)t}$$

We find that a conjugate complex pair of frequencies, $\mathbf{s}_1 = \sigma + j\omega$ and $\mathbf{s}_2 = \mathbf{s}_1^* = \sigma - j\omega$, is also required to describe the exponentially damped sinusoid. In general, neither σ nor ω is zero, and we see that the exponentially varying sinusoidal waveform is the general case; the constant, sinusoidal, and exponential waveforms are special cases.

As numerical illustrations, we should now recognize at sight the complex frequencies associated with these voltages:

$$v(t) = 100 \qquad \mathbf{s} = 0$$

$$v(t) = 5e^{-2t} \qquad \mathbf{s} = -2 + j0$$

$$v(t) = 2 \sin 500t \qquad \begin{cases} \mathbf{s}_1 = j500 \\ \mathbf{s}_2 = \mathbf{s}_1^* = -j500 \end{cases}$$

$$v(t) = 4e^{-3t} \sin (6t + 10°) \qquad \begin{cases} \mathbf{s}_1 = -3 + j6 \\ \mathbf{s}_2 = \mathbf{s}_1^* = -3 - j6 \end{cases}$$

The reverse type of example is also worth considering. Given a complex frequency or a pair of conjugate complex frequencies, we must be able to identify the nature of the function with which they are associated. The most special case, $\mathbf{s} = 0$, defines a constant or dc function. With reference to the defining functional form of Eq. (6), it is apparent that the constant $\mathbf{K}$ must be real if the function is to be real.

Let us next consider real values of $\mathbf{s}$. A positive real value, such as $\mathbf{s} = 5 + j0$, identifies an exponentially increasing function $\mathbf{K}e^{+5t}$, where again $\mathbf{K}$ must be real if the function is to be a physical one. A negative real value for $\mathbf{s}$, such as $\mathbf{s} = -5 + j0$, refers to an exponentially decreasing function $\mathbf{K}e^{-5t}$.

A purely imaginary value of $\mathbf{s}$, such as $j10$, can never be associated with a purely real quantity. The functional form is $\mathbf{K}e^{j10t}$, which can also be written as $\mathbf{K}(\cos 10t + j \sin 10t)$; it obviously possesses both real and imaginary parts. Each part is sinusoidal. In order to construct a real function, it is necessary to consider conjugate values of $\mathbf{s}$, such as $\mathbf{s}_{1,2} = \pm j10$, with which must be associated conjugate values of $\mathbf{K}$. Loosely speaking, however, we may identify either of the complex frequencies $\mathbf{s}_1 = +j10$ or $\mathbf{s}_2 = -j10$ with a sinusoidal voltage at the radian frequency of 10 rad/s. The presence of the conjugate complex frequency is understood. The amplitude and phase angle of the sinusoidal voltage will depend on the choice of $\mathbf{K}$ for each of the two frequencies. Thus, selecting $\mathbf{s}_1 = j10$ and $\mathbf{K}_1 = 6 - j8$, where

$$v(t) = \mathbf{K}_1 e^{\mathbf{s}_1 t} + \mathbf{K}_2 e^{\mathbf{s}_2 t} \qquad \mathbf{s}_2 = \mathbf{s}_1^* \qquad \text{and} \qquad \mathbf{K}_2 = \mathbf{K}_1^*$$

we obtain the real sinusoid $20 \cos (10t - 53.1°)$.

In a similar manner, a general value for $\mathbf{s}$, such as $3 - j5$, can be associated with a real quantity only if it is accompanied by its conjugate, $3 + j5$. Speaking loosely again, we may think of either of these two conjugate frequencies as describing an exponentially increasing sinusoidal function, $e^{3t} \cos 5t$; the specific amplitude and phase angle will again depend on the specific values of the conjugate complex $\mathbf{K}$'s.

By now we should have achieved some appreciation of the physical nature of the complex frequency $\mathbf{s}$; in general, it describes an exponentially varying sinusoid. The real part of $\mathbf{s}$ is associated with the exponential variation; if it is negative, the function decays as t increases; if it is positive, the function increases; and if it is zero, the sinusoidal amplitude is constant. The larger the *magnitude* of the real part of $\mathbf{s}$, the greater is the rate of exponential increase or decrease. The imaginary part of $\mathbf{s}$ describes the sinusoidal variation; it is specifically the radian frequency. A large magnitude for the imaginary part of $\mathbf{s}$ indicates a more rapidly changing function of time. Thus, larger magnitudes for the real part of $\mathbf{s}$, the imaginary part of $\mathbf{s}$, or the magnitude of $\mathbf{s}$ indicate a more rapidly varying function.

It is customary to use the letter σ to designate the real part of $\mathbf{s}$, and ω (*not* $j\omega$) to designate the imaginary part:

$$\mathbf{s} = \sigma + j\omega \qquad (7)$$

The radian frequency is sometimes referred to as the "real frequency," but this terminology can be very confusing when we find that we must then say that "the real frequency is the imaginary part of the complex frequency"! When we need to be specific, we shall call **s** the complex frequency, σ the neper frequency, ω the radian frequency, and f the cyclic frequency; when no confusion seems likely, it is permissible to use "frequency" to refer to any of these four quantities. The neper frequency is measured in nepers per second, radian frequency is measured in radians per second, and complex frequency **s** is measured in units which are variously termed complex nepers per second or complex radians per second.

12-1. Identify all the complex frequencies present in the real time functions: (a) $(2e^{-100t} + e^{-200t}) \sin 2000t$; (b) $(2 - e^{-10t}) \cos (4t + \phi)$; (c) $e^{-10t} \cos 10t \sin 40t$.

Ans: $-100 + j2000, -100 - j2000, -200 + j2000, -200 - j2000$ s^{-1}; $j4, -j4, -10 + j4, -10 - j4$ s^{-1}; $-10 + j30, -10 - j30, -10 + j50, -10 - j50$ s^{-1}

12-2. Use real constants A, B, C, ϕ, and so forth, to construct the general form of the real function of time for a current having components at these frequencies: (a) $0, 10, -10$ s^{-1}; (b) $-5, j8, -5 - j8$ s^{-1}; (c) $-20, 20, -20 + j20, 20 - j20$ s^{-1}.

Ans: $A + Be^{10t} + Ce^{-10t}$; $Ae^{-5t} + B \cos (8t + \phi_1) + Ce^{-5t} \cos (8t + \phi_2)$; $Ae^{-20t} + Be^{20t} + Ce^{-20t} \cos (20t + \phi_1) + De^{20t} \cos (20t + \phi_2)$

12-3

The damped sinusoidal forcing function

We have devoted enough time to the definition and introductory interpretation of complex frequency; now it is time to put this concept to work and become familiar with it by seeing what it will do and how it is used.

The general exponentially varying sinusoid, which we may represent for the moment as the voltage

$$v(t) = V_m e^{\sigma t} \cos (\omega t + \theta) \tag{8}$$

is expressible in terms of the complex frequency **s** by making use of Euler's identity as before:

$$v(t) = \text{Re} \, (V_m e^{\sigma t} e^{j(\omega t + \theta)}) \tag{9}$$

or

$$v(t) = \text{Re} \, (V_m e^{\sigma t} e^{j(-\omega t - \theta)}) \tag{10}$$

Either representation is suitable, and the two expressions should remind us that a pair of conjugate complex frequencies is associated with a sinusoid or an exponentially damped sinusoid. Equation (9) is more directly related to the given damped sinusoid, and we shall concern ourselves principally with it. Collecting factors, to get

$$v(t) = \text{Re} \, (V_m e^{j\theta} e^{(\sigma + j\omega)t})$$

we now substitute $\mathbf{s} = \sigma + j\omega$ and obtain

$$v(t) = \text{Re} \, (V_m e^{j\theta} e^{\mathbf{s}t}) \tag{11}$$

Before we apply a forcing function of this form to any circuit, we should note the resemblance of this last representation of the damped sinusoid to the corresponding representation of the *undamped* sinusoid that we studied back in Sec. 8-3,

$$\mathrm{Re}\ (V_m e^{j\theta} e^{j\omega t})$$

The only difference is that we now have **s** where we previously had $j\omega$. Instead of restricting ourselves to sinusoidal forcing functions and their radian frequencies, we have now extended our notation to include the damped sinusoidal forcing function at a complex frequency. It should be no surprise at all to see later in this section and in the following section that we shall develop a *frequency-domain* description of the exponentially damped sinusoid in exactly the same way that we did for the sinusoid; we shall simply omit the Re notation and suppress e^{st}.

We are now ready to apply the exponentially damped sinusoid, as given by Eq. (8), (9), (10), or (11), to an electrical network. The forced response—say, a current in some branch of the network—is the desired response. Since the forced response has the form of the forcing function, its integral, and its derivatives, the response may be assumed to be

$$i(t) = I_m e^{\sigma t} \cos\ (\omega t + \phi)$$

or

$$i(t) = \mathrm{Re}\ (I_m e^{j\phi} e^{st})$$

where the complex frequency of the source and the response must be identical.

If we now recall that the real part of a complex forcing function produces the real part of the response while the imaginary part of the complex forcing function causes the imaginary part of the response, then we are again led to the application of a *complex* forcing function to our network. We shall obtain a complex response whose real part is the desired real response. Actually, we shall work with the Re notation omitted, but we should realize that it may be reinserted at any time and that it *must* be reinserted whenever we desire the time-domain response. Thus, given the real forcing function

$$v(t) = \mathrm{Re}\ (V_m e^{j\theta} e^{st})$$

we apply the complex forcing function $e^{j\theta} e^{st}$; the resultant forced response $e^{j\phi} e^{st}$ is complex, and it must have as its real part the desired time-domain forced response

$$i(t) = \mathrm{Re}\ (I_m e^{j\phi} e^{st})$$

The solution of our circuit analysis problem consists of the determination of the unknown response amplitude I_m and phase angle ϕ.

Before we actually carry out the details of an analysis problem and see how exactly the procedure follows that which was used in the sinusoidal analysis, it is worthwhile to outline the steps of the basic method. We first characterize the circuit with a set of loop or nodal integrodifferential equations. The given forcing functions, in complex form, and the assumed forced responses, also in complex form, are then substituted in the equations and the indicated integrations and differentiations are performed. Each term in every equation will then contain the same factor e^{st}. We therefore divide throughout by this factor, or "suppress e^{st}," understanding that it must be reinserted if a time-domain description of any response function is desired. With the Re symbol and the e^{st} factor gone, we have converted all the voltages and currents from the time

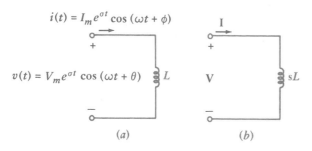

Figure 12-2

(*a*) The time-domain inductor voltage and current are related by $v = L\,di/dt$. (*b*) The (complex) frequency-domain inductor voltage and current are related by $\mathbf{V} = \mathbf{s}L\mathbf{I}$.

and the current as

$$i(t) = \mathrm{Re}\,(I_m e^{j\phi} e^{\mathbf{s}t}) = \mathrm{Re}\,(\mathbf{I}e^{\mathbf{s}t})$$

then the substitution of these expressions in the defining equation of an inductor,

$$v(t) = L\frac{di(t)}{dt}$$

leads to

$$\mathrm{Re}\,(\mathbf{V}e^{\mathbf{s}t}) = \mathrm{Re}\,(\mathbf{s}L\mathbf{I}e^{\mathbf{s}t})$$

We now drop Re, thus considering the complex response to a complex forcing function, and suppress the superfluous factor $e^{\mathbf{s}t}$:

$$\mathbf{V} = \mathbf{s}L\mathbf{I}$$

The ratio of the complex voltage to the complex current is once again the impedance. Since it depends in general upon the complex frequency $\mathbf{s}$, this functional dependence is sometimes indicated by writing

$$\mathbf{Z}(\mathbf{s}) = \frac{\mathbf{V}}{\mathbf{I}} = \mathbf{s}L$$

In a similar manner, the admittance of an inductor L is

$$\mathbf{Y}(\mathbf{s}) = \frac{1}{\mathbf{s}L}$$

We shall still call $\mathbf{V}$ and $\mathbf{I}$ *phasors*. Each of these complex quantities has an amplitude and phase angle which, along with a specific complex-frequency value, enable us to characterize the exponentially varying sinusoidal waveform completely. The phasor is still a frequency-domain description, but its application is not limited to the realm of radian frequencies. The frequency-domain equivalent of Fig. 12-2*a* is shown in Fig. 12-2*b*; phasor currents, phasor voltages, and impedances or admittances are now used.

The impedances of a resistor and a capacitor at a complex frequency $\mathbf{s}$ are obtained by a similar sequence of steps. Without going through the details, the results for all three of the elements are shown in the following table:

	R	L	C
$\mathbf{Z}(\mathbf{s})$	R	$\mathbf{s}L$	$\dfrac{1}{\mathbf{s}C}$
$\mathbf{Y}(\mathbf{s})$	$\dfrac{1}{R}$	$\dfrac{1}{\mathbf{s}L}$	$\mathbf{s}C$

Example 12-2 Reconsider the series *RLC* example (Example 12-1 and Fig. 12-1) in the frequency domain by utilizing $\mathbf{Z(s)}$ to find the current.

Solution: The source voltage

$$v(t) = 60e^{-2t} \cos (4t + 10°)$$

is transformed to the phasor voltage

$$\mathbf{V} = 60\underline{/10°}$$

at a frequency $\mathbf{s} = -2 + j4$ s^{-1}, and a phasor current $\mathbf{I}$ is assumed. The impedance of each element at $\mathbf{s} = -2 + j4$ is next determined. For the resistor, $\mathbf{Z}_R(\mathbf{s}) = 2$. The inductor impedance becomes $\mathbf{Z}_L(\mathbf{s}) = \mathbf{s}L = (-2 + j4)3 = -6 + j12$, while $\mathbf{Z}_C = 1/\mathbf{s}C = 1/[(-2 + j4)0.1] = -1 - j2$. These values are placed on a frequency-domain diagram, Fig. 12-3. The unknown current is now easily obtained by dividing the phasor voltage by the sum of the three impedances:

$$\mathbf{I} = \frac{60\underline{/10°}}{2 + (-6 + j12) + (-1 - j2)} = \frac{60\underline{/10°}}{-5 + j10}$$

$$= 5.37\underline{/-106.6°} \text{ A}$$

Thus, the previous result is obtained, but much more directly and rapidly. ∎

Figure 12-3

The frequency-domain equivalent of the series *RLC* circuit shown in Fig. 12-1.

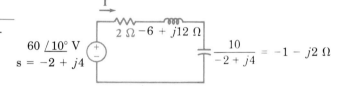

It is hardly necessary to say that all the techniques which we have used in the past to simplify frequency-domain analysis, such as mesh analysis, nodal analysis, superposition, source transformations, duality, and Thévenin's and Norton's theorems, are still valid and useful. We illustrate by making use of Thévenin's theorem.

Example 12-3 Find the Thévenin equivalent of the network shown in Fig. 12-4a.

Figure 12-4

(a) A given two-terminal network.
(b) The frequency-domain Thévenin equivalent.

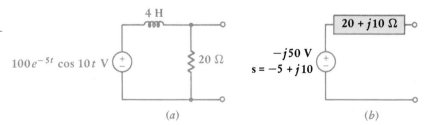

Solution: The complex frequency is $\mathbf{s} = -5 + j10$, and the frequency-domain source voltage is $\mathbf{V}_s = 100\underline{/0°}$ V. Hence, the inductor impedance is

$$\mathbf{Z}_L(\mathbf{s}) = 4(-5 + j10) = -20 + j40$$

and the resistor impedance is $\mathbf{Z}_R(\mathbf{s}) = 20$.

The Thévenin impedance is the parallel equivalent of $\mathbf{Z}_L$ and $\mathbf{Z}_R$:

$$\mathbf{Z}_{th} = \frac{20(-20 + j40)}{20 - 20 + j40} = 20 + j10$$

The open-circuit voltage is found by voltage division:

$$\mathbf{V}_{oc} = 100\underline{/0°} \, \frac{20}{20 - 20 + j40} = -j50$$

Thus, the frequency-domain Thévenin-equivalent network is that shown in Fig. 12-4b. The return to the time domain may be taken after some desired forced response has been determined in the frequency domain. For example, if another 4-H inductor is placed across the open circuit, the frequency-domain current is

$$\mathbf{I} = \frac{-j50}{20 + j10 - 20 + j40} = -1\,\text{A}$$

corresponding to the time-domain current

$$i(t) = -e^{-5t}\cos 10t \,\text{A} \qquad\qquad ■$$

Drill Problems

12-5. For the circuit of Fig. 12-5, find, in terms of $\mathbf{I}_s$: (a) $\mathbf{V}_1(\mathbf{s})$; (b) $\mathbf{V}_2(\mathbf{s})$.
Ans: $100\mathbf{I}_s(\mathbf{s} + 20)/(\mathbf{s}^2 + 20\mathbf{s} + 500)$; $100\mathbf{s}\mathbf{I}_s/(\mathbf{s}^2 + 20\mathbf{s} + 500)$

12-6. Find the phasor voltage $\mathbf{V}_2$ in Fig. 12-5 if $\mathbf{I}_s = 3\underline{/110°}$ A and $\mathbf{s}$ equals (a) 0; (b) -10 Np/s; (c) $j10$ s^{-1}; (d) $-j10$ s^{-1}; (e) $-10 + j10$ s^{-1}.
Ans: 0; $7.5\underline{/-70°}$ V; $6.71\underline{/173.4°}$ V; $6.71\underline{/46.6°}$ V; $14.14\underline{/-115°}$ V

12-7. The time-domain expression for the source current in Fig. 12-5 is $i_s(t) = 3e^{-5t}\cos(10t + 18°)$ A. Find v_2 at t equal to (a) 0; (b) 0.1 s.
Ans: -4.55 V; -5.96 V

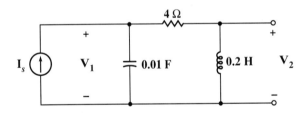

Figure 12-5

See Drill Probs. 12-5 to 12-7.

In Sec. 9-5 we considered the forced response of a circuit as a function of the radian frequency ω. We represented such quantities as impedance, admittance, specific voltages or currents, voltage and current gain, and transfer impedances and admittances also as functions of ω. Then we located the poles and zeros of these quantities and sketched the responses as functions of ω. Before we discuss the more general problem of frequency response as a function of the complex frequency $\mathbf{s}$ in the next section, let us devote a little time to the simpler problem of frequency response as a function of σ. This is a subject which is very useful in synthesizing RC and RL circuits, a popular topic in more advanced courses.

As a simple example we may select the series RL circuit excited by the frequency-domain voltage source $V_m\underline{/0°}$. The current is obtained as a function

12-5

Frequency response as a function of σ

of **s** by dividing the source voltage by the input impedance:

$$\mathbf{I} = \frac{V_m \underline{/0°}}{R + \mathbf{s}L}$$

We now set $\omega = 0$, $\mathbf{s} = \sigma + j0$, thus restricting ourselves to time-domain sources of the form

$$v_s = V_m e^{\sigma t}$$

and thus

$$I = \frac{V_m}{R + \sigma L}$$

or

$$I = \frac{V_m}{L} \frac{1}{\sigma + R/L} \qquad (13)$$

Transforming to the time domain,

$$i(t) = \frac{V_m}{R + \sigma L} e^{\sigma t} \qquad (14)$$

The necessary information about this forced response, however, is all contained in the frequency-domain description of Eq. (13). As the neper frequency σ varies, a qualitative description of the response is easily provided. When σ is a large negative number, corresponding to a rapidly decreasing exponential function, the current response of Eq. (13) is negative and relatively small in amplitude; it is, of course, also a rapidly decreasing (in magnitude) exponential function. As σ increases, becoming a smaller negative number, the magnitude of the negative response increases. We can obtain an arbitrarily large response by letting σ be sufficiently close to $-R/L$, a pole of the response. Thus, if $V_m = 1$ V, $L = 1$ H, and $R = 1$ Ω, the pole is located at $\sigma = -1$ Np/s, and the amplitude of the current response, $V_m/(R + \sigma L) = 1/(\sigma + 1)$, will be greater than 1 000 000 A in magnitude whenever σ is within 10^{-6} Np/s of -1 Np/s, or $0 < |\sigma + 1| < 10^{-6}$.

As σ continues to increase, the next noteworthy point occurs when $\sigma = 0$. Since $v_s = V_m$, we are now faced with the dc case, and the forced response is obviously V_m/R, which agrees with Eq. (13). Positive values of σ must all provide positive amplitude responses, the larger amplitudes arising when σ is smaller. Finally, an infinite value of σ provides a zero-amplitude response, and thus establishes a zero. The only critical frequencies are the pole at $\sigma = -R/L$ and the zero at $\sigma = \pm\infty$.

This information may be presented quite easily by plotting $|I|$, the current magnitude, as a function of σ, as shown in Fig. 12-6. In preparation for the

Figure 12-6

A plot of the magnitude of **I** versus the neper frequency σ for a series RL circuit excited by an exponential voltage source $V_m e^{\sigma t}$. The only pole is located at $\sigma = -R/L$; the zero is located at $\sigma = \pm\infty$.

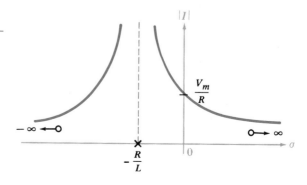

more general response information with changing **s** in the next section, we show the *magnitude* of I versus σ; the phase angle (not shown) is, of course, either 0° or 180°.

Let us now turn our attention to the two critical frequencies of this forced response. The only finite critical frequency is the pole at $\sigma = -R/L$. Equation (13) implies that the application of a finite-amplitude forcing function at this frequency will cause a response of infinite amplitude. This is inconveniently large, of course, as well as being mathematically unsuitable. A much more informative approach to an understanding of the nature of the pole is based on a consideration of the natural response. For this series RL circuit, an appropriate form for the natural response is

$$i_n(t) = I_m e^{-Rt/L} \tag{15}$$

where I_m is the amplitude of the natural response at $t = 0$. We are not applying any forcing function at all, and the natural response is therefore the complete response. If we now interpret this response as the *forced response* produced by a zero-amplitude forcing function, we find that the zero-amplitude forcing function produces a nonzero-amplitude forced response. Instead of applying a forcing function of nonzero amplitude at the frequency of the pole and obtaining a response of infinite amplitude, we apply a forcing function of zero amplitude and see a response of finite amplitude. Moreover, the form of this forced response is identical with the form of the natural response

Let us emphasize this idea by considering a numerical example.

Example 12-4 Once again consider the simple series circuit with $L = 1$ H, $R = 1$ Ω, and a pole at $\sigma = -1$ Np/s. The forcing function is $v_s = V_m e^{\sigma t}$, and we ask what amplitude V_m must have to cause a current of 1-A amplitude to be present. That is, we desire the response $i(t) = 1e^{\sigma t}$ A, at $\mathbf{s} = \sigma + j0$.

Solution: Using Eq. (13) again with $I = 1$, $R = 1$, and $L = 1$, we find that

$$1 = \frac{V_m}{\sigma + 1}$$

or $$V_m = \sigma + 1$$

This is the amplitude of the forcing function that is required to maintain a current of 1-A amplitude in the circuit. We note that a zero-amplitude forcing function, or a short circuit, is sufficient at the frequency of the pole.

■

The relationship between the frequencies of the poles and the form of the natural response arises in several of the examples discussed in this section, but we shall not discuss it thoroughly until the following section.

Example 12-5 As a second example, let us consider the circuit drawn in Fig. 12-7a. The exponential current source $i_s(t) = 4e^{\sigma t}$ A is applied to the RC circuit, and the desired response is the output voltage $v_o(t)$.

Solution: The input impedance of the network to the right of the source is

$$Z_{in}(\sigma) = 5 \left\| \left(5 + \frac{1}{0.01\sigma} \right) \right. = \frac{5(5 + 100/\sigma)}{5 + 5 + 100/\sigma}$$

Figure 12-7

(a) An RC circuit excited by an exponential current source. (b) The magnitude of the voltage response $|V_o|$ possesses a pole at $\sigma = -10$ and a zero at $\sigma = -20$ Np/s.

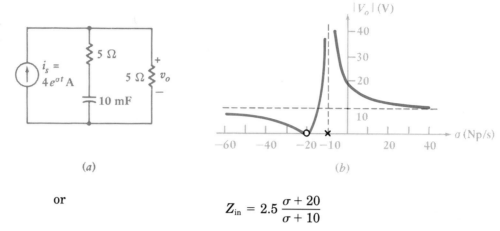

(a)

(b)

or

$$Z_{in} = 2.5 \frac{\sigma + 20}{\sigma + 10}$$

Therefore, the frequency-domain output voltage is

$$V_o = I_s Z_{in} = \frac{10(\sigma + 20)}{\sigma + 10} \qquad (16)$$

This last form of the forced response, written as a constant times the ratio of factors having the form $(\sigma + \sigma_1)$, is obviously well suited for the quick determination of the poles and zeros of the response. The voltage response indicates a pole at $\sigma = -10$ and a zero at $\sigma = -20$; infinite frequency is not a critical frequency. The response magnitude $|V_o|$ is plotted as a function of frequency in Fig. 12-7b. The pole at $\sigma = -10$ indicates that this forcing function, having an amplitude of 4 A, can cause an output voltage of arbitrarily large amplitude if the frequency is brought sufficiently close to -10 Np/s. Alternatively, we can provide an output of some given finite amplitude by using a vanishingly small input amplitude for which σ approaches -10 Np/s.

The zero at $\sigma = -20$ Np/s shows that no finite amplitude of the source at this frequency is capable of yielding any output voltage.

Finally, our desired time-domain output voltage is

$$v_o(t) = \frac{10(\sigma + 20)}{\sigma + 10} e^{\sigma t} \qquad (17)$$

∎

Our last example leads to a more complicated response curve.

Example 12-6 We wish to obtain a plot of $|I(\sigma)|$ versus σ for the circuit shown in Fig. 12-8a.

Solution: The current is easily found:

$$I = \frac{100}{6 + \sigma + 5/\sigma}$$

or

$$I = 100 \frac{\sigma}{(\sigma + 1)(\sigma + 5)} \qquad (18)$$

The response curve may be obtained by indicating the locations of all poles and zeros on the σ axis, and placing vertical asymptotes at the poles. When this is done, we find that a relative minimum of the forced response must

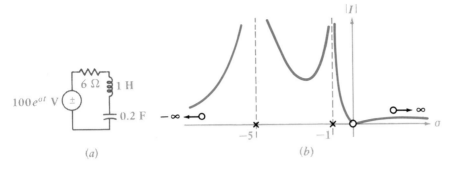

Figure 12-8

(a) A series RLC circuit is driven by an exponential forcing function. (b) The resultant current-magnitude response curve shows zeros at $\sigma = 0$ and $\sigma = \pm\infty$, and poles at $\sigma = -5$ and -1 Np/s.

exist between the two poles, and a relative maximum must be present at some frequency greater than zero. By differentiation, the locations of this minimum and maximum may be determined as $\sigma = -\sqrt{5}$ and $\sigma = \sqrt{5}$, respectively. The value of the response at the relative minimum turns out to be 65.5, and that of the relative maximum, 9.55. The response curve is shown in Fig. 12-8b.

To emphasize the relationship between the time domain and frequency domain once again, the forced response at $\sigma = -3$ is found from Eq. (18) to be 75 A; hence, excitation of the network by the forcing function $v(t) = 100e^{-3t}$ produces the forced current response $i(t) = 75e^{-3t}$. ∎

The two poles occurring in this last example may again be used to construct the natural response,

$$i_n = Ae^{-t} + Be^{-5t}$$

The values of these two natural frequencies may be compared with those obtained using the methods of Sec. 6-6 for a series RLC circuit. That is, we let

$$i_n = A_1 e^{s_1 t} + A_2 e^{s_2 t}$$

where

$$s_{1,2} = -\frac{R}{2L} \pm \sqrt{\left(\frac{R}{2L}\right)^2 - \frac{1}{LC}} = -1 \text{ and } -5$$

Thus, as the frequency of a nonzero forcing function approaches either of these two frequencies, $\sigma = -1$ or $\sigma = -5$, an arbitrarily large forced response results; a zero-amplitude forcing function may be associated with a finite-amplitude forced response which can be interpreted as the natural response of the circuit.

Drill Problems

12-8. Find V_{out}/V_{in} as a function of σ for the circuit of Fig. 12-9a, and then: (a) determine all the critical frequencies for the expression. (b) Evaluate V_{out}/V_{in} at $\sigma = -200, -80, -40,$ and 0 Np/s, and sketch $|V_{out}/V_{in}|$ versus σ.
Ans: $-\infty, -100, -10$ Np/s; 0.0526, -0.714; -0.556, 1.000

12-9. Find an expression for the input impedance seen by the source as a function of σ for the network of Fig. 12-9a, and then: (a) specify all the critical frequencies. (b) Evaluate Z_{in} at $\sigma = -105, -95, -50,$ and 10 Np/s, and then sketch $|Z_{in}|$ versus σ. *Ans:* $-100, -10, -90, 0$ Np/s; 6.03, -17.89, 20.0, 44.0 kΩ

Figure 12-9

See Drill Probs. 12-8 to
12-10.

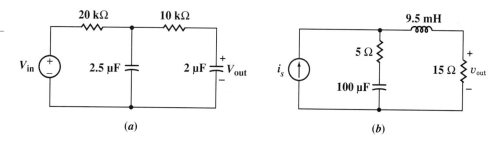

(a)   (b)

12-10. Let $i_s(t) = 5e^{-1000t}$ mA in Fig. 12-9b, and then find $v_{out}(t)$ at t equal to
(a) -1 ms; (b) 0; (c) 1 ms.   *Ans:* -2.04 V; -0.750 V; -0.276 V

12-11. Specify an admittance $Y_1(\sigma)$ that has zeros at $\sigma = -2$ and -4 Np/s, and
poles at $\sigma = -1$ and -3 Np/s. Specify the amplitude by causing $Y_1(\sigma)$ to approach
5 mS as $\sigma \to \pm \infty$. An admittance of 10 mS is now placed in series with $Y_1(\sigma)$.
Find all the critical frequencies of this series combination.

Ans: -3.22, -1.451, -2.00, -4.00 Np/s

12-6

The complex-frequency plane

Now that we have considered the forced response of a circuit as ω varies (with
$\sigma = 0$) and as σ varies (with $\omega = 0$), we are prepared to develop a more general
graphical presentation by graphing quantities as functions of $\mathbf{s}$; that is, we wish
to show the response simultaneously as functions of both σ and ω.

Such a graphical portrayal of the forced response as a function of the
complex frequency $\mathbf{s}$ is a useful, enlightening technique in the analysis of
circuits, as well as in the design or synthesis of circuits. After we have developed
the concept of the complex-frequency plane, or $\mathbf{s}$ plane, we shall see how quickly
the behavior of a circuit can be approximated from a graphical representation
of its critical frequencies in this $\mathbf{s}$ plane. The converse procedure is also very
useful: if we are given a desired response curve (the frequency response of a
filter, for example), it will be possible to decide upon the necessary location of
its poles and zeros in the $\mathbf{s}$ plane and then to synthesize the filter. This synthesis
problem is a subject for detailed study in subsequent courses and is not one
which we shall more than briefly examine here. The $\mathbf{s}$ plane is also the basic
tool with which the possible presence of undesired oscillations is investigated
in feedback amplifiers and automatic control systems.

Let us develop a method of obtaining circuit response as a function of $\mathbf{s}$ by
extending the methods we have been using to find the response as a function
of either σ or ω. To review these methods, we shall obtain the input or driving-
point impedance of a network composed of a 3-Ω resistor in series with a 4-H
inductor. As a function of $\mathbf{s}$, we have

$$\mathbf{Z(s)} = 3 + 4\mathbf{s}$$

If we wish to obtain a graphical interpretation of the impedance variation with
σ, we let $\mathbf{s} = \sigma + j0$:

$$\mathbf{Z}(\sigma) = 3 + 4\sigma$$

and recognize a zero at $\sigma = -\frac{3}{4}$ and a pole at infinity. These critical frequencies
are marked on a σ axis, and after identifying the value of $\mathbf{Z}(\sigma)$ at some con-
venient noncritical frequency [perhaps $\mathbf{Z}(0) = 3$], it is easy to sketch $|\mathbf{Z}(\sigma)|$
versus σ.

In order to plot the response as a function of the radian frequency ω, we let $\mathbf{s} = 0 + j\omega$:

$$\mathbf{Z}(j\omega) = 3 + j4\omega$$

and then obtain both the magnitude and phase angle of $\mathbf{Z}(j\omega)$ as functions of ω:

$$|\mathbf{Z}(j\omega)| = \sqrt{9 + 16\omega^2}$$

$$\text{ang } \mathbf{Z}(j\omega) = \tan^{-1}\frac{4\omega}{3}$$

The magnitude function shows a single pole at infinity and a minimum at $\omega = 0$; it can be sketched readily as a curve of $|\mathbf{Z}(j\omega)|$ versus ω. The phase angle is an inverse tangent function, zero at $\omega = 0$ and $\pm 90°$ at $\omega = \pm\infty$; it is also easily presented as a plot of ang $\mathbf{Z}(j\omega)$ versus ω.

In graphing the response $\mathbf{Z}(j\omega)$ as a function of ω, two 2-dimensional plots are required, magnitude and phase angle as functions of ω. When exponential excitation is assumed, we could present all the information on a single two-dimensional graph by allowing both positive and negative values of $\mathbf{Z}(\sigma)$ versus σ. However, we chose to plot the magnitude of $\mathbf{Z}(\sigma)$ in order that our sketches would compare more closely with those depicting the magnitude of $\mathbf{Z}(j\omega)$. The phase angle ($\pm 180°$ only) of $\mathbf{Z}(\sigma)$ was largely ignored. The important point to note is that there is only one independent variable, σ in the case of exponential excitation and ω in the sinusoidal case. Now let us consider what alternatives are available to us if we wish to plot a response as a function of $\mathbf{s}$.

The complex frequency $\mathbf{s}$ requires two parameters, σ and ω, for its complete specification. The response is also a complex function, and we must therefore consider sketching both the magnitude and phase angle as functions of $\mathbf{s}$. Either of these quantities—for example, the magnitude—is a function of the two parameters σ and ω, and we can plot it in two dimensions only as a family of curves, such as magnitude versus ω, with σ as the parameter. Conversely, we could also show the magnitude versus σ, with ω as the parameter. Such a family of curves represents a tremendous amount of work, however, and this is what we are trying to avoid; it is also questionable whether we could ever draw any useful conclusions from the family of curves even after they were obtained.

A better method of representing the magnitude of some complex response graphically involves using a *three*-dimensional model. Although such a model is difficult to draw on a two-dimensional sheet of paper, we shall find that the model is not difficult to visualize; most of the drawing will be done mentally, since in one's head few supplies are needed and construction, correction, and erasures are quickly accomplished. Let us think of a σ axis and a $j\omega$ axis, perpendicular to each other, laid out on a horizontal surface such as the floor. The floor now represents a *complex-frequency plane*, or $\mathbf{s}$ *plane*, as sketched in Fig. 12-10. To each point in this plane there corresponds exactly one value of $\mathbf{s}$, and with each value of $\mathbf{s}$ we may associate a single point in this complex plane.

Since we are already quite familiar with the type of time-domain function associated with a particular value of the complex frequency $\mathbf{s}$, it is now possible to associate the functional form of a forcing function or forced response with a specific region in the $\mathbf{s}$ plane. The origin, for example, represents a dc quantity. Points lying on the σ axis represent exponential functions, decaying for $\sigma < 0$,

Figure 12-10

The complex-frequency plane, or **s** plane.

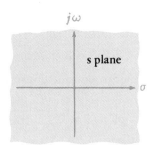

increasing for $\sigma > 0$. Pure sinusoids are associated with points on the positive or negative $j\omega$ axis. The right half of the **s** plane, usually referred to simply as the RHP, contains points describing frequencies with positive real parts and thus corresponds to time-domain quantities which are exponentially increasing sinusoids, except on the σ axis. Correspondingly, points in the left half of the **s** plane (LHP) describe the frequencies of exponentially decreasing sinusoids, again with the exception of the negative σ axis. Figure 12-11 summarizes the relationship between the time domain and the various regions of the **s** plane.

Let us now return to our search for an appropriate method of representing a response graphically as a function of the complex frequency **s**. The magnitude of the response may be represented by constructing, say, a plaster model whose height above the floor at every point corresponds to the magnitude of the response at that value of **s**. In other words, we have added a third axis, perpen-

Figure 12-11

The nature of the time-domain function is sketched in the region of the complex-frequency plane to which it corresponds.

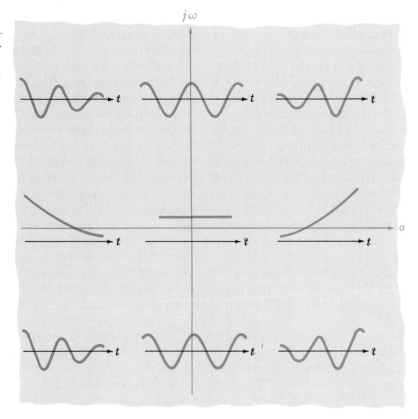

dicular to both the σ axis and the $j\omega$ axis and passing through the origin; this axis is labeled $|\mathbf{Z}|$, $|\mathbf{Y}|$, $|\mathbf{V}_2/\mathbf{V}_1|$, or with another appropriate symbol. The response magnitude is determined for every value of $\mathbf{s}$, and the resultant plot is a surface lying above (or just touching) the $\mathbf{s}$ plane.

We may try out these preliminary ideas by seeing what one such plaster model might look like.

Example 12-7 Let us consider the admittance of the series combination of a 1-H inductor and a 3-Ω resistor.

Solution: The admittance of these two series elements is

$$\mathbf{Y}(\mathbf{s}) = \frac{1}{\mathbf{s}+3}$$

In terms of both σ and ω, we have, as the magnitude,

$$|\mathbf{Y}(\mathbf{s})| = \frac{1}{\sqrt{(\sigma+3)^2 + \omega^2}}$$

When $\mathbf{s} = -3 + j0$, the response magnitude is infinite; and when $\mathbf{s}$ is infinite, the magnitude of $\mathbf{Y}(\mathbf{s})$ is zero. Thus our model must have infinite height over the point $(-3 + j0)$, and it must have zero height at all points infinitely far away from the origin. A cutaway view of such a model is shown in Fig. 12-12a .

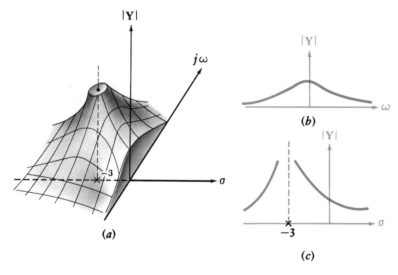

(a)

(b)

(c)

Figure 12-12

(a) A cutaway view of a plaster model whose top surface represents $|\mathbf{Y}(\mathbf{s})|$ for the series combination of a 1-H inductor and a 3-Ω resistor. (b) $|\mathbf{Y}(\mathbf{s})|$ as a function of ω. (c) $|\mathbf{Y}(\mathbf{s})|$ as a function of σ.

Once the model is constructed, it is simple to visualize the variation of $|\mathbf{Y}|$ as a function of ω (with $\sigma = 0$) by cutting the model with a perpendicular plane containing the $j\omega$ axis. The model shown in Fig. 12-12a happens to be cut along this plane, and the desired plot of $|\mathbf{Y}|$ versus ω can be seen; the curve is also drawn in Fig. 12-12b. In a similar manner, a vertical plane containing the σ axis enables us to obtain $|\mathbf{Y}|$ versus σ (with $\omega = 0$), as shown in Fig. 12-12c. ∎

How might we obtain some qualitative response information without doing all this work? After all, most of us have neither the time nor the inclination to be good plasterers, and some more practical method is needed. Let us visualize the **s** plane once again as the floor and then imagine a larger rubber sheet laid on it. We now fix our attention on all the poles and zeros of the response. At each zero, the response is zero, the height of the sheet must be zero, and we therefore tack the sheet to the floor. At the value of **s** corresponding to each pole, we may prop up the sheet with a thin vertical rod. Zeros and poles at infinity must be treated by using a large-radius clamping ring or a high circular fence, respectively. If we have used an infinitely large, weightless, perfectly elastic sheet, tacked down with vanishingly small tacks, and propped up with infinitely long, zero-diameter rods, then the rubber sheet assumes a height which is exactly proportional to the magnitude of the response. Less accurate rubber-sheet models may actually be constructed in the laboratory, but their main advantage lies in the ease by which their construction may be visualized from a knowledge of the pole-zero locations of the response.

These comments may be illustrated by considering the configuration of the poles and zeros, sometimes called a *pole-zero constellation*, which locates all the critical frequencies of some frequency-domain quantity—say, the impedance $\mathbf{Z}(\mathbf{s})$. A pole-zero constellation for an impedance is shown in Fig. 12-13a. If we visualize a rubber-sheet model, tacked down at $\mathbf{s} = -2 + j0$ and propped up at $\mathbf{s} = -1 + j5$ and at $\mathbf{s} = -1 - j5$, we should see a terrain whose distinguishing features are two mountains and one conical crater or depression. The portion of the model for the upper LHP is shown in Fig. 12-13b.

Figure 12-13

(a) The pole-zero constellation of some impedance $\mathbf{Z}(\mathbf{s})$. (b) A portion of the rubber-sheet model of the magnitude of $\mathbf{Z}(\mathbf{s})$.

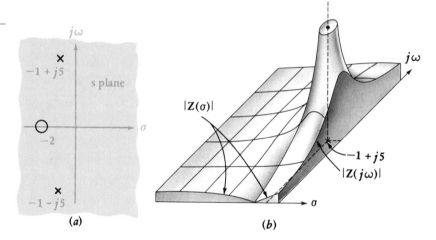

Let us now build up the expression for $\mathbf{Z}(\mathbf{s})$ which leads to this pole-zero configuration. The zero requires a factor of $(\mathbf{s} + 2)$ in the numerator, and the two poles require the factors $(\mathbf{s} + 1 - j5)$ and $(\mathbf{s} + 1 + j5)$ in the denominator. Except for a multiplying constant k, we now know the form of $\mathbf{Z}(\mathbf{s})$:

$$\mathbf{Z}(\mathbf{s}) = k \, \frac{\mathbf{s} + 2}{(\mathbf{s} + 1 - j5)(\mathbf{s} + 1 + j5)}$$

or
$$\mathbf{Z}(\mathbf{s}) = k \, \frac{\mathbf{s} + 2}{\mathbf{s}^2 + 2\mathbf{s} + 26} \tag{19}$$

Let us select k by assuming a single additional fact about $\mathbf{Z}(\mathbf{s})$: let $\mathbf{Z}(0) = 1$. By direct substitution in Eq. (19), we find that k is 13, and therefore

$$\mathbf{Z(s)} = 13\,\frac{\mathbf{s} + 2}{\mathbf{s}^2 + 2\mathbf{s} + 26} \tag{20}$$

The plots $|\mathbf{Z}(\sigma)|$ versus σ and $|\mathbf{Z}(j\omega)|$ versus ω may be obtained exactly from Eq. (20), but the general form of the function is apparent from the pole-zero configuration and the rubber-sheet analogy. Portions of these two curves appear at the sides of the model shown in Fig. 12-13b.

Thus far, we have been using the **s** plane and the rubber-sheet model to obtain *qualitative* information about the variation of the *magnitude* of the frequency-domain function with frequency. It is possible, however, to get *quantitative* information concerning the variation of both the *magnitude* and *phase angle*. The method provides us with a powerful new tool.

Consider the representation of a complex frequency in polar form, as suggested by an arrow drawn from the origin of the **s** plane to the complex frequency under consideration. The length of the arrow is the magnitude of the frequency, and the angle that the arrow makes with the positive direction of the σ axis is the angle of the complex frequency. The frequency $\mathbf{s}_1 = -3 + j4 = 5/\underline{126.9°}$ is indicated in Fig. 12-14a.

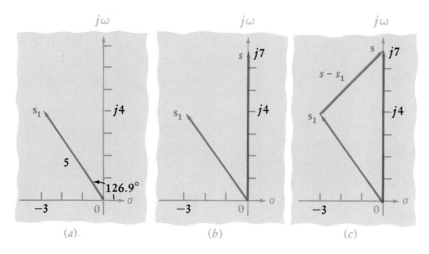

Figure 12-14

(a) The complex frequency $\mathbf{s}_1 = -3 + j4$ is indicated by drawing an arrow from the origin to $\mathbf{s}_1$. (b) The frequency $\mathbf{s} = j7$ is also represented vectorially. (c) The difference $\mathbf{s} - \mathbf{s}_1$ is represented by the vector drawn from $\mathbf{s}_1$ to $\mathbf{s}$.

It is also necessary to represent the difference between two values of **s** as an arrow or vector on the complex plane. Let us select a value of **s** that corresponds to a sinusoid $\mathbf{s} = j7$ and indicate it also as a vector, as shown in Fig. 12-14b. The *difference* $\mathbf{s} - \mathbf{s}_1$ is seen to be the vector drawn from the last-named point $\mathbf{s}_1$ to the first-named point $\mathbf{s}$; the vector $\mathbf{s} - \mathbf{s}_1$ is drawn in Fig. 12-14c. Note that $\mathbf{s}_1 + (\mathbf{s} - \mathbf{s}_1) = \mathbf{s}$. Numerically, $\mathbf{s} - \mathbf{s}_1 = j7 - (-3 + j4) = 3 + j3 = 4.24/\underline{45°}$, and this value agrees with the graphical difference.

Let us see how this graphical interpretation of the difference $(\mathbf{s} - \mathbf{s}_1)$ enables us to determine frequency response. Consider the admittance

$$\mathbf{Y(s)} = \mathbf{s} + 2$$

This expression may be interpreted as the difference between some frequency of interest **s** and a zero location. Thus, the zero is present at $\mathbf{s}_2 = -2 + j0$, and the factor $\mathbf{s} + 2$, which may be written as $\mathbf{s} - \mathbf{s}_2$, is represented by the vector drawn from the zero location $\mathbf{s}_2$ to the frequency **s** at which the response is desired. If the sinusoidal response is desired, **s** must lie on the $j\omega$ axis, as illustrated in Fig. 12-15a. The magnitude of $\mathbf{s} + 2$ may now be visualized as ω

Figure 12-15

(a) The vector representing the admittance $\mathbf{Y}(\mathbf{s}) = \mathbf{s} + 2$ is shown for $\mathbf{s} = j\omega$.
(b) Sketches of $|\mathbf{Y}(j\omega)|$ and ang $\mathbf{Y}(j\omega)$ as they might be obtained from the performance of the vector as $\mathbf{s}$ moves up or down the $j\omega$ axis from the origin.

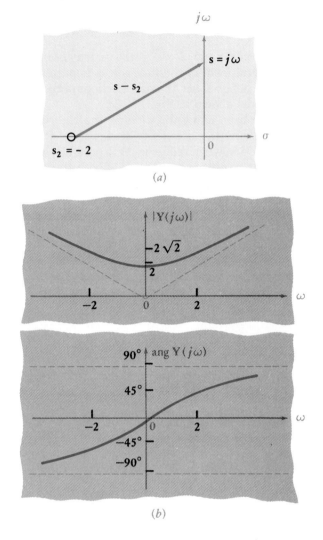

varies from zero to infinity. When $\mathbf{s}$ is zero, the vector has a magnitude of 2 and an angle of 0°. Thus $\mathbf{Y}(0) = 2$. As ω increases, the magnitude increases, slowly at first, and then almost linearly with ω; the phase angle increases almost linearly at first, and then gradually approaches 90° as ω becomes infinite. At $\omega = 7$, $\mathbf{Y}(j7)$ has a magnitude of $\sqrt{2^2 + 7^2}$ and a phase angle of $\tan^{-1}(3.5)$. The magnitude and phase of $\mathbf{Y}(\mathbf{s})$ are sketched as functions of ω in Fig. 12-15b.

Let us now construct a more realistic example by considering a frequency-domain function given by the quotient of two factors,

$$\mathbf{V}(\mathbf{s}) = \frac{\mathbf{s} + 2}{\mathbf{s} + 3}$$

We again select a value of $\mathbf{s}$ which corresponds to sinusoidal excitation and draw the vectors $\mathbf{s} + 2$ and $\mathbf{s} + 3$, the first from the zero to the chosen point on the $j\omega$ axis and the second from the pole to the chosen point. The two vectors are sketched in Fig. 12-16a. The quotient of these two vectors has a magnitude equal to the quotient of the magnitudes and a phase angle equal to the difference of the numerator and denominator phase angles. An investigation of the varia-

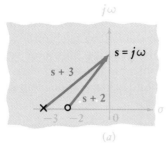

Figure 12-16

(a) Vectors are drawn from the two critical frequencies of the voltage response $V(s) = (s + 2)/(s + 3)$.
(b) Sketches of the magnitude and the phase angle of $V(j\omega)$ as obtained from the quotient of the two vectors shown in part a.

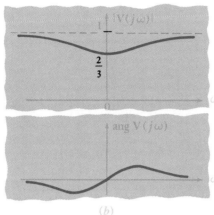

tion of the magnitude of $V(s)$ versus ω is made by allowing s to move from the origin up the $j\omega$ axis and considering the ratio of the distance from the zero to $s = j\omega$ and the distance from the pole to the same point on the $j\omega$ axis. The ratio evidently is $\frac{2}{3}$ at $\omega = 0$ and approaches unity as ω becomes infinite. A consideration of the difference of the two phase angles shows that ang $V(j\omega)$ is $0°$ at $\omega = 0$; increases at first as ω increases, since the angle of the vector $s + 2$ is greater than that of $s + 3$; and then decreases with a further increase in ω, finally approaching $0°$ at infinite frequency, where both vectors possess $90°$ angles. These results are sketched in Fig. 12-16b. Although no quantitative markings are present on these sketches, it is important to note that they could be obtained easily. For example, the complex response at $s = j4$ must be given by the ratio

$$V(j4) = \frac{\sqrt{4 + 16} \,\underline{/\tan^{-1}\left(\frac{4}{2}\right)}}{\sqrt{9 + 16} \,\underline{/\tan^{-1}\left(\frac{4}{3}\right)}}$$

$$= \sqrt{\tfrac{20}{25}} \,\underline{/\tan^{-1} 2 - \tan^{-1}\left(\tfrac{4}{3}\right)}$$

$$= 0.894 \,\underline{/10.3°}$$

In designing circuits to produce some desired response, the behavior of the vectors drawn from the respective critical frequencies to a general point on the $j\omega$ axis is an important aid. For example, if it were necessary to increase the hump in the phase response of Fig. 12-16b, we can see that we would have to provide a greater difference in the angles of the two vectors. This may be achieved in Fig. 12-16a either by moving the zero closer to the origin or by locating the pole farther from the origin, or both.

The ideas we have been discussing to help in the graphical determination of the magnitude and angular variation of some frequency-domain function with frequency will be needed in the following chapter when we investigate the frequency performance of highly selective filters, or resonant circuits. These concepts are fundamental in obtaining a quick, clear understanding of the behavior of electrical networks and other engineering systems. The procedure is briefly summarized as follows:

1 Draw the pole-zero constellation of the frequency-domain function under consideration in the **s** plane, and locate a test point corresponding to the frequency at which the function is to be evaluated.
2 Draw an arrow from each pole and zero to the test point.
3 Determine the length of each pole arrow and each zero arrow, and the value of each pole-arrow angle and each zero-arrow angle.
4 Divide the product of the zero-arrow lengths by the product of the pole-arrow lengths. This quotient is the magnitude of the frequency-domain function for the assumed frequency of the test point [within a multiplying constant, since $\mathbf{F}(\mathbf{s})$ and $k\mathbf{F}(\mathbf{s})$ have the same pole-zero constellations].
5 Subtract the sum of the pole-arrow angles from the sum of the zero-arrow angles. The resultant difference is the angle of the frequency-domain function, evaluated at the frequency of the test point. The angle does not depend upon the value of the real multiplying constant k.

Drill Problems

12-12. The parallel combination of 0.25 mH and 5 Ω is in series with the parallel combination of 40 μF and 5 Ω. (*a*) Find $\mathbf{Z}_{in}(\mathbf{s})$, the input impedance of the series combination. (*b*) Specify all the zeros of $\mathbf{Z}_{in}(\mathbf{s})$. (*c*) Specify all the poles of $\mathbf{Z}_{in}(\mathbf{s})$. (*d*) Draw the pole-zero configuration.   *Ans:* $5(\mathbf{s}^2 + 10\ 000\mathbf{s} + 10^8)/(\mathbf{s}^2 + 25\ 000\mathbf{s} + 10^8)$ Ω; $-5 \pm j8.66$ complex krad/s; $-5, -20$ krad/s

12-13. Three pole-zero constellations are shown in Fig. 12-17. Each applies to a voltage gain $\mathbf{G}_V$. Obtain an expression for each gain that is a ratio of polynomials in **s**.   *Ans:* $(15\mathbf{s}^2 + 45\mathbf{s})/(\mathbf{s}^2 + 6\mathbf{s} + 8)$; $(2\mathbf{s}^3 + 22\mathbf{s}^2 + 88\mathbf{s} + 120)/(\mathbf{s}^2 + 4\mathbf{s} + 8)$; $(3\mathbf{s}^2 + 27)/(\mathbf{s}^2 + 2\mathbf{s})$

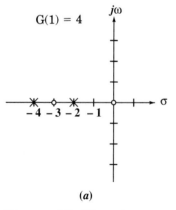

(*a*)

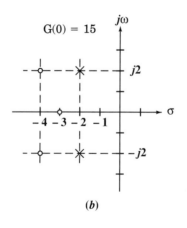

(*b*)

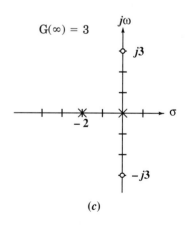

(*c*)

Figure 12-17

See Drill Prob. 12-13.

12-14. A plaster model of an impedance shows zeros at $\mathbf{s} = -1 \pm j5$ s^{-1} and poles at $\mathbf{s} = -3 \pm j4$ s^{-1}. If the height of the model is 6 cm at the origin, find the height at $\mathbf{s}$ equal to (a) -2 s^{-1}; (b) $j2$ s^{-1}; (c) $-2 + j2$ s^{-1}; (d) ∞.

Ans: 8.82 cm; 5.33 cm; 9.48 cm; 5.77 cm

12-15. The pole-zero configuration for an admittance $\mathbf{Y}(\mathbf{s})$ has one pole at $\mathbf{s} = -10 + j0$ s^{-1} and one zero at $\mathbf{s} = z_1 + j0$, where $z_1 < 0$. Let $\mathbf{Y}(0) = 0.1$ S. Find the value of z_1 if (a) ang $\mathbf{Y}(j5) = 20°$; (b) $|\mathbf{Y}(j5)| = 0.2$ S.

Ans: -4.73 Np/s; -2.50 Np/s

12-7

Natural response and the s plane

There is a tremendous amount of information contained in the pole-zero plot of some forced response in the s plane. We shall find out how a complete current response, natural plus forced, produced by an arbitrary forcing function can be quickly written from the pole-zero configuration of the forced current response and from the initial conditions; the method is similarly effective in finding the complete voltage response produced by an arbitrary source.

Let us introduce the method by considering the simplest example, a series RL circuit as shown in Fig. 12-18. A general voltage source $v_s(t)$ causes the current $i(t)$ to flow after closure of the switch at $t = 0$. The complete response $i(t)$ for $t > 0$ is composed of a natural response and a forced response:

$$i(t) = i_n(t) + i_f(t) \tag{21}$$

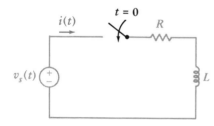

Figure 12-18

An example which illustrates the determination of the complete response through a knowledge of the critical frequencies of the impedance faced by the source.

We may find the forced response by working in the frequency domain, assuming, of course, that $v_s(t)$ has a functional form which we can transform to the frequency domain; if $v_s(t) = 1/(1 + t^2)$, for example, we must proceed as best we can from the basic differential equation for the circuit. For the circuit of Fig. 12-18, we have

$$\mathbf{I}_f(\mathbf{s}) = \frac{\mathbf{V}_s}{R + \mathbf{s}L}$$

or

$$\mathbf{I}_f(\mathbf{s}) = \frac{1}{L} \frac{\mathbf{V}_s}{\mathbf{s} + R/L} \tag{22}$$

and $i_f(t)$ is obtained by replacing $\mathbf{s}$, L, and R by their values, reinserting $\mathbf{e}^{\mathbf{s}t}$, and taking the real part. The answer may even be obtained as a function of a general ω, σ, R, and L if desired.

Now let us consider the natural response. Of course, we know that the form will be a decaying exponential with the time constant L/R, but we may pretend that we are finding it for the first time. The form of the natural, or *source-free*, response is, by definition, independent of the forcing function; the forcing function contributes, along with the other initial conditions, only to the magnitude of the natural response. To find the proper form, replace all independent sources by their internal impedances; here, $v_s(t)$ is replaced by a short circuit. Now let us try to obtain this natural response as a limiting case of the forced response. We return to the frequency-domain expression of Eq. (22) and obediently set $\mathbf{V}_s = 0$. On the surface, it appears that $\mathbf{I}(\mathbf{s})$ must also be zero, but this is not necessarily true if we are operating at a complex frequency that is a simple pole of $\mathbf{I}(\mathbf{s})$. That is, the denominator and the numerator may both be zero so that $\mathbf{I}(\mathbf{s})$ need not be zero.

Let us inspect this new idea from a slightly different vantage point. We fix our attention on the ratio of the desired forced response to the forcing function. Let us designate this ratio in general by $\mathbf{H}(\mathbf{s})$ and call it a *transfer function*. Here,

$$\frac{\mathbf{I}_f(\mathbf{s})}{\mathbf{V}_s} = \mathbf{H}(\mathbf{s}) = \frac{1}{L(\mathbf{s} + R/L)}$$

In this example, the transfer function is the input admittance faced by $\mathbf{V}_s$. We seek the natural, or source-free, response by setting $\mathbf{V}_s = 0$. However, $\mathbf{I}_f(\mathbf{s}) = \mathbf{V}_s\mathbf{H}(\mathbf{s})$, and if $\mathbf{V}_s = 0$, a nonzero value for the current can be obtained only by operating at a pole of $\mathbf{H}(\mathbf{s})$. The poles of the transfer function therefore assume a special significance.

Returning to our series RL circuit, we see that the pole of the transfer function occurs when the operating frequency is $\mathbf{s} = -R/L + j0$. A finite current at this frequency thus represents the natural response

$$\mathbf{I}(\mathbf{s}) = A \quad \text{at} \quad \mathbf{s} = -\frac{R}{L} + j0$$

where A is an unknown constant. Transforming this natural response to the time domain, we find

$$i_n(t) = \text{Re}\,(Ae^{-Rt/L})$$

or

$$i_n(t) = Ae^{-Rt/L}$$

To complete this example, the total response is then

$$i(t) = Ae^{-Rt/L} + i_f(t)$$

and A may be determined once the initial conditions are specified for this circuit.

Now let us generalize these results. Figures 12-19a and b show single sources connected to networks containing no independent sources. The desired response, which might be some current $\mathbf{I}_1(\mathbf{s})$ or some voltage $\mathbf{V}_2(\mathbf{s})$, may be

Figure 12-19

The poles of the response, $\mathbf{I}_1(\mathbf{s})$ or $\mathbf{V}_2(\mathbf{s})$, produced by a voltage source $\mathbf{V}_s$ (*a*) or a current source $\mathbf{I}_s$ (*b*), determine the form of the natural response, $i_{1n}(t)$ or $v_{2n}(t)$, that occurs when $\mathbf{V}_s$ is replaced by a short circuit or $\mathbf{I}_s$ by an open circuit and some initial energy is available.

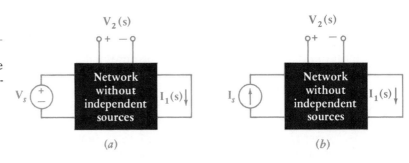

expressed by a transfer function that displays all the critical frequencies. To be specific, we select the response $\mathbf{V}_2(\mathbf{s})$ in Fig. 12-19a:

$$\frac{\mathbf{V}_2(\mathbf{s})}{\mathbf{V}_s} = \mathbf{H}(\mathbf{s}) = k\,\frac{(\mathbf{s} - \mathbf{s}_1)(\mathbf{s} - \mathbf{s}_3)\;\cdot\;\cdot\;\cdot}{(\mathbf{s} - \mathbf{s}_2)(\mathbf{s} - \mathbf{s}_4)\;\cdot\;\cdot\;\cdot} \qquad (23)$$

The poles of $\mathbf{H}(\mathbf{s})$ occur at $\mathbf{s} = \mathbf{s}_2,\ \mathbf{s}_4,\ \ldots$, and thus a finite voltage $\mathbf{V}_2(\mathbf{s})$ at each of these frequencies must be a possible functional form for the natural response. Therefore, we think of a zero-volt source (which is just a short circuit) applied to the input terminals; the natural response which occurs when the input terminals are short-circuited thus must have the form

$$v_{2n}(t) = \mathbf{A}_2 e^{\mathbf{s}_2 t} + \mathbf{A}_4 e^{\mathbf{s}_4 t} + \cdot\;\cdot\;\cdot$$

where each $\mathbf{A}$ must be evaluated in terms of the initial conditions (including the initial value of any voltage source applied at the input terminals).

To find the form of the natural response $i_{1n}(t)$ in Fig. 12-19a, we should determine the poles of the transfer function, $\mathbf{H}(\mathbf{s}) = \mathbf{I}_1(\mathbf{s})/\mathbf{V}_s$. The transfer functions applying to the situations depicted in Fig. 12-19b would obviously be $\mathbf{I}_1(\mathbf{s})/\mathbf{I}_s$ and $\mathbf{V}_2(\mathbf{s})/\mathbf{I}_s$, and their poles then determine the natural responses $i_{1n}(t)$ and $v_{2n}(t)$, respectively.

If the natural response is desired for a network that does not contain any independent sources, then a source $\mathbf{V}_s$ or $\mathbf{I}_s$ may be inserted at any convenient point, restricted only by the condition that the original network is obtained when the source is killed. The corresponding transfer function is then determined and its poles specify the natural frequencies. Note that the same frequencies must be obtained for any of the many source locations possible. If the network already contains a source, that source may be set equal to zero and another source inserted at a more convenient point.

Before we illustrate this method with several examples, completeness requires us to acknowledge two very special cases that might arise. One occurs when the network in Fig. 12-19a or b contains two or more parts that are isolated from each other. For example, we might have the parallel combination of three networks: R_1 in series with C, R_2 in series with L, and a short circuit. Obviously, a voltage source in series with R_1 and C cannot produce any current in R_2 and L; that transfer function would be zero. To find the form of the natural response of the inductor voltage, for example, the voltage source must be installed in the R_2L network. A case of this type can often be recognized by an inspection of the network before a source is installed; but if it is not, then a transfer function equal to zero will be obtained. When $\mathbf{H}(\mathbf{s}) = 0$, we obtain no information about the frequencies characterizing the natural response, and a more suitable location for the source must be used. One network of this type is present in the problems at the end of the chapter, but we forget which problem it is.

The other case involves multiple poles in the transfer function. That is, in Eq. (22) we might find that $\mathbf{s}_2$, $\mathbf{s}_4$, and $\mathbf{s}_{10}$ are identical. Transfer functions of this form are handled so much more easily by Laplace transform techniques that we shall postpone consideration of all problems of this type until Chap. 19. However, it is interesting to note that the critically damped RLC circuits of Chap. 6 possess a double pole, and the form of their natural response provides a clue by which curious readers might extrapolate and guess the results that we shall obtain later.

Let us now try out these techniques on two examples.

Example 12-8 Our first example involves a circuit that is source-free (Fig. 12-20). We seek expressions for i_1 and i_2 for $t > 0$, given the initial conditions $i_1(0) = i_2(0) = 11$ A.

Figure 12-20

A circuit for which the natural responses i_1 and i_2 are desired.

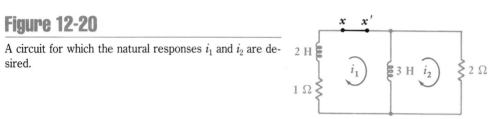

Solution: Let us install a voltage source $\mathbf{V}_s$ between points x and x' and find the transfer function $\mathbf{H(s)} = \mathbf{I_1(s)}/\mathbf{V}_s$, which also happens to be the input admittance seen by the voltage source. We have

$$\mathbf{I_1(s)} = \frac{\mathbf{V}_s}{2\mathbf{s} + 1 + 6\mathbf{s}/(3\mathbf{s} + 2)} = \frac{(3\mathbf{s} + 2)\mathbf{V}_s}{6\mathbf{s}^2 + 13\mathbf{s} + 2}$$

or

$$\mathbf{H(s)} = \frac{\mathbf{I_1(s)}}{\mathbf{V}_s} = \frac{\frac{1}{2}(\mathbf{s} + \frac{2}{3})}{(\mathbf{s} + 2)(\mathbf{s} + \frac{1}{6})}$$

Thus, i_1 must be of the form

$$i_1(t) = Ae^{-2t} + Be^{-t/6}$$

The solution is completed by using the given initial conditions to establish the values of A and B. Although we have been able to obtain the *form* of the solution rather rapidly, the procedure for evaluating the unknown constants in this second-order system is similar to that used for the *RLC* circuits back in Chap. 6; any saving in effort here must come later with a study of the Laplace transform.

For practice, however, let us complete the solution for i_1. Since $i_1(0)$ is given as 11 A,

$$11 = A + B$$

The necessary additional equation is obtained by writing the KVL equation around the perimeter of our circuit:

$$1i_1 + 2\frac{di_1}{dt} + 2i_2 = 0$$

and solving for the derivative:

$$\left.\frac{di_1}{dt}\right|_{t=0} = -\frac{1}{2}[2i_2(0) + 1i_1(0)] = -\frac{22 + 11}{2} = -2A - \frac{1}{6}B$$

Thus, $A = 8$ and $B = 3$, and so the desired solution is

$$i_1(t) = 8e^{-2t} + 3e^{-t/6}$$

The natural frequencies constituting i_2 are the same as those of i_1, and a similar procedure used to evaluate the arbitrary constants leads to

$$i_2(t) = 12e^{-2t} - e^{-t/6}$$

■

Example 12-9 As our last example, we shall find the complete response $v(t)$ of the circuit shown in Fig. 12-21. The make-before-break switch ensures that all currents and voltages to the right of the switch are initially zero. At $t = 0$ the switch moves up, and the voltage across the 3-Ω resistor is to be found for $t > 0$.

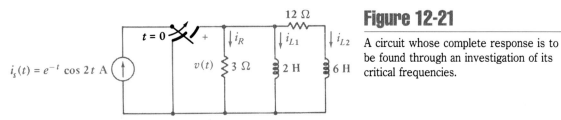

Figure 12-21

A circuit whose complete response is to be found through an investigation of its critical frequencies.

Solution: This response is composed of both a forced and a natural response:

$$v(t) = v_f(t) + v_n(t)$$

Each may be found through a knowledge of the pole-zero configuration of the transfer function, $\mathbf{H}(\mathbf{s}) = \mathbf{V}(\mathbf{s})/\mathbf{I}_s$, which is also the input impedance of the portion of the network to the right of the switch. We have

$$\mathbf{V}(\mathbf{s}) = \frac{\mathbf{I}_s}{\frac{1}{3} + 1/2\mathbf{s} + 1/(6\mathbf{s} + 12)}$$

or, after combining and factoring,

$$\mathbf{H}(\mathbf{s}) = \frac{\mathbf{V}(\mathbf{s})}{\mathbf{I}_s} = \frac{3\mathbf{s}(\mathbf{s} + 2)}{(\mathbf{s} + 1)(\mathbf{s} + 3)} \qquad (24)$$

The form of the natural response may now be written as

$$v_n(t) = Ae^{-t} + Be^{-3t}$$

In order to find the forced response, the frequency-domain current source $\mathbf{I}_s(\mathbf{s}) = 1$ at $\mathbf{s} = -1 + j2$ may be multiplied by the transfer function, evaluated at $\mathbf{s} = -1 + j2$:

$$\mathbf{V}(\mathbf{s}) = \mathbf{I}_s(\mathbf{s})\mathbf{H}(\mathbf{s}) = 3\frac{(-1 + j2)(1 + j2)}{j2(2 + j2)}$$

and thus

$$\mathbf{V}(\mathbf{s}) = 1.875\sqrt{2}\ \underline{/45°}$$

Transforming to the time domain, we have

$$v_f(t) = 1.875\sqrt{2}e^{-t} \cos{(2t + 45°)}$$

The complete response is therefore

$$v(t) = Ae^{-t} + Be^{-3t} + 1.875\sqrt{2}e^{-t} \cos{(2t + 45°)}$$

Since both inductor currents are initially zero, the initial source current of 1 A must flow through the 3-Ω resistor. Thus

$$v(0) = 3 = A + B + \frac{1.875\sqrt{2}}{\sqrt{2}} \qquad (25)$$

Again it is necessary to differentiate and then to obtain an initial condition for dv/dt. We first find

$$\frac{dv}{dt}\bigg|_{t=0} = 1.875\sqrt{2}\left(-\frac{2}{\sqrt{2}} - \frac{1}{\sqrt{2}}\right) - A - 3B$$

or

$$\frac{dv}{dt}\bigg|_{t=0} = -5.625 - A - 3B \qquad (26)$$

The initial value of this rate of change is obtained by analyzing the circuit. However, those rates of change which are most easily found are the derivatives of the inductor currents, for $v = L\,di/dt$, and the initial values of the inductor voltages should not be difficult to find. We therefore express the response $v(t)$ in terms of the resistor current:

$$v(t) = 3i_R$$

and then apply Kirchhoff's current law:

$$v(t) = 3i_s - 3i_{L1} - 3i_{L2}$$

Now we may take the derivative:

$$\frac{dv}{dt} = 3\frac{di_s}{dt} - 3\frac{di_{L1}}{dt} - 3\frac{di_{L2}}{dt}$$

Differentiation of the source function and evaluation at $t = 0$ provide a value of -3 V/s for the first term; the second term is numerically $\frac{3}{2}$ of the initial voltage across the 2-H inductor, or -4.5 V/s; and the last term is -1.5 V/s. Thus,

$$\frac{dv}{dt}\bigg|_{t=0} = -9$$

and we may now use Eqs. (24) and (25) to determine the unknown amplitudes:

$$A = 0 \qquad B = 1.125$$

The complete response is therefore

$$v(t) = 1.125e^{-3t} + 1.875\sqrt{2}e^{-t}\cos(2t + 45°)$$ ∎

The process which we must pursue to evaluate the amplitude coefficients of the natural response is a detailed one, except in those cases where the initial values of the desired response and its derivatives are obvious. However, we should not lose sight of the ease and rapidity with which the *form* of the natural response can be obtained.

Drill Problems

12-16. (*a*) If a current source $i_1(t) = u(t)$ A is present at *a-b* in Fig. 12-22 with the arrow entering *a*, find $\mathbf{H}(\mathbf{s}) = \mathbf{V}_{cd}/\mathbf{I}_1$, and specify the natural frequencies present in $v_{cd}(t)$. (*b*) Repeat for the voltage source $v_{ab}(t) = u(t)$ V, $\mathbf{H}(\mathbf{s}) = \mathbf{V}_{cd}/\mathbf{V}_{ab}$, and $v_{cd}(t)$.

Ans: 120s/(s + 20 000) Ω, −20 000 s⁻¹; s/(s + 50 000) Ω, −50 000 s⁻¹

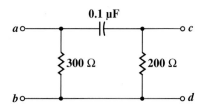

Figure 12-22

See Drill Prob. 12-16.

12-17. Find the natural frequencies present in the following natural response for the circuit of Fig. 12-23: (a) $i_1(t)$, if a source $v_s(t)$ is suddenly connected in place of the short circuit at a-b; (b) $v_2(t)$, if a source $i_s(t)$ is suddenly connected between c and d.   *Ans:* $-0.375 \pm j0.331$ s^{-1}; $-0.375 \pm j0.331$ s^{-1}

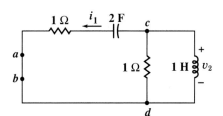

Figure 12-23

See Drill Prob. 12-17.

12-18. In Fig. 12-24, the transfer function $\mathbf{I}_2/\mathbf{V}_1$ is given by $8(\mathbf{s}+5)/(\mathbf{s}+20)$. (a) If $v_1(t) = 10e^{-8t}$ V, find the forced response, $i_{2f}(t)$. (b) If $v_1(t) = 20$ V, find the forced response $i_{2f}(t)$. (c) If $v_1(t) = 10e^{-8t}u(t)$ V and $i_2(0^+) = 50$ A, find the complete response $i_2(t)$.   *Ans:* $-20e^{-8t}$ A; 40 A; $(70e^{-20t} - 20e^{-8t})u(t)$ A

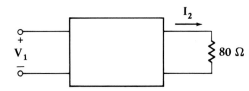

Figure 12-24

See Drill Prob. 12-18.

Much of the discussion in this chapter has been related to the poles and zeros of a transfer function. We have located them on the complex-frequency plane, we have used them to express transfer functions as ratios of factors or polynomials in **s**, we have calculated forced responses from them, and in the preceding section we have used their poles to establish the form of the natural response.

Now let us see how we might determine a network that can provide a desired transfer function. We consider only a small part of the general problem, working with a transfer function of the form $\mathbf{H}(\mathbf{s}) = \mathbf{V}_{out}(\mathbf{s})/\mathbf{V}_{in}(\mathbf{s})$, as indicated in Fig. 12-25. For simplicity, we restrict $\mathbf{H}(\mathbf{s})$ to critical frequencies on the negative σ axis (including the origin). Thus, we shall consider transfer functions such as

$$\mathbf{H}_1(\mathbf{s}) = \frac{10(\mathbf{s}+2)}{\mathbf{s}+5} \tag{27}$$

or

$$\mathbf{H}_2(\mathbf{s}) = \frac{-5\mathbf{s}}{(\mathbf{s}+8)^2} \tag{28}$$

or

$$\mathbf{H}_3(\mathbf{s}) = 0.1\mathbf{s}(\mathbf{s}+2) \tag{29}$$

12-8

A technique for synthesizing the voltage ratio H(s) = $\mathbf{V}_{out}/\mathbf{V}_{in}$

Figure 12-25

Given $\mathbf{H(s)} = \mathbf{V}_{out}/\mathbf{V}_{in}$, we seek a network having a specified $\mathbf{H(s)}$.

Let us begin by finding the voltage gain of the network of Fig. 12-26, which contains an ideal op-amp. The voltage between the two input terminals of the op-amp is essentially zero, and the input impedance of the op-amp is essentially infinite. We therefore may set the sum of the currents entering the inverting input terminal equal to zero:

$$\frac{\mathbf{V}_{in}}{\mathbf{Z}_1} + \frac{\mathbf{V}_{out}}{\mathbf{Z}_f} = 0$$

or

$$\frac{\mathbf{V}_{out}}{\mathbf{V}_{in}} = - \frac{\mathbf{Z}_f}{\mathbf{Z}_1} \tag{30}$$

Figure 12-26

For an ideal op-amp, $\mathbf{H(s)} = \mathbf{V}_{out}/\mathbf{V}_{in} = -\mathbf{Z}_f/\mathbf{Z}_1$.

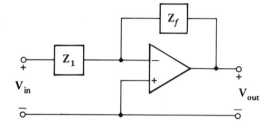

If $\mathbf{Z}_f$ and $\mathbf{Z}_1$ are both resistances, the circuit acts as an inverting amplifier, or possibly an attenuator. Our present interest, however, lies with those cases in which one of these impedances is a resistance while the other is an RC network.

In Fig. 12-27a, we let $\mathbf{Z}_1 = R_1$, while $\mathbf{Z}_f$ is the parallel combination of R_f and C_f. Therefore,

$$\mathbf{Z}_f = \frac{R_f/sC_f}{R_f + (1/sC_f)} = \frac{R_f}{1 + sC_fR_f} = \frac{1/C_f}{s + (1/R_fC_f)}$$

and

$$\mathbf{H(s)} = \frac{\mathbf{V}_{out}}{\mathbf{V}_{in}} = - \frac{\mathbf{Z}_f}{\mathbf{Z}_1} = - \frac{1/R_1C_f}{s + (1/R_fC_f)} \tag{31}$$

We have a transfer function with a single (finite) critical frequency, a pole at $\mathbf{s} = -1/R_fC_f$.

Figure 12-27

(a) The transfer function $\mathbf{H(s)} = \mathbf{V}_{out}/\mathbf{V}_{in}$ has a pole at $\mathbf{s} = -1/R_fC_f$. (b) Here, there is a zero at $\mathbf{s} = -1/R_1C_1$.

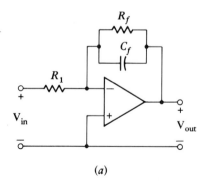

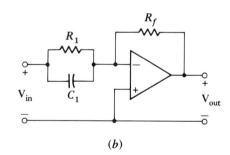

(a)

(b)

Moving on to Fig. 12-27b, we now let $\mathbf{Z}_f$ be resistive while $\mathbf{Z}_1$ is an RC parallel combination:

$$\mathbf{Z}_1 = \frac{1/C_1}{\mathbf{s} + (1/R_1C_1)}$$

and

$$\mathbf{H}(\mathbf{s}) = \frac{\mathbf{V}_{out}}{\mathbf{V}_{in}} = -\frac{\mathbf{Z}_f}{\mathbf{Z}_1} = -R_fC_1 \left(\mathbf{s} + \frac{1}{R_1C_1} \right) \qquad (32)$$

The only finite critical frequency is a zero at $\mathbf{s} = -1/R_1C_1$.

For our ideal op-amps, the output or Thévenin impedance is zero and therefore $\mathbf{V}_{out}$ and $\mathbf{V}_{out}/\mathbf{V}_{in}$ are not functions of any load $\mathbf{Z}_L$ that may be placed across the output terminals. This includes the input to another op-amp, as well, and therefore we may connect circuits having poles and zeros at specified locations in cascade, where the output of one op-amp is connected directly to the input of the next, and thus generates any desired transfer function.

Let us try a numerical example.

Example 12-10   To illustrate these ideas, we shall synthesize a circuit that will yield the transfer function

$$\mathbf{H}(\mathbf{s}) = \frac{\mathbf{V}_{out}}{\mathbf{V}_{in}} = \frac{10(\mathbf{s} + 2)}{\mathbf{s} + 5}$$

Solution: The pole at $\mathbf{s} = -5$ may be obtained by a network of the form of Fig. 12-27a. Calling this network A, we have $1/R_{fA}C_{fA} = 5$. We arbitrarily select $R_{fA} = 100$ kΩ; therefore $C_{fA} = 2$ μF. For this portion of the complete circuit,

$$\mathbf{H}_A(\mathbf{s}) = -\frac{1/R_{1A}C_{fA}}{\mathbf{s} + (1/R_{fA}C_{fA})} = -\frac{5 \times 10^5/R_{1A}}{\mathbf{s} + 5}$$

Next, we consider the zero at $\mathbf{s} = -2$. From Fig. 12-27b, $1/R_{1B}C_{1B} = 2$, and, with $R_{1B} = 100$ kΩ, we have $C_{1B} = 5$ μF. Thus

$$\mathbf{H}_B(\mathbf{s}) = -R_{fB}C_{1B} \left(\mathbf{s} + \frac{1}{R_{1B}C_{1B}} \right)$$

$$= -5 \times 10^{-6}R_{fB}(\mathbf{s} + 2)$$

and

$$\mathbf{H}(\mathbf{s}) = \mathbf{H}_A(\mathbf{s})\mathbf{H}_B(\mathbf{s}) = 2.5 \frac{R_{fB}}{R_{1A}} \frac{\mathbf{s} + 2}{\mathbf{s} + 5}$$

We complete the design by letting $R_{fB} = 100$ kΩ and $R_{1A} = 25$ kΩ. The result is shown in Fig. 12-28. The capacitors in this circuit are fairly large, but

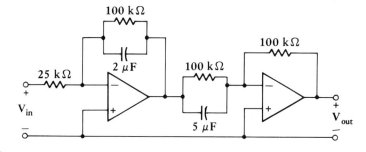

Figure 12-28

This network contains two ideal op-amps and gives the voltage transfer function $\mathbf{H}(\mathbf{s}) = \mathbf{V}_{out}/\mathbf{V}_{in} = 10(\mathbf{s} + 2)/(\mathbf{s} + 5)$.

this is a direct consequence of the low frequencies selected for the pole and zero of $\mathbf{H(s)}$. If $\mathbf{H(s)}$ were changed to $10(\mathbf{s} + 2000)/(\mathbf{s} + 5000)$, we could use 2- and 5-nF values.   ■

Drill Problem

12-19. Specify suitable element values for $\mathbf{Z}_1$ and $\mathbf{Z}_f$ in each of three cascaded stages to realize the transfer function $\mathbf{H(s)} = -20\mathbf{s}^2/(\mathbf{s} + 1000)$.

Ans: 1 μF $\parallel$ ∞, 100 kΩ; 1 μF $\parallel$ ∞, 100 kΩ; 100 kΩ $\parallel$ 10 nF, 50 kΩ

Problems

1 Let the real part of the complex time-varying current $\mathbf{i}(t)$ be $i(t)$. Find (*a*) $i_x(t)$ if $\mathbf{i}_x(t) = (4 - j7)e^{(-3+j15)t}$; (*b*) $i_y(t)$ if $\mathbf{i}_y(t) = (4 + j7)e^{-3t}(\cos 15t - j \sin 15t)$; (*c*) $i_A(0.4)$ if $\mathbf{i}_A(t) = \mathbf{K}_A e^{\mathbf{s}_A t}$, where $\mathbf{K}_A = 5 - j8$ and $\mathbf{s}_A = -1.5 + j12$; (*d*) $i_B(0.4)$ if $\mathbf{i}_B(t) = \mathbf{K}_B e^{\mathbf{s}_B t}$, where $\mathbf{K}_B$ is the conjugate of $\mathbf{K}_A$ and $\mathbf{s}_B$ is the conjugate of $\mathbf{s}_A$.

2 (*a*) Find v_x for $t < 0$ for the circuit of Fig. 12-29. (*b*) Find v_x for $t > 0$. (*c*) List all the complex frequencies found in v_x for $t > 0$.

Figure 12-29

See Prob. 2.

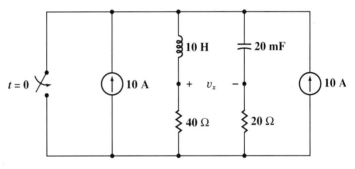

3 Let $i_s(t) = 2u(t)$ A in the circuit shown in Fig. 12-30. (*a*) Find $v_x(t)$. (*b*) List all the complex frequencies present in $v_x(t)$.

Figure 12-30

See Prob. 3.

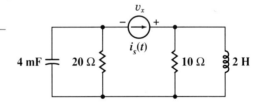

4 If a complex time-varying voltage is given as $\mathbf{v}_s(t) = (20 - j30)e^{(-2+j50)t}$ V, find (*a*) $\mathbf{v}_s(0.1)$ in polar form; (*b*) Re $[\mathbf{v}_s(t)]$; (*c*) Re $[\mathbf{v}_s(0.1)]$; (*d*) $\mathbf{s}$; (*e*) $\mathbf{s}^*$.

5 A 20-Ω resistor, an 80-mH inductor, and a charged 0.1-μF capacitor are in series with an open switch. The switch closes suddenly at $t = 0$. Find the complex frequencies present in the current response.

6 Refer to Fig. 12-31 and find (*a*) $\mathbf{Z}_{in}(\mathbf{s})$ as a ratio of two polynomials in $\mathbf{s}$; (*b*) $\mathbf{Z}_{in}(-80)$; (*c*) $\mathbf{Z}_{in}(j80)$; (*d*) the admittance of the parallel RL branch, $\mathbf{Y}_{RL}(\mathbf{s})$, as a ratio of polynomials in $\mathbf{s}$. (*e*) Repeat for $\mathbf{Y}_{RC}(\mathbf{s})$. (*f*) Show that $\mathbf{Z}_{in}(\mathbf{s}) = (\mathbf{Y}_{RL} + \mathbf{Y}_{RC})/\mathbf{Y}_{RL}\mathbf{Y}_{RC}$.

Figure 12-31

See Prob. 6.

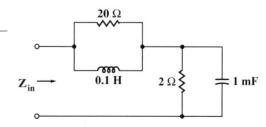

7 (*a*) Find $\mathbf{Z}_{in}(\mathbf{s})$ for the network of Fig. 12-32 as a ratio of two polynomials in **s**. (*b*) Find $\mathbf{Z}_{in}(j8)$ in rectangular form. (*c*) Find $\mathbf{Z}_{in}(-2 + j6)$ in polar form. (*d*) To what value should the 16-Ω resistor be changed in order that $\mathbf{Z}_{in} = 0$ at $\mathbf{s} = -5 + j0$? (*e*) To what value should the 16-Ω resistor be changed in order that $\mathbf{Z}_{in} = \infty$ at $\mathbf{s} = -5 + j0$?

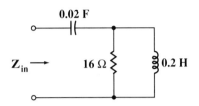

Figure 12-32

See Prob. 7.

8 (*a*) Let $v_s = 10e^{-2t} \cos(10t + 30°)$ V in the circuit of Fig. 12-33, and work in the frequency domain to find $\mathbf{I}_x$. (*b*) Find $i_x(t)$.

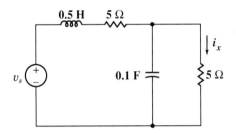

Figure 12-33

See Prob. 8.

9 Let $i_{s1} = 20e^{-3t} \cos 4t$ A and $i_{s2} = 30e^{-3t} \sin 4t$ A in the circuit of Fig. 12-34. (*a*) Work in the frequency domain to find $\mathbf{V}_x$. (*b*) Find $v_x(t)$.

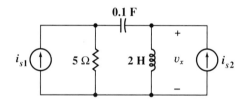

Figure 12-34

See Prob. 9.

10 The series combination of a 5-Ω resistance and a 0.2-F capacitance is in parallel with the series combination of a 2-Ω resistance and a 5-H inductance. (*a*) Find the input admittance, $\mathbf{Y}_1(\mathbf{s})$, of this parallel combination as a ratio of two polynomials in **s**. (*b*) Identify all the poles and zeros of $\mathbf{Y}_1(\mathbf{s})$. (*c*) Identify all the poles of the input admittance obtained if a 10-Ω resistance is connected in parallel with $\mathbf{Y}_1(\mathbf{s})$. (*d*) Identify all the zeros of the input admittance obtained if a 10-Ω resistance is connected in series with $\mathbf{Y}_1(\mathbf{s})$.

11 A network is composed of a 1-μF capacitor in parallel with the series combination of a 20-mH inductor and a 500-Ω resistor. (*a*) Find $\mathbf{Z}_{in}(\mathbf{s})$ for the network as a ratio of two polynomials in **s**. (*b*) Let $\mathbf{s} = \sigma + j0$ and find the negative value of σ, $-20 < \sigma < -10$ kNp/s, at which $|\mathbf{Z}_{in}(\sigma)|$ is a minimum. (*c*) Again let $\mathbf{s} = \sigma + j0$ and find the negative value of σ, $\sigma < -30$ kNp/s, at which $|\mathbf{Z}_{in}(\sigma)|$ is a maximum.

12 For the network shown in Fig. 12-35: (*a*) determine $|\mathbf{Z}_{in}(\sigma)|$ as a ratio of polynomials in σ; (*b*) determine all the poles and zeros of $|\mathbf{Z}_{in}(\sigma)|$; (*c*) sketch $|\mathbf{Z}_{in}(\sigma)|$ versus σ.

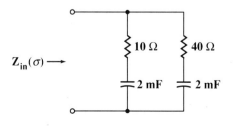

Figure 12-35

See Prob. 12.

13 (a) Find $Z_{in}(\sigma)$ as a function of σ for the network of Fig. 12-36 and express it as a constant times a ratio of polynomials in σ. (b) Find all the zeros and poles of $Z_{in}(\sigma)$. (c) Sketch $|Z_{in}(\sigma)|$ versus σ. (d) Sketch ang $Z_{in}(\sigma)$ versus σ.

Figure 12-36

See Prob. 13.

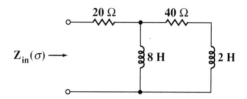

14 An admittance $Y(s)$ has zeros at $s = 0$ and $s = -10$, and poles at $s = -5$ and -20 s^{-1}. If $Y(s) \rightarrow 12$ S as $s \rightarrow \infty$, find (a) $Y(j10)$; (b) $Y(-j10)$; (c) $Y(-15)$; (d) the poles and zeros of $5 + Y(s)$.

15 (a) Find $Z_{in}(s)$ for the network shown in Fig. 12-37. (b) Find all the critical frequencies of $Z_{in}(s)$. (c) Sketch $|Z_{in}(\sigma)|$ versus σ. (d) Repeat parts a to c if a 2-Ω resistor is connected across the input terminals.

Figure 12-37

See Prob. 15.

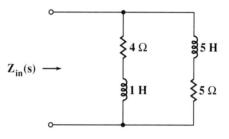

16 The pole-zero configuration of $H(s) = V_2(s)/V_1(s)$ is shown in Fig. 12-38. Let $H(0) = 1$. Sketch $|H(s)|$ versus: (a) σ if $\omega = 0$; (b) ω if $\sigma = 0$. (c) Find $|H(j\omega)|_{max}$.

Figure 12-38

See Prob. 16.

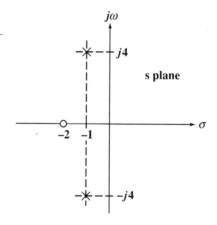

17 Given the voltage gain $H(s) = (10s^2 + 55s + 75)/(s^2 + 16)$: (a) indicate the critical frequencies on the s plane; (b) calculate $H(0)$ and $H(\infty)$. (c) If a scale model of $|H(s)|$ has a height of 3 cm at the origin, how high is it at $s = j3$? (d) Roughly sketch $|H(\sigma)|$ versus σ and $|H(j\omega)|$ versus ω.

18 The pole-zero constellation shown in Fig. 12-39 applies to a current gain $H(s) = I_{out}/I_{in}$. Let $H(-2) = 6$. (a) Express $H(s)$ as a ratio of polynomials in s. (b) Find $H(0)$ and $H(\infty)$. (c) Determine the magnitude and direction of each arrow from a critical frequency to $s = j2$.

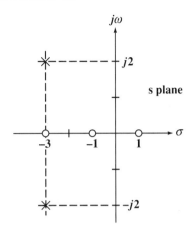

Figure 12-39
See Prob. 18.

19 Find $\mathbf{H}(\mathbf{s}) = \mathbf{V}_{out}/\mathbf{V}_{in}$ for the network of Fig. 12-40 and locate all its critical frequencies.

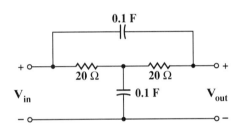

Figure 12-40
See Prob. 19.

20 Let $\mathbf{H}(\mathbf{s}) = 100(\mathbf{s} + 2)/(\mathbf{s}^2 + 2\mathbf{s} + 5)$ and (a) show the pole-zero plot for $\mathbf{H}(\mathbf{s})$; (b) find $\mathbf{H}(j\omega)$; (c) find $|\mathbf{H}(j\omega)|$; (d) sketch $|\mathbf{H}(j\omega)|$ versus ω; (e) find ω_{max}, the frequency at which $|\mathbf{H}(j\omega)|$ is a maximum.

21 The three-element network shown in Fig. 12-41 has an input impedance $\mathbf{Z}_A(\mathbf{s})$ that has a zero at $\mathbf{s} = -10 + j0$. If a 20-Ω resistor is placed in series with the network, the zero of the new impedance shifts to $\mathbf{s} = -3.6 + j0$. Find R and C.

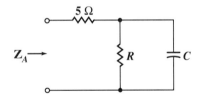

Figure 12-41
See Prob. 21.

22 Let $\mathbf{Z}_{in}(\mathbf{s}) = (5\mathbf{s} + 20)/(\mathbf{s} + 2)$ Ω for the network shown in Fig. 12-42. Find (a) the voltage $v_{ab}(t)$ between the open-circuited terminals if $v_{ab}(0) = 25$ V; (b) the current $i_{ab}(t)$ in a short circuit between terminals a and b if $i_{ab}(0) = 3$ A.

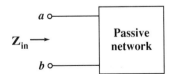

Figure 12-42
See Probs. 22 and 23.

23 Let $\mathbf{Z}_{in}(\mathbf{s}) = 5(\mathbf{s}^2 + 4\mathbf{s} + 20)/(\mathbf{s} + 1)$ Ω for the passive network of Fig. 12-42. Find $i_a(t)$, the instantaneous current entering terminal a, given $v_{ab}(t)$ equal to (a) $160e^{-6t}$ V; (b) $160e^{-6t}u(t)$ V, with $i_a(0) = 0$ and $di_a/dt = 32$ A/s at $t = 0$.

24 (a) Determine $\mathbf{H}(\mathbf{s}) = \mathbf{I}_C/\mathbf{I}_s$ for the circuit shown in Fig. 12-43. (b) Find the poles of $\mathbf{H}(\mathbf{s})$. (c) Find α, ω_0, and ω_d for the RLC circuit. (d) Determine the forced response $i_{C,f}(t)$

Figure 12-43

See Prob. 24.

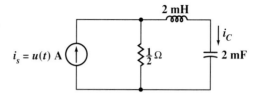

completely. (e) Give the form of the natural response $i_{C,n}(t)$. (f) Determine values for $i_C(0^+)$ and di_C/dt at $t = 0^+$. (g) Write the complete response, $i_C(t)$.

25 For the circuit of Fig. 12-44: (a) find the poles of $\mathbf{H}(\mathbf{s}) = \mathbf{I}_{in}/\mathbf{V}_{in}$. (b) Let $i_1(0^+) = 5$ A and $i_2(0^+) = 2$ A, and find $i_{in}(t)$ if $v_{in}(t) = 500u(t)$ V.

Figure 12-44

See Prob. 25.

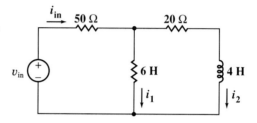

26 (a) Find $\mathbf{H}(\mathbf{s}) = \mathbf{V}(\mathbf{s})/\mathbf{I}_s(\mathbf{s})$ for the circuit of Fig. 12-45. Find $v(t)$ if $i_s(t)$ equals (b) $2u(t)$ A; (c) $4e^{-10t}$ A; (d) $4e^{-10t}u(t)$ A.

Figure 12-45

See Prob. 26.

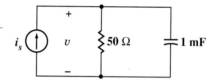

27 For the circuit shown in Fig. 12-46: (a) find $\mathbf{H}(\mathbf{s}) = \mathbf{V}_{C2}/\mathbf{V}_s$; (b) let $v_{C1}(0^+) = 0$ and $v_{C2}(0^+) = 0$, and find $v_{C2}(t)$ if $v_s(t) = u(t)$ V.

Figure 12-46

See Prob. 27.

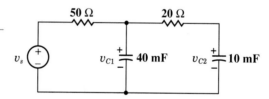

28 Refer to Fig. 12-47 and find the impedance $\mathbf{Z}_{in}(\mathbf{s})$ seen by the source. Use this expression to help determine $v_{in}(t)$ for $t > 0$.

Figure 12-47

See Prob. 28.

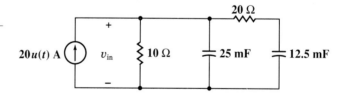

29 For the circuit shown in Fig. 12-48, determine (a) the transfer function $\mathbf{V}_2/\mathbf{I}_{s1}$, if $\mathbf{I}_{s1}$ is in parallel with the inductor with its arrow directed upward; (b) the transfer function $\mathbf{V}_2/\mathbf{I}_{s2}$, if $\mathbf{I}_{s2}$ is in parallel with the capacitor with its arrow directed upward; (c) the transfer function $\mathbf{V}_2/\mathbf{V}_{s1}$, if $\mathbf{V}_{s1}$ is in series with the inductor with its positive reference on top. (d) Specify the form of the natural response $v_{2n}(t)$.

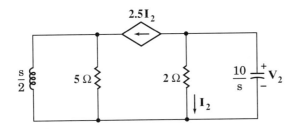

Figure 12-48

See Prob. 29.

30 In the op-amp circuit of Fig. 12-27a, let R_f = 20 kΩ, and then specify values for R_1 and C_f so that $\mathbf{H(s)} = \mathbf{V}_{out}/\mathbf{V}_{in}$ equals (a) −50; (b) −10³/(s + 10⁴); (c) −10⁴/(s + 10³); (d) 100/(s + 10⁵), using two stages.

31 In the circuit of Fig. 12-27b, let R_f = 20 kΩ, and then specify values for R_1 and C_1 so that $\mathbf{H(s)} = \mathbf{V}_{out}/\mathbf{V}_{in}$ equals (a) −50; (b) −10⁻³(s + 10⁴); (c) −10⁻⁴(s + 10³); (d) 10⁻³(s + 10⁵), using two stages.

32 Find $\mathbf{H(s)} = \mathbf{V}_{out}/\mathbf{V}_{in}$ as a ratio of polynomials in s for the op-amp circuit of Fig. 12-26, given the impedance values (in Ω): (a) $\mathbf{Z}_1(\mathbf{s})$ = 10³ + (10⁸/s), $\mathbf{Z}_f(\mathbf{s})$ = 5000; (b) $\mathbf{Z}_1(\mathbf{s})$ = 5000, $\mathbf{Z}_f(\mathbf{s})$ = 10³ + (10⁸/s); (c) $\mathbf{Z}_1(\mathbf{s})$ = 10³ + (10⁸/s), $\mathbf{Z}_f(\mathbf{s})$ = 10⁴ + (10⁸/s).

33 Use several op-amps in cascade to realize the transfer function $\mathbf{H(s)} = \mathbf{V}_{out}/\mathbf{V}_{in}$ = −10⁻⁴s(s + 10²)/(s + 10³). Use only 10-kΩ resistors, open circuits, or short circuits, but specify all capacitance values.

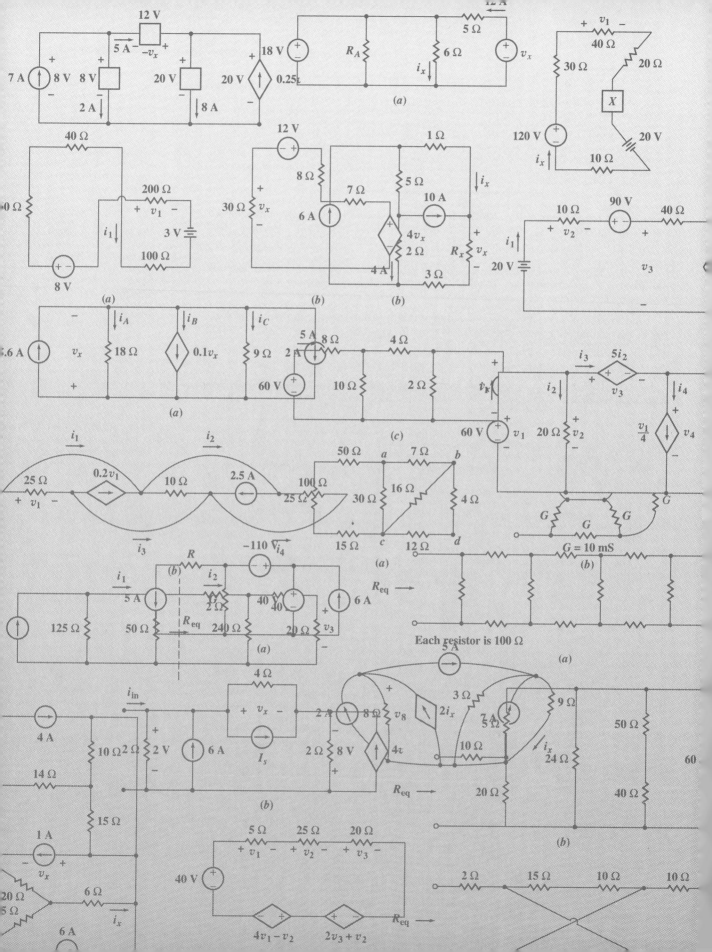

Frequency Response

Frequency response is a subject that has come up for consideration several times before. In Chap. 9 we discussed plots of admittance, impedance, current, and voltage as functions of ω, and the pole-zero concept was introduced as an aid in constructing and interpreting response curves. Response as a function of the neper frequency σ was discussed from the same standpoint in the last chapter. At that time, we also broadened our concept of frequency and introduced the complex frequency $\mathbf{s}$ and the $\mathbf{s}$ plane. We found that a plot of the critical frequencies of a response on the complex-frequency plane enabled us to tie together the forced response and the natural response; the critical frequencies themselves presented us almost directly with the form of the natural response, and the visualization of a three-dimensional rubber-sheet model or the performance of vectors drawn from each critical frequency to some test frequency gave us valuable information concerning the variation of the forced response with frequency.

In this chapter, we shall concentrate again on the forced response, particularly its variation with the radian frequency ω.

Why should we be so interested in the response to sinusoidal forcing functions when we so seldom encounter them in practice as such? The electric power industry is an exception, for the sinusoidal waveform appears throughout, although it is sometimes necessary to consider other frequencies introduced by the nonlinearity of some devices. But in most other electrical systems, the forcing functions and responses are not sinusoidal. In any system in which *information* is to be transmitted, the sinusoid by itself is almost valueless; it contains no information because its future values are exactly predictable from its past values. Moreover, once one period has been completed, any periodic nonsinusoidal waveform also contains no additional information.

Sinusoidal analysis, however, provides us with the response of a network as a function of ω, and the later work in Chaps. 18 and 19 will develop methods of determining network response to aperiodic signals (which can have a high information content) from the known sinusoidal frequency response.

The frequency response of a network provides useful information in its own right, however. Let us suppose that a certain forcing function is found to contain sinusoidal components having frequencies within the range of 10 to 100 Hz. Now let us imagine that this forcing function is applied to a network which has the property that all sinusoidal voltages with frequencies from zero to 200 Hz

13-1

Introduction

applied at the input terminals appear doubled in magnitude at the output terminals, with no change in phase angle. The output function is therefore an undistorted facsimile of the input function, but with twice the amplitude. If, however, the network has a frequency response such that the magnitudes of input sinusoids between 10 and 50 Hz are multiplied by a different factor than are those between 50 and 100 Hz, then the output would in general be distorted; it would no longer be a magnified version of the input. This distorted output might be desirable in some cases and undesirable in others. That is, the network frequency response might be chosen deliberately to reject some frequency components of the forcing function, or to emphasize others.

Such behavior is characteristic of tuned circuits or resonant circuits, as we shall see in this chapter. In discussing resonance we shall be able to apply all the methods we have discussed in presenting frequency response.

13-2

Parallel resonance

In this section we shall begin the study of a very important phenomenon which may occur in circuits containing both inductors and capacitors. The phenomenon is called *resonance,* and it may be loosely described as the condition existing in any physical system when a fixed-amplitude sinusoidal forcing function produces a response of maximum amplitude. However, we often speak of resonance as occurring even when the forcing function is not sinusoidal. The resonant system may be electrical, mechanical, hydraulic, acoustic, or some other kind, but we shall restrict our attention, for the most part, to electrical systems. We shall define resonance more exactly in what follows.

Resonance is a familiar phenomenon. Jumping up and down on the bumper of an automobile, for example, can put the vehicle into rather large oscillatory motion if the jumping is done *at the proper frequency* (about one jump per second), and if the shock absorbers are somewhat decrepit. However, if the jumping frequency is increased or decreased, the vibrational response of the automobile will be considerably less than it was before. A further illustration is furnished in the case of an opera singer who is able to shatter crystal goblets by means of a well-formed note *at the proper frequency.* In each of these examples, we are thinking of frequency as being adjusted until resonance occurs; it is also possible to adjust the size, shape, and material of the mechanical object being vibrated, but this may not be so easily accomplished physically.

The condition of resonance may or may not be desirable, depending upon the purpose which the physical system is to serve. In the automotive example, a large amplitude of vibration may help to separate locked bumpers, but it would be somewhat disagreeable at 65 mi/h (105 km/h).

Let us now define resonance more carefully. In a two-terminal electrical network containing at least one inductor and one capacitor, we define *resonance* as the condition which exists when the input impedance of the network is purely resistive. Thus,

> a network is *in resonance* (or *resonant*) when the voltage and current at the network input terminals are in phase.

We shall also find that a maximum-amplitude response is produced in the network when it is in the resonant condition or *almost in the resonant condition.*

We first apply the definition of resonance to the parallel *RLC* network shown in Fig. 13-1. In many practical situations, this circuit is a very good

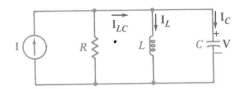

Figure 13-1

The parallel combination of a resistor, an inductor, and a capacitor, often referred to as a *parallel resonant circuit*.

approximation to the circuit we might build in the laboratory by connecting a physical inductor in parallel with a physical capacitor, where this parallel combination is driven by an energy source having a very high output impedance. The admittance offered to the ideal current source is

$$\mathbf{Y} = \frac{1}{R} + j\left(\omega C - \frac{1}{\omega L}\right) \tag{1}$$

and thus resonance occurs when

$$\omega C - \frac{1}{\omega L} = 0$$

The resonant condition may be achieved by adjusting L, C, or ω; we shall devote our attention to the case for which ω is the variable.[1] Hence, the resonant frequency ω_0 is

$$\omega_0 = \frac{1}{\sqrt{LC}} \quad \text{rad/s} \tag{2}$$

or

$$f_0 = \frac{1}{2\pi\sqrt{LC}} \quad \text{Hz} \tag{3}$$

This resonant frequency ω_0 is identical to the resonant frequency defined in Eq. (10), Chap. 6.

The pole-zero configuration of the admittance function can also be used to considerable advantage here. Given $\mathbf{Y}(\mathbf{s})$,

$$\mathbf{Y}(\mathbf{s}) = \frac{1}{R} + \frac{1}{\mathbf{s}L} + \mathbf{s}C$$

or

$$\mathbf{Y}(\mathbf{s}) = C\,\frac{\mathbf{s}^2 + \mathbf{s}/RC + 1/LC}{\mathbf{s}} \tag{4}$$

we may display the zeros of $\mathbf{Y}(\mathbf{s})$ by factoring the numerator:

$$\mathbf{Y}(\mathbf{s}) = C\,\frac{(\mathbf{s} + \alpha - j\omega_d)(\mathbf{s} + \alpha + j\omega_d)}{\mathbf{s}}$$

where α and ω_d represent the same quantities that they did when we discussed the natural response of the parallel RLC circuit in Sec. 6-2. That is, α is the exponential damping coefficient,

$$\alpha = \frac{1}{2RC}$$

[1] The techniques by which the applied frequency may be varied systematically are part of the SPICE program and are described in Sec. A5-9 of Appendix 5.

and ω_d is the natural resonant frequency (*not* the resonant frequency ω_0),

$$\omega_d = \sqrt{\omega_0^2 - \alpha^2}$$

The pole-zero constellation shown in Fig. 13-2a follows directly from the factored form.

Figure 13-2

(a) The pole-zero constellation of the input admittance of a parallel resonant circuit is shown on the **s** plane; $\omega_0^2 = \alpha^2 + \omega_d^2$. (b) The pole-zero constellation of the input impedance.

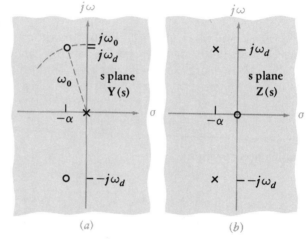

In view of the relationship among α, ω_d, and ω_0, it is apparent that the distance from the origin of the **s** plane to one of the admittance zeros is numerically equal to ω_0. Given the pole-zero configuration, the resonant frequency may therefore be obtained by purely graphical methods. We merely swing an arc, using the origin of the **s** plane as a center, through one of the zeros. The intersection of this arc and the positive $j\omega$ axis locates the point $\mathbf{s} = j\omega_0$. It is evident that ω_0 is slightly greater than the natural resonant frequency ω_d, but their ratio approaches unity as the ratio of ω_d to α increases.

Next let us examine the magnitude of the response, the voltage $\mathbf{V}(\mathbf{s})$ indicated in Fig. 13-1, as the frequency of the forcing function is varied. If we assume a constant-amplitude sinusoidal current source, the voltage response is proportional to the input impedance. This response can therefore be obtained from the pole-zero plot of the impedance $\mathbf{Z}(\mathbf{s})$, shown in Fig. 13-2b. The response obviously starts at zero, reaches a maximum value in the vicinity of the natural resonant frequency, and then drops again to zero as ω becomes infinite. The frequency response is sketched in Fig. 13-3. The maximum value of the response

Figure 13-3

The magnitude of the voltage response of a parallel resonant circuit is shown as a function of frequency.

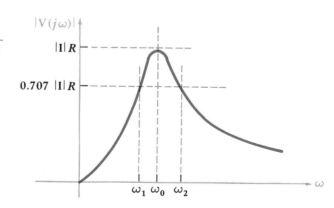

is indicated as R times the amplitude of the source current, implying that the maximum magnitude of the circuit impedance is R; moreover, the response maximum is shown to occur *exactly* at the resonant frequency ω_0. The two frequencies, ω_1 and ω_2, which we shall later use as a measure of the width of the response curve, are also identified. Let us first show that the maximum impedance magnitude is R and that this maximum occurs at resonance.

The admittance, as specified by Eq. (1), possesses a constant conductance and a susceptance which has a minimum magnitude (zero) at resonance. The minimum admittance magnitude therefore occurs at resonance, and it is $1/R$. Hence, the maximum impedance magnitude is R, and it occurs at resonance.

At the resonant frequency, therefore, the voltage across the parallel resonant circuit of Fig. 13-1 is simply $\mathbf{I}R$, and we see that the *entire* source current $\mathbf{I}$ flows through the resistor. However, current is also present in L and C. For the former, $\mathbf{I}_{L,0} = \mathbf{I}R/j\omega_0 L$, and the capacitor current at resonance is $\mathbf{I}_{C,0} = j\omega_0 CR\mathbf{I}$. Since $1/\omega_0 C = \omega_0 L$ at resonance, we find that

$$\mathbf{I}_{C,0} = -\mathbf{I}_{L,0} = j\omega_0 CR\mathbf{I} \tag{5}$$

and
$$\mathbf{I}_{C,0} + \mathbf{I}_{L,0} = \mathbf{I}_{LC} = 0$$

The maximum value of the response magnitude and the frequency at which it occurs are not always found so easily. In less standard resonant circuits, we may find it necessary to express the magnitude of the response in analytical form, usually as the square root of the sum of the real part squared and the imaginary part squared; then we should differentiate this expression with respect to frequency, equate the derivative to zero, solve for the frequency of maximum response, and finally substitute this frequency in the magnitude expression to obtain the maximum-amplitude response. This procedure may be carried out for this simple case merely as a corroborative exercise; but, as we have seen, it is not necessary.

It should be emphasized that, although the height of the response curve of Fig. 13-3 depends only upon the value of R for constant-amplitude excitation, the width of the curve or the steepness of the sides depends upon the other two element values also. We shall shortly relate the "width of the response curve" to a more carefully defined quantity, the bandwidth, but it will be helpful to express this relationship in terms of a very important parameter, the *quality factor Q*.[2]

We shall find that the sharpness of the response curve of any resonant circuit is determined by the maximum amount of energy that can be stored in the circuit, compared with the energy that is lost during one complete period of the response. We define Q as

$$\boxed{Q = \text{quality factor} = 2\pi \frac{\text{maximum energy stored}}{\text{total energy lost per period}}} \tag{6}$$

The proportionality constant 2π is included in the definition in order to simplify the more useful expressions for Q which we shall now obtain. Since energy can be stored only in the inductor and the capacitor, and can be lost only in the resistor, we may express Q in terms of the instantaneous energy associated

---

[2] This Q should not be confused with Q for charge or Q for reactive power.

with each of the reactive elements and the average power dissipated in the resistor:

$$Q = 2\pi \frac{[w_L(t) + w_C(t)]_{\text{max}}}{P_R T}$$

where T is the period of the sinusoidal frequency at which Q is being evaluated.

Now let us apply this definition to the parallel RLC circuit of Fig. 13-1 and determine the value of Q at the resonant frequency. This value of Q is denoted by Q_0. We select the current forcing function

$$i(t) = I_m \cos \omega_0 t$$

and obtain the corresponding voltage response at resonance,

$$v(t) = Ri(t) = RI_m \cos \omega_0 t$$

Then the energy stored in the capacitor is

$$w_C(t) = \frac{1}{2} C v^2 = \frac{I_m^2 R^2 C}{2} \cos^2 \omega_0 t$$

The instantaneous energy stored in the inductor becomes

$$w_L(t) = \frac{1}{2} L i_L^2 = \frac{1}{2} L \left(\frac{1}{L} \int_0^t v \, dt \right)^2$$

Thus

$$w_L(t) = \frac{I_m^2 R^2 C}{2} \sin^2 \omega_0 t$$

The total *instantaneous* stored energy is therefore constant:

$$w(t) = w_L(t) + w_C(t) = \frac{I_m^2 R^2 C}{2}$$

and this constant value must also be the maximum value. In order to find the energy lost in the resistor in one period, we take the average power absorbed by the resistor,

$$P_R = \tfrac{1}{2} I_m^2 R$$

and multiply by one period, obtaining

$$P_R T = \frac{1}{2f_0} I_m^2 R$$

We thus find the quality factor at resonance:

$$Q_0 = 2\pi \frac{I_m^2 R^2 C/2}{I_m^2 R/2f_0}$$

or
$$Q_0 = 2\pi f_0 RC = \omega_0 RC \tag{7}$$

This equation [as well as the expressions in Eq. (8)] holds only for the simple parallel RLC circuit of Fig. 13-1. Equivalent expressions for Q_0 which are often quite useful may be obtained by simple substitution:

$$Q_0 = R\sqrt{\frac{C}{L}} = \frac{R}{X_{C,0}} = \frac{R}{X_{L,0}} \tag{8}$$

It is apparent that Q_0 is a dimensionless constant which is a function of all three circuit elements in the parallel resonant circuit. The concept of Q, however, is not limited to electric circuits or even to electrical systems; it is useful in describing any resonant phenomenon. For example, let us consider a bouncing golf ball. If we assume a weight W and release the golf ball from a height h_1 above a very hard (lossless) horizontal surface, then the ball rebounds to some lesser height h_2. The energy stored initially is Wh_1, and the energy lost in one period is $W(h_1 - h_2)$. The Q_0 is therefore

$$Q_0 = 2\pi \frac{h_1 W}{(h_1 - h_2)W} = \frac{2\pi h_1}{h_1 - h_2}$$

A perfect golf ball would rebound to its original height and have an infinite Q_0; a more typical value is 35. It should be noted that the Q in this mechanical example has been calculated from the natural response and not from the forced response. The Q of an electric circuit may also be determined from a knowledge of the natural response, as illustrated by Eqs. (10) and (11) in the following discussion.

Another useful interpretation of Q is obtained when we inspect the inductor and capacitor currents at resonance, as given by Eq. (5),

$$\mathbf{I}_{C,0} = -\mathbf{I}_{L,0} = j\omega_0 C R \mathbf{I} \tag{9}$$

Note that each is Q_0 times the source current in amplitude and that each is 180° out of phase with the other. Thus, if we apply 2 mA at the resonant frequency to a parallel resonant circuit with a Q_0 of 50, we find 2 mA in the resistor, and 100 mA in both the inductor and the capacitor. A parallel resonant circuit can therefore act as a current amplifier, but not, of course, as a power amplifier, since it is a passive network.

Let us now relate to each other the various parameters which we have associated with a parallel resonant circuit. The three parameters α, ω_d, and ω_0 were introduced much earlier in connection with the natural response. Resonance, by definition, is fundamentally associated with the forced response, since it is defined in terms of a purely resistive input impedance, a sinusoidal steady-state concept. The two most important parameters of a resonant circuit are perhaps the resonant frequency ω_0 and the quality factor Q_0. Both the exponential damping coefficient and the natural resonant frequency may be expressed in terms of ω_0 and Q_0:

$$\alpha = \frac{1}{2RC} = \frac{1}{2(Q_0/\omega_0 C)C}$$

or

$$\alpha = \frac{\omega_0}{2Q_0} \tag{10}$$

and

$$\omega_d = \sqrt{\omega_0^2 - \alpha^2}$$

or

$$\omega_d = \omega_0 \sqrt{1 - \left(\frac{1}{2Q_0}\right)^2} \tag{11}$$

For future reference it may be helpful to note one additional relationship involving ω_0 and Q_0. The quadratic factor appearing in the numerator of Eq. (4),

$$s^2 + \frac{1}{RC}s + \frac{1}{LC}$$

may be written in terms of α and ω_0:

$$\mathbf{s}^2 + 2\alpha\mathbf{s} + \omega_0^2$$

In the field of system theory or automatic control theory, it is traditional to write this factor in a slightly different form that utilizes the dimensionless parameter ζ (zeta), called the *damping factor*:

$$\mathbf{s}^2 + 2\zeta\omega_0\mathbf{s} + \omega_0^2$$

Comparison of these expressions enables us to relate ζ to our other parameters:

$$\zeta = \frac{\alpha}{\omega_0} = \frac{1}{2Q_0} \tag{12}$$

Let us evaluate some of these parameters for a simple parallel resonant circuit.

Example 13-1   Calculate numerical values of ω_0, α, ω_d, and R for a parallel resonant circuit having $L = 2.5$ mH, $Q_0 = 5$, and $C = 0.01$ μF.

Solution: From Eq. (2), we see that $\omega_0 = 1/\sqrt{LC} = 200$ krad/s, while $f_0 = \omega_0/2\pi = 31.8$ kHz.

The value of α may be obtained quickly by using Eq. (10), $\alpha = \omega_0/2Q_0 = 2 \times 10^5/(2 \times 5) = 2 \times 10^4$ Np/s. Now we may make use of our old friend from Chap. 6, $\omega_d = \sqrt{\omega_0^2 - \alpha^2}$, to find that $\omega_d = \sqrt{(2 \times 10^5)^2 - (2 \times 10^4)^2} = 199.00$ krad/s. Finally, we need a value for the parallel resistance, and Eq. (7) gives us the answer: $Q_0 = \omega_0 RC$, $R = Q_0/\omega_0 C = 5/(2 \times 10^5 \times 10^{-8}) = 2.50$ kΩ.   ∎

Now let us interpret Q_0 in terms of the pole-zero locations of the admittance $\mathbf{Y}(\mathbf{s})$ of the parallel RLC circuit. We shall keep ω_0 constant; this may be done, for example, by changing R while holding L and C constant. As Q_0 is increased, the relationships relating α, Q_0, and ω_0 indicate that the two zeros must move closer to the $j\omega$ axis. These relationships also show that the zeros must simultaneously move away from the σ axis. The exact nature of the movement becomes clearer when we remember that the point at which $\mathbf{s} = j\omega_0$ could be located on the $j\omega$ axis by swinging an arc, centered at the origin, through one of the zeros and over to the positive $j\omega$ axis; since ω_0 is to be held constant, the radius must be constant, and the zeros must therefore move along this arc toward the positive $j\omega$ axis as Q_0 increases.

The two zeros are indicated in Fig. 13-4, and the arrows show the path they take as R increases. When R is infinite, Q_0 is also infinite, and the two zeros are found at $\mathbf{s} = \pm j\omega_0$ on the $j\omega$ axis. As R decreases, the zeros move toward the σ axis along the circular locus, joining to form a double zero on the σ axis at $\mathbf{s} = -\omega_0$ when $R = \frac{1}{2}\sqrt{L/C}$ or $Q_0 = \frac{1}{2}$. This condition may be recalled as that for critical damping, so that $\omega_d = 0$ and $\alpha = \omega_0$. Lower values of R and lower values of Q_0 cause the zeros to separate and move in opposite directions on the negative σ axis, but these low values of Q_0 are not really typical of resonant circuits and we need not track them any further.

Later, we shall use the criterion $Q_0 \geq 5$ to describe a high-Q circuit. When $Q_0 = 5$, the zeros are located at $\mathbf{s} = -0.1\omega_0 \pm j0.995\omega_0$, and thus ω_0 and ω_d differ by only one-half of 1 percent.

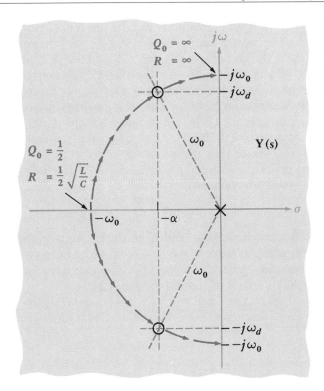

Figure 13-4

The two zeros of the admittance $\mathbf{Y}(\mathbf{s})$, located at $\mathbf{s} = -\alpha \pm j\omega_d$, provide a semicircular locus as R increases from $\frac{1}{2}\sqrt{L/C}$ to ∞.

13-1. Apply the definition of resonance to the network shown in Fig. 13-5 and find (a) the resonant (radian) frequency; (b) the value of $\mathbf{Z}_{in}$ at resonance.

Ans: 100.79 rad/s; $128.0 + j0\ \Omega$

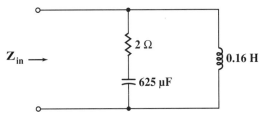

Figure 13-5

See Drill Prob. 13-1.

13-2. A parallel resonant circuit is composed of the elements $R = 8\ k\Omega$, $L = 50$ mH, and $C = 80$ nF. Find (a) ω_0; (b) Q_0; (c) ω_d; (d) α; (e) ζ.

Ans: 15.811 krad/s; 10.12; 15.792 krad/s; 781 Np/s; 0.0494

13-3. Find the values of R, L, and C in a parallel resonant circuit for which $\omega_0 = 1000$ rad/s, $\omega_d = 998$ rad/s, and $\mathbf{Y}_{in} = 1$ mS at resonance.

Ans: 1000 Ω; 126.4 mH; 7.91 μF

We continue our discussion of parallel resonance by defining half-power frequencies and bandwidth, and then we shall make good use of these new concepts in obtaining approximate response data for high-Q circuits.

The "width" of a resonance response curve, such as the one shown in Fig. 13-3, may now be defined more carefully and related to Q_0. Let us first define the two *half-power frequencies* ω_1 and ω_2 as those frequencies at which the magnitude of the input admittance of a parallel resonant circuit is greater than the magnitude at resonance by a factor of $\sqrt{2}$. Since the response curve of

13-3

More about parallel resonance

Fig. 13-3 displays the voltage produced across the parallel circuit by a sinusoidal current source as a function of frequency, the half-power frequencies also locate those points at which the voltage response is $1/\sqrt{2}$, or 0.707, times its maximum value. A similar relationship holds for the impedance magnitude. We shall select ω_1 as the *lower half-power frequency* and ω_2 as the *upper half-power frequency*. These names arise from the fact that a voltage which is 0.707 times the resonant voltage is equivalent to a squared voltage which is *one-half* the squared voltage at resonance. Thus, at the half-power frequencies, the resistor absorbs one-half the power that it does at resonance.

The (half-power) bandwidth of a resonant circuit is defined as the difference of these two half-power frequencies.

$$\mathcal{B} = \omega_2 - \omega_1 \tag{13}$$

We think of this bandwidth as the "width" of the response curve, even though the curve actually extends from $\omega = 0$ to $\omega = \infty$. More exactly, the half-power bandwidth is measured by that portion of the response curve which is equal to or greater than 70.7 percent of the maximum value.

Now let us express the bandwidth $\mathcal{B}$ in terms of Q_0 and the resonant frequency. In order to do so, we first express the admittance of the parallel RLC circuit,

$$\mathbf{Y} = \frac{1}{R} + j\left(\omega C - \frac{1}{\omega L}\right)$$

in terms of Q_0:

$$\mathbf{Y} = \frac{1}{R} + j\frac{1}{R}\left(\frac{\omega\omega_0 CR}{\omega_0} - \frac{\omega_0 R}{\omega\omega_0 L}\right)$$

or

$$\mathbf{Y} = \frac{1}{R}\left[1 + jQ_0\left(\frac{\omega}{\omega_0} - \frac{\omega_0}{\omega}\right)\right] \tag{14}$$

We note again that the magnitude of the admittance at resonance is $1/R$, and then realize that an admittance magnitude of $\sqrt{2}/R$ can occur only when a frequency is selected such that the imaginary part of the bracketed quantity has a magnitude of unity. Thus

$$Q_0\left(\frac{\omega_2}{\omega_0} - \frac{\omega_0}{\omega_2}\right) = 1 \quad \text{and} \quad Q_0\left(\frac{\omega_1}{\omega_0} - \frac{\omega_0}{\omega_1}\right) = -1$$

Solving, we have

$$\omega_1 = \omega_0\left[\sqrt{1 + \left(\frac{1}{2Q_0}\right)^2} - \frac{1}{2Q_0}\right] \tag{15}$$

$$\omega_2 = \omega_0\left[\sqrt{1 + \left(\frac{1}{2Q_0}\right)^2} + \frac{1}{2Q_0}\right] \tag{16}$$

Although these expressions are somewhat unwieldy, their difference provides a very simple formula for the bandwidth:

$$\mathcal{B} = \omega_2 - \omega_1 = \frac{\omega_0}{Q_0} \tag{17}$$

Equations (15) and (16) may be multiplied by each other to show that ω_0 is exactly equal to the geometric mean of the half-power frequencies:

$$\omega_0^2 = \omega_1\omega_2$$

or

$$\omega_0 = \sqrt{\omega_1\omega_2}$$

Circuits possessing a higher Q_0 have a narrower bandwidth, or a sharper response curve; they have greater *frequency selectivity,* or higher quality (factor).

Many resonant circuits are deliberately designed to have a large Q_0 in order to take advantage of the narrow bandwidth and high frequency selectivity associated with such circuits. When Q_0 is larger than about 5, it is possible to make some very useful approximations in the expressions for the upper and lower half-power frequencies and in the general expressions for the response in the neighborhood of resonance. Let us arbitrarily refer to a high-Q circuit as one for which Q_0 is equal to or greater than 5. The pole-zero configuration of $\mathbf{Y(s)}$ for a parallel RLC circuit having a Q_0 of about 5 is shown in Fig. 13-6.

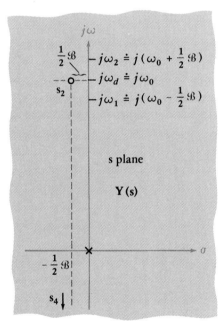

Figure 13-6

The pole-zero constellation of $\mathbf{Y(s)}$ for a parallel RLC circuit. The two zeros are exactly $\frac{1}{2}\mathcal{B}$ Np/s (or rad/s) to the left of the $j\omega$ axis and approximately $j\omega_0$ rad/s (or Np/s) from the σ axis. The upper and lower half-power frequencies are separated exactly $\mathcal{B}$ rad/s, and each is approximately $\frac{1}{2}\mathcal{B}$ rad/s away from the resonant frequency and the natural resonant frequency.

Since

$$\alpha = \frac{\omega_0}{2Q_0}$$

then

$$\alpha = \tfrac{1}{2}\mathcal{B}$$

and the locations of the two zeros may be approximated:

$$\mathbf{s}_{2,4} = -\alpha \pm j\omega_d \doteq -\tfrac{1}{2}\mathcal{B} \pm j\omega_0$$

Moreover, the locations of the two half-power frequencies (on the positive $j\omega$ axis) may also be determined in a concise approximate form:

$$\omega_{1,2} = \omega_0\left[\sqrt{1+\left(\frac{1}{2Q_0}\right)^2} \mp \frac{1}{2Q_0}\right] \doteq \omega_0\left(1 \mp \frac{1}{2Q_0}\right)$$

or
$$\omega_{1,2} \doteq \omega_0 \mp \tfrac{1}{2}\mathcal{B} \tag{18}$$

In a high-Q circuit, therefore, each half-power frequency is located approximately one-half bandwidth from the resonant frequency; this is indicated in Fig. 13-6.

The approximate relationships for ω_1 and ω_2 in Eq. (18) may be added to each other to show that ω_0 is approximately equal to the arithmetic mean of ω_1 and ω_2 in high-Q circuits:

$$\omega_0 \doteq \tfrac{1}{2}(\omega_1 + \omega_2)$$

Now let us visualize a test point slightly above $j\omega_0$ on the $j\omega$ axis. In order to determine the admittance offered by the parallel RLC network at this frequency, we construct the three vectors from the critical frequencies to the test point. If the test point is close to $j\omega_0$, then the vector from the pole is approximately $j\omega_0$ and that from the lower zero is nearly $j2\omega_0$. The admittance is therefore given approximately by

$$\mathbf{Y(s)} \doteq C\frac{(j2\omega_0)(\mathbf{s} - \mathbf{s}_2)}{j\omega_0} \doteq 2C(\mathbf{s} - \mathbf{s}_2) \tag{19}$$

where C is the capacitance, as shown in Eq. (4). In order to determine a useful approximation for the vector $(\mathbf{s} - \mathbf{s}_2)$, let us consider an enlarged view of that portion of the $\mathbf{s}$ plane in the neighborhood of the zero $\mathbf{s}_2$ (Fig. 13-7).

Figure 13-7

An enlarged portion of the pole-zero constellation for $\mathbf{Y(s)}$ of a high-Q_0 parallel RLC circuit.

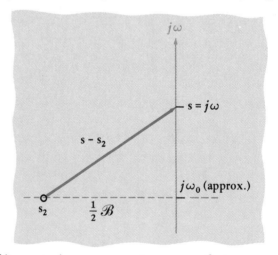

In terms of its cartesian components, we see that

$$\mathbf{s} - \mathbf{s}_2 \doteq \tfrac{1}{2}\mathcal{B} + j(\omega - \omega_0)$$

where this expression would be exact if ω_0 were replaced by ω_d. We now substitute this equation in the approximation for $\mathbf{Y(s)}$, Eq. (19), and factor out $\tfrac{1}{2}\mathcal{B}$:

$$\mathbf{Y(s)} \doteq 2C\left(\tfrac{1}{2}\mathcal{B}\right)\left(1 + j\frac{\omega - \omega_0}{\tfrac{1}{2}\mathcal{B}}\right)$$

or
$$\mathbf{Y}(\mathbf{s}) \doteq \frac{1}{R}\left(1 + j\,\frac{\omega - \omega_0}{\frac{1}{2}\mathcal{B}}\right) \tag{20}$$

The fraction $(\omega - \omega_0)/(\frac{1}{2}\mathcal{B})$ may be interpreted as the "number of half-bandwidths off resonance" and abbreviated by N. Thus,

$$\mathbf{Y}(\mathbf{s}) \doteq \frac{1}{R}(1 + jN) \tag{21}$$

where

$$N = \frac{\omega - \omega_0}{\frac{1}{2}\mathcal{B}} \qquad \text{(number of half-bandwidths off resonance)} \tag{22}$$

At the upper half-power frequency, $\omega_2 \doteq \omega_0 + \frac{1}{2}\mathcal{B}$, $N = +1$, and we are one half-bandwidth above resonance. For the lower half-power frequency, $\omega_1 \doteq \omega_0 - \frac{1}{2}\mathcal{B}$, so that $N = -1$, locating us one half-bandwidth below resonance.

Equation (21) is much easier to use than the exact relationships we have had heretofore. It shows that the magnitude of the admittance is

$$|\mathbf{Y}(j\omega)| \doteq \frac{1}{R}\sqrt{1 + N^2} \tag{23}$$

while the angle of $\mathbf{Y}(j\omega)$ is given by the inverse tangent of N:

$$\text{ang } \mathbf{Y}(j\omega) \doteq \tan^{-1} N \tag{24}$$

Example 13-2 As an example of the use of these approximations, let us determine the approximate value of the admittance of a parallel RLC network for which $R = 40$ kΩ, $L = 1$ H, and $C = \frac{1}{64}$ μF.

Solution: We find that $Q_0 = 5$, $\omega_0 = 8$ krad/s, $\mathcal{B} = 1.6$ krad/s, and $\frac{1}{2}\mathcal{B} = 0.8$ krad/s.

Let us evaluate the admittance at $\omega = 8.2$ krad/s. Thus,

$$N = \frac{8.2 - 8}{0.8} = 0.25$$

and we see that we are operating 0.25 half-bandwidth above resonance. Next, we calculate

$$\text{ang } \mathbf{Y} \doteq \tan^{-1} 0.25 = 14.04°$$

and
$$|\mathbf{Y}| \doteq 25\sqrt{1 + (0.25)^2} = 25.77 \ \mu\text{S}$$

An exact calculation of the admittance shows that

$$\mathbf{Y}(j8200) = 25.75\underline{/13.87°} \ \mu\text{S}$$

The approximate method therefore leads to values of admittance magnitude and angle that are quite accurate for this example. ∎

Our intention is to use these approximations for high-Q circuits near resonance. We have already agreed that we shall let "high-Q" imply $Q_0 \geq 5$, but how near is "near"? It can be shown that the error in magnitude or phase is less than 5 percent if $Q_0 \geq 5$ and $0.9\omega_0 \leq \omega \leq 1.1\omega_0$. Although this narrow band of frequencies may seem to be prohibitively small, it is usually more than sufficient to contain the range of frequencies in which we are most interested.

For example, an AM home or car radio usually contains a circuit tuned to a resonant frequency of 455 kHz with a half-power bandwidth of 10 kHz. This circuit must then have a value of 45.5 for Q_0, and the half-power frequencies are about 450 and 460 kHz. Our approximations, however, are valid from 409.5 to 500.5 kHz (with errors less than 5 percent), a range which covers essentially all the peaked portion of the response curve; only in the remote "tails" of the response curve do the approximations lead to unreasonably large errors.[3]

Let us conclude our coverage of the parallel resonant circuit by reviewing the various conclusions we have reached. The resonant frequency ω_0 is the frequency at which the imaginary part of the input admittance becomes zero, or the angle of the admittance becomes zero. Then, $\omega_0 = 1/\sqrt{LC}$. The circuit's figure of merit Q_0 is defined as 2π times the ratio of the maximum energy stored in the circuit to the energy lost each period in the circuit. From this definition, we find that $Q_0 = \omega_0 RC$. The two half-power frequencies ω_1 and ω_2 are defined as the frequencies at which the admittance magnitude is $\sqrt{2}$ times the minimum admittance magnitude. These are also the frequencies at which the voltage response is 70.7 percent of the maximum response. The exact and approximate (for high Q_0) expressions for these two frequencies are

$$\omega_{1,2} = \omega_0 \left[\sqrt{1 + \left(\frac{1}{2Q_0}\right)^2} \mp \frac{1}{2Q_0} \right] \doteq \omega_0 \mp \frac{1}{2}\mathcal{B}$$

where $\mathcal{B}$ is the difference between the upper and the lower half-power frequencies. This half-power bandwidth is given by

$$\mathcal{B} = \omega_2 - \omega_1 = \frac{\omega_0}{Q_0}$$

The input admittance may also be expressed in approximate form for high-Q circuits:

$$\mathbf{Y} \doteq \frac{1}{R}(1 + jN) = \frac{1}{R}\sqrt{1 + N^2}\ \underline{/\tan^{-1}N}$$

where

$$N = \frac{\omega - \omega_0}{\frac{1}{2}\mathcal{B}}$$

The approximation is valid for frequencies which do not differ from the resonant frequency by more than one-tenth of the resonant frequency.

Drill Problems

13-4. A marginally high-Q parallel resonant circuit has $f_0 = 440$ Hz with $Q_0 = 6$. Use Eqs. (15) and (16) to obtain accurate values for (a) f_1; (b) f_2. Now use Eq. (18) to calculate approximate values for (c) f_1; (d) f_2.

Ans: 478.2 Hz; 404.9 Hz; 476.7 Hz; 403.3 Hz

13-5. A parallel resonant circuit contains element values $L = 1$ mH, $C = 1$ nF, $R = 20$ kΩ. Use the high-Q approximations to find $\mathbf{Z}_{in}$ at ω equal to (a) 1020 krad/s; (b) 1040 krad/s; (c) 960 krad/s.

Ans: 15.62$\underline{/-38.7°}$ kΩ; 10.60$\underline{/-58.0°}$ kΩ; 10.60$\underline{/58.0°}$ kΩ

13-6. Find accurate values for the three parts of Drill Prob. 13-5.

Ans: 15.677$\underline{/-38.38°}$ kΩ; 10.748$\underline{/-57.49°}$ kΩ; 10.443$\underline{/58.52°}$ kΩ

---

[3] At frequencies remote from resonance, we are often satisfied with very rough results; greater accuracy is not always necessary.

Although we probably find less use for the series RLC circuit than we do for the parallel RLC circuit, it is still worthy of our attention. We shall consider the circuit shown in Fig. 13-8. It should be noted that the various circuit elements are given the subscript s (for series) for the time being in order to avoid confusing them with the parallel elements when the circuits are compared.

Our discussion of parallel resonance occupied two sections of considerable length. We could now give the series RLC circuit the same kind of treatment, but it is much cleverer to avoid such needless repetition and use the duality principle. For simplicity, let us concentrate on the conclusions presented in the last paragraph of the preceding section on parallel resonance. The important results are contained there, and the use of dual language enables us to transcribe this paragraph to present the important results for the series RLC circuit.

13-4

Series resonance

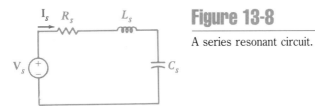

Figure 13-8

A series resonant circuit.

Let us conclude our coverage of the series resonant circuit by summarizing the various conclusions we have reached. The resonant frequency ω_{0s} is the frequency at which the imaginary part of the input impedance becomes zero or the angle of the impedance becomes zero. Then, $\omega_{0s} = 1/\sqrt{L_s C_s}$. The circuit's figure of merit Q_{0s} is defined as 2π times the ratio of the maximum energy stored in the circuit to the energy lost each period in the circuit. From this definition, we find that $Q_{0s} = \omega_{0s} L_s / R_s$. The two half-power frequencies ω_{1s} and ω_{2s} are defined as the frequencies at which the impedance magnitude is $\sqrt{2}$ times the minimum impedance magnitude. These are also the frequencies at which the voltage response is 70.7 percent of the maximum response. The exact and approximate (for high Q_{0s}) expressions for these two frequencies are

$$\omega_{1s,2s} = \omega_{0s}\left[\sqrt{1 + \left(\frac{1}{2Q_{0s}}\right)^2} \mp \frac{1}{2Q_{0s}}\right] \doteq \omega_{0s} \mp \frac{1}{2}\mathcal{B}_s$$

where $\mathcal{B}_s$ is the difference between the upper and the lower half-power frequencies. This half-power bandwidth is given by

$$\mathcal{B}_s = \omega_{2s} - \omega_{1s} = \frac{\omega_{0s}}{Q_{0s}}$$

The input impedance may also be expressed in approximate form for high-Q_s circuits:

$$\mathbf{Z}_s \doteq R_s(1 + jN_s) = R_s\sqrt{1 + N_s^2}\ \underline{/\tan^{-1}N_s}$$

where

$$N_s = \frac{\omega - \omega_{0s}}{\frac{1}{2}\mathcal{B}_s}$$

The approximation is valid for frequencies which do not differ from the resonant frequency by more than one-tenth of the resonant frequency.

The series resonant circuit is characterized by a low impedance at resonance, whereas the parallel resonant circuit produces a high resonant impedance. The latter circuit provides inductor currents and capacitor currents at resonance which have amplitudes Q_0 times as great as the source current; the series resonant circuit provides inductor voltages and capacitor voltages which are greater than the source voltage by the factor Q_{0s}. The series circuit thus provides voltage amplification at resonance.

A comparison of our results for series and parallel resonance, as well as the exact and approximate expressions we have developed, appears in Table 13-1.

From this point on, we shall no longer identify series resonant circuits by use of the subscript s, unless clarity requires it.

Table 13-1

A short summary of resonance

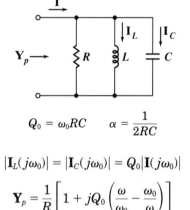

$$Q_0 = \omega_0 RC \qquad \alpha = \frac{1}{2RC}$$

$$|\mathbf{I}_L(j\omega_0)| = |\mathbf{I}_C(j\omega_0)| = Q_0|\mathbf{I}(j\omega_0)|$$

$$\mathbf{Y}_p = \frac{1}{R}\left[1 + jQ_0\left(\frac{\omega}{\omega_0} - \frac{\omega_0}{\omega}\right)\right]$$

$$Q_0 = \frac{\omega_0 L}{R} \qquad \alpha = \frac{R}{2L}$$

$$|\mathbf{V}_L(j\omega_0)| = |\mathbf{V}_C(j\omega_0)| = Q_0|\mathbf{V}(j\omega_0)|$$

$$\mathbf{Z}_s = R\left[1 + jQ_0\left(\frac{\omega}{\omega_0} - \frac{\omega_0}{\omega}\right)\right]$$

Exact expressions

$$\omega_0 = \frac{1}{\sqrt{LC}} = \sqrt{\omega_1\omega_2} \qquad \omega_d = \sqrt{\omega_0^2 - \alpha^2} = \omega_0\sqrt{1 - \left(\frac{1}{2Q_0}\right)^2}$$

$$\omega_{1,2} = \omega_0\left[\sqrt{1 + \left(\frac{1}{2Q_0}\right)^2} \mp \frac{1}{2Q_0}\right]$$

$$N = \frac{\omega - \omega_0}{\frac{1}{2}\mathcal{B}} \qquad \mathcal{B} = \omega_2 - \omega_1 = \frac{\omega_0}{Q_0} = 2\alpha$$

Approximate expressions

$(Q_0 \geq 5 \qquad 0.9\omega_0 \leq |\omega| \leq 1.1\omega_0)$

$$\omega_d \doteq \omega_0 \qquad \omega_{1,2} \doteq \omega_0 \mp \tfrac{1}{2}\mathcal{B}$$

$$\omega_0 \doteq \tfrac{1}{2}(\omega_1 + \omega_2)$$

$$\mathbf{Y}_p \doteq \frac{\sqrt{1 + N^2}}{R} \underline{/\tan^{-1} N}$$

$$\mathbf{Z}_s \doteq R\sqrt{1 + N^2} \underline{/\tan^{-1} N}$$

13-7. A series resonant circuit has a bandwidth of 100 Hz and contains a 20-mH inductance and a 2-μF capacitance. Determine (a) f_0; (b) Q_0; (c) $\mathbf{Z}_{\text{in}}$ at resonance; (d) f_2.   *Ans: 796 Hz; 7.96; 12.57 + j0 Ω; 846 Hz (approx.)*

13-8. The voltage $v_s = 100 \cos \omega t$ mV is applied to a series resonant circuit composed of a 10-Ω resistance, a 0.2-μF capacitance, and a 2-mH inductance. Use both exact and approximate methods to calculate the current amplitude if ω equals (a) 48 krad/s; (b) 55 krad/s.   *Ans: 7.745, 7.809 mA; 4.640, 4.472 mA*

The parallel and series *RLC* circuits of the previous two sections represent *idealized* resonant circuits; they are no more than useful, approximate representations of a physical circuit which might be constructed by combining a coil of wire, a carbon resistor, and a tantalum capacitor in parallel or series. The degree of accuracy with which the idealized model fits the actual circuit depends on the operating frequency range, the Q of the circuit, the materials present in the physical elements, the element sizes, and many other factors. We are not studying the techniques for determining the best model of a given physical circuit, for this requires some knowledge of electromagnetic field theory and the properties of materials; we are, however, concerned with the problem of reducing a more complicated model to one of the two simpler models with which we are more familiar.

The network shown in Fig. 13-9a is a reasonably accurate model for the parallel combination of a physical inductor, capacitor, and resistor. The resistor labeled R_1 is a hypothetical resistor that is included to account for the ohmic losses, core losses, and radiation losses of the physical coil. The losses in the dielectric within the physical capacitor, as well as the resistance of the physical resistor in the given *RLC* circuit, are accounted for by the resistor labeled R_2. In this model, there is no way to combine elements and produce a simpler model which is equivalent to the original model *for all frequencies*. We shall show, however, that a simpler equivalent may be constructed which is valid over a frequency band which is usually large enough to include all frequencies of interest. The equivalent will take the form of the network shown in Fig. 13-9b.

13-5

Other resonant forms

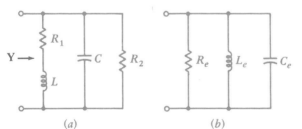

(a)   (b)

Figure 13-9

(a) A useful model of a physical network which consists of a physical inductor, capacitor, and resistor in parallel. (b) A network which can be equivalent to part *a* over a narrow frequency band.

Before we learn how to develop such an equivalent circuit, let us first consider the given circuit, Fig. 13-9a. The resonant radian frequency for this network is *not* $1/\sqrt{LC}$, although if R_1 is sufficiently small it may be very close to this value. The definition of resonance is unchanged, and we may determine the resonant frequency by setting the imaginary part of the input admittance equal to zero:

$$\text{Im} \left[\mathbf{Y}(j\omega) \right] = \text{Im} \left(\frac{1}{R_2} + j\omega C + \frac{1}{R_1 + j\omega L} \right) = 0$$

Thus,

$$C = \frac{L}{R_1^2 + \omega^2 L^2}$$

and

$$\omega_0 = \sqrt{\frac{1}{LC} - \left(\frac{R_1}{L}\right)^2}$$

We note that ω_0 is less than $1/\sqrt{LC}$, but sufficiently small values of the ratio R_1/L may result in a negligible difference between ω_0 and $1/\sqrt{LC}$.

The maximum magnitude of the input impedance also deserves consideration. It is *not* R_2, and it does *not* occur at ω_0 (or at $\omega = 1/\sqrt{LC}$). The proof of these statements will not be shown, because the expressions soon become algebraically cumbersome; the theory, however, is straightforward. Let us be content with a numerical example.

Example 13-3   We select the simple values $R_1 = 2\ \Omega$, $L = 1$ H, $C = \frac{1}{8}$ F, and $R_2 = 3\ \Omega$ for Fig. 13-9a, and seek to determine the resonant frequency.

Solution: Substituting the appropriate values in the last equation, we find

$$\omega_0 = \sqrt{8 - 2^2} = 2\ \text{rad/s}$$

and this enables us to calculate the input admittance,

$$\mathbf{Y} = \frac{1}{3} + j2\left(\frac{1}{8}\right) + \frac{1}{2 + j(2)(1)} = \frac{1}{3} + \frac{1}{4} = 0.583\ \text{S}$$

and then the input impedance at resonance:

$$\mathbf{Z}(j2) = \frac{1}{0.583} = 1.714\ \Omega$$

At the frequency which would be the resonant frequency if R_1 were zero,

$$\frac{1}{\sqrt{LC}} = 2.83\ \text{rad/s}$$

the input impedance is

$$\mathbf{Z}(j2.83) = 1.947\underline{/-13.26°}\ \Omega$$

However, the frequency at which the maximum impedance magnitude occurs, indicated by ω_m, is found to be

$$\omega_m = 3.26\ \text{rad/s}$$

and the impedance having the maximum magnitude is

$$\mathbf{Z}(j3.26) = 1.980\underline{/-21.4°}\ \Omega \qquad\qquad ■$$

The impedance magnitude at resonance and the maximum magnitude differ by about 13 percent. Although it is true that such an error may be neglected occasionally in practice, it is too large to neglect on an exam in class. The later work in this section will show that the Q of the inductor-resistor combination at 2 rad/s is unity; this low value accounts for the 13 percent discrepancy.

In order to transform the given circuit of Fig. 13-9a into an equivalent of the form of that shown in Fig. 13-9b, we must discuss the Q of a simple series or parallel combination of a resistor and a reactor (inductor or capacitor). We first consider the series circuit shown in Fig. 13-10a. The Q of this network is again defined as 2π times the ratio of the maximum stored energy to the energy lost each period, but the Q may be evaluated at any frequency we choose. In other words, Q is a function of ω. It is true that we shall choose to evaluate it at a frequency which is, or apparently is, the resonant frequency of some network of which the series arm is a part. This frequency, however, is not known until a more complete circuit is available. The eager reader is encouraged to show that the Q of this series arm is $|X_s|/R_s$, whereas the Q of the parallel network of Fig. 13-10b is $R_p/|X_p|$.

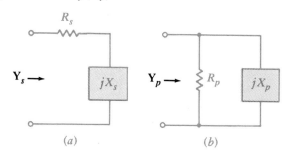

(a)   (b)

Figure 13-10

(a) A series network which consists of a resistance R_s and an inductive or capacitive reactance X_s may be transformed into (b) a parallel network such that $\mathbf{Y}_s = \mathbf{Y}_p$ at one specific frequency. The reverse transformation is equally possible.

Let us now carry out the details necessary to find values for R_p and X_p so that the parallel network of Fig. 13-10b is equivalent to the series network of Fig. 13-10a at some single specific frequency. We equate $\mathbf{Y}_s$ and $\mathbf{Y}_p$,

$$\mathbf{Y}_s = \frac{1}{R_s + jX_s} = \frac{R_s - jX_s}{R_s^2 + X_s^2} = \mathbf{Y}_p = \frac{1}{R_p} - j\frac{1}{X_p}$$

and obtain

$$R_p = \frac{R_s^2 + X_s^2}{R_s} \qquad X_p = \frac{R_s^2 + X_s^2}{X_s^2}$$

Dividing these two expressions, we find

$$\frac{R_p}{X_p} = \frac{X_s}{R_s}$$

It follows that the Q's of the series and parallel networks must be equal:

$$Q_p = Q_s = Q$$

The transformation equations may therefore be simplified:

$$R_p = R_s (1 + Q^2) \qquad\qquad (25)$$

$$X_p = X_s \left(1 + \frac{1}{Q^2}\right) \qquad\qquad (26)$$

It is apparent that R_s and X_s may also be found if R_p and X_p are the given values; the transformation in either direction may be performed.

If $Q \geq 5$, little error is introduced by using the approximate relationships

$$R_p \doteq Q^2 R_s \qquad\qquad (27)$$

$$X_p \doteq X_s \qquad (C_p \doteq C_s \quad \text{or} \quad L_p \doteq L_s) \qquad\qquad (28)$$

Example 13-4   Let us try out some of these relationships by finding the parallel equivalent of the series combination of a 100-mH inductor and a 5-Ω resistor. We shall perform the transformation at a frequency of 1000 rad/s, a value selected because it is approximately the resonant frequency of the network (not shown) of which this series arm is a part.

Solution: We find that X_s is 100 Ω and Q is 20. Since the Q is sufficiently high, we use Eqs. (27) and (28) to obtain

$$R_p \doteq Q^2 R_s = 2000 \; \Omega \qquad L_p \doteq L_s = 100 \text{ mH}$$

The conclusion is that a 100-mH inductor in series with a 5-Ω resistor provides essentially the same input impedance as does a 100-mH inductor in parallel with a 2000-Ω resistor at the frequency 1000 rad/s. In order to check the accuracy of the equivalence, let us evaluate the input impedance for each network at 1000 rad/s. We find

$$\mathbf{Z}_s(j1000) = 5 + j100 = 100.1\underline{/87.1°}$$

$$\mathbf{Z}_p(j1000) = \frac{2000(j100)}{2000 + j100} = 99.9\underline{/87.1°}$$

and conclude that the approximation is exceedingly accurate at the transformation frequency. The accuracy at 900 rad/s is also reasonably good, because

$$\mathbf{Z}_s(j900) = 90.1\underline{/86.8°}$$

$$\mathbf{Z}_p(j900) = 89.9\underline{/87.4°}$$

If this inductor and series resistor had been used as part of a series RLC circuit for which the resonant frequency was 1000 rad/s, then the half-power bandwidth would have been

$$\mathcal{B} = \frac{\omega_0}{Q_0} = \frac{1000}{20} = 50$$

and the frequency of 900 rad/s would have represented a frequency that was 4 half-bandwidths off resonance. Thus the equivalent networks that we evaluated in this example would have been adequate for reproducing essentially all the peaked portion of the response curve.   ∎

As a further example of the replacement of a more complicated resonant circuit by an equivalent series or parallel RLC circuit, let us consider a problem in electronic instrumentation. The simple series RLC network in Fig. 13-11a is excited by a sinusoidal voltage source at the resonant frequency. The effective value of the source voltage is 0.5 V, and we wish to measure the effective value of the voltage across the capacitor with an electronic voltmeter (VM) having an internal resistance of 100 000 Ω. That is, an equivalent representation of the voltmeter is an ideal voltmeter in parallel with a 100-kΩ resistor.

Before the voltmeter is connected, we find that the resonant frequency is 10^5 rad/s, $Q_0 = 50$, the current is 25 mA, and the rms capacitor voltage is 25 V. As indicated at the end of Sec. 13-4, this voltage is Q_0 times the applied voltage. Thus, if the voltmeter were ideal, it would read 25 V when connected across the capacitor.

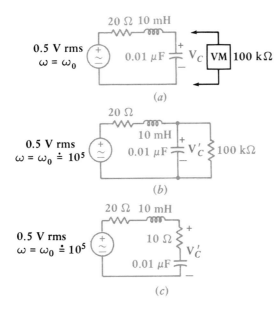

Figure 13-11

(*a*) A given series resonant circuit in which the capacitor voltage is to be measured by a non-ideal electronic voltmeter. (*b*) The effect of the voltmeter is included in the circuit; it reads V_C'. (*c*) A series resonant circuit is obtained when the parallel RC network in part *b* is replaced by the series RC network which is equivalent at 10^5 rad/s.

However, when the actual voltmeter is connected, the circuit shown in Fig. 13-11*b* results. In order to obtain a series RLC circuit, it is now necessary to replace the parallel RC network by a series RC network. Let us assume that the Q of this RC network is sufficiently high that the equivalent series capacitor will be the same as the given parallel capacitor. We do this in order to approximate the resonant frequency of the final series RLC circuit. Thus, if the series RLC circuit also contains a 0.01-μF capacitor, the resonant frequency remains 10^5 rad/s. We need to know this estimated resonant frequency in order to calculate the Q of the parallel RC network; it is

$$Q = \frac{R_p}{|X_p|} = \omega R_p C_p = 10^5(10^5)(10^{-8}) = 100$$

Since this value is greater than 5, our vicious circle of assumptions is justified, and the equivalent series RC network consists of the capacitor

$$C_s = 0.01 \ \mu\text{F}$$

and the resistor

$$R_s \doteq \frac{R_p}{Q^2} = 10 \ \Omega$$

Hence, the equivalent circuit of Fig. 13-11*c* is obtained. The resonant Q of this circuit is now only 33.3, and thus the voltage across the capacitor in the circuit of Fig. 13-11*c* is $16\frac{2}{3}$ V. But we need to find $|\mathbf{V}_C|$, the voltage across the series RC combination; we obtain

$$|\mathbf{V}_C| = \frac{0.5}{30}|10 - j1000| = 16.67 \text{ V}$$

The capacitor voltage and $|\mathbf{V}_C|$ are essentially equal, since the voltage across the 10-Ω resistor is quite small.

The final conclusion must be that an apparently good voltmeter may produce a severe effect on the response of a high-Q resonant circuit. A similar effect may occur when a nonideal ammeter is inserted in the circuit.

We conclude this section with a technical fable.

Once upon a time there was a student whom we shall call Pat, who had a professor identified simply as Dr. Noe.

In the laboratory one afternoon, Dr. Noe gave Pat three practical circuit devices: a resistor, an inductor, and a capacitor, having nominal element values of 20 Ω, 20 mH, and 1 μF. The student was asked to connect a variable-frequency voltage source to the series combination of these three elements, to measure the resultant voltage across the resistor as a function of frequency, and then to calculate numerical values for the resonant frequency, the Q at resonance, and the half-power bandwidth. The student was also asked to predict the results of the experiment before making the measurements.

Pat, who was not one of the more spectacular performers in the class, formed a mental picture of the equivalent circuit for this problem that was like the circuit of Fig. 13-12a, and then calculated:

$$f_0 = \frac{1}{2\pi\sqrt{LC}} = \frac{1}{2\pi\sqrt{20 \times 10^{-3} \times 10^{-6}}} = 1125 \text{ Hz}$$

$$Q_0 = \frac{\omega_0 L}{R} = 7.07$$

$$\mathcal{B} = \frac{f_0}{Q_0} = 159 \text{ Hz}$$

Figure 13-12

(a) A first model for a 20-mH inductor, a 1-μF capacitor, and a 20-Ω resistor in series with a voltage generator. (b) An improved model in which more accurate values are used and the losses in the inductor and capacitor are acknowledged. (c) The final model also contains the output resistance of the voltage source.

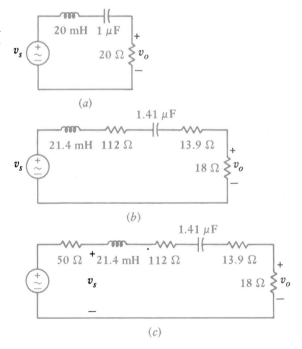

(a)

(b)

(c)

Next, Pat made the measurements that Dr. Noe requested, compared them with the predicted values, and then felt a strong urge to transfer to the business school. The results were

$$f_0 = 1000 \text{ Hz} \qquad Q_0 = 0.625 \qquad \mathcal{B} = 1600 \text{ Hz}$$

Even Pat knew that discrepancies of this magnitude could not be characterized as being "within engineering accuracy" or "due to meter errors." Sadly, the results were handed to the professor.

Remembering many past errors in judgment, some of which were even self-made, Dr. Noe smiled benevolently and called Pat's attention to the Q-meter (or impedance bridge) which is present in most well-equipped laboratories, and suggested that it might be used to find out what these practical circuit elements really looked like at some convenient frequency near resonance, say, 1000 Hz.

Upon doing so, Pat discovered that the resistor had a measured value of 18 Ω and the inductor was 21.4 mH with a Q of 1.2, while the capacitor had a capacitance of 1.41 μF and a dissipation factor (the reciprocal of Q) equal to 0.123.

So, with the hope that springs eternal within the human breast, Pat reasoned that a better model for the practical inductor would be 21.4 mH in series with $\omega L/Q = 112$ Ω, while a more appropriate model for the capacitor would be 1.41 μF in series with $1/\omega CQ = 13.9$ Ω. Using these data, Pat prepared the modified circuit model shown as Fig. 13-12b and calculated a new set of predicted values:

$$f_0 = \frac{1}{2\pi\sqrt{21.4 \times 10^{-3} \times 1.41 \times 10^{-6}}} = 916 \text{ Hz}$$

$$Q_0 = \frac{2\pi \times 916 \times 21.4 \times 10^{-3}}{143.9} = 0.856$$

$$\mathcal{B} = \frac{916}{0.856} = 1070 \text{ Hz}$$

Since these results were much closer to the measured values, Pat was much happier. Dr. Noe, however, being a stickler for detail, looked askance at the differences in the predicted and measured values for both Q_0 and the bandwidth. "Have you," Dr. Noe asked, "given any consideration to the output impedance of the voltage source?" "Not yet," said Pat, trotting back to the laboratory bench.

It turned out that the output impedance in question was 50 Ω, and so Pat added this value to the circuit diagram, as shown in Fig. 13-12c. Using the new equivalent resistance value of 193.9 Ω, improved values for Q_0 and $\mathcal{B}$ were then obtained:

$$Q_0 = 0.635 \qquad \mathcal{B} = 1442 \text{ Hz}$$

Since all the theoretical and experimental values now agreed within 10 percent, Pat was once again suffused with the joys of studying electrical engineering. Dr. Noe simply nodded her head agreeably as she moralized:

**When using real devices,
Watch the models that you choose;
Think well before you calculate,
And mind your Z's and Q's!**

Drill Problems

13-9. At $\omega = 1000$ rad/s, find parallel networks that are equivalent to the series combinations in Figs. 13-13a and b, and find series equivalents for the parallel networks of Figs. 13-13c and d.

Ans: 8 H, 640 kΩ; 0.5 μF, 100 kΩ; 5 H, 250 Ω; 1 μF, 33.3 Ω

Figure 13-13

See Drill Prob. 13-9.

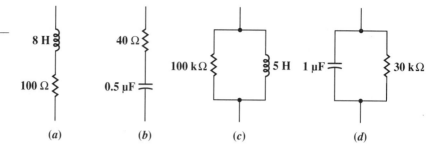

(a)   (b)   (c)   (d)

13-10. At a frequency of 50 krad/s, find the Q of (a) a 0.1-μF capacitor in series with a resistance of 2 Ω; (b) an 80-kΩ resistance in parallel with the network of part a; (c) a 2-Ω resistance in series with the network of part b.

Ans: 100; 80; 44.4

13-11. The series combination of 10 Ω and 10 nF is in parallel with the series combination of 20 Ω and 10 mH. (a) Find the approximate resonant frequency of the parallel network. (b) Find the Q of the RC branch. (c) Find the Q of the RL branch. (d) Find the three-element equivalent of the original network.

Ans: 10^5 rad/s; 100; 50; 10 nF $\|$ 10 mH $\|$ 33.3 kΩ

13-6

Scaling

Some of the examples and problems which we have been solving have involved circuits containing passive element values ranging around a few ohms, a few henrys, and a few farads. The applied frequencies were a few radians per second. These particular numerical values were used not because they are those commonly met in practice, but because arithmetic manipulations are so much easier than they would be if it were necessary to carry along various powers of 10 throughout the calculations. The scaling procedures that will be discussed in this section enable us to analyze networks composed of practical-sized elements by scaling the element values to permit more convenient numerical calculations. We shall consider both magnitude scaling and frequency scaling.

Let us select the parallel resonant circuit shown in Fig. 13-14a as our example. The impractical element values lead to the unlikely response curve drawn as Fig. 13-14b; the maximum impedance is 2.5 Ω, the resonant frequency is 1 rad/s, Q_0 is 5, and the bandwidth is 0.2 rad/s. These numerical values are much more characteristic of the electrical analog of some mechanical system than they are of any basically electrical device. We have convenient numbers with which to calculate, but an impractical circuit to construct.

Figure 13-14

(a) A parallel resonant circuit used as an example to illustrate magnitude and frequency scaling. (b) The magnitude of the input impedance is shown as a function of frequency.

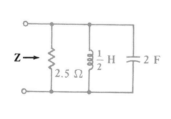

(a)

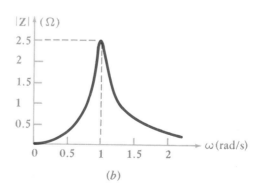

(b)

Let us assume that our goal is to scale this network in such a way as to provide an impedance maximum of 5000 Ω at a resonant frequency of 5×10^6 rad/s, or 796 kHz. In other words, we may use the same response curve shown in Fig. 13-14b if every number on the ordinate scale is increased by a factor of 2000 and every number on the abscissa scale is increased by a factor of 5×10^6. We shall treat this as two problems: (1) scaling in magnitude by a factor of 2000 and (2) scaling in frequency by a factor of 5×10^6.

Magnitude scaling is defined as the process by which the *impedance* of a two-terminal network is increased by a factor of K_m, the frequency remaining constant. The factor K_m is real and positive; it may be greater or smaller than unity. We shall understand that the shorter statement "the network is scaled in magnitude by a factor of 2" indicates that the impedance of the new network is to be *twice* that of the old network at any frequency. Let us now determine how we must scale each type of passive element. To increase the input impedance of a network by a factor of K_m, it is sufficient to increase the impedance of each element in the network by this same factor. Thus, a resistance R must be replaced by a resistance $K_m R$. Each inductance must also exhibit an impedance which is K_m times as great at any frequency. In order to increase an impedance sL by a factor of K_m when s remains constant, the inductance L must be replaced by an inductance $K_m L$. In a similar manner, each capacitance C must be replaced by a capacitance C/K_m. In summary, these changes will produce a network which is scaled in magnitude by a factor of K_m:

$$\left.\begin{array}{c} R \rightarrow K_m R \\ L \rightarrow K_m L \\ C \rightarrow \dfrac{C}{K_m} \end{array}\right\} \quad \text{magnitude scaling}$$

When each element in the network of Fig. 13-14a is scaled in magnitude by a factor of 2000, the network shown in Fig. 13-15a results. The response curve shown in Fig. 13-15b indicates that no change in the previously drawn response curve need be made other than a change in the scale of the ordinate.

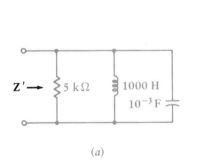

(a)

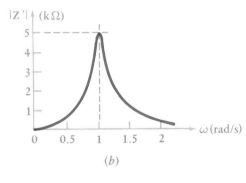

(b)

Figure 13-15

(a) The network of Fig. 13-14a after being scaled in magnitude by a factor $K_m = 2000$. (b) The corresponding response curve.

Let us now take this new network and scale it in frequency. We define *frequency scaling* as the process by which the frequency at which any impedance occurs is increased by a factor of K_f. Again, we shall make use of the shorter expression "the network is scaled in frequency by a factor of 2" to indicate that the same impedance is now obtained at a frequency twice as great. Frequency scaling is accomplished by scaling each passive element in frequency. It is apparent that no resistor is affected. The impedance of any inductor is sL, and if this same impedance is to be obtained at a frequency K_f times as great, then the inductance L must be replaced by an inductance of L/K_f. Similarly, a capacitance C is to be replaced by a capacitance C/K_f. Thus, if a network is to be scaled in frequency by a factor of K_f, then the changes necessary in each passive element are

$$\left.\begin{array}{c} R \to R \\[6pt] L \to \dfrac{L}{K_f} \\[6pt] C \to \dfrac{C}{K_f} \end{array}\right\} \quad \text{frequency scaling}$$

When each element of the magnitude-scaled network of Fig. 13-15a is scaled in frequency by a factor of 5×10^6, the network of Fig. 13-16a is obtained. The corresponding response curve is shown in Fig. 13-16b.

Figure 13-16

(a) The network of Fig. 13-15a after being scaled in frequency by a factor $K_f = 5 \times 10^6$. (b) The corresponding response curve.

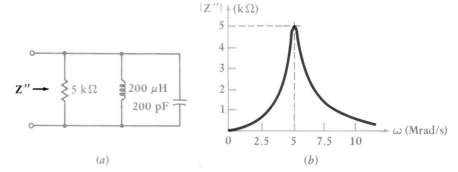

The circuit elements in this last network have values which are easily achieved in physical circuits; the network can actually be built and tested. It follows that, if the original network of Fig. 13-14a were actually an analog of some mechanical resonant system, we could have scaled this analog in both magnitude and frequency in order to achieve a network which we might construct in the laboratory; tests that are expensive or inconvenient to run on the mechanical system could then be made on the scaled electrical system, and the results should then be "unscaled" and converted into mechanical units to complete the analysis.

The effect of magnitude scaling or frequency scaling on the pole-zero constellation of an impedance is not very difficult to ascertain, but its determination offers such an excellent opportunity to review the meaning and significance of a pole-zero plot in the s plane that it forms the basis for one of the drill problems at the end of this section.

An impedance which is given as a function of s may also be scaled in magnitude or frequency, and this may be done without any knowledge of the specific elements out of which the two-terminal network is composed. In order to scale $\mathbf{Z}(\mathbf{s})$ in magnitude, the definition of magnitude scaling shows that it is necessary only to multiply $\mathbf{Z}(\mathbf{s})$ by K_m in order to obtain the magnitude-scaled impedance. Thus, the impedance of the parallel resonant circuit shown in Fig. 13-14a is

$$\mathbf{Z}(\mathbf{s}) = \frac{\mathbf{s}}{2\mathbf{s}^2 + 0.4\mathbf{s} + 2}$$

or
$$\mathbf{Z(s)} = (0.5)\frac{\mathbf{s}}{(\mathbf{s} + 0.1 + j0.995)(\mathbf{s} + 0.1 - j0.995)}$$

The impedance $\mathbf{Z'(s)}$ of the magnitude-scaled network is

$$\mathbf{Z'(s)} = K_m \mathbf{Z(s)}$$

If we again select $K_m = 2000$, we have

$$\mathbf{Z'(s)} = (1000)\frac{\mathbf{s}}{(\mathbf{s} + 0.1 + j0.995)(\mathbf{s} + 0.1 - j0.995)}$$

If $\mathbf{Z'(s)}$ is now to be scaled in frequency by a factor of 5×10^6, then $\mathbf{Z''(s)}$ and $\mathbf{Z'(s)}$ are to provide identical values of impedance if $\mathbf{Z''(s)}$ is evaluated at a frequency K_f times that at which $\mathbf{Z'(s)}$ is evaluated. After some careful cerebral activity, this conclusion may be stated concisely in functional notation:

$$\mathbf{Z''(s)} = \mathbf{Z'}\left(\frac{\mathbf{s}}{K_f}\right)$$

Note that we obtain $\mathbf{Z''(s)}$ by replacing every $\mathbf{s}$ in $\mathbf{Z'(s)}$ by $\mathbf{s}/K_f$. The analytic expression for the impedance of the network shown in Fig. 13-16a must therefore be

$$\mathbf{Z''(s)} = (1000)\frac{\mathbf{s}/(5 \times 10^6)}{[\mathbf{s}/(5 \times 10^6) + 0.1 + j0.995][\mathbf{s}/(5 \times 10^6) + 0.1 - j0.995]}$$

or
$$\mathbf{Z''(s)} = \frac{5 \times 10^9 \times \mathbf{s}}{(\mathbf{s} + 0.5 \times 10^6 + j4.975 \times 10^6)(\mathbf{s} + 0.5 \times 10^6 - j4.975 \times 10^6)}$$

Although scaling is a process normally applied to passive elements, dependent sources may also be scaled in magnitude and frequency. We assume that the output of any source is given as $k_x v_x$ or $k_y i_y$, where k_x has the dimensions of an admittance for a dependent current source and is dimensionless for a dependent voltage source, whereas k_y has the dimensions of ohms for a dependent voltage source and is dimensionless for a dependent current source. If the network containing the dependent source is scaled in magnitude by K_m, then it is necessary only to treat k_x or k_y as if it were the type of element consistent with its dimensions. That is, if k_x (or k_y) is dimensionless, it is left unchanged; if it is an admittance, it is divided by K_m; and if it is an impedance, it is multiplied by K_m. Frequency scaling does not affect the dependent sources.

Drill Problems

13-12. A parallel resonant circuit is defined by $C = 0.01$ F, $\mathcal{B} = 2.5$ rad/s, and $\omega_0 = 20$ rad/s. Find the values of R and L if the network is scaled in (a) magnitude by a factor of 800; (b) frequency by a factor of 10^4; (c) magnitude by a factor of 800 and frequency by a factor of 10^4.
Ans: 32 kΩ, 200 H; 40 Ω, 25 μH; 32 kΩ, 20 mH

13-13. The pole-zero plot of a certain $\mathbf{H(s)}$ shows a zero at $\mathbf{s} = -6 + j10$ and a magnitude of 80 at the origin. The network to which it applies is now scaled in magnitude by $K_m = 4$ and in frequency by $K_f = 20$. Specify the location of the zero and the magnitude at the origin if $\mathbf{H(s)}$ is (a) an impedance; (b) a voltage gain; (c) an admittance; (d) a current gain.
Ans: $-120 + j200$, 320; $-120 + j200$, 80; $-120 + j200$, 20; $-120 + j200$, 80

13-14. Scale the network shown in Fig. 13-17 by $K_m = 20$ and $K_f = 50$. (a) Give parameter values for all three elements, and (b) find $\mathbf{Z}_{in}(\mathbf{s})$ for the scaled network.　　　*Ans:* 50 μF, 0.01$\mathbf{V}_1$, 0.2 H; $(0.2\mathbf{s}^2 - 40\mathbf{s} + 20\,000)/\mathbf{s}$

Figure 13-17

See Drill Prob. 13-14.

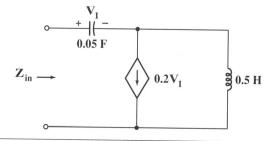

13-7

Bode diagrams

This chapter is concerned with frequency response, as the title promises, and we have already investigated several topics that we illustrated with plots showing certain responses as functions of frequency. Curves of this type first began to appear back in Chap. 9 when the sinusoidal forcing function was introduced, and then reappeared in Chap. 12 as we sought to understand complex frequency.

In this section we shall discover a quick method of obtaining an approximate picture of the amplitude and phase variation of a given transfer function as functions of ω. Accurate curves may, of course, be plotted after calculating values with a programmable calculator or a computer; curves may also be produced directly on the computer. Our object here, however, is to obtain a better picture of the response than we could visualize from a pole-zero plot, but yet not mount an all-out computational offensive.

The approximate response curve we construct is called an *asymptotic plot,* or a *Bode plot,* or a *Bode diagram,* after its developer, Hendrik W. Bode,[4] who was an electrical engineer and mathematician with the Bell Telephone Laboratories. Both the magnitude and phase curves are shown using a logarithmic frequency scale for the abscissa, and the magnitude itself is also shown in logarithmic units called *decibels* (dB). We define the value of $|\mathbf{H}(j\omega)|$ in dB as follows:

$$H_{dB} = 20 \log |\mathbf{H}(j\omega)| \tag{29}$$

where the common logarithm (base 10) is used. A different definition is used for power ratios, but we shall not need it here. The inverse operation is

$$|\mathbf{H}(j\omega)| = 10^{(H_{dB}/20)} \tag{30}$$

Before we actually begin a detailed discussion of the technique for drawing Bode diagrams, it will help to gain some feeling for the size of the decibel unit, to learn a few of its important values, and to recall some of the properties of the logarithm. Since $\log 1 = 0$, $\log 2 = 0.301\,03$, and $\log 10 = 1$, we note the correspondences:

$$|\mathbf{H}(j\omega)| = 1 \Leftrightarrow H_{dB} = 0$$

$$|\mathbf{H}(j\omega)| = 2 \Leftrightarrow H_{dB} \doteq 6 \text{ dB}$$

$$|\mathbf{H}(j\omega)| = 10 \Leftrightarrow H_{dB} = 20 \text{ dB}$$

[4] Sounds like "OK."

An increase of $|\mathbf{H}(j\omega)|$ by a factor of 10 corresponds to an increase in H_{dB} by 20 dB. Moreover, $\log 10^n = n$, and thus $10^n \Leftrightarrow 20n$ dB, so that 1000 corresponds to 60 dB, while 0.01 is represented as -40 dB. Using only the values already given, we may also find that $20 \log 5 = 20 \log \frac{10}{2} = 20 \log 10 - 20 \log 2 = 20 - 6 = 14$ dB, and thus $5 \Leftrightarrow 14$ dB. Also, $\log \sqrt{x} = \frac{1}{2} \log x$, and therefore $\sqrt{2} \Leftrightarrow 3$ dB and $1/\sqrt{2} \Leftrightarrow -3$ dB.[5]

We shall write our transfer functions in terms of $\mathbf{s}$, substituting $\mathbf{s} = j\omega$ when we are ready to find the magnitude or phase angle.

Our next step is to factor $\mathbf{H}(\mathbf{s})$ to display its poles and zeros. We first consider a zero at $\mathbf{s} = -a$, written in a standardized form as

$$\mathbf{H}(\mathbf{s}) = 1 + \frac{\mathbf{s}}{a}$$

The Bode diagram for this function consists of the two asymptotic curves approached by H_{dB} for very large and very small values of ω. Thus,

$$|\mathbf{H}(j\omega)| = \left|1 + \frac{j\omega}{a}\right| = \sqrt{1 + \frac{\omega^2}{a^2}}$$

and

$$H_{dB} = 20 \log \left|1 + \frac{j\omega}{a}\right| = 20 \log \sqrt{1 + \frac{\omega^2}{a^2}}$$

When $\omega \ll a$,

$$H_{dB} \doteq 20 \log 1 = 0 \qquad (\omega \ll a)$$

This simple asymptote is shown in Fig. 13-18. It is drawn as a solid line for $\omega < a$, and as a broken line for $\omega > a$. When $\omega \gg a$,

$$H_{dB} \doteq 20 \log \frac{\omega}{a} \qquad (\omega \gg a)$$

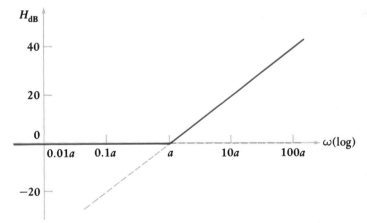

Figure 13-18

The Bode amplitude plot for $\mathbf{H}(\mathbf{s}) = 1 + \mathbf{s}/a$ consists of the low- and high-frequency asymptotes, shown in solid color. They intersect on the abscissa at the corner frequency.

At $\omega = a$, $H_{dB} = 0$; at $\omega = 10a$, $H_{dB} = 20$ dB; and at $\omega = 100a$, $H_{dB} = 40$ dB. Thus, the value of H_{dB} increases 20 dB for every 10-fold increase in frequency. The asymptote thus has a slope of 20 dB/decade. Since H_{dB} increases by 6 dB when ω doubles, an alternate value for the slope is 6 dB/octave. The high-frequency asymptote is also shown in Fig. 13-18, a solid line for $\omega > a$, and a

[5] Note that we are being slightly dishonest by using $20 \log 2 = 6$ dB rather than 6.02 dB.

broken line for $\omega < a$. Note that the two asymptotes intersect at $\omega = a$, the frequency of the zero. This frequency is also described as the *corner, break,* 3-dB, or *half-power frequency.*

The Bode plot represents the response in terms of two asymptotes, both straight lines, and both easily drawn.

Now let us see how much error is embodied in our asymptotic response curve. At the corner frequency,

$$H_{\mathrm{dB}} = 20 \log \sqrt{1 + \frac{a^2}{a^2}} = 3 \text{ dB}$$

as compared with an asymptotic value of 0 dB. At $\omega = 0.5a$, we have

$$H_{\mathrm{dB}} = 20 \log \sqrt{1.25} \doteq 1 \text{ dB}$$

Thus, the exact response is represented by a smooth curve that lies 3 dB above the asymptotic response at $\omega = a$, and 1 dB above it at $\omega = 0.5a$ (and also at $\omega = 2a$). This information can always be used to smooth out the corner if a more exact result is desired.

We now need the phase response for the simple zero,

$$\text{ang } \mathbf{H}(j\omega) = \text{ang}\left(1 + \frac{j\omega}{a}\right) = \tan^{-1}\frac{\omega}{a}$$

This expression is also represented by its asymptotes, although three straight line segments are required. For $\omega \ll a$, ang $\mathbf{H}(j\omega) \doteq 0°$, and we use this as our asymptote when $\omega < 0.1a$:

$$\text{ang } \mathbf{H}(j\omega) = 0° \qquad (\omega < 0.1a)$$

At the high end, $\omega \gg a$, we have ang $\mathbf{H}(j\omega) \doteq 90°$, and we use this above $\omega = 10a$:

$$\text{ang } \mathbf{H}(j\omega) = 90° \qquad (\omega > 10a)$$

Since the angle is 45° at $\omega = a$, we now construct the straight line asymptote extending from 0° at $\omega = 0.1a$, through 45° at $\omega = a$, to 90° at $\omega = 10a$. This straight line has a slope of 45°/decade. It is shown as a solid curve in Fig. 13-19, while the exact angle response is shown as a broken line. The maximum

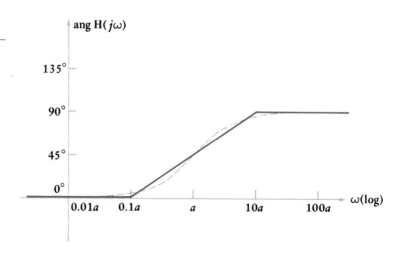

Figure 13-19

The asymptotic angle response for $\mathbf{H}(\mathbf{s}) = 1 + \mathbf{s}/a$ is shown as the three straight line segments in solid color. The endpoints of the ramp are 0° at $0.1a$ and 90° at $10a$.

differences between the asymptotic and true responses are $\pm 5.71°$ at $\omega = 0.1a$ and $10a$. Errors of $\pm 5.29°$ also occur at $\omega = 0.394a$ and $2.54a$; the error is zero at $\omega = 0.159a$, a, and $6.31a$.

We next consider a simple pole,

$$\mathbf{H(s)} = \frac{1}{1 + \mathbf{s}/a}$$

Since this is the reciprocal of a zero, the logarithmic operation leads to a Bode plot which is the negative of that obtained previously. The amplitude is 0 dB up to $\omega = a$, and then the slope is -20 dB/decade for $\omega > a$. The angle plot is $0°$ for $\omega < 0.1a$, $-90°$ for $\omega > 10a$, and $-45°$ at $\omega = a$, and it has a slope of $-45°$/decade when $0.1a < \omega < 10a$.

Another term that can appear in $\mathbf{H(s)}$ is a factor of $\mathbf{s}$ in the numerator or denominator. If $\mathbf{H(s)} = \mathbf{s}$, then

$$H_{dB} = 20 \log |\omega|$$

Thus, we have an infinite straight line passing through 0 dB at $\omega = 1$ and having a slope everywhere of 20 dB/decade. This is shown in Fig. 13-20a. If the $\mathbf{s}$ factor occurs in the denominator, a straight line is obtained having a slope of -20 dB/decade and passing through 0 dB at $\omega = 1$, as shown in Fig. 13-20b.

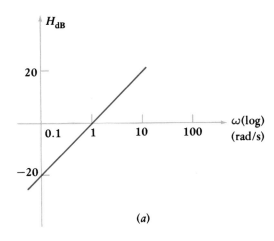

(a)

Figure 13-20

The asymptotic diagrams are shown for (a) $\mathbf{H(s)} = \mathbf{s}$ and (b) $\mathbf{H(s)} = 1/\mathbf{s}$. Both are infinitely long straight lines passing through 0 dB at $\omega = 1$ and having slopes of ± 20 dB/decade.

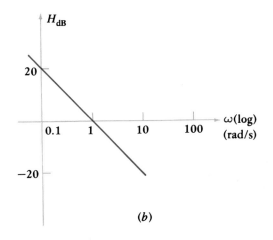

(b)

Another simple term found in $\mathbf{H}(\mathbf{s})$ is the multiplying constant K. This yields a Bode plot which is a horizontal straight line lying $20 \log |K|$ dB above the abscissa. It will actually be below the abscissa if $|K| < 1$.

Let us now consider an example in which several of these different factors are combined in $\mathbf{H}(\mathbf{s})$.

Example 13-5   We seek the Bode plot of the input impedance of the network shown in Fig. 13-21.

Figure 13-21

If $\mathbf{H}(\mathbf{s})$ is selected as $\mathbf{Z}_{\text{in}}(\mathbf{s})$ for this network, then the Bode plot for H_{dB} is as shown in Fig. 13-22b.

$$H(s) = Z_{\text{in}}(s) \rightarrow$$

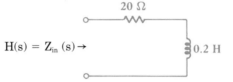

Solution: We have the input impedance,

$$\mathbf{Z}_{\text{in}}(\mathbf{s}) = \mathbf{H}(\mathbf{s}) = 20 + 0.2\mathbf{s}$$

Putting this in standard form, we obtain

$$\mathbf{H}(\mathbf{s}) = 20 \left(1 + \frac{\mathbf{s}}{100} \right)$$

The two factors constituting $\mathbf{H}(\mathbf{s})$ are a zero at $\omega = 100$ and a constant, equivalent to $20 \log 20 = 26$ dB. Each of these is sketched lightly in Fig. 13-22a. Since we are working with the logarithm of $\mathbf{H}(\mathbf{s})$, we next add together the Bode plots corresponding to the individual factors. The resultant magnitude plot appears as Fig. 13-22b. No attempt has been made to smooth out the corner with a +3-dB correction at $\omega = 100$ rad/s.

Figure 13-22

(a) The Bode plots for the factors of $\mathbf{H}(\mathbf{s})$ = $20(1 + \mathbf{s}/100)$ are sketched individually. (b) The composite Bode plot is shown as the sum of the plots of part a.

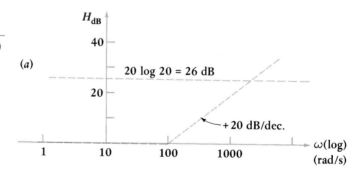

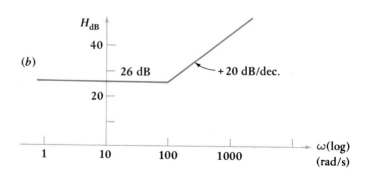

The Bode phase plot is merely the angle corresponding to the zero at $\omega = 100$. We therefore have $0°$ below $\omega = 10$, $90°$ above $\omega = 1000$, $45°$ at $\omega = 100$, and a $45°$/decade slope for $10 < \omega < 1000$. The result is shown as Fig. 13-23. ∎

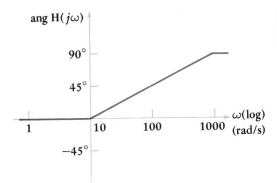

Figure 13-23

The Bode phase plot for $\mathbf{H(s)} = 20(1 + \mathbf{s}/100)$.

Example 13-6 As a second example, we choose a circuit similar to an amplifier with both low- and high-frequency limitations. We need the Bode plot for the gain of the circuit shown in Fig. 13-24.

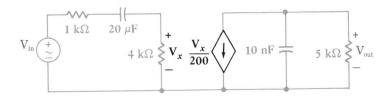

Figure 13-24

If $\mathbf{H(s)} = \mathbf{V}_{out}/\mathbf{V}_{in}$, this amplifier is found to have the Bode amplitude plot shown in Fig. 13-25b, and the phase plot shown in Fig. 13-26.

Solution: We work from left to right through the circuit and write the expression for the voltage gain,

$$\mathbf{H(s)} = \frac{\mathbf{V}_{out}}{\mathbf{V}_{in}} = \frac{4000}{5000 + 10^6/20\mathbf{s}}\left(-\frac{1}{200}\right)\frac{(10^8/\mathbf{s})5000}{5000 + 10^8/\mathbf{s}}$$

$$= \frac{8 \times 10^4\mathbf{s}}{10^6 + 10^5\mathbf{s}}\left(-\frac{1}{200}\right)\frac{5 \times 10^{11}}{10^8 + 5000\mathbf{s}}$$

which simplifies to

$$\mathbf{H(s)} = \frac{-2\mathbf{s}}{(1 + \mathbf{s}/10)(1 + \mathbf{s}/20\,000)}$$

We see a constant, $20 \log |-2| = 6$ dB, poles at $\omega = 10$ and $\omega = 20\,000$, and a linear factor $\mathbf{s}$. Each of these is sketched in Fig. 13-25a, and the four sketches are added to give the Bode magnitude plot in Fig. 13-25b. ∎

Before we construct the phase plot for this amplifier, let us take a few moments to investigate several of the details of the magnitude plot.

First, it is wise not to rely too heavily on graphical addition of the individual magnitude plots. Instead, the exact value of the combined magnitude plot may be found easily at selected points by considering the asymptotic value of each factor of $\mathbf{H(s)}$ at the point in question. For example, in the flat region of Fig. 13-25b between $\omega = 10$ and $\omega = 20\,000$, we are below the corner at $\omega = 20\,000$,

Figure 13-25

(a) Individual Bode magnitude sketches are made for the factors (-2), (s), $(1 + s/10)^{-1}$, and $(1 + s/20\,000)^{-1}$. (b) The four separate plots of part a are added to give the Bode magnitude plot for the amplifier of Fig. 13-24.

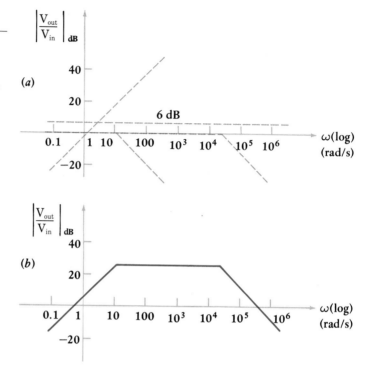

and so we represent $(1 + s/20\,000)$ by 1; but we are above $\omega = 10$, so $(1 + s/10)$ is represented as $\omega/10$. Hence,

$$H_{dB} = 20 \log \left| \frac{-2\omega}{(\omega/10)1} \right|$$

$$= 20 \log 20 = 26 \text{ dB} \qquad (10 < \omega < 20\,000)$$

We might also wish to know the frequency at which the asymptotic response crosses the abscissa at the high end. The two corners are expressed here as $\omega/10$ and $\omega/20\,000$; thus

$$H_{dB} = 20 \log \left| \frac{-2\omega}{(\omega/10)(\omega/20\,000)} \right| = 20 \log \left| \frac{400\,000}{\omega} \right|$$

Since $H_{dB} = 0$ at the abscissa crossing, $400\,000/\omega = 1$, and $\omega = 400\,000$ rad/s.

Many times we may not need an accurate Bode plot drawn on printed semilog paper. Instead we construct a rough logarithmic frequency axis on simple lined paper. After selecting the interval for a decade—say, a distance L extending from $\omega = \omega_1$ to $\omega = 10\omega_1$ (where ω_1 is usually an integral power of 10)—we let x locate the distance that ω lies to the right of ω_1, so that $x/L = \log(\omega/\omega_1)$. Of particular help is the knowledge that $x = 0.3L$ when $\omega = 2\omega_1$, $x = 0.6L$ at $\omega = 4\omega_1$, and $x = 0.7L$ at $\omega = 5\omega_1$.

Now let us finish up the example of Fig. 13-24.

Example 13-6 (cont.) Draw the phase plot for $\mathbf{H(s)} = -2\mathbf{s}/(1 + \mathbf{s}/10)(1 + \mathbf{s}/20\,000)$.

Solution: We now inspect $\mathbf{H}(j\omega)$:

$$\mathbf{H}(j\omega) = \frac{-j2\omega}{(1 + j\omega/10)(1 + j\omega/20\,000)} \qquad (31)$$

The angle of the numerator is a constant, $-90°$, and the remaining factors are represented as the sum of the angles contributed by simple poles at $\omega = 10$ and $\omega = 20\,000$. These three terms appear as broken-line asymptotic curves in Fig. 13-26, and their sum is shown as the solid curve. An equivalent representation is obtained if the curve is shifted upward by $360°$.

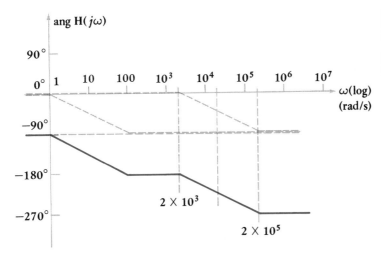

Figure 13-26

The solid curve displays the asymptotic phase response of the amplifier shown in Fig. 13-24.

Exact values can also be obtained for the asymptotic phase response. For example, at $\omega = 10^4$ rad/s, the angle in Fig. 13-26 is obtained from the numerator and denominator terms in Eq. (31). The numerator angle is $-90°$. The angle for the pole at $\omega = 10$ is $-90°$, since ω is greater than 10 times the corner frequency. Between 0.1 and 10 times the corner frequency, we recall that the slope is $-45°$/decade for a simple pole. For the pole at 20 000 rad/s, we therefore calculate the angle, $-45° \log(\omega/0.1a) = -45° \log[10\,000/(0.1 \times 20\,000)] = -31.5°$. The algebraic sum of these three contributions is $-90° - 90° - 31.5° = -211.5°$, a value which appears to be moderately near the asymptotic phase curve of Fig. 13-26. ∎

The zeros and poles that we have been considering are all first-order terms, such as $\mathbf{s}^{\pm1}$, $(1 + 0.2\mathbf{s})^{\pm1}$, and so forth. We may extend our analysis to higher-order poles and zeros very easily, however. A term $\mathbf{s}^{\pm n}$ yields a magnitude response that passes through $\omega = 1$ with a slope of $\pm20n$ dB/decade; the phase response is a constant angle of $(\pm90°)n$. Also, a multiple zero, $(1 + \mathbf{s}/a)^n$, must represent the sum of n of the magnitude-response curves, or n of the phase-response curves of the simple zero. We therefore obtain an asymptotic magnitude plot that is 0 dB for $\omega < a$ and has a slope of $20n$ dB/decade when $\omega > a$; the error is $-3n$ dB at $\omega = a$, and $-n$ dB at $\omega = 0.5a$ and $2a$. The phase plot is $0°$ for $\omega < 0.1a$, $(90°)n$ for $\omega > 10a$, $(45°)n$ at $\omega = a$, and a straight line with a slope of $(45°)n$/decade for $0.1a < \omega < 10a$, and it has errors as large as $(\pm5.71°)n$ at two frequencies.

The asymptotic magnitude and phase curves associated with a factor such as $(1 + \mathbf{s}/20)^{-3}$ may be drawn quickly, but the relatively large errors associated with the higher powers should be kept in mind.

The last type of factor that we need to consider represents a conjugate complex pair of poles or zeros. We adopt the following as the standard form for a pair of zeros:

$$\mathbf{H(s)} = 1 + 2\zeta\left(\frac{\mathbf{s}}{\omega_0}\right) + \left(\frac{\mathbf{s}}{\omega_0}\right)^2$$

The quantity ζ is the damping factor introduced in Sec. 13-2, and we shall see shortly that ω_0 is the corner frequency of the asymptotic response.[6] If $\zeta = 1$, we see that $\mathbf{H(s)} = 1 + 2(\mathbf{s}/\omega_0) + (\mathbf{s}/\omega_0)^2 = (1 + \mathbf{s}/\omega_0)^2$, a second-order zero such as we have just considered. If $\zeta > 1$, then $\mathbf{H(s)}$ may be factored to show two simple zeros. Thus, if $\zeta = 1.25$, then $\mathbf{H(s)} = 1 + 2.5(\mathbf{s}/\omega_0) + (\mathbf{s}/\omega_0)^2 = (1 + \mathbf{s}/2\omega_0)(1 + \mathbf{s}/0.5\omega_0)$, and we again have a familiar situation.

A new case arises when $0 \le \zeta < 1$. There is no need to find values for the conjugate complex pair of roots. Instead, we determine the low- and high-frequency asymptotic values for both the magnitude and phase response, and then apply a correction that depends on the value of ζ.

For the magnitude response, we have

$$H_{\text{dB}} = 20 \log |\mathbf{H}(j\omega)| = 20 \log \left| 1 + j2\zeta\left(\frac{\omega}{\omega_0}\right) - \left(\frac{\omega}{\omega_0}\right)^2 \right| \qquad (32)$$

When $\omega \ll \omega_0$, $H_{\text{dB}} \doteq 20 \log |1| = 0$ dB. This is the low-frequency asymptote. Next, if $\omega \gg \omega_0$, only the squared term is important, and $H_{\text{dB}} \doteq 20 \log |-(\omega/\omega_0)^2| = 40 \log (\omega/\omega_0)$. We have a slope of $+40$ dB/decade. This is the high-frequency asymptote, and the two asymptotes intersect at 0 dB, $\omega = \omega_0$. The solid curve in Fig. 13-27 shows this asymptotic representation of the magnitude response. However, a correction must be applied in the neighborhood of the corner frequency. We let $\omega = \omega_0$ in Eq. (32) and have

$$H_{\text{dB}} = 20 \log \left| j2\zeta\left(\frac{\omega}{\omega_0}\right) \right| = 20 \log (2\zeta)$$

If $\zeta = 1$, a limiting case, the correction is $+6$ dB; for $\zeta = 0.5$, no correction is required; and if $\zeta = 0.1$, the correction is -14 dB. Knowing this one correction value is often sufficient to draw a satisfactory asymptotic magnitude response. Figure 13-27 shows more accurate curves for $\zeta = 1$, 0.5, 0.25, and 0.1, as calculated from Eq. (31). For example, if $\zeta = 0.25$, then the exact value of H_{dB} at $\omega = 0.5\omega_0$ is

$$H_{\text{dB}} = 20 \log |1 + j0.25 - 0.25| = 20 \log \sqrt{0.75^2 + 0.25^2} = -2.0 \text{ dB}$$

The negative peaks do not show a minimum value exactly at $\omega = \omega_0$, as we can see by the curve for $\zeta = 0.5$. The valley is always found at a slightly lower frequency.

If $\zeta = 0$, then $\mathbf{H}(j\omega_0) = 0$ and $H_{\text{dB}} = -\infty$. Bode plots are not usually drawn for this situation.

Our last task is to draw the asymptotic phase response for $\mathbf{H}(j\omega) = 1 + j2\zeta(\omega/\omega_0) - (\omega/\omega_0)^2$. Below $\omega = 0.1\omega_0$, we let ang $\mathbf{H}(j\omega) = 0°$; above $\omega = 10\omega_0$, we have ang $\mathbf{H}(j\omega) \doteq \text{ang}\,[-(\omega/\omega_0)^2] = 180°$. At the corner frequency, ang $\mathbf{H}(j\omega_0) = \text{ang}\,(j\,2\zeta) = 90°$. In the interval, $0.1\omega_0 < \omega < 10\omega_0$, we begin with the straight

---

[6] It is also the resonant frequency if $\zeta = 0$.

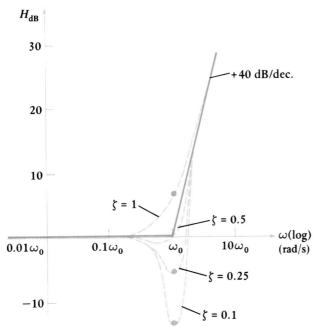

Figure 13-27

Bode amplitude plots are shown for $\mathbf{H(s)} = 1 + 2\zeta(\mathbf{s}/\omega_0) + (\mathbf{s}/\omega_0)^2$ for several values of the damping factor ζ.

line shown as a solid curve in Fig. 13-28. It extends from $(0.1\omega_0, 0°)$, through $(\omega_0, 90°)$, and terminates at $(10\omega_0, 180°)$; it has a slope of 90°/decade.

We must now provide some correction to this basic curve for various values of ζ. From Eq. (32), we have

$$\text{ang } \mathbf{H}(j\omega) = \tan^{-1}\frac{2\zeta(\omega/\omega_0)}{1-(\omega/\omega_0)^2}$$

One accurate value above and one below $\omega = \omega_0$ may be sufficient to give an approximate shape to the curve. If we take $\omega = 0.5\omega_0$, we find ang $\mathbf{H}(j0.5\omega_0) = \tan^{-1}(4\zeta/3)$, while the angle is $180° - \tan^{-1}(4\zeta/3)$ at $\omega = 2\omega_0$. Phase curves are shown as broken lines in Fig. 13-28 for $\zeta = 1, 0.5, 0.25,$ and 0.1; heavy dots identify accurate values at $\omega = 0.5\omega_0$ and $\omega = 2\omega_0$.

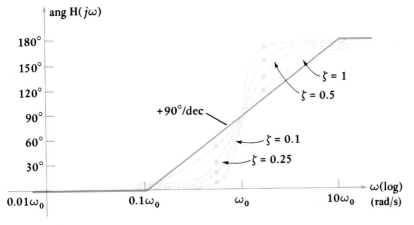

Figure 13-28

The straight line approximation to the phase characteristic for $\mathbf{H}(j\omega) = 1 + j2\zeta(\omega/\omega_0) - (\omega/\omega_0)^2$ is shown as a solid curve, and the true phase response is shown for $\zeta = 1, 0.5, 0.25,$ and 0.1 as broken lines.

If the quadratic factor appears in the denominator, both the magnitude and phase curves are the negatives of those just discussed.

We conclude with an example that contains both linear and quadratic factors.

Example 13-7 Construct the Bode plot for the transfer function

$$\mathbf{H(s)} = \frac{100\,000\,\mathbf{s}}{(\mathbf{s} + 1)(10\,000 + 20\mathbf{s} + \mathbf{s}^2)}$$

Solution: Let's consider the quadratic factor first and arrange it in a form such that we can see the value of ζ. We begin by dividing the second-order factor by its constant term, 10 000:

$$\mathbf{H(s)} = \frac{10\mathbf{s}}{(1 + \mathbf{s})(1 + 0.002\mathbf{s} + 0.0001\mathbf{s}^2)}$$

An inspection of the $\mathbf{s}^2$ term next shows that $\omega_0 = \sqrt{1/0.0001} = 100$. Then the linear term of the quadratic is written to display the factor 2, the factor $(\mathbf{s}/\omega_0)$, and finally the factor ζ:

$$\mathbf{H(s)} = \frac{10\mathbf{s}}{(1 + \mathbf{s})[1 + 2(0.1)(\mathbf{s}/100) + (\mathbf{s}/100)^2]}$$

We see that $\zeta = 0.1$.

The asymptotes of the magnitude-response curve are sketched in lightly in Fig. 13-29: 20 dB for the factor of 10, an infinite straight line through $\omega = 1$ with a +20 dB/decade slope for the $\mathbf{s}$ factor, a corner at $\omega = 1$ for the simple pole, and a corner at $\omega = 100$ with a slope of −40 dB/decade for the second-order term in the denominator. Adding these four curves and supplying a correction of +14 dB for the quadratic factor leads to the heavy curve of Fig. 13-29.

Figure 13-29

The Bode magnitude plot of the transfer function $\mathbf{H(s)} = \dfrac{100\,000\mathbf{s}}{(\mathbf{s} + 1)(10\,000 + 20\mathbf{s} + \mathbf{s}^2)}$

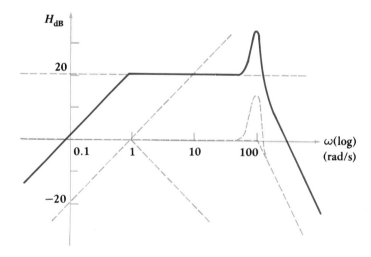

The phase response contains three components: +90° for the factor $\mathbf{s}$; 0° for $\omega < 0.1$, −90° for $\omega > 10$, and −45°/decade for the simple pole at $\omega = 1$; and 0° for $\omega < 10$, −180° for $\omega > 1000$, and −90°/decade for the quadratic factor. The addition of these three asymptotes plus some improvement for $\zeta = 0.1$ is shown as the solid curve in Fig. 13-30. ∎

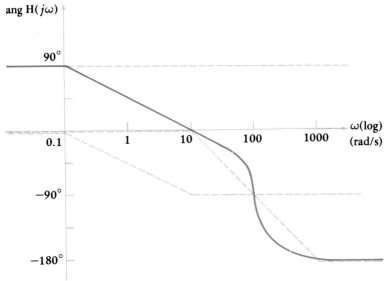

Figure 13-30

The Bode phase plot of the transfer function $\mathbf{H}(\mathbf{s}) =$
$$\frac{100\,000\mathbf{s}}{(\mathbf{s} + 1)(10\,000 + 20\mathbf{s} + \mathbf{s}^2)}$$

Drill Problems

13-15. Calculate H_{dB} at $\omega = 146$ rad/s if $\mathbf{H}(\mathbf{s})$ equals (a) $20/(\mathbf{s} + 100)$; (b) $20(\mathbf{s} + 100)$; (c) $20\mathbf{s}$. Calculate $|H(j\omega)|$ if H_{dB} equals (d) 29.2 dB; (e) -15.6 dB; (f) -0.318 dB.   *Ans:* -18.94 dB; 71.0 dB; 69.3 dB; 28.8; 0.1660; 0.964

13-16. Construct a Bode magnitude plot for $\mathbf{H}(\mathbf{s})$ equal to (a) $50/(\mathbf{s} + 100)$; (b) $(\mathbf{s} + 10)/(\mathbf{s} + 100)$; (c) $(\mathbf{s} + 10)/\mathbf{s}$.
Ans: (a) -6 dB, $\omega < 100$; -20 dB/decade, $\omega > 100$; (b) -20 dB, $\omega < 10$; $+20$ dB/decade, $10 < \omega < 100$; 0 dB, $\omega > 100$; (c) 0 dB, $\omega > 10$; -20 dB/decade, $\omega < 10$

13-17. Draw the Bode phase plot for each $\mathbf{H}(\mathbf{s})$ in Drill Prob. 13-16.
Ans: (a) $0°$, $\omega < 10$; $-45°$/decade, $10 < \omega < 100$; $-90°$, $\omega > 1000$; (b) $0°$, $\omega < 1$; $+45°$/decade, $1 < \omega < 10$; $45°$, $10 < \omega < 100$; $-45°$/decade, $100 < \omega < 1000$; $0°$, $\omega > 1000$; (c) $-90°$, $\omega < 1$; $+45°$/decade, $1 < \omega < 100$; $0°$, $\omega > 100$

13-18. If $\mathbf{H}(\mathbf{s}) = 1000\mathbf{s}^2/(\mathbf{s}^2 + 5\mathbf{s} + 100)$, sketch the Bode amplitude plot and calculate a value for (a) ω when $H_{dB} = 0$; (b) H_{dB} at $\omega = 1$; (c) H_{dB} as $\omega \to \infty$.
Ans: 0.316 rad/s; 20 dB; 60 dB

Problems

1 Find the resonant frequency of the two-terminal network shown in Fig. 13-31.

2 Let $R = 1$ MΩ, $L = 1$ H, $C = 1$ μF, and $\mathbf{I} = 10\underline{/0°}$ μA in the circuit of Fig. 13-1. (a) Find ω_0 and Q_0. (b) Plot $|\mathbf{V}|$ as a function of ω, $995 < \omega < 1005$ rad/s.

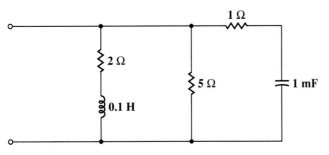

Figure 13-31

See Prob. 1.

3 For the network shown in Fig. 13-32, find (a) the resonant frequency ω_0; (b) $\mathbf{Z}_{in}(j\omega_0)$.

Figure 13-32

See Probs. 3 and 42.

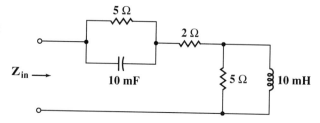

4 A parallel resonant circuit has impedance poles at $\mathbf{s} = -50 \pm j1000$ s^{-1}, and a zero at the origin. If $C = 1 \ \mu F$: (a) find L and R; (b) calculate $\mathbf{Z}$ at $\omega = 1000$ rad/s.

5 A parallel resonant circuit has parameter values of $\alpha = 80$ Np/s and $\omega_d = 1200$ rad/s. If the impedance at $\mathbf{s} = -2\alpha + j\omega_d$ has a magnitude of 400 Ω, calculate Q_0, R, L, and C.

6 Design a parallel resonant circuit for an AM radio so that a variable inductor can adjust the resonant frequency over the AM broadcast band, 535 to 1605 kHz, with $Q_0 = 45$ at one end of the band and $Q_0 \leq 45$ throughout the band. Let $R = 20$ kΩ, and specify values for C, L_{min}, and L_{max}.

7 (a) Find $\mathbf{Y}_{in}$ for the network shown in Fig. 13-33. (b) Determine ω_0 and $\mathbf{Z}_{in}(j\omega_0)$ for the network.

Figure 13-33

See Prob. 7.

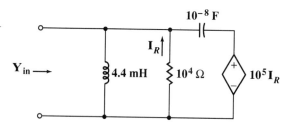

8 A parallel resonant circuit has $\omega_0 = 1000$ rad/s, $Q_0 = 80$, and $C = 0.2 \ \mu F$. (a) Find R and L. (b) Use approximate methods to plot $|\mathbf{Z}|$ versus ω.

9 Use the exact relationships to find R, L, and C for a parallel resonant circuit that has $\omega_1 = 103$ rad/s, $\omega_2 = 118$ rad/s, and $|\mathbf{Z}(j105)| = 10 \ \Omega$.

10 Let $\omega_0 = 30$ krad/s, $Q_0 = 10$, and $R = 600 \ \Omega$ for a certain parallel resonant circuit. (a) Find the bandwidth. (b) Calculate N at $\omega = 28$ krad/s. (c) Use approximate methods to determine $\mathbf{Z}_{in}(j28\ 000)$. (d) Find the true value of $\mathbf{Z}_{in}(j28\ 000)$. (e) State the percentage error incurred by using the approximate relationships to calculate $|\mathbf{Z}_{in}|$ and ang $\mathbf{Z}_{in}$ at 28 krad/s.

11 A parallel resonant circuit is resonant at 400 Hz with $Q_0 = 8$ and $R = 500 \ \Omega$. If a current of 2 mA is applied to the circuit, use approximate methods to find the cyclic frequency of the current if (a) the voltage across the circuit has a magnitude of 0.5 V; (b) the resistor current has a magnitude of 0.5 mA.

12 A parallel resonant circuit has $\omega_0 = 1$ Mrad/s and $Q_0 = 10$. Let $R = 5$ kΩ and find (a) L; (b) the frequency above ω_0 at which $|\mathbf{Z}_{in}| = 2$ kΩ; (c) the frequency at which ang $\mathbf{Z}_{in} = -30°$.

13 Use good approximations on the circuit of Fig. 13-34 to (a) find ω_0; (b) calculate $\mathbf{V}_1$ at the resonant frequency; (c) calculate $\mathbf{V}_1$ at a frequency that is 15 krad/s above resonance.

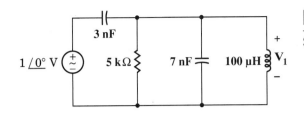

Figure 13-34

See Prob. 13.

14 (*a*) Apply the definition of resonance to find ω_0 for the network of Fig. 13-35. (*b*) Find $\mathbf{Z}_{in}(j\omega_0)$.

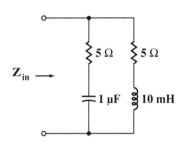

Figure 13-35

See Prob. 14.

15 A parallel resonant circuit is characterized by $f_0 = 1000$ Hz, $Q_0 = 40$, and $|\mathbf{Z}_{in}(j\omega_0)|$ = 2 kΩ. Use the approximate relationships to find (*a*) $\mathbf{Z}_{in}$ at 1010 Hz; (*b*) the frequency range over which the approximations are reasonably accurate.

16 (*a*) Use approximate techniques to plot $|\mathbf{V}_{out}|$ versus ω for the circuit shown in Fig. 13-36. (*b*) Find an exact value for $\mathbf{V}_{out}$ at $\omega = 9$ rad/s.

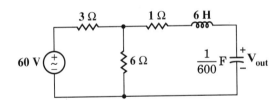

Figure 13-36

See Prob. 16.

17 A series resonant network consists of a 50-Ω resistor, a 4-mH inductor, and a 0.1-μF capacitor. Calculate values for (*a*) ω_0; (*b*) f_0; (*c*) Q_0; (*d*) $\mathcal{B}$; (*e*) ω_1; (*f*) ω_2; (*g*) $\mathbf{Z}_{in}$ at 45 krad/s; (*h*) the ratio of magnitudes of the capacitor impedance to the resistor impedance at 45 krad/s.

18 After deriving $\mathbf{Z}_{in}(\mathbf{s})$ in Fig. 13-37, find: (*a*) ω_0; (*b*) Q_0.

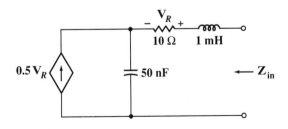

Figure 13-37

See Prob. 18.

19 Inspect the circuit of Fig. 13-38, noting the amplitude of the source voltage. Now decide whether you would be willing to put your bare hands across the capacitor if the circuit were actually built in the lab. Plot $|\mathbf{V}_C|$ versus ω to justify your answer.

Figure 13-38

See Prob. 19.

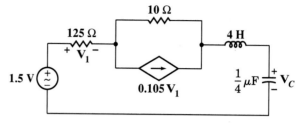

20 A certain series resonant circuit has $f_0 = 500$ Hz, $Q_0 = 10$, and $X_L = 500\ \Omega$ at resonance. (*a*) Find R, L, and C. (*b*) If a source $\mathbf{V}_s = 1\underline{/0°}$ V is connected in series with the circuit, find exact values for $|\mathbf{V}_C|$ at $f = 450$, 500, and 550 Hz.

21 A three-element network has an input impedance $\mathbf{Z}(\mathbf{s})$ that shows poles at $\mathbf{s} = 0$ and infinity, and a pair of zeros at $\mathbf{s} = -20\,000 \pm j80\,000$ s^{-1}. Specify the three element values if $\mathbf{Z}_{in}(-10\,000) = -20 + j0\ \Omega$.

22 Work Drill Prob. 13-8*a* again with a 1-kΩ resistor connected in parallel with the capacitor.

23 Make a few reasonable approximations on the network of Fig. 13-39 and obtain values for ω_0, Q_0, $\mathscr{B}$, $\mathbf{Z}_{in}(j\omega_0)$, and $\mathbf{Z}_{in}(j99\,000)$.

Figure 13-39

See Probs. 23 and 24.

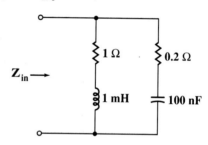

24 What value of resistance should be connected across the input of the network in Fig. 13-39 to cause it to have a Q_0 of 50?

25 Refer to the network shown in Fig. 13-40 and use approximate techniques to determine the minimum magnitude of $\mathbf{Z}_{in}$ and the frequency at which it occurs.

Figure 13-40

See Prob. 25.

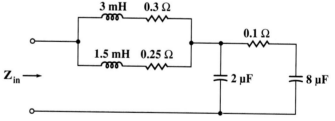

26 For the circuit of Fig. 13-41: (*a*) prepare an approximate response curve of $|\mathbf{V}|$ versus ω, and (*b*) calculate the exact value of $\mathbf{V}$ at $\omega = 50$ rad/s.

Figure 13-41

See Prob. 26.

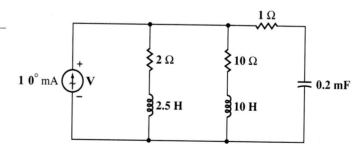

27 (*a*) Use approximate methods to calculate $|\mathbf{V}_x|$ at $\omega = 2000$ rad/s for the circuit of Fig. 13-42. (*b*) Obtain the exact value of $|\mathbf{V}_x \, (j2000)|$.

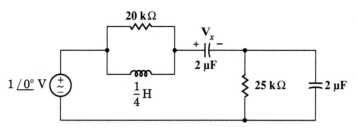

Figure 13-42

See Prob. 27.

28 The filter shown in Fig. 13-43*a* has the response curve shown in Fig. 13-43*b*. (*a*) Scale the filter so that it operates between a 50-Ω source and a 50-Ω load and has a cutoff frequency of 20 kHz. (*b*) Draw the new response curve.

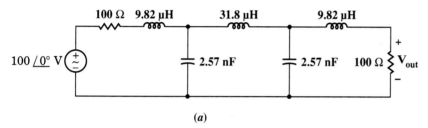

(*a*)

Figure 13-43

See Prob. 28.

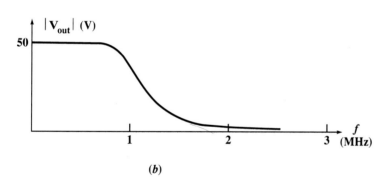

(*b*)

29 (*a*) Find $\mathbf{Z}_{in}(\mathbf{s})$ for the network shown in Fig. 13-44. (*b*) Write an expression for $\mathbf{Z}_{in}(\mathbf{s})$ after it has been scaled by $K_m = 2$, $K_f = 5$. (*c*) Scale the elements in the network by $K_m = 2$, $K_f = 5$, and draw the new network.

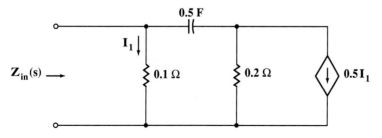

Figure 13-44

See Prob. 29.

30 (*a*) Use good approximations to find ω_0 and Q_0 for the circuit of Fig. 13-45. (*b*) Scale the network to the right of the source so that it is resonant at 1 Mrad/s. (*c*) Specify ω_0 and $\mathcal{B}$ for the scaled circuit.

Figure 13-45

See Prob. 30.

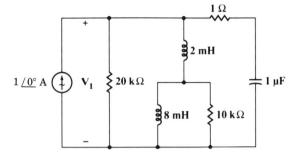

31 (a) Draw the new configuration for Fig. 13-46 after the network is scaled by $K_m = 250$ and $K_f = 400$. (b) Determine the Thévenin equivalent of the scaled network at $\omega = 1$ krad/s.

Figure 13-46

See Prob. 31.

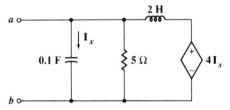

32 A network composed entirely of ideal R's, L's, and C's has a pair of input terminals to which a sinusoidal current source $\mathbf{I}_s$ is connected, and a pair of open-circuited output terminals at which a voltage $\mathbf{V}_{out}$ is defined. If $\mathbf{I}_s = 1\underline{/0°}$ A at $\omega = 50$ rad/s, then $\mathbf{V}_{out} = 30\underline{/25°}$ V. Specify $\mathbf{V}_{out}$ for each condition described as follows. If it is impossible to determine the value of $\mathbf{V}_{out}$, write OTSK.[7] (a) $\mathbf{I}_s = 2\underline{/0°}$ A at $\omega = 50$ rad/s; (b) $\mathbf{I}_s = 2\underline{/40°}$ A at $\omega = 50$ rad/s; (c) $\mathbf{I}_s = 2\underline{/40°}$ A at 200 rad/s; (d) the network is scaled by $K_m = 30$, $\mathbf{I}_s = 2\underline{/40°}$ A, $\omega = 50$ rad/s; (e) $K_m = 30$, $K_f = 4$, $\mathbf{I}_s = 2\underline{/40°}$ A, $\omega = 200$ rad/s.

33 Find H_{dB} if $\mathbf{H}(\mathbf{s})$ equals (a) 0.2; (b) 50; (c) $12/(\mathbf{s} + 2) + 26/(\mathbf{s} + 20)$ for $\mathbf{s} = j10$. Find $|\mathbf{H}(\mathbf{s})|$ if H_{dB} equals (d) 37.6 dB; (e) -8 dB; (f) 0.01 dB.

34 Draw the Bode amplitude plot for (a) $20(\mathbf{s} + 1)/(\mathbf{s} + 100)$; (b) $2000(\mathbf{s} + 1)\mathbf{s}/(\mathbf{s} + 100)^2$; (c) $\mathbf{s} + 45 + 200/\mathbf{s}$.

35 For Fig. 13-47, prepare Bode amplitude and phase plots for transfer function, $\mathbf{H}(\mathbf{s}) = \mathbf{V}_C/\mathbf{I}_s$.

Figure 13-47

See Prob. 35.

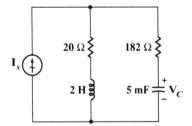

36 (a) Using an origin at $\omega = 1$, $H_{dB} = 0$, construct the Bode amplitude plot for $\mathbf{H}(\mathbf{s}) = 5 \times 10^8 \mathbf{s}(\mathbf{s} + 100)/[(\mathbf{s} + 20)(\mathbf{s} + 1000)^3]$. (b) Give the coordinates for all corners and all intercepts on the Bode plot. (c) Give the exact value of $20 \log |\mathbf{H}(j\omega)|$ for each corner frequency in part b.

37 (a) Construct a Bode phase plot for $\mathbf{H}(\mathbf{s}) = 5 \times 10^8 \mathbf{s}(\mathbf{s} + 100)/[(\mathbf{s} + 20)(\mathbf{s} + 1000)^3]$. Place the origin at $\omega = 1$, ang $= 0°$. (b) Give the coordinates for all points on the phase plot at which the slope changes. (c) Give the exact value of ang $\mathbf{H}(j\omega)$ for each frequency listed in part b.

[7] Only The Shadow Knows.

38 (a) Make a Bode magnitude plot for the transfer function $\mathbf{H(s)} = 1 + 20/\mathbf{s} + 400/\mathbf{s}^2$. (b) Compare the Bode plot and exact values at $\omega = 5$ and 100 rad/s.

39 (a) Find $\mathbf{H(s)} = \mathbf{V}_R/\mathbf{V}_s$ for the circuit shown in Fig. 13-48. (b) Draw Bode amplitude and phase plots for $\mathbf{H(s)}$. (c) Calculate the exact values of H_{dB} and ang $\mathbf{H}(j\omega)$ at $\omega = 20$ rad/s.

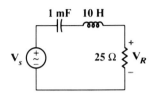

Figure 13-48

See Prob. 39.

40 Construct the Bode amplitude plot for the transfer function $\mathbf{H(s)} = \mathbf{V}_{out}/\mathbf{V}_{in}$ of the network shown in Fig. 13-49.

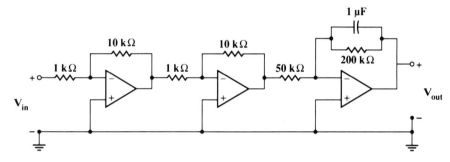

Figure 13-49

See Prob. 40.

41 For the network of Fig. 13-50: (a) find $\mathbf{H(s)} = \mathbf{V}_{out}/\mathbf{V}_{in}$; (b) draw the Bode amplitude plot for H_{dB}; (c) draw the Bode phase plot for $\mathbf{H}(j\omega)$.

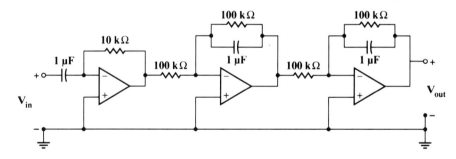

Figure 13-50

See Prob. 41.

42 (SPICE) Apply a sinusoidal voltage source, $20\underline{/0°}$ V, to the network of Fig. 13-32, and use SPICE to calculate the magnitude of the input current at the frequencies, $f = 10, 11, 12, \ldots, 20, 21$, and 22 Hz.

43 (SPICE) Use a SPICE program to determine the magnitude of $\mathbf{V}_{in}$ in Fig. 13-51 at frequencies obtained by dividing the frequency decade $500 \leq f \leq 5000$ Hz into 10 equal logarithmic intervals, 500, 629.5, 792.4, ... Hz.

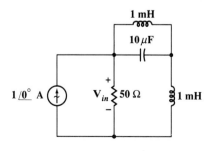

Figure 13-51

See Prob. 43.

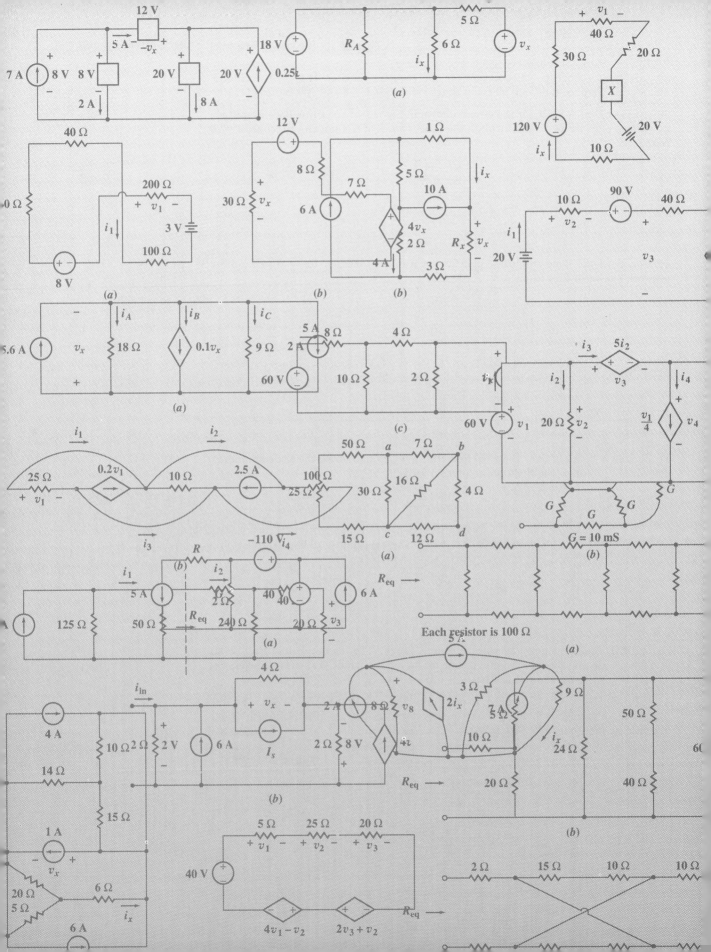

Part Five:
Two-Port Networks

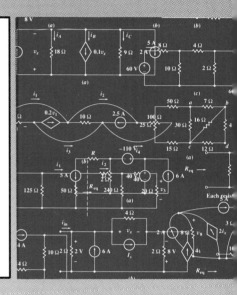

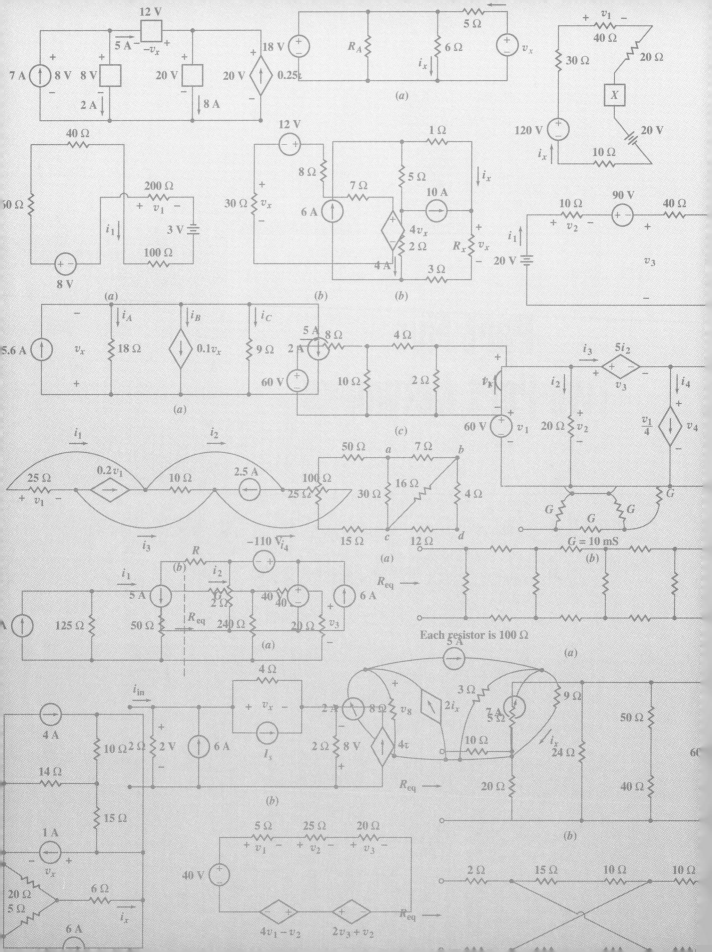

Magnetically Coupled Circuits

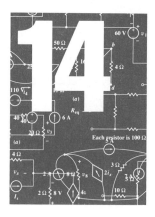

Several hundred pages ago the inductor was introduced as a circuit element and defined in terms of the voltage across it and the time rate of change of the current through it. Strictly speaking, our definition was of *self-inductance,* but, loosely speaking, *inductance* is the commonly used term. Now we need to consider mutual inductance, a property which is associated mutually with two or more coils that are physically close together. A circuit element called the "mutual inductor" does not exist; furthermore, mutual inductance is not a property which is associated with a single pair of terminals, but instead is defined with reference to two pairs of terminals.

Mutual inductance results through the presence of a common magnetic flux which links two coils. It may be defined in terms of this common magnetic flux, just as we might have defined self-inductance in terms of the magnetic flux about the single coil. However, we have agreed to confine our attention to circuit concepts, and such quantities as magnetic flux and flux linkages are mentioned only in passing; we cannot define these quantities easily or accurately at this time, and we must accept them as nebulous concepts which are useful in establishing some background only.

The physical device whose operation is based inherently on mutual inductance is the transformer. The 60-Hz power system uses many transformers, ranging in size from the dimensions of a living room to those of the living room wastebasket. They are used to change the amplitude of the voltage, increasing it for more economical transmission, and then decreasing it for safer operation of home or industrial electrical equipment. Most radios contain one or more transformers, as do television receivers, hi-fi systems, some telephones, automobiles, and electrified railroads (BART and Lionel both).

We first define mutual inductance and study the methods whereby its effects are included in the circuit equations. We conclude with a study of the important characteristics of a linear transformer and an important approximation to a good iron-core transformer, which is known as an *ideal transformer.*

14-2

Mutual inductance

When we defined inductance, we did so by specifying the relationship between the terminal voltage and current,

$$v(t) = L \frac{di(t)}{dt}$$

where the passive sign convention is assumed. We learned, however, that the physical basis for such a current-voltage characteristic rests upon two things: the production of a magnetic flux by a current, the flux being proportional to the current in linear inductors; and the production of a voltage by the time-varying magnetic field, the voltage being proportional to the time rate of change of the magnetic field or the magnetic flux. The proportionality between voltage and time rate of change of current thus becomes evident.

Mutual inductance results from a slight extension of this same argument. A current flowing in one coil establishes a magnetic flux about that coil and also about a second coil which is in its vicinity. The time-varying flux surrounding the second coil produces a voltage across the terminals of this second coil; this voltage is proportional to the time rate of change of the current flowing through the first coil. Figure 14-1a shows a simple model of two coils L_1 and L_2, sufficiently close together that the flux produced by a current $i_1(t)$ flowing through L_1 establishes an open-circuit voltage $v_2(t)$ across the terminals of L_2.

Figure 14-1

(a) A current i_1 at L_1 produces an open-circuit voltage v_2 at L_2. (b) A current i_2 at L_2 produces an open-circuit voltage v_1 at L_1.

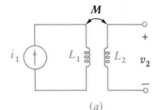
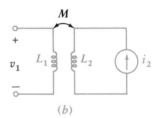

(a)   (b)

Without considering the proper algebraic sign for the relationship at this time, we define the *coefficient of mutual inductance*, or simply *mutual inductance*, M_{21},

$$v_2(t) = M_{21} \frac{di_1(t)}{dt} \qquad (1)$$

The order of the subscripts on M_{21} indicates that a voltage response is produced at L_2 by a current source at L_1. If the system is reversed, as indicated in Fig. 14-1b, and a voltage response is produced at L_1 by a current source at L_2, then we have

$$v_1(t) = M_{12} \frac{di_2(t)}{dt} \qquad (2)$$

Two coefficients of mutual inductance are not necessary, however; we shall use energy relationships a little later to prove that M_{12} and M_{21} are equal. Thus, $M_{12} = M_{21} = M$. The existence of mutual coupling between two coils is indicated by a double-headed arrow, as shown in Figs. 14-1a and b.

Mutual inductance is measured in henrys and, like resistance, inductance, and capacitance, is always positive.[1] The voltage $M\, di/dt$, however, may appear as either a positive or a negative quantity in the same way that $v = -Ri$ is useful.

---

[1] Mutual inductance is not universally assumed to be positive. It is particularly convenient to allow it to "carry its own sign" when three or more coils are involved and each coil interacts with each other coil. We shall restrict our attention to the more important simple case of two coils.

The inductor is a two-terminal element, and we are able to use the passive sign convention in order to select the correct sign for the voltage $L\, di/dt$, $j\omega L\mathbf{I}$, or $\mathbf{s}L\mathbf{I}$. If the current enters the terminal at which the positive voltage reference is located, then the positive sign is used. Mutual inductance, however, cannot be treated in exactly the same way, because four terminals are involved. The choice of a correct sign is established by use of one of several possibilities that include the "dot convention," or an extension of the dot convention which involves the use of a larger variety of special symbols; or by an examination of the particular way in which each coil is wound. We shall use the dot convention and merely look briefly at the physical construction of the coils; the use of other special symbols is not necessary when only two coils are coupled.

The dot convention makes use of a large dot placed at one end of each of the two coils which are mutually coupled. We determine the sign of the mutual voltage as follows:

> A current entering the dotted terminal of one coil produces an open-circuit voltage between the terminals of the second coil which is sensed in the direction indicated by a positive voltage reference at the dotted terminal of this second coil.

Thus, in Fig. 14-2a, i_1 enters the dotted terminal of L_1, v_2 is sensed positively at the dotted terminal of L_2, and $v_2 = M\, di_1/dt$. We have found previously that it is often not possible to select voltages or currents throughout a circuit so that the passive sign convention is everywhere satisfied; the same situation arises with mutual coupling. For example, it may be more convenient to represent v_2 by a positive voltage reference at the undotted terminal, as shown in Fig. 14-2b; then $v_2 = -M\, di_1/dt$. Currents which enter the dotted terminal are also not always available, as indicated by Figs. 14-2c and d. We note then that:

> A current entering the *undotted* terminal of one coil provides a voltage which is positively sensed at the *undotted* terminal of the second coil.

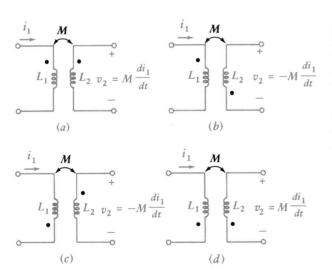

(a)   (b)   (c)   (d)

Figure 14-2

Current entering the dotted terminal of one coil produces a voltage which is sensed positively at the dotted terminal of the second coil. Current entering the undotted terminal of one coil produces a voltage which is sensed positively at the undotted terminal of the second coil.

We have as yet considered only a mutual voltage present across an *open-circuited* coil. In general, a nonzero current will be flowing in each of the two coils, and a mutual voltage will be produced in each coil because of the current flowing in the other coil. *This mutual voltage is present independently of and in addition to any voltage of self-induction.* In other words, the voltage across the terminals of L_1 will be composed of two terms, $L_1\,di_1/dt$ and $M\,di_2/dt$, each carrying a sign depending on the current directions, the assumed voltage sense, and the placement of the two dots. In the portion of a circuit drawn in Fig. 14-3a, currents i_1 and i_2 are shown, each arbitrarily assumed entering the dotted terminal. The voltage across L_1 is thus composed of two parts,

$$v_1 = L_1\frac{di_1}{dt} + M\frac{di_2}{dt}$$

as is the voltage across L_2,

$$v_2 = L_2\frac{di_2}{dt} + M\frac{di_1}{dt}$$

Figure 14-3

(a) Since the pairs v_1, i_1 and v_2, i_2 each satisfy the passive sign convention, the voltages of self-induction are both positive; since i_1 and i_2 each enter dotted terminals, and since v_1 and v_2 are both positively sensed at the dotted terminals, the voltages of mutual induction are also both positive. (b) Since the pairs v_1, i_1 and v_2, i_2 are not sensed according to the passive sign convention, the voltages of self-induction are both negative; since i_1 enters the dotted terminal and v_2 is positively sensed at the dotted terminal, the mutual term of v_2 is positive; and since i_2 enters the undotted terminal and v_1 is positively sensed at the undotted terminal, the mutual term of v_1 is also positive.

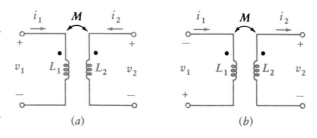

In Fig. 14-3b the currents and voltages are not selected with the object of obtaining all positive terms for v_1 and v_2. By inspecting only the reference symbols for i_1 and v_1, it is apparent that the passive sign convention is not satisfied and the sign of $L_1\,di_1/dt$ must therefore be negative. An identical conclusion is reached for the term $L_2\,di_2/dt$. The mutual term of v_2 is signed by inspecting the direction of i_1 and v_2; since i_1 enters the dotted terminal and v_2 is sensed positive at the dotted terminal, the sign of $M\,di_1/dt$ must be positive. Finally, i_2 enters the undotted terminal of L_2, and v_1 is sensed positive at the undotted terminal of L_1; hence, the mutual portion of v_1, $M\,di_2/dt$, must also be positive. Thus, we have

$$v_1 = -L_1\frac{di_1}{dt} + M\frac{di_2}{dt} \qquad v_2 = -L_2\frac{di_2}{dt} + M\frac{di_1}{dt}$$

The same considerations lead to identical choices of signs for excitation at a complex frequency **s**,

$$\mathbf{V}_1 = -\mathbf{s}L_1\mathbf{I}_1 + \mathbf{s}M\mathbf{I}_2 \qquad \mathbf{V}_2 = -\mathbf{s}L_2\mathbf{I}_2 + \mathbf{s}M\mathbf{I}_1$$

or in the sinusoidal steady state where $\mathbf{s} = j\omega$,

$$\mathbf{V}_1 = -j\omega L_1 \mathbf{I}_1 + j\omega M \mathbf{I}_2 \qquad \mathbf{V}_2 = -j\omega L_2 \mathbf{I}_2 + j\omega M \mathbf{I}_1$$

   Before we apply the dot convention to the analysis of a numerical example, we can gain a more complete understanding of the dot symbolism by looking at the physical basis for the convention. The meaning of the dots is now interpreted in terms of magnetic flux. Two coils are shown wound on a cylindrical form in Fig. 14-4, and the direction of each winding is evident. Let us assume that the current i_1 is positive and increasing with time. The magnetic flux that i_1 produces within the form has a direction which may be found by the right-hand rule: when the right hand is wrapped around the coil with the fingers pointing in the direction of current flow, the thumb indicates the direction of the flux within the coil. Thus i_1 produces a flux which is directed downward; since i_1 is increasing with time, the flux, which is proportional to i_1, is also increasing with time. Turning now to the second coil, let us also think of i_2 as positive and increasing; the application of the right-hand rule shows that i_2 also produces a magnetic flux which is directed downward and is increasing. In other words, the assumed currents i_1 and i_2 produce *additive* fluxes.

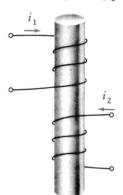

Figure 14-4

The physical construction of two mutually coupled coils. From a consideration of the direction of magnetic flux produced by each coil, it is shown that dots may be placed either on the upper terminal of each coil or on the lower terminal of each coil.

   The voltage across the terminals of any coil results from the time rate of change of the flux within that coil. The voltage across the terminals of the first coil is therefore greater with i_2 flowing than it would be if i_2 were zero. Thus i_2 induces a voltage in the first coil which has the same sense as the self-induced voltage in that coil. The sign of the self-induced voltage is known from the passive sign convention, and the sign of the mutual voltage is thus obtained.

   The dot convention merely enables us to suppress the physical construction of the coils by placing a dot at one terminal of each coil such that currents entering dot-marked terminals produce additive fluxes. It is apparent that there are always two possible locations for the dots, because both dots may always be moved to the other ends of the coils and additive fluxes will still result.

   Let us see how easy this procedure is by attacking a numerical example.

Example 14-1   Refer to the circuit shown in Fig. 14-5. We desire the ratio of the output voltage across the 400-Ω resistor to the source voltage.

   Solution: Two conventional mesh currents are established, and KVL is applied to each mesh. In the left mesh, the sign of the mutual term is determined by applying the dot convention. Since $\mathbf{I}_2$ enters the undotted

Figure 14-5

A circuit containing mutual inductance in which the voltage ratio V_2/V_1 is desired.

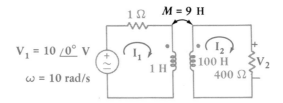

terminal of L_2, the mutual voltage across L_1 must have the positive reference at the undotted terminal. Thus,

$$I_1(1 + j10) - j90I_2 = 10$$

The sign of the mutual term in the second mesh is determined in a similar manner. Since I_1 enters the dot-marked terminal, the mutual term in the right mesh has its ($+$) reference at the dotted terminal of the 100-H inductor. Thus, we may write

$$I_2(400 + j1000) - j90I_1 = 0$$

The two equations may be solved by determinants (or a simple elimination of I_1):

$$I_2 = \frac{\begin{vmatrix} 1 + j10 & 10 \\ -j90 & 0 \end{vmatrix}}{\begin{vmatrix} 1 + j10 & -j90 \\ -j90 & 400 + j1000 \end{vmatrix}}$$

or

$$I_2 = 0.1724\underline{/-16.7°}\text{ A}$$

Thus,

$$\frac{V_2}{V_1} = \frac{400(0.1724\underline{/-16.7°})}{10} = 6.90\underline{/-16.7°}$$

∎

The output voltage is greater in magnitude than the input voltage, and this example shows that a voltage gain is possible with mutual coupling just as it is in a resonant circuit. The voltage gain available in this circuit, however, is present over a relatively wide frequency range; in a moderately high-Q resonant circuit, the voltage step-up is proportional to the Q and occurs only over a frequency range which is inversely proportional to Q. Let us see if this point can be made clear by reference to the complex plane.

We find $I_2(\mathbf{s})$ for this particular circuit,

$$I_2(\mathbf{s}) = \frac{\begin{vmatrix} 1 + \mathbf{s} & V_1 \\ -9\mathbf{s} & 0 \end{vmatrix}}{\begin{vmatrix} 1 + \mathbf{s} & -9\mathbf{s} \\ -9\mathbf{s} & 400 + 100\mathbf{s} \end{vmatrix}}$$

and thus obtain the ratio of output voltage to input voltage as a function of $\mathbf{s}$,

$$\frac{V_2}{V_1} = \frac{3600\mathbf{s}}{19\mathbf{s}^2 + 500\mathbf{s} + 400} = (189.5)\frac{\mathbf{s}}{(\mathbf{s} + 0.826)(\mathbf{s} + 25.5)}$$

The pole-zero plot of this transfer function is shown in Fig. 14-6. The location of the pole at $\mathbf{s} = -25.5$ is distorted in order to show both poles clearly; the ratio of the distances of the two poles from the origin is actually about 30:1. An

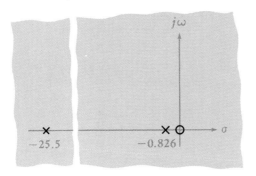

Figure 14-6

A pole-zero plot of the transfer function $\mathbf{V}_2/\mathbf{V}_1$ for the circuit shown in Fig. 14-5. The plot is useful in showing that the magnitude of the transfer function is relatively large from $\omega = 1$ or 2 to $\omega = 15$ or 20 rad/s.

inspection of this plot shows that the transfer function is zero at zero frequency but that as soon as ω is greater than, say, 1 or 2 the ratio of the distances from the zero and the pole nearer the origin is essentially unity. Thus the voltage gain is affected only by the distant pole on the negative σ axis, and this distance does not increase appreciably until ω approaches 15 or 20.

These tentative conclusions are verified by the response curve of Fig. 14-7, which shows that the magnitude of the voltage gain is greater than 0.707 times its maximum value from $\omega = 0.78$ to $\omega = 27$.

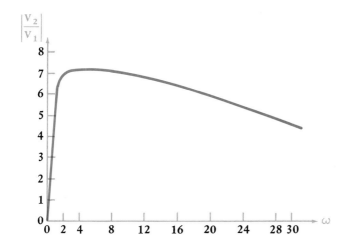

Figure 14-7

The voltage gain $|\mathbf{V}_2/\mathbf{V}_1|$ of the circuit shown in Fig. 14-5 is plotted as a function of ω. The voltage gain is greater than 5 from about $\omega = 0.75$ to approximately $\omega = 28$ rad/s.

The circuit is still passive, except for the voltage source, and the voltage gain must not be mistakenly interpreted as a power gain. At $\omega = 10$, the voltage gain is 6.90, but the ideal voltage source, having a terminal voltage of 10 V, delivers a total power of 8.07 W, of which only 5.94 W reaches the 400-Ω resistor. The ratio of the output power to the source power, which we may define as the power gain, is thus 0.736.

Let us consider briefly one additional example.

Example 14-2 Our object is to write a correct set of equations for the circuit illustrated in Fig. 14-8.

Solution: The circuit contains three meshes, and three mesh currents are assigned. Applying Kirchhoff's voltage law to the first mesh, a positive sign

Figure 14-8

A three-mesh circuit with mutual coupling may be analyzed most easily by using loop or mesh currents.

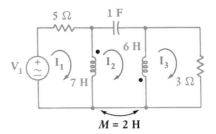

for the mutual term is assured by selecting $(\mathbf{I}_3 - \mathbf{I}_2)$ as the current through the second coil. Thus,

$$5\mathbf{I}_1 + 7\mathbf{s}(\mathbf{I}_1 - \mathbf{I}_2) + 2\mathbf{s}(\mathbf{I}_3 - \mathbf{I}_2) = \mathbf{V}_1$$

or
$$(5 + 7\mathbf{s})\mathbf{I}_1 - 9\mathbf{s}\mathbf{I}_2 + 2\mathbf{s}\mathbf{I}_3 = \mathbf{V}_1 \tag{3}$$

The second mesh requires two self-inductance terms and two mutual-inductance terms; the equation cannot be written carelessly. We obtain

$$7\mathbf{s}(\mathbf{I}_2 - \mathbf{I}_1) + 2\mathbf{s}(\mathbf{I}_2 - \mathbf{I}_3) + \frac{1}{\mathbf{s}}\mathbf{I}_2 + 6\mathbf{s}(\mathbf{I}_2 - \mathbf{I}_3) + 2\mathbf{s}(\mathbf{I}_2 - \mathbf{I}_1) = 0$$

or
$$-9\mathbf{s}\mathbf{I}_1 + \left(17\mathbf{s} + \frac{1}{\mathbf{s}}\right)\mathbf{I}_2 - 8\mathbf{s}\mathbf{I}_3 = 0 \tag{4}$$

Finally, for the third mesh,

$$6\mathbf{s}(\mathbf{I}_3 - \mathbf{I}_2) + 2\mathbf{s}(\mathbf{I}_1 - \mathbf{I}_2) + 3\mathbf{I}_3 = 0$$

or
$$2\mathbf{s}\mathbf{I}_1 - 8\mathbf{s}\mathbf{I}_2 + (3 + 6\mathbf{s})\mathbf{I}_2 = 0 \tag{5}$$

Equations (3) to (5) may be solved by any of the conventional methods. ■

Drill Problems

14-1. In the circuit of Fig. 14-9, let $v_s = 20e^{-1000t}$ V. Write an appropriate mesh equation for the (a) left mesh; (b) right mesh.
Ans: $20e^{-1000t} = 3i_1 + 0.002\, di_1/dt - 0.003\, di_2/dt;\ 10i_2 + 0.005\, di_2/dt - 0.003\, di_1/dt = 0$

Figure 14-9

See Drill Probs. 14-1 to 14-3.

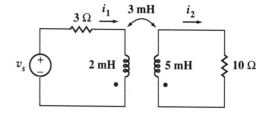

14-2. In the circuit of Fig. 14-9, let $\mathbf{V}_s = 20$ V and $\mathbf{s} = -1000 + j0$ s^{-1}. Write an appropriate mesh equation in terms of the phasor currents $\mathbf{I}_1$ and $\mathbf{I}_2$ for the (a) left mesh; (b) right mesh. Solve these two equations for (c) $\mathbf{I}_1$; (d) $\mathbf{I}_2$.
Ans: $20 = 3\mathbf{I}_1 - 2\mathbf{I}_1 + 3\mathbf{I}_2;\ 10\mathbf{I}_2 - 5\mathbf{I}_2 + 3\mathbf{I}_1 = 0;\ -25$ A; 15 A

14-3. A 15-mH inductor is inserted in series with the 10-Ω resistor of Fig. 14-9. Find all the critical frequencies of $\mathbf{H}(\mathbf{s}) = \mathbf{V}_s(\mathbf{s})/\mathbf{I}_1(\mathbf{s})$.
Ans: -455 Np/s; -2130 Np/s; -500 Np/s; $\pm\infty$ Np/s

Let us now consider the energy stored in a pair of mutually coupled inductors. The results will be useful in several different ways. We shall first justify our assumption that $M_{12} = M_{21}$, and we may then determine the maximum possible value of the mutual inductance between two given inductors.

The pair of coupled coils shown in Fig. 14-3a has currents, voltages, and polarity dots indicated. In order to show that $M_{12} = M_{21}$ we begin by letting all currents and voltages be zero, thus establishing zero initial energy storage in the network. We then open-circuit the right-hand terminal pair and increase i_1 from zero to some constant value I_1 at time $t = t_1$. The power entering the network from the left at any instant is

$$v_1 i_1 = L_1 \frac{di_1}{dt} i_1$$

and that entering from the right is

$$v_2 i_2 = 0$$

since $i_2 = 0$.

The energy stored within the network when $i_1 = I_1$ is thus

$$\int_0^{t_1} v_1 i_1 \, dt = \int_0^{I_1} L_1 i_1 \; di_1 = \frac{1}{2} L_1 I_1^2$$

We now hold i_1 constant, $i_1 = I_1$, and let i_2 change from zero at $t = t_1$ to some constant value I_2 at $t = t_2$. The energy delivered from the right-hand source is thus

$$\int_{t_1}^{t_2} v_2 i_2 \, dt = \int_0^{I_2} L_2 i_2 \, di_2 = \frac{1}{2} L_2 I_2^2$$

However, even though the value of i_1 remains constant, the left-hand source also delivers energy to the network during this time interval:

$$\int_{t_1}^{t_2} v_1 i_1 \, dt = \int_{t_1}^{t_2} M_{12} \frac{di_2}{dt} i_1 \, dt = M_{12} I_1 \int_0^{I_2} di_2 = M_{12} I_1 I_2$$

The total energy stored in the network when both i_1 and i_2 have reached constant values is

$$W_{\text{total}} = \tfrac{1}{2} L_1 I_1^2 + \tfrac{1}{2} L_2 I_2^2 + M_{12} I_1 I_2$$

Now, we may establish the same final currents in this network by allowing the currents to reach their final values in the reverse order, that is, first increasing i_2 from zero to I_2 and then holding i_2 constant while i_1 increases from zero to I_1. If the total energy stored is calculated for this experiment, the result is found to be

$$W_{\text{total}} = \tfrac{1}{2} L_1 I_1^2 + \tfrac{1}{2} L_2 I_2^2 + M_{21} I_1 I_2$$

The only difference is the interchange of the mutual inductances M_{21} and M_{12}. The initial and final conditions in the network are the same, however, and the two values of the stored energy must be identical. Thus,

$$M_{12} = M_{21} = M$$

and
$$W = \tfrac{1}{2} L_1 I_1^2 + \tfrac{1}{2} L_2 I_2^2 + M I_1 I_2 \tag{6}$$

14-3

Energy considerations

If one current enters a dot-marked terminal while the other leaves a dot-marked terminal, the sign of the mutual energy term is reversed:

$$W = \tfrac{1}{2}L_1I_1^2 + \tfrac{1}{2}L_2I_2^2 - MI_1I_2 \tag{7}$$

Although Eqs. (6) and (7) were derived by treating the final values of the two currents as constants, it is apparent that these "constants" may have any value, and the energy expressions correctly represent the energy stored when the *instantaneous* values of i_1 and i_2 are I_1 and I_2, respectively. In other words, lowercase symbols might just as well be used:

$$w(t) = \tfrac{1}{2}L_1[i_1(t)]^2 + \tfrac{1}{2}L_2[i_2(t)]^2 \pm M[i_1(t)][i_2(t)] \tag{8}$$

The only assumption upon which Eq. (8) is based is the logical establishment of a zero-energy reference level when both currents are zero.

Equation (8) may now be used to establish an upper limit for the value of M. Since $w(t)$ represents the energy stored within a *passive* network, it cannot be negative for any values of i_1, i_2, L_1, L_2, or M. Let us assume first that i_1 and i_2 are either both positive or both negative; their product is therefore positive. From Eq. (8), the only case in which the energy could possibly be negative is

$$w = \tfrac{1}{2}L_1i_1^2 + \tfrac{1}{2}L_2i_2^2 - Mi_1i_2$$

which we may write, by completing the square, as

$$w = \tfrac{1}{2}(\sqrt{L_1}\,i_1 - \sqrt{L_2}\,i_2)^2 + \sqrt{L_1L_2}\,i_1i_2 - Mi_1i_2$$

Now the energy cannot be negative; the right-hand side of this equation therefore cannot be negative. The first term, however, may be as small as zero, and thus the sum of the last two terms cannot be negative. Hence,

$$\sqrt{L_1L_2} \geq M$$

or
$$M \leq \sqrt{L_1L_2} \tag{9}$$

There is, therefore, an upper limit to the possible magnitude of the mutual inductance; it can be no larger than the geometric mean of the inductances of the two coils between which the mutual inductance exists. Although we have derived this inequality on the assumption that i_1 and i_2 carried the same algebraic sign, a similar development is possible if the signs are opposite; it is necessary only to select the positive sign in Eq. (8).

We might also have demonstrated the truth of inequality (9) from a physical consideration of the magnetic coupling; if we think of i_2 as being zero and the current i_1 as establishing the magnetic flux linking both L_1 and L_2, it is apparent that the flux within L_2 cannot be greater than the flux within L_1, which represents the total flux. Qualitatively, then, there is an upper limit to the magnitude of the mutual inductance possible between two given inductors. For example, if $L_1 = 1$ H and $L_2 = 10$ H, then $M \leq 3.16$ H.

The degree to which M approaches its maximum value is exactly described by the coefficient of coupling. We define the *coefficient of coupling*, symbolized as k:

$$k = \frac{M}{\sqrt{L_1L_2}} \tag{10}$$

It is evident that

$$0 \leq k \leq 1$$

The larger values of the coefficient of coupling are obtained with coils which are physically closer, which are wound or oriented to provide a larger common magnetic flux, or which are provided with a common path through a material which serves to concentrate and localize the magnetic flux (a high-permeability material). Coils having a coefficient of coupling close to unity are said to be *tightly coupled.*

Drill Problems

14-4. In Fig. 14-3a, let $L_1 = 0.4$ H, $L_2 = 2.5$ H, $k = 0.6$, and $i_1 = 4i_2 = 20 \cos (500t - 20°)$ mA. Evaluate the following quantities at $t = 0$: (a) i_2; (b) v_1; (c) the total energy stored in the system.   *Ans:* 4.70 mA; 1.881 V; 151.2 μJ

14-5. Let $i_s = 2 \cos 10t$ A in Fig. 14-10, and find the total energy stored in the passive network at $t = 0$ if $k = 0.6$ and terminals x and y are (a) left open-circuited; (b) short-circuited.   *Ans:* 0.8 J; 0.512 J

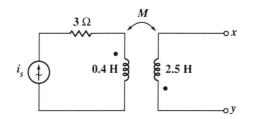

Figure 14-10

See Drill Prob. 14-5.

We are now ready to apply our knowledge of magnetic coupling to an analytical description of the performance of two specific practical devices, each of which may be represented by a model containing mutual inductance. Both of the devices are *transformers,* a term which we may define as a network containing two or more coils which are deliberately coupled magnetically. In this section we shall consider the linear transformer, which happens to be an excellent model for the practical linear transformer used at radio frequencies, or higher frequencies. In the following section we shall consider the ideal transformer, which is an idealized unity-coupled model of a physical transformer that has a core made of some magnetic material, usually an iron alloy.

14-4

The linear transformer

In Fig. 14-11 a transformer is shown with two mesh currents identified. The first mesh, usually containing the source, is called the *primary,* while the second mesh, usually containing the load, is known as the *secondary.* The inductors labeled L_1 and L_2 are also referred to as the primary and secondary, respectively, of the transformer. We shall assume that the transformer is linear. This implies that magnetic material causing a nonlinear flux-versus-current relationship is not employed. Without such material, however, it is difficult to achieve a coefficient of coupling greater than a few tenths. The two resistors serve to account for the resistance of the wire out of which the primary and secondary coils are wound, and any other losses.

In many applications, the linear transformer is used with a tuned, or resonant, secondary; the primary winding is also often operated in a resonant

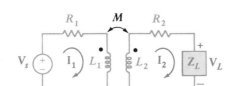

Figure 14-11

A linear transformer containing a source in the primary circuit and a load in the secondary circuit. Resistance is also included in both the primary and the secondary.

condition by replacing the ideal voltage source by a current source in parallel with a large resistance and a capacitance. The analysis of such a single-tuned or double-tuned circuit is a fairly lengthy process, and we shall not undertake it at this time. It may be pointed out, however, that the secondary response is characterized by the familiar resonance curve for relatively small coefficients of coupling, but that a greater control of the shape of the response curve as a function of frequency becomes possible for larger coefficients of coupling. Response curves which possess flatter tops and sharper drops on each side may be achieved in the double-tuned circuit.

Consider the input impedance offered at the terminals of the primary circuit. The two mesh equations are

$$\mathbf{V}_s = \mathbf{I}_1(R_1 + \mathbf{s}L_1) - \mathbf{I}_2\mathbf{s}M \tag{11}$$

$$0 = -\mathbf{I}_1\mathbf{s}M + \mathbf{I}_2(R_2 + \mathbf{s}L_2 + \mathbf{Z}_L) \tag{12}$$

We may simplify by defining

$$\mathbf{Z}_{11} = R_1 + \mathbf{s}L_1 \qquad \mathbf{Z}_{22} = R_2 + \mathbf{s}L_2 + \mathbf{Z}_L$$

and thus

$$\mathbf{V}_s = \mathbf{I}_1\mathbf{Z}_{11} - \mathbf{I}_2\mathbf{s}M \tag{13}$$

$$0 = -\mathbf{I}_1\mathbf{s}M + \mathbf{I}_2\mathbf{Z}_{22} \tag{14}$$

Solving the second equation for $\mathbf{I}_2$ and inserting the result in the first equation enables us to find the input impedance,

$$\mathbf{Z}_{\text{in}} = \frac{\mathbf{V}_s}{\mathbf{I}_1} = \mathbf{Z}_{11} - \frac{\mathbf{s}^2 M^2}{\mathbf{Z}_{22}} \tag{15}$$

Before manipulating this expression any further, we can draw several exciting conclusions. In the first place, this result is independent of the location of the dots on either winding, for if either dot is moved to the other end of the coil, the result is a change in sign of each term involving M in Eqs. (11) to (14). This same effect could be obtained by replacing M by $(-M)$, and such a change cannot affect the input impedance, as Eq. (15) demonstrates. We also may note in Eq. (15) that the input impedance is simply $\mathbf{Z}_{11}$ if the coupling is reduced to zero. As the coupling is increased from zero, the input impedance differs from $\mathbf{Z}_{11}$ by an amount $-\mathbf{s}^2 M^2/\mathbf{Z}_{22}$, termed the *reflected impedance*. The nature of this change is more evident if examined under steady-state sinusoidal operation. Letting $\mathbf{s} = j\omega$,

$$\mathbf{Z}_{\text{in}}(j\omega) = \mathbf{Z}_{11}(j\omega) + \frac{\omega^2 M^2}{R_{22} + jX_{22}}$$

and rationalizing the reflected impedance,

$$\mathbf{Z}_{\text{in}} = \mathbf{Z}_{11} + \frac{\omega^2 M^2 R_{22}}{R_{22}^2 + X_{22}^2} + \frac{-j\omega^2 M^2 X_{22}}{R_{22}^2 + X_{22}^2}$$

Since $\omega^2 M^2 R_{22}/(R_{22}^2 + X_{22}^2)$ must be positive, it is evident that the presence of the secondary increases the losses in the primary circuit. In other words, the presence of the secondary might be accounted for in the primary circuit by increasing the value of R_1. Moreover, the reactance which the secondary *reflects* into the primary circuit has a sign which is opposite to that of X_{22}, the net reactance around the secondary loop. This reactance X_{22} is the sum of ωL_2 and

X_L; it is necessarily positive for inductive loads and either positive or negative for capacitive loads, depending on the magnitude of the load reactance.

Let us consider the effects of this reflected reactance and resistance by considering the special case in which both the primary and the secondary are identical series resonant circuits. Thus, $R_1 = R_2 = R$, $L_1 = L_2 = L$, and the load impedance $\mathbf{Z}_L$ is produced by a capacitance C, identical to a capacitance inserted in series in the primary circuit. The series resonant frequency of either the primary or the secondary alone is thus $\omega_0 = 1/\sqrt{LC}$. At this resonant frequency, the net secondary reactance is zero, the net primary reactance is zero, and no reactance is reflected into the primary by the secondary. The input impedance is therefore a pure resistance, and a resonant condition is present. At a slightly higher frequency, the net primary and secondary reactances are both inductive, and the reflected reactance is therefore capacitive. If the magnetic coupling is sufficiently large, the input impedance may once again be a pure resistance; another resonant condition is achieved, but at a frequency slightly higher than ω_0. A similar condition will also occur at a frequency slightly below ω_0. Each circuit alone is then capacitive, the reflected reactance is inductive, and cancellation may again occur.

Although we are leaping at conclusions, it seems possible that these three adjacent resonances, each similar to a series resonance, may permit a relatively large primary current to flow from the voltage source. The large primary current in turn provides a large induced voltage in the secondary circuit, and a large secondary current. The large secondary current is present over a band of frequencies extending from slightly below to slightly above ω_0, and thus a maximum response is achieved over a wider range of frequencies than is possible in a simple resonant circuit. Such a response curve is obviously desirable if the primary source is some intelligence signal containing energy distributed throughout a band of frequencies, rather than at a single frequency. Such signals are present in AM and FM radio, television, telemetry, radar, and all other communication systems.

It is often convenient to replace a transformer by an equivalent network in the form of a T or Π. If we separate the primary and secondary resistances from the transformer, only the pair of mutually coupled inductors remains, as shown in Fig. 14-12. Note that the two lower terminals of the transformer are con-

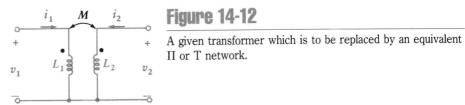

Figure 14-12

A given transformer which is to be replaced by an equivalent Π or T network.

nected together to form a three-terminal network. We do this because both of our equivalent networks are also three-terminal networks. The differential equations describing this circuit are, once again,

$$v_1 = L_1 \frac{di_1}{dt} + M \frac{di_2}{dt} \qquad (16)$$

and

$$v_2 = M \frac{di_1}{dt} + L_2 \frac{di_2}{dt} \qquad (17)$$

The form of these two equations is familiar and may be easily interpreted in terms of mesh analysis. Let us select a clockwise i_1 and a counterclockwise i_2 so that i_1 and i_2 are exactly identifiable with the currents in Fig. 14-12. The terms $M\,di_2/dt$ in Eq. (16) and $M\,di_1/dt$ in Eq. (17) indicate that the two meshes must then have a common *self*-inductance M. Since the total inductance around the left-hand mesh is L_1, a self-inductance of $L_1 - M$ must be inserted in the first mesh, but not in the second mesh. Similarly, a self-inductance of $L_2 - M$ is required in the second mesh, but not in the first mesh. The resultant equivalent network is shown in Fig. 14-13. The equivalence is guaranteed by the identical pairs of equations relating v_1, i_1, v_2, and i_2 for the two networks.

Figure 14-13

The T equivalent of the transformer shown in Fig. 14-12.

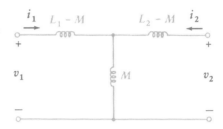

If either of the dots on the windings of the given transformer is placed on the opposite end of its coil, the sign of the mutual terms in Eqs. (16) and (17) will be negative. This is analogous to replacing M by $-M$, and such a replacement in the network of Fig. 14-13 leads to the correct equivalent for this case. The three self-inductance values are now $L_1 + M$, $-M$, and $L_2 + M$.

The inductances in the T equivalent are all self-inductances; no mutual inductance is present. It is possible that negative values of inductance may be obtained for the equivalent circuit, but this is immaterial if our only desire is a mathematical analysis; the actual construction of the equivalent network is, of course, impossible in any form involving a negative inductance. However, there are times when procedures for synthesizing networks to provide a desired transfer function lead to circuits containing a T network having a negative inductance; this network may then be realized by use of an appropriate linear transformer.

Example 14-3　Find the T equivalent of the linear transformer shown in Fig. 14-14a.

Solution: We identify $L_1 = 30$ mH, $L_2 = 60$ mH, and $M = 40$ mH, and note that the dots are both at the upper terminals, as they are in the basic circuit of Fig. 14-12.

Hence, $L_1 - M = -10$ mH is in the upper left arm, $L_2 - M = 20$ mH is at the upper right, and the center stem contains $M = 40$ mH. The complete equivalent T is shown in Fig. 14-14b.

Figure 14-14

(a) A linear transformer used as an example. (b) The T-equivalent network of the transformer.

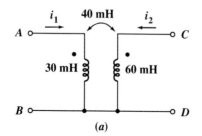

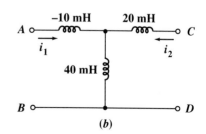

To demonstrate the equivalence, let us leave terminals C and D open-circuited and apply $v_{AB} = 10 \cos 100t$ V to the input in Fig. 14-14a. Thus,

$$i_1 = \frac{1}{30 \times 10^{-3}} \int 10 \cos 100t \, dt = 3.33 \sin 100t \qquad \text{V}$$

and

$$v_{CD} = M \frac{di_1}{dt} = 40 \times 10^{-3} \times 3.33 \times 100 \cos 100t$$

$$= 13.33 \cos 100t \qquad \text{V}$$

Applying the same voltage in the T equivalent, we find that

$$i_1 = \frac{1}{(-10 + 40)10^{-3}} \int 10 \cos 100t \, dt = 3.33 \sin 100t \qquad \text{V}$$

once again. Also, the voltage at C and D is equal to the voltage across the 40-mH inductor. Thus,

$$v_{CD} = 40 \times 10^{-3} \times 3.33 \times 100 \cos 100t = 13.33 \cos 100t \qquad \text{V}$$

and the two networks yield equal results. ∎

The equivalent Π network is not obtained as easily. It is more complicated, and it is not used as much. We develop it by solving Eq. (17) for di_2/dt and substituting the result in Eq. (16):

$$v_1 = L_1 \frac{di_1}{dt} + \frac{M}{L_2} v_2 - \frac{M^2}{L_2} \frac{di_1}{dt}$$

or

$$\frac{di_1}{dt} = \frac{L_2}{L_1 L_2 - M^2} v_1 - \frac{M}{L_1 L_2 - M^2} v_2$$

If we now integrate from 0 to t, we obtain

$$i_1 - i_1(0)u(t) = \frac{L_2}{L_1 L_2 - M^2} \int_0^t v_1 \, dt - \frac{M}{L_1 L_2 - M^2} \int_0^t v_2 \, dt \qquad (18)$$

In a similar fashion, we also have

$$i_2 - i_2(0)u(t) = \frac{-M}{L_1 L_2 - M^2} \int_0^t v_1 \, dt + \frac{L_1}{L_1 L_2 - M^2} \int_0^t v_2 \, dt \qquad (19)$$

Equations (18) and (19) may be interpreted as a pair of nodal equations. A step-current source must be installed at each node in order to provide the proper initial conditions. The factors multiplying each integral are evidently the inverses of certain equivalent inductances. Thus, the second coefficient in Eq. (18), $M/(L_1 L_2 - M^2)$, is $1/L_B$, or the reciprocal of the inductance extending between nodes 1 and 2, as shown on the equivalent Π network, Fig. 14-15. So

$$L_B = \frac{L_1 L_2 - M^2}{M}$$

The first coefficient in Eq. (18), $L_2/(L_1 L_2 - M^2)$, is $1/L_A + 1/L_B$. Thus,

$$\frac{1}{L_A} = \frac{L_2}{L_1 L_2 - M^2} - \frac{M}{L_1 L_2 - M^2}$$

or

$$L_A = \frac{L_1 L_2 - M^2}{L_2 - M}$$

Figure 14-15

The Π network which is equivalent to the transformer shown in Fig. 14-12.

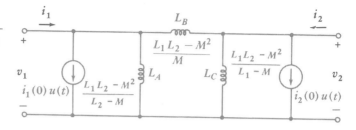

Finally,

$$L_C = \frac{L_1 L_2 - M^2}{L_1 - M}$$

No magnetic coupling is present among the inductors in the equivalent Π, and the initial currents *in the three self-inductances* are zero.

We may compensate for a reversal of either dot in the given transformer by merely changing the sign of M in the equivalent network. Also, just as we found in the equivalent T, negative self-inductances may appear in the equivalent Π network.

Example 14-4　Let us again return to the transformer of Fig. 14-14a, this time finding the equivalent Π. We assume zero initial currents.

Solution: We first evaluate the term $L_1 L_2 - M^2$, obtaining $30 \times 10^{-3} \times 60 \times 10^{-3} - (40 \times 10^{-3})^2 = 2 \times 10^{-4}$ H². Thus, $L_A = (L_1 L_2 - M^2)/(L_2 - M) = 2 \times 10^{-4}/(20 \times 10^{-3}) = 10$ mH, $L_C = (L_1 L_2 - M^2)/(L_1 - M) = -20$ mH, and $L_B = (L_1 L_2 - M^2)/M = 5$ mH. The equivalent Π network is shown in Fig. 14-16.

Figure 14-16

The Π equivalent of the linear transformer shown in Fig. 14-14a. It is assumed that $i_1(0) = 0$ and $i_2(0) = 0$.

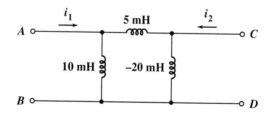

If we again check our result by letting $v_{AB} = 10 \cos 100t$ V with C-D open-circuited, the output voltage is quickly obtained by voltage division:

$$v_{CD} = \frac{-20 \times 10^{-3}}{5 \times 10^{-3} - 20 \times 10^{-3}} 10 \cos 100t = 13.33 \cos 100t \quad \text{V} \qquad \blacksquare$$

Drill Problems

14-6. Element values for a certain linear transformer are $R_1 = 3\ \Omega$, $R_2 = 6\ \Omega$, $L_1 = 2$ mH, $L_2 = 10$ mH, and $M = 4$ mH. If $\omega = 5000$ rad/s, find $\mathbf{Z}_{\text{in}}$ for $\mathbf{Z}_L$ equal to (a) $10\ \Omega$; (b) $j20\ \Omega$; (c) $10 + j20\ \Omega$; (d) $-j20\ \Omega$.
　　　　Ans: $5.32 + j2.74\ \Omega$; $3.49 + j4.33\ \Omega$; $4.24 + j4.57\ \Omega$; $5.56 - j2.82\ \Omega$

14-7. (a) If the networks of Figs. 14-17a and b are equivalent, specify values for $L_x, L_y,$ and L_z. (b) Repeat if the dot on the secondary in Fig. 14-17b is located at the bottom of the coil.　　　　*Ans:* $-1.5, 2.5, 3.5$ H; $5.5, 9.5, -3.5$ H

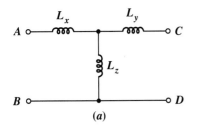

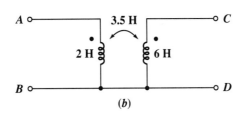

Figure 14-17

See Drill Prob. 7.

(a)

(b)

14-8. Determine equivalent T's for both dot locations in a lossless linear transformer for which $L_1 = 4$ mH, $L_2 = 18$ mH, and $M = 8$ mH, and then use the T's to find the three equivalent input inductances obtained when the secondary is open-circuited, short-circuited, and connected in parallel with the primary.

Ans: 4, 0.444, 1.333 mH; 4, 0.444, 0.211 mH

An ideal transformer is a useful approximation of a very tightly coupled transformer in which the coefficient of coupling is almost unity and both the primary and secondary inductive reactances are extremely large in comparison with the terminating impedances. These characteristics are closely approached by most well-designed iron-core transformers over a reasonable range of frequencies for a reasonable range of terminal impedances. The approximate analysis of a circuit containing an iron-core transformer may be achieved very simply by replacing that transformer by an ideal transformer; the ideal transformer may be thought of as a first-order model of an iron-core transformer.

One new concept arises with the ideal transformer: the *turns ratio a*. The self-inductance of either the primary or the secondary coil is proportional to the square of the number of turns of wire forming the coil. This relationship is valid only if all the flux established by the current flowing in the coil links all the turns. In order to develop this result logically, it is necessary to utilize magnetic field concepts, a subject which is not included in our discussion of circuit analysis. However, a qualitative argument may suffice. If a current I flows through a coil of N turns, then N times the magnetic flux of a single-turn coil will be produced. If we think of the N turns as being coincident, then all the flux certainly links all the turns. As the current and flux change with time, a voltage is then induced *in each turn* which is N times larger than that caused by a single-turn coil. Finally, the voltage induced *in the N-turn coil* must be N^2 times the single-turn voltage. Thus, the proportionality between inductance and the square of the numbers of turns arises. It follows that

$$\frac{L_2}{L_1} = \frac{N_2^2}{N_1^2} = a^2$$

where

$$a = \frac{N_2}{N_1}$$

Figure 14-18 shows an ideal transformer to which a secondary load is connected. The ideal nature of the transformer is established by several conventions: the use of the vertical lines between the two coils to indicate the iron laminations present in many iron-core transformers, the unity value of the

14-5

The ideal transformer

Figure 14-18

An ideal transformer is connected to a general load impedance.

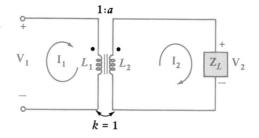

coupling coefficient, and the presence of the symbol $1:a$, suggesting a turns ratio of N_1 to N_2.

Let us analyze this transformer in the sinusoidal steady state in order to interpret our assumptions in the simplest context. The two mesh equations are

$$\mathbf{V}_1 = \mathbf{I}_1 j\omega L_1 - \mathbf{I}_2 j\omega M \tag{20}$$

$$0 = -\mathbf{I}_1 j\omega M + \mathbf{I}_2(\mathbf{Z}_L + j\omega L_2) \tag{21}$$

We first determine the input impedance of an ideal transformer. Although we shall let the self-inductance of each winding become infinite, the input impedance will remain finite. By solving Eq. (21) for $\mathbf{I}_2$ and substituting in Eq. (20), we obtain

$$\mathbf{V}_1 = \mathbf{I}_1 j\omega L_1 + \mathbf{I}_1 \frac{\omega^2 M^2}{\mathbf{Z}_L + j\omega L_2}$$

and
$$\mathbf{Z}_{\text{in}} = \frac{\mathbf{V}_1}{\mathbf{I}_1} = j\omega L_1 + \frac{\omega^2 M^2}{\mathbf{Z}_L + j\omega L_2}$$

Since $k = 1$, $M^2 = L_1 L_2$ and

$$\mathbf{Z}_{\text{in}} = j\omega L_1 + \frac{\omega^2 L_1 L_2}{\mathbf{Z}_L + j\omega L_2}$$

We now must let both L_1 and L_2 tend to infinity. Their ratio, however, remains finite, as specified by the turns ratio. Thus,

$$L_2 = a^2 L_1$$

and
$$\mathbf{Z}_{\text{in}} = j\omega L_1 + \frac{\omega^2 a^2 L_1^2}{\mathbf{Z}_L + j\omega a^2 L_1}$$

Now if we let L_1 become infinite, both of the terms on the right-hand side of the preceding expression become infinite, and the result is indeterminate. It is necessary to first combine these two terms:

$$\mathbf{Z}_{\text{in}} = \frac{j\omega L_1 \mathbf{Z}_L - \omega^2 a^2 L_1^2 + \omega^2 a^2 L_1^2}{\mathbf{Z}_L + j\omega a^2 L_1} \tag{22}$$

or
$$\mathbf{Z}_{\text{in}} = \frac{j\omega L_1 \mathbf{Z}_L}{\mathbf{Z}_L + j\omega a^2 L_1} \tag{23}$$

Now as L_1 becomes infinite, it is apparent that $\mathbf{Z}_{\text{in}}$ becomes

$$\mathbf{Z}_{\text{in}} = \frac{\mathbf{Z}_L}{a^2} \tag{24}$$

for finite $\mathbf{Z}_L$.

This result has some interesting implications, and at least one of them appears to contradict one of the characteristics of the linear transformer. It should not, of course, since the linear transformer represents the more general case. The input impedance of the ideal transformer is an impedance which is proportional to the load impedance, the proportionality constant being the reciprocal of the square of the turns ratio. In other words, if the load impedance is a capacitive impedance, then the input impedance is a capacitive impedance. In the linear transformer, however, the reflected impedance suffered a sign change in its reactive part; a capacitive load led to an inductive contribution to the input impedance. The explanation of this occurrence is achieved by first realizing that $\mathbf{Z}_L/a^2$ is *not* the reflected impedance, although it is often loosely called by that name. The true reflected impedance is infinite in the ideal transformer; otherwise it could not "cancel" the infinite impedance of the primary inductance. This cancellation occurs in the numerator of Eq. (22). The impedance $\mathbf{Z}_L/a^2$ represents a small term which is the amount by which an exact cancellation does not occur. The true reflected impedance in the ideal transformer *does* change sign in its reactive part; as the primary and secondary inductances become infinite, however, the effect of the infinite primary-coil reactance and the infinite, but negative, reflected reactance of the secondary coil is one of cancellation.

The first important characteristic of the ideal transformer is therefore its ability to change the magnitude of an impedance, or to change impedance level. An ideal transformer having 100 primary turns and 10 000 secondary turns has a turns ratio of 10 000/100, or 100. Any impedance placed across the secondary then appears at the primary terminals reduced in magnitude by a factor of 100^2, or 10 000. A 20 000-Ω resistor looks like 2 Ω, a 200-mH inductor looks like 20 μH, and a 100-pF capacitor looks like 1 μF. If the primary and secondary windings are interchanged, then $a = 0.01$ and the load impedance is apparently increased in magnitude. In practice, this exact change in magnitude does not always occur, for we must remember that as we took the last step in our derivation and allowed L_1 to become infinite in Eq. (23), it was necessary to neglect $\mathbf{Z}_L$ in comparison with $j\omega L_1$. Since L_2 can never be infinite, it is evident that the ideal transformer model will become invalid if the load impedances are very large.

A practical example of the use of an iron-core transformer as a device for changing impedance level is in the coupling of an audio power amplifier to a speaker system. In order to achieve maximum power transfer, we know that the resistance of the load should be equal to the internal resistance of the source; the speaker usually has an impedance magnitude (often assumed to be a resistance) of only a few ohms, while the power amplifier possesses an internal resistance of several thousand ohms. An ideal transformer is required in which $N_2 < N_1$. For example, if the amplifier (or generator) internal impedance is 4000 Ω and the speaker impedance is 8 Ω, then we desire that

$$\mathbf{Z}_g = 4000 = \frac{\mathbf{Z}_L}{a^2} = \frac{8}{a^2}$$

or

$$a = \frac{1}{22.4}$$

and thus

$$\frac{N_1}{N_2} = 22.4$$

There is also a simple relationship between the primary and secondary currents $\mathbf{I}_1$ and $\mathbf{I}_2$ in an ideal transformer. From Eq. (21),

$$\frac{\mathbf{I}_2}{\mathbf{I}_1} = \frac{j\omega M}{\mathbf{Z}_L + j\omega L_2}$$

We allow L_2 to become infinite, and then it follows that

$$\frac{\mathbf{I}_2}{\mathbf{I}_1} = \frac{j\omega M}{j\omega L_2} = \sqrt{\frac{L_1}{L_2}}$$

or

$$\frac{\mathbf{I}_2}{\mathbf{I}_1} = \frac{1}{a} \tag{25}$$

The ratio of the primary and secondary currents is the turns ratio. If we have $N_2 > N_1$, then $a > 1$, and it is apparent that the larger current flows in the winding with the smaller number of turns. In other words,

$$N_1\mathbf{I}_1 = N_2\mathbf{I}_2$$

It should also be noted that the current ratio is the negative of the turns ratio if either current is reversed or if either dot location is changed.

In our example in which an ideal transformer was used to change the impedance level to efficiently match a speaker to a power amplifier, an rms current of 50 mA at 1000 Hz in the primary causes an rms current of 1.12 A at 1000 Hz in the secondary. The power delivered to the speaker is $(1.12)^2(8)$, or 10 W, and the power delivered to the transformer by the power amplifier is $(0.05)^2(4000)$, or 10 W. The result is comforting, since the ideal transformer contains neither an active device which can generate power nor any resistor which can absorb power.

Since the power delivered to the ideal transformer is identical with that delivered to the load, whereas the primary and secondary currents are related by the turns ratio, it is obvious that the primary and secondary voltages must also be related to the turns ratio. If we define the secondary voltage, or load voltage, as

$$\mathbf{V}_2 = \mathbf{I}_2\mathbf{Z}_L$$

and the primary voltage as the voltage across L_1, then

$$\mathbf{V}_1 = \mathbf{I}_1\mathbf{Z}_{\text{in}} = \mathbf{I}_1\frac{\mathbf{Z}_L}{a^2}$$

The ratio of the two voltages then becomes

$$\frac{\mathbf{V}_2}{\mathbf{V}_1} = a^2\frac{\mathbf{I}_2}{\mathbf{I}_1}$$

or

$$\frac{\mathbf{V}_2}{\mathbf{V}_1} = a = \frac{N_2}{N_1} \tag{26}$$

The ratio of the secondary to the primary voltage is equal to the turns ratio. This ratio may also be negative if either voltage is reversed or either dot location is changed.

Combining the voltage and current ratios, Eqs. (25) and (26),

$$\mathbf{V}_2\mathbf{I}_2 = \mathbf{V}_1\mathbf{I}_1$$

and we see that the primary and secondary complex voltamperes are equal. The magnitude of this product is usually specified as a maximum allowable value on power transformers. If the load has a phase angle θ, or

$$\mathbf{Z}_L = |\mathbf{Z}_L| \underline{/\theta}$$

then $\mathbf{V}_2$ leads $\mathbf{I}_2$ by an angle θ. Moreover, the input impedance is $\mathbf{Z}_L/a^2$, and thus $\mathbf{V}_1$ also leads $\mathbf{I}_1$ by the same angle θ. If we let the voltage and current represent rms values, then $|\mathbf{V}_2| \, |\mathbf{I}_2| \cos \theta$ must equal $|\mathbf{V}_1| \, |\mathbf{I}_1| \cos \theta$, and all the power delivered to the primary terminals reaches the load; none is absorbed by or delivered to the ideal transformer.

The characteristics of the ideal transformer which we have obtained have all been determined by frequency-domain analysis. They are certainly true in the sinusoidal steady state, but we have no reason to believe that they are correct for the complete response. Actually, they are applicable in general, and the demonstration that this statement is true is much simpler than the frequency-domain analysis we have just completed. Our analysis, however, has served to point out the specific approximations which must be made on a more exact model of an actual transformer in order to obtain an ideal transformer. For example, we have seen that the reactance of the secondary winding must be much greater in magnitude than the impedance of any load which is connected to the secondary. Some feeling for those operating conditions under which a transformer ceases to behave as an ideal transformer is thus achieved.

Let us now determine how the time-domain quantities v_1 and v_2 are related in the ideal transformer. Returning to the circuit shown in Fig. 14-12 and the two equations, (16) and (17), describing it, we may solve the second equation for di_2/dt and substitute in the first equation:

$$v_1 = L_1 \frac{di_1}{dt} + \frac{M}{L_2} v_2 - \frac{M^2}{L_2} \frac{di_1}{dt}$$

However, for unity coupling, $M^2 = L_1 L_2$, and thus

$$v_1 = \frac{M}{L_2} v_2 = \sqrt{\frac{L_1}{L_2}} v_2 = \frac{1}{a} v_2$$

The relationship between primary and secondary voltage is thus found to apply to the complete time-domain response.

An expression relating primary and secondary current in the time domain is most quickly obtained by dividing Eq. (16) throughout by L_1,

$$\frac{v_1}{L_1} = \frac{di_1}{dt} + \frac{M}{L_1} \frac{di_2}{dt} = \frac{di_1}{dt} + a \frac{di_2}{dt}$$

and then invoking one of the hypotheses underlying the ideal transformer: L_1 must be infinite. If we assume that v_1 is not infinite, then

$$\frac{di_1}{dt} = -a \frac{di_2}{dt}$$

Integrating,

$$i_1 = -a i_2 + A$$

where A is a constant of integration which does not vary with time. Thus, if we neglect any direct currents in the two windings and fix our attention only on the time-varying portion of the response, then

$$i_1 = -ai_2$$

The minus sign arises, of course, from the placement of the dots and selection of the current directions in Fig. 14-12.

The same current and voltage relationships are therefore obtained in the time domain as were obtained previously in the frequency domain, provided that dc components are ignored. The time-domain results are more general, but they have been obtained by a less informative process.

The characteristics of the ideal transformer which we have established may be utilized to simplify circuits in which ideal transformers appear. Let us assume, for purposes of illustration, that everything to the left of the primary terminals has been replaced by its Thévenin equivalent, as has the network to the right of the secondary terminals. We thus consider the circuit shown in Fig. 14-19. Excitation at any complex frequency **s** is assumed.

Figure 14-19

The networks connected to the primary and secondary terminals of an ideal transformer are represented by their Thévenin equivalents.

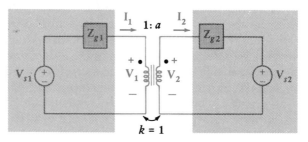

Thévenin's or Norton's theorem may now be used to achieve an equivalent circuit which does not contain a transformer. For example, let us determine the Thévenin equivalent of the network to the left of the secondary terminals. Open-circuiting the secondary, $\mathbf{I}_2 = 0$ and therefore $\mathbf{I}_1 = 0$ (remember that L_1 is infinite). No voltage appears across $\mathbf{Z}_{g1}$, and thus $\mathbf{V}_1 = \mathbf{V}_{s1}$ and $\mathbf{V}_{2oc} = a\mathbf{V}_{s1}$. The Thévenin impedance is obtained by killing $\mathbf{V}_{s1}$ and utilizing the square of the turns ratio, being careful to use the reciprocal turns ratio, since we are looking in at the secondary terminals. Thus, $\mathbf{Z}_{th2} = \mathbf{Z}_{g1}a^2$. As a check on our equivalent, let us also determine the short-circuit secondary current $\mathbf{I}_{2sc}$. With the secondary short-circuited, the primary generator faces an impedance of $\mathbf{Z}_{g1}$, and, thus, $\mathbf{I}_1 = \mathbf{V}_{s1}/\mathbf{Z}_{g1}$. Therefore, $\mathbf{I}_{2sc} = \mathbf{V}_{s1}/a\mathbf{Z}_{g1}$. The ratio of the open-circuit voltage to the short-circuit current is $a^2\mathbf{Z}_{g1}$, as it should be. The Thévenin equivalent of the transformer and the primary circuit is shown in the circuit of Fig. 14-20.

Figure 14-20

The Thévenin equivalent of the network to the left of the secondary terminals in Fig. 14-19 is used to simplify that circuit.

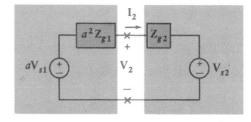

Each primary voltage may therefore be multiplied by the turns ratio, each primary current divided by the turns ratio, and each primary impedance multiplied by the square of the turns ratio; and then these modified voltages, currents, and impedances replace the given voltages, currents, and impedances plus the transformer. If either dot is interchanged, the equivalent may be obtained by using the negative of the turns ratio.

Note that this equivalence, as illustrated by Fig. 14-20, is possible only if the network connected to the two primary terminals, and that connected to the two secondary terminals, can be replaced by their Thévenin equivalents. That is, each must be a two-terminal network. For example, if we cut the two primary leads at the transformer, the circuit must be divided into two separate networks; there can be no element or network bridging across the transformer between primary and secondary, such as those shown in Fig. 14-52 or Fig. 14-54 in the problems at the end of this chapter.

A similar analysis of the transformer and the secondary network shows that everything to the right of the primary terminals may be replaced by an identical network without the transformer, each voltage being divided by a, each current being multiplied by a, and each impedance being divided by a^2. A reversal of either winding corresponds to the use of a turns ratio of $-a$.

Example 14-5 As a simple example of this application of equivalent circuits, consider the circuit given in Fig. 14-21 . We seek the equivalent circuit in which the transformer and the secondary circuit are replaced, and also that in which the transformer and the primary circuit are replaced.

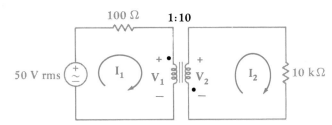

Figure 14-21

A simple circuit in which a resistive load is matched to the source impedance by means of an ideal transformer.

Solution: Let $a = 10$. The input impedance is $10\,000/(10)^2$, or $100\ \Omega$. Thus $I_1 = 0.25$ A, $V_1 = 25$ V, and the source delivers 12.5 W, of which 6.25 W is dissipated in the internal resistance of the source and 6.25 W is delivered to the load. This is the condition for maximum power transfer to the load.

If the secondary circuit and the ideal transformer are removed by the use of the Thévenin equivalent, the simplified circuit of Fig. 14-22a is obtained. The primary current and voltage are now immediately evident.

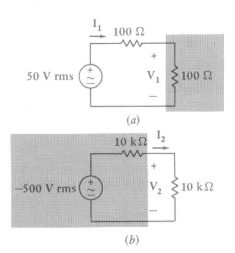

(a)

(b)

Figure 14-22

The circuit of Fig. 14-21 is simplified by replacing (a) the transformer and secondary circuit by the Thévenin equivalent or (b) the transformer and primary circuit by the Thévenin equivalent.

If, instead, the network to the left of the secondary terminals is replaced by its Thévenin equivalent, the simpler circuit of Fig. 14-22*b* is obtained. The presence of the minus sign on the equivalent source should be verified. The corresponding Norton equivalents may also be obtained easily. ∎

Drill Problems

14-9. Let $N_1 = 1000$ turns and $N_2 = 5000$ turns in the ideal transformer shown in Fig. 14-23. If $\mathbf{Z}_L = 500 - j400 \ \Omega$, find the average power delivered to $\mathbf{Z}_L$ for (*a*) $\mathbf{I}_2 = 1.4\underline{/20°}$ A rms; (*b*) $\mathbf{V}_2 = 900\underline{/40°}$ V rms; (*c*) $\mathbf{V}_1 = 80\underline{/100°}$ V rms; (*d*) $\mathbf{I}_1 = 6\underline{/45°}$ A rms; (*e*) $\mathbf{V}_s = 200\underline{/0°}$ V rms.

Ans: 980 W; 988 W; 195.1 W; 720 W; 692 W

Figure 14-23

See Drill Prob. 14-9.

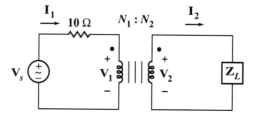

14-10. Find the average power delivered to each of the four resistors in the circuit of Fig. 14-24.   *Ans:* 191.4 W; 73.2 W; 61.0 W; 549 W

Figure 14-24

See Drill Prob. 14-10.

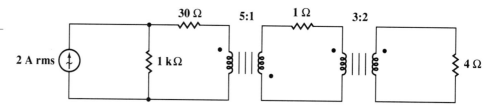

Problems

1 The physical construction of three pairs of coupled coils is shown in Fig. 14-25. Show the two different possible locations for the two dots on each pair of coils.

Figure 14-25

See Prob. 1.

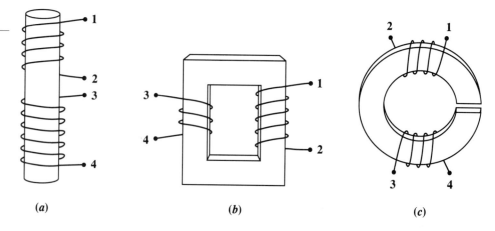

    (*a*)   (*b*)   (*c*)

2 Let $i_{s1}(t) = 4t$ A and $i_{s2}(t) = 10t$ A in the circuit shown in Fig. 14-26. Find (*a*) v_{AG}; (*b*) v_{CG}; (*c*) v_{BG}.

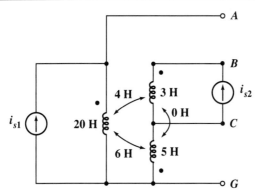

Figure 14-26

See Prob. 2.

3 In the circuit shown in Fig. 14-27, find the average power absorbed by (*a*) the source; (*b*) each of the two resistors; (*c*) each of the two inductances; (*d*) the mutual inductance.

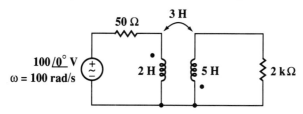

Figure 14-27

See Probs. 3 and 5.

4 (*a*) Write a set of mesh equations in terms of $\mathbf{I}_1(\mathbf{s})$, $\mathbf{I}_2(\mathbf{s})$, and $\mathbf{I}_3(\mathbf{s})$ for the circuit shown in Fig. 14-28. (*b*) Find $\mathbf{I}_3$ if $\mathbf{s} = -1$ Np/s.

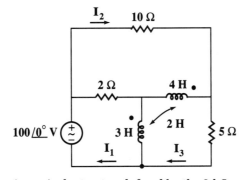

Figure 14-28

See Prob. 4.

5 (*a*) Find the Thévenin equivalent network faced by the 2-kΩ resistor in the circuit of Prob. 3. (*b*) What is the maximum average power that can be drawn from the network by an optimum value of $\mathbf{Z}_L$ (instead of 2 kΩ)?

6 Refer to the network shown in Fig. 14-29 and (*a*) write two equations giving $v_A(t)$ and $v_B(t)$ as functions of $i_1(t)$ and $i_2(t)$ for the network of Fig. 14-29*a*; (*b*) write two equations giving $\mathbf{V}_1(\mathbf{s})$ and $\mathbf{V}_2(\mathbf{s})$ as functions of $\mathbf{I}_A(\mathbf{s})$ and $\mathbf{I}_B(\mathbf{s})$ for the network of Fig. 14-29*b*.

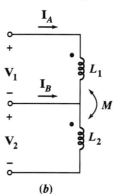

Figure 14-29

See Prob. 6.

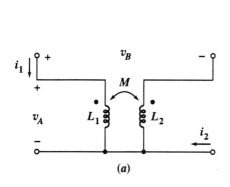

(*a*)

(*b*)

7 Find $i_C(t)$ for $t > 0$ in the circuit of Fig. 14-30 if $v_s(t) = 10t^2u(t)/(t^2 + 0.01)$ V.

Figure 14-30

See Prob. 7.

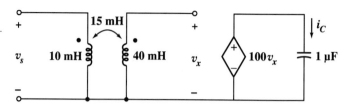

8 (a) Find $\mathbf{Z}_{in}(\mathbf{s})$ for the network of Fig. 14-31. (b) List all the critical frequencies of $\mathbf{Z}_{in}(\mathbf{s})$. (c) Find $\mathbf{Z}_{in}(j\omega)$ for $\omega = 50$ rad/s.

Figure 14-31

See Prob. 8.

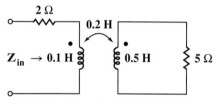

9 Note that there is no mutual coupling between the 5-H and 6-H inductors in the circuit of Fig. 14-32. (a) Write a set of equations in terms of $\mathbf{I}_1(\mathbf{s})$, $\mathbf{I}_2(\mathbf{s})$, and $\mathbf{I}_3(\mathbf{s})$. (b) Find $\mathbf{I}_3(\mathbf{s})$ if $\mathbf{s} = -2$ Np/s.

Figure 14-32

See Prob. 9.

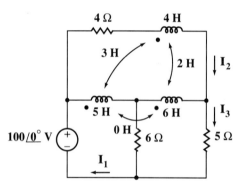

10 Let $i_{s1} = 2 \cos 10t$ A and $i_{s2} = 1.2 \cos 10t$ A in Fig. 14-33. Find (a) $v_1(t)$; (b) $v_2(t)$; (c) the average power being supplied by each source.

Figure 14-33

See Prob. 10.

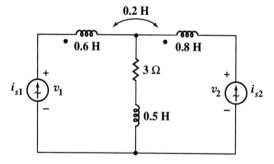

11 Find $\mathbf{I}_L$ in the circuit shown in Fig. 14-34.

Figure 14-34

See Prob. 11.

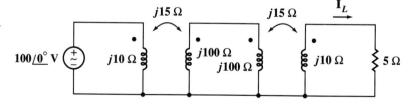

12 Let $i_s = 2 \cos 10t$ A in the circuit of Fig. 14-35. Find the total energy stored at $t = 0$ if (a) a-b is open-circuited as shown; (b) a-b is short-circuited.

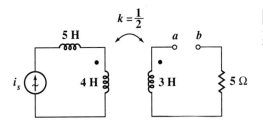

Figure 14-35

See Prob. 12.

13 Let $\mathbf{V}_s = 12/\underline{0°}$ V rms in the linear transformer of Fig. 14-36. With $\omega = 100$ rad/s, find the average power supplied to the 24-Ω resistor as a function of k.

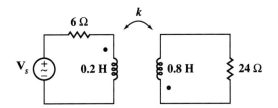

Figure 14-36

See Prob. 13.

14 If $i_1 = 2 \cos 500t$ A in the network of Fig. 14-37, find the value of the maximum energy stored in the network.

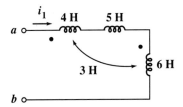

Figure 14-37

See Prob. 14.

15 Let $\omega = 100$ rad/s in the circuit of Fig. 14-38 and find the average power: (a) delivered to the 10-Ω load; (b) delivered to the 20-Ω load; (c) generated by the source.

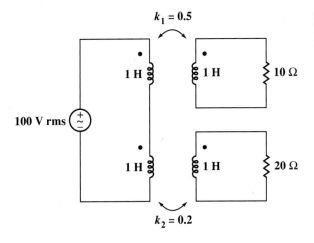

Figure 14-38

See Probs. 15 and 36.

16 Use the equivalent T to help determine the input impedance $\mathbf{Z}(\mathbf{s})$ for the network shown in Fig. 14-39.

Figure 14-39

See Prob. 16.

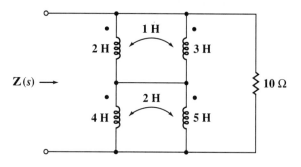

17 Let $\omega = 1000$ rad/s for the circuit of Fig. 14-40 and determine the value of the ratio $\mathbf{V}_2/\mathbf{V}_s$ if (a) $L_1 = 1$ mH, $L_2 = 25$ mH, and $k = 1$; (b) $L_1 = 1$ H, $L_2 = 25$ H, and $k = 0.99$; (c) $L_1 = 1$ H, $L_2 = 25$ H, and $k = 1$.

Figure 14-40

See Prob. 17.

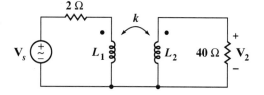

18 A load $\mathbf{Z}_L$ is connected to the secondary of a linear transformer that is characterized by inductances $L_1 = 1$ H and $L_2 = 4$ H and a unity coefficient of coupling. If $\omega = 1000$ rad/s, find the equivalent series network (R, L, and C values) seen at the input terminals if $\mathbf{Z}_L$ is represented by (a) 100 Ω; (b) 0.1 H; (c) 10 μF.

19 (a) An inductance bridge used on the coupled coils of Fig. 14-41 measures the following values under short-circuit or open-circuit conditions: $L_{AB,CD\ oc} = 10$ mH, $L_{CD,\ AB\ oc} = 5$ mH, $L_{AB,CD\ sc} = 8$ mH. Find k. (b) Assuming dots at A and D, and with $i_1 = 5$ A, what value should i_2 have in order for 100 mJ to be stored in the system?

Figure 14-41

See Prob. 19.

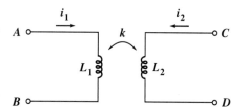

20 Find the equivalent inductance seen at terminals 1 and 2 in the network of Fig. 14-42 if the following terminals are connected together: (a) none; (b) A to B; (c) B to C; (d) A to C.

Figure 14-42

See Prob. 20.

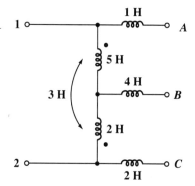

21 (*a*) Refer to Fig. 14-43 and use the equivalent T to help find the ratio $\mathbf{I}_L(\mathbf{s})/\mathbf{V}_s(\mathbf{s})$. (*b*) Set $v_s(t) = 100u(t)$ V and find $i_L(t)$. [Hint: You may wish to write the two differential equations for the circuit to help find di_L/dt at $t = 0^+$.]

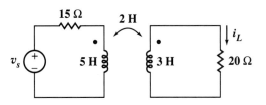

Figure 14-43

See Prob. 21.

22 A linear transformer has $L_1 = 6$ H, $L_2 = 12$ H, and $M = 5$ H. Find the eight different values of L_{in} that can be obtained for the eight different possible methods for obtaining a two-terminal network (single inductances, series and parallel combinations, short-circuited transformers, various dot combinations). Show each network and give its L_{in}.

23 Find $\mathbf{H}(\mathbf{s}) = \mathbf{V}_o/\mathbf{V}_s$ for the circuit shown in Fig. 14-44.

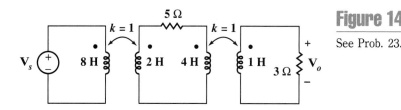

Figure 14-44

See Prob. 23.

24 Let $\mathbf{V}_s = 100\underline{/0°}$ V rms and $\omega = 100$ rad/s in the circuit of Fig. 14-45. Find the Thévenin equivalent of the network: (*a*) to the right of terminals a and b; (*b*) to the left of terminals c and d.

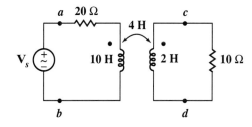

Figure 14-45

See Prob. 24.

25 Repeat Prob. 24 if L_1 is increased to 125 H, L_2 is increased to 20 H, and M is increased until $k = 1$.

26 Let $\mathbf{V}_s = 50\underline{/0°}$ V rms with $\omega = 100$ rad/s in the circuit of Fig. 14-46. Find the average power delivered to each resistor.

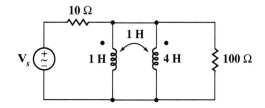

Figure 14-46

See Prob. 26.

27 (*a*) What is the maximum value of average power that can be delivered to R_L in the circuit shown in Fig. 14-47? (*b*) Let $R_L = 100$ Ω and connect a 40-Ω resistor between the upper terminals of the primary and secondary. Find P_L.

Figure 14-47

See Prob. 27.

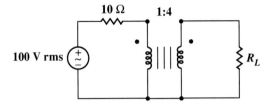

28 Find the average power delivered to the 8-Ω load in the circuit of Fig. 14-48 if c equals (*a*) 0; (*b*) 0.04 S; (*c*) -0.04 S.

Figure 14-48

See Prob. 28.

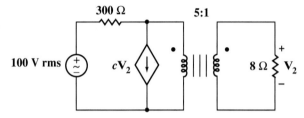

29 Find the Thévenin equivalent at terminals a and b for the network shown in Fig. 14-49.

Figure 14-49

See Prob. 29.

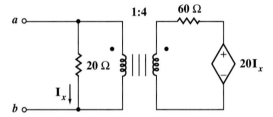

30 Select values for a and b in the circuit of Fig. 14-50 so that the ideal source supplies 1000 W, half of which is delivered to the 100-Ω load.

Figure 14-50

See Prob. 30.

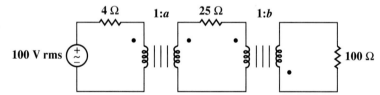

31 For the circuit shown in Fig. 14-51, find (*a*) $\mathbf{I}_1$; (*b*) $\mathbf{I}_2$; (*c*) $\mathbf{I}_3$; (*d*) $P_{25\Omega}$; (*e*) $P_{2\Omega}$; (*f*) $P_{3\Omega}$.

Figure 14-51

See Prob. 31.

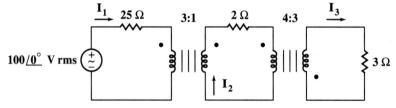

32 Find $\mathbf{V}_2$ in the circuit of Fig. 14-52.

Figure 14-52

See Prob. 32.

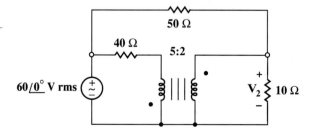

33 Find the power being dissipated in each resistor in the circuit of Fig. 14-53.

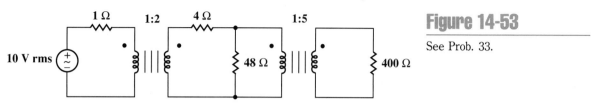

Figure 14-53

See Prob. 33.

34 Find $\mathbf{I}_x$ in the circuit of Fig. 14-54.

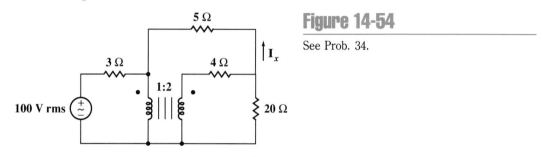

Figure 14-54

See Prob. 34.

35 (*a*) Find the average power delivered to each 10-Ω resistor in the circuit shown in Fig. 14-55. (*b*) Repeat after connecting *A* to *C* and *B* to *D*.

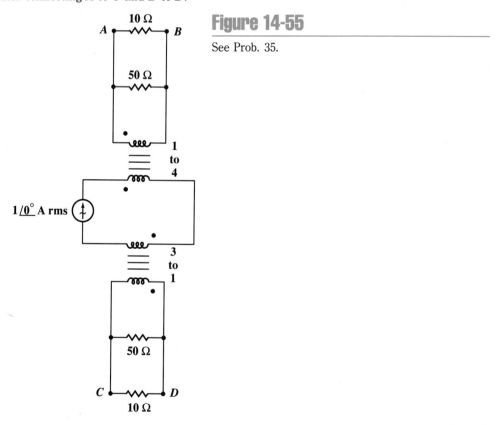

Figure 14-55

See Prob. 35.

36 (SPICE) Let $\omega = 100$ rad/s in the circuit shown in Fig. 14-38. Use a SPICE analysis to find the rms current amplitude that leaves the upper source terminal if the circuit is modified by placing a small 10^{-6}-Ω resistor in series with the source and (*a*) connecting the lower terminals of the two secondary coils and the lower terminal of the source together; (*b*) same as *a*, but also install a 1-μF capacitor in parallel with the 10-Ω load.

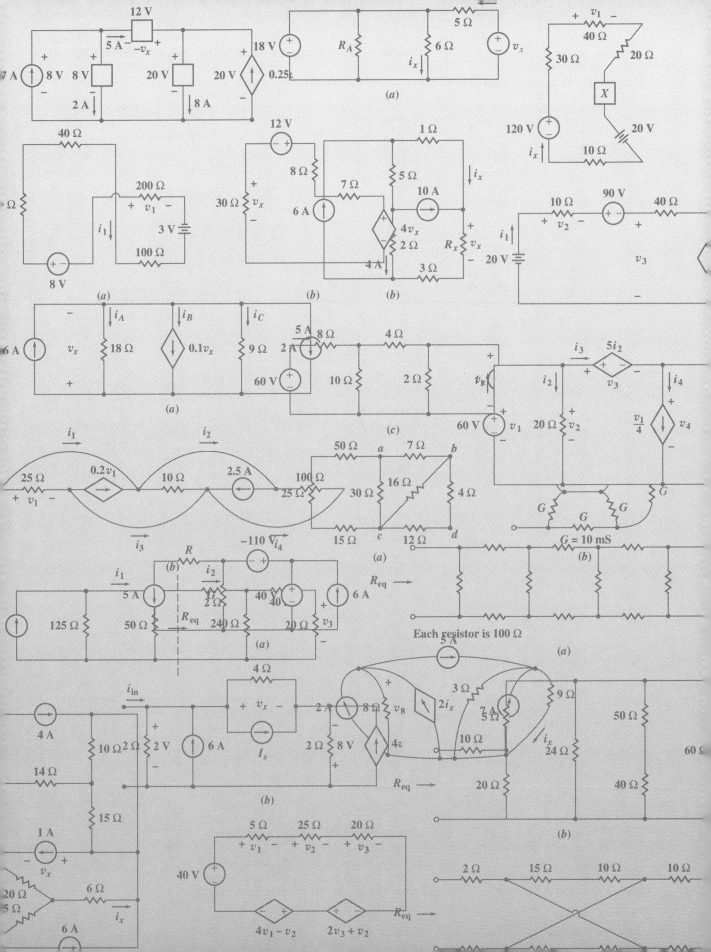

General Two-Port Networks

A general network having two pairs of terminals, one perhaps labeled the "input terminals" and the other the "output terminals," is a very important building block in electronic systems, communication systems, automatic control systems, transmission and distribution systems, or other systems in which an electrical signal or electric energy enters the input terminals, is acted upon by the network, and leaves via the output terminals. The output terminal pair may very well connect with the input terminal pair of another network. A pair of terminals at which a signal may enter or leave a network is called a *port,* and a network having only one such pair of terminals is called a *one-port network,* or simply a *one-port.* No connections may be made to any other nodes internal to the one-port, and it is therefore evident that i_a must equal i_b in the one-port shown in Fig. 15-1a. When more than one pair of terminals is present, the network is known as a *multiport network.* The two-port network to which this chapter is principally devoted is shown in Fig. 15-1b. The currents in the two leads making up each port must be equal, and it follows that $i_a = i_b$ and $i_c = i_d$ in the two-port shown in Fig. 15-1b. Sources and loads must be connected directly across the two terminals of a port if the methods of this chapter are to be used. In other words, each port can be connected only to a one-port network or to a port of another multiport network. For example, no device may be connected between terminals a and c of the two-port network in Fig. 15-1b. If such a circuit must be analyzed, general loop or nodal equations should usually be written.

The special methods of analysis which are developed for two-port networks, or simply two-ports, emphasize the current and voltage relationships at the terminals of the networks and suppress the specific nature of the currents and voltages within the networks. Our introductory study should serve to acquaint us with a number of important parameters and their use in simplifying and systematizing linear two-port network analysis.

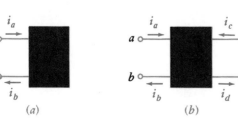

(a) i_a i_b

(b) i_a i_c a b c d i_b i_d

Figure 15-1

(*a*) A one-port network. (*b*) A two-port network.

15-2

One-port networks

Some of the introductory study of one- and two-port networks is accomplished best by using a generalized network notation and the abbreviated nomenclature for determinants introduced in Appendix 1. Thus, if we write a set of loop equations for a passive network,

$$\mathbf{Z}_{11}\mathbf{I}_1 + \mathbf{Z}_{12}\mathbf{I}_2 + \mathbf{Z}_{13}\mathbf{I}_3 + \cdots + \mathbf{Z}_{1N}\mathbf{I}_N = \mathbf{V}_1$$
$$\mathbf{Z}_{21}\mathbf{I}_1 + \mathbf{Z}_{22}\mathbf{I}_2 + \mathbf{Z}_{23}\mathbf{I}_3 + \cdots + \mathbf{Z}_{2N}\mathbf{I}_N = \mathbf{V}_2$$
$$\mathbf{Z}_{31}\mathbf{I}_1 + \mathbf{Z}_{32}\mathbf{I}_2 + \mathbf{Z}_{33}\mathbf{I}_3 + \cdots + \mathbf{Z}_{3N}\mathbf{I}_N = \mathbf{V}_3 \qquad (1)$$
$$\cdots\cdots\cdots\cdots\cdots\cdots\cdots\cdots$$
$$\mathbf{Z}_{N1}\mathbf{I}_1 + \mathbf{Z}_{N2}\mathbf{I}_2 + \mathbf{Z}_{N3}\mathbf{I}_3 + \cdots + \mathbf{Z}_{NN}\mathbf{I}_N = \mathbf{V}_N$$

then the coefficient of each current will be an impedance $\mathbf{Z}_{ij}(\mathbf{s})$, and the circuit determinant, or determinant of the coefficients, is

$$\Delta_{\mathbf{Z}} = \begin{vmatrix} \mathbf{Z}_{11} & \mathbf{Z}_{12} & \mathbf{Z}_{13} & \cdots & \mathbf{Z}_{1N} \\ \mathbf{Z}_{21} & \mathbf{Z}_{22} & \mathbf{Z}_{23} & \cdots & \mathbf{Z}_{2N} \\ \mathbf{Z}_{31} & \mathbf{Z}_{32} & \mathbf{Z}_{33} & \cdots & \mathbf{Z}_{3N} \\ \cdots & \cdots & \cdots & \cdots & \cdots \\ \mathbf{Z}_{N1} & \mathbf{Z}_{N2} & \mathbf{Z}_{N3} & \cdots & \mathbf{Z}_{NN} \end{vmatrix} \qquad (2)$$

where N loops have been assumed, the currents appear in subscript order in each equation, and the order of the equations is the same as that of the currents. We also assume that KVL is applied so that the sign of each $\mathbf{Z}_{ii}$ term ($\mathbf{Z}_{11}$, $\mathbf{Z}_{22}$, ..., $\mathbf{Z}_{NN}$) is positive; the sign of any $\mathbf{Z}_{ij}$ ($i \neq j$) or mutual term may be either positive or negative, depending on the reference directions assigned to $\mathbf{I}_i$ and $\mathbf{I}_j$.

If there are dependent sources within the network, then it is possible that all the coefficients in the loop equations may not be resistances or impedances, as we saw in Chap. 2. Even so, we shall continue to refer to the circuit determinant as $\Delta_{\mathbf{Z}}$.

The use of minor notation (Appendix 1) enables the input or driving-point impedance at the terminals of a *one-port* network to be expressed very concisely. The result is also applicable to a *two-port* network if one of the two ports is terminated in a passive impedance, including an open or a short circuit.

Let us suppose that the one-port network shown in Fig. 15-2a is composed entirely of passive elements and dependent sources; linearity is also assumed. An ideal voltage source $\mathbf{V}_1$ is connected to the port, and the source current is identified as the current in loop 1. By the familiar procedure, then,

$$\mathbf{I}_1 = \frac{\begin{vmatrix} \mathbf{V}_1 & \mathbf{Z}_{12} & \mathbf{Z}_{13} & \cdots & \mathbf{Z}_{1N} \\ 0 & \mathbf{Z}_{22} & \mathbf{Z}_{23} & \cdots & \mathbf{Z}_{2N} \\ 0 & \mathbf{Z}_{32} & \mathbf{Z}_{33} & \cdots & \mathbf{Z}_{3N} \\ \cdots & \cdots & \cdots & \cdots & \cdots \\ 0 & \mathbf{Z}_{N2} & \mathbf{Z}_{N3} & \cdots & \mathbf{Z}_{NN} \end{vmatrix}}{\begin{vmatrix} \mathbf{Z}_{11} & \mathbf{Z}_{12} & \mathbf{Z}_{13} & \cdots & \mathbf{Z}_{1N} \\ \mathbf{Z}_{21} & \mathbf{Z}_{22} & \mathbf{Z}_{23} & \cdots & \mathbf{Z}_{2N} \\ \mathbf{Z}_{31} & \mathbf{Z}_{32} & \mathbf{Z}_{33} & \cdots & \mathbf{Z}_{3N} \\ \cdots & \cdots & \cdots & \cdots & \cdots \\ \mathbf{Z}_{N1} & \mathbf{Z}_{N2} & \mathbf{Z}_{N3} & \cdots & \mathbf{Z}_{NN} \end{vmatrix}}$$

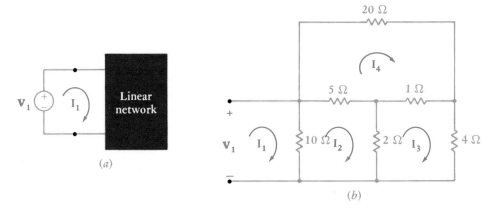

Figure 15-2

(a) An ideal voltage source V_1 is connected to the single port of a linear one-port network containing no independent sources; $Z_{in} = \Delta_z/\Delta_{11}$. (b) A resistive one-port used as an example. (c) A one-port containing a dependent source used as an example.

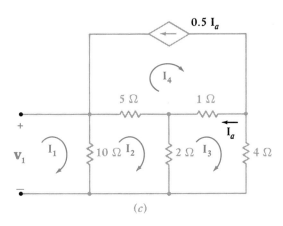

or, more concisely,

$$I_1 = \frac{V_1 \Delta_{11}}{\Delta_z}$$

Thus,

$$Z_{in} = \frac{V_1}{I_1} = \frac{\Delta_z}{\Delta_{11}} \qquad (3)$$

Example 15-1 Let us use this result to calculate the input impedance for the one-port resistive network shown in Fig. 15-2b.

Solution: We first assign the four mesh currents as shown and write the circuit determinant by inspection:

$$\Delta_z = \begin{vmatrix} 10 & -10 & 0 & 0 \\ -10 & 17 & -2 & -5 \\ 0 & -2 & 7 & -1 \\ 0 & -5 & -1 & 26 \end{vmatrix}$$

Its value is 9680 Ω^4. Eliminating the first row and first column, we have

$$\Delta_{11} = \begin{vmatrix} 17 & -2 & -5 \\ -2 & 7 & -1 \\ -5 & -1 & 26 \end{vmatrix} = 2778 \ \Omega^3$$

Thus, Eq. (3) provides the value of the input impedance,

$$\mathbf{Z}_{in} = \frac{9680}{2778} = 3.48 \ \Omega$$

∎

Now let us include a dependent source, Fig. 15-2c.

Example 15-2   Find the input impedance of the network shown in Fig. 15-2c.

Solution: The four mesh equations are written in terms of the four assigned mesh currents:

$$10\mathbf{I}_1 - 10\mathbf{I}_2 = \mathbf{V}_1$$

$$-10\mathbf{I}_1 + 17\mathbf{I}_2 - 2\mathbf{I}_3 - 5\mathbf{I}_4 = 0$$

$$-2\mathbf{I}_2 + 7\mathbf{I}_3 - \mathbf{I}_4 = 0$$

$$\mathbf{I}_4 = -0.5\mathbf{I}_a = -0.5(\mathbf{I}_4 - \mathbf{I}_3) \quad \text{or} \quad -0.5\mathbf{I}_3 + 1.5\mathbf{I}_4 = 0$$

and

$$\Delta_\mathbf{Z} = \begin{vmatrix} 10 & -10 & 0 & 0 \\ -10 & 17 & -2 & -5 \\ 0 & -2 & 7 & -1 \\ 0 & 0 & -0.5 & 1.5 \end{vmatrix} = 590 \ \Omega^3$$

while

$$\Delta_{11} = \begin{vmatrix} 17 & -2 & -5 \\ -2 & 7 & -1 \\ 0 & -0.5 & 1.5 \end{vmatrix} = 159 \ \Omega^2$$

giving

$$\mathbf{Z}_{in} = \frac{590}{159} = 3.71 \ \Omega$$

∎

We may also select a similar procedure using nodal equations, yielding the input admittance:

$$\mathbf{Y}_{in} = \frac{1}{\mathbf{Z}_{in}} = \frac{\Delta_\mathbf{Y}}{\Delta_{11}} \tag{4}$$

where Δ_{11} now refers to the minor of $\Delta_\mathbf{Y}$.

Example 15-3   Use Eq. (4) to again determine the input impedance of the network shown in Fig. 15-2b.

Solution: For the example of Fig. 15-2b, we order the node voltages $\mathbf{V}_1$, $\mathbf{V}_2$, and $\mathbf{V}_3$ from left to right, select the reference at the bottom node, and write the system admittance matrix by inspection:

$$\Delta_\mathbf{Y} = \begin{vmatrix} 0.35 & -0.2 & -0.05 \\ -0.2 & 1.7 & -1 \\ -0.05 & -1 & 1.3 \end{vmatrix} = 0.347 \ S^3$$

$$\Delta_{11} = \begin{vmatrix} 1.7 & -1 \\ -1 & 1.3 \end{vmatrix} = 1.21 \ S^2$$

so that

$$\mathbf{Y}_{in} = \frac{0.347}{1.21} = 0.287 \text{ S}$$

which corresponds to

$$\mathbf{Z}_{in} = \frac{1}{0.287} = 3.48 \text{ } \Omega$$

once again. ∎

Problems 7 and 8 at the end of the chapter give one-ports that can be built using operational amplifiers. These problems illustrate that *negative* resistances may be obtained from networks whose only passive circuit elements are resistors, and that inductors may be simulated with only resistors and capacitors.

Drill Problems

15-1. Find the input impedance of the network shown in Fig. 15-3 if it is formed into a one-port network by breaking it at terminals: (*a*) *a* and *a′*; (*b*) *b* and *b′*; (*c*) *c* and *c′*.   *Ans:* 9.47 Ω; 10.63 Ω; 7.58 Ω

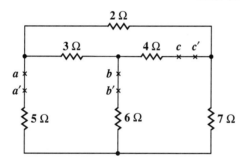

Figure 15-3

See Drill Prob. 15-1.

15-2. Write a set of nodal equations for the circuit of Fig. 15-4, calculate $\Delta_\mathbf{Y}$, and then find the input admittance seen between (*a*) node 1 and the reference node; (*b*) node 2 and the reference.   *Ans:* 10.68 S; 13.80 S

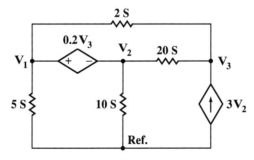

Figure 15-4

See Drill Prob. 15-2.

Let us now turn our attention to two-port networks. We shall assume in all that follows that the network is composed of linear elements and contains no independent sources; dependent sources *are* permissible. Further conditions will also be placed on the network in some special cases.

We shall consider the two-port as it is shown in Fig. 15-5; the voltage and current at the input terminals are $\mathbf{V}_1$ and $\mathbf{I}_1$, and $\mathbf{V}_2$ and $\mathbf{I}_2$ are specified at the

15-3

Admittance parameters

Figure 15-5

A general two-port with terminal voltages and currents specified. The two-port is composed of linear elements, possibly including dependent sources, but not containing any independent sources.

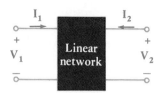

output port. The directions of $\mathbf{I}_1$ and $\mathbf{I}_2$ are both customarily selected as *into* the network at the upper conductors (and out at the lower conductors). Since the network is linear and contains no independent sources within it, $\mathbf{I}_1$ may be considered to be the superposition of two components, one caused by $\mathbf{V}_1$ and the other by $\mathbf{V}_2$. When the same argument is applied to $\mathbf{I}_2$, we may begin with the set of equations

$$\mathbf{I}_1 = \mathbf{y}_{11}\mathbf{V}_1 + \mathbf{y}_{12}\mathbf{V}_2 \tag{5}$$

$$\mathbf{I}_2 = \mathbf{y}_{21}\mathbf{V}_1 + \mathbf{y}_{22}\mathbf{V}_2 \tag{6}$$

where the $\mathbf{y}$'s are no more than proportionality constants, or unknown coefficients, for the present. However, it is obvious that their dimensions must be A/V, or S. They are called the $\mathbf{y}$ parameters, and they are defined by Eqs. (5) and (6).

The $\mathbf{y}$ parameters, as well as other sets of parameters we shall define later in the chapter, are represented concisely as matrices.[1] Here, we define the (2×1) column matrix $[\mathbf{I}]$,

$$[\mathbf{I}] = \begin{bmatrix} \mathbf{I}_1 \\ \mathbf{I}_2 \end{bmatrix} \tag{7}$$

the (2×2) square matrix of the $\mathbf{y}$ parameters,

$$[\mathbf{y}] = \begin{bmatrix} \mathbf{y}_{11} & \mathbf{y}_{12} \\ \mathbf{y}_{21} & \mathbf{y}_{22} \end{bmatrix} \tag{8}$$

and the (2×1) column matrix $[\mathbf{V}]$,

$$[\mathbf{V}] = \begin{bmatrix} \mathbf{V}_1 \\ \mathbf{V}_2 \end{bmatrix} \tag{9}$$

Thus, we may write the matrix equation $[\mathbf{I}] = [\mathbf{y}][\mathbf{V}]$, or

$$\begin{bmatrix} \mathbf{I}_1 \\ \mathbf{I}_2 \end{bmatrix} = \begin{bmatrix} \mathbf{y}_{11} & \mathbf{y}_{12} \\ \mathbf{y}_{21} & \mathbf{y}_{22} \end{bmatrix} \begin{bmatrix} \mathbf{V}_1 \\ \mathbf{V}_2 \end{bmatrix}$$

and matrix multiplication of the right-hand side gives us the equality

$$\begin{bmatrix} \mathbf{I}_1 \\ \mathbf{I}_2 \end{bmatrix} = \begin{bmatrix} \mathbf{y}_{11}\mathbf{V}_1 + \mathbf{y}_{12}\mathbf{V}_2 \\ \mathbf{y}_{21}\mathbf{V}_1 + \mathbf{y}_{22}\mathbf{V}_2 \end{bmatrix}$$

These (2×1) matrices must be equal, element by element, and thus we are led to the defining equations, (5) and (6).

---

[1] Our use of matrices is quite elementary. A brief discussion of the necessary techniques appears in Appendix 2.

The most useful and informative way to attach a physical meaning to the **y** parameters is through a direct inspection of Eqs. (5) and (6). Consider Eq. (5), for example; if we let $\mathbf{V}_2$ be zero, then we see that $\mathbf{y}_{11}$ must be given by the ratio of $\mathbf{I}_1$ to $\mathbf{V}_1$. We therefore describe $\mathbf{y}_{11}$ as the admittance measured at the input terminals with the output terminals *short-circuited* ($\mathbf{V}_2 = 0$). Since there can be no question which terminals are short-circuited, $\mathbf{y}_{11}$ is best described as the *short-circuit input admittance*. Alternatively, we might describe $\mathbf{y}_{11}$ as the reciprocal of the input impedance measured with the output terminals short-circuited, but a description as an admittance is obviously more direct. It is not the *name* of the parameter that is important; rather, it is the conditions which must be applied to Eq. (5) or (6), and hence to the network, that are most meaningful; when the conditions are determined, the parameter can be found directly from an analysis of the circuit (or by experiment on the physical circuit). Each of the **y** parameters may be described as a current-voltage ratio with either $\mathbf{V}_1 = 0$ (the input terminals short-circuited) or $\mathbf{V}_2 = 0$ (the output terminals short-circuited):

$$\mathbf{y}_{11} = \left. \frac{\mathbf{I}_1}{\mathbf{V}_1} \right|_{\mathbf{V}_2 = 0} \tag{10}$$

$$\mathbf{y}_{12} = \left. \frac{\mathbf{I}_1}{\mathbf{V}_2} \right|_{\mathbf{V}_1 = 0} \tag{11}$$

$$\mathbf{y}_{21} = \left. \frac{\mathbf{I}_2}{\mathbf{V}_1} \right|_{\mathbf{V}_2 = 0} \tag{12}$$

$$\mathbf{y}_{22} = \left. \frac{\mathbf{I}_2}{\mathbf{V}_2} \right|_{\mathbf{V}_1 = 0} \tag{13}$$

Because each parameter is an admittance which is obtained by short-circuiting either the output or the input port, the **y** parameters are known as the *short-circuit admittance parameters*. The specific name of $\mathbf{y}_{11}$ is the *short-circuit input admittance*, $\mathbf{y}_{22}$ is the *short-circuit output admittance*, and $\mathbf{y}_{12}$ and $\mathbf{y}_{21}$ are the *short-circuit transfer admittances*.

Example 15-4   Find the four short-circuit admittance parameters for the resistive two-port shown in Fig. 15-6a.

Solution: The values of the parameters may be easily established by applying Eqs. (10) to (13), which we obtained directly from the defining equations, (5) and (6). To determine $\mathbf{y}_{11}$, we short-circuit the output and find the ratio of $\mathbf{I}_1$ to $\mathbf{V}_1$. This may be done by letting $\mathbf{V}_1 = 1$ V, for then $\mathbf{y}_{11} = \mathbf{I}_1$. By inspection of Fig. 15-6a, it is apparent that 1 V applied at the input with the output short-circuited will cause an input current of $(\frac{1}{5} + \frac{1}{10})$, or 0.3 A. Hence,

$$\mathbf{y}_{11} = 0.3 \text{ S}$$

In order to find $\mathbf{y}_{12}$, we short-circuit the input terminals and apply 1 V at the output terminals. The input current flows through the short circuit and is $\mathbf{I}_1 = -\frac{1}{10}$ A. Thus

$$\mathbf{y}_{12} = -0.1 \text{ S}$$

Figure 15-6

(a) A resistive two-port. (b) The resistive two-port is terminated with specific one-ports.

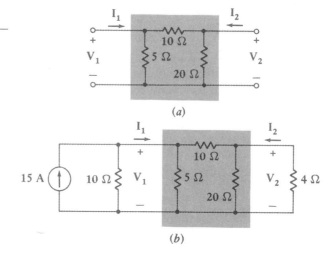

(a)

(b)

By similar methods,

$$\mathbf{y}_{21} = -0.1 \text{ S} \qquad \mathbf{y}_{22} = 0.15 \text{ S}$$

The describing equations for this two-port in terms of the admittance parameters are, therefore,

$$\mathbf{I}_1 = 0.3\mathbf{V}_1 - 0.1\mathbf{V}_2 \tag{14}$$

$$\mathbf{I}_2 = -0.1\mathbf{V}_1 + 0.15\mathbf{V}_2 \tag{15}$$

and
$$[\mathbf{y}] = \begin{bmatrix} 0.3 & -0.1 \\ -0.1 & 0.15 \end{bmatrix} \quad \text{(all S)} \qquad \blacksquare$$

It is not necessary to find these parameters one at a time by using Eqs. (10) to (13), however. We may find them all at once.

Example 15-5   Assign node voltages $\mathbf{V}_1$ and $\mathbf{V}_2$ in the two-port of Fig. 15-6a and write the expressions for $\mathbf{I}_1$ and $\mathbf{I}_2$ in terms of them.

Solution: We have

$$\mathbf{I}_1 = \frac{\mathbf{V}_1}{5} + \frac{\mathbf{V}_1 - \mathbf{V}_2}{10} = 0.3\mathbf{V}_1 - 0.1\mathbf{V}_2$$

and
$$\mathbf{I}_2 = \frac{\mathbf{V}_2 - \mathbf{V}_1}{10} + \frac{\mathbf{V}_2}{20} = -0.1\mathbf{V}_1 + 0.15\mathbf{V}_2$$

These equations are identical with Eqs. (14) and (15), and the four **y** parameters may be read from them directly. $\blacksquare$

In general, it is easier to use Eq. (10), (11), (12), or (13) when only one parameter is desired. If we need all of them, however, it is usually easier to assign $\mathbf{V}_1$ and $\mathbf{V}_2$ to the input and output nodes, to assign other node-to-reference voltages at any interior nodes, and then to carry through with the general solution.

In order to see what use might be made of such a system of equations, let us now terminate each port with some specific one-port network. The simple example of Fig. 15-6b shows a practical current source connected to the input port and a resistive load to the output port. A relationship must now exist between $\mathbf{V}_1$ and $\mathbf{I}_1$ that is independent of the two-port network. This relationship may be determined solely from this external circuit. If we apply KCL (or write a single nodal equation) at the input,

$$\mathbf{I}_1 = 15 - 0.1\mathbf{V}_1$$

For the output, Ohm's law yields

$$\mathbf{I}_2 = -0.25\mathbf{V}_2$$

Substituting these expressions for $\mathbf{I}_1$ and $\mathbf{I}_2$ in Eqs. (14) and (15), we have

$$15 = 0.4\mathbf{V}_1 - 0.1\mathbf{V}_2$$
$$0 = -0.1\mathbf{V}_1 + 0.4\mathbf{V}_2$$

from which are obtained

$$\mathbf{V}_1 = 40 \text{ V} \qquad \mathbf{V}_2 = 10 \text{ V}$$

The input and output currents are also easily found:

$$\mathbf{I}_1 = 11 \text{ A} \qquad \mathbf{I}_2 = -2.5 \text{ A}$$

and the complete terminal characteristics of this resistive two-port are then known.

The advantages of two-port analysis do not show up very strongly for such a simple example, but it should be apparent that once the $\mathbf{y}$ parameters are determined for a more complicated two-port, the performance of the two-port for different terminal conditions is easily determined; it is necessary only to relate $\mathbf{V}_1$ to $\mathbf{I}_1$ at the input and $\mathbf{V}_2$ to $\mathbf{I}_2$ at the output.

In the example just concluded, $\mathbf{y}_{12}$ and $\mathbf{y}_{21}$ were both found to be -0.1 S. It is not difficult to show that this equality is also obtained if three general impedances $\mathbf{Z}_A$, $\mathbf{Z}_B$, and $\mathbf{Z}_C$ are contained in this Π network. It is somewhat more difficult to determine the specific conditions which are necessary in order that $\mathbf{y}_{12} = \mathbf{y}_{21}$, but the use of determinant notation is of some help. Let us see if the relationships of Eqs. (10) to (13) can be expressed in terms of the impedance determinant and its minors.

Since our concern is with the two-port and not with the specific networks with which it is terminated, we shall let $\mathbf{V}_1$ and $\mathbf{V}_2$ be represented by two ideal voltage sources. Equation (10) is applied by letting $\mathbf{V}_2 = 0$ (thus short-circuiting the output) and finding the input admittance. The network now, however, is simply a one-port, and the input impedance of a one-port was found in the previous section. We select loop 1 to include the input terminals, and let $\mathbf{I}_1$ be that loop's current; we identify $(-\mathbf{I}_2)$ as the loop current in loop 2 and assign the remaining loop currents in any convenient manner. Thus,

$$\mathbf{Z}_{in}\bigg|_{\mathbf{V}_2=0} = \frac{\Delta_\mathbf{z}}{\Delta_{11}}$$

and, therefore,

$$\mathbf{y}_{11} = \frac{\Delta_{11}}{\Delta_\mathbf{z}} \tag{16}$$

Similarly,

$$\mathbf{y}_{22} = \frac{\Delta_{22}}{\Delta_{\mathbf{z}}} \tag{17}$$

In order to find $\mathbf{y}_{12}$, we let $\mathbf{V}_1 = 0$ and find $\mathbf{I}_1$ as a function of $\mathbf{V}_2$. We find that $\mathbf{I}_1$ is given by the ratio

$$\mathbf{I}_1 = \cfrac{\begin{vmatrix} 0 & \mathbf{Z}_{12} & \cdots & \mathbf{Z}_{1N} \\ -\mathbf{V}_2 & \mathbf{Z}_{22} & \cdots & \mathbf{Z}_{2N} \\ 0 & \mathbf{Z}_{32} & \cdots & \mathbf{Z}_{3N} \\ \cdots & \cdots & \cdots & \cdots \\ 0 & \mathbf{Z}_{N2} & \cdots & \mathbf{Z}_{NN} \end{vmatrix}}{\begin{vmatrix} \mathbf{Z}_{11} & \mathbf{Z}_{12} & \cdots & \mathbf{Z}_{1N} \\ \mathbf{Z}_{21} & \mathbf{Z}_{22} & \cdots & \mathbf{Z}_{2N} \\ \mathbf{Z}_{31} & \mathbf{Z}_{32} & \cdots & \mathbf{Z}_{3N} \\ \cdots & \cdots & \cdots & \cdots \\ \mathbf{Z}_{N1} & \mathbf{Z}_{N2} & \cdots & \mathbf{Z}_{NN} \end{vmatrix}}$$

Thus,

$$\mathbf{I}_1 = -\frac{(-\mathbf{V}_2)\Delta_{21}}{\Delta_{\mathbf{z}}}$$

and

$$\mathbf{y}_{12} = \frac{\Delta_{21}}{\Delta_{\mathbf{z}}} \tag{18}$$

In a similar manner, we may show that

$$\mathbf{y}_{21} = \frac{\Delta_{12}}{\Delta_{\mathbf{z}}} \tag{19}$$

The equality of $\mathbf{y}_{12}$, and $\mathbf{y}_{21}$ is thus contingent on the equality of the two minors of $\Delta_{\mathbf{z}}$, Δ_{12}, and Δ_{21}. These two minors are

$$\Delta_{21} = \begin{vmatrix} \mathbf{Z}_{12} & \mathbf{Z}_{13} & \mathbf{Z}_{14} & \cdots & \mathbf{Z}_{1N} \\ \mathbf{Z}_{32} & \mathbf{Z}_{33} & \mathbf{Z}_{34} & \cdots & \mathbf{Z}_{3N} \\ \mathbf{Z}_{42} & \mathbf{Z}_{43} & \mathbf{Z}_{44} & \cdots & \mathbf{Z}_{4N} \\ \cdots & \cdots & \cdots & \cdots & \cdots \\ \mathbf{Z}_{N2} & \mathbf{Z}_{N3} & \mathbf{Z}_{N4} & \cdots & \mathbf{Z}_{NN} \end{vmatrix}$$

and

$$\Delta_{12} = \begin{vmatrix} \mathbf{Z}_{21} & \mathbf{Z}_{23} & \mathbf{Z}_{24} & \cdots & \mathbf{Z}_{2N} \\ \mathbf{Z}_{31} & \mathbf{Z}_{33} & \mathbf{Z}_{34} & \cdots & \mathbf{Z}_{3N} \\ \mathbf{Z}_{41} & \mathbf{Z}_{43} & \mathbf{Z}_{44} & \cdots & \mathbf{Z}_{4N} \\ \cdots & \cdots & \cdots & \cdots & \cdots \\ \mathbf{Z}_{N1} & \mathbf{Z}_{N3} & \mathbf{Z}_{N4} & \cdots & \mathbf{Z}_{NN} \end{vmatrix}$$

Their equality is shown by first interchanging the rows and columns of one minor, say, Δ_{21}, an operation which any college algebra book proves is valid, and then letting every mutual impedance $\mathbf{Z}_{ij}$ be replaced by $\mathbf{Z}_{ji}$. Thus, we set

$$\mathbf{Z}_{12} = \mathbf{Z}_{21} \qquad \mathbf{Z}_{23} = \mathbf{Z}_{32} \qquad \text{etc.}$$

This equality of $\mathbf{Z}_{ij}$ and $\mathbf{Z}_{ji}$ is certainly obvious for the three familiar passive elements, the resistor, capacitor, and inductor, and it is also true for mutual inductance, as we proved in the preceding chapter. However, it is not true for every type of device which we may wish to include inside a two-port network. Specifically, it is not true in general for a dependent source, and it is not true for the gyrator, a useful model for Hall-effect devices and for waveguide sections containing ferrites. Over a narrow range of radian frequencies, the gyrator provides an additional phase shift of $180°$ for a signal passing from the output to the input over that for a signal in the forward direction, and thus $\mathbf{y}_{12} = -\mathbf{y}_{21}$. A common type of passive element leading to the inequality of $\mathbf{Z}_{ij}$ and $\mathbf{Z}_{ji}$, however, is a nonlinear element.

Any device for which $\mathbf{Z}_{ij} = \mathbf{Z}_{ji}$ is called a *bilateral element*, and a circuit which contains only bilateral elements is called a *bilateral circuit*. We have therefore shown that an important property of a bilateral two-port is

$$\mathbf{y}_{12} = \mathbf{y}_{21}$$

and this property is glorified by stating it as the *reciprocity theorem*:

> In any passive linear bilateral network, if the single voltage source $\mathbf{V}_x$ in branch x produces the current response $\mathbf{I}_y$ in branch y, then the removal of the voltage source from branch x and its insertion in branch y will produce the current response $\mathbf{I}_y$ in branch x.

A simple way of stating the theorem is to say that the interchange of an ideal voltage source and an ideal ammeter in any passive, linear, bilateral circuit will not change the ammeter reading.

If we had been working with the admittance determinant of the circuit and had proved that the minors Δ_{21} and Δ_{12} of the admittance determinant Δ_Y were equal, then we should have obtained the reciprocity theorem in its dual form:

> In any passive linear bilateral network, if the single current source $\mathbf{I}_x$ between nodes x and x' produces the voltage response $\mathbf{V}_y$ between nodes y and y', then the removal of the current source from nodes x and x' and its insertion between nodes y and y' will produce the voltage response $\mathbf{I}_x$ between nodes x and x'.

In other words, the interchange of an ideal current source and an ideal voltmeter in any passive linear bilateral circuit will not change the voltmeter reading.

Two-ports containing dependent sources receive emphasis in the following section.

Drill Problems

15-3. By applying the appropriate 1-V sources and short circuits to the circuit shown in Fig. 15-7, find (*a*) $\mathbf{y}_{11}$; (*b*) $\mathbf{y}_{21}$; (*c*) $\mathbf{y}_{22}$; (*d*) $\mathbf{y}_{12}$.
     Ans: 0.1192 S; −0.1115 S; 0.1269 S; −0.1115 S

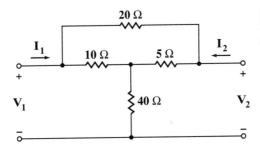

Figure 15-7

See Drill Probs. 15-3 and 15-4.

15-4. In the circuit of Fig. 15-7, let I_1 and I_2 represent ideal current sources. Assign the node voltage V_1 at the input, V_2 at the output, and V_x from the central node to the reference node. Write three nodal equations, eliminate V_x to obtain two equations, and then rearrange these equations into the form of Eqs. (5) and (6) so that all four **y** parameters may be read directly from the equations.

$$Ans: \begin{bmatrix} 0.1192 & -0.1115 \\ -0.1115 & 0.1269 \end{bmatrix} \text{ (all S)}$$

15-5. Find [**y**] for the two-port shown in Fig. 15-8.

$$Ans: \begin{bmatrix} 0.6 & 0 \\ -0.2 & 0.2 \end{bmatrix} \text{ (all S)}$$

Figure 15-8

See Drill Prob. 15-5.

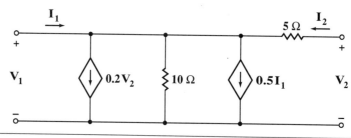

15-4

Some equivalent networks

In analyzing electronic circuits, it is usually necessary to replace the active device (and perhaps some of its associated passive circuitry) with an equivalent two-port containing only three or four impedances. The validity of the equivalent may be restricted to small signal amplitudes and a single frequency, or perhaps a limited range of frequencies. The equivalent is also a linear approximation of a nonlinear circuit. However, if we are faced with a network containing a number of resistors, capacitors, and inductors, plus a transistor labeled 2N3823, then we cannot analyze the circuit by any of the techniques we have studied previously; the transistor must first be replaced by a linear model, just as we replaced the op-amp by a linear model. The **y** parameters provide one such model in the form of a two-port network that is often used at high frequencies. Another common linear model for a transistor appears in Sec. 15-6.

The two basic equations which determine the short-circuit admittance parameters,

$$I_1 = y_{11}V_1 + y_{12}V_2 \tag{20}$$

$$I_2 = y_{21}V_1 + y_{22}V_2 \tag{21}$$

have the form of a pair of nodal equations written for a circuit containing two nonreference nodes. The determination of an equivalent circuit that leads to Eqs. (20) and (21) is made more difficult by the inequality, in general, of y_{12} and y_{21}; it helps to resort to a little trickery in order to obtain a pair of equations that possess equal mutual coefficients. Let us both add and subtract $y_{12}V_1$ [the term we would like to see present on the right side of Eq. (21)]:

$$I_2 = y_{12}V_1 + y_{22}V_2 + (y_{21} - y_{12})V_1 \tag{22}$$

or

$$I_2 - (y_{21} - y_{12})V_1 = y_{12}V_1 + y_{22}V_2 \tag{23}$$

The right-hand sides of Eqs. (20) and (23) now show the proper symmetry for a bilateral circuit; the left-hand side of Eq. (23) may be interpreted as the algebraic sum of two current sources, one an independent source $\mathbf{I}_2$ entering node 2, and the other a dependent source $(\mathbf{y}_{21} - \mathbf{y}_{12})\mathbf{V}_1$ leaving node 2.

Let us now "read" the equivalent network from Eqs. (20) and (23). We first provide a reference node, and then a node labeled $\mathbf{V}_1$ and one labeled $\mathbf{V}_2$. From Eq. (20), we establish the current $\mathbf{I}_1$ flowing into node 1, we supply a mutual admittance $(-\mathbf{y}_{12})$ between nodes 1 and 2, and we supply an admittance of $(\mathbf{y}_{11} + \mathbf{y}_{12})$ between node 1 and the reference node. With $\mathbf{V}_2 = 0$, the ratio of $\mathbf{I}_1$ to $\mathbf{V}_1$ is then $\mathbf{y}_{11}$, as it should be. Now consider Eq. (23); we cause the current $\mathbf{I}_2$ to flow into the second node, we cause the current $(\mathbf{y}_{21} - \mathbf{y}_{12})\mathbf{V}_1$ to leave the node, we note that the proper admittance $(-\mathbf{y}_{12})$ exists between the nodes, and we complete the circuit by installing the admittance $(\mathbf{y}_{22} + \mathbf{y}_{12})$ from node 2 to the reference node. The completed circuit is shown in Fig. 15-9a.

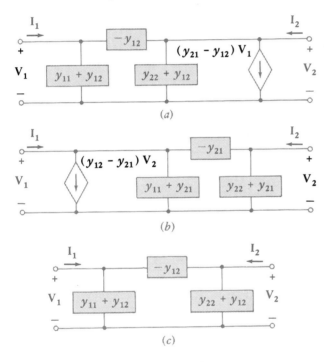

Figure 15-9

(a, b) Two-ports which are equivalent to any general linear two-port. The dependent source in part a depends on $\mathbf{V}_1$, and that in part b depends on $\mathbf{V}_2$. (c) An equivalent for a bilateral network.

Another form of equivalent network is obtained by subtracting and adding $\mathbf{y}_{21}\mathbf{V}_2$ in Eq. (20); this equivalent circuit is shown in Fig. 15-9b.

If the two-port is bilateral, then $\mathbf{y}_{12} = \mathbf{y}_{21}$, and either of the equivalents reduces to a simple passive Π network. The dependent source disappears. This equivalent of the bilateral two-port is shown in Fig. 15-9c.

There are several uses to which these equivalent circuits may be put. In the first place, we have succeeded in showing that an equivalent of any complicated linear two-port *exists*. It does not matter how many nodes or loops are contained within the network; the equivalent is no more complex than the circuits of Fig. 15-9. One of these may be much simpler to use than the given circuit if we are interested only in the terminal characteristics of the given network.

The three-terminal network shown in Fig. 15-10a is often referred to as a Δ of impedances, while that in Fig. 15-10b is called a Y. One network may be

Figure 15-10

The three-terminal Δ network (a) and the three-terminal Y network (b) are equivalent if the six impedances satisfy the conditions of the Y-Δ (or Π-T) transformation, Eqs. (24) to (29).

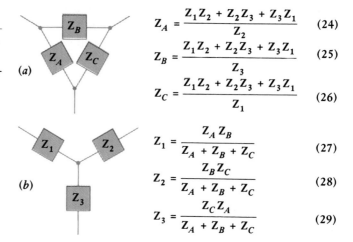

$$Z_A = \frac{Z_1 Z_2 + Z_2 Z_3 + Z_3 Z_1}{Z_2} \quad (24)$$

$$Z_B = \frac{Z_1 Z_2 + Z_2 Z_3 + Z_3 Z_1}{Z_3} \quad (25)$$

$$Z_C = \frac{Z_1 Z_2 + Z_2 Z_3 + Z_3 Z_1}{Z_1} \quad (26)$$

$$Z_1 = \frac{Z_A Z_B}{Z_A + Z_B + Z_C} \quad (27)$$

$$Z_2 = \frac{Z_B Z_C}{Z_A + Z_B + Z_C} \quad (28)$$

$$Z_3 = \frac{Z_C Z_A}{Z_A + Z_B + Z_C} \quad (29)$$

replaced by the other if certain specific relationships between the impedances are satisfied, and these interrelationships may be established by use of the **y** parameters. We find that

$$\mathbf{y}_{11} = \frac{1}{\mathbf{Z}_A} + \frac{1}{\mathbf{Z}_B} = \frac{1}{\mathbf{Z}_1 + \mathbf{Z}_2 \mathbf{Z}_3/(\mathbf{Z}_2 + \mathbf{Z}_3)}$$

$$\mathbf{y}_{12} = \mathbf{y}_{21} = -\frac{1}{\mathbf{Z}_B} = \frac{-\mathbf{Z}_3}{\mathbf{Z}_1 \mathbf{Z}_2 + \mathbf{Z}_2 \mathbf{Z}_3 + \mathbf{Z}_3 \mathbf{Z}_1}$$

$$\mathbf{y}_{22} = \frac{1}{\mathbf{Z}_C} + \frac{1}{\mathbf{Z}_B} = \frac{1}{\mathbf{Z}_2 + \mathbf{Z}_1 \mathbf{Z}_3/(\mathbf{Z}_1 + \mathbf{Z}_3)}$$

These equations may be solved for $\mathbf{Z}_A$, $\mathbf{Z}_B$, and $\mathbf{Z}_C$ in terms of $\mathbf{Z}_1$, $\mathbf{Z}_2$, and $\mathbf{Z}_3$:

$$\mathbf{Z}_A = \frac{\mathbf{Z}_1 \mathbf{Z}_2 + \mathbf{Z}_2 \mathbf{Z}_3 + \mathbf{Z}_3 \mathbf{Z}_1}{\mathbf{Z}_2} \quad (24)$$

$$\mathbf{Z}_B = \frac{\mathbf{Z}_1 \mathbf{Z}_2 + \mathbf{Z}_2 \mathbf{Z}_3 + \mathbf{Z}_3 \mathbf{Z}_1}{\mathbf{Z}_3} \quad (25)$$

$$\mathbf{Z}_C = \frac{\mathbf{Z}_1 \mathbf{Z}_2 + \mathbf{Z}_2 \mathbf{Z}_3 + \mathbf{Z}_3 \mathbf{Z}_1}{\mathbf{Z}_1} \quad (26)$$

or, for the inverse relationships:

$$\mathbf{Z}_1 = \frac{\mathbf{Z}_A \mathbf{Z}_B}{\mathbf{Z}_A + \mathbf{Z}_B + \mathbf{Z}_C} \quad (27)$$

$$\mathbf{Z}_2 = \frac{\mathbf{Z}_B \mathbf{Z}_C}{\mathbf{Z}_A + \mathbf{Z}_B + \mathbf{Z}_C} \quad (28)$$

$$\mathbf{Z}_3 = \frac{\mathbf{Z}_C \mathbf{Z}_A}{\mathbf{Z}_A + \mathbf{Z}_B + \mathbf{Z}_C} \quad (29)$$

These equations enable us to transform easily between the equivalent Y and Δ networks, a process known as the Y-Δ transformation (or Π-T transformation if the networks are drawn in the forms of those letters). In going from Y to Δ, Eqs. (24) to (26), first find the value of the common numerator as the sum of the products of the impedances in the Y taken two at a time. Each impedance in the Δ is then found by dividing the numerator by the impedance of that element in the Y which has no common node with the desired Δ element. Conversely, given the Δ, first take the sum of the three impedances around the Δ; then divide the product of the two Δ impedances having a common node with the desired Y element by that sum.

These transformations are often useful in simplifying passive networks, particularly resistive ones, thus avoiding the need for any mesh or nodal analysis.

Example 15-6 Find the input resistance of the circuit shown in Fig. 15-11a.

Solution: We first make a Δ-Y transformation on the upper Δ appearing in Fig. 15-11a. The sum of the three resistances forming this Δ is $1 + 4 + 3 = 8\ \Omega$. The product of the two resistors connected to the top node is $1 \times 4 = 4\ \Omega^2$. Thus, the upper resistor of the Y is $\frac{4}{8}$, or $\frac{1}{2}\ \Omega$. Repeating this procedure for the other two resistors, we obtain the network shown in Fig. 15-11b.

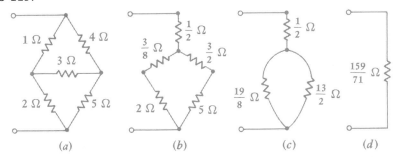

(a)   (b)   (c)   (d)

Figure 15-11

(a) A given resistive network whose input resistance is desired. (b) The upper Δ is replaced by an equivalent Y. (c, d) Series and parallel combinations give the equivalent input resistance $\frac{159}{71}\ \Omega$.

We next make the series and parallel combinations indicated, obtaining in succession Figs. 15-11c and d. Thus, the input resistance of the circuit in Fig. 15-11a is found to be $\frac{159}{71}$, or $2.24\ \Omega$. ∎

Now let us tackle a slightly more complicated example, shown as Fig. 15-12. We note that the circuit contains a dependent source, and thus the Y-Δ transformation is not applicable.

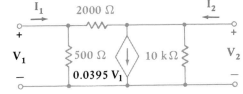

Figure 15-12

The linear equivalent circuit of a transistor in common-emitter configuration with resistive feedback between collector and base. It is used as a two-port example.

Example 15-7 The circuit shown in Fig. 15-12 may be considered to be an approximate linear equivalent of a transistor amplifier in which the emitter terminal is the bottom node, the base terminal is the upper input node, and the collector terminal is the upper output node. A 2000-Ω resistor is connected between collector and base for some special application and makes the analysis of the circuit more difficult. Determine the **y** parameters for this circuit.

Solution: There are several ways we might think about this circuit. If we recognize it as being in the form of the equivalent circuit shown in Fig.

15-9a, then we may immediately determine the values of the **y** parameters. If recognition is not immediate, then the **y** parameters may be determined for the two-port by applying the relationships of Eqs. (10) to (13). We also might avoid any use of two-port analysis methods and write equations directly for the circuit as it stands.

Let us compare the network with the equivalent circuit of Fig. 15-9a. Thus, we first obtain

$$\mathbf{y}_{12} = -\tfrac{1}{2000} = -0.5 \text{ mS}$$

and hence,

$$\mathbf{y}_{11} = \tfrac{1}{500} - (-\tfrac{1}{2000}) = 2.5 \text{ mS}$$

Then

$$\mathbf{y}_{22} = \tfrac{1}{10\,000} - (-\tfrac{1}{2000}) = 0.6 \text{ mS}$$

and

$$\mathbf{y}_{21} = 0.0395 + (-\tfrac{1}{2000}) = 39 \text{ mS}$$

The following equations must then apply:

$$\mathbf{I}_1 = 2.5\mathbf{V}_1 - 0.5\mathbf{V}_2 \tag{30}$$

$$\mathbf{I}_2 = 39\mathbf{V}_1 + 0.6\mathbf{V}_2 \tag{31}$$

where we are now using units of mA, V, and mS or kΩ. ∎

Now let us make use of Eqs. (30) and (31) by analyzing the performance of this two-port under several different operating conditions. We first provide a current source of $1/\underline{0^\circ}$ mA at the input and connect a 0.5-kΩ (2-mS) load to the output. The terminating networks are therefore both one-ports and give us the following specific information relating $\mathbf{I}_1$ to $\mathbf{V}_1$ and $\mathbf{I}_2$ to $\mathbf{V}_2$:

$$\mathbf{I}_1 = 1 \text{ (for any } \mathbf{V}_1) \qquad \mathbf{I}_2 = -2\mathbf{V}_2$$

We now have four equations in the four variables, $\mathbf{V}_1, \mathbf{V}_2, \mathbf{I}_1,$ and $\mathbf{I}_2$. Substituting the two one-port relationships in Eqs. (30) and (31), we obtain two equations relating $\mathbf{V}_1$ and $\mathbf{V}_2$:

$$1 = 2.5\mathbf{V}_1 - 0.5\mathbf{V}_2 \qquad 0 = 39\mathbf{V}_1 + 2.6\mathbf{V}_2$$

Solving, we find that

$$\mathbf{V}_1 = 0.1 \text{ V} \qquad \mathbf{V}_2 = -1.5 \text{ V}$$

$$\mathbf{I}_1 = 1 \text{ mA} \qquad \mathbf{I}_2 = 3 \text{ mA}$$

These four values apply to the two-port operating with a prescribed input $(\mathbf{I}_1 = 1 \text{ mA})$ and a specified load $(R_L = 0.5 \text{ kΩ})$.

The performance of an amplifier is often described by giving a few specific values. Let us calculate four of these values for this two-port with its terminations. We shall define and evaluate the voltage gain, the current gain, the power gain, and the input impedance.

The *voltage gain* $\mathbf{G}_V$ is

$$\mathbf{G}_V = \frac{\mathbf{V}_2}{\mathbf{V}_1}$$

From the numerical results, it is easy to see that $\mathbf{G}_V = -15$.

The *current gain* $\mathbf{G}_I$ is defined as

$$\mathbf{G}_I = \frac{\mathbf{I}_2}{\mathbf{I}_1}$$

and we have

$$\mathbf{G}_I = 3$$

Let us define and calculate the *power gain* G_P for an assumed sinusoidal excitation. We have

$$G_P = \frac{P_{\text{out}}}{P_{\text{in}}} = \frac{\text{Re}\,[-\frac{1}{2}\mathbf{V}_2\mathbf{I}_2^*]}{\text{Re}\,[\frac{1}{2}\mathbf{V}_1\mathbf{I}_1^*]} = 45$$

The device might be termed either a voltage, a current, or a power amplifier, since all the gains are greater than unity. If the 2-kΩ resistor were removed, the power gain would rise to 354.

The input and output impedances of the amplifier are often desired in order that maximum power transfer may be achieved to or from an adjacent two-port. We define the *input impedance* $\mathbf{Z}_{\text{in}}$ as the ratio of input voltage to current:

$$\mathbf{Z}_{\text{in}} = \frac{\mathbf{V}_1}{\mathbf{I}_1} = 0.1\,\text{k}\Omega$$

This is the impedance offered to the current source when the 500-Ω load is connected to the output. (With the output short-circuited, the input impedance is necessarily $1/\mathbf{y}_{11}$, or 400 Ω.)

It should be noted that the input impedance *cannot* be determined by replacing every source by its internal impedance and then combining resistances or conductances. In the given circuit, such a procedure would yield a value of 416 Ω. The error, of course, comes from treating the dependent source as an independent source. If we think of the input impedance as being numerically equal to the input voltage produced by an input current of 1 A, the application of the 1-A source produces some input voltage $\mathbf{V}_1$, and the strength of the dependent source $(0.0395\mathbf{V}_1)$ cannot be zero. We should recall that when we obtain the Thévenin equivalent impedance of a circuit containing a dependent source along with one or more independent sources, we must replace the independent sources by short circuits or open circuits, but a dependent source must not be killed. Of course, if the voltage or current on which the dependent source depends is zero, then the dependent source will itself be inactive; occasionally a circuit may be simplified by recognizing such an occurrence.

Besides $\mathbf{G}_V$, $\mathbf{G}_I$, G_P, and $\mathbf{Z}_{\text{in}}$, there is one other performance parameter that is quite useful. This is the *output impedance* $\mathbf{Z}_{\text{out}}$, and it is determined for a different circuit configuration.

The output impedance is just another term for the Thévenin impedance appearing in the Thévenin equivalent circuit of that portion of the network faced by the load. In our circuit, which we have assumed is driven by a $1\underline{/0°}$-mA current source, we therefore replace this independent source by an open circuit, leave the dependent source alone, and seek the *input* impedance seen looking to the left from the output terminals (with the load removed). Thus, we define

$$\mathbf{Z}_{\text{out}} = \mathbf{V}_2 \Big|_{\mathbf{I}_2 = 1\text{ A with all other independent sources killed and } R_L \text{ removed}}$$

We therefore remove the load resistor, apply $1\underline{/0°}$ mA (since we are working in V, mA, and kΩ) at the output terminals, and determine $\mathbf{V}_2$. We place these requirements on Eqs. (30) and (31), and obtain

$$0 = 2.5\mathbf{V}_1 - 0.5\mathbf{V}_2 \qquad 1 = 39\mathbf{V}_1 + 0.6\mathbf{V}_2$$

Solving,

$$\mathbf{V}_2 = 0.1190 \text{ V}$$

and thus

$$\mathbf{Z}_{\text{out}} = 0.1190 \text{ k}\Omega$$

An alternative procedure might be to find the open-circuit output voltage and the short-circuit output current. That is, the Thévenin impedance is the output impedance:

$$\mathbf{Z}_{\text{out}} = \mathbf{Z}_{\text{th}} = -\frac{\mathbf{V}_{2\text{oc}}}{\mathbf{I}_{2\text{sc}}}$$

Carrying out this procedure, we first rekindle the independent source so that $\mathbf{I}_1 = 1$ mA, and then open-circuit the load so that $\mathbf{I}_2 = 0$. We have

$$1 = 2.5\mathbf{V}_1 - 0.5\mathbf{V}_2 \qquad 0 = 39\mathbf{V}_1 + 0.6\mathbf{V}_2$$

and thus

$$\mathbf{V}_{2\text{oc}} = -1.857 \text{ V}$$

Next, we apply short-circuit conditions by setting $\mathbf{V}_2 = 0$ and again let $\mathbf{I}_1 = 1$ mA. We find that

$$\mathbf{I}_1 = 1 = 2.5\mathbf{V}_1 - 0 \qquad \mathbf{I}_2 = 39\mathbf{V}_1 + 0$$

and thus

$$\mathbf{I}_{2\text{oc}} = 15.6 \text{ mA}$$

The assumed directions of $\mathbf{V}_1$ and $\mathbf{I}_2$ therefore result in a Thévenin or output impedance

$$\mathbf{Z}_{\text{out}} = -\frac{\mathbf{V}_{2\text{oc}}}{\mathbf{I}_{2\text{sc}}} = -\frac{-1.857}{15.6} = 0.1190 \text{ k}\Omega$$

as before.

We now have enough information to enable us to draw the Thévenin or Norton equivalent of the two-port of Fig. 15-12 when it is driven by a $1\underline{/0°}$-mA current source and terminated in a 500-Ω load. Thus, the Norton equivalent presented to the load must contain a current source equal to the short-circuit current $\mathbf{I}_{2\text{sc}}$ in parallel with the output impedance; this equivalent is shown in Fig. 15-13a. Also, the Thévenin equivalent offered to the $1\underline{/0°}$-mA input source must consist solely of the input impedance, as drawn in Fig. 15-13b.

Figure 15-13

(a) The Norton equivalent of the network (Fig. 15-12) to the left of the output terminals, $\mathbf{I}_1 = 1\underline{/0°}$ mA. (b) The Thévenin equivalent of that portion of the network to the right of the input terminals, if $\mathbf{I}_2 = -2\mathbf{V}_2$ mA.

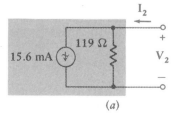
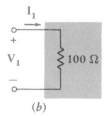

(a)   (b)

Before leaving the **y** parameters, we should recognize their usefulness in describing the parallel connection of two-ports, as indicated in Fig. 15-14. When we first defined a port in Sec. 15-1, we noted that the currents entering and leaving the two terminals of a port had to be equal, and there could be no external connections made that bridged between ports. Apparently the parallel connection shown in Fig. 15-14 violates this condition. However, if each two-

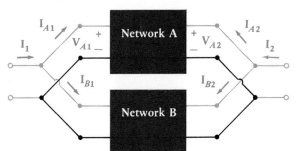

Figure 15-14

The parallel connection of two two-port networks. If both inputs and outputs have the same reference node, then $[\mathbf{y}] = [\mathbf{y}_A] + [\mathbf{y}_B]$.

port has a reference node that is common to its input and output port, and if the two-ports are connected in parallel so that they have a common reference node, then all ports remain ports after the connection. Thus, for the A network,

$$[\mathbf{I}_A] = [\mathbf{y}_A][\mathbf{V}_A]$$

where

$$[\mathbf{I}_A] = \begin{bmatrix} \mathbf{I}_{A1} \\ \mathbf{I}_{A2} \end{bmatrix} \quad \text{and} \quad [\mathbf{V}_A] = \begin{bmatrix} \mathbf{V}_{A1} \\ \mathbf{V}_{A2} \end{bmatrix}$$

and for the B network

$$[\mathbf{I}_B] = [\mathbf{y}_B][\mathbf{V}_B]$$

But

$$[\mathbf{V}_A] = [\mathbf{V}_B] = [\mathbf{V}] \quad \text{and} \quad [\mathbf{I}] = [\mathbf{I}_A] + [\mathbf{I}_B]$$

Thus,

$$[\mathbf{I}] = ([\mathbf{y}_A] + [\mathbf{y}_B])[\mathbf{V}]$$

and we see that each **y** parameter of the parallel network is given as the sum of the corresponding parameters of the individual networks,

$$[\mathbf{y}] = [\mathbf{y}_A] + [\mathbf{y}_B] \tag{32}$$

This may obviously be extended to any number of two-ports connected in parallel.

Drill Problems

15-6. A linear two-port (such as that shown in Fig. 15-5) is described by the equations $\mathbf{I}_1 = 0.25\mathbf{V}_1 - 0.4\mathbf{V}_2$ and $\mathbf{I}_2 = 40\mathbf{V}_1 + 0.5\mathbf{V}_2$, where all admittance values are in mS. An ideal source $\mathbf{I}_s$, directed upward, in parallel with 2500 Ω, is connected to the input port, and a load $R_L = 500\ \Omega$ is connected to the output. Find values for (a) $\mathbf{G}_V$; (b) $\mathbf{G}_I$; (c) G_P. *Ans: -16; 4.81; 77.0*

15-7. With the input and output one-ports in place for the two-port of Drill Prob. 15-6, calculate (a) $\mathbf{Z}_{\text{in}}$; (b) $\mathbf{Z}_{\text{out}}$. *Ans: 0.1504 kΩ; $39.8\ \Omega$*

15-8. Find $[\mathbf{y}]$ and $\mathbf{Z}_{\text{out}}$ for the terminated two-port shown in Fig. 15-15.

$$Ans: \begin{bmatrix} 2 \times 10^{-4} & -10^{-4} \\ -4 \times 10^{-3} & 20.3 \times 10^{-3} \end{bmatrix} \text{(S)}; 51.1\ \Omega$$

Figure 15-15

See Drill Prob. 15-8.

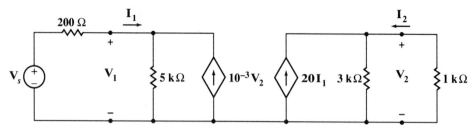

15-9. Use Δ-Y and Y-Δ transformations to determine R_{in} for the network shown in (a) Fig. 15-16a; (b) Fig. 15-16b.   *Ans:* 11.43 Ω; 1.311 Ω

Figure 15-16

See Drill Prob. 15-9.

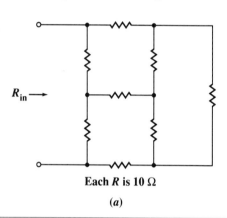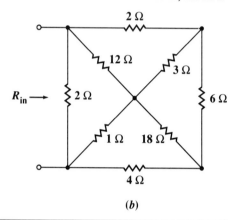

Each R is 10 Ω

(a)   (b)

15-5

Impedance parameters

The concept of two-port parameters has been introduced in terms of the short-circuit admittance parameters. There are other sets of parameters, however, and each set is associated with a particular class of networks for which its use provides the simplest analysis. We shall consider three other types of parameters, the open-circuit impedance parameters, which are the subject of this section; and the hybrid and the transmission parameters, which are discussed in following sections.

We begin again with a general linear two-port that does not contain any independent sources; the currents and voltages are assigned as before (Fig. 15-5). Now let us consider the voltage V_1 as the response produced by two current sources I_1 and I_2. We thus write for V_1

$$V_1 = z_{11}I_1 + z_{12}I_2 \tag{33}$$

and for V_2

$$V_2 = z_{21}I_1 + z_{22}I_2 \tag{34}$$

or

$$[V] = \begin{bmatrix} V_1 \\ V_2 \end{bmatrix} = [z][I] = \begin{bmatrix} z_{11} & z_{12} \\ z_{21} & z_{22} \end{bmatrix}\begin{bmatrix} I_1 \\ I_2 \end{bmatrix} \tag{35}$$

Of course, in using these equations it is not necessary that I_1 and I_2 be current sources; nor is it necessary that V_1 and V_2 be voltage sources. In general, we may have any networks terminating the two-port at either end. As the equations are written, we probably think of V_1 and V_2 as given quantities, or independent variables, and I_1 and I_2 as unknowns, or dependent variables. The six ways in which two equations may be written to relate these four quantities define the different systems of parameters. We study the four most important of these six systems of parameters.

The most informative description of the **z** parameters, defined in Eqs. (33) and (34), is obtained by setting each of the currents equal to zero. Thus

$$\mathbf{z}_{11} = \frac{\mathbf{V}_1}{\mathbf{I}_1}\bigg|_{\mathbf{I}_2=0} \tag{36}$$

$$\mathbf{z}_{12} = \frac{\mathbf{V}_1}{\mathbf{I}_2}\bigg|_{\mathbf{I}_1=0} \tag{37}$$

$$\mathbf{z}_{21} = \frac{\mathbf{V}_2}{\mathbf{I}_1}\bigg|_{\mathbf{I}_2=0} \tag{38}$$

$$\mathbf{z}_{22} = \frac{\mathbf{V}_2}{\mathbf{I}_2}\bigg|_{\mathbf{I}_1=0} \tag{39}$$

Since zero current results from an open-circuit termination, the **z** parameters are known as the *open-circuit impedance parameters*. They are easily related to the short-circuit admittance parameters by solving Eqs. (33) and (34) for $\mathbf{I}_1$ and $\mathbf{I}_2$:

$$\mathbf{I}_1 = \frac{\begin{vmatrix} \mathbf{V}_1 & \mathbf{z}_{12} \\ \mathbf{V}_2 & \mathbf{z}_{22} \end{vmatrix}}{\begin{vmatrix} \mathbf{z}_{11} & \mathbf{z}_{12} \\ \mathbf{z}_{21} & \mathbf{z}_{22} \end{vmatrix}}$$

or

$$\mathbf{I}_1 = \left(\frac{\mathbf{z}_{22}}{\mathbf{z}_{11}\mathbf{z}_{22} - \mathbf{z}_{12}\mathbf{z}_{21}}\right)\mathbf{V}_1 + \left(-\frac{\mathbf{z}_{12}}{\mathbf{z}_{11}\mathbf{z}_{22} - \mathbf{z}_{12}\mathbf{z}_{21}}\right)\mathbf{V}_2$$

Using determinant notation, and being careful that the subscript is a lowercase **z**, we assume that $\Delta_{\mathbf{z}} \neq 0$ and obtain

$$\mathbf{y}_{11} = \frac{\Delta_{11}}{\Delta_{\mathbf{z}}} = \frac{\mathbf{z}_{22}}{\Delta_{\mathbf{z}}} \qquad \mathbf{y}_{12} = -\frac{\Delta_{21}}{\Delta_{\mathbf{z}}} = -\frac{\mathbf{z}_{12}}{\Delta_{\mathbf{z}}}$$

and from solving for $\mathbf{I}_2$,

$$\mathbf{y}_{21} = -\frac{\Delta_{12}}{\Delta_{\mathbf{z}}} = -\frac{\mathbf{z}_{21}}{\Delta_{\mathbf{z}}} \qquad \mathbf{y}_{22} = \frac{\Delta_{22}}{\Delta_{\mathbf{z}}} = \frac{\mathbf{z}_{11}}{\Delta_{\mathbf{z}}}$$

In a similar manner, the **z** parameters may be expressed in terms of the admittance parameters. Transformations of this nature are possible between any of the various parameter systems, and quite a collection of occasionally useful formulas may be obtained. Transformations between the **y** and **z** parameters (as well as the **h** and **t** parameters which we will consider in the following sections) are given in Table 15-1 as a helpful reference.

If the two-port is a bilateral network, reciprocity is present; it is easy to show that this results in the equality of $\mathbf{z}_{12}$ and $\mathbf{z}_{21}$.

Equivalent circuits may again be obtained from an inspection of Eqs. (33) and (34); their construction is facilitated by adding and subtracting either $\mathbf{z}_{12}\mathbf{I}_1$ in Eq. (34) or $\mathbf{z}_{21}\mathbf{I}_2$ in Eq. (33). Each of these equivalent circuits contains a dependent voltage source.

Table 15-1

Transformations between $\mathbf{y}$, $\mathbf{z}$, $\mathbf{h}$, and $\mathbf{t}$ parameters

		y		**z**		**h**		**t**	
y		$\mathbf{y}_{11}$	$\mathbf{y}_{12}$	$\dfrac{\mathbf{z}_{22}}{\Delta_{\mathbf{z}}}$	$\dfrac{-\mathbf{z}_{12}}{\Delta_{\mathbf{z}}}$	$\dfrac{1}{\mathbf{h}_{11}}$	$\dfrac{-\mathbf{h}_{12}}{\mathbf{h}_{11}}$	$\dfrac{\mathbf{t}_{22}}{\mathbf{t}_{12}}$	$\dfrac{-\Delta_{\mathbf{t}}}{\mathbf{t}_{12}}$
		$\mathbf{y}_{21}$	$\mathbf{y}_{22}$	$\dfrac{-\mathbf{z}_{21}}{\Delta_{\mathbf{z}}}$	$\dfrac{\mathbf{z}_{11}}{\Delta_{\mathbf{z}}}$	$\dfrac{\mathbf{h}_{21}}{\mathbf{h}_{11}}$	$\dfrac{\Delta_{\mathbf{h}}}{\mathbf{h}_{11}}$	$\dfrac{-1}{\mathbf{t}_{12}}$	$\dfrac{\mathbf{t}_{11}}{\mathbf{t}_{12}}$
z		$\dfrac{\mathbf{y}_{22}}{\Delta_{\mathbf{y}}}$	$\dfrac{-\mathbf{y}_{12}}{\Delta_{\mathbf{y}}}$	$\mathbf{z}_{11}$	$\mathbf{z}_{12}$	$\dfrac{\Delta_{\mathbf{h}}}{\mathbf{h}_{22}}$	$\dfrac{\mathbf{h}_{12}}{\mathbf{h}_{22}}$	$\dfrac{\mathbf{t}_{11}}{\mathbf{t}_{21}}$	$\dfrac{\Delta_{\mathbf{t}}}{\mathbf{t}_{21}}$
		$\dfrac{-\mathbf{y}_{21}}{\Delta_{\mathbf{y}}}$	$\dfrac{\mathbf{y}_{11}}{\Delta_{\mathbf{y}}}$	$\mathbf{z}_{21}$	$\mathbf{z}_{22}$	$\dfrac{-\mathbf{h}_{21}}{\mathbf{h}_{22}}$	$\dfrac{1}{\mathbf{h}_{22}}$	$\dfrac{1}{\mathbf{t}_{21}}$	$\dfrac{\mathbf{t}_{22}}{\mathbf{t}_{21}}$
h		$\dfrac{1}{\mathbf{y}_{11}}$	$\dfrac{-\mathbf{y}_{12}}{\mathbf{y}_{11}}$	$\dfrac{\Delta_{\mathbf{z}}}{\mathbf{z}_{22}}$	$\dfrac{\mathbf{z}_{12}}{\mathbf{z}_{22}}$	$\mathbf{h}_{11}$	$\mathbf{h}_{12}$	$\dfrac{\mathbf{t}_{12}}{\mathbf{t}_{22}}$	$\dfrac{\Delta_{\mathbf{t}}}{\mathbf{t}_{22}}$
		$\dfrac{\mathbf{y}_{21}}{\mathbf{y}_{11}}$	$\dfrac{\Delta_{\mathbf{y}}}{\mathbf{y}_{11}}$	$\dfrac{-\mathbf{z}_{21}}{\mathbf{z}_{22}}$	$\dfrac{1}{\mathbf{z}_{22}}$	$\mathbf{h}_{21}$	$\mathbf{h}_{22}$	$\dfrac{-1}{\mathbf{t}_{22}}$	$\dfrac{\mathbf{t}_{21}}{\mathbf{t}_{22}}$
t		$\dfrac{-\mathbf{y}_{22}}{\mathbf{y}_{21}}$	$\dfrac{-1}{\mathbf{y}_{21}}$	$\dfrac{\mathbf{z}_{11}}{\mathbf{z}_{21}}$	$\dfrac{\Delta_{\mathbf{z}}}{\mathbf{z}_{21}}$	$\dfrac{-\Delta_{\mathbf{h}}}{\mathbf{h}_{21}}$	$\dfrac{-\mathbf{h}_{11}}{\mathbf{h}_{21}}$	$\mathbf{t}_{11}$	$\mathbf{t}_{12}$
		$\dfrac{-\Delta_{\mathbf{y}}}{\mathbf{y}_{21}}$	$\dfrac{-\mathbf{y}_{11}}{\mathbf{y}_{21}}$	$\dfrac{1}{\mathbf{z}_{21}}$	$\dfrac{\mathbf{z}_{22}}{\mathbf{z}_{21}}$	$\dfrac{-\mathbf{h}_{22}}{\mathbf{h}_{21}}$	$\dfrac{-1}{\mathbf{h}_{21}}$	$\mathbf{t}_{21}$	$\mathbf{t}_{22}$

For all parameter sets: $\Delta_{\mathbf{p}} = \mathbf{p}_{11}\mathbf{p}_{22} - \mathbf{p}_{12}\mathbf{p}_{21}$.

Let us leave the derivation of such an equivalent to some leisure moment, such as during the next hourly exam, and consider next an example of a rather general nature. Can we construct a general Thévenin equivalent of the two-port, as viewed from the output terminals? It is necessary first to assume a specific input circuit configuration, and we shall select an independent voltage source $\mathbf{V}_s$ (positive sign at top) in series with a generator impedance $\mathbf{Z}_g$. Thus

$$\mathbf{V}_s = \mathbf{V}_1 + \mathbf{I}_1 \mathbf{Z}_g$$

Combining this result with Eqs. (33) and (34), we may eliminate $\mathbf{V}_1$ and $\mathbf{I}_1$ and obtain

$$\mathbf{V}_2 = \frac{\mathbf{z}_{21}}{\mathbf{z}_{11} + \mathbf{Z}_g}\mathbf{V}_s + \left(\mathbf{z}_{22} - \frac{\mathbf{z}_{12}\mathbf{z}_{21}}{\mathbf{z}_{11} + \mathbf{Z}_g}\right)\mathbf{I}_2$$

The Thévenin equivalent circuit may be drawn directly from this equation; it is shown in Fig. 15-17. The output impedance, expressed in terms of the $\mathbf{z}$ parameters, is

$$\mathbf{Z}_{\text{out}} = \mathbf{z}_{22} - \frac{\mathbf{z}_{12}\mathbf{z}_{21}}{\mathbf{z}_{11} + \mathbf{Z}_g}$$

Figure 15-17

The Thévenin equivalent of a general two-port, as viewed from the output terminals, expressed in terms of the open-circuit impedance parameters.

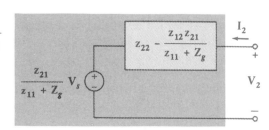

If the generator impedance is zero, the simpler expression

$$\mathbf{Z}_{\text{out}} = \frac{\mathbf{z}_{11}\mathbf{z}_{22} - \mathbf{z}_{12}\mathbf{z}_{21}}{\mathbf{z}_{11}} = \frac{\Delta_z}{\Delta_{22}} = \frac{1}{\mathbf{y}_{22}} \qquad (\mathbf{Z}_g = 0)$$

is obtained. For this special case, the output *admittance* is identical to $\mathbf{y}_{22}$, as indicated by the basic relationship of Eq. (13).

Example 15-8   Given the set of impedance parameters

$$[\mathbf{z}] = \begin{bmatrix} 10^3 & 10 \\ -10^6 & 10^4 \end{bmatrix} \qquad (\text{all } \Omega)$$

which is representative of a transistor operating in the common-emitter configuration, we desire the voltage, current, and power gains, as well as the input and output impedances. We shall consider the two-port as driven by an ideal sinusoidal voltage source $\mathbf{V}_s$ in series with a 500-Ω resistor and terminated in a 10-kΩ load resistor.

    Solution: The two describing equations for the two-port are

$$\mathbf{V}_1 = 10^3\mathbf{I}_1 + 10\mathbf{I}_2 \tag{40}$$

$$\mathbf{V}_2 = -10^6\mathbf{I}_1 + 10^4\mathbf{I}_2 \tag{41}$$

and the characterizing equations of the input and output networks are

$$\mathbf{V}_s = 500\mathbf{I}_1 + \mathbf{V}_1 \tag{42}$$

$$\mathbf{V}_2 = -10^4\mathbf{I}_2 \tag{43}$$

From these last four equations, we may easily obtain expressions for $\mathbf{V}_1$, $\mathbf{I}_1$, $\mathbf{V}_2$, and $\mathbf{I}_2$ in terms of $\mathbf{V}_s$:

$$\mathbf{V}_1 = 0.75\mathbf{V}_s \qquad \mathbf{I}_1 = \frac{\mathbf{V}_s}{2000}$$

$$\mathbf{V}_2 = -250\mathbf{V}_s \qquad \mathbf{I}_2 = \frac{\mathbf{V}_s}{40}$$

From this information, it is simple to determine the voltage gain,

$$\mathbf{G}_V = \frac{\mathbf{V}_2}{\mathbf{V}_1} = -333$$

the current gain,

$$\mathbf{G}_I = \frac{\mathbf{I}_2}{\mathbf{I}_1} = 50$$

the power gain,

$$\mathbf{G}_P = \frac{\text{Re}\,[-\tfrac{1}{2}\mathbf{V}_2\mathbf{I}_2^*]}{\text{Re}\,[\tfrac{1}{2}\mathbf{V}_1\mathbf{I}_1^*]} = 16\,670$$

and the input impedance,

$$\mathbf{Z}_{\text{in}} = \frac{\mathbf{V}_1}{\mathbf{I}_1} = 1500\ \Omega$$

The output impedance may be obtained by referring to Fig. 15-17:

$$\mathbf{Z}_{\text{out}} = \mathbf{z}_{22} - \frac{\mathbf{z}_{12}\mathbf{z}_{21}}{\mathbf{z}_{11} + \mathbf{Z}_g} = 16.67 \text{ k}\Omega$$

In accordance with the predictions of the maximum power transfer theorem, the power gain reaches a maximum value when $\mathbf{Z}_L = \mathbf{Z}_{\text{out}} = 16.67$ kΩ; that maximum value is 17 045.   ■

The **y** parameters are useful when two-ports are interconnected in parallel, and, in a dual manner, the **z** parameters simplify the problem of a series connection of networks, shown in Fig. 15-18. Note that the series connection is

Figure 15-18

The series connection of two two-port networks is made by connecting the four common reference nodes together; then $[\mathbf{z}] = [\mathbf{z}_A] + [\mathbf{z}_B]$.

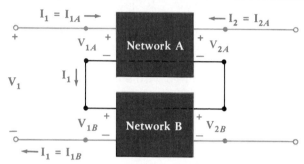

not the same as the cascade connection that we shall discuss later in connection with the transmission parameters. If each two-port has a common reference node for its input and output, and if the references are connected together as indicated in Fig. 15-18, then $\mathbf{I}_1$ flows through the input ports of the two networks in series. A similar statement holds for $\mathbf{I}_2$. Thus, ports remain ports after the interconnection. It follows that $[\mathbf{I}] = [\mathbf{I}_A] = [\mathbf{I}_B]$ and

$$[\mathbf{V}] = [\mathbf{V}_A] + [\mathbf{V}_B] = [\mathbf{z}_A][\mathbf{I}_A] + [\mathbf{z}_B][\mathbf{I}_B]$$

$$= ([\mathbf{z}_A] + [\mathbf{z}_B])[\mathbf{I}] = [\mathbf{z}][\mathbf{I}]$$

where

$$[\mathbf{z}] = [\mathbf{z}_A] + [\mathbf{z}_B] \tag{44}$$

so that $\mathbf{z}_{11} = \mathbf{z}_{11A} + \mathbf{z}_{11B}$, and so forth.

Drill Problems

15-10. Find [**z**] for the two-port shown in (*a*) Fig. 15-19*a*; (*b*) Fig. 15-19*b*.

$$Ans: \begin{bmatrix} 45 & 25 \\ 25 & 75 \end{bmatrix} (\Omega); \begin{bmatrix} 21.2 & 11.76 \\ 11.76 & 67.6 \end{bmatrix} (\Omega)$$

15-11. Find [**z**] for the two-port shown in Fig. 15-19*c*.

$$Ans: \begin{bmatrix} 70 & 100 \\ 50 & 150 \end{bmatrix} (\Omega)$$

15-12. The open-circuit impedance determinant of the amplifier shown in Fig. 15-19*d* is $[\mathbf{z}] = \begin{bmatrix} 20 & 3 \\ 100 & 5 \end{bmatrix}$ (Ω). Find (*a*) $\mathbf{G}_I$; (*b*) $\mathbf{G}_V$; (*c*) G_P; (*d*) $\mathbf{Z}_{\text{in}}$; (*e*) $\mathbf{Z}_{\text{out}}$.
Ans: -0.952; 5.56; 5.29; 17.14 Ω; 0.714 Ω

Figure 15-19

See Drill Probs. 15-10 to 15-12.

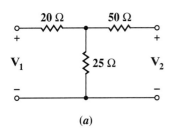

(a)

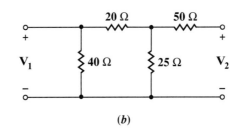

(b)

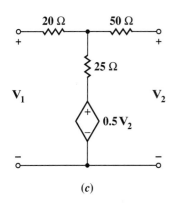

(c)

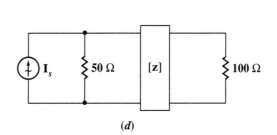

(d)

15-13. Find the **z** parameters of a two-port network for which (a) $[\mathbf{y}] = \begin{bmatrix} 4 & -1 \\ 1 & 20 \end{bmatrix}$ (mS); (b) $\mathbf{V}_1 = \mathbf{V}_2 + \mathbf{I}_2$ and $\mathbf{I}_1 = \mathbf{I}_2 + \mathbf{V}_1 + \mathbf{V}_2$ (V and A).

$Ans:\begin{bmatrix} 247 & 12.35 \\ -12.35 & 49.4 \end{bmatrix} (\Omega); \begin{bmatrix} 0.5 & 0 \\ 0.5 & -1 \end{bmatrix} (\Omega)$

The use of the hybrid parameters is well suited to transistor circuits, because these parameters are among the most convenient to measure experimentally for a transistor. The difficulty in measuring, say, the open-circuit impedance parameters arises when a parameter such as $\mathbf{z}_{21}$ must be measured. A known sinusoidal current is easily supplied at the input terminals, but because of the exceedingly high output impedance of the transistor circuit, it is difficult to open-circuit the output terminals and yet supply the necessary dc biasing voltages and measure the sinusoidal output voltage. A short-circuit current measurement at the output terminals is much simpler to instrument.

The hybrid parameters are defined by writing the pair of equations relating $\mathbf{V}_1$, $\mathbf{I}_1$, $\mathbf{V}_2$, and $\mathbf{I}_2$ as if $\mathbf{V}_1$ and $\mathbf{I}_2$ were the independent variables:

15-6

Hybrid parameters

$$\mathbf{V}_1 = \mathbf{h}_{11}\mathbf{I}_1 + \mathbf{h}_{12}\mathbf{V}_2 \qquad (45)$$

$$\mathbf{I}_2 = \mathbf{h}_{21}\mathbf{I}_1 + \mathbf{h}_{22}\mathbf{V}_2 \qquad (46)$$

or

$$\begin{bmatrix} \mathbf{V}_1 \\ \mathbf{I}_2 \end{bmatrix} = [\mathbf{h}] \begin{bmatrix} \mathbf{I}_1 \\ \mathbf{V}_2 \end{bmatrix} \qquad (47)$$

The nature of the parameters is made clear by first setting $\mathbf{V}_2 = 0$. We see that

$$\mathbf{h}_{11} = \left.\frac{\mathbf{V}_1}{\mathbf{I}_1}\right|_{\mathbf{V}_2=0} = \text{short-circuit input impedance}$$

$$\mathbf{h}_{21} = \left.\frac{\mathbf{I}_2}{\mathbf{I}_1}\right|_{\mathbf{V}_2=0} = \text{short-circuit forward current gain}$$

Letting $\mathbf{I}_1 = 0$, we obtain

$$\mathbf{h}_{12} = \left.\frac{\mathbf{V}_1}{\mathbf{V}_2}\right|_{\mathbf{I}_1=0} = \text{open-circuit reverse voltage gain}$$

$$\mathbf{h}_{22} = \left.\frac{\mathbf{I}_2}{\mathbf{V}_2}\right|_{\mathbf{I}_1=0} = \text{open-circuit output admittance}$$

Since the parameters represent an impedance, an admittance, a voltage gain, and a current gain, it is understandable that they are called the "hybrid" parameters.

The subscript designations for these parameters are often simplified when they are applied to transistors. Thus, $\mathbf{h}_{11}$, $\mathbf{h}_{12}$, $\mathbf{h}_{21}$, and $\mathbf{h}_{22}$ become $\mathbf{h}_i$, $\mathbf{h}_r$, $\mathbf{h}_f$, and $\mathbf{h}_o$, respectively, where the subscripts denote input, reverse, forward, and output.

Example 15-9   In order to illustrate the ease with which these parameters may be evaluated, consider the bilateral resistive circuit drawn in Fig. 15-20. We seek [**h**].

Figure 15-20

A bilateral network for which the **h** parameters are found: $\mathbf{h}_{12} = -\mathbf{h}_{21}$.

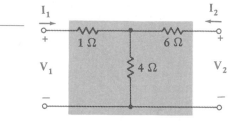

Solution: With the output short-circuited ($\mathbf{V}_2 = 0$), the application of a 1-A source at the input ($\mathbf{I}_1 = 1$ A) produces an input voltage of 3.4 V ($\mathbf{V}_1 = 3.4$ V); hence, $\mathbf{h}_{11} = 3.4\ \Omega$. Under these same conditions, the output current is easily obtained by current division, $\mathbf{I}_2 = -0.4$ A; thus, $\mathbf{h}_{21} = -0.4$.

The remaining two parameters are obtained with the input open-circuited ($\mathbf{I}_1 = 0$). Let us apply a voltage of 1 V at the output terminals ($\mathbf{V}_2 = 1$ V). The response at the input terminals is 0.4 V ($\mathbf{V}_1 = 0.4$ V), and thus $\mathbf{h}_{12} = 0.4$. The current delivered by this source at the output terminals is 0.1 A ($\mathbf{I}_2 = 0.1$ A), and therefore $\mathbf{h}_{22} = 0.1$ S.

We therefore have $[\mathbf{h}] = \begin{bmatrix} 3.4\ \Omega & 0.4 \\ -0.4 & 0.1\ \text{S} \end{bmatrix}$. It is a consequence of the reciprocity theorem that $\mathbf{h}_{12} = -\mathbf{h}_{21}$ for a bilateral network.   ∎

The circuit shown in Fig. 15-21 is a direct translation of the two defining equations, (45) and (46). The first represents KVL about the input loop, while

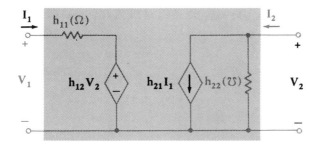

Figure 15-21

The four **h** parameters are referred to a two-port. The pertinent equations are $V_1 = h_{11}I_1 + h_{12}V_2$, $I_2 = h_{21}I_1 + h_{22}V_2$.

the second is obtained from KCL at the upper output node. This circuit is also a popular transistor equivalent circuit. Let us assume some reasonable values for the common-emitter configuration: $h_{11} = 1200\ \Omega$, $h_{12} = 2 \times 10^{-4}$, $h_{21} = 50$, $h_{22} = 50 \times 10^{-6}$ S, a voltage generator of $1\underline{/0°}$ mV in series with 800 Ω, and a 5-kΩ load. For the input,

$$10^{-3} = (1200 + 800)I_1 + 2 \times 10^{-4}V_2$$

and at the output,

$$I_2 = -2 \times 10^{-4}V_2 = 50I_1 + 50 \times 10^{-6}V_2$$

Solving,

$$I_1 = 0.510\ \mu A \qquad V_1 = 0.592\ mV$$

$$I_2 = 20.4\ \mu A \qquad V_2 = -102\ mV$$

Through the transistor we have a current gain of 40, a voltage gain of -172, and a power gain of 6880. The input impedance to the transistor is 1160 Ω, and a few more calculations show that the output impedance is 22.2 kΩ.

Hybrid parameters may be added directly when two-ports are connected in series at the input and in parallel at the output. This is called a series-parallel interconnection, and it is not used very often.

Drill Problems

15-14. Find [**h**] for the two-port shown in (a) Fig. 15-22a; (b) Fig. 15-22b.

$$Ans: \begin{bmatrix} 20\ \Omega & 1 \\ -1 & 0.025\ S \end{bmatrix}; \begin{bmatrix} 8\ \Omega & 0.8 \\ -0.8 & 0.02\ S \end{bmatrix}$$

Figure 15-22

See Drill Prob. 15-14.

(a)

(b)

15-15. If $[\mathbf{h}] = \begin{bmatrix} 5\ \Omega & 2 \\ -0.5 & 0.1\ S \end{bmatrix}$, find (a) [**y**]; (b) [**z**].

$$Ans: \begin{bmatrix} 0.2 & -0.4 \\ -0.1 & 0.3 \end{bmatrix} (all\ S); \begin{bmatrix} 15 & 20 \\ 5 & 10 \end{bmatrix} (all\ \Omega)$$

15-16. Find $[\mathbf{h}]$ at $\omega = 100$ Mrad/s for the high-frequency transistor equivalent circuit of Fig. 15-23.

$$Ans: \begin{bmatrix} 1.109\,\underline{/-56.3^\circ}\,\text{k}\Omega & 0.277\,\underline{/33.7^\circ} \\ 5.55\,\underline{/-59.2^\circ} & 1.532\,\underline{/38.9^\circ}\,\text{mS} \end{bmatrix}$$

Figure 15-23

See Drill Prob. 15-16.

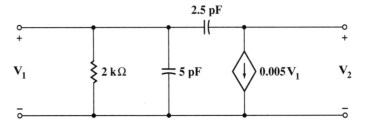

15-17. Find $\mathbf{G}_V$ and $\mathbf{Z}_{\text{out}}$ for an amplifier having $[\mathbf{h}] = \begin{bmatrix} 5\,\Omega & 0.5 \\ 10 & 2\,\text{S} \end{bmatrix}$, if it is driven by a source having an internal resistance of 10 Ω and is terminated in a load resistance $R_L = 1\,\Omega$. *Ans:* -1.000; 0.6 Ω

15-7

Transmission parameters

The last two-port parameters that we shall consider are called the **t** *parameters*, the **ABCD** *parameters*, or simply the *transmission parameters*. They are defined by

$$\mathbf{V}_1 = \mathbf{t}_{11}\mathbf{V}_2 - \mathbf{t}_{12}\mathbf{I}_2 \tag{48}$$

$$\mathbf{I}_1 = \mathbf{t}_{21}\mathbf{V}_2 - \mathbf{t}_{22}\mathbf{I}_2 \tag{49}$$

or

$$\begin{bmatrix} \mathbf{V}_1 \\ \mathbf{I}_1 \end{bmatrix} = [\mathbf{t}] \begin{bmatrix} \mathbf{V}_2 \\ -\mathbf{I}_2 \end{bmatrix} \tag{50}$$

where $\mathbf{V}_1$, $\mathbf{V}_2$, $\mathbf{I}_1$, and $\mathbf{I}_2$ are defined as usual (Fig. 15-5). The minus signs that appear in Eqs. (48) and (49) should be associated with the output current, as $(-\mathbf{I}_2)$. Thus, both $\mathbf{I}_1$ and $-\mathbf{I}_2$ are directed to the right, the direction of energy or signal transmission.

Other widely used nomenclature for this set of parameters is

$$\begin{bmatrix} \mathbf{t}_{11} & \mathbf{t}_{12} \\ \mathbf{t}_{21} & \mathbf{t}_{22} \end{bmatrix} = \begin{bmatrix} \mathbf{A} & \mathbf{B} \\ \mathbf{C} & \mathbf{D} \end{bmatrix} \tag{51}$$

Note that there are no minus signs in the **t** or **ABCD** matrices.

Looking again at Eqs. (48) to (50), we see that the quantities on the left, often thought of as the given or independent variables, are the input voltage and current, $\mathbf{V}_1$ and $\mathbf{I}_1$; the dependent variables, $\mathbf{V}_2$ and $\mathbf{I}_2$, are the output quantities. Thus, the transmission parameters provide a direct relationship between input and output. Their major use arises in transmission-line analysis and in cascaded networks.

Let us find the **t** parameters for the bilateral resistive two-port of Fig. 15-24a. To illustrate one possible procedure for finding a single parameter, consider

$$\mathbf{t}_{12} = \left. \frac{\mathbf{V}_1}{-\mathbf{I}_2} \right|_{\mathbf{V}_2=0}$$

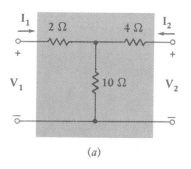 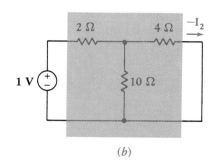

Figure 15-24

(a) A two-port resistive network for which the **t** parameters are to be found. (b) To find $\mathbf{t}_{12}$, set $\mathbf{V}_2 = 1$ V; then $\mathbf{t}_{12} = 1/(-\mathbf{I}_2) = 6.8\ \Omega$.

We therefore short-circuit the output ($\mathbf{V}_2 = 0$) and set $\mathbf{V}_1 = 1$ V, as shown in Fig. 15-24b. Note that we cannot set the denominator equal to unity by placing a 1-A current source at the output; we already have a short circuit there. The equivalent resistance offered to the 1-V source is $R_{\text{eq}} = 2 + (4 \parallel 10)\ \Omega$, and we then use current division to get

$$-\mathbf{I}_2 = \frac{1}{2 + (4 \parallel 10)} \times \frac{10}{10 + 4} = \frac{5}{34}\ \text{A}$$

Hence,

$$\mathbf{t}_{12} = \frac{1}{-\mathbf{I}_2} = \frac{34}{5} = 6.8\ \Omega$$

If it is necessary to find all four parameters, we write any convenient pair of equations using all four terminal quantities, $\mathbf{V}_1$, $\mathbf{V}_2$, $\mathbf{I}_1$, and $\mathbf{I}_2$. From Fig. 15-24a, we have two mesh equations:

$$\mathbf{V}_1 = 12\mathbf{I}_1 + 10\mathbf{I}_2 \qquad (52)$$

$$\mathbf{V}_2 = 10\mathbf{I}_1 + 14\mathbf{I}_2 \qquad (53)$$

Solving Eq. (53) for $\mathbf{I}_1$, we get

$$\mathbf{I}_1 = 0.1\mathbf{V}_2 - 1.4\mathbf{I}_2$$

so that $\mathbf{t}_{21} = 0.1$ S and $\mathbf{t}_{22} = 1.4$. Substituting the expression for $\mathbf{I}_1$ in Eq. (52), we find

$$\mathbf{V}_2 = 12(0.1\mathbf{V}_2 - 1.4\mathbf{I}_2) + 10\mathbf{I}_2 = 1.2\mathbf{V}_2 - 6.8\mathbf{I}_2$$

and $\mathbf{t}_{11} = 1.2$ and $\mathbf{t}_{12} = 6.8\ \Omega$, once again.

For reciprocal networks, the determinant of the **t** matrix is equal to unity:

$$\Delta_{\mathbf{t}} = \mathbf{t}_{11}\mathbf{t}_{22} - \mathbf{t}_{12}\mathbf{t}_{21} = 1 \qquad (54)$$

In the resistive example of Fig. 15-24, $\Delta_{\mathbf{t}} = 1.2 \times 1.4 - 6.8 \times 0.1 = 1$. Good!

We conclude our two-port discussion by connecting two two-ports in cascade, as illustrated for two networks in Fig. 15-25. Terminal voltages and

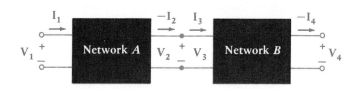

Figure 15-25

When two-port networks A and B are cascaded, the t-parameter matrix for the combined network is given by the matrix product, $[\mathbf{t}] = [\mathbf{t}_A][\mathbf{t}_B]$.

currents are indicated for each two-port, and the corresponding **t**-parameter relationships are, for network A,

$$\begin{bmatrix} \mathbf{V}_1 \\ \mathbf{I}_1 \end{bmatrix} = [\mathbf{t}_A] \begin{bmatrix} \mathbf{V}_2 \\ -\mathbf{I}_2 \end{bmatrix} = [\mathbf{t}_A] \begin{bmatrix} \mathbf{V}_3 \\ \mathbf{I}_3 \end{bmatrix}$$

and for network B,

$$\begin{bmatrix} \mathbf{V}_3 \\ \mathbf{I}_3 \end{bmatrix} = [\mathbf{t}_B] \begin{bmatrix} \mathbf{V}_4 \\ -\mathbf{I}_4 \end{bmatrix}$$

Combining these results, we have

$$\begin{bmatrix} \mathbf{V}_1 \\ \mathbf{I}_1 \end{bmatrix} = [\mathbf{t}_A][\mathbf{t}_B] \begin{bmatrix} \mathbf{V}_4 \\ -\mathbf{I}_4 \end{bmatrix}$$

Therefore, the **t** parameters for the cascaded networks are found by the matrix product,

$$[\mathbf{t}] = [\mathbf{t}_A] [\mathbf{t}_B]$$

This product is *not* obtained by multiplying corresponding elements in the two matrices. If necessary, review the correct procedure for matrix multiplication in Appendix 2.

Example 15-10   As an example of such a calculation, consider the cascaded connection shown in Fig. 15-26. We desire the **t** parameters for the cascaded networks.

Figure 15-26

A cascaded connection for which $[\mathbf{t}] =$

$$\begin{bmatrix} 1.2 & 6.8 \\ 0.1 & 1.4 \end{bmatrix} \begin{bmatrix} 1.2 & 13.6 \\ 0.05 & 1.4 \end{bmatrix} = \begin{bmatrix} 1.78 & 25.84 \\ 0.19 & 3.32 \end{bmatrix}.$$

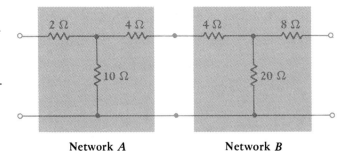

Network A Network B

Solution: Network A is the two-port of Fig. 15-24, and, therefore

$$[\mathbf{t}_A] = \begin{bmatrix} 1.2 & 6.8\,\Omega \\ 0.1\,\text{S} & 1.4 \end{bmatrix}$$

while network B has resistance values twice as large, so that

$$[\mathbf{t}_B] = \begin{bmatrix} 1.2 & 13.6\,\Omega \\ 0.05\,\text{S} & 1.4 \end{bmatrix}$$

For the combined network,

$$[\mathbf{t}] = [\mathbf{t}_A][\mathbf{t}_B] = \begin{bmatrix} 1.2 & 6.8 \\ 0.1 & 1.4 \end{bmatrix} \begin{bmatrix} 1.2 & 13.6 \\ 0.05 & 1.4 \end{bmatrix}$$

$$= \begin{bmatrix} 1.2 \times 1.2 + 6.8 \times 0.05 & 1.2 \times 13.6 + 6.8 \times 1.4 \\ 0.1 \times 1.2 + 1.4 \times 0.05 & 0.1 \times 13.6 + 1.4 \times 1.4 \end{bmatrix}$$

and $$[\mathbf{t}] = \begin{bmatrix} 1.78 & 25.84\,\Omega \\ 0.19\,\text{S} & 3.32 \end{bmatrix}$$ ∎

15-18. Find [**t**] for the two-port shown in (a) Fig. 15-27a; (b) Fig. 15-27b.

Drill Problems

$$Ans: \begin{bmatrix} 1.6 & 25.2\,\Omega \\ 0.1\,\text{S} & 2.2 \end{bmatrix}; \begin{bmatrix} 1 & 10\,\Omega \\ 0.1\,\text{S} & 2.2 \end{bmatrix}$$

Figure 15-27

See Drill Prob. 15-18.

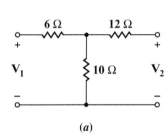
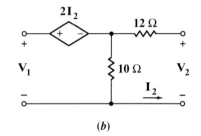

(a)　　　　　　　　　　(b)

15-19. Given $[\mathbf{t}] = \begin{bmatrix} 3.2 & 8\,\Omega \\ 0.2\,\text{S} & 4 \end{bmatrix}$, find (a) [**z**]; (b) [**t**] for two identical networks in cascade; (c) [**z**] for two identical networks in cascade.

$$Ans: \begin{bmatrix} 16 & 56 \\ 5 & 20 \end{bmatrix}(\Omega); \begin{bmatrix} 11.84 & 57.6\,\Omega \\ 1.44\,\text{S} & 17.6 \end{bmatrix}; \begin{bmatrix} 8.22 & 87.1 \\ 0.694 & 12.22 \end{bmatrix}(\Omega)$$

---

1 Find Δ_z for the network shown in Fig. 15-28, and then use it as a help in finding the power generated by a 100-V dc source inserted in the outside branch of mesh: (a) 1; (b) 2; (c) 3.

Problems

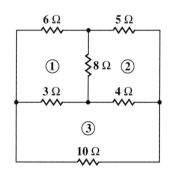

Figure 15-28

See Prob. 1.

2 Find Δ_Y for the network shown in Fig. 15-29, and then use it as a help in finding the power generated by a 10-A dc source inserted between the reference node and node: (a) 1; (b) 2; (c) 3.

Figure 15-29

See Prob. 2.

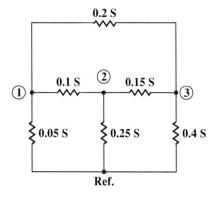

3 The resistance matrix of a certain one-port network is given as Fig. 15-30. Find R_{in} for a source inserted only in mesh 1.

Figure 15-30

See Prob. 3.

$$[\mathbf{R}] = \begin{bmatrix} 3 & -1 & -2 & 0 \\ -1 & 4 & 1 & 3 \\ -2 & 2 & 5 & 2 \\ 0 & 3 & -2 & 6 \end{bmatrix} \ (\Omega)$$

4 Find the Thévenin equivalent impedance $\mathbf{Z}_{th}(\mathbf{s})$ for the one-port of Fig. 15-31.

Figure 15-31

See Prob. 4.

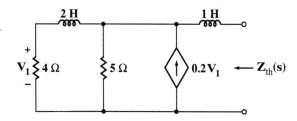

5 Find $\mathbf{Z}_{in}$ for the one-port shown in Fig. 15-32 by (a) finding Δ_Z; (b) finding Δ_Y and $\mathbf{Y}_{in}$ first, and then $\mathbf{Z}_{in}$.

Figure 15-32

See Prob. 5.

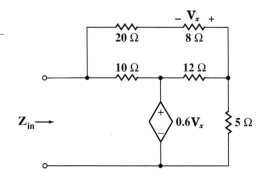

6 Find the output impedance for the network of Fig. 15-33, as a function of **s**.

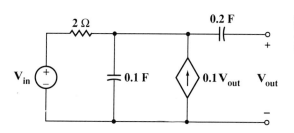

Figure 15-33

See Prob. 6.

7 If the op-amp shown in Fig. 15-34 is assumed to be ideal ($R_i = \infty$, $R_o = 0$, and $A = \infty$), find R_{in}.

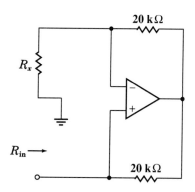

Figure 15-34

See Prob. 7.

8 (a) If both the op-amps shown in the circuit of Fig. 15-35 are assumed to be ideal ($R_i = \infty$, $R_o = 0$, and $A = \infty$), find $\mathbf{Z}_{in}$. (b) Let $R_1 = 4$ kΩ, $R_2 = 10$ kΩ, $R_3 = 10$ kΩ, $R_4 = 1$ kΩ, and $C = 200$ pF, and show that $\mathbf{Z}_{in} = j\omega L_{in}$, where $L_{in} = 0.8$ mH.

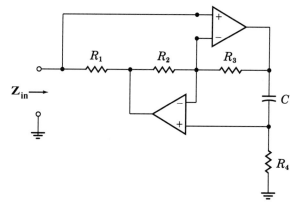

Figure 15-35

See Prob. 8.

9 Find $\mathbf{y}_{11}$ and $\mathbf{y}_{12}$ for the two-port shown in Fig. 15-36.

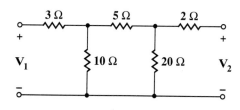

Figure 15-36

See Prob. 9.

10 If the two-port shown in Fig. 15-37 has the parameter values $y_{11} = 10$, $y_{12} = -5$, $y_{21} = 50$, and $y_{22} = 20$, all in mS, find V_1 and V_2 when $V_s = 100$ V, $R_s = 25$ Ω, and $R_L = 100$ Ω.

Figure 15-37

See Prob. 10.

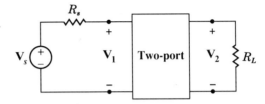

11 Find the four **y** parameters for the network of Fig. 15-38.

Figure 15-38

See Prob. 11.

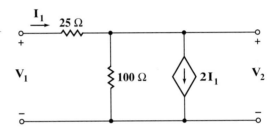

12 Find [**y**] for the two-port shown in Fig. 15-39.

Figure 15-39

See Prob. 12.

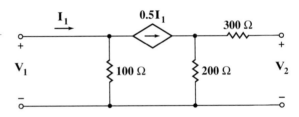

13 Let $[\mathbf{y}] = \begin{bmatrix} 0.1 & -0.0025 \\ -8 & 0.05 \end{bmatrix}$ (S) for the two-port of Fig. 15-40. (a) Find values for the ratios V_2/V_1, I_2/I_1, and V_1/I_1. (b) Remove the 5-Ω resistor, set the 1-V source equal to zero, and find V_2/I_2.

Figure 15-40

See Prob. 13.

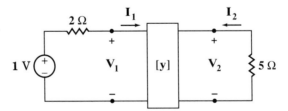

14 The admittance parameters of a certain two-port are $[\mathbf{y}] = \begin{bmatrix} 10 & -5 \\ -20 & 2 \end{bmatrix}$ (mS). Find the new [**y**] if a 100-Ω resistor is connected: (a) in series with one of the input leads; (b) in series with one of the output leads.

15 Complete the table given as part of Fig. 15-41, and also give values for the **y** parameters.

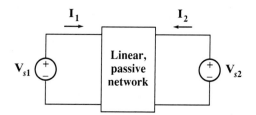

Figure 15-41

See Prob. 15.

	V_{s1} (V)	V_{s2} (V)	I_1 (A)	I_2 (A)
Exp't #1	100	50	5	−32.5
Exp't #2	50	100	−20	−5
Exp't #3	20	0		
Exp't #4			5	0
Exp't #5			5	15

16 Find R_{in} for the one-port shown in Fig. 15-42 by using Y-Δ and Δ-Y transformations.

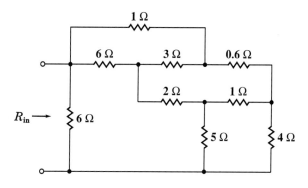

Figure 15-42

See Prob. 16.

17 Use Y-Δ and Δ-Y transformations to find the input resistance of the one-port shown in Fig. 15-43.

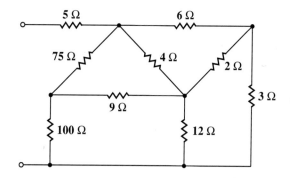

Figure 15-43

See Prob. 17.

18 Find $\mathbf{Z}_{in}$ for the network of Fig. 15-44.

Figure 15-44

See Prob. 18.

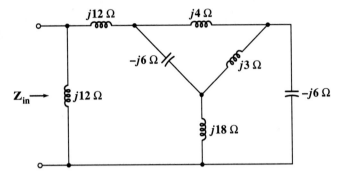

19 Let $[\mathbf{y}] = \begin{bmatrix} 0.4 & -0.002 \\ -5 & 0.04 \end{bmatrix}$ (S) for the two-port of Fig. 15-45, and find (a) $\mathbf{G}_V$; (b) $\mathbf{G}_I$; (c) G_P; (d) $\mathbf{Z}_{\text{in}}$; (e) $\mathbf{Z}_{\text{out}}$.

Figure 15-45

See Prob. 19.

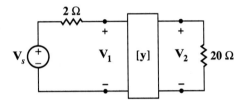

20 Let $[\mathbf{y}] = \begin{bmatrix} 0.1 & -0.05 \\ -0.5 & 0.2 \end{bmatrix}$ (S) for the two-port of Fig. 15-46. Find (a) $\mathbf{G}_V$; (b) $\mathbf{G}_I$; (c) G_P; (d) $\mathbf{Z}_{\text{in}}$; (e) $\mathbf{Z}_{\text{out}}$. (f) If the reverse voltage gain $\mathbf{G}_{V,\text{rev}}$ is defined as $\mathbf{V}_1/\mathbf{V}_2$ with $\mathbf{V}_s = 0$ and R_L removed, calculate $\mathbf{G}_{V,\text{rev}}$. (g) If the insertion power gain G_{ins} is defined as the ratio of $P_{5\Omega}$ with the two-port in place to $P_{5\Omega}$ with the two-port replaced by jumpers connecting each input terminal to the corresponding output terminal, calculate G_{ins}.

Figure 15-46

See Prob. 20.

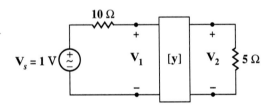

21 (a) Draw an equivalent circuit in the form of Fig. 15-9b for which $[\mathbf{y}] = \begin{bmatrix} 1.5 & -1 \\ 4 & 3 \end{bmatrix}$ (mS). (b) If two of these two-ports are connected in parallel, draw the new equivalent circuit and show that $[\mathbf{y}]_{\text{new}} = 2[\mathbf{y}]$.

22 (a) Find $[\mathbf{y}]_a$ for the two-port of Fig. 15-47a. (b) Find $[\mathbf{y}]_b$ for Fig. 15-47b. (c) Draw the network that is obtained when these two-ports are connected in parallel, and show that $[\mathbf{y}]$ for this network is equal to $[\mathbf{y}]_a + [\mathbf{y}]_b$.

Figure 15-47

See Prob. 22.

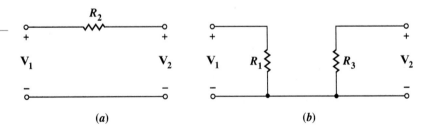

(a)   (b)

23 Find [**z**] for the two-port shown in Fig. 15-48.

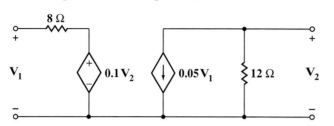

Figure 15-48

See Prob. 23.

24 (*a*) Find [**z**] for the two-port of Fig. 15-49. (*b*) If $\mathbf{I}_1 = \mathbf{I}_2 = 1$ A, find the voltage gain $\mathbf{G}_V$.

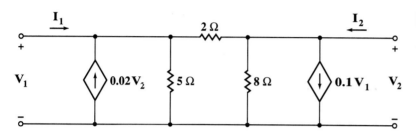

Figure 15-49

See Prob. 24.

25 A certain two-port is described by $[\mathbf{z}] = \begin{bmatrix} 4 & 1.5 \\ 10 & 3 \end{bmatrix}$ (Ω). The input consists of a source $\mathbf{V}_s$ in series with 5 Ω, while the output is $R_L = 2$ Ω. Find (*a*) $\mathbf{G}_I$; (*b*) $\mathbf{G}_V$; (*c*) G_P; (*d*) $\mathbf{Z}_{\text{in}}$; (*e*) $\mathbf{Z}_{\text{out}}$.

26 Let $[\mathbf{z}] = \begin{bmatrix} 1000 & 100 \\ -2000 & 400 \end{bmatrix}$ (Ω) for the two-port of Fig. 15-50. Find the average power delivered to the (*a*) 200-Ω resistor; (*b*) 500-Ω resistor; (*c*) two-port.

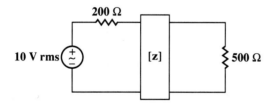

Figure 15-50

See Prob. 26.

27 Find the four **z** parameters at $\omega = 10^8$ rad/s for the transistor high-frequency equivalent circuit shown in Fig. 15-51.

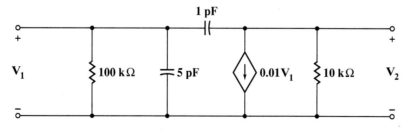

Figure 15-51

See Prob. 27.

28 A two-port for which $[\mathbf{z}] = \begin{bmatrix} 20 & 2 \\ 40 & 10 \end{bmatrix}$ (Ω) is driven by a source $\mathbf{V}_s = 100\underline{/0°}$ V in series with 5 Ω, and terminated in a 25-Ω resistor. Find the Thévenin-equivalent circuit presented to the 25-Ω resistor.

29 The **h** parameters for a certain two-port are $[\mathbf{h}] = \begin{bmatrix} 9\,\Omega & -2 \\ 20 & 0.2\,\mathrm{S} \end{bmatrix}$. Find the new $[\mathbf{h}]$ that results if a 1-Ω resistor is connected in series with (a) the input; (b) the output.

30 Find $\mathbf{Z}_{\mathrm{in}}$ and $\mathbf{Z}_{\mathrm{out}}$ for a two-port driven by a source having $R_s = 100\ \Omega$ and terminated with $R_L = 500\ \Omega$, if $[\mathbf{h}] = \begin{bmatrix} 100\,\Omega & 0.01 \\ 20 & 1\,\mathrm{mS} \end{bmatrix}$.

31 Refer to the two-port shown in Fig. 15-52 and find (a) $\mathbf{h}_{12}$; (b) $\mathbf{z}_{12}$; (c) $\mathbf{y}_{12}$.

Figure 15-52

See Prob. 31.

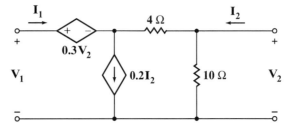

32 Let $\mathbf{h}_{11} = 1\ \mathrm{k}\Omega$, $\mathbf{h}_{12} = -1$, $\mathbf{h}_{21} = 4$, and $\mathbf{h}_{22} = 500\ \mu\mathrm{S}$ for the two-port shown in Fig. 15-53. Find the average power delivered to (a) $R_s = 200\ \Omega$; (b) $R_L = 1\ \mathrm{k}\Omega$; (c) the entire two-port.

Figure 15-53

See Prob. 32.

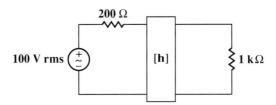

33 (a) Find $[\mathbf{h}]$ for the two-port of Fig. 15-54. (b) Find $\mathbf{Z}_{\mathrm{out}}$ if the input contains $\mathbf{V}_s$ in series with $R_s = 200\ \Omega$.

Figure 15-54

See Prob. 33.

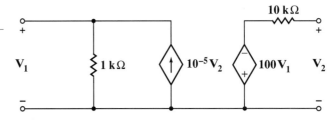

34 Find $[\mathbf{y}]$, $[\mathbf{z}]$, and $[\mathbf{h}]$ for both of the two-ports shown in Fig. 15-55. If any parameter is infinite, skip that parameter set.

Figure 15-55

See Prob. 34.

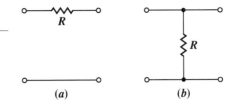

35 Given $[\mathbf{y}] = \begin{bmatrix} 1 & -2 \\ 3 & 4 \end{bmatrix}$, $[\mathbf{b}] = \begin{bmatrix} 4 & 6 \\ -1 & 5 \end{bmatrix}$, $[\mathbf{c}] = \begin{bmatrix} 3 & 2 & 4 & -1 \\ -2 & 3 & 5 & 0 \end{bmatrix}$, and $[\mathbf{d}] =$

$$\begin{bmatrix} 1 & 2 & -1 \\ 3 & 0 & 5 \\ -2 & -3 & 1 \\ 4 & -4 & 2 \end{bmatrix}, \text{ calculate: } (a)\ [\mathbf{y}][\mathbf{b}];\ (b)\ [\mathbf{b}][\mathbf{y}];\ (c)\ [\mathbf{b}][\mathbf{c}];\ (d)\ [\mathbf{c}][\mathbf{d}];\ (e)\ [\mathbf{y}][\mathbf{b}][\mathbf{c}][\mathbf{d}].$$

36 (a) Find [**t**] for the two-port shown in Fig. 15-56. (b) Calculate $\mathbf{Z}_{\text{out}}$ for this two-port if $R_s = 15\ \Omega$ for the source.

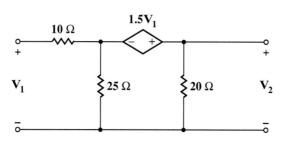

Figure 15-56

See Prob. 36.

37 Find [**t**] for the two-port shown in Fig. 15-57.

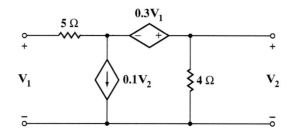

Figure 15-57

See Prob. 37.

38 (a) Find $[\mathbf{t}]_A$, $[\mathbf{t}]_B$, and $[\mathbf{t}]_C$ for the cascaded two-ports of Fig. 15-58. (b) Find [**t**] for the six-resistor two-port.

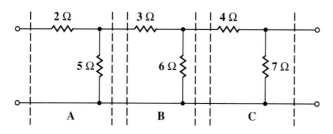

Figure 15-58

See Prob. 38.

39 (a) Find $[\mathbf{t}]_A$ for the single 2-Ω resistor of Fig. 15-59. (b) Show that [**t**] for a single 10-Ω resistor can be obtained by $([\mathbf{t}]_A)^5$.

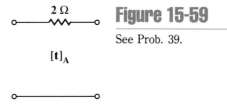

Figure 15-59

See Prob. 39.

40 (a) Find $[\mathbf{t}]_a$, $[\mathbf{t}]_b$, and $[\mathbf{t}]_c$ for the networks shown in Figs. 15-60a, b, and c. (b) By using the rules for interconnecting two-ports in cascade, find [**t**] for the network of Fig. 15-60d.

Figure 15-60

See Prob. 40.

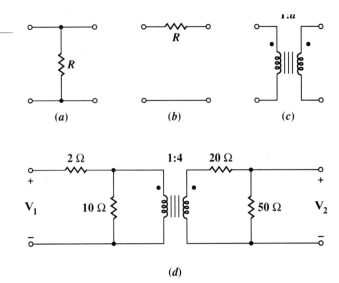

(a)   (b)   (c)

(d)

41 (a) Find [**t**] for the two-port shown in Fig. 15-61. (b) Use the techniques of cascading two-ports to find [**t**]$_{new}$ if a 20-Ω resistor is connected across the output.

Figure 15-61

See Prob. 41.

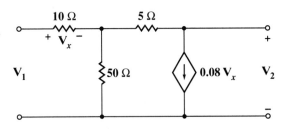

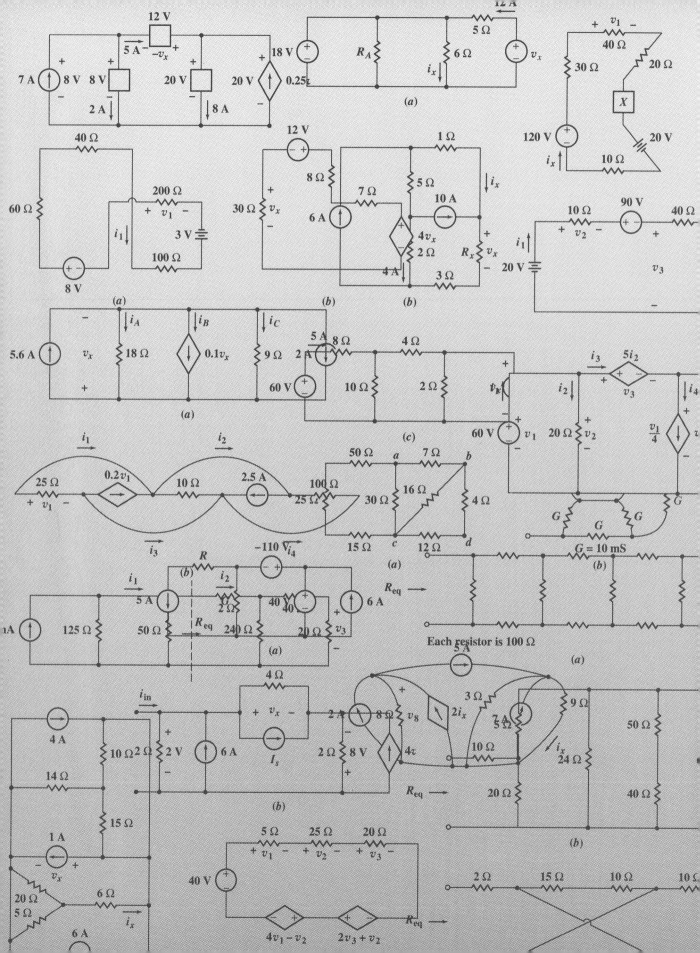

Part Six:
Signal Analysis

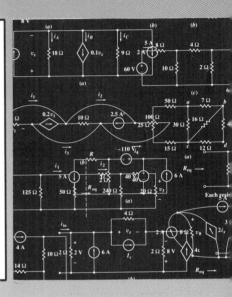

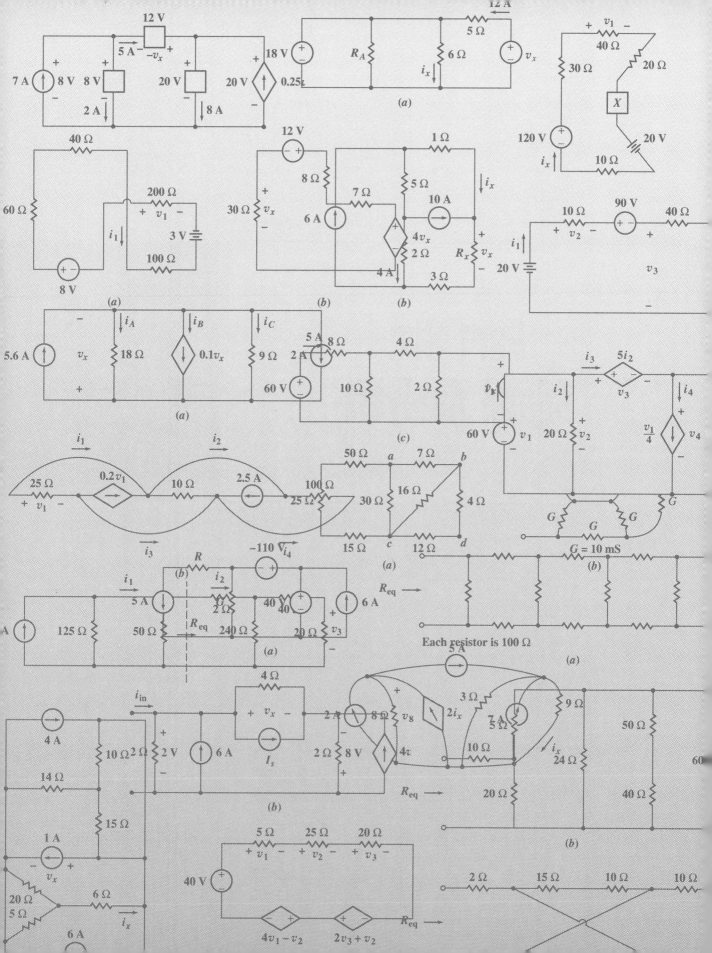

State-Variable Analysis

Up to this point, we have seen several different methods by which circuits might be analyzed. The resistive circuit came first, and for it we wrote a set of algebraic equations, often cast in the form of nodal or mesh equations. However, we also found that we could choose other, more convenient voltage or current variables after drawing an appropriate tree for the network. The tree sprouts up again in this chapter, in the selection of circuit variables.

We then added inductors and capacitors to our networks, and this produced equations containing derivatives and integrals with respect to time. Except for simple first- and second-order systems that either were source-free or contained only dc sources, we did not attempt solving these equations. The results we obtained were found by time-domain methods.

Sinusoidal forcing functions led us to the use of phasors and complex algebraic equations in the frequency domain. We were still able to use the techniques we developed for resistive circuits, even though the presence of complex quantities made their use somewhat more cumbersome.

In this last major section of the text, we shall begin to look at more general and more powerful methods of circuit analysis. We return to the time domain in this chapter as we introduce the use of state variables. Once again we shall not obtain many explicit solutions for circuits of even moderate complexity, but we shall write sets of equations compatible with programs available for digital computers. These programs make use of numerical analysis techniques, and we shall indicate the procedure toward the end of the chapter. SPICE is also applicable, but when it is used, the procedures of state-variable analysis are unnecessary.

Some of these equations will be written to describe circuits we have already analyzed by other methods; now we shall solve them by means of techniques to be developed in this chapter. As we progress through these sections it may seem that the solutions are being made more complicated than they have been in the past, and the reader may feel that the old ways were best. But we must remember, continually, that these new methods of characterizing circuits by means of state variables will allow us to analyze circuits whose complexity makes our old methods of solution completely ineffectual. Patience is the watchword!

In Chap. 17 we return to the frequency domain once again and consider methods of describing periodic functions in terms of their sinusoidal components, a process known as *Fourier analysis*. The final two chapters are devoted to Fourier transforms and Laplace transforms, each of which provides yet another link between the time domain and the frequency domain.

Now let's see what state variables are, and how they are used.

16-2

State variables and normal-form equations

State-variable analysis, or state-space analysis, as it is sometimes called, is a procedure that can be applied both to linear and, with some modifications, to nonlinear circuits, as well as to circuits containing time-varying parameters, such as the capacitance $C = 50 \cos 20t$ pF. Our attention, however, will be restricted to time-invariant linear circuits

We introduce some of the ideas underlying state variables by looking at a general *RLC* circuit that first appeared back in Chap. 3 as Fig. 3-18. It is redrawn as Fig. 16-1. When we first wrote equations for this circuit, we arbitrarily chose to use nodal analysis, the two dependent variables being the node voltages at the central and right nodes. We also could have opted for mesh analysis and used two mesh currents as the variables. Finally, we could have drawn a tree first and then selected a set of tree-branch voltages or link currents as the dependent variables. It is possible that each approach could lead to a different number of variables, although two seems to be the most likely number for this circuit.

Figure 16-1

A four-node *RLC* circuit that we saw first as Fig. 3-18.

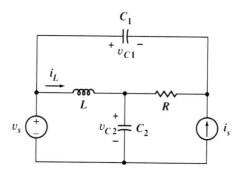

The set of variables we shall select in state-variable analysis is a hybrid set that may include both currents and voltages. They are the *inductor currents* and the *capacitor voltages*. Each of these quantities may be used directly to express the energy stored in the inductor or capacitor at any instant of time. That is, they collectively describe the *energy state* of the system, and for that reason, they are called the *state variables*.

Let us try to write a set of equations for the circuit of Fig. 16-1 in terms of the state variables i_L, v_{C1}, and v_{C2}, as defined on the circuit diagram. The method we use will be outlined more formally in the following section, but for the present let us try to use KVL once for each inductor, and KCL once for each capacitor.

Beginning with the inductor, we set the sum of voltages around the lower left mesh equal to zero:

$$Li_L' + v_{C2} - v_s = 0 \tag{1}$$

We presume that the source voltage v_s and the source current i_s are known, and we therefore have one equation in terms of our chosen state variables.

Next, we consider the capacitor C_1. Since the left terminal of C_1 is also one terminal of a voltage source, it will become part of a supernode. Therefore, we select the right terminal of C_1 as the node to which we apply KCL. The downward current through the capacitor branch is $C_1v'_{C1}$, the upward source current is i_s, and the current in R is obtained by noting that the voltage across R, positive reference on the left, is $(v_{C2} - v_s + v_{C1})$ and, therefore, the current to the right in R is $(v_{C2} - v_s + v_{C1})/R$. Thus,

$$C_1v'_{C1} + \frac{1}{R}(v_{C2} - v_s + v_{C1}) + i_s = 0 \tag{2}$$

Again we have been able to write an equation without introducing any new variables, although we might not have been able to express the current through R directly in terms of the state variables if the circuit had been any more complicated.

Finally, we apply KCL to the upper terminal of C_2:

$$C_2v'_{C2} - i_L + \frac{1}{R}(v_{C2} - v_s + v_{C1}) = 0 \tag{3}$$

Equations (1) to (3) are written solely in terms of the three state variables, the known element values, and the two known forcing functions. They are not, however, written in the standardized form which state-variable analysis demands. The state equations are said to be in *normal form* when the derivative of each state variable is expressed as a linear combination of all the state variables and forcing functions. The ordering of the equations defining the derivatives and the order in which the state variables appear in every equation must be the same. Let us arbitrarily select the order i_L, v_{C1}, v_{C2}, and rewrite Eq. (1) as

$$i'_L = -\frac{1}{L}v_{C2} + \frac{1}{L}v_s \tag{4}$$

Then Eq. (2) is rewritten as

$$v'_{C1} = -\frac{1}{RC_1}v_{C1} - \frac{1}{RC_1}v_{C2} + \frac{1}{RC_1}v_s - \frac{1}{C_1}i_s \tag{5}$$

while Eq. (3) becomes

$$v'_{C2} = \frac{1}{C_2}i_L - \frac{1}{RC_2}v_{C1} - \frac{1}{RC_2}v_{C2} + \frac{1}{RC_2}v_s \tag{6}$$

Note that these equations define, in order, i'_L, v'_{C1}, and v'_{C2}, with a corresponding order of variables on the right sides, i_L, v_{C1}, and v_{C2}. The forcing functions come last and may be written in any convenient order.

As another example of the determination of a set of normal-form equations, we look at the circuit shown in Fig. 16-2a. Since the circuit has one capacitor and one inductor, we expect two state variables, the capacitor voltage and the inductor current. To facilitate writing the normal-form equations, let us construct a tree for this circuit which follows all the rules for tree construction first presented back in Sec. 2-7, and in addition requires all capacitors to be located in the tree and all inductors in the cotree. This is usually possible and it leads to a *normal tree*. In those few exceptional cases where it is not possible to draw a normal tree, we use a slightly different method which will be considered at the end of Sec. 16-3. Here, we are able to place C in the tree and L and i_s in the cotree, as shown in Fig. 16-2b. This is the only normal tree possible for

Figure 16-2

(*a*) An *RLC* circuit requiring two state variables. (*b*) A normal tree showing the state variables v_C and i_L. (*c*) The current in every link and the voltage across every tree branch is expressed in terms of the state variables.

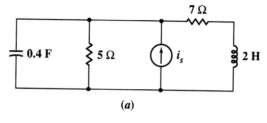

(*a*)

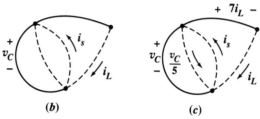

(*b*)   (*c*)

this circuit. The source quantities and the state variables are indicated on the tree and cotree.

Next, we determine the current in every link and the voltage across each tree branch in terms of the state variables. For a simple circuit such as this, it is possible to do this by beginning with any resistor for which either the current or the voltage is obvious. The results are shown on the tree in Fig. 16-2*c*.

We may now write the normal-form equations by invoking KCL at the upper terminal of the capacitor:

$$0.4v_C' + 0.2v_C - i_s + i_L = 0$$

or, in normal form,

$$v_C' = -0.5v_C - 2.5i_L + 2.5i_s \qquad (7)$$

Around the outer loop, we have

$$2i_L' - v_C + 7i_L = 0$$

or $$i_L' = 0.5v_C - 3.5i_L \qquad (8)$$

Equations (7) and (8) are the desired normal-form equations. Their solution will yield all the information necessary for a complete analysis of the given circuit. Of course, explicit expressions for the state variables can be obtained only if a specific function is given for $i_s(t)$. For example, it will be shown later that if

$$i_s(t) = 12 + 3.2e^{-2t}u(t) \qquad A \qquad (9)$$

then

$$v_C(t) = 35 + (10e^{-t} - 12e^{-2t} + 2e^{-3t})u(t) \qquad V \qquad (10)$$

and $$i_L(t) = 5 + (2e^{-t} - 4e^{-2t} + 2e^{-3t})u(t) \qquad A \qquad (11)$$

The solutions, however, are far from obvious, and we shall develop the technique for obtaining them from the normal-form equations in Sec. 16-7.

Now let us see if we can organize the procedure we have been following to obtain the set of normal-form equations. This is the subject of the next section.

16-1. Write a set of normal-form equations for the circuit shown in Fig. 16-3.
Order the state variables as i_{L1}, i_{L2}, and v_C.

Ans: $i'_{L1} = -0.8i_{L1} + 0.8i_{L2} - 0.2v_C + 0.2v_s$; $i'_{L2} = 2i_{L1} - 2i_{L2}$; $v'_C = 10i_{L1}$

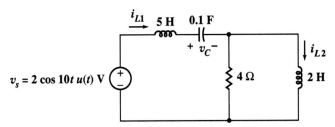

Figure 16-3

See Drill Prob. 16-1.

In the two examples considered in the previous section, the methods whereby we obtained a set of normal-form equations may have seemed to be more of an art than a science. In order to bring a little order into our chaos, let's try to follow the procedure used when we were studying nodal analysis, mesh analysis, and the use of trees in general loop and general nodal analysis. We seek a set of guidelines that will systematize the procedure. Then we will apply these rules to three new examples, each a little more involved than the preceding one.

Here are the six steps that we have been following:

1 *Establish a normal tree.* Place capacitors and voltage sources in the tree, and inductors and current sources in the cotree; and place control voltages in the tree and control currents in the cotree if possible. More than one normal tree may be possible. Certain types of networks do not permit any normal tree to be drawn; these exceptions are considered at the end of this section.

2 *Assign voltage and current variables.* Assign a voltage (with polarity reference) to every capacitor and a current (with arrow) to every inductor; these voltages and currents are the state variables. Indicate the voltage across every tree branch and the current through every link in terms of the source voltages, the source currents, and the state variables, if possible; otherwise, assign a new voltage or current variable to that resistive tree branch or link.

3 *Write the C equations.* Use KCL to write one equation for each capacitor. Set Cv'_C equal to the sum of link currents obtained by considering the node (or supernode) at either end of the capacitor. The supernode is identified as the set of all tree branches connected to that terminal of the capacitor. Do not introduce any new variables.

4 *Write the L equations.* Use KVL to write one equation for each inductor. Set Li'_L equal to the sum of tree-branch voltages obtained by considering the single closed path consisting of the link in which L lies and a convenient set of tree branches. Do not introduce any new variables.

5 *Write the R equations (if necessary).* If any new voltage variables were assigned to resistors in step 2, use KCL to set v_R/R equal to a sum of link currents. If any new current variables were assigned to resistors in step 2, use KVL to set $i_R R$ equal to a sum of tree-branch voltages. Solve these resistor equations simultaneously to obtain explicit expressions for each v_R and i_R in terms of the state variables and source quantities.

16-3

Writing a set of normal-form equations

6 *Write the normal-form equations.* Substitute the expressions for each v_R and i_R into the equations obtained in steps 3 and 4, thus eliminating all resistor variables. Put the resultant equations in normal form.

Example 16-1   As an example of the use of these rules, obtain the normal-form equations for the circuit of Fig. 16-4a, a four-node circuit containing two capacitors, one inductor, and two independent sources.

Figure 16-4

(a) A given circuit for which normal-form equations are to be written. (b) Only one normal tree is possible. (c) Tree-branch voltages and link currents are assigned.

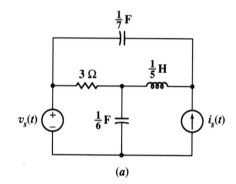

(a)

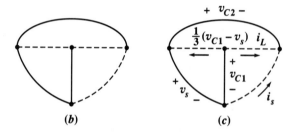

(b)   (c)

Solution: Following step 1, we draw a normal tree. Note that only one such tree is possible here, as shown in Fig. 16-4b, since the two capacitors and the voltage source must be in the tree and the inductor and the current source must be in the cotree.

We next define the voltage across the $\frac{1}{6}$-F capacitor as v_{C1}, the voltage across the $\frac{1}{7}$-F capacitor as v_{C2}, and the current through the inductor as i_L. The source voltage is indicated across its tree branch, the source current is marked on its link, and only the resistor link remains without an assigned variable. The current directed to the left through that link is obviously the voltage $v_{C1} - v_s$ divided by 3 Ω, and we thus find it unnecessary to introduce any additional variables. These tree-branch voltages and link currents are shown in Fig. 16-4c.

Two equations must be written for step 3. For the $\frac{1}{6}$-F capacitor, we apply KCL to the central node:

$$\frac{v_{C1}'}{6} + i_L + \frac{1}{3}(v_{C1} - v_s) = 0$$

while the right-hand node is most convenient for the $\frac{1}{7}$-F capacitor:

$$\frac{v_{C2}'}{7} + i_L + i_s = 0$$

Moving on to step 4, KVL is applied to the inductor link and the entire tree in this case:

$$\frac{i'_L}{5} - v_{C2} + v_s - v_{C1} = 0$$

Since there were no new variables assigned to the resistor, we skip step 5 and simply rearrange the three preceding equations to obtain the desired normal-form equations,

$$v'_{C1} = -2v_{C1} - 6i_L + 2v_s \tag{12}$$

$$v'_{C2} = -7i_L - 7i_s \tag{13}$$

$$i'_L = 5v_{C1} + 5v_{C2} - 5v_s \tag{14}$$

The state variables have arbitrarily been ordered as v_{C1}, v_{C2}, and i_L. If the order i_L, v_{C1}, v_{C2} had been selected instead, the three normal-form equations would be

$$i'_L = 5v_{C1} + 5v_{C2} - 5v_s$$

$$v'_{C1} = -6i_L - 2v_{C1} + 2v_s$$

$$v'_{C2} = -7i_L - 7i_s$$

Note that the ordering of the terms on the right-hand sides of the equations has been changed to agree with the ordering of the equations. ∎

Example 16-2 Now let us write a set of normal-form equations for the circuit of Fig. 16-5a.

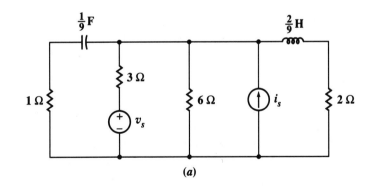

(a)

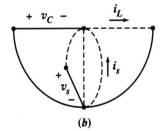

(b)

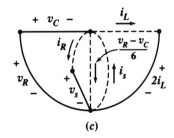

(c)

Figure 16-5

(a) The given circuit. (b) One of many possible normal trees. (c) The assigned voltages and currents.

Solution: This circuit contains several resistors, and this time it will be necessary to introduce resistor variables. Many different normal trees can be constructed for this network, and it is often worthwhile to sketch several possible trees to see whether resistor voltage and current variables can be avoided by a judicious selection. The tree we shall use is shown in Fig. 16-5b. The state variables v_C and i_L and the forcing functions v_s and i_s are indicated on the graph.

Although we might study this circuit and the tree for a few minutes and arrive at a method to avoid the introduction of any new variables, let us plead temporary stupidity and assign the tree-branch voltage v_R and link current i_R for the 1-Ω and 3-Ω branches, respectively. The voltage across the 2-Ω resistor is easily expressed as $2i_L$, while the downward current through the 6-Ω resistor becomes $(v_R - v_C)/6$. All the link currents and tree-branch voltages are marked on Fig. 16-5c.

The capacitor equation may now be written,

$$\frac{v_C'}{9} = i_R + i_L - i_s + \frac{v_R - v_C}{6} \tag{15}$$

and that for the inductor is

$$\tfrac{2}{9} i_L' = -v_C + v_R - 2i_L \tag{16}$$

We begin step 5 with the 1-Ω resistor. Since it is in the tree, we must equate its current to a sum of link currents. Both terminals of the capacitor collapse into a supernode, and we have

$$\frac{v_R}{1} = i_s - i_R - i_L - \frac{v_R - v_C}{6}$$

The 3-Ω resistor is next, and its voltage may be written as

$$3i_R = -v_C + v_R - v_s$$

These last two equations must now be solved simultaneously for v_R and i_R in terms of i_L, v_C, and the two forcing functions. Doing so, we find that

$$v_R = \frac{v_C}{3} - \frac{2}{3} i_L + \frac{2}{3} i_s + \frac{2}{9} v_s$$

$$i_R = -\frac{2}{9} v_C - \frac{2}{9} i_L + \frac{2}{9} i_s - \frac{7}{27} v_s$$

Finally, these results are substituted in Eqs. (15) and (16), and the normal-form equations for Fig. 16-5 are obtained:

$$v_C' = -3v_C + 6i_L - 2v_s - 6i_s \tag{17}$$

$$i_L' = -3v_C - 12i_L + v_s + 3i_s \tag{18}$$

■

Up to now we have discussed only those circuits in which the voltage and current sources were independent sources. At this time we shall look at a circuit containing a dependent source.

Example 16-3 Let us insert a dependent voltage source in series with the $\frac{1}{7}$-F capacitor of Fig. 16-4a, as shown in Fig. 16-6a. We seek the set of normal-form equations for this circuit.

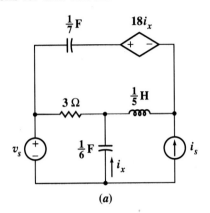

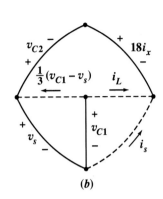

Figure 16-6

(a) A circuit containing a dependent voltage source. (b) The normal tree for this circuit with tree-branch voltages and link currents assigned.

(a)   (b)

Solution: The only possible normal tree is that shown in Fig. 16-6b, and it might be noted that it was not possible to place the controlling current i_x in a link. The tree-branch voltages and link currents are shown on the linear graph, and they are unchanged from the earlier example except for the additional source voltage, $18i_x$.

For the $\frac{1}{6}$-F capacitor, we again find that

$$\frac{v'_{C1}}{6} + i_L + \frac{1}{3}(v_{C1} - v_s) = 0 \tag{19}$$

Letting the dependent voltage source shrink into a supernode, we also find the relationship unchanged for the $\frac{1}{7}$-F capacitor,

$$\frac{v'_{C2}}{7} + i_L + i_s = 0 \tag{20}$$

Our previous result for the inductor changes, however, because there is an added branch in the tree:

$$\frac{i'_L}{5} - 18i_x - v_{C2} + v_s - v_{C1} = 0$$

Finally, we must write a control equation that expresses i_x in terms of our tree-branch voltages and link currents. It is

$$i_x = i_L + \frac{v_{C1} - v_s}{3}$$

and thus the inductor equation becomes

$$\frac{i'_L}{5} - 18i_L - 6v_{C1} + 6v_s - v_{C2} + v_s - v_{C1} = 0 \tag{21}$$

When Eqs. (19) to (21) are written in normal form, we have

$$v'_{C1} = -2v_{C1} - 6i_L + 2v_s \tag{22}$$

$$v'_{C2} = -7i_L - 7i_s \qquad (23)$$

$$i'_L = 35v_{C1} + 5v_{C2} + 90i_L - 35v_s \qquad (24)$$

∎

Our equation-writing process must be modified slightly if we cannot construct a normal tree for the circuit, and there are two types of networks for which this occurs. One contains a loop in which every element is a capacitor or voltage source, thus making it impossible to place them all in the tree.[1] The other occurs if there is a node or supernode that is connected to the remainder of the circuit only by inductors and current sources.[2]

When either of these events occurs, we meet the challenge by leaving a capacitor out of the tree in one case and omitting an inductor from the cotree in the other. We then are faced with a *capacitor* in a link for which we must specify a *current*, or an *inductor*, in the tree which requires a *voltage*. The capacitor current may be expressed as the capacitance times the time derivative of the voltage across it, as defined by a sequence of tree-branch voltages; and the inductor voltage is given by the inductance times the derivative of the current entering or leaving the node or supernode at either terminal of the inductor.

Example 16-4 As an example of a circuit in which an inductor must be placed in the tree, let us write a set of normal-form equations for the network shown in Fig. 16-7a.

Figure 16-7

(a) A circuit for which a normal tree cannot be drawn. (b) A tree is constructed in which one inductor must be a tree branch.

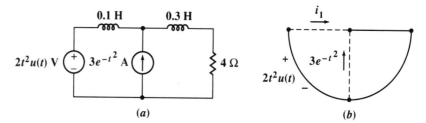

(a)   (b)

Solution: The only elements connected to the upper central node are the two inductors and the current source, and we therefore must place one inductor in the tree, as illustrated by Fig. 16-7b. The two forcing functions and the single state variable, the current i_1, are shown on the graph. Note that the current in the upper right-hand inductor is known in terms of i_1 and $i_s(t)$. Thus we could not specify the energy states of the two inductors independently of each other, and this system requires only the single state variable i_1. Of course, if we had placed the 0.1-H inductor in the tree (instead of the 0.3-H inductor), the current in the right-hand inductor would have ended up as the single state variable.

We must still assign voltages to the remaining two tree branches in Fig. 16-7b. Since the current directed to the right in the 0.3-H inductor is $i_1 + 3e^{-t^2}$, the voltages appearing across the inductor and the 4-Ω resistor are $0.3\,d(i_1 + 3e^{-t^2})/dt = 0.3i'_1 - 1.8te^{-t^2}$ and $4i_1 + 12e^{-t^2}$, respectively.

---

[1] Remember, a tree, by definition, contains no closed paths (loops).
[2] And every node must have at least one tree branch connected to it.

The single normal-form equation is obtained in step 4 of our procedure:

$$0.1i_1' + 0.3i_1' - 1.8te^{-t^2} + 4i_1 + 12e^{-t^2} - 2t^2u(t) = 0$$

or

$$i_1' = -10i_1 + e^{-t^2}(4.5t - 30) + 5t^2u(t) \qquad (25)$$

Note that one of the terms on the right-hand side of the equation is proportional to the derivative of one of the source functions. ∎

In this example, the two inductors and the current source were all connected to a common node. With circuits containing more branches and more nodes, the connection may be to a supernode. As an indication of such a network, Fig. 16-8 shows a slight rearrangement of two elements appearing in Fig. 16-7a. Note that, once again, the energy states of the two inductors cannot be specified independently; with one specified, and the source current given, the other is then specified also. The method of obtaining the normal-form equation is the same.

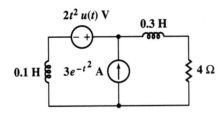

Figure 16-8

The circuit of Fig. 16-7a is redrawn in such a way as to cause the supernode containing the voltage source to be connected to the remainder of the network only through the two inductors and the current source.

The exception created by a loop of capacitors and voltage sources is treated by a similar (dual) procedure, and the wary student should have only the most minor troubles. Be sure to try Drill Prob. 16-4.

16-2. Write normal-form equations for the circuit of Fig. 16-9a.

Ans: $i_1' = -20\,000i_1 - 20\,000i_5 + 1000\cos 2t\,u(t)$; $i_5' = -4000i_1 - 14\,000i_5 + 200\cos 2t\,u(t)$ [using $i_1 \downarrow$ and $i_5 \downarrow$]

Drill Problems

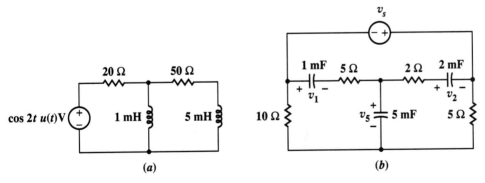

Figure 16-9

(a) See Drill Prob. 16-2.
(b) See Drill Prob. 16-3.

(a)   (b)

16-3. Write the normal-form equations for the circuit of Fig. 16-9b using the state-variable order v_1, v_2, v_5.

Ans: $v_1' = -160v_1 - 100v_2 - 60v_5 - 140v_s$; $v_2' = -50v_1 - 125v_2 + 75v_5 - 75v_s$; $v_5' = -12v_1 + 30v_2 - 42v_5 + 2v_s$

16-4. Find the normal-form equation for the circuit shown in Fig. 16-10 using the state variable: (a) v_{C1}; (b) v_{C2}.

Ans: $v_{C1}' = -6.67v_{C1} + 6.67v_s + 0.333v_s'$; $v_{C2}' = -6.67v_{C2} + 0.667v_s'$

Figure 16-10

See Drill Prob. 16-4.

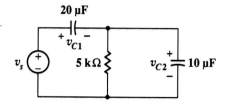

16-4

The use of matrix notation

In the examples that we studied in the previous two sections, the state variables selected were the capacitor voltages and the inductor currents, except for the final case, where it was not possible to draw a normal tree and only one inductor current could be selected as a state variable. As the number of inductors and capacitors in a network increases, it is apparent that the number of state variables will increase. More complicated circuits thus require a greater number of state equations, each of which contains a greater array of state variables on the right-hand side. Not only does the solution of such a set of equations require computer assistance,[3] but also the sheer effort of writing down all the equations can easily lead to writer's cramp.

In this section we shall establish a useful symbolic notation that minimizes the equation-writing effort.

Let us introduce this method by recalling the normal-form equations that we obtained for the circuit of Fig. 16-6, which contained two independent sources, v_s and i_s. The results were given as Eqs. (22) to (24) of Sec. 16-3:

$$v'_{C1} = -2v_{C1} - 6i_L + 2v_s \tag{22}$$

$$v'_{C2} = -7i_L - 7i_s \tag{23}$$

$$i'_L = 35v_{C1} + 5v_{C2} + 90i_L - 35v_s \tag{24}$$

Two of our state variables are voltages, one is a current, one source function is a voltage, one is a current, and the units associated with the constants on the right-hand sides of these equations have the dimensions of ohms, or siemens, or they are dimensionless.

To avoid notational problems in a more generalized treatment, we shall use the letter q to denote a state variable, a to indicate a constant multiplier of q, and f to represent the entire forcing function appearing on the right-hand side of an equation. Thus, Eqs. (22) to (24) become

$$q'_1 = a_{11}q_1 + a_{12}q_2 + a_{13}q_3 + f_1 \tag{26}$$

$$q'_2 = a_{21}q_1 + a_{22}q_2 + a_{23}q_3 + f_2 \tag{27}$$

$$q'_3 = a_{31}q_1 + a_{32}q_2 + a_{33}q_3 + f_3 \tag{28}$$

where

$q_1 = v_{C1}$	$a_{11} = -2$	$a_{12} = 0$	$a_{13} = -6$	$f_1 = 2v_s$
$q_2 = v_{C2}$	$a_{21} = 0$	$a_{22} = 0$	$a_{23} = -7$	$f_2 = -7i_s$
$q_3 = i_L$	$a_{31} = 35$	$a_{32} = 5$	$a_{33} = 90$	$f_3 = -35v_s$

---

[3] Although SPICE can be used to analyze any of the circuits we meet in this chapter, other computer programs are used specifically to solve normal-form equations.

We now turn to the use of matrices and linear algebra to simplify our equations and further generalize the methods.[4] We first define a matrix $\mathbf{q}$ which we call the *state vector*:

$$\mathbf{q}(t) = \begin{bmatrix} q_1(t) \\ q_2(t) \\ \cdot \\ \cdot \\ \cdot \\ q_n(t) \end{bmatrix} \tag{29}$$

The derivative of a matrix is obtained by taking the derivative of each element of the matrix. Thus,

$$\mathbf{q}'(t) = \begin{bmatrix} q_1'(t) \\ q_2'(t) \\ \cdot \\ \cdot \\ \cdot \\ q_n'(t) \end{bmatrix}$$

We shall represent all matrices and vectors in this chapter by lowercase boldface letters, such as $\mathbf{q}$ or $\mathbf{q}(t)$, with the single exception of the identity matrix $\mathbf{I}$, to be defined in Sec. 16-6. The elements of any matrix are scalars, and they are symbolized by lowercase italic letters, such as q_1 or $q_1(t)$.

We also define the set of forcing functions, $f_1, f_2, \ldots, f_n$, as a matrix $\mathbf{f}$ and call it the *forcing-function matrix*:

$$\mathbf{f}(t) = \begin{bmatrix} f_1(t) \\ f_2(t) \\ \cdot \\ \cdot \\ \cdot \\ f_n(t) \end{bmatrix} \tag{30}$$

Now let us turn our attention to the coefficients a_{ij}, which represent elements in the $(n \times n)$ square matrix $\mathbf{a}$,

$$\mathbf{a} = \begin{bmatrix} a_{11} & a_{12} & \cdots & a_{1n} \\ a_{21} & a_{22} & \cdots & a_{2n} \\ \cdot & \cdot & \cdot & \cdot \\ a_{n1} & a_{n2} & \cdots & a_{nn} \end{bmatrix} \tag{31}$$

The matrix $\mathbf{a}$ is termed the *system matrix*.

Using the matrices we have defined in the preceding paragraphs, we can combine these results to obtain a concise, compact representation of the state equations,

$$\mathbf{q}' = \mathbf{aq} + \mathbf{f} \tag{32}$$

---

[4] Some knowledge of matrix notation will make the remainder of this chapter easier to assimilate, but the lack of such knowledge should not prevent continued study; just proceed more slowly, more carefully, and more doggedly.

The matrices $\mathbf{q}'$ and $\mathbf{f}$, and the matrix product $\mathbf{aq}$, are all $(n \times 1)$ column matrices.

The advantages of this representation are obvious, because a system of 100 equations in 100 state variables has exactly the same form as one equation in one state variable.

For the example of Eqs. (22) to (24), the four matrices in Eq. (32) may be written out explicitly as

$$\begin{bmatrix} v'_{C1} \\ v'_{C2} \\ i'_L \end{bmatrix} = \begin{bmatrix} -2 & 0 & -6 \\ 0 & 0 & -7 \\ 35 & 5 & 90 \end{bmatrix} \begin{bmatrix} v_{C1} \\ v_{C2} \\ i_L \end{bmatrix} + \begin{bmatrix} 2v_s \\ -7i_s \\ -35v_s \end{bmatrix} \tag{33}$$

Everyone except full-blooded matrix experts should take a few minutes off to expand Eq. (33) and then check the results with Eqs. (22) to (24); three identical equations should result.

What now is our status with respect to state-variable analysis? Given a circuit, we should be able to construct a normal tree, specify a set of state variables, order them as the state vector, write a set of normal-form equations, and finally specify the system matrix and the forcing-function vector from the equations.

The next problem facing us is to obtain the explicit functions of time that the state variables represent.

Drill Problem

16-5. (a) Using the state vector $\mathbf{q} = \begin{bmatrix} i \\ v \end{bmatrix}$, determine the system matrix and the forcing-function vector for the circuit of Fig. 16-11. (b). Repeat for the state vector $\mathbf{q} = \begin{bmatrix} v \\ i \end{bmatrix}$.

$$Ans: \begin{bmatrix} -3.33 & 0.333 \\ -1.667 & -0.833 \end{bmatrix}, \begin{bmatrix} 43.3u(t) \\ -8.33u(t) \end{bmatrix}; \begin{bmatrix} -0.833 & -1.667 \\ 0.333 & -3.33 \end{bmatrix}, \begin{bmatrix} -8.33u(t) \\ 43.3u(t) \end{bmatrix}$$

Figure 16-11

See Drill Prob. 16-5.

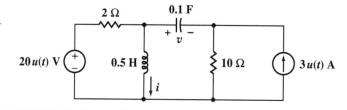

16-5

Solution of the first-order equation

The matrix equation representing the set of normal-form equations for a general nth-order system was obtained as Eq. (32) in the previous section, and it is repeated here for convenience as Eq. (34):

$$\mathbf{q}' = \mathbf{aq} + \mathbf{f} \tag{34}$$

The $(n \times n)$ system matrix $\mathbf{a}$ is composed of constant elements for our time-invariant circuits, and $\mathbf{q}'$, $\mathbf{q}$, and $\mathbf{f}$ are all $(n \times 1)$ column matrices. We need to solve this matrix equation for $\mathbf{q}$, whose elements are $q_1, q_2, \ldots, q_n$. Each must be found as a function of time. Remember that these are the state variables, the collection of which enables us to specify every voltage and current in the given circuit.

Probably the simplest way of approaching this problem is to recall the method by which we solved the corresponding first-order (scalar) equation back in Sec. 5-4. We will quickly repeat that process, but, as we do so, we should keep in mind the fact that we are next going to extend the procedure to a matrix equation.

If each matrix in Eq. (34) has only one row and one column, then we may write the matrix equation as

$$[q_1'(t)] = [a_{11}]\,[q_1(t)] + [f_1(t)]$$
$$= [a_{11}q_1(t)] + [f_1(t)]$$
$$= [a_{11}q_1(t) + f_1(t)]$$

and therefore, we have the first-order equation

$$q_1'(t) = a_{11}q_1(t) + f_1(t) \tag{35}$$

or
$$q_1'(t) - a_{11}q_1(t) = f_1(t) \tag{36}$$

Equation (36) has the same form as Eq. (2) of Sec. 5-4, and we therefore proceed with a similar method of solution by multiplying each side of the equation by the integrating factor $e^{-ta_{11}}$:

$$e^{-ta_{11}}q_1'(t) - e^{-ta_{11}}a_{11}q_1(t) = e^{-ta_{11}}f_1(t)$$

The left-hand side of this equation is again an exact derivative, and so we have

$$\frac{d}{dt}[e^{-ta_{11}}q_1(t)] = e^{-ta_{11}}f_1(t) \tag{37}$$

The order in which the various factors in Eq. (37) have been written may seem a little strange, because a term which is the product of a constant and a time function is usually written with the constant appearing first and the time function following it. In scalar equations, multiplication is commutative, and so the order in which the factors appear is of no consequence. But in the matrix equations which we will be considering next, the corresponding factors will be matrices, and matrix multiplication is *not* commutative. That is, we know that

$$\begin{bmatrix} 2 & 4 \\ 6 & 8 \end{bmatrix}\begin{bmatrix} a & b \\ c & d \end{bmatrix} = \begin{bmatrix} (2a+4c) & (2b+4d) \\ (6a+8c) & (6b+8d) \end{bmatrix}$$

while

$$\begin{bmatrix} a & b \\ c & d \end{bmatrix}\begin{bmatrix} 2 & 4 \\ 6 & 8 \end{bmatrix} = \begin{bmatrix} (2a+6b) & (4a+8b) \\ (2c+6d) & (4c+8d) \end{bmatrix}$$

Different results are obtained, and we therefore need to be careful later about the order in which we write matrix factors.

Continuing with Eq. (37), let us integrate each side with respect to time from $-\infty$ to a general time t:

$$e^{-ta_{11}}q_1(t) = \int_{-\infty}^{t} e^{-za_{11}}f_1(z)\,dz \tag{38}$$

where z is simply a dummy variable of integration, and where we have assumed that $e^{-ta_{11}}q_1(t)$ approaches zero as t approaches $-\infty$. We now multiply (premultiply, if this were a matrix equation) each side of Eq. (38) by the exponential factor $e^{ta_{11}}$, obtaining

$$q_1(t) = e^{ta_{11}} \int_{-\infty}^{t} e^{-za_{11}} f_1(z) \, dz \qquad (39)$$

which is the desired expression for the single unknown state variable.

In many circuits, however, particularly those in which switches are present and the circuit is reconfigured at some instant of time (often $t = 0$), we do not know the forcing function or the normal-form equation prior to that instant. We therefore incorporate all of the past history in an integral from $-\infty$ to that instant, here assumed to be $t = 0$, by letting $t = 0$ in Eq. (39):

$$q_1(0) = \int_{-\infty}^{0} e^{-za_{11}} f_1(z) \, dz$$

We then use this initial value in the general solution for $q_1(t)$:

$$q_1(t) = e^{ta_{11}} q_1(0) + e^{ta_{11}} \int_{0}^{t} e^{-za_{11}} f_1(z) \, dz \qquad (40)$$

This last expression shows that the state-variable time function may be interpreted as the sum of two terms. The first is the response that would arise if the forcing function were zero [$f_1(t) = 0$], and in the language of state-variable analysis it is called the *zero-input response*. It has the *form* of the natural response, although it may not have the same amplitude as the term we have been calling the natural response. The zero-input response also is the solution to the homogeneous normal-form equation, obtained by letting $f_1(t) = 0$ in Eq. (36).

The second part of the solution would represent the complete response if $q_1(0)$ were zero, and it is termed the *zero-state response*. We shall see in a following example that what we have termed the forced response appears as a part of the zero-state response.

Let us use Eqs. (39) and (40) on two first-order examples.

Example 16-5 The first example is shown in Fig. 16-12. We seek $v_C(t)$.

Figure 16-12

A first-order circuit for which $v_C(t)$ is to be found through the methods of state-variable analysis.

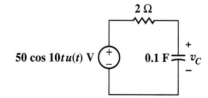

Solution: The normal-form equation is easily found to be

$$v_C' = -5v_C + 250 \cos 10t \, u(t)$$

Thus, $a_{11} = -5$, $f_1(t) = 250 \cos 10t \, u(t)$, and we may substitute directly in Eq. (39) to obtain the solution:

$$v_C(t) = e^{-5t} \int_{-\infty}^{t} e^{5z} 250 \cos 10z \, u(z) \, dz$$

The unit-step function inside the integral may be replaced by $u(t)$ outside the integral if the lower limit is changed to zero:

$$v_C(t) = e^{-5t} u(t) \int_{0}^{t} e^{5z} 250 \cos 10z \, dz$$

Integrating, we have

$$v_C(t) = e^{-5t}u(t)\left[\frac{250e^{5z}}{5^2 + 10^2}(5\cos 10z + 10\sin 10z)\right]_0^t$$

or $\qquad v_C(t) = [-10e^{-5t} + 10(\cos 10t + 2\sin 10t)]u(t) \qquad (41)$

The same result may be obtained through the use of Eq. (40). Since $v_C(0) = 0$, we have

$$v_C(t) = e^{-5t}(0) + e^{-5t}\int_0^t e^{5z}250\cos 10z \, u(z) \, dz$$

$$= e^{-5t}u(t)\int_0^t e^{5z}250\cos 10z \, dz$$

and this leads to the same solution as before, of course. However, we also see that the entire solution for $v_C(t)$ is the zero-state response, and there is no zero-input response. It is interesting to note that, if we had solved this problem by the methods of Chap. 5, we would have set up a natural response,

$$v_{C,n}(t) = Ae^{-5t}$$

and computed the forced response by frequency-domain methods,

$$V_{C,f} = \frac{50}{2 - j1}(-j1) = 10 - j20$$

so that

$$v_{C,f}(t) = 10\cos 10t + 20\sin 10t$$

Thus,

$$v_C(t) = Ae^{-5t} + 10(\cos 10t + 2\sin 10t) \qquad (t > 0)$$

and the application of the initial condition, $v_C(0) = 0$, leads to $A = -10$ and an expression identical to Eq. (41) once again. Looking at the partial responses obtained in the two methods, we therefore find that

$$v_{C,\text{zero-input}} = 0$$

$$v_{C,\text{zero-state}} = [-10e^{-5t} + 10(\cos 10t + 2\sin 10t)]u(t)$$

$$v_{C,n} = -10e^{-5t}$$

$$v_{C,f} = 10(\cos 10t + 2\sin 10t) \qquad\qquad ■$$

Now let us set up an example in which a zero-input response is present.

Example 16-6 The switch in the circuit of Fig. 16-13 is thrown at $t = 0$, and the form of the circuit changes at that instant. Find $i_L(t)$ for $t > 0$.

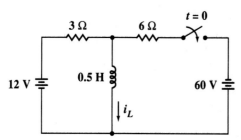

Figure 16-13

A first-order example in which the form of the circuit changes at $t = 0$.

Solution: We represent everything before $t = 0$ by the statement that $i_L(0) = 4$ A, and we obtain the normal-form equation for the circuit in the configuration it has *after* $t = 0$. It is

$$i_L'(t) = -4i_L(t) - 24$$

This time, we must use Eq. (40), since we do not have a single normal-form equation that is valid for all time. The result is

$$i_L(t) = e^{-4t}(4) + e^{-4t} \int_0^t e^{-z}(-24)\,dz$$

or
$$i_L(t) = 4e^{-4t} - 6(1 - e^{-4t}) \qquad (t > 0)$$

The several components of the response are now identified:

$$i_{L,\text{zero-input}} = 4e^{-4t} \qquad (t > 0)$$

$$i_{L,\text{zero-state}} = -6(1 - e^{-4t}) \qquad (t > 0)$$

while our earlier analytical methods would have led to

$$i_{L,n} = 10e^{-4t} \qquad (t > 0)$$

$$i_{L,f} = -6 \qquad (t > 0) \qquad \blacksquare$$

These first-order networks certainly do not require the use of state variables for their analyses. However, the method by which we solved the single normal-form equation does offer a few clues to how the nth-order solution might be obtained. We follow this exciting trail in the next section.

Drill Problem

16-6. (*a*) Write the normal-form equation for the circuit of Fig. 16-14. Find $v_C(t)$ for $t > 0$ by (*b*) Eq. (39); (*c*) Eq. 40.

Ans: $v_C' = -4v_C - 120 + 200u(t)$; $20 - 50e^{-4t}$ V; $20 - 50e^{-4t}$ V

Figure 16-14

See Drill Prob. 16-6.

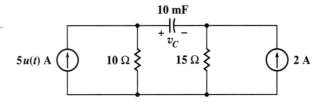

16-6

The solution of the matrix equation

The general matrix equation for the nth-order system that we now wish to solve is given as Eq. (34) in Sec. 16-4,

$$\mathbf{q}' = \mathbf{aq} + \mathbf{f} \qquad (34)$$

where $\mathbf{a}$ is an $(n \times n)$ square matrix of constants and the other three matrices are all $(n \times 1)$ column matrices whose elements are, in general, time functions. In the most general case, all the matrices would be composed of time functions.

In this section we shall obtain the matrix solution for this equation. In the following section we will interpret our results and indicate how we might obtain a useful solution for $\mathbf{q}$.

We begin by subtracting the matrix product $\mathbf{aq}$ from each side of Eq. (34):

$$\mathbf{q}' - \mathbf{aq} = \mathbf{f} \qquad (42)$$

Recalling our case of the integrating factor $e^{-ta_{11}}$ in the first-order case, let us premultiply each side of Eq. (42) by e^{-ta}:

$$e^{-ta}\mathbf{q}' - e^{-ta}\mathbf{aq} = e^{-ta}\mathbf{f} \tag{43}$$

Although the presence of a matrix as an exponent may seem to be somewhat strange, the function e^{-ta} may be defined in terms of its infinite power series expansion in t,

$$e^{-ta} = \mathbf{I} - t\mathbf{a} + \frac{t^2}{2!}(\mathbf{a})^2 - \frac{t^3}{3!}(\mathbf{a})^3 + \cdots \tag{44}$$

We identify $\mathbf{I}$ as the $(n \times n)$ identity matrix,

$$\mathbf{I} = \begin{bmatrix} 1 & 0 & \cdots & 0 \\ 0 & 1 & \cdots & 0 \\ \cdots\cdots\cdots\cdots\cdots \\ 0 & 0 & \cdots & 1 \end{bmatrix}$$

such that

$$\mathbf{Ia} = \mathbf{aI} = \mathbf{a}$$

The products $(\mathbf{a})^2$, $(\mathbf{a})^3$, and so forth in Eq. (44) may be obtained by repeated multiplication of the matrix $\mathbf{a}$ by itself, and therefore each term in the expansion is again an $(n \times n)$ matrix. Thus, it is apparent that e^{-ta} is also an $(n \times n)$ square matrix, but its elements are all functions of time in general.

Again following the first-order procedure, we would now like to show that the left-hand side of Eq. (43) is equal to the time derivative of $e^{-ta}\mathbf{q}$. Since this is a product of two functions of time, we have

$$\frac{d}{dt}(e^{-ta}\mathbf{q}) = e^{-ta}\frac{d}{dt}(\mathbf{q}) + \left[\frac{d}{dt}(e^{-ta})\right]\mathbf{q}$$

The derivative of e^{-ta} is obtained by again considering the infinite series of Eq. (44), and we find that it is given by $-\mathbf{a}e^{-ta}$. The series expansion can also be used to show that $-\mathbf{a}e^{-ta} = -e^{-ta}\mathbf{a}$. Thus

$$\frac{d}{dt}(e^{-ta}\mathbf{q}) = e^{-ta}\mathbf{q}' - e^{-ta}\mathbf{aq}$$

and Eq. (43) helps to simplify this expression into the form

$$\frac{d}{dt}(e^{-ta}\mathbf{q}) = e^{-ta}\mathbf{f}$$

Multiplying by dt and integrating from $-\infty$ to t, we have

$$e^{-ta}\mathbf{q} = \int_{-\infty}^{t} e^{-za}\mathbf{f}(z)\,dz \tag{45}$$

To solve for $\mathbf{q}$, we must premultiply the left-hand side of Eq. (45) by the matrix inverse of e^{-ta}. That is, any square matrix $\mathbf{b}$ has an inverse $\mathbf{b}^{-1}$ such that $\mathbf{b}^{-1}\mathbf{b} = \mathbf{b}\,\mathbf{b}^{-1} = \mathbf{I}$. In this case, another power series expansion shows that the inverse of e^{-ta} is e^{ta}, or

$$e^{ta}e^{-ta}\mathbf{q} = \mathbf{Iq} = \mathbf{q}$$

and we may therefore write our solution as

$$\mathbf{q} = e^{t\mathbf{a}} \int_{-\infty}^{t} e^{-z\mathbf{a}}\mathbf{f}(z)\, dz \qquad (46)$$

In terms of the initial value of the state vector,

$$\mathbf{q} = e^{t\mathbf{a}}\mathbf{q}(0) + e^{t\mathbf{a}} \int_{0}^{t} e^{-z\mathbf{a}}\mathbf{f}(z)\, dz \qquad (47)$$

The function $e^{t\mathbf{a}}$ is a very important quantity in state-space analysis. It is called the *state-transition matrix,* for it describes how the state of the system changes from its zero state to its state at time t. Equations (46) and (47) are the nth-order matrix equations that correspond to the first-order results we numbered (39) and (40). Although these expressions represent the "solutions" for $\mathbf{q}$, the fact that we can express $e^{t\mathbf{a}}$ and $e^{-t\mathbf{a}}$ only as infinite series is a serious deterrent to our making any effective use of these results. We would have infinite power series in t for each $q_i(t)$, and while a computer might find this procedure compatible with its memory and computational speed, we would probably have other, more pressing chores to handle than carrying out the summation by ourselves.

We therefore shall obtain a more satisfactory representation for $e^{-t\mathbf{a}}$ and $e^{t\mathbf{a}}$ in the next section. Be patient.

Drill Problem

16-7. Let $\mathbf{a} = \begin{bmatrix} 0 & 2 \\ -1 & 1 \end{bmatrix}$ and $t = 0.1$ s. Use the power series expansion to find the (a) matrix $e^{-t\mathbf{a}}$; (b) matrix $e^{t\mathbf{a}}$; (c) value of the determinant of $e^{-t\mathbf{a}}$; (d) value of the determinant of $e^{t\mathbf{a}}$; (e) product of the last two results.

Ans: $\begin{bmatrix} 0.9903 & -0.1897 \\ 0.0948 & 0.8955 \end{bmatrix}$; $\begin{bmatrix} 0.9897 & 0.2097 \\ -0.1048 & 1.0945 \end{bmatrix}$; 0.9048; 1.1052; 1.0000

16-7

A further look at the state-transition matrix

In this section we seek a more satisfactory representation for $e^{t\mathbf{a}}$ and $e^{-t\mathbf{a}}$. If the system matrix $\mathbf{a}$ is an $(n \times n)$ square matrix, then each of these exponentials is an $(n \times n)$ square matrix of time functions, and one of the consequences of a theorem developed in linear algebra, known as the Cayley-Hamilton theorem, shows that such a matrix may be expressed as an $(n-1)$st-degree polynomial in the matrix $\mathbf{a}$. That is,

$$e^{t\mathbf{a}} = u_0\mathbf{I} + u_1\mathbf{a} + u_2(\mathbf{a})^2 + \cdots + u_{n-1}(\mathbf{a})^{n-1} \qquad (48)$$

where each of the u_i is a scalar function of time that is still to be determined; the $\mathbf{a}^i$ are constant $(n \times n)$ matrices. The theorem also states that Eq. (48) remains an equality if $\mathbf{I}$ is replaced by unity and $\mathbf{a}$ is replaced by any one of the scalar roots s_i of the nth-degree scalar equation,

$$\det(\mathbf{a} - s\mathbf{I}) = 0 \qquad (49)$$

The expression $\det(\mathbf{a} - s\mathbf{I})$ indicates the determinant of the matrix $(\mathbf{a} - s\mathbf{I})$. This determinant is an nth-degree polynomial in s. We shall assume that the n roots are all different. Equation (49) is called the *characteristic equation* of the matrix $\mathbf{a}$, and the values of s which are the roots of the equation are known as the *eigenvalues* of $\mathbf{a}$.

These values of s are identical with the natural frequencies that we dealt with most recently in Chap. 12 as poles of an appropriate transfer function. That is, if our state variable is $v_{C1}(t)$, then the poles of $\mathbf{H(s)} = \mathbf{V}_{C1}(s)/\mathbf{I}_s$ or of $\mathbf{V}_{C1}(s)/\mathbf{V}_s$ are also the eigenvalues of the characteristic equation.

Thus, this is the procedure we shall follow to obtain a simpler form for $e^{t\mathbf{a}}$:

> 1 Given $\mathbf{a}$, form the matrix $(\mathbf{a} - s\mathbf{I})$.
> 2 Set the determinant of this square matrix equal to zero.
> 3 Solve the resultant nth-degree polynomial for its n roots, $s_1, s_2,$ $\ldots, s_n$.
> 4 Write the n scalar equations of the form
> $$e^{ts_i} = u_0 + u_1 s_i + \cdots + u_{n-1} s_i^{n-1} \tag{50}$$
> 5 Solve for the n time functions $u_0, u_1, \ldots, u_{n-1}$.
> 6 Substitute these time functions in Eq. (48) to obtain the $(n \times n)$ matrix $e^{t\mathbf{a}}$.

To illustrate this procedure, let us use the system matrix that corresponds to Eqs. (7) and (8) of Sec. 16-2 and to the circuit shown in Fig. 16-2a:

$$\mathbf{a} = \begin{bmatrix} -0.5 & -2.5 \\ 0.5 & -3.5 \end{bmatrix}$$

Therefore,

$$(\mathbf{a} - s\mathbf{I}) = \begin{bmatrix} -0.5 & -2.5 \\ 0.5 & -3.5 \end{bmatrix} - s\begin{bmatrix} 1 & 0 \\ 0 & 1 \end{bmatrix} = \begin{bmatrix} (-0.5 - s) & -2.5 \\ 0.5 & (-3.5 - s) \end{bmatrix}$$

and the expansion of the corresponding (2×2) determinant gives

$$\det \begin{bmatrix} (-0.5 - s) & -2.5 \\ 0.5 & (-3.5 - s) \end{bmatrix} = \begin{vmatrix} (-0.5 - s) & -2.5 \\ 0.5 & (-3.5 - s) \end{vmatrix}$$

$$= (-0.5 - s)(-3.5 - s) + 1.25$$

so that

$$\det (\mathbf{a} - s\mathbf{I}) = s^2 + 4s + 3$$

The roots of this polynomial are $s_1 = -1$ and $s_2 = -3$, and we substitute each of these values in Eq. (50), obtaining the two equations

$$e^{-t} = u_0 - u_1 \quad \text{and} \quad e^{-3t} = u_0 - 3u_1$$

Subtracting, we find that

$$u_1 = 0.5e^{-t} - 0.5e^{-3t}$$

and, therefore,

$$u_0 = 1.5e^{-t} - 0.5e^{-3t}$$

Note that each u_i has the general form of the natural response.

When these two functions are installed in Eq. (48), so that

$$e^{t\mathbf{a}} = (1.5e^{-t} - 0.5e^{-3t})\mathbf{I} + (0.5e^{-t} - 0.5e^{-3t})\begin{bmatrix} -0.5 & -2.5 \\ 0.5 & -3.5 \end{bmatrix}$$

we may carry out the indicated operations to find our desired expression for $e^{t\mathbf{a}}$,

$$e^{t\mathbf{a}} = \begin{bmatrix} (1.25e^{-t} - 0.25e^{-3t}) & (-1.25e^{-t} + 1.25e^{-3t}) \\ (0.25e^{-t} - 0.25e^{-3t}) & (-0.25e^{-t} + 1.25e^{-3t}) \end{bmatrix} \tag{51}$$

Having $e^{t\mathbf{a}}$, we form $e^{-t\mathbf{a}}$ by replacing each t in Eq. (51) by $-t$. To complete this example, we may identify $\mathbf{q}$ and $\mathbf{f}$ from Eqs. (7), (8), and (9):

$$\mathbf{q} = \begin{bmatrix} v_C \\ i_L \end{bmatrix}$$

$$\mathbf{f} = \begin{bmatrix} 30 + 80e^{-2t}u(t) \\ 0 \end{bmatrix} \tag{52}$$

Since part of the forcing-function vector is present for $t < 0$, we may as well use Eq. (46) to solve for the state vector:

$$\mathbf{q} = e^{t\mathbf{a}} \int_{-\infty}^{t} e^{-z\mathbf{a}}\mathbf{f}(z)\,dz \tag{46}$$

The matrix product $e^{-z\mathbf{a}}\mathbf{f}(z)$ is next formed by replacing t by $-z$ in Eq. (51) and then postmultiplying by Eq. (52) with z replacing t. With only moderate labor, we find that

$$e^{-z\mathbf{a}}\mathbf{f} = \begin{bmatrix} 37.5e^{z} - 7.5e^{3z} + (10e^{-z} - 2e^{z})u(z) \\ 7.5e^{z} - 7.5e^{3z} + (2e^{-z} - 2e^{z})u(z) \end{bmatrix}$$

Integrating the first two terms of each element from $-\infty$ to t and the last two terms from 0 to t, we have

$$\int_{-\infty}^{t} e^{-z\mathbf{a}}\mathbf{f}\,dz = \begin{bmatrix} 35.5e^{t} - 2.5e^{3t} - 10e^{-t} + 12 \\ 5.5e^{t} - 2.5e^{3t} - 2e^{-t} + 4 \end{bmatrix} \quad (t > 0)$$

Finally, we must premultiply this matrix by Eq. (51) and this matrix multiplication involves a large number of scalar multiplications and algebraic additions of like time functions. The result is the desired state vector

$$\mathbf{q} = \begin{bmatrix} 35 + 10e^{-t} - 12e^{-2t} + 2e^{-3t} \\ 5 + 2e^{-t} - 4e^{-2t} + 2e^{-3t} \end{bmatrix} \quad (t > 0)$$

These are the expressions that were given offhandedly in Eqs. (10) and (11) at the end of Sec. 16-2.

We now have a general technique that we could apply to higher-order problems. Such a procedure is theoretically possible, but the labor involved soon becomes monumental. Instead, we utilize computer programs that work directly from the normal-form equations and the initial values of the state variables. Knowing $\mathbf{q}(0)$, we compute $\mathbf{f}(0)$ and solve the equations for $\mathbf{q}'(0)$. Then $\mathbf{q}(\Delta t)$ may be approximated from the initial value and its derivative: $\mathbf{q}(\Delta t) \doteq \mathbf{q}(0) + \mathbf{q}'(0)\,\Delta t$. With this value for $\mathbf{q}(\Delta t)$, the normal-form equations are used to determine a value for $\mathbf{q}'(\Delta t)$, and the process moves along by time increments of Δt. Smaller values for Δt lead to greater accuracy for $\mathbf{q}(t)$ at any time $t > 0$, but with attendant penalties in computational time and computer storage requirements.

With this example, we complete our introduction to the subject of state variables, and it is well to ask what our accomplishments have been. First, and perhaps most important, we have learned some of the terms and ideas of this branch of system analysis, and this should make future study of this field more meaningful and pleasurable.

Another accomplishment is the general solution for the first-order case in matrix form.

We have also obtained the matrix solution for the general case. This introduction to the use of matrices in circuit and system analysis is a tool which becomes increasingly necessary in more advanced work in these areas.

Finally, we have indicated how a numerical solution might be obtained using numerical methods, hopefully by a digital computer. We extend our capabilities to include forcing functions having periodic nature in the next chapter.

16-8. Given $\mathbf{q} = \begin{bmatrix} v_C \\ i_L \end{bmatrix}$, $\mathbf{a} = \begin{bmatrix} 0 & -27 \\ \frac{1}{3} & -10 \end{bmatrix}$, $\mathbf{f} = \begin{bmatrix} 243u(t) \\ 40u(t) \end{bmatrix}$, and $\mathbf{q}(0) = \begin{bmatrix} 150 \\ 5 \end{bmatrix}$, use **Drill Problem**
a Δt of 0.001 s and calculate: (a) $v_C(0.001)$; (b) $i_L(0.001)$; (c) $v_C'(0.001)$; (d) $i_L'(0.001)$; (e) $v_C(0.002)$; (f) $i_L(0.002)$.

Ans: 150.108 V; 5.04 A; 106.92 V/s; 39.636 A/s; 150.2149 V; 5.0796 A

---

1 Using the order i_1, i_2, i_3, write the following equations as a set of normal-form equa- **Problems**
tions: $-2i_1' - 6i_3' = 5 + 2 \cos 10t - 3i_1 + 2i_2$, $4i_2 = 0.05i_1' - 0.15i_2' + 0.25i_3'$, $i_2 = -2i_1 - 5i_3 + 0.4 \int_0^t (i_1 - i_3)\, dt + 8$.

2 Given the two linear differential equations $x' + y' = x + y + 1$ and $x' - 2y' = 2x - y - 1$: (a) write the two normal-form equations, using the order x, y; (b) obtain a single differential equation involving only $x(t)$ and its derivatives. (c) If $x(0) = 2$ and $y(0) = -5$, find $x'(0)$, $x''(0)$, and $x'''(0)$.

3 Write the following equations in normal form, using the order x, y, z: $x' - 2y - 3z' = f_1(t)$, $2x' + 5z = 3$, $z' - 2y' - x = 0$.

4 If $x' = -2x - 3y + 4$ and $y' = 5x - 6y + 7$, let $x(0) = 2$ and $y(0) = \frac{1}{3}$, and find (a) $x''(0)$; (b) $y''(0)$; (c) $y'''(0)$.

5 If $v_s = 100 \cos 120 \pi t$ V, write a set of normal-form equations for the circuit shown in Fig. 16-15. Use i_L and v_C as the state variables.

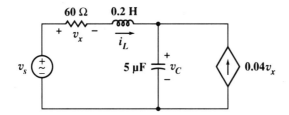

Figure 16-15

See Prob. 5.

6 (a) Draw a normal tree for the circuit of Fig. 16-16 and assign the necessary state variables (for uniformity, put + references at the top or left side of an element, and direct arrows down or to the right). (b) Specify every link current and tree-branch voltage in terms of the sources, element values, and state variables. (c) Write the normal-form equations using the order i_L, v_x, v_{2F}.

Figure 16-16

See Prob. 6.

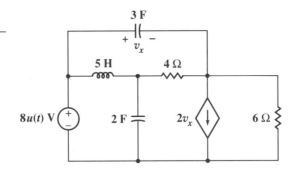

7 Write a set of normal-form equations for each circuit shown in Fig. 16-17. Use the state-variable order given below the circuit.

Figure 16-17

See Probs. 7, 8, and 14.

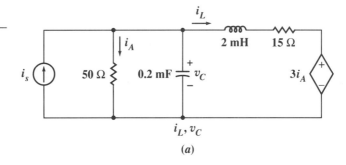

(a)

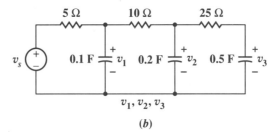

(b)

8 A 12.5-Ω resistor is placed in series with the 0.2-mF capacitor in the circuit of Fig. 16-17a. Using one current and one voltage, in that order, write a set of normal-form equations for this circuit.

9 (a) Write normal-form equations in the state variables v_1, v_2, i_1, and i_2 for Fig. 16-18. (b) Repeat if the current source is replaced with a 3-Ω resistor.

Figure 16-18

See Prob. 9.

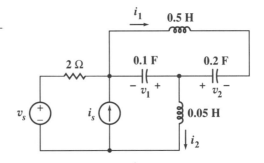

10 Write a set of normal-form equations in the order i_L, v_C for each circuit shown in Fig. 16-19.

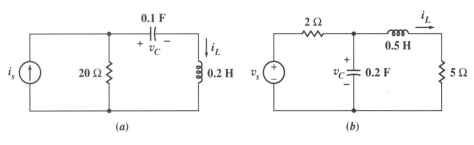

Figure 16-19

See Prob. 10.

(a)   (b)

11 Write a set of normal-form equations for the circuit shown in Fig. 16-20. Use the variable order v, i.

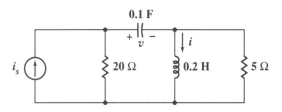

Figure 16-20

See Prob. 11.

12 If $v_s = 5 \sin 2t \, u(t)$ V, write a set of normal-form equations for the circuit of Fig. 16-21.

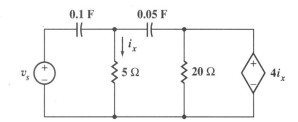

Figure 16-21

See Prob. 12.

13 (a) Given $\mathbf{q} = \begin{bmatrix} i_{L1} \\ i_{L2} \\ v_C \end{bmatrix}$, $\mathbf{a} = \begin{bmatrix} -1 & -2 & -3 \\ 4 & -5 & 6 \\ 7 & -8 & -9 \end{bmatrix}$, and $\mathbf{f} = \begin{bmatrix} 2t \\ 3t^2 \\ 1+t \end{bmatrix}$, write out the three normal-form equations. (b) If $\mathbf{q} = \begin{bmatrix} i_L \\ v_C \end{bmatrix}$, $\mathbf{a} = \begin{bmatrix} 0 & -6 \\ 4 & 0 \end{bmatrix}$, and $\mathbf{f} = \begin{bmatrix} 1 \\ 0 \end{bmatrix}$, show the state variables and the elements (with their values) on the skeleton circuit diagram of Fig. 16-22.

Figure 16-22

See Prob. 13.

14 Replace the capacitors in the circuit of Fig. 16-17b by 0.1-H, 0.2-H, and 0.5-H inductors and find $\mathbf{a}$ and $\mathbf{f}$ if i_{L1}, i_{L2}, and i_{L3} are the state variables.

15 Given the state vector $\mathbf{q} = \begin{bmatrix} v_{C1} \\ v_{C2} \\ i_L \end{bmatrix}$, let the system matrix $\mathbf{a} = \begin{bmatrix} -3 & 3 & 0 \\ 2 & -1 & 0 \\ 1 & 4 & -2 \end{bmatrix}$, and the forcing function vector $\mathbf{f} = \begin{bmatrix} 10 \\ 0 \\ 0 \end{bmatrix}$. Other voltages and currents in the circuit appear in the vector $\mathbf{w} = \begin{bmatrix} v_{o1} \\ v_{o2} \\ i_{R1} \\ i_{R2} \end{bmatrix}$, and they are related to the state vector by $\mathbf{w} = \mathbf{bq} + \mathbf{d}$, where

$\mathbf{b} = \begin{bmatrix} 1 & 2 & 0 \\ 0 & 0 & -1 \\ 0 & -1 & 2 \\ 1 & 3 & 0 \end{bmatrix}$ and $\mathbf{d} = \begin{bmatrix} 0 \\ 2 \\ -2 \\ 0 \end{bmatrix}$. Write out the set of equations giving v'_{o1}, v'_{o2}, i'_{R1},

and i'_{R2} as functions of the state variables.

16 Let $\mathbf{q} = \begin{bmatrix} q_1 \\ q_2 \\ q_3 \end{bmatrix}$, $\mathbf{a} = \begin{bmatrix} -3 & 1 & 2 \\ -2 & -2 & 1 \\ -1 & 3 & 0 \end{bmatrix}$, $\mathbf{f} = \begin{bmatrix} \cos 2\pi t \\ \sin 2\pi t \\ 0 \end{bmatrix}$, $\mathbf{y} = \begin{bmatrix} y_1 \\ y_2 \\ y_3 \\ y_4 \end{bmatrix}$, $\mathbf{b} =$

$\begin{bmatrix} 1 & 2 & 0 & 3 \\ 0 & -1 & 1 & 1 \\ 2 & -1 & -1 & 3 \end{bmatrix}$, and $\mathbf{d} = \begin{bmatrix} 2 \\ 1 \\ 3 \end{bmatrix}$. Then if $\mathbf{q}' = \mathbf{aq} + \mathbf{f}$ and $\mathbf{q} = \mathbf{by} + \mathbf{d}$, determine

$\mathbf{q}(0)$ and $\mathbf{q}'(0)$ if $\mathbf{y}(0) = \begin{bmatrix} 10 \\ -10 \\ -5 \\ 5 \end{bmatrix}$.

17 Find $\mathbf{a}$ and $\mathbf{f}$ for the circuit shown in Fig. 16-23 if $\mathbf{q} = \begin{bmatrix} v_C \\ i_L \end{bmatrix}$.

Figure 16-23

See Prob. 17.

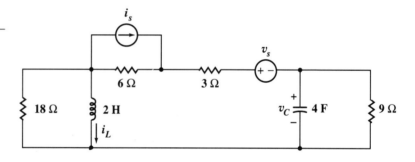

18 For the circuit shown in Fig. 16-24: (a) write the normal-form equation for i, $t > 0$; (b) solve this equation for i; (c) identify the zero-state and zero-input responses; (d) find i by the methods of Chap. 5 and specify the natural and forced responses.

Figure 16-24

See Prob. 18.

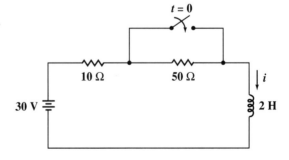

19 Let $v_s = 2tu(t)$ V in the circuit shown in Fig. 16-25. Find $i_2(t)$.

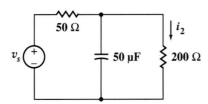

Figure 16-25

See Prob. 19.

20 (a) Use state-variable methods to find $i_L(t)$ for all t in the circuit shown in Fig. 16-26. (b) Identify the zero-input, zero-state, natural, and forced responses.

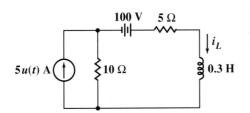

Figure 16-26

See Prob. 20.

21 Let $v_s = 100[u(t) - u(t - 0.5)] \cos \pi t$ V in Fig. 16-27. Find v_C for $t > 0$.

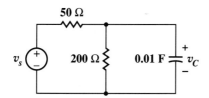

Figure 16-27

See Prob. 21.

22 (a) Write the normal-form equation for the circuit shown in Fig. 16-28. (b) Find $v(t)$ for $t < 0$ and $t > 0$. (c) Identify the forced response, the natural response, the zero-state response, and the zero-input response.

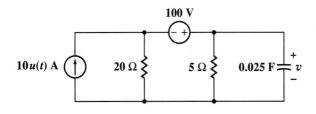

Figure 16-28

See Prob. 22.

23 Let $v_s = 90e^{-t}[u(t) - u(t - 0.5)]$ V in the circuit of Fig. 16-29. Find $i_L(t)$.

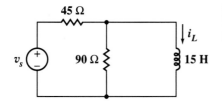

Figure 16-29

See Prob. 23.

24 Given the system matrix $\mathbf{a} = \begin{bmatrix} -8 & 5 \\ 10 & -10 \end{bmatrix}$, let $t = 10$ ms and use the infinite power series for the exponential to find (a) $e^{-t\mathbf{a}}$; (b) $e^{t\mathbf{a}}$; (c) $e^{-t\mathbf{a}}e^{t\mathbf{a}}$.

25 (a) If $\mathbf{q} = \begin{bmatrix} x \\ y \\ z \end{bmatrix}$, $\mathbf{f} = \begin{bmatrix} u(t) \\ \cos t \\ -u(t) \end{bmatrix}$, and $\mathbf{a} = \begin{bmatrix} 1 & 2 & -1 \\ 0 & -1 & 3 \\ -2 & -3 & -1 \end{bmatrix}$, write the set of normal-form

equations. (b) If $\mathbf{q}(0) = \begin{bmatrix} 2 \\ -3 \\ 1 \end{bmatrix}$, estimate $\mathbf{q}(0.1)$ using $\Delta t = 0.1$. (c) Repeat for $\Delta t = 0.05$.

26 Determine the eigenvalues of the matrix $\mathbf{a} = \begin{bmatrix} -1 & 2 & 3 \\ 0 & -1 & 2 \\ 3 & 1 & -1 \end{bmatrix}$. As a help, one root

is near $s = -3.5$.

27 Using the method described by Eqs. (48) to (50), find $e^{t\mathbf{a}}$ if $\mathbf{a} = \begin{bmatrix} -3 & 2 \\ 1 & -4 \end{bmatrix}$.

28 (a) Find the normal-form equations for the circuit of Fig. 16-30. Let $\mathbf{q} = \begin{bmatrix} i \\ v \end{bmatrix}$.
(b) Find the eigenvalues of $\mathbf{a}$. (c) Determine u_0 and u_1. (d) Specify $e^{t\mathbf{a}}$. (e) Use $e^{t\mathbf{a}}$ to find $\mathbf{q}$ for $t > 0$.

Figure 16-30

See Prob. 28.

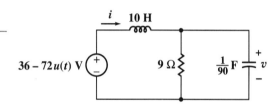

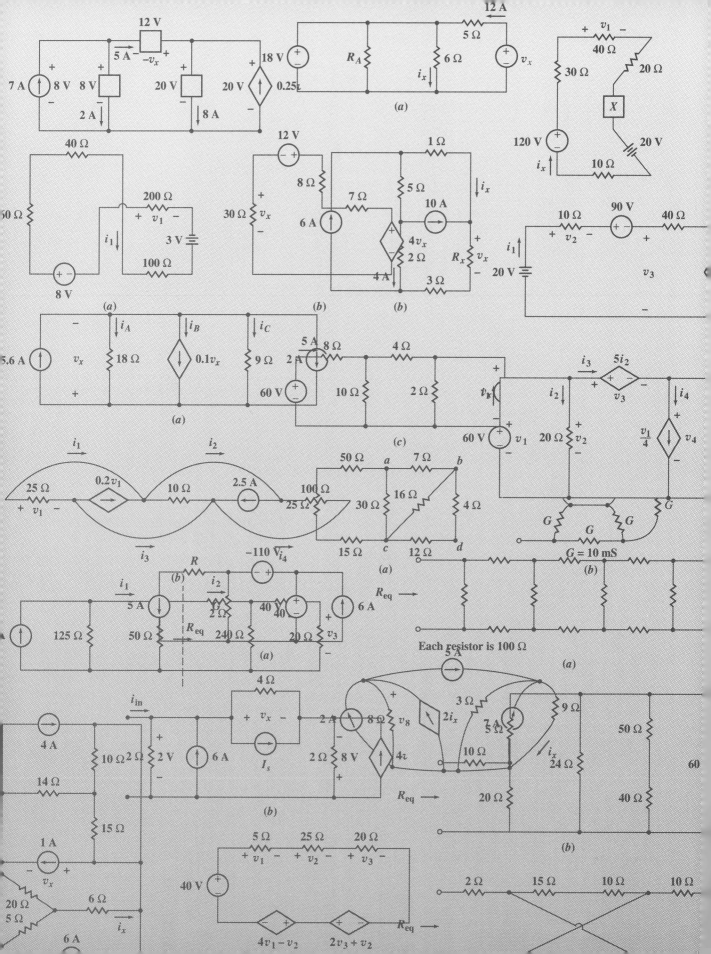

Fourier Analysis

In this chapter we continue our introduction to circuit analysis by studying periodic functions in both the time and frequency domains.

We know that the complete response of a linear circuit to an arbitrary forcing function is composed of the sum of a forced response and a natural response. The natural response was initially considered in Chaps. 4 to 6, but with few exceptions, only simple series or parallel RL, RC, and RLC circuits were examined. Later, the complex-frequency concept in Chap. 12 provided us with a general method of obtaining the natural response; we discovered that we could write the form of the natural response after locating the poles of an appropriate transfer function of the network. Thus, we found a powerful general method for determining the natural response.

A second general method for finding the natural response (and also the forced response) is based on state-variable theory, a subject we just met in Chap. 16, and one which will no doubt return for a more extensive treatment in a future course.

Now let us consider our status with respect to the forced response. We are able to find the forced response in any purely resistive linear circuit, regardless of the nature of the forcing function, but this can hardly be classed as a scientific breakthrough. If the circuit contains energy-storage elements, then we can certainly find the forced response for those circuits and forcing functions to which we can apply the impedance concept, that is, whenever the forcing function is direct current, exponential, sinusoidal, or exponentially varying sinusoidal in form.

In this chapter we consider forcing functions which are *periodic* and have functional natures which satisfy certain mathematical restrictions that are characteristic of any function which we can generate in the laboratory. Any such function may be represented as the sum of an infinite number of sine and cosine functions which are harmonically related. Therefore, since the forced response to each sinusoidal component can be determined easily by sinusoidal steady-state analysis, the response of the linear network to the general periodic forcing function may be obtained by superposing the partial responses.

Some feeling for the validity of representing a general periodic function by an infinite sum of sine and cosine functions may be gained by considering a simple example. Let us first assume a cosine function of radian frequency ω_0,

$$v_1(t) = 2 \cos \omega_0 t$$

17-1

Introduction

where

$$\omega_0 = 2\pi f_0$$

and the period T is

$$T = \frac{1}{f_0} = \frac{2\pi}{\omega_0}$$

Although T does not usually carry a zero subscript, it is the period of the fundamental frequency. The *harmonics* of this sinusoid have frequencies $n\omega_0$, where ω_0 is the fundamental frequency and $n = 1, 2, 3, \ldots$. The frequency of the first harmonic is the fundamental frequency.

Next let us select a third-harmonic voltage

$$v_{3a}(t) = \cos 3\omega_0 t$$

The fundamental $v_1(t)$, the third harmonic $v_{3a}(t)$, and the sum of these two waves are shown as functions of time in Fig. 17-1a. It should be noted that the sum is periodic, with period $T = 2\pi/\omega_0$.

Figure 17-1

Several of the infinite number of different wave-forms which may be obtained by combining a fundamental and a third harmonic. The fundamental is $v_1 = 2 \cos \omega_0 t$, and the third harmonic is (a) $v_{3a} = \cos 3\omega_0 t$; (b) $v_{3b} = 1.5 \cos 3\omega_0 t$; (c) $v_{3c} = \sin 3\omega_0 t$.

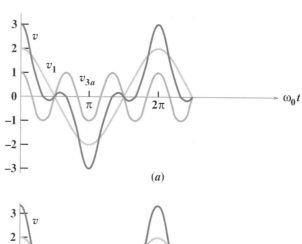

(a)

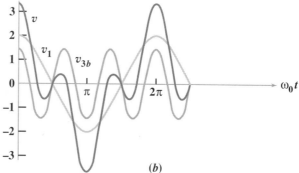

(b)

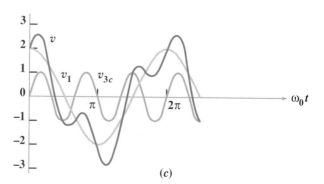

(c)

The form of the resultant periodic function changes as the phase and amplitude of the third-harmonic component change. Thus, Fig. 17-1b shows the effect of combining $v_1(t)$ and a third harmonic of slightly larger amplitude,

$$v_{3b}(t) = 1.5 \cos 3\omega_0 t$$

By shifting the phase of the third harmonic, to give

$$v_{3c}(t) = \sin 3\omega_0 t$$

the sum, shown in Fig. 17-1c, takes on a still different character. In all cases, the period of the resultant waveform is the same as the period of the fundamental waveform. The nature of the waveform depends on the amplitude and phase of every possible harmonic component, and we shall find that we are able to generate waveforms which have extremely nonsinusoidal characteristics by an appropriate combination of sinusoidal functions.

After we have become familiar with the use of the sum of an infinite number of sine and cosine functions to represent a periodic waveform, we shall consider the frequency-domain representation of a general nonperiodic waveform in the next chapter.

17-1. Let a third-harmonic voltage be added to the fundamental to yield $v = 2 \cos \omega_0 t + V_{m3} \sin 3\omega_0 t$, the waveform shown in Fig. 17-1c for $V_{m3} = 1$. (a) Find the value of V_{m3} so that $v(t)$ will have zero slope at $\omega_0 t = 2\pi/3$. (b) Evaluate $v(t)$ at $\omega_0 t = 2\pi/3$.

Drill Problem

Ans: 0.577; -1.000

17-2
Trigonometric form of the Fourier series

We first consider a *periodic* function $f(t)$, defined in Sec. 10-3 by the functional relationship

$$f(t) = f(t + T)$$

where T is the period. We further assume that the function $f(t)$ satisfies the following properties:

1. $f(t)$ is single-valued everywhere; that is, $f(t)$ satisfies the mathematical definition of a function.
2. The integral $\int_{t_0}^{t_0+T} |f(t)|\, dt$ exists (i.e., is not infinite) for any choice of t_0.
3. $f(t)$ has a finite number of discontinuities in any one period.
4. $f(t)$ has a finite number of maxima and minima in any one period.

We shall consider $f(t)$ to represent a voltage or current waveform, and any voltage or current waveform which we can actually produce must satisfy these conditions. Certain mathematical functions which we might hypothesize may not satisfy these conditions, but we shall assume that these four conditions are always satisfied.

Given such a periodic function $f(t)$, the Fourier theorem[1] states that $f(t)$ may be represented by the infinite series

$$f(t) = a_0 + a_1 \cos \omega_0 t + a_2 \cos 2\omega_0 t + \cdots$$
$$+ b_1 \sin \omega_0 t + b_2 \sin 2\omega_0 t + \cdots$$
$$= a_0 + \sum_{n=1}^{\infty} (a_n \cos n\omega_0 t + b_n \sin n\omega_0 t) \qquad (1)$$

[1] Jean Baptiste Joseph Fourier published this theorem in 1822. Some rather unbelievable pronunciations of this French name come from American students; it should rhyme with "poor pay."

where the fundamental frequency ω_0 is related to the period T by

$$\omega_0 = \frac{2\pi}{T}$$

and where a_0, a_n, and b_n are constants which depend upon n and $f(t)$. Equation (1) is the trigonometric form of the *Fourier series for f(t)*, and the process of determining the values of the constants a_0, a_n, and b_n is called *Fourier analysis*. Our object is not the proof of this theorem, but only a simple development of the procedures of Fourier analysis and a feeling that the theorem is plausible.

Before we discuss the evaluation of the constants appearing in the Fourier series, let us collect a set of useful trigonometric integrals. We let both n and k represent any element of the set of integers 1, 2, 3, ... In the following integrals, 0 and T are used as the integration limits, but it is understood that any interval of one period is equally correct. Since the average value of a sinusoid over one period is zero,

$$\int_0^T \sin n\omega_0 t \, dt = 0 \tag{2}$$

and
$$\int_0^T \cos n\omega_0 t \, dt = 0 \tag{3}$$

It is also a simple matter to show that the following three definite integrals are zero:

$$\int_0^T \sin k\omega_0 t \cos n\omega_0 t \, dt = 0 \tag{4}$$

$$\int_0^T \sin k\omega_0 t \sin n\omega_0 t \, dt = 0 \qquad (k \neq n) \tag{5}$$

$$\int_0^T \cos k\omega_0 t \cos n\omega_0 t \, dt = 0 \qquad (k \neq n) \tag{6}$$

Those cases which are excepted in Eqs. (5) and (6) are also easily evaluated; we obtain

$$\int_0^T \sin^2 n\omega_0 t \, dt = \frac{T}{2} \tag{7}$$

$$\int_0^T \cos^2 n\omega_0 t \, dt = \frac{T}{2} \tag{8}$$

The evaluation of the unknown constants in the Fourier series may now be accomplished readily. We first attack a_0. If we integrate each side of Eq. (1), over a full period, we obtain

$$\int_0^T f(t) \, dt = \int_0^T a_0 \, dt + \int_0^T \sum_{n=1}^{\infty} (a_n \cos n\omega_0 t + b_n \sin n\omega_0 t) \, dt$$

But every term in the summation is of the form of Eq. (2) or (3), and thus

$$\int_0^T f(t) \, dt = a_0 T$$

or
$$a_0 = \frac{1}{T} \int_0^T f(t) \, dt \tag{9}$$

This constant a_0 is simply the average value of $f(t)$ over a period, and we therefore describe it as the dc component of $f(t)$.

To evaluate one of the cosine coefficients—say, a_k, the coefficient of $\cos k\omega_0 t$—we first multiply each side of Eq. (1) by $\cos k\omega_0 t$ and then integrate both sides of the equation over a full period:

$$\int_0^T f(t)\cos k\omega_0 t\, dt = \int_0^T a_0 \cos k\omega_0 t\, dt$$

$$+ \int_0^T \sum_{n=1}^{\infty} a_n \cos k\omega_0 t \cos n\omega_0 t\, dt$$

$$+ \int_0^T \sum_{n=1}^{\infty} b_n \cos k\omega_0 t \sin n\omega_0 t\, dt$$

From Eqs. (3), (4), and (6) we note that every term on the right-hand side of this equation is zero except for the single a_n term where $k = n$. We evaluate that term using Eq. (8), and in so doing we find a_k, or a_n:

$$a_n = \frac{2}{T}\int_0^T f(t)\cos n\omega_0 t\, dt \tag{10}$$

This result is *twice* the average value of the product $f(t)\cos n\omega_0 t$ over a period.

In a similar way, we obtain b_k by multiplying by $\sin k\omega_0 t$, integrating over a period, noting that all but one of the terms on the right-hand side are zero, and performing that single integration by Eq. (7). The result is

$$b_n = \frac{2}{T}\int_0^T f(t)\sin n\omega_0 t\, dt \tag{11}$$

which is *twice* the average value of $f(t)\sin n\omega_0 t$ over a period.

Equations (9) to (11) now enable us to determine values for a_0 and all the a_n and b_n in the Fourier series, Eq. (1):

$$f(t) = a_0 + \sum_{n=1}^{\infty}(a_n \cos n\omega_0 t + b_n \sin n\omega_0 t) \tag{1}$$

$$\omega_0 = \frac{2\pi}{T} = 2\pi f_0$$

$$a_0 = \frac{1}{T}\int_0^T f(t)\, dt \tag{9}$$

$$a_n = \frac{2}{T}\int_0^T f(t)\cos n\omega_0 t\, dt \tag{10}$$

$$b_n = \frac{2}{T}\int_0^T f(t)\sin n\omega_0 t\, dt \tag{11}$$

Let us consider a numerical example.

Example 17-1 The "half-sinusoidal" waveform shown in Fig. 17-2a represents the voltage response obtained at the output of a half-wave rectifier circuit, a nonlinear circuit whose purpose is to convert a sinusoidal input voltage to a (pulsating) dc output voltage. Find the Fourier series representation of this waveform.

Figure 17-2

(a) The output of a half-wave rectifier to which a sinusoidal input is applied. (b) The discrete line spectrum of the waveform in part a.

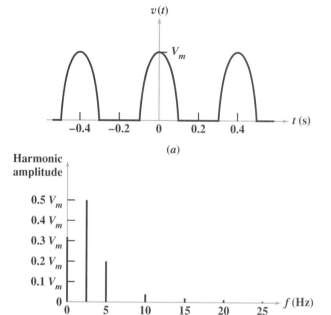

(a)

(b)

Solution: In order to represent this voltage as a Fourier series, we must first determine the period and then express the graphical voltage as an analytical function of time. From the graph, the period is seen to be

$$T = 0.4 \text{ s}$$

and thus

$$f_0 = 2.5 \text{ Hz}$$

and

$$\omega_0 = 5\pi \text{ rad/s}$$

With these three quantities determined, we now seek an appropriate expression for $f(t)$ or $v(t)$ which is valid throughout the period. Obtaining this equation or set of equations proves to be the most difficult part of Fourier analysis for many students. The source of the difficulty is apparently either the inability to recognize the given curve, carelessness in determining multiplying constants within the functional expression, or a negligence which results in not writing the complete expression. In this example, the statement of the problem implies that the functional form is a sinusoid whose amplitude is V_m. The radian frequency has already been determined as 5π, and only the positive portion of the cosine wave is present. The functional expression over the period $t = 0$ to $t = 0.4$ is therefore

$$v(t) = \begin{cases} V_m \cos 5\pi t & 0 \le t \le 0.1 \\ 0 & 0.1 \le t \le 0.3 \\ V_m \cos 5\pi t & 0.3 \le t \le 0.4 \end{cases}$$

It is evident that the choice of the period extending from $t = -0.1$ to $t = 0.3$ will result in fewer equations and, hence, fewer integrals:

$$v(t) = \begin{cases} V_m \cos 5\pi t & -0.1 \le t \le 0.1 \\ 0 & 0.1 \le t \le 0.3 \end{cases} \tag{12}$$

This form is preferable, although either description will yield the correct results.

The zero-frequency component is easily obtained:

$$a_0 = \frac{1}{0.4} \int_{-0.1}^{0.3} v(t)\, dt$$

$$= \frac{1}{0.4} \left[\int_{-0.1}^{0.1} V_m \cos 5\pi t\, dt + \int_{0.1}^{0.3} (0)\, dt \right]$$

and
$$a_0 = \frac{V_m}{\pi} \tag{13}$$

Notice that integration over an entire period must be broken up into subintervals of the period, in each of which the functional form of $v(t)$ is known.

The amplitude of a general cosine term is

$$a_n = \frac{2}{0.4} \int_{-0.1}^{0.1} V_m \cos 5\pi t \cos 5\pi n t\, dt$$

The form of the function we obtain upon integrating is different when n is unity than it is for any other choice of n. If $n = 1$, we have

$$a_1 = 5V_m \int_{-0.1}^{0.1} \cos^2 5\pi t\, dt = \frac{V_m}{2} \tag{14}$$

whereas if n is not equal to unity, we find

$$a_n = 5V_m \int_{-0.1}^{0.1} \cos 5\pi t \cos 5\pi n t\, dt$$

$$= 5V_m \int_{-0.1}^{0.1} \frac{1}{2} \left[\cos 5\pi(1+n)t + \cos 5\pi(1-n)t \right] dt$$

or
$$a_n = \frac{2V_m}{\pi} \frac{\cos(\pi n/2)}{1-n^2} \qquad (n \ne 1) \tag{15}$$

Some of the details of the integration have been left out for those who prefer to work out the small, tedious steps for themselves. It should be pointed out, incidentally, that the expression for a_n when $n \ne 1$ will yield the correct result for $n = 1$ in the limit as $n \to 1$.

A similar integration shows that $b_n = 0$ for any value of n, and the Fourier series thus contains no sine terms. The Fourier series is therefore obtained from Eqs. (13), (14), and (15):

$$v(t) = \frac{V_m}{\pi} + \frac{V_m}{2} \cos 5\pi t + \frac{2V_m}{3\pi} \cos 10\pi t - \frac{2V_m}{15\pi} \cos 20\pi t$$

$$+ \frac{2V_m}{35\pi} \cos 30\pi t - \cdots \tag{16}$$

In Fig. 17-2a, $v(t)$ is shown graphically as a function of time; in Eq. (12), $v(t)$ is expressed as an analytical function of time. Either of these representations is a time-domain representation. Equation (16), the Fourier series representation of $v(t)$, is also a time-domain expression, but it may be transformed easily into a frequency-domain representation. For example, we could locate in the **s** plane the points that represent the frequencies present in Eq. (16). The result would be a mark at the origin and symmetrical marks on the positive and negative $j\omega$ axis. A more customary method of presenting this information, and one which shows the amplitude of each frequency component, is by a *line spectrum*. A line spectrum for Eq. (16) is shown in Fig. 17-2b; the amplitude of each frequency component is indicated by the length of the vertical line located at the corresponding frequency. We also speak of this spectrum as a *discrete* spectrum, because any finite frequency interval contains only a finite number of frequency components.

One note of caution must be injected. The example we have considered contains no sine terms, and the amplitude of the nth harmonic is therefore $|a_n|$. If b_n is not zero, then the amplitude of the component at a frequency $n\omega_0$ must be $\sqrt{a_n^2 + b_n^2}$. This is the general quantity which we must show in a line spectrum. When we discuss the complex form of the Fourier series, we shall see that this amplitude is obtained more directly.

In addition to the amplitude spectrum, we also may construct a discrete *phase spectrum*. At any frequency $n\omega_0$, we combine the cosine and sine terms to determine the phase angle ϕ_n:

$$a_n \cos n\omega_0 t + b_n \sin n\omega_0 t = \sqrt{a_n^2 + b_n^2} \cos\left(n\omega_0 t + \tan^{-1}\frac{-b_n}{a_n}\right)$$

$$= \sqrt{a_n^2 + b_n^2} \cos(n\omega_0 t + \phi_n)$$

or
$$\phi_n = \tan^{-1}\frac{-b_n}{a_n}$$

In Eq. (16), $\phi_n = 0°$ or $180°$ for *every* n.

The Fourier series obtained for this example includes no sine terms and no odd harmonics (except the fundamental) among the cosine terms. It is possible to anticipate the absence of certain terms in a Fourier series, before any integrations are performed, by an inspection of the symmetry of the given time function. We shall investigate the use of symmetry in the following section.

Drill Problems

17-2. A periodic waveform $f(t)$ is described as follows: $f(t) = -4, 0 < t < 0.3$; $f(t) = 6, 0.3 < t < 0.4$; $f(t) = 0, 0.4 < t < 0.5$; $T = 0.5$. Evaluate: (a) a_0; (b) a_3; (c) b_1.
    Ans: $-1.200; 1.383; -4.44$

17-3. Write the Fourier series for the three voltage waveforms shown in Fig. 17-3.

$$Ans: \frac{4}{\pi}\left(\sin \pi t + \frac{1}{3}\sin 3\pi t + \frac{1}{5}\sin 5\pi t + \cdots\right) V;$$

$$\frac{4}{\pi}\left(\cos \pi t - \frac{1}{3}\cos 3\pi t + \frac{1}{5}\cos 5\pi t - \cdots\right) V;$$

$$\frac{8}{\pi^2}\left(\sin \pi t - \frac{1}{9}\sin 3\pi t + \frac{1}{25}\sin 5\pi t - \cdots\right) V$$

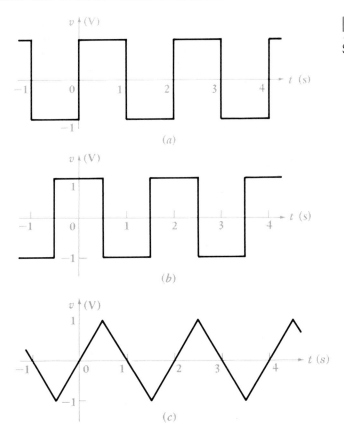

Figure 17-3

See Drill Probs. 17-3 and 17-7.

The two types of symmetry which are most readily recognized are *even-function symmetry* and *odd-function symmetry*, or simply *even symmetry* and *odd symmetry*. We say that $f(t)$ possesses the property of even symmetry if

$$f(t) = f(-t) \tag{17}$$

Such functions as t^2, $\cos 3t$, $\ln(\cos t)$, $\sin^2 7t$, and a constant C all possess even symmetry; the replacement of t by $(-t)$ does not change the value of any of these functions. This type of symmetry may also be recognized graphically, for if $f(t) = f(-t)$ then mirror symmetry exists about the $f(t)$ axis. The function shown in Fig. 17-4a possesses even symmetry; if the figure were to be folded along the $f(t)$ axis, then the portions of the graph for positive and negative time would fit exactly, one on top of the other.

We define odd symmetry by stating that if odd symmetry is a property of $f(t)$, then

$$f(t) = -f(-t) \tag{18}$$

In other words, if t is replaced by $(-t)$, then the negative of the given function is obtained; for example, t, $\sin t$, $t \cos 70t$, $t\sqrt{1 + t^2}$, and the function sketched in Fig. 17-4b are all odd functions and possess odd symmetry. The graphical characteristics of odd symmetry are apparent if the portion of $f(t)$ for $t > 0$ is rotated about the positive t axis and the resultant figure is then rotated about

17-3

The use of symmetry

Figure 17-4

(a) A waveform showing even symmetry. (b) A waveform showing odd symmetry.

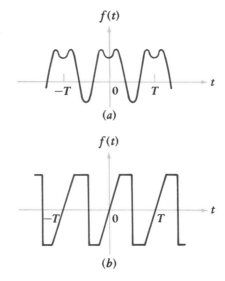

the $f(t)$ axis; the two curves will fit exactly, one on top of the other. That is, we now have symmetry about the origin, rather than about the $f(t)$ axis as we did for even functions.

Having definitions for even and odd symmetry, we should note that the product of two functions with even symmetry, or of two functions with odd symmetry, yields a function with even symmetry. Furthermore, the product of an even and an odd function gives a function with odd symmetry.

Now let us investigate the effect that even symmetry produces in a Fourier series. If we think of the expression which equates an even function $f(t)$ and the sum of an infinite number of sine and cosine functions, then it is apparent that the sum must also be an even function. A sine wave, however, is an odd function, and no sum of sine waves can produce any even function other than zero (which is both even and odd). It is thus plausible that the Fourier series of any even function is composed of only a constant and cosine functions. Let us now show carefully that $b_n = 0$. We have

$$b_n = \frac{2}{T} \int_{-T/2}^{T/2} f(t) \sin n\omega_0 t \, dt$$

$$= \frac{2}{T} \left[\int_{-T/2}^{0} f(t) \sin n\omega_0 t \, dt + \int_{0}^{T/2} f(t) \sin n\omega_0 t \, dt \right]$$

Now let us replace the variable t in the first integral by $-\tau$, or $\tau = -t$, and make use of the fact that $f(t) = f(-t) = f(\tau)$:

$$b_n = \frac{2}{T} \left[\int_{T/2}^{0} f(-\tau) \sin(-n\omega_0 \tau)(-d\tau) + \int_{0}^{T/2} f(t) \sin n\omega_0 t \, dt \right]$$

$$= \frac{2}{T} \left[-\int_{0}^{T/2} f(\tau) \sin n\omega_0 \tau \, d\tau + \int_{0}^{T/2} f(t) \sin n\omega_0 t \, dt \right]$$

But the symbol we use to identify the variable of integration cannot affect the value of the integral. Thus,

$$\int_0^{T/2} f(\tau) \sin n\omega_0 \tau \, d\tau = \int_0^{T/2} f(t) \sin n\omega_0 t \, dt$$

and $\qquad\qquad b_n = 0 \qquad$ (even sym.) $\qquad\qquad$ (19)

No sine terms are present. Therefore, if $f(t)$ shows even symmetry, then $b_n = 0$; conversely, if $b_n = 0$, then $f(t)$ must have even symmetry.

A similar examination of the expression for a_n leads to an integral over the *half period* extending from $t = 0$ to $t = \frac{1}{2}T$:

$$a_n = \frac{4}{T} \int_0^{T/2} f(t) \cos n\omega_0 t \, dt \qquad \text{(even sym.)} \qquad\qquad (20)$$

The fact that a_n may be obtained for an even function by taking "twice the integral over half the range" should seem logical.

A function having odd symmetry can contain no constant term or cosine terms in its Fourier expansion. Let us prove the second part of this statement. We have

$$a_n = \frac{2}{T} \int_{-T/2}^{T/2} f(t) \cos n\omega_0 t \, dt$$

$$= \frac{2}{T} \left[\int_{-T/2}^{0} f(t) \cos n\omega_0 t \, dt + \int_0^{T/2} f(t) \cos n\omega_0 t \, dt \right]$$

and we now let $t = -\tau$ in the first integral:

$$a_n = \frac{2}{T} \left[\int_{T/2}^{0} f(-\tau) \cos (-n\omega_0 \tau) (-d\tau) + \int_0^{T/2} f(t) \cos n\omega_0 t \, dt \right]$$

$$= \frac{2}{T} \left[\int_0^{T/2} f(-\tau) \cos n\omega_0 \tau \, d\tau + \int_0^{T/2} f(t) \cos n\omega_0 t \, dt \right]$$

But $f(-\tau) = -f(\tau)$, and therefore

$$a_n = 0 \qquad \text{(odd sym.)} \qquad\qquad (21)$$

A similar, but simpler, proof shows that

$$a_0 = 0 \qquad \text{(odd sym.)}$$

With odd symmetry, therefore, $a_n = 0$ and $a_0 = 0$; conversely, if $a_n = 0$ and $a_0 = 0$, odd symmetry is present.

The values of b_n may again be obtained by integrating over half the range:

$$b_n = \frac{4}{T} \int_0^{T/2} f(t) \sin n\omega_0 t \, dt \qquad \text{(odd sym.)} \qquad\qquad (22)$$

Examples of even and odd symmetry were afforded by Drill Prob. 17-3, preceding this section. In both parts a and b, a square wave of the same amplitude and period is the given function. The time origin, however, is selected to provide odd symmetry in part a and even symmetry in part b, and the resultant series contain, respectively, only sine terms and only cosine terms. It is also noteworthy that the point at which $t = 0$ could be selected to provide neither even nor odd symmetry; the determination of the coefficients of the terms in the Fourier series would then take at least twice as long.

The Fourier series for both of these square waves have one other interesting characteristic: neither contains any even *harmonics.*[2] That is, the only frequency components present in the series have frequencies which are odd multiples of the fundamental frequency; a_n and b_n are zero for even values of n. This result is caused by another type of symmetry, called half-wave symmetry. We shall say that $f(t)$ possesses *half-wave symmetry* if

$$f(t) = -f(t - \tfrac{1}{2}T)$$

or the equivalent expression,

$$f(t) = -f(t + \tfrac{1}{2}T)$$

Except for a change of sign, each half cycle is like the adjacent half cycles. Half-wave symmetry, unlike even and odd symmetry, is not a function of the choice of the point $t = 0$. Thus, we can state that the square wave (Fig. 17-3a or b) shows half-wave symmetry. Neither waveform shown in Fig. 17-4 has half-wave symmetry, but the two somewhat similar functions plotted in Fig. 17-5 do possess half-wave symmetry.

Figure 17-5

(*a*) A waveform somewhat similar to the one shown in Fig. 17-4*a*, but possessing half-wave symmetry. (*b*) A waveform somewhat similar to the one shown in Fig. 17-4*b*, but possessing half-wave symmetry.

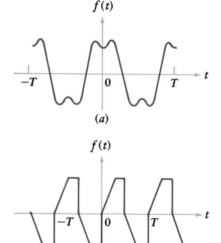

It may be shown that the Fourier series of any function which has half-wave symmetry contains only odd harmonics. Let us consider the coefficients a_n. We have again

$$a_n = \frac{2}{T} \int_{-T/2}^{T/2} f(t) \cos n\omega_0 t \, dt$$

$$= \frac{2}{T} \left[\int_{-T/2}^{0} f(t) \cos n\omega_0 t \, dt + \int_{0}^{T/2} f(t) \cos n\omega_0 t \, dt \right]$$

---

[2] Constant vigilance is required to avoid confusion between an even function and an even harmonic, or between an odd function and an odd harmonic. For example, b_{10} is the coefficient of an even harmonic, and it is zero if $f(t)$ is an even function.

which we may represent as

$$a_n = \frac{2}{T}(I_1 + I_2)$$

Now we substitute the new variable $\tau = t + \frac{1}{2}T$ in the integral I_1:

$$I_1 = \int_0^{T/2} f\left(\tau - \frac{1}{2}T\right) \cos n\omega_0\left(\tau - \frac{1}{2}T\right) d\tau$$

$$= \int_0^{T/2} -f(\tau)\left(\cos n\omega_0\tau \cos\frac{n\omega_0 T}{2} + \sin n\omega_0\tau \sin\frac{n\omega_0 T}{2}\right) d\tau$$

But $\omega_0 T$ is 2π, and thus

$$\sin\frac{n\omega_0 T}{2} = \sin n\pi = 0$$

Hence

$$I_1 = -\cos n\pi \int_0^{T/2} f(\tau) \cos n\omega_0\tau \, d\tau$$

After noting the form of I_2, we therefore may write

$$a_n = \frac{2}{T}(1 - \cos n\pi) \int_0^{T/2} f(t) \cos n\omega_0 t \, dt$$

The factor $(1 - \cos n\pi)$ indicates that a_n is zero if n is even. Thus,

$$a_n = \begin{cases} \dfrac{4}{T}\displaystyle\int_0^{T/2} f(t) \cos n\omega_0 t \, dt & n \text{ odd} \\ 0 & n \text{ even} \end{cases} \qquad \text{(half-wave sym.)} \qquad (23)$$

A similar investigation shows that b_n is also zero for all even n, and therefore

$$b_n = \begin{cases} \dfrac{4}{T}\displaystyle\int_0^{T/2} f(t) \sin n\omega_0 t \, dt & n \text{ odd} \\ 0 & n \text{ even} \end{cases} \qquad \text{(half-wave sym.)} \qquad (24)$$

It should be noted that half-wave symmetry may be present in a waveform which also shows odd symmetry or even symmetry. The waveform sketched in Fig. 17-5a, for example, possesses both even symmetry and half-wave symmetry. When a waveform possesses half-wave symmetry and either even or odd symmetry, then it is possible to reconstruct the waveform if the function is known over any quarter-period interval. The value of a_n or b_n may also be found by integrating over any quarter period. Thus,

$$\left.\begin{aligned} a_n &= \frac{8}{T}\int_0^{T/4} f(t) \cos n\omega_0 t \, dt & n \text{ odd} \\ a_n &= 0 & n \text{ even} \\ b_n &= 0 & \text{all } n \end{aligned}\right\} \quad \begin{matrix} \text{(half-wave and} \\ \text{even sym.)} \end{matrix} \qquad (25)$$

$$a_n = 0 \qquad\qquad\qquad \text{all } n$$

$$\left.\begin{array}{l} b_n = \dfrac{8}{T}\displaystyle\int_0^{T/4} f(t)\sin n\omega_0 t\, dt \qquad n \text{ odd} \\[4mm] b_n = 0 \qquad\qquad\qquad\qquad n \text{ even} \end{array}\right\} \begin{array}{c}\text{(half-wave and}\\ \text{odd sym.)}\end{array} \qquad (26)$$

It is always worthwhile spending a few moments investigating the symmetry of a function for which a Fourier series is to be determined.

Drill Problems

17-4. Sketch each of the functions described, state whether or not even symmetry, odd symmetry, and half-wave symmetry are present, and give the period: (*a*) $v = 0$, $-2 < t < 0$ and $2 < t < 4$; $v = 5$, $0 < t < 2$; $v = -5$, $4 < t < 6$; repeats; (*b*) $v = 10$, $1 < t < 3$; $v = 0$, $3 < t < 7$; $v = -10$, $7 < t < 9$; repeats; (*c*) $v = 8t$, $-1 < t < 1$; $v = 0$, $1 < t < 3$; repeats.

Ans: No, no, yes, 8; no, no, no, 8; no, yes, no, 4

17-5. Determine the Fourier series for the waveforms of Drill Prob. 17-4*a* and *b*.

$$\textit{Ans: } \sum_{n=1(\text{odd})}^{\infty} \frac{10}{n\pi}\left(\sin\frac{n\pi}{2}\cos\frac{n\pi t}{4} + \sin\frac{n\pi t}{4}\right);$$

$$\sum_{n=1}^{\infty} \frac{10}{n\pi}\left[\left(\sin\frac{3n\pi}{4} - 3\sin\frac{n\pi}{4}\right)\cos\frac{n\pi t}{4} + \left(\cos\frac{n\pi}{4} - \cos\frac{3n\pi}{4}\right)\sin\frac{n\pi t}{4}\right]$$

17-4

Complete response to periodic forcing functions

Through the use of the Fourier series, we may now express an arbitrary periodic forcing function as the sum of an infinite number of sinusoidal forcing functions. The forced response to each of these functions may be determined by conventional steady-state analysis, and the form of the natural response may be determined from the poles of an appropriate network transfer function. The initial conditions existing throughout the network, including the initial value of the forced response, enable the amplitude of the natural response to be selected; and then the complete response is obtained as the sum of the forced and natural responses. Let us illustrate this general procedure by a specific example.

Example 17-2   Find the periodic response obtained when the square wave of Fig. 17-6*a*, including its dc component, is applied to the series *RL* circuit shown in Fig. 17-6*b*. The forcing function is applied at $t = 0$, and the current is the desired response. Its initial value is zero.

Figure 17-6

(*a*) A square-wave voltage forcing function. (*b*) The forcing function of part *a* is applied to this series *RL* circuit at $t = 0$; the complete response $i(t)$ is desired.

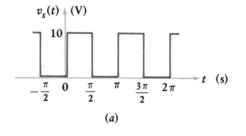

(*a*)

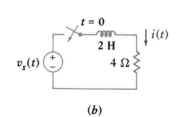

(*b*)

Solution: The forcing function has a fundamental frequency $\omega_0 = 2$ rad/s, and its Fourier series may be written down by comparison with the Fourier series, which was developed for the waveform of Fig. 17-3a in the solution of Drill Prob. 17-3.

$$v_s(t) = 5 + \frac{20}{\pi} \sum_{n=1(\text{odd})}^{\infty} \frac{\sin 2nt}{n}$$

We shall find the forced response for the nth harmonic by working in that frequency domain. Thus,

$$v_{sn}(t) = \frac{20}{\pi n} \sin 2nt$$

and

$$\mathbf{V}_{sn} = \frac{20}{\pi n}(-j1)$$

The impedance offered by the RL circuit at this frequency is

$$\mathbf{Z}_n = 4 + j(2n)2 = 4 + j4n$$

and thus the component of the forced response at this frequency is

$$\mathbf{I}_{fn} = \frac{\mathbf{V}_{sn}}{\mathbf{Z}_n} = \frac{-j5}{\pi n(1 + jn)}$$

Transforming to the time domain, we have

$$i_{fn}(t) = \frac{5}{\pi n} \frac{1}{\sqrt{1 + n^2}} \cos\left(2nt - 90° - \tan^{-1} n\right)$$

$$= \frac{5}{\pi(1 + n^2)} \left(\frac{\sin 2nt}{n} - \cos 2nt\right)$$

Since the response to the dc component is obviously 1.25 A, the forced response may be expressed as the summation

$$i_f(t) = 1.25 + \frac{5}{\pi} \sum_{n=1(\text{odd})}^{\infty} \left[\frac{\sin 2nt}{n(1 + n^2)} - \frac{\cos 2nt}{1 + n^2}\right]$$

The familiar natural response of this simple circuit is the single exponential term [characterizing the single pole of the transfer function, $\mathbf{I}_f/\mathbf{V}_s = 1/(4 + 2\mathbf{s})$]

$$i_n(t) = Ae^{-2t}$$

The complete response is therefore the sum

$$i(t) = i_f(t) + i_n(t)$$

and, since $i(0) = 0$, it is necessary to select A so that

$$A = -i_f(0)$$

Letting $t = 0$, we find that $i_f(0)$ is given by

$$i_f(0) = 1.25 - \frac{5}{\pi} \sum_{n=1(\text{odd})}^{\infty} \frac{1}{1 + n^2}$$

Although we could express A in terms of this summation, it is more convenient to use the numerical value of the summation. The sum of the first five terms of $\Sigma \, 1/(1 + n^2)$ is 0.671, the sum of the first ten terms is 0.695, the sum of the first twenty terms is 0.708, and the exact sum[3] is 0.720 to three significant figures. Thus

$$A = -1.25 + \frac{5}{\pi}(0.720) = -0.104$$

and

$$i(t) = -0.104e^{-2t} + 1.25 + \frac{5}{\pi} \sum_{n=1\,(\text{odd})}^{\infty} \left[\frac{\sin 2nt}{n(1 + n^2)} - \frac{\cos 2nt}{1 + n^2} \right]$$

In obtaining this solution, we have had to use many of the most general concepts introduced in this and the preceding 16 chapters. Some we did not have to use because of the simple nature of this particular circuit, but their places in the general analysis were indicated. In this sense, we may look upon the solution of this problem as a significant achievement in our introductory study of circuit analysis. In spite of this glorious feeling of accomplishment, however, it must be pointed out that the complete response, as obtained in Example 17-2 in analytical form, is not of much value as it stands; it furnishes no clear picture of the nature of the response. What we really need is a sketch of $i(t)$ as a function of time. This may be obtained by a laborious calculation at a sufficient number of instants of time; an available digital computer or a programmable hand calculator can be of great assistance here. The sketch may be approximated by the graphical addition of the natural response, the dc term, and the first few harmonics; this is an unrewarding task. When all is said and done, the most informative solution of this problem is probably obtained by making a repeated transient analysis. That is, the form of the response can certainly be calculated in the interval from $t = 0$ to $t = \pi/2$ s; it is an exponential rising toward 2.5 A. After determining the value at the end of this first interval, we have an initial condition for the next $(\pi/2)$-second interval. The process is repeated until the response assumes a generally periodic nature. The method is eminently suitable to this example, for there is negligible change in the current waveform in the successive periods $\pi/2 < t < 3\pi/2$ and $3\pi/2 < t < 5\pi/2$. The complete current response is sketched in Fig. 17-7. ∎

Figure 17-7

The initial portion of the complete response of the circuit of Fig. 17-6b to the forcing function of Fig. 17-6a.

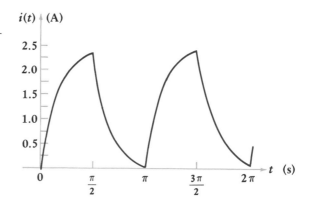

[3] The sum of this series is known in closed form:

$$\sum_{n=1\,(\text{odd})}^{\infty} \frac{1}{1 + n^2} = \frac{\pi}{4} \tanh \frac{\pi}{2}$$

17-6. Use the methods of Chap. 5 to determine the value of the current sketched in Fig. 17-7 at t equal to (a) $\pi/2$; (b) π; (c) $3\pi/2$.

Ans: 2.392 A; 0.1034 A; 2.396 A

In obtaining a frequency spectrum, we have seen that the amplitude of each frequency component depends on both a_n and b_n; that is, the sine term and the cosine term both contribute to the amplitude. The exact expression for this amplitude is $\sqrt{a_n^2 + b_n^2}$. It is also possible to obtain the amplitude directly by using a form of Fourier series in which each term is a cosine function with a phase angle; the amplitude and phase angle are functions of $f(t)$ and n.

An even more convenient and concise form of the Fourier series is obtained if the sines and cosines are expressed as exponential functions with complex multiplying constants.

Let us first take the trigonometric form of the Fourier series:

$$f(t) = a_0 + \sum_{n=1}^{\infty} (a_n \cos n\omega_0 t + b_n \sin n\omega_0 t)$$

and then substitute the exponential forms for the sine and cosine. After rearranging,

$$f(t) = a_0 + \sum_{n=1}^{\infty} \left(e^{jn\omega_0 t} \frac{a_n - jb_n}{2} + e^{-jn\omega_0 t} \frac{a_n + jb_n}{2} \right)$$

We now define a complex constant $\mathbf{c}_n$:

$$\mathbf{c}_n = \tfrac{1}{2}(a_n - jb_n) \qquad (n = 1, 2, 3, \ldots) \tag{27}$$

The values of a_n, b_n, and $\mathbf{c}_n$ all depend on n and $f(t)$. Suppose we now replace n by $(-n)$; how do the values of the constants change? The coefficients a_n and b_n are defined by Eqs. (10) and (11), and it is evident that

$$a_{-n} = a_n$$

but

$$b_{-n} = -b_n$$

From Eq. (27), then,

$$\mathbf{c}_{-n} = \tfrac{1}{2}(a_n + jb_n) \qquad (n = 1, 2, 3, \ldots) \tag{28}$$

Thus,

$$\mathbf{c}_n = \mathbf{c}_{-n}^*$$

We also let

$$\mathbf{c}_0 = a_0$$

We may therefore express $f(t)$ as

$$f(t) = \mathbf{c}_0 + \sum_{n=1}^{\infty} \mathbf{c}_n e^{jn\omega_0 t} + \sum_{n=1}^{\infty} \mathbf{c}_{-n} e^{-jn\omega_0 t}$$

or

$$f(t) = \sum_{n=0}^{\infty} \mathbf{c}_n e^{jn\omega_0 t} + \sum_{n=1}^{\infty} \mathbf{c}_{-n} e^{-jn\omega_0 t}$$

17-5

Complex form of the Fourier series

Finally, instead of summing the second series over the positive integers from 1 to ∞, let us sum over the negative integers, from -1 to $-\infty$:

$$f(t) = \sum_{n=0}^{\infty} \mathbf{c}_n e^{jn\omega_0 t} + \sum_{n=-1}^{-\infty} \mathbf{c}_n e^{jn\omega_0 t}$$

or

$$f(t) = \sum_{n=-\infty}^{\infty} \mathbf{c}_n e^{jn\omega_0 t} \qquad (29)$$

By agreement, a summation from $-\infty$ to ∞ is understood to include a term for $n = 0$.

Equation (29) is the complex form of the Fourier series for $f(t)$; its conciseness is one of the most important reasons for its use. In order to obtain the expression by which a particular complex coefficient $\mathbf{c}_n$ may be evaluated, we substitute Eqs. (10) and (11) in Eq. (27):

$$\mathbf{c}_n = \frac{1}{T} \int_{-T/2}^{T/2} f(t) \cos n\omega_0 t \, dt - j \frac{1}{T} \int_{-T/2}^{T/2} f(t) \sin n\omega_0 t \, dt$$

and then we use the exponential equivalents of the sine and cosine and simplify:

$$\mathbf{c}_n = \frac{1}{T} \int_{-T/2}^{T/2} f(t) e^{-jn\omega_0 t} \, dt \qquad (30)$$

Thus, a single concise equation serves to replace the two equations required for the trigonometric form of the Fourier series. Instead of evaluating two integrals to find the Fourier coefficients, only one integration is required; moreover, it is almost always a simpler integration. It should be noted that the integral of Eq. (30) contains the multiplying factor $1/T$, whereas the integrals for a_n and b_n both contain the factor $2/T$.

Collecting the two basic relationships for the exponential form of the Fourier series, we have

$$\boxed{\begin{aligned} f(t) &= \sum_{n=-\infty}^{\infty} \mathbf{c}_n e^{jn\omega_0 t} \qquad &(29) \\[1em] \mathbf{c}_n &= \frac{1}{T} \int_{-T/2}^{T/2} f(t) e^{-jn\omega_0 t} \, dt \qquad &(30) \end{aligned}}$$

where $\omega_0 = 2\pi/T$ as usual.

The amplitude of the component of the exponential Fourier series at $\omega = n\omega_0$, where $n = 0, \pm 1, \pm 2, \ldots$, is $|\mathbf{c}_n|$. We may plot a discrete frequency spectrum giving $|\mathbf{c}_n|$ versus $n\omega_0$ or nf_0, using an abscissa that shows both positive and negative values; and when we do this, the graph is symmetrical about the origin, since Eqs. (27) and (28) show that $|\mathbf{c}_n| = |\mathbf{c}_{-n}|$.

We note also from Eqs. (29) and (30) that the amplitude of the sinusoidal component at $\omega = n\omega_0$, where $n = 1, 2, 3, \ldots$, is $\sqrt{a_n^2 + b_n^2} = 2|\mathbf{c}_n| = 2|\mathbf{c}_{-n}| = |\mathbf{c}_n| + |\mathbf{c}_{-n}|$.

For the dc component, $a_0 = \mathbf{c}_0$.

The exponential Fourier coefficients, given by Eq. (30), are also affected by the presence of certain symmetries in $f(t)$. Thus, appropriate expressions for $\mathbf{c}_n$ are

$$\mathbf{c}_n = \frac{2}{T} \int_0^{T/2} f(t) \cos n\omega_0 t \, dt \qquad \text{(even sym.)} \qquad (31)$$

$$\mathbf{c}_n = \frac{-j2}{T} \int_0^{T/2} f(t) \sin n\omega_0 t \, dt \qquad \text{(odd sym.)} \qquad (32)$$

$$\mathbf{c}_n = \begin{cases} \dfrac{2}{T} \int_0^{T/2} f(t) e^{-jn\omega_0 t} \, dt & (n \text{ odd, half-wave sym.}) \qquad (33a) \\[2mm] 0 & (n \text{ even, half-wave sym.}) \qquad (33b) \end{cases}$$

$$\mathbf{c}_n = \begin{cases} \dfrac{4}{T} \int_0^{T/4} f(t) \cos n\omega_0 t \, dt & (n \text{ odd, half-wave and even}) \qquad (34a) \\[2mm] 0 & (n \text{ even, half-wave and even}) \qquad (34b) \end{cases}$$

$$\mathbf{c}_n = \begin{cases} \dfrac{-j4}{T} \int_0^{T/4} f(t) \sin n\omega_0 t \, dt & (n \text{ odd, half-wave and odd}) \qquad (35a) \\[2mm] 0 & (n \text{ even, half-wave and odd}) \qquad (35b) \end{cases}$$

Example 17-3 As a simple example of an exponential Fourier series, let us determine $\mathbf{c}_n$ for the square wave of Fig. 17-3b.

Solution: This square wave possesses both even and half-wave symmetry. If we ignore this symmetry and use the general equation (30), we let $T = 2$, $\omega_0 = 2\pi/2 = \pi$, and have

$$\mathbf{c}_n = \frac{1}{T} \int_{-T/2}^{T/2} f(t) e^{-jn\omega_0 t} \, dt$$

$$= \frac{1}{2} \left[\int_{-1}^{-0.5} -e^{-jn\pi t} \, dt + \int_{-0.5}^{0.5} e^{-jn\pi t} \, dt - \int_{0.5}^{1} e^{-jn\pi t} \, dt \right]$$

$$= \frac{1}{2} \left[\frac{-1}{-jn\pi} (e^{-jn\pi t})_{-1}^{-0.5} + \frac{1}{-jn\pi} (e^{-jn\pi t})_{-0.5}^{0.5} + \frac{-1}{-jn\pi} (e^{-jn\pi t})_{0.5}^{1} \right]$$

$$= \frac{1}{j2n\pi} (e^{jn\pi/2} - e^{jn\pi} - e^{-jn\pi/2} + e^{jn\pi/2} + e^{-jn\pi} - e^{-jn\pi/2})$$

$$= \frac{1}{j2n\pi} (2e^{jn\pi/2} - 2e^{-jn\pi/2}) = \frac{2}{n\pi} \sin \frac{n\pi}{2}$$

We thus find that $\mathbf{c}_0 = 0$, $\mathbf{c}_1 = 2/\pi$, $\mathbf{c}_2 = 0$, $\mathbf{c}_3 = -2/3\pi$, $\mathbf{c}_4 = 0$, $\mathbf{c}_5 = 2/5\pi$, and so forth. These values agree with the trigonometric Fourier series given as the answer we obtained in Drill Prob. 17-3 for the waveform shown in Fig. 17-3b if we remember that $a_n = 2\mathbf{c}_n$ when $b_n = 0$.

Utilizing the symmetry of the waveform (even and half-wave), there is less work when we apply Eqs. (34a) and (34b), leading to

$$\mathbf{c}_n = \frac{4}{T} \int_0^{T/4} f(t) \cos n\omega_0 t \, dt$$

$$= \frac{4}{2} \int_0^{0.5} \cos n\pi t \, dt = \frac{2}{n\pi} (\sin n\pi)_0^{0.5}$$

$$= \begin{cases} \dfrac{2}{n\pi} \sin \dfrac{n\pi}{2} & (n \text{ odd}) \\[2mm] 0 & (n \text{ even}) \end{cases}$$

These results are the same as those we just obtained when we did not take the symmetry of the waveform into account. ∎

Now let us consider a more difficult, more interesting example.

Example 17-4   Our function $f(t)$ is a train of rectangular pulses of amplitude V_0 and duration τ, recurring periodically every T seconds, as shown in Fig. 17-8a. We seek the exponential Fourier series for $f(t)$.

Figure 17-8

(a) A periodic sequence of rectangular pulses. (b) The corresponding discrete line spectrum for $|c_n|$, $f = nf_0$, $n = 0$, ± 1, ± 2, ... (c) $\sqrt{a_n^2 + b_n^2}$ versus $f = nf_0$, $n = 0, 1, 2, ...$

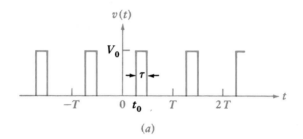

(a)

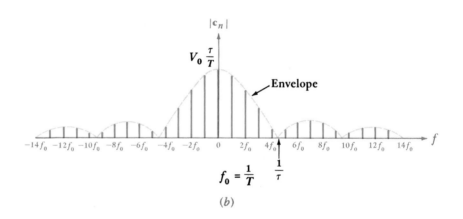

(b)

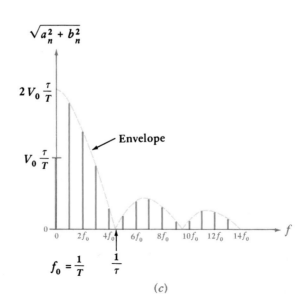

(c)

Solution: The fundamental frequency is $f_0 = 1/T$. No symmetry is present, and the value of a general complex coefficient is found from Eq. (30):

$$\mathbf{c}_n = \frac{1}{T}\int_{-T/2}^{T/2} f(t)e^{-jn\omega_0 t}\,dt = \frac{V_0}{T}\int_{t_0}^{t_0+\tau} e^{-jn\omega_0 t}\,dt$$

$$= \frac{V_0}{-jn\omega_0 T}\left(e^{-jn\omega_0(t_0+\tau)} - e^{-jn\omega_0 t_0}\right)$$

$$= \frac{2V_0}{n\omega_0 T}e^{-jn\omega_0(t_0+\tau/2)}\sin\left(\frac{1}{2}n\omega_0\tau\right)$$

$$= \frac{V_0\tau}{T}\,\frac{\sin\left(\frac{1}{2}n\omega_0\tau\right)}{\frac{1}{2}n\omega_0\tau}\,e^{-jn\omega_0(t_0+\tau/2)}$$

The magnitude of $\mathbf{c}_n$ is therefore

$$|\mathbf{c}_n| = \frac{V_0\tau}{T}\left|\frac{\sin\left(\frac{1}{2}n\omega_0\tau\right)}{\frac{1}{2}n\omega_0\tau}\right| \qquad (36)$$

and the angle of $\mathbf{c}_n$ is

$$\text{ang }\mathbf{c}_n = -n\omega_0\left(t_0 + \frac{\tau}{2}\right) \qquad \text{(possibly plus 180°)} \qquad (37)$$

Equations (36) and (37) represent our solution to this exponential Fourier series problem. ∎

The trigonometric factor in Eq. (36) occurs frequently in modern communication theory, and it is called the *sampling function*. The "sampling" refers to the time function of Fig. 17-8a from which the sampling function is derived. The product of this sequence of pulses and any other function $f(t)$ represents *samples* of $f(t)$ every T seconds if τ is small and $V_0 = 1$. We define

$$\text{Sa}(x) = \frac{\sin x}{x}$$

Because of the way in which it helps to determine the amplitude of the various frequency components in $f(t)$, it is worth our while to discover the important characteristics of this function. First, we note that $\text{Sa}(x)$ is zero whenever x is an integral multiple of π; that is,

$$\text{Sa}(n\pi) = 0 \qquad n = 1, 2, 3, \ldots$$

When x is zero, the function is indeterminate, but it is easy to show that its value is unity:

$$\text{Sa}(0) = 1$$

The magnitude of $\text{Sa}(x)$ therefore decreases from unity at $x = 0$ to zero at $x = \pi$. As x increases from π to 2π, $|\text{Sa}(x)|$ increases from zero to a maximum less than unity, and then decreases to zero once again. As x continues to increase, the successive maxima continually become smaller because the numerator of $\text{Sa}(x)$ cannot exceed unity and the denominator is continually increasing. Also, $\text{Sa}(x)$ shows even symmetry.

Now let us construct the line spectrum. We first consider $|\mathbf{c}_n|$, writing Eq. (36) in terms of the fundamental cyclic frequency f_0:

$$|\mathbf{c}_n| = \frac{V_0\tau}{T}\left|\frac{\sin{(n\pi f_0\tau)}}{n\pi f_0\tau}\right| \tag{38}$$

The amplitude of any $\mathbf{c}_n$ is obtained from Eq. (38) by using the known values τ and $T = 1/f_0$ and selecting the desired value of n, $n = 0, \pm 1, \pm 2, \ldots$. Instead of evaluating Eq. (38) at these discrete frequencies, let us sketch the *envelope* of $|\mathbf{c}_n|$ by considering the frequency nf_0 to be a continuous variable. That is, f, which is nf_0, can actually take on only the discrete values of the harmonic frequencies 0, $\pm f_0$, $\pm 2f_0$, $\pm 3f_0$, and so forth, but we may think of n for the moment as a continuous variable. When f is zero, $|\mathbf{c}_n|$ is evidently $V_0\tau/T$, and when f has increased to $1/\tau$, $|\mathbf{c}_n|$ is zero. The resultant envelope is sketched as the broken line of Fig. 17-8b. The line spectrum is then obtained by simply erecting a vertical line at each harmonic frequency, as shown in the sketch. The amplitudes shown are those of the $\mathbf{c}_n$. The particular case sketched applies to the case where $\tau/T = 1/(1.5\pi) = 0.212$. In this example, it happens that there is no harmonic exactly at that frequency at which the envelope amplitude is zero; another choice of τ or T could produce such an occurrence, however.

In Fig. 17-8c, the amplitude of the sinusoidal component is plotted as a function of frequency. Note again that $a_0 = \mathbf{c}_0$ and $\sqrt{a_n^2 + b_n^2} = |\mathbf{c}_n| + |\mathbf{c}_{-n}|$.

There are several observations and conclusions which we may make about the line spectrum of a periodic sequence of rectangular pulses, as given in Fig. 17-8c. With respect to the envelope of the discrete spectrum, it is evident that the "width" of the envelope depends upon τ, and not upon T. As a matter of fact, the shape of the envelope is not a function of T. It follows that the bandwidth of a filter which is designed to pass the periodic pulses is a function of the pulse width τ, but not of the pulse period T; an inspection of Fig. 17-8c indicates that the required bandwidth is about $1/\tau$ Hz. If the pulse period T is increased (or the pulse repetition frequency f_0 is decreased), the bandwidth $1/\tau$ does not change, but the number of spectral lines between zero frequency and $1/\tau$ Hz increases, albeit discontinuously; the amplitude of each line is inversely proportional to T. Finally, a shift in the time origin does not change the line spectrum; that is, $|\mathbf{c}_n|$ is not a function of t_0. The relative phases of the frequency components do change with the choice of t_0.

Drill Problem

17-7. Determine the general coefficient $\mathbf{c}_n$ in the complex Fourier series for the waveform shown in Fig.: (*a*) 17-3a; (*b*) 17-3c.

> *Ans:* $-j\,2/(n\pi)$ for n odd, 0 for n even; $-j\,[4/(n^2\pi^2)]\sin{n\pi/2}$ for all n

Problems

1 Let $v(t) = 3 - 3\cos{(100\pi t - 40°)} + 4\sin{(200\pi t - 10°)} + 2.5\cos{300\pi t}$ V. Find (*a*) V_{av}; (*b*) V_{eff}; (*c*) T; (*d*) $v(18\text{ ms})$.

2 (*a*) Make a sketch of the voltage waveform $v(t) = 2\cos{2\pi t} + 1.8\sin{4\pi t}$ in the interval $0 < t < T$. (*b*) Find the maximum value of $v(t)$ in this interval. (*c*) Find the magnitude of the most negative value of $v(t)$ in this interval.

3 The waveform shown in Fig. 17-9 is periodic with $T = 10$ s. Find (*a*) the average value; (*b*) the effective value; (*c*) the value of a_3.

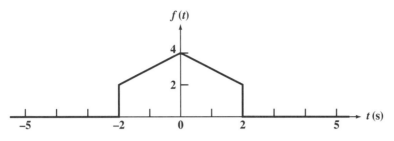

Figure 17-9

See Prob. 3.

4 For the periodic waveform illustrated in Fig. 17-10, find (*a*) T; (*b*) f_0; (*c*) ω_0; (*d*) a_0; (*e*) b_2.

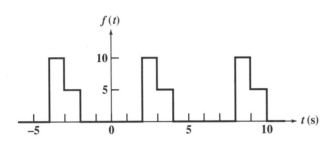

Figure 17-10

See Probs. 4 and 5.

5 Find a_3, b_3, and $\sqrt{a_3^2 + b_3^2}$ for the waveform shown in Fig. 17-10.

6 Obtain the trigonometric form of the Fourier series, give the value of T, and determine the average value for each of these periodic functions of time: (*a*) $3.8 \cos^2 80\pi t$; (*b*) $3.8 \cos^3 80\pi t$; (*c*) $3.8 \cos 79\pi t - 3.8 \sin 80\pi t$.

7 A periodic function of time with $T = 2$ s has the following values: $f(t) = 0$, $-1 < t < 0$; $f(t) = 1$, $0 < t < t_1$; and $f(t) = 0$, $t_1 < t < 1$. (*a*) What value of t_1 will maximize b_4? (*b*) Find $b_{4,\text{max}}$.

8 Let an electrical signal be described by $g(t) = -5 + 8 \cos 10t - 5 \cos 15t + 3 \cos 20t - 8 \sin 10t - 4 \sin 15t + 2 \sin 20t$. Find (*a*) the period of $g(t)$; (*b*) the bandwidth (in hertz) of the signal; (*c*) the average value of $g(t)$; (*d*) the effective value of $g(t)$; (*e*) the discrete amplitude and phase spectra of the signal.

9 The waveform of Example 17-1 (shown as Fig. 17-2) is the output of a half-wave rectifier. If the half sinusoids occupy all the intervals $-0.5 < t < -0.3$, $-0.3 < t < -0.1$, $-0.1 < t < 0.1$, and so forth, then the output is that of a full-wave rectifier. Find the trigonometric Fourier series for this case.

10 (*a*) Specify the types of symmetry present in the waveform of Fig. 17-11. (*b*) Which of the a_n, b_n, or a_0 are zero? (*c*) Calculate a_1, b_1, a_2, b_2, a_3, and b_3.

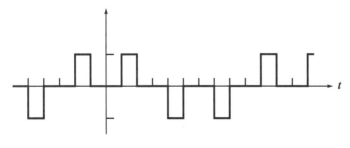

Figure 17-11

See Prob. 10.

11 The periodic function $y(t)$ is known to have odd symmetry and the amplitude spectrum shown in Fig. 17-12. If all the a_n and b_n are nonnegative: (*a*) determine the Fourier series for $y(t)$; (*b*) find the effective value of $y(t)$; (*c*) calculate the value of $y(0.2 \text{ ms})$.

Figure 17-12

See Prob. 11.

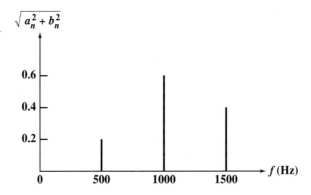

12 Use the waveform given for $f(t)$ over the interval $0 < t < 3$ in Fig. 17-13 to sketch a new function $g(t)$ that is equal to $f(t)$ for $0 < t < 3$ but also has (*a*) $T = 6$ and even symmetry; (*b*) $T = 6$ and odd symmetry; (*c*) $T = 12$, and even and half-wave symmetry; (*d*) $T = 12$, and odd and half-wave symmetry. (*e*) Evaluate a_5 and b_5 for each case.

Figure 17-13

See Prob. 12.

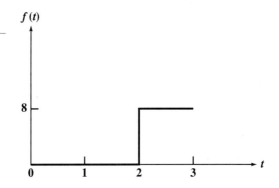

13 The waveform shown in Fig. 17-14 repeats every 4 ms. (*a*) Find the dc component a_0. (*b*) Specify the values of a_1 and b_1. (*c*) Specify a function $f_x(t)$ that equals $f(t)$ in the 4-ms interval shown, but has a period of 8 ms and shows even symmetry. (*d*) Find a_1 and b_1 for $f_x(t)$.

Figure 17-14

See Prob. 13.

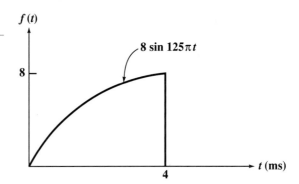

14 Make use of symmetry as much as possible to obtain numerical values for a_0, a_n, and b_n, $1 \le n \le 10$, for the waveform shown in Fig. 17-15.

15 A function $f(t)$ has both odd and half-wave symmetry. The period is 8 ms. It is also known that $f(t) = 10^3 t$, $0 < t < 1$ ms, and $f(t) = 0$, $1 < t < 2$ ms. Find values for b_n, $1 \le n \le 5$.

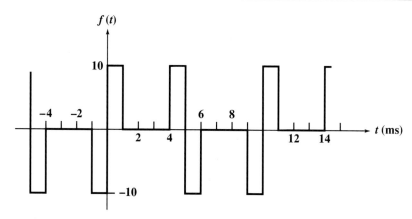

Figure 17-15

See Prob. 14.

16 A portion of $f(t)$ is shown in Fig. 17-16. Show $f(t)$ over the interval $0 < t < 8$ s if $f(t)$ has (a) odd symmetry and $T = 4$ s; (b) even symmetry and $T = 4$ s; (c) odd and half-wave symmetry and $T = 8$ s; (d) even and half-wave symmetry and $T = 8$ s.

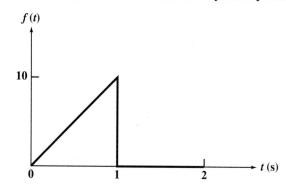

Figure 17-16

See Prob. 16.

17 Replace the square wave of Fig. 17-6a with that shown in Fig. 17-17 and repeat the analysis of Sec. 17-4 to obtain a new expression for (a) $i_f(t)$; (b) $i(t)$.

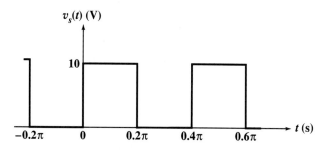

Figure 17-17

See Probs. 17 through 19.

18 The waveform for $v_s(t)$ shown in Fig. 17-17 is applied to the circuit of Fig. 17-6b. Use the standard methods of transient analysis to calculate $i(t)$ at t equal to (a) 0.2π s; (b) 0.4π s; (c) 0.6π s.

19 An ideal voltage source v_s, an open switch, a 2-Ω resistor, and a 2-F capacitor are in series. The source voltage is shown in Fig. 17-17. The switch closes at $t = 0$ and the capacitor voltage is the desired response. (a) Work in the frequency domain of the nth harmonic to find the forced response as a trigonometric Fourier series. (b) Specify the functional form of the natural response. (c) Determine the complete response.

20 Let $T = 6$ ms for the periodic waveform shown in Fig. 17-18. Find $\mathbf{c}_3$, $\mathbf{c}_{-3}$, $|\mathbf{c}_3|$, a_3, b_3, and $\sqrt{a_3^2 + b_3^2}$.

Figure 17-18

See Prob. 20.

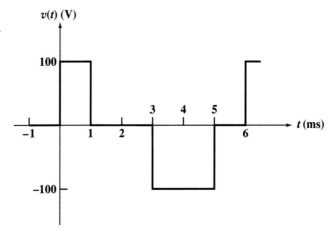

21 (a) Find the complex Fourier series for the periodic waveform shown in Fig. 17-19. (b) Give numerical values for c_n, $n = 0$, ±1, and ±2.

Figure 17-19

See Prob. 21.

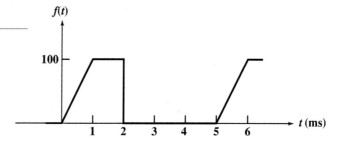

22 The pulses shown in Fig. 17-8a have an amplitude of 8 V, a duration of 0.2 μs, and a repetition rate of 6000 pulses per second. (a) Find the frequency at which the envelope of the frequency spectrum has an amplitude of zero. (b) Determine the frequency separation of the spectral lines. (c) Find $|c_n|$ for that spectral component closest to 20 kHz. (d) . . . closest to 2 MHz. (e) Specify the nominal bandwidth that an amplifier should have to transmit this pulse train with reasonable fidelity. (f) State the number of spectral components in the frequency range $2 < \omega < 2.2$ Mrad/s. (g) Calculate the amplitude of c_{227} and state its frequency.

23 A voltage waveform has a period $T = 5$ ms and complex coefficient values: $c_0 = 1$, $c_1 = 0.2 - j0.2$, $c_2 = 0.5 + j0.25$, $c_3 = -1 - j2$, and $c_n = 0$ for $|n| \geq 4$. (a) Find $v(t)$. (b) Calculate $v(1$ ms$)$.

24 A pulse sequence has a period of 5 μs, an amplitude of unity for $-0.6 < t < -0.4\ \mu$s and for $0.4 < t < 0.6\ \mu$s, and zero amplitude elsewhere in the period interval. This series of pulses might represent the decimal number 3 being transmitted in binary form by a digital computer. (a) Find c_n. (b) Evaluate c_4. (c) Evaluate c_0. (d) Find $|c_n|_{max}$. (e) Find N so that $|c_n| \leq 0.01|c_n|_{max}$ for all $n > N$. (f) What bandwidth is required to transmit this portion of the spectrum?

25 Let a periodic voltage $v_s(t) = 40$ V for $0 < t < \frac{1}{96}$ s, and 0 for $\frac{1}{96} < t < \frac{1}{16}$ s. If $T = \frac{1}{16}$ s, find (a) $\mathbf{c}_3$; (b) the power delivered to the load in the circuit of Fig. 17-20.

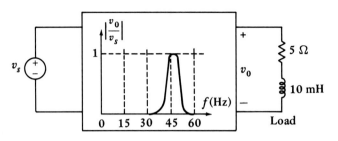

Figure 17-20

See Prob. 25.

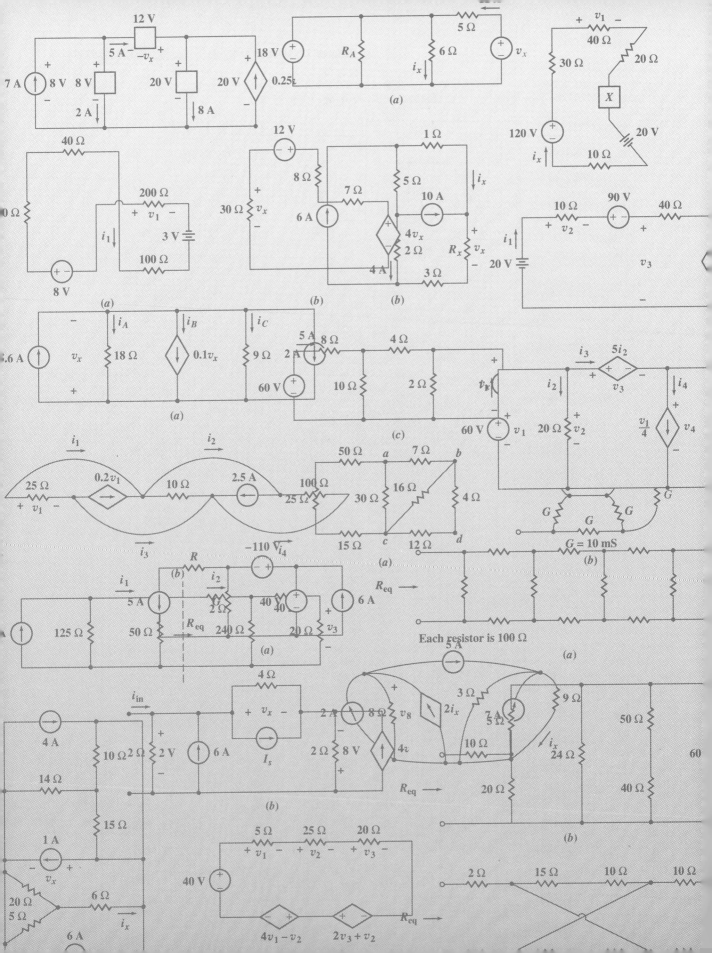

Fourier Transforms

In this chapter we begin to consider the use of transform methods in studying the behavior of linear circuits through a discussion of the Fourier transform. The Laplace transform is the subject of the next chapter.

The Fourier transform and the Laplace transform are operations that convert a function of time into a function of $j\omega$ (Fourier transform) or $\mathbf{s}$ (Laplace transform). Given a suitable function of time, there is one and only one Fourier transform that corresponds to it, and a similar unique correspondence exists for the Laplace transform. Moreover, a one-to-one relationship also exists for the inverse transform; that is, given a suitable function of $j\omega$, or $\mathbf{s}$ for the Laplace transform, there is one and only one function of time that corresponds to it.

The Fourier and Laplace transforms are *integral transforms* which are extremely important in the study of many types of engineering systems, including linear electric circuits. Up to now we have not really needed them, because most of our circuits, along with their forcing and response functions, have been rather simple. The use of relatively powerful integral-transform techniques on them could be likened to using a diesel locomotive to crack walnuts; we might lose sight of the meat. But now we have progressed to a point where our forcing functions, and perhaps our circuits, are getting too complicated for the tools we have developed so far. We want to look at forcing functions that are *not* periodic, and in later studies at (random) forcing functions or signals that are not expressible at all as time functions. This desire is often gratified through use of the integral-transform methods which now burst upon us.

18-1
Introduction

Let us proceed to define the Fourier transform by first recalling the spectrum of the periodic train of rectangular pulses we obtained at the end of Chap. 17. That was a *discrete* line spectrum, which is the type that we must always obtain for periodic functions of time. The spectrum was discrete in the sense that it was not a smooth or continuous function of frequency; instead, it had nonzero values only at specific frequencies.

There are many important forcing functions, however, that are not periodic functions of time, such as a single rectangular pulse, a step function, a ramp function, or a rather strange type of function called the *impulse function* to be defined later in this chapter. Frequency spectra may be obtained for such

18-2
Definition of the Fourier transform

nonperiodic functions, but they will be *continuous* spectra in which some energy, in general, may be found in any nonzero frequency interval, no matter how small.

We shall develop this concept by beginning with a periodic function and then letting the period become infinite. Our experience with the periodic rectangular pulses at the end of Chap. 17 should indicate that the envelope will decrease in amplitude without otherwise changing shape, and that more and more frequency components will be found in any given frequency interval. In the limit, we should expect an envelope of vanishingly small amplitude, filled with an infinite number of frequency components separated by vanishingly small frequency intervals. The number of frequency components between, say, 0 and 100 Hz becomes infinite, but the amplitude of each one approaches zero. At first thought, a spectrum of zero amplitude is a puzzling concept. We know that the line spectrum of a periodic forcing function shows the amplitude of each frequency component. But what does the zero-amplitude continuous spectrum of a nonperiodic forcing function signify? That question will be answered in the following section; now we proceed to carry out the limiting procedure just suggested.

We begin with the exponential form of the Fourier series:

$$f(t) = \sum_{n=-\infty}^{\infty} \mathbf{c}_n e^{jn\omega_0 t} \tag{1}$$

where

$$\mathbf{c}_n = \frac{1}{T} \int_{-T/2}^{T/2} f(t) e^{-jn\omega_0 t} \, dt \tag{2}$$

and

$$\omega_0 = \frac{2\pi}{T} \tag{3}$$

We now let

$$T \to \infty$$

and thus, from Eq. (3), ω_0 must become vanishingly small. We represent this limit by a differential:

$$\omega_0 \to d\omega$$

Thus

$$\frac{1}{T} = \frac{\omega_0}{2\pi} \to \frac{d\omega}{2\pi} \tag{4}$$

Finally, the frequency of any "harmonic" $n\omega_0$ must now correspond to the general frequency variable which describes the continuous spectrum. In other words, n must tend to infinity as ω_0 approaches zero, so that the product is finite:

$$n\omega_0 \to \omega \tag{5}$$

When these four limiting operations are applied to Eq. (2), we find that $\mathbf{c}_n$ must approach zero, as we had previously presumed. If we multiply each side of Eq. (2) by the period T and then undertake the limiting process, a nontrivial result is obtained:

$$\mathbf{c}_n T \to \int_{-\infty}^{\infty} f(t) e^{-j\omega t} \, dt$$

The right-hand side of this expression is a function of ω (and *not* of t), and we represent it by $\mathbf{F}(j\omega)$:

$$\mathbf{F}(j\omega) = \int_{-\infty}^{\infty} f(t)e^{-j\omega t}\, dt \tag{6}$$

Now let us apply the limiting process to Eq. (1). We begin by multiplying and dividing the summation by T,

$$f(t) = \sum_{n=-\infty}^{\infty} \mathbf{c}_n T e^{jn\omega_0 t} \frac{1}{T}$$

next replacing $\mathbf{c}_n T$ by the new quantity $\mathbf{F}(j\omega)$, and then making use of expressions (4) and (5). In the limit, the summation becomes an integral, and

$$f(t) = \frac{1}{2\pi} \int_{-\infty}^{\infty} \mathbf{F}(j\omega)e^{j\omega t}\, d\omega \tag{7}$$

Equations (6) and (7) are collectively called the *Fourier transform pair*. The function $\mathbf{F}(j\omega)$ is the *Fourier transform* of $f(t)$, and $f(t)$ is the *inverse* Fourier transform of $\mathbf{F}(j\omega)$.

This transform-pair relationship is most important! We should memorize it, draw arrows pointing to it, and mentally keep it on the conscious level henceforth and forevermore.[1] We emphasize the importance of these relations by repeating them in boxed form:

$$\boxed{\begin{aligned} \mathbf{F}(j\omega) &= \int_{-\infty}^{\infty} e^{-j\omega t} f(t)\, dt \\[2mm] f(t) &= \frac{1}{2\pi} \int_{-\infty}^{\infty} e^{j\omega t} \mathbf{F}(j\omega)\, d\omega \end{aligned}} \tag{8a} \tag{8b}$$

The exponential terms in these two equations carry opposite signs for the exponents. To keep them straight, it may help to note that the positive sign is associated with the expression for $f(t)$, as it is with the complex Fourier series, Eq. (1).

It is appropriate to raise one question at this time. For the Fourier transform relationships of Eq. (8), can we obtain the Fourier transform of *any* arbitrarily chosen $f(t)$? It turns out that the answer is affirmative for essentially any voltage or current that we can actually produce. A sufficient condition for the existence of $\mathbf{F}(j\omega)$ is that

$$\int_{-\infty}^{\infty} |f(t)|\, dt < \infty$$

This condition is not necessary, however, because some functions that do not meet it still have a Fourier transform; the step function is one such example. Furthermore, we shall see later that $f(t)$ does not even need to be nonperiodic in order to have a Fourier transform; the Fourier series representation for a periodic time function is just a special case of the more general Fourier transform representation.

As we indicated earlier, the Fourier transform-pair relationship is unique. For a given $f(t)$ there is one specific $\mathbf{F}(j\omega)$; and for a given $\mathbf{F}(j\omega)$ there is one specific $f(t)$.

---

[1] Future used-car dealers and politicians may forget it.

Example 18-1   Use the Fourier transform to obtain the continuous spectrum of a single rectangular pulse. Specifically, we select that pulse in Fig. 17-8a (repeated as Fig. 18-1a) which occurs in the interval $t_0 < t < t_0 + \tau$.

Solution: The pulse is described by

$$f(t) = \begin{cases} V_0 & t_0 < t < t_0 + \tau \\ 0 & t < t_0 \text{ and } t > t_0 + \tau \end{cases}$$

The Fourier transform of $f(t)$ is found from Eq. (8a):

$$\mathbf{F}(j\omega) = \int_{t_0}^{t_0+\tau} V_0 e^{-j\omega t}\, dt$$

and this may be easily integrated and simplified:

$$\mathbf{F}(j\omega) = V_0 \tau \, \frac{\sin \frac{1}{2}\omega\tau}{\frac{1}{2}\omega\tau}\, e^{-j\omega(t_0+\tau/2)}$$

The magnitude of $\mathbf{F}(j\omega)$ yields the continuous frequency spectrum, and it is obviously of the form of the sampling function. The value of $\mathbf{F}(0)$ is $V_0\tau$. The shape of the spectrum is identical with the envelope in Fig. 18-1b. A plot of $|\mathbf{F}(j\omega)|$ as a function of ω does *not* indicate the magnitude of the voltage present at any given frequency. What is it, then? Examination of Eq. (7) shows that, if $f(t)$ is a voltage waveform, then $\mathbf{F}(j\omega)$ is dimensionally "volts per unit frequency," a concept that may be strange to many of us. In order to understand this a little better, we next look into some of the properties of $\mathbf{F}(j\omega)$. ∎

Figure 18-1

(a) A periodic sequence of rectangular pulses. (b) The corresponding discrete line spectrum for $|c_n|$.

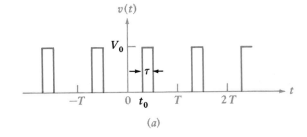

(a)

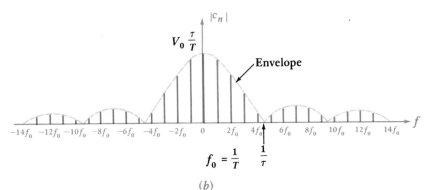

(b)

18-1. If $f(t) = -10$ V, $-0.2 < t < -0.1$ s, $f(t) = 10$ V, $0.1 < t < 0.2$ s, and $f(t) = 0$ for all other t, evaluate $\mathbf{F}(j\omega)$ for ω equal to (a) 0; (b) 10π rad/s; (c) -10π rad/s; (d) 15π rad/s; (e) -20π rad/s.

Ans: 0; $j1.273$ V/(rad/s); $-j1.273$ V/(rad/s); $-j0.424$ V/(rad/s); 0

18-2. If $\mathbf{F}(j\omega) = -10$ V/(rad/s) for $-4 < \omega < -2$ rad/s, $+10$ V/(rad/s) for $2 < \omega < 4$ rad/s, and 0 for all other ω, find the numerical value of $f(t)$ at t equal to (a) 10^{-4} s; (b) 10^{-2} s; (c) $\pi/4$ s; (d) $\pi/2$ s; (e) π s.

Ans: $j1.9099 \times 10^{-3}$ V; $j0.1910$ V; $j4.05$ V; $-j4.05$ V; 0

Our object in this section is to establish several of the mathematical properties of the Fourier transform and, even more important, to understand its physical significance.

We begin by using Euler's identity to replace $e^{-j\omega t}$ in Eq. (8a):

$$\mathbf{F}(j\omega) = \int_{-\infty}^{\infty} f(t) \cos \omega t \, dt - j \int_{-\infty}^{\infty} f(t) \sin \omega t \, dt \qquad (9)$$

Since $f(t)$, $\cos \omega t$, and $\sin \omega t$ are all real functions of time, both the integrals in Eq. (9) are real functions of ω. Thus, by letting

$$\mathbf{F}(j\omega) = A(\omega) + jB(\omega) = |\mathbf{F}(j\omega)|e^{j\phi(\omega)} \qquad (10)$$

we have

$$A(\omega) = \int_{-\infty}^{\infty} f(t) \cos \omega t \, dt \qquad (11)$$

$$B(\omega) = -\int_{-\infty}^{\infty} f(t) \sin \omega t \, dt \qquad (12)$$

$$|\mathbf{F}(j\omega)| = \sqrt{A^2(\omega) + B^2(\omega)} \qquad (13)$$

and

$$\phi(\omega) = \tan^{-1} \frac{B(\omega)}{A(\omega)} \qquad (14)$$

18-3

Some properties of the Fourier transform

Replacing ω by $-\omega$ shows that $A(\omega)$ and $|\mathbf{F}(j\omega)|$ are both even functions of ω, while $B(\omega)$ and $\phi(\omega)$ are both odd functions of ω.

Now, if $f(t)$ is an even function of t, then the integrand of Eq. (12) is an odd function of t, and the symmetrical limits force $B(\omega)$ to be zero; thus, if $f(t)$ is even, its Fourier transform $\mathbf{F}(j\omega)$ is a real, even function of ω, and the phase function $\phi(\omega)$ is zero or π for all ω. However, if $f(t)$ is an odd function of t, then $A(\omega) = 0$ and $\mathbf{F}(j\omega)$ is both odd and a pure imaginary function of ω; $\phi(\omega)$ is $\pm\pi/2$. In general, however, $\mathbf{F}(j\omega)$ is a complex function of ω.

Finally, we note that the replacement of ω by $-\omega$ in Eq. (9) forms the conjugate of $\mathbf{F}(j\omega)$. Thus,

$$\mathbf{F}(-j\omega) = A(\omega) - jB(\omega) = \mathbf{F}^*(j\omega)$$

and we have

$$\mathbf{F}(j\omega)\mathbf{F}(-j\omega) = \mathbf{F}(j\omega)\mathbf{F}^*(j\omega) = A^2(\omega) + B^2(\omega) = |\mathbf{F}(j\omega)|^2$$

With these basic mathematical properties of the Fourier transform in mind, we are now ready to consider its physical significance. Let us suppose that $f(t)$ is either the voltage across or the current through a 1-Ω resistor, so that $f^2(t)$ is the instantaneous power delivered to the 1-Ω resistor by $f(t)$. Integrating this power over all time, we obtain the total energy delivered by $f(t)$ to the 1-Ω resistor,

$$W_{1\Omega} = \int_{-\infty}^{\infty} f^2(t)\, dt \tag{15}$$

Now let us resort to a little trickery. Thinking of the integrand in Eq. (15) as $f(t)$ times itself, we replace one of those functions by Eq. (8b):

$$W_{1\Omega} = \int_{-\infty}^{\infty} f(t) \left[\frac{1}{2\pi} \int_{-\infty}^{\infty} e^{j\omega t} \mathbf{F}(j\omega)\, d\omega \right] dt$$

Since $f(t)$ is not a function of the variable of integration ω, we may move it inside the bracketed integral and then interchange the order of integration:

$$W_{1\Omega} = \frac{1}{2\pi} \int_{-\infty}^{\infty} \left[\int_{-\infty}^{\infty} \mathbf{F}(j\omega) e^{j\omega t} f(t)\, dt \right] d\omega$$

Next we shift $\mathbf{F}(j\omega)$ outside the inner integral, causing that integral to become $\mathbf{F}(-j\omega)$:

$$W_{1\Omega} = \frac{1}{2\pi} \int_{-\infty}^{\infty} \mathbf{F}(j\omega)\mathbf{F}(-j\omega)\, d\omega = \frac{1}{2\pi} \int_{-\infty}^{\infty} |\mathbf{F}(j\omega)|^2\, d\omega$$

Collecting these results,

$$\int_{-\infty}^{\infty} f^2(t)\, dt = \frac{1}{2\pi} \int_{-\infty}^{\infty} |\mathbf{F}(j\omega)|^2\, d\omega \tag{16}$$

Equation (16) is a very useful expression known as Parseval's theorem.[2] This theorem, along with Eq. (15), tells us that the energy associated with $f(t)$ can be obtained either from an integration over all time in the time domain or by $1/(2\pi)$ times an integration over all (radian) frequency in the frequency domain.

Parseval's theorem also leads us to a greater understanding and interpretation of the meaning of the Fourier transform. Consider a voltage $v(t)$ with Fourier transform $\mathbf{F}_v(j\omega)$ and 1-Ω energy $W_{1\Omega}$:

$$W_{1\Omega} = \frac{1}{2\pi} \int_{-\infty}^{\infty} |\mathbf{F}_v(j\omega)|^2\, d\omega = \frac{1}{\pi} \int_0^{\infty} |\mathbf{F}_v(j\omega)|^2\, d\omega$$

where the rightmost equality follows from the fact that $|\mathbf{F}_v(j\omega)|^2$ is an even function of ω. Then, since $\omega = 2\pi f$, we can write

$$W_{1\Omega} = \int_{-\infty}^{\infty} |\mathbf{F}_v(j\omega)|^2\, df = 2 \int_0^{\infty} |\mathbf{F}_v(j\omega)|^2\, df \tag{17}$$

Figure 18-2 illustrates a typical plot of $|\mathbf{F}_v(j\omega)|^2$ as a function of both ω and f. If we divide the frequency scale up into vanishingly small increments df, Eq. (17) shows us that the area of a differential slice under the $|\mathbf{F}_v(j\omega)|^2$ curve, having a width df, is $|\mathbf{F}_v(j\omega)|^2\, df$. This area is shown shaded. The sum of all such areas, as f ranges from minus to plus infinity, is the total 1-Ω energy contained in $v(t)$. Thus, $|\mathbf{F}_v(j\omega)|^2$ is the (1-Ω) *energy density* or energy per unit bandwidth (J/Hz) of $v(t)$, and this energy density is always a real, even,

---

[2] Marc Antoine Parseval-Deschenes was a rather obscure French mathematician, geographer, and occasional poet who published these results in 1805, seventeen years before Fourier published his theorem.

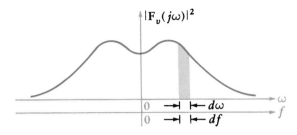

Figure 18-2

The area of the slice $|\mathbf{F}_v(j\omega)|^2$ is the 1-Ω energy associated with $v(t)$ lying in the bandwidth df.

nonnegative function of ω. By integrating $|\mathbf{F}_v(j\omega)|^2$ over an appropriate frequency interval, we are able to calculate that portion of the total energy lying within the chosen interval. Note that the energy density is not a function of the phase of $\mathbf{F}_v(j\omega)$, and thus there are an infinite number of time functions and Fourier transforms that possess identical energy-density functions.

Example 18-2   As an example of an energy-density calculation, let us assume that the one-sided [i.e., $v(t) = 0$ for $t < 0$] exponential pulse

$$v(t) = 4e^{-3t}u(t)$$

is applied to the input of an ideal bandpass filter.[3] We let the filter passband be defined by $1 < |f| < 2$ Hz, and calculate the total output energy.

Solution: We call the filter output voltage $v_o(t)$. The energy in $v_o(t)$ will therefore be equal to the energy of that part of $v(t)$ having frequency components in the intervals $1 < f < 2$ and $-2 < f < -1$. We determine the Fourier transform of $v(t)$,

$$\mathbf{F}_v(j\omega) = 4\int_{-\infty}^{\infty} e^{-j\omega t}e^{-3t}u(t)\,dt$$

$$= 4\int_0^{\infty} e^{-(3+j\omega)t}\,dt = \frac{4}{3+j\omega}$$

and then we may calculate the total 1-Ω energy in the input signal by either

$$W_{1\Omega} = \frac{1}{2\pi}\int_{-\infty}^{\infty} |\mathbf{F}_v(j\omega)|^2\,d\omega$$

$$= \frac{8}{\pi}\int_{-\infty}^{\infty} \frac{d\omega}{9+\omega^2} = \frac{16}{\pi}\int_0^{\infty} \frac{d\omega}{9+\omega^2} = \frac{8}{3} \text{ J}$$

or

$$W_{1\Omega} = \int_{-\infty}^{\infty} v^2(t)\,dt = 16\int_0^{\infty} e^{-6t}\,dt = \frac{8}{3} \text{ J}$$

The total energy in $v_o(t)$, however, is smaller:

$$W_{o1} = \frac{1}{2\pi}\int_{-4\pi}^{-2\pi} \frac{16\,d\omega}{9+\omega^2} + \frac{1}{2\pi}\int_{2\pi}^{4\pi} \frac{16\,d\omega}{9+\omega^2}$$

$$= \frac{16}{\pi}\int_{2\pi}^{4\pi} \frac{d\omega}{9+\omega^2} = \frac{16}{3\pi}\left(\tan^{-1}\frac{4\pi}{3} - \tan^{-1}\frac{2\pi}{3}\right) = 0.358 \text{ J} \qquad \blacksquare$$

[3] An ideal bandpass filter is a two-port network which allows all those frequency components of the input signal for which $\omega_1 < |\omega| < \omega_2$ to pass unattenuated from the input terminals to the output terminals; all components having frequencies outside this so-called *passband* are completely attenuated.

In general, we see that an ideal bandpass filter enables us to remove energy from prescribed frequency ranges while still retaining the energy contained in other frequency ranges. The Fourier transform helps us to describe the filtering action quantitatively without actually evaluating $v_o(t)$, although we shall see later that the Fourier transform can also be used to obtain the expression for $v_o(t)$ if we wish to do so.

Drill Problems

18-3. If $i(t) = 10e^{20t}[u(t + 0.1) - u(t - 0.1)]$ A, find (a) $\mathbf{F}_i(j0)$; (b) $\mathbf{F}_i(j10)$; (c) $A_i(10)$; (d) $B_i(10)$; (e) $\phi_i(10)$.

   Ans: 3.63 A/(rad/s); 3.33$\underline{/-31.7°}$ A/(rad/s); 2.83 A/(rad/s); -1.749 A/(rad/s); $-31.7°$

18-4. Find the 1-Ω energy associated with the current $i(t) = 20e^{-10t}u(t)$ A in the interval: (a) $-0.1 < t < 0.1$ s; (b) $-10 < \omega < 10$ rad/s; (c) $10 < \omega < \infty$ rad/s.   *Ans:* 17.29 J; 10 J; 5 J

18-4

The unit-impulse function

Before continuing our discussion of the Fourier transform, we need to pause briefly to define a new singularity function called the *unit-impulse* or *delta function*. We shall see that the unit-impulse function enables us to confirm our previous statement that periodic, as well as nonperiodic, time functions possess Fourier transforms. In addition, our analysis of general *RLC* circuits can be enhanced by utilization of the unit impulse. We have been avoiding the possibility that the voltage across a capacitor or the current through an inductor might change by a finite amount in zero time, for in these cases the capacitor current or the inductor voltage would have had to assume infinite values. Although this is not physically possible, it is mathematically possible. Thus, if a step voltage $V_0 u(t)$ is applied directly across an uncharged capacitor C, the time constant is zero, since R_{eq} is zero and C is finite. This means that a charge CV_0 must be established on the capacitor in a vanishingly small time. We might say, with more insight than rigor, that "infinite current flowing for zero time produces a finite charge" on the capacitor. This type of phenomenon can be described through use of the unit-impulse function.

We shall define the unit impulse as a function of time which is zero when its argument, generally $(t - t_0)$, is less than zero; which is also zero when its argument is greater than zero; which is infinite when its argument is equal to zero; and which has unit area. Mathematically, the defining statements are

$$\delta(t - t_0) = 0 \qquad t \neq t_0 \tag{18}$$

and

$$\int_{-\infty}^{\infty} \delta(t - t_0)\, dt = 1 \tag{19}$$

where the symbol δ (curly delta) is used to represent the unit impulse. In view of the functional values expressed by Eq. (18), it is apparent that the limits on the integral in Eq. (19) may be any values which are less than t_0 and greater than t_0. In particular, we may let t_0^- and t_0^+ represent values of time which are arbitrarily close to t_0 and then express Eq. (19) as

$$\int_{t_0^-}^{t_0^+} \delta(t - t_0)\, dt = 1 \tag{20}$$

For the most part, we shall concern ourselves with signals that have only a single discontinuity, and we shall select our time scale so that the switching operation occurs at $t = 0$. For this special case, the defining equations are

$$\delta(t) = 0 \qquad t \neq 0 \tag{21}$$

and
$$\int_{-\infty}^{\infty} \delta(t) \, dt = 1 \tag{22}$$

or
$$\int_{0^-}^{0^+} \delta(t) \, dt = 1 \tag{23}$$

The unit impulse may be multiplied by a constant also; this cannot, of course, affect Eq. (18) or (21), because the value must still be zero when the argument is not zero. However, multiplication of any of the integral expressions by a constant shows that the area under the impulse is now equal to the constant multiplying factor; this area is called the *strength* of the impulse. Thus, the impulse $5\delta(t)$ has a strength of 5, and the impulse $-10\delta(t - 2)$ has a strength of -10. If the unit impulse is multiplied by a function of time, then the strength of the impulse must be the value of that function at the time for which the impulse *argument* is zero. In other words, the strength of the impulse $e^{-t/2}\delta(t - 2)$ is $e^{-2/2} = 0.368$, and the strength of the impulse $\sin(5\pi t + \pi/4)\,\delta(t)$ is 0.707. It is therefore possible to write the following integrals[4] which make the same statement in mathematical form:

$$\int_{-\infty}^{\infty} f(t)\delta(t) \, dt = f(0) \tag{24}$$

or
$$\int_{-\infty}^{\infty} f(t)\,\delta(t - t_0)\, dt = f(t_0) \tag{25}$$

The graphical symbol for an impulse is shown in Fig. 18-3, where $f(t) = 4\delta(t + 1) - 3\delta(t)$ is plotted as a function of time. It is customary to indicate the strength of the impulse in parentheses adjacent to the impulse. Note that no attempt should be made to indicate the strength of an impulse by adjusting its amplitude; each spike has infinite amplitude, and all impulses should be drawn as arrows having the same convenient amplitude. Positive and negative impulses should be drawn above and below the time axis, respectively. In order to avoid confusion with the ordinate, the lines and arrowheads used to form the impulses are drawn thicker than are the axes.

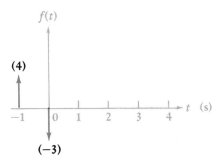

Figure 18-3

A positive and negative impulse are plotted graphically as functions of time. The strengths of the impulses are 4 and -3, respectively, and thus $f(t) = 4\delta(t + 1) - 3\delta(t)$.

We shall find it convenient to be familiar with several other interpretations of the unit impulse. We seek graphical forms which do not have infinite amplitudes, but which will approximate an impulse as the amplitude increases. Let us first consider a rectangular pulse, such as that shown in Fig. 18-4. The pulse width is selected as Δ and its amplitude as $1/\Delta$, thus forcing the area of the pulse to be unity, regardless of the magnitude of Δ. As Δ decreases, the amplitude

---

[4] They are called *sifting integrals,* because the integral sifts out a particular value of $f(t)$.

Figure 18-4

A rectangular pulse of unit area which approaches a unit impulse as $\Delta \to 0$.

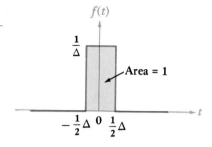

$f(t)$

$\dfrac{1}{\Delta}$

Area = 1

$-\dfrac{1}{2}\Delta$ 0 $\dfrac{1}{2}\Delta$

t

$1/\Delta$ increases, and the rectangular pulse becomes a better approximation to a unit impulse. The response of a circuit element to a unit impulse may be determined by finding its response to this rectangular pulse and then letting Δ approach zero. However, since the impulse response is easily found, we should also realize that this response may in itself be an acceptable approximation to the response produced by a narrow rectangular pulse.

A triangular pulse, sketched in Fig. 18-5, may also be used to approximate the unit impulse. Since we again prefer unit area, a pulse of amplitude $1/\Delta$ must possess an overall width of 2Δ. As Δ approaches zero, the triangular pulse approaches the unit impulse.

Figure 18-5

A triangular pulse of unit area which approaches a unit impulse as $\Delta \to 0$.

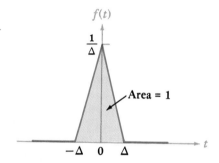

$f(t)$

$\dfrac{1}{\Delta}$

Area = 1

$-\Delta$ 0 Δ

t

There are many other pulse shapes which, in the limit, approach the unit impulse, but we shall take the decaying exponential as our last limiting form. We first construct such a waveform with unit area by finding the area under the general exponential described by

$$f(t) = \begin{cases} 0 & t < 0 \\ Ae^{-t/\tau} & t > 0 \end{cases}$$

or

$$f(t) = Ae^{-t/\tau}u(t)$$

Thus

$$\text{Area} = \int_0^{\infty} Ae^{-t/\tau}\,dt = -\tau Ae^{-t/\tau}\bigg|_0^{\infty} = \tau A$$

and we must set $A = 1/\tau$. The time constant will be very short, and we shall again use Δ to represent this short time. Thus, the exponential function

$$f(t) = \frac{1}{\Delta}e^{-t/\Delta}u(t)$$

approaches the unit impulse as $\Delta \to 0$. This representation of the unit impulse indicates that the exponential decay of a current or voltage in a circuit approaches an impulse (but not necessarily a unit impulse) as the time constant is reduced.

For our final interpretation of the unit impulse, let us try to establish a relationship with the unit-step function. The function shown in Fig. 18-6a is almost a unit step; however, Δ seconds are required for it to complete the linear change from zero to unit amplitude. Beneath this modified unit-step function, Fig. 18-6b shows its derivative; since the linear portion of the modified step rises at the rate of 1 unit every Δ seconds, the derivative must be a rectangular pulse of amplitude $1/\Delta$ and width Δ. This, however, was the first function we

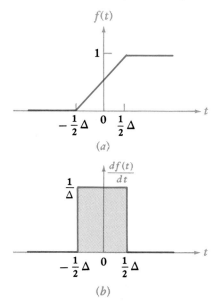

Figure 18-6

(a) A modified unit-step function; the transition from zero to unity is linear over a Δ-second time interval. (b) The derivative of the modified unit step. As $\Delta \to 0$, the graph of part a becomes the unit step and that of part b approaches the unit impulse.

considered as an approximation to the unit impulse, and we know that it approaches the unit impulse as Δ approaches zero. But the modified unit step approaches the unit step itself as Δ approaches zero, and we conclude that the unit impulse may be regarded as the time derivative of the unit-step function.[5] Mathematically,

$$\delta(t) = \frac{du(t)}{dt} \tag{26}$$

and conversely,

$$u(t) = \int_{0^-}^{t} \delta(t)\,dt \qquad t > 0 \tag{27}$$

where the lower limit may in general be any value of t less than zero. Either Eq. (26) or Eq. (27) could be used as the definition of the unit impulse if we wished.

We thus see another method suggesting itself for the determination of the response to a unit impulse. If we can find the unit-step response, then the linear

---

[5] The fact that the derivative of the unit step does not exist at the point of the discontinuity creates quite a skeptical attitude in many mathematicians; nevertheless, the impulse is a useful analytical function.

nature of our circuits requires the response to a unit impulse to be the derivative of the response to the unit step. From the opposite point of view, if the response to a unit impulse is known, then the integral of this response from $-\infty$ to t must be the unit-step response.

Let us apply Eq. (27) to show how a current impulse can place an immediate charge on a capacitor. Since the strength of an impulse may be interpreted as the area under the impulse, a current impulse must have a strength that has the units of current times time, or charge. Thus, we apply the impulse

$$i(t) = Q_0 \delta(t - t_0)$$

to a capacitor C. Assuming the passive sign convention, the capacitor voltage is

$$v(t) = \frac{1}{C}\int_{-\infty}^{t} i(t)\,dt = \frac{Q_0}{C}\int_{-\infty}^{t} \delta(t - t_0)\,dt = \frac{Q_0}{C} u(t - t_0)$$

We see that the capacitor voltage jumps discontinuously from zero to Q_0/C at $t = t_0$.

Drill Problems

18-5. Calculate the strength of an impulse defined by (a) the limit as $a \to 0$ of $\dfrac{5x}{x^2 + a^2}\,[u(x) - u(x - a)]$; ($b$) $2e^{-x}\ln(x + 2)\delta(x - 1)$; ($c$) part of $f'(t)$ if $f(t) = 5[\sin(t + 1)]\,u(t)$.

$\qquad\qquad$ *Ans:* 1.733; 0.808; 4.21

18-6. Evaluate: (a) $\displaystyle\int_{-\infty}^{\infty} 8e^{-2\pi x}\delta(x - 0.2)\,dx$; ($b$) $\displaystyle\int_{-1}^{1} 2\cos 4\pi x\,[\delta(x) - \delta(x - 0.1)]\,dx$; ($c$) $\displaystyle\int_{-0.1}^{0.8} \pi t\,\delta(\sin 4\pi t)\,dt$.

$\qquad\qquad$ *Ans:* 2.28; 1.382; 4.71

18-5

Fourier transform pairs for some simple time functions

We now seek the Fourier transform of the unit impulse $\delta(t - t_0)$. That is, we are interested in the spectral properties or frequency-domain description of this singularity function. If we use the notation $\mathscr{F}\{\ \}$ to symbolize "Fourier transform of $\{\ \}$," then

$$\mathscr{F}\{\delta(t - t_0)\} = \int_{-\infty}^{\infty} e^{-j\omega t}\delta(t - t_0)\,dt$$

From our earlier discussion of this type of integral, and by Eq. (25) in particular, we have

$$\mathscr{F}\{\delta(t - t_0)\} = e^{-j\omega t_0} = \cos\omega t_0 - j\sin\omega t_0 \qquad (28)$$

This complex function of ω leads to the 1-Ω energy-density function,

$$|\mathscr{F}\{\delta(t - t_0)\}|^2 = \cos^2\omega t_0 + \sin^2\omega t_0 = 1$$

This remarkable result says that the (1-Ω) energy per unit bandwidth is unity *at all frequencies,* and that the total energy in the unit impulse is infinitely large.[6] No wonder, then, that we must conclude that the unit impulse is "impractical" in the sense that it cannot be generated in the laboratory. Moreover, even

---

[6] Note, for example, from Fig. 18-4 that the total energy in the unit impulse is

$$\lim_{\Delta \to 0}\int_{-\Delta/2}^{\Delta/2} \left(\frac{1}{\Delta}\right)^2 dt = \lim_{\Delta \to 0}\frac{1}{\Delta} = \infty$$

if one were available to us, it must appear distorted after being subjected to the finite bandwidth of any practical laboratory instrument.

Since there is a unique one-to-one correspondence between a time function and its Fourier transform, we can say that the inverse Fourier transform of $e^{-j\omega t_0}$ is $\delta(t - t_0)$. Utilizing the symbol $\mathcal{F}^{-1}\{\ \}$ for the inverse transform, we have

$$\mathcal{F}^{-1}\{e^{-j\omega t_0}\} = \delta(t - t_0)$$

Thus, we now know that

$$\frac{1}{2\pi}\int_{-\infty}^{\infty} e^{j\omega t}e^{-j\omega t_0}\, d\omega = \delta(t - t_0)$$

even though we would fail in an attempt at the direct evaluation of this improper integral. Symbolically, we may write

$$\delta(t - t_0) \Leftrightarrow e^{-j\omega t_0} \tag{29}$$

where $\Leftrightarrow$ indicates that the two functions constitute a Fourier transform pair.

Continuing with our consideration of the unit-impulse function, let us consider a Fourier transform in that form,

$$\mathbf{F}(j\omega) = \delta(\omega - \omega_0)$$

which is a unit impulse *in the frequency domain* located at $\omega = \omega_0$. Then $f(t)$ must be

$$f(t) = \mathcal{F}^{-1}\{\mathbf{F}(j\omega)\} = \frac{1}{2\pi}\int_{-\infty}^{\infty} e^{j\omega t}\delta(\omega - \omega_0)\, d\omega = \frac{1}{2\pi}e^{j\omega_0 t}$$

where we have used the sifting property of the unit impulse. Thus we may now write

$$\frac{1}{2\pi}e^{j\omega_0 t} \Leftrightarrow \delta(\omega - \omega_0)$$

or

$$e^{j\omega_0 t} \Leftrightarrow 2\pi\delta(\omega - \omega_0) \tag{30}$$

Also, by a simple sign change we obtain

$$e^{-j\omega_0 t} \Leftrightarrow 2\pi\delta(\omega + \omega_0) \tag{31}$$

Clearly, the time function is complex in both expressions (30) and (31), and does not exist in the real world of the laboratory. Time functions such as $\cos\omega_0 t$, for example, can be produced with laboratory equipment, but a function like $e^{-j\omega_0 t}$ cannot.

However, we know that

$$\cos\omega_0 t = \tfrac{1}{2}e^{j\omega_0 t} + \tfrac{1}{2}e^{-j\omega_0 t}$$

and it is easily seen from the definition of the Fourier transform that

$$\mathcal{F}\{f_1(t)\} + \mathcal{F}\{f_2(t)\} = \mathcal{F}\{f_1(t) + f_2(t)\} \tag{32}$$

Therefore,

$$\mathcal{F}\{\cos\omega_0 t\} = \mathcal{F}\{\tfrac{1}{2}e^{j\omega_0 t}\} + \mathcal{F}\{\tfrac{1}{2}e^{-j\omega_0 t}\}$$

$$= \pi\delta(\omega - \omega_0) + \pi\delta(\omega + \omega_0)$$

which indicates that the frequency-domain description of $\cos\omega_0 t$ shows a *pair* of impulses, located at $\omega = \pm\omega_0$. This should not be a great surprise, for in our first discussion of complex frequency in Sec. 12-2, we noted that a sinusoidal function of time was always represented by a pair of imaginary frequencies located at $\mathbf{s} = \pm j\omega_0$. We have, therefore,

$$\cos \omega_0 t \Leftrightarrow \pi[\delta(\omega + \omega_0) + \delta(\omega - \omega_0)] \tag{33}$$

Before establishing the Fourier transforms of any more time functions, we should understand just where we are heading, and why. Thus far we have determined several Fourier transform pairs. As we build up our knowledge of such pairs, they, in turn, can be used to obtain more pairs, and eventually we will have a catalog of most of the familiar time functions encountered in circuit analysis, along with the corresponding Fourier transforms. Thus we will have not only the time-domain descriptions of such functions, but their frequency-domain descriptions as well. Then, just as the use of phasor transforms simplified the determination of the steady-state sinusoidal response, we shall find that the use of the Fourier transforms of various forcing functions can simplify the determination of the complete response, both the natural and the forced components. When we extend our thinking to the use of the Laplace transform in the following chapter, we shall even be able to account for the troublesome initial conditions that have plagued us in the past. With these thoughts in mind, let us look at just a few more transform pairs with the goal of listing our findings in a form that will be useful for quick reference later.

The first forcing function that we considered many chapters ago was a dc voltage or current. To find the Fourier transform of a constant function of time, $f(t) = K$, our first inclination might be to substitute this constant in the defining equation for the Fourier transform and evaluate the resulting integral. If we did, we would find ourselves with an indeterminate expression on our hands. Fortunately, however, we have already solved this problem, for from expression (31),

$$e^{-j\omega_0 t} \Leftrightarrow 2\pi\delta(\omega + \omega_0)$$

We see that, if we simply let $\omega_0 = 0$, then the resulting transform pair is

$$1 \Leftrightarrow 2\pi\delta(\omega) \tag{34}$$

from which it follows that

$$K \Leftrightarrow 2\pi K\delta(\omega) \tag{35}$$

and our problem is solved. The frequency spectrum of a constant function of time consists only of a component at $\omega = 0$, which we knew all along.

As another example, let us obtain the Fourier transform of a singularity function known as the *signum function,* sgn (t), defined by

$$\text{sgn}\,(t) = \begin{cases} -1 & t < 0 \\ 1 & t > 0 \end{cases} \tag{36}$$

or　　　　　　　　　$\text{sgn}\,(t) = u(t) - u(-t)$

Again, if we should try to substitute this time function in the defining equation for the Fourier transform, we would face an indeterminate expression upon substitution of the limits of integration. This same problem will arise every time we attempt to obtain the Fourier transform of a time function that does

Chapter 18: Fourier Transforms

not approach zero as $|t|$ approaches infinity. Eventually, we will avoid this situation by using the *Laplace transform,* because this transform will be seen to contain a built-in convergence factor that will cure many of the inconvenient ills associated with the evaluation of certain Fourier transforms.

The signum function under consideration can be written as

$$\text{sgn}\,(t) = \lim_{a\to 0}\,[e^{-at}u(t) - e^{at}u(-t)]$$

Notice that the expression within the brackets *does* approach zero as $|t|$ gets very large. Using the definition of the Fourier transform, we obtain

$$\mathscr{F}\{\text{sgn}\,(t)\} = \lim_{a\to 0}\left[\int_0^\infty e^{-j\omega t}e^{-at}\,dt - \int_{-\infty}^0 e^{-j\omega t}e^{at}\,dt\right]$$

$$= \lim_{a\to 0}\frac{-j2\omega}{\omega^2 + a^2} = \frac{2}{j\omega}$$

The real component is zero, since sgn (t) is an odd function of t. Thus,

$$\text{sgn}\,(t) \Leftrightarrow \frac{2}{j\omega} \tag{37}$$

As a final example in this section, let us look at the familiar unit-step function, $u(t)$. Making use of our work on the signum function in the preceding paragraphs, we represent the unit step by

$$u(t) = \tfrac{1}{2} + \tfrac{1}{2}\,\text{sgn}\,(t)$$

and obtain the Fourier transform pair

$$u(t) \Leftrightarrow \left[\pi\delta(\omega) + \frac{1}{j\omega}\right] \tag{38}$$

Table 18-1 presents the conclusions drawn from the examples discussed in this section, along with a few others that have not been detailed here.

Example 18-3   Use Table 18-1 to find the Fourier transform of the time function $3e^{-t}\cos 4t\,u(t)$.

Solution: From the next to the last entry in the table, we have

$$e^{-\alpha t}\cos\omega_d t\,u(t) \Leftrightarrow \frac{\alpha + j\omega}{(\alpha + j\omega)^2 + \omega_d^2}$$

We therefore identify α as 1 and ω_d as 4, and have

$$\mathbf{F}(j\omega) = (3)\frac{1 + j\omega}{(1 + j\omega)^2 + 16} \qquad\blacksquare$$

Drill Problems

18-7. Evaluate the Fourier transform at $\omega = 12$ for the time function: (a) $4u(t) - 10\delta(t)$; (b) $5e^{-8t}u(t)$; (c) $4\cos 8t\,u(t)$; (d) -4 sgn (t).
Ans: $10.01\underline{/-178.1°}$; $0.347\underline{/-56.3°}$; $-j0.6$; $j0.667$

18-8. Find $f(t)$ at $t = 2$ if $\mathbf{F}(j\omega)$ is (a) $5e^{-j3\omega} - j(4/\omega)$; (b) $8[\delta(\omega - 3) + \delta(\omega + 3)]$; (c) $(8/\omega)\sin 5\omega$.
Ans: 2.00; 2.45; 0.800

Table 18-1

Some familiar Fourier transform pairs

$f(t)$	$f(t)$	$\mathcal{F}\{f(t)\} = \mathbf{F}(j\omega)$	$\lvert\mathbf{F}(j\omega)\rvert$
	$\delta(t - t_0)$	$e^{-j\omega t_0}$	
	$e^{j\omega_0 t}$	$2\pi\delta(\omega - \omega_0)$	
	$\cos \omega_0 t$	$\pi[\delta(\omega + \omega_0) + \delta(\omega - \omega_0)]$	
	1	$2\pi\delta(\omega)$	
	$\mathrm{sgn}(t)$	$\dfrac{2}{j\omega}$	
	$u(t)$	$\pi\delta(\omega) + \dfrac{1}{j\omega}$	
	$e^{-\alpha t}u(t)$	$\dfrac{1}{\alpha + j\omega}$	
	$e^{-\alpha t}\cos \omega_d t \cdot u(t)$	$\dfrac{\alpha + j\omega}{(\alpha + j\omega)^2 + \omega_d^2}$	
	$u(t + \tfrac{1}{2}T) - u(t - \tfrac{1}{2}T)$	$T\dfrac{\sin \frac{\omega T}{2}}{\frac{\omega T}{2}}$	

In Sec. 18-2 we remarked that we would be able to show that periodic time functions, as well as nonperiodic functions, possess Fourier transforms. Since a promise made is a debt unpaid, let us now establish this fact on a rigorous basis. Consider a periodic time function $f(t)$ with period T and Fourier series expansion, as outlined by Eqs. (1), (2), and (3),

$$f(t) = \sum_{n=-\infty}^{\infty} \mathbf{c}_n e^{jn\omega_0 t} \tag{1}$$

$$\mathbf{c}_n = \frac{1}{T} \int_{-T/2}^{T/2} f(t) e^{-jn\omega_0 t}\, dt \tag{2}$$

and

$$\omega_0 = \frac{2\pi}{T} \tag{3}$$

Bearing in mind that the Fourier transform of a sum is just the sum of the transforms of the terms in the sum, and that $\mathbf{c}_n$ is not a function of time, we can write

$$\mathcal{F}\{f(t)\} = \mathcal{F}\left\{ \sum_{n=-\infty}^{\infty} \mathbf{c}_n e^{jn\omega_0 t} \right\} = \sum_{n=-\infty}^{\infty} \mathbf{c}_n \mathcal{F}\{e^{jn\omega_0 t}\}$$

After obtaining the transform of $e^{jn\omega_0 t}$ from expression (30), we have

$$f(t) \Leftrightarrow 2\pi \sum_{n=-\infty}^{\infty} \mathbf{c}_n \delta(\omega - n\omega_0) \tag{39}$$

This shows that $f(t)$ has a discrete spectrum consisting of impulses located at points on the ω axis given by $\omega = n\omega_0$, $n = \ldots, -2, -1, 0, 1, \ldots$. The strength of each impulse is 2π times the value of the corresponding Fourier coefficient appearing in the complex form of the Fourier series expansion for $f(t)$.

As a check on our work, let us see whether the inverse Fourier transform of the right side of expression (39) is once again $f(t)$. This inverse transform can be written as

$$\mathcal{F}^{-1}\{\mathbf{F}(j\omega)\} = \frac{1}{2\pi} \int_{-\infty}^{\infty} e^{j\omega t}\left[2\pi \sum_{n=-\infty}^{\infty} \mathbf{c}_n \delta(\omega - n\omega_0) \right] d\omega \stackrel{?}{=} f(t)$$

Since the exponential term does not contain the index of summation n, we can interchange the order of the integration and summation operations:

$$\mathcal{F}^{-1}\{\mathbf{F}(j\omega)\} = \sum_{n=-\infty}^{\infty} \int_{-\infty}^{\infty} \mathbf{c}_n e^{j\omega t} \delta(\omega - n\omega_0)\, d\omega \stackrel{?}{=} f(t)$$

Because it is not a function of the variable of integration, $\mathbf{c}_n$ can be treated as a constant. Then, using the sifting property of the impulse, we obtain

$$\mathcal{F}^{-1}\{\mathbf{F}(j\omega)\} = \sum_{n=-\infty}^{\infty} \mathbf{c}_n e^{jn\omega_0 t} \stackrel{?}{=} f(t)$$

which is exactly the same as Eq. (1), the complex Fourier series expansion for $f(t)$. The question marks in the preceding equations can now be removed, and the existence of the Fourier transform for a periodic time function is established. This should come as no great surprise, however. In the last section we evaluated the Fourier transform of a cosine function, which is certainly periodic, although we made no direct reference to its periodicity. However, we did use a back-handed approach in getting the transform. But now we have a mathematical tool by which the transform can be obtained more directly. To demonstrate this procedure, consider $f(t) = \cos \omega_0 t$ once more. First we evaluate the Fourier coefficients c_n:

$$\mathbf{c}_n = \frac{1}{T} \int_{-T/2}^{T/2} \cos \omega_0 t \, e^{-jn\omega_0 t} \, dt = \begin{cases} \frac{1}{2} & n = \pm 1 \\ 0 & \text{otherwise} \end{cases}$$

Then

$$\mathscr{F}\{f(t)\} = 2\pi \sum_{n=-\infty}^{\infty} \mathbf{c}_n \delta(\omega - n\omega_0)$$

This expression has values that are nonzero only when $n = \pm 1$, and it follows, therefore, that the entire summation reduces to

$$\mathscr{F}\{\cos \omega_0 t\} = \pi[\delta(\omega - \omega_0) + \delta(\omega + \omega_0)]$$

which is precisely the expression that we obtained before. What a relief!

Drill Problem

18-9. Find (a) $\mathscr{F}\{5 \sin^2 3t\}$; (b) $\mathscr{F}\{A \sin \omega_0 t\}$; (c) $\mathscr{F}\{6 \cos (8t + 0.1\pi)\}$.
Ans: $2.5\pi[2\delta(\omega) + \delta(\omega + 6) + \delta(\omega - 6)]$; $j\pi A[\delta(\omega + \omega_0) - \delta(\omega - \omega_0)]$; $[18.85\underline{/18°}]\delta(\omega - 8) + [18.85\underline{/-18°}]\delta(\omega + 8)$

18-7

Convolution and circuit response in the time domain

Before continuing our discussion of the Fourier transform, let us note again where all this is leading. Our goal is a technique for simplifying those problems in linear circuit analysis that involve the determination of explicit expressions for response functions caused by the application of one or more forcing functions. We accomplish this by utilizing a transfer function called the *system function* of the circuit. It turns out that this system function is the Fourier transform of the unit-impulse response of the circuit.

The specific analytical technique that we will use requires the evaluation of the Fourier transform of the forcing function, the multiplication of this transform by the system function to obtain the transform of the response function, and then the inverse-transform operation to obtain that response function. By these means some relatively complicated integral expressions will be reduced to simple functions of ω, and the mathematical operations of integration and differentiation will be replaced by the simpler operations of algebraic multiplication and division. With these remarks in mind, let us now proceed to examine the unit-impulse response of a circuit and eventually establish its relation to the system function. Then we can look at some specific analysis problems.

Consider a linear electrical network N, without initial stored energy, to which a forcing function $x(t)$ is applied. At some point in this circuit, a response function $y(t)$ is present. We show this in block diagram form in Fig. 18-7a along with general sketches of typical time functions. The forcing function is arbitrarily shown to exist only in the interval $a < t < b$. Thus, $y(t)$ exists only for $t > a$.

The question that we now wish to answer is this: If we know the form of $x(t)$, then how is $y(t)$ described? To answer this question, it is obvious that we need to know something about N. Suppose, therefore, that our knowledge of N consists of knowing its response when the forcing function is a unit impulse.

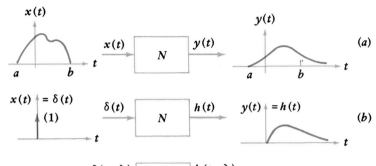

Figure 18-7

A conceptual development of the convolution integral,

$$y(t) = \int_{-\infty}^{\infty} x(\lambda)h(t - \lambda)\, d\lambda$$

That is, we are assuming that we know $h(t)$, the response function resulting when a unit impulse is supplied as the forcing function at $t = 0$, as shown in Fig. 18-7b. The function $h(t)$ is commonly called the unit-impulse response function, or the *impulse response*. This is a very important descriptive property for an electric circuit.

Instead of applying the unit impulse at time $t = 0$, let us now suppose that it is applied at time $t = \lambda$ (lambda). We see that the only change in the output is a time delay. Thus, the output becomes $h(t - \lambda)$ when the input is $\delta(t - \lambda)$, as shown in Fig. 18-7c. Next, suppose that the input impulse has some strength other than unity. Specifically, let the strength of the impulse be numerically equal to the value of $x(t)$ when $t = \lambda$. This value $x(\lambda)$ is a constant; we know that the multiplication of a single forcing function in a linear circuit by a constant simply causes the response to change proportionately. Thus, if the input is changed to $x(\lambda)\delta(t - \lambda)$, then the response becomes $x(\lambda)h(t - \lambda)$, as shown in Fig. 18-7d.

Now let us sum this latest input over all possible values of λ and use the result as a forcing function for N. Linearity decrees that the output must be equal to the sum of the responses resulting from the use of all possible values of λ. Loosely speaking, the integral of the input produces the integral of the output, as shown in Fig. 18-7e. But what is the input now? Given the sifting property of the unit impulse, we see that the input is simply $x(t)$, the original input.

Our question is now answered. When $x(t)$, the input to N, is known, and when $h(t)$, the impulse response of N, is known, then $y(t)$, the output or response function, is expressed by

$$y(t) = \int_{-\infty}^{\infty} x(\lambda)h(t - \lambda)\, d\lambda \tag{40}$$

as shown in Fig. 18-7*f*. This important relationship is known far and wide as the *convolution integral*. In words, this last equation states that *the output is equal to the input convolved with the impulse response*. It is often abbreviated by means of

$$y(t) = x(t) * h(t) \tag{41}$$

where the asterisk is read "convolved with."

Equation (40) sometimes appears in a slightly different but equivalent form. If we let $z = t - \lambda$, $d\lambda = -dz$, the expression for $y(t)$ becomes

$$y(t) = \int_{\infty}^{-\infty} -x(t - z)h(z)\, dz = \int_{-\infty}^{\infty} x(t - z)h(z)\, dz$$

and since the symbol that we use for the variable of integration is unimportant, we can modify Eq. (40) to write

$$y(t) = x(t) * h(t) = \int_{-\infty}^{\infty} x(z)h(t - z)\, dz = \int_{-\infty}^{\infty} x(t - z)h(z)\, dz \tag{42}$$

The two forms of the convolution integral given by Eq. (42) are worth memorizing.

The result that we have in Eq. (42) is very general. It applies to any linear system. However, we are usually interested in *physically realizable* systems, those that *do* exist or *could* exist, and such systems have a property that modifies the convolution integral slightly. That is, the response of the system cannot begin before the forcing function is applied. In particular, $h(t)$ is the response of the system resulting from the application of a unit impulse at $t = 0$. Therefore, $h(t)$ cannot exist for $t < 0$. It follows that, in the second integral of Eq. (42), the integrand is zero when $z < 0$; in the first integral, the integrand is zero when $(t - z)$ is negative, or when $z > t$. Therefore, for realizable systems the limits of integration change in the convolution integrals:

$$y(t) = x(t) * h(t) = \int_{-\infty}^{t} x(z)h(t - z)\, dz = \int_{0}^{\infty} x(t - z)h(z)\, dz \tag{43}$$

Equations (42) and (43) are both valid, but the latter is more specific when we are speaking of *realizable* linear systems.

Before discussing the significance of the impulse response of a circuit any further, let us consider a numerical example which will give us some insight into just how the convolution integral can be evaluated. Although the expression itself is simple enough, the evaluation is sometimes troublesome, especially with regard to the values used as the limits of integration.

Suppose that the input is a rectangular voltage pulse that starts at $t = 0$, has a duration of 1 s, and is 1 V in amplitude:

$$x(t) = v_i(t) = u(t) - u(t - 1)$$

Suppose also that the impulse response of this circuit is known to be an exponential function of the form:[7]

$$h(t) = 2e^{-t}u(t)$$

---

[7] A description of one possible circuit to which this impulse response might apply is developed in Prob. 35.

We wish to evaluate the output voltage $v_o(t)$, and we can write the answer immediately in integral form,

$$y(t) = v_o(t) = v_i(t) * h(t) = \int_0^\infty v_i(t-z)h(z)\,dz$$

$$= \int_0^\infty [u(t-z) - u(t-z-1)][2e^{-z}u(z)]\,dz$$

Obtaining this expression for $v_o(t)$ is simple enough, but the presence of the many unit-step functions tends to make its evaluation confusing. Careful attention must be paid to the determination of those portions of the range of integration in which the integrand is zero.

Let us use some graphical assistance to help us understand what the convolution integral says. We begin by drawing several z axes lined up one above the other, as shown in Fig. 18-8. We know what $v_i(t)$ looks like, and so we know what $v_i(z)$ looks like also; this is plotted as Fig. 18-8a. The function $v_i(-z)$ is simply $v_i(z)$ run backward with respect to z, or rotated about the ordinate axis; it is shown in Fig. 18-8b. Next we wish to represent $v_i(t-z)$, which is $v_i(-z)$ after it is shifted to the right by an amount $z = t$ as shown in Fig. 18-8c. On the next z axis, in Fig. 18-8d, the hypothetical impulse response is plotted.

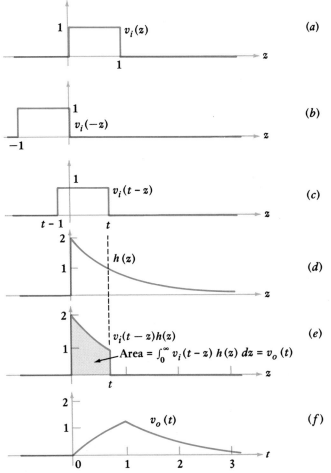

Figure 18-8

Graphical concepts in evaluating a convolution integral.

Finally, we multiply the two functions $v_i(t - z)$ and $h(z)$. The result is shown in Fig. 18-8e. Since $h(z)$ does not exist prior to $t = 0$ and $v_i(t - z)$ does not exist for $z > t$, we notice that the product of these two functions has nonzero values only in the interval $0 < z < t$ for the case shown where $t < 1$; when $t > 1$, the nonzero values for the product are obtained in the interval $(t - 1) < z < t$. The *area* under the product curve (shown shaded in the figure) is numerically equal to the value of v_o corresponding to the specific value of t selected in Fig. 18-8c. As t increases from zero to unity, the area under the product curve continues to rise, and thus $v_o(t)$ continues to rise. But as t increases beyond $t = 1$, the area under the product curve, which is equal to $v_o(t)$, starts decreasing and approaches zero. For $t < 0$, the curves representing $v_i(t - z)$ and $h(z)$ do not overlap at all, and so the area under the product curve is obviously zero. Now let us use these graphical concepts to obtain an explicit expression for $v_o(t)$.

For values of t that lie between zero and unity, we must integrate from $z = 0$ to $z = t$; for values of t that exceed unity, the range of integration is $t - 1 < z < t$. Thus, we may write

$$
v_o(t) = \begin{cases} 0 & t < 0 \\ \int_0^t 2e^{-z}\,dz = 2(1 - e^{-t}) & 0 < t < 1 \\ \int_{t-1}^t 2e^{-z}\,dz = 2(e - 1)e^{-t} & t > 1 \end{cases}
$$

This function is shown plotted versus the time variable t in Fig. 18-8f, and our solution is completed.

As a second convolution example, let us apply a unit-step function, $x(t) = u(t)$, as the input to a system whose impulse response is $h(t) = u(t) - 2u(t - 1) + u(t - 2)$. These two functions of time are shown in Fig. 18-9a and b. We

Figure 18-9

Sketches of (a) the input signal $x(t) = u(t)$ and (b) the unit-impulse response $h(t) = u(t) - 2u(t - 1) + u(t - 2)$, for a linear system.

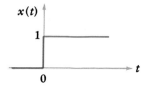

(a)

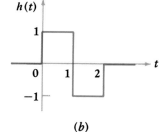

(b)

arbitrarily choose to evaluate the first integral of Eq. (43),

$$
y(t) = \int_{-\infty}^t x(z)h(t - z)\,dz
$$

and prepare a sequence of sketches to help select the correct limits of integration. Figure 18-10 shows these functions in order: the input $x(z)$ as a function of z; the impulse response $h(z)$; the curve of $h(-z)$, which is just $h(z)$ rotated about the vertical axis; and $h(t - z)$, obtained by sliding $h(-z)$ to the right t units. For this sketch, we have selected t in the range $0 < t < 1$.

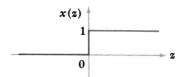

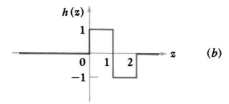

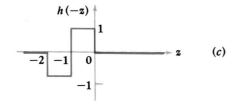

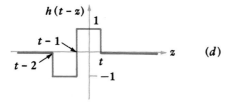

Figure 18-10

(a) The input signal and (b) the impulse response are plotted as functions of z. (c) $h(-z)$ is obtained by flipping $h(z)$ about the vertical axis, and (d) $h(t-z)$ results when $h(-z)$ is slid t units to the right.

It is now easy to visualize the product of the first graph, $x(z)$, and the last, $h(t-z)$, for the various ranges of t. When t is less than zero, there is no overlap, and

$$y(t) = 0 \qquad t < 0$$

For the case sketched in Fig. 18-10d, the curves overlap from $z = 0$ to $z = t$, and each is unity in value. Thus,

$$y(t) = \int_0^t 1 \times 1 \, dz = t \qquad 0 < t < 1$$

When t lies between 1 and 2, $h(t-z)$ has slid far enough to the right to bring under the step function that part of the negative square wave extending from $z = 0$ to $z = t-1$. We have

$$y(t) = \int_0^{t-1} 1 \times (-1) \, dz + \int_{t-1}^t 1 \times 1 \, dz$$

$$= -z \Big|_{z=0}^{z=t-1} + z \Big|_{z=t-1}^{z=t}$$

Therefore,

$$y(t) = -(t-1) + t - (t-1) = 2 - t \qquad 1 < t < 2$$

Finally, when t is greater than 2, $h(t - z)$ has slid far enough to the right so that it lies entirely to the right of $z = 0$. The intersection with the unit step is complete, and

$$y(t) = \int_{t-2}^{t-1} 1 \times (-1)\, dz + \int_{t-1}^{t} 1 \times 1\, dz$$

$$= -z \Big|_{z=t-2}^{z=t-1} + z \Big|_{z=t-1}^{z=t}$$

or $$y(t) = -(t - 1) + (t - 2) + t - (t - 1) = 0 \qquad t > 2$$

These four segments of $y(t)$ are collected as a continuous curve in Fig. 18-11.

Figure 18-11

The result of convolving the $x(t)$ and $h(t)$ shown in Fig. 18-9.

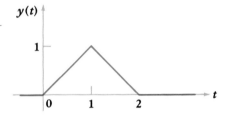

There is a great deal of information and technique wrapped up in these two examples. The drill problems that follow offer an opportunity to make sure that the procedure is understood sufficiently well to make it worthwhile to pass on to new material. In other words, pay attention to the old Chinese proverb: I hear, I forget; I see, I remember; I do, I understand.

Drill Problems

18-10. The impulse response of a network is given by $h(t) = 5u(t - 1)$. If an input signal $x(t) = 2[u(t) - u(t - 3)]$ is applied, determine the output $y(t)$ at t equal to (a) -0.5; (b) 0.5; (c) 1.5; (d) 2.5; (e) 3.5; (f) 4.5.

Ans: 0; 0; 5; 15; 25; 30

18-11. Let $h(t) = 0.5[u(t) - u(t - 0.7)]$ and $x(t) = 0.8e^{-2t}$. Find $y(t)$ at t equal to (a) -0.2; (b) 0.5; (c) 0.7; (d) 1.   Ans: 0; 0.1264; 0.1507; 0.0827

18-8

The system function and response in the frequency domain

In the previous section the problem of determining the output of a physical system in terms of the input and the impulse response was solved by using the convolution integral and working entirely in the time domain. The input, the output, and the impulse response are all time functions. Now let us see whether some analytical simplification can be wrought by working with frequency-domain descriptions of these three functions.

To do this we examine the Fourier transform of the system output. Assuming arbitrarily that the input and output are voltages, we apply the basic definition of the Fourier transform and express the output by the convolution integral of Eq. (42):

$$\mathscr{F}\{v_o(t)\} = \mathbf{F}_o(j\omega) = \int_{-\infty}^{\infty} e^{-j\omega t}\left[\int_{-\infty}^{\infty} v_i(t - z)h(z)\, dz\right] dt$$

where we again assume no initial energy storage. At first glance this expression may seem rather formidable, but it can be reduced to a result that is surprisingly simple. We may move the exponential term inside the inner integral because it does not contain the variable of integration z. Next we reverse the order of integration, obtaining

$$\mathbf{F}_o(j\omega) = \int_{-\infty}^{\infty} \left[\int_{-\infty}^{\infty} e^{-j\omega t} v_i(t-z) h(z) \, dt \right] dz$$

Since it is not a function of t, we can extract $h(z)$ from the inner integral and simplify the integration with respect to t by a change of variable, $t - z = x$:

$$\mathbf{F}_o(j\omega) = \int_{-\infty}^{\infty} h(z) \left[\int_{-\infty}^{\infty} e^{-j\omega(x+z)} v_i(x) \, dx \right] dz$$

$$= \int_{-\infty}^{\infty} e^{-j\omega z} h(z) \left[\int_{-\infty}^{\infty} e^{-j\omega x} v_i(x) \, dx \right] dz$$

But now the sum is starting to break through, for the inner integral is merely the Fourier transform of $v_i(t)$. Furthermore, it contains no z terms and can be treated as a constant in any integration involving z. Thus, we can move this transform, $\mathbf{F}_i(j\omega)$, completely outside all the integral signs:

$$\mathbf{F}_o(j\omega) = \mathbf{F}_i(j\omega) \int_{-\infty}^{\infty} e^{-j\omega z} h(z) \, dz$$

Finally, the remaining integral exhibits our old friend once more, another Fourier transform! This one is the Fourier transform of the impulse response, which we shall designate by the notation $\mathbf{H}(j\omega)$. Therefore, all our work has boiled down to the simple result:[8]

$$\mathbf{F}_o(j\omega) = \mathbf{F}_i(j\omega) \mathbf{H}(j\omega) = \mathbf{F}_i(j\omega) \mathcal{F}\{h(t)\} \tag{44}$$

This is another important result: it defines the *system function* $\mathbf{H}(j\omega)$ as the ratio of the Fourier transform of the response function to the Fourier transform of the forcing function. Moreover, the system function and the impulse response constitute a Fourier transform pair:

$$h(t) \Leftrightarrow \mathbf{H}(j\omega) \tag{45}$$

The development in the preceding paragraph also serves to prove the general statement that the Fourier transform of the convolution of two time functions is the product of their Fourier transforms,

$$\boxed{\mathcal{F}\{f(t) * g(t)\} = \mathbf{F}_f(j\omega) \mathbf{F}_g(j\omega)} \tag{46}$$

To recapitulate, if we know the Fourier transforms of the forcing function and the impulse response, then the Fourier transform of the response function can be obtained as their product. The result is a description of the response function in the frequency domain; if we wish to do so, we can obtain the time-domain description of the response function by taking the inverse Fourier transform. Thus we see that the process of convolution in the time domain is equivalent to the relatively simple operation of multiplication in the frequency domain. This is one fact that makes the use of integral transforms so attractive.

The foregoing comments might make us wonder once again why we would ever choose to work in the time domain at all, but we must always remember

---

[8] Not to be confused with Boyle's law in your physics classes!

that we seldom get something for nothing. A poet once said, "Our sincerest laughter/with some pain is fraught."[9] The pain herein is the occasional difficulty in obtaining the inverse Fourier transform of a response function, for reasons of mathematical complexity. On the other hand, a modern digital computer can convolve two time functions with magnificent celerity. For that matter, it can also obtain an FFT (fast Fourier transform)[10] quite rapidly. Consequently there is no clear-cut advantage between working in the time domain and in the frequency domain. A decision must be made each time a new problem arises; it should be based on the given information available and on the computational facilities at hand.

Now let us attempt a frequency-domain analysis of the problem that we worked in the preceding section with the convolution integral. We had a forcing function of the form

$$v_i(t) = u(t) - u(t-1)$$

and a unit-impulse response defined by

$$h(t) = 2e^{-t}u(t)$$

We first obtain the corresponding Fourier transforms. The forcing function is the difference between two unit-step functions. These two functions are identical, except that one is initiated 1 s after the other. We shall evaluate the response due to $u(t)$; the response due to $u(t-1)$ is the same, but delayed in time by 1 s. The difference between these two partial responses will be the total response due to $v_i(t)$.

The Fourier transform of $u(t)$ was obtained in Sec. 18-5:

$$\mathcal{F}\{u(t)\} = \pi\delta(\omega) + \frac{1}{j\omega}$$

The system function is obtained by taking the Fourier transform of $h(t)$, listed in Table 18-1,

$$\mathcal{F}\{h(t)\} = \mathbf{H}(j\omega) = \mathcal{F}\{2e^{-t}u(t)\} = \frac{2}{1+j\omega}$$

The inverse transform of the product of these two functions yields that component of $v_o(t)$ caused by $u(t)$,

$$v_{o1}(t) = \mathcal{F}^{-1}\left\{\frac{2\pi\delta(\omega)}{1+j\omega} + \frac{2}{j\omega(1+j\omega)}\right\}$$

Using the sifting property of the unit impulse, the inverse transform of the first term is just a constant equal to unity. Thus,

$$v_{o1}(t) = 1 + \mathcal{F}^{-1}\left\{\frac{2}{j\omega(1+j\omega)}\right\}$$

The second term contains a product of terms in the denominator, each of the form $(\alpha + j\omega)$, and its inverse transform is found most easily by making

[9] A cultural message from P. B. Shelley, "To a Skylark," 1821.
[10] The fast Fourier transform is a type of *discrete* Fourier transform, which is a numerical approximation to the (continuous) Fourier transform we have been considering. Many references concerning it are given in G. D. Bergland, "A guided tour of the fast Fourier transform," *IEEE Spectrum*, vol. 6, no. 7, July 1969, pp. 41–52.

use of the partial-fraction expansion that we all should have seen in our introductory calculus course. Let us select a technique for obtaining a partial-
fraction expansion that has one big advantage—it always works, although
faster methods are usually available for most situations.[11] We assign an
unknown quantity in the numerator of each fraction, here two in number,

$$\frac{2}{j\omega(1+j\omega)} = \frac{A}{j\omega} + \frac{B}{1+j\omega}$$

and then substitute a corresponding number of simple values for $j\omega$. Here we
let $j\omega = 1$:

$$1 = A + \frac{B}{2}$$

and then let $j\omega = -2$:

$$1 = -\frac{A}{2} - B$$

This leads to $A = 2$ and $B = -2$. Thus,

$$\mathcal{F}^{-1}\left\{\frac{2}{j\omega(1+j\omega)}\right\} = \mathcal{F}^{-1}\left\{\frac{2}{j\omega} - \frac{2}{1+j\omega}\right\} = \text{sgn}\,(t) - 2e^{-t}u(t)$$

so that

$$v_{o1}(t) = 1 + \text{sgn}\,(t) - 2e^{-t}u(t)$$
$$= 2u(t) - 2e^{-t}u(t)$$
$$= 2(1 - e^{-t})u(t)$$

It follows that $v_{o2}(t)$, the component of $v_o(t)$ produced by $u(t-1)$, is

$$v_{o2}(t) = 2(1 - e^{-(t-1)})u(t-1)$$

Therefore,

$$v_o(t) = v_{o1}(t) + v_{o2}(t)$$
$$= 2(1-e^{-t})u(t) - 2(1-e^{-t+1})u(t-1)$$

The discontinuities at $t = 0$ and $t = 1$ dictate a separation into three time
intervals:

$$v_o(t) = \begin{cases} 0 & t < 0 \\ 2(1-e^{-t}) & 0 < t < 1 \\ 2(e-1)e^{-t} & t > 1 \end{cases}$$

This is the same result we obtained by working this problem with the convolution integral in the time domain.

It appears that some ease of solution is realized by working this particular
problem in the frequency domain rather than in the time domain. We are
able to trade a relatively complicated time-domain integration for a simple
multiplication of two Fourier transforms in the frequency domain, and (this is

[11] See that old calculus text!

a most noteworthy point) we did not have to pay the price of having a difficult Fourier transform to invert in determining $v_o(t)$.

Drill Problems

18-12. The impulse response of a certain linear network is $h(t) = 6e^{-20t}u(t)$. The input signal is $3e^{-6t}u(t)$ V. Find (a) $\mathbf{H}(j\omega)$; (b) $\mathbf{V}_i(j\omega)$; (c) $\mathbf{V}_o(j\omega)$; (d) $v_o(0.1)$; (e) $v_o(0.3)$; (f) $v_{o,\max}$.

     Ans: $6/(20 + j\omega)$; $3/(6 + j\omega)$; $18/[(20 + j\omega)(3 + j\omega)]$; 0.532 V; 0.209 V; 0.5372 V

18-13. A certain linear system has the system function $(10 - j2\omega)/(4 + j\omega)$. Work in the frequency domain to find the time-domain output if the input voltage is (a) a unit impulse; (b) a unit-step function; (c) $2e^{-5t}u(t)$ V.

     Ans: $-2\delta(t) + 18e^{-4t}u(t)$ V; $(2.5 - 4.5e^{-4t})u(t)$ V; $(36e^{-4t} - 40e^{-5t})u(t)$ V

18-9

The physical significance of the system function

In this section we shall try to connect several aspects of the Fourier transform with work we completed in earlier chapters.

     Given a general linear two-port network N without any initial energy storage, we assume sinusoidal forcing and response functions, arbitrarily assumed to be voltages, as shown in Fig. 18-12. We let the input voltage be simply $A \cos(\omega_x t + \theta)$, and the output can be described in general terms as $v_o(t) = B \cos(\omega_x t + \phi)$, where the amplitude B and phase angle ϕ are functions of ω_x. In phasor form, we can write the forcing and response functions as $\mathbf{V}_i = Ae^{j\theta}$ and $\mathbf{V}_o = Be^{j\phi}$. The ratio of the phasor response to the phasor forcing function is a complex number that is a function of ω_x:

$$\frac{\mathbf{V}_o}{\mathbf{V}_i} = \mathbf{G}(\omega_x) = \frac{B}{A}e^{j(\phi - \theta)}$$

where B/A is the amplitude of $\mathbf{G}$ and $\phi - \theta$ is its phase angle. This transfer function $\mathbf{G}(\omega_x)$ could be obtained in the laboratory by varying ω_x over a large range of values and measuring the amplitude B/A and phase $\phi - \theta$ for each value of ω_x. If we then plotted each of these parameters as a function of frequency, the resultant pair of curves would completely describe the transfer function.

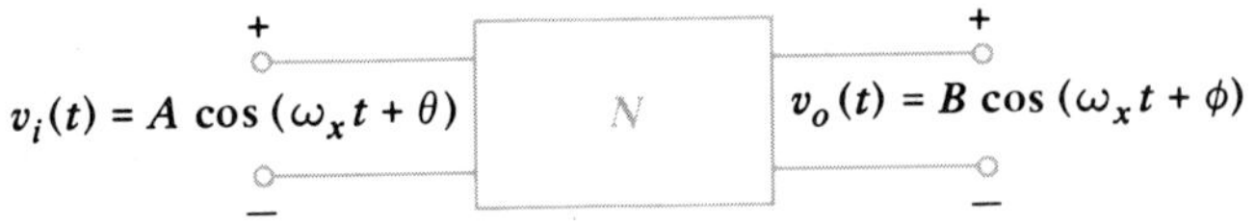

Figure 18-12

Sinusoidal analysis can be used to determine the transfer function $\mathbf{H}(j\omega_x) = (B/A)e^{j(\phi - \theta)}$, where B and ϕ are functions of ω_x.

Now let us hold these comments in the backs of our minds for a moment as we consider a slightly different aspect of the same analysis problem.

     For the circuit with sinusoidal input and output shown in Fig. 18-12, what is the system function $\mathbf{H}(j\omega)$? To answer this question, we begin with the definition of $\mathbf{H}(j\omega)$ as the ratio of the Fourier transforms of the output and the input. Both of these time functions involve the functional form $\cos(\omega_x t + \beta)$, whose Fourier transform we have not evaluated as yet, although we can handle $\cos \omega_x t$. The transform we need is

$$\mathscr{F}\{\cos(\omega_x t + \beta)\} = \int_{-\infty}^{\infty} e^{-j\omega t} \cos(\omega_x t + \beta)\, dt$$

If we make the substitution $\omega_x t + \beta = \omega_x \tau$, then

$$\mathscr{F}\{\cos(\omega_x t + \beta)\} = \int_{-\infty}^{\infty} e^{-j\omega\tau + j\omega\beta/\omega_x} \cos \omega_x \tau \, d\tau$$

$$= e^{j\omega\beta/\omega_x} \mathscr{F}\{\cos \omega_x t\}$$

$$= \pi e^{j\omega\beta/\omega_x} [\delta(\omega - \omega_x) + \delta(\omega + \omega_x)]$$

This is a new Fourier transform pair,

$$\cos(\omega_x t + \beta) \Leftrightarrow \pi e^{j\omega\beta/\omega_x} [\delta(\omega - \omega_x) + \delta(\omega + \omega_x)] \tag{47}$$

which we can now use to evaluate the desired system function,

$$\mathbf{H}(j\omega) = \frac{\mathscr{F}\{B\cos(\omega_x t + \phi)\}}{\mathscr{F}\{A\cos(\omega_x t + \theta)\}}$$

$$= \frac{\pi B e^{j\omega\phi/\omega_x}[\delta(\omega - \omega_x) + \delta(\omega + \omega_x)]}{\pi A e^{j\omega\theta/\omega_x}[\delta(\omega - \omega_x) + \delta(\omega + \omega_x)]}$$

$$= \frac{B}{A} e^{j\omega(\phi - \theta)/\omega_x}$$

Now we recall the expression for $\mathbf{G}(\omega_x)$,

$$\mathbf{G}(\omega_x) = \frac{B}{A} e^{j(\phi - \theta)}$$

where B and ϕ were evaluated at $\omega = \omega_x$, and we see that evaluating $\mathbf{H}(j\omega)$ at $\omega = \omega_x$ gives

$$\mathbf{H}(\omega_x) = \mathbf{G}(\omega_x) = \frac{B}{A} e^{j(\phi - \theta)}$$

Since there is nothing special about the x subscript, we conclude that the system function and the transfer function are identical:

$$\mathbf{H}(j\omega) = \mathbf{G}(\omega) \tag{48}$$

The fact that one argument is ω while the other is indicated by $j\omega$ is immaterial and arbitrary; the j merely makes possible a more direct comparison between the Fourier and Laplace transforms.

Equation (48) represents a direct connection between Fourier transform techniques and sinusoidal steady-state analysis. Our previous work on steady-state sinusoidal analysis using phasors was but a special case of the more general techniques of Fourier transform analysis. It was "special" in the sense that the inputs and outputs were sinusoids, whereas the use of Fourier transforms and system functions enables us to handle nonsinusoidal forcing functions and responses.

Thus, to find the system function $\mathbf{H}(j\omega)$ for a network, all we need to do is to determine the corresponding sinusoidal transfer function as a function of ω (or $j\omega$).

Example 18-4   As an example of the application of these profound generalities, let us look at the simple *RL* series circuit of Fig. 18-13a. We seek the voltage across the inductor when the input voltage is a simple exponentially decaying pulse.

Figure 18-13

(a) The response $v_o(t)$ caused by $v_i(t)$ is desired. (b) The system function $\mathbf{H}(j\omega)$ may be determined by sinusoidal steady-state analysis: $\mathbf{H}(j\omega) = \mathbf{V}_o/\mathbf{V}_i$.

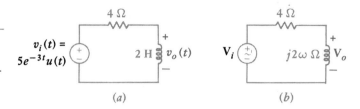

(a)   (b)

Solution: We need the system function; but it is not necessary to apply an impulse, find the impulse response, and then determine its inverse transform. Instead we use Eq. (48) to obtain the system function $\mathbf{H}(j\omega)$ by assuming that the input and output voltages are both sinusoids described by their corresponding phasors, as shown in Fig. 18-13b. Using voltage division, we have

$$\mathbf{H}(j\omega) = \frac{\mathbf{V}_o}{\mathbf{V}_i} = \frac{j2\omega}{4 + j2\omega}$$

The transform of the forcing function is

$$\mathscr{F}\{v_i(t)\} = \frac{5}{3 + j\omega}$$

and thus the transform of $v_o(t)$ is given as

$$\mathscr{F}\{v_o(t)\} = \mathbf{H}(j\omega)\mathscr{F}\{v_i(t)\}$$

$$= \frac{j2\omega}{4 + j2\omega}\frac{5}{3 + j\omega}$$

$$= \frac{15}{3 + j\omega} - \frac{10}{2 + j\omega}$$

where the partial fractions appearing in the last step help to determine the inverse Fourier transform

$$v_o(t) = \mathscr{F}^{-1}\left\{\frac{15}{3 + j\omega} - \frac{10}{2 + j\omega}\right\}$$

$$= 15e^{-3t}u(t) - 10e^{-2t}u(t)$$

$$= 5(3e^{-3t} - 2e^{-2t})u(t)$$

Our problem is completed without fuss, convolution, or differential equations. ∎

Returning again to Eq. (48), the identity between the system function $\mathbf{H}(j\omega)$ and the sinusoidal steady-state transfer function $\mathbf{G}(\omega)$, we may now consider the system function as the ratio of the output phasor to the input phasor. Suppose that we hold the input-phasor amplitude at unity and the phase angle at zero. Then the output phasor is $\mathbf{H}(j\omega)$. Under these conditions, if we record the output amplitude and phase as functions of ω, for all ω, we have recorded the system function $\mathbf{H}(j\omega)$ as a function of ω, for all ω. We thus have examined the system response under the condition that an infinite number of sinusoids, all with unity amplitude and zero phase, were successively applied at the input. Now suppose that our input is a single unit impulse, and look at the impulse response $h(t)$. Is the information we examine really any different from that we just obtained? The Fourier transform of the unit impulse is a constant equal to unity, indicating that all frequency components are present, all with the same magnitude, and all with zero phase. Our system response is the sum of the responses to all these components. The result might be viewed at the output on a cathode-ray oscilloscope. It is evident that the system function and the impulse-response function contain equivalent information regarding the response of the system.

We therefore have two different methods of describing the response of a system to a general forcing function; one is a time-domain description, and the other a frequency-domain description. Working in the time domain, we convolve the forcing function with the impulse response of the system to obtain the response function. As we saw when we first considered convolution, this procedure may be interpreted by thinking of the input as a continuum of impulses of different strengths and times of application; the output which results is a continuum of impulse responses.

In the frequency domain, however, we determine the response by multiplying the Fourier transform of the forcing function by the system function. In this case we interpret the transform of the forcing function as a frequency spectrum, or a continuum of sinusoids. Multiplying this by the system function, we obtain the response function, also as a continuum of sinusoids.

Whether we choose to think of the output as a continuum of impulse responses or as a continuum of sinusoidal responses, the linearity of the network and the superposition principle enable us to determine the total output as a time function by summing over all frequencies (the inverse Fourier transform), or as a frequency function by summing over all time (the Fourier transform).

Unfortunately, both of these techniques have some difficulties or limitations associated with their use. In using convolution, the integral itself can often be rather difficult to evaluate when complicated forcing functions or impulse response functions are present. Furthermore, from the experimental point of view, we cannot really measure the impulse response of a system because we cannot actually generate an impulse. Even if we approximate the impulse by a narrow high-amplitude pulse, we would probably drive our system into saturation and out of its linear operating range.

With regard to the frequency domain, we encounter one absolute limitation in that we may easily hypothesize forcing functions that we would like to apply theoretically that do not possess Fourier transforms. Moreover, if we wish to find the time-domain description of the response function, we must evaluate an inverse Fourier transform, and some of these inversions can be extremely difficult.

Finally, neither of these techniques offers a very convenient method of handling initial conditions.

The greatest benefits derived from the use of the Fourier transform arise through the abundance of useful information it provides about the spectral properties of a signal, particularly the energy or power per unit bandwidth.

Most of the difficulties and limitations associated with the Fourier transform are overcome with the use of the Laplace transform. We shall see that the Laplace transform is defined in such a way that there is a built-in convergence factor that enables transforms to be determined for a much wider range of input time functions than the Fourier transform can accommodate. Moreover, we shall find that initial conditions may be handled in a manner that may make us wonder why this marvelous technique was not introduced before the last chapter. Finally, the spectral information that is so readily available in the Fourier transform can also be obtained from the Laplace transform, at least for most of the time functions that arise in real engineering problems.

Well, why *has* this all been withheld until now? The best answer is probably that these powerful techniques can overcomplicate the solution of simple problems and tend to obscure the physical interpretation of the performance of the simpler networks. For example, if we are interested only in the forced response, then there is little point in using the Laplace transform and obtaining both the forced and natural response after laboring through a difficult inverse transform operation.

So much for generalities; on to Laplace.

Drill Problem

18-14. Use Fourier-transform techniques on the circuit of Fig. 18-14 to find $i_1(t)$ at $t = 1.5$ ms if i_s equals (a) $\delta(t)$ A; (b) $u(t)$ A; (c) $\cos 500t$ A.

Ans: -141.7 A; 0.683 A; 0.308 A

Figure 18-14

See Drill Prob. 18-14.

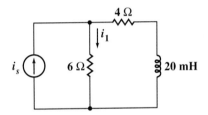

Problems

1 Given the time function $f(t) = 5[u(t + 3) + u(t + 2) - u(t - 2) - u(t - 3)]$: (a) sketch $f(t)$; (b) use the definition of the Fourier transform to find $\mathbf{F}(j\omega)$.

2 Use the defining equations for the Fourier transform to find $\mathbf{F}(j\omega)$ if $f(t)$ equals (a) $e^{-at}u(t)$, $a > 0$; (b) $e^{-a(t-t_0)}u(t - t_0)$, $a > 0$; (c) $te^{-at}u(t)$, $a > 0$.

3 Find the Fourier transform of the single triangular pulse in Fig. 18-15.

Figure 18-15

See Prob. 3.

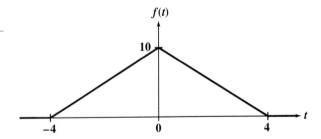

4 Find the Fourier transform of the single sinusoidal pulse in Fig. 18-16.

5 Let $f(t) = (8 \cos t) [u(t + 0.5\pi) - u(t - 0.5\pi)]$. Calculate $\mathbf{F}(j\omega)$ for ω equal to (a) 0; (b) 0.8; (c) 3.1.

6 Use the defining equations for the inverse Fourier transform to find $f(t)$, and then evaluate it at $t = 0.8$ for $\mathbf{F}(j\omega)$ equal to (a) $4[u(\omega + 2) - u(\omega - 2)]$; (b) $4e^{-2|\omega|}$; (c) $(4 \cos \pi\omega) [u(\omega + 0.5) - u(\omega - 0.5)]$.

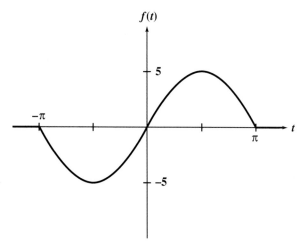

$f(t)$

Figure 18-16

See Prob. 4.

7 Given the voltage $v(t) = 20e^{1.5t}u(-t - 2)$ V, find (a) $\mathbf{F}_v(j0)$; (b) $A_v(2)$; (c) $B_v(2)$; (d) $|\mathbf{F}_v(j2)|$; (e) $\phi_v(2)$.

8 Let $i(t)$ be the time-varying current through a 4-Ω resistor. If the magnitude of the Fourier transform of $i(t)$ is known to be $|\mathbf{I}(j\omega)| = (3\cos 10\omega)[u(\omega + 0.05\pi) - u(\omega - 0.05\pi)]$ A/(rad/s), find (a) the total energy present in the signal; (b) the frequency ω_x such that half the total energy lies in the range $|\omega| < \omega_x$.

9 Let $f(t) = 10te^{-4t}u(t)$, and find (a) the 1-Ω energy represented by that signal; (b) $|\mathbf{F}(j\omega)|$; (c) the energy density at $\omega = 0$ and $\omega = 4$ rad/s.

10 If $v(t) = 8e^{-2|t|}$ V, find (a) the 1-Ω energy associated with this signal; (b) $|\mathbf{F}_v(j\omega)|$; (c) the frequency range $|\omega| < \omega_1$ in which 90 percent of the 1-Ω energy lies.

11 Find the strength of the impulse that is defined by (a) the limit as $a \to 0$ of $(1/a)[\cos(\pi t/2a)][u(t + a) - u(t - a)]$; (b) the limit as $a \to 0$ of $(\pi/a)[u(t + a) + u(t + 0.4a) - u(t - 0.4a) - u(t - a)]$; (c) $(\tan t)[\delta(t - 1)][u(t + 1.5) - u(t - 1.5)]$.

12 (a) Find $i_R(t)$, $i_C(t)$, and $i_s(t)$ for the circuit shown in Fig. 18-17. Express all currents in terms of singularity functions. (b) Specify numerical values for the differences $i_R(0^+) - i_R(0^-)$, $i_C(0^+) - i_C(0^-)$, $i_s(0^+) - i_s(0^-)$, and $q_C(0^+) - q_C(0^-)$.

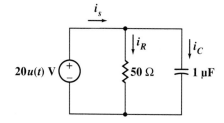

Figure 18-17

See Prob. 12.

13 (a) Find $v_2(t)$, $i_6(t)$, $v_5(t)$, and $v_s(t)$ for the circuit shown in Fig. 18-18. Each answer should be expressed in terms of singularity functions. (b) Show that $v_5(t) = 6 \, di_6/dt$. Remember that $(d/dt)(uv) = u \, dv/dt + v \, du/dt$.

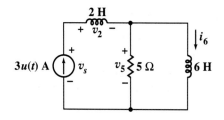

Figure 18-18

See Prob. 13.

14 (*a*) Let $i_s = 2\delta(t)$ A in the circuit shown in Fig. 18-19, and find $i_{3,a}(t)$. (*b*) Find $i_{3,b}(t)$ if $i_s = 2u(t)$ A. (*c*) Show that $i_{3,a}(t) = (d/dt)[i_{3,b}(t)]$.

Figure 18-19

See Prob. 14.

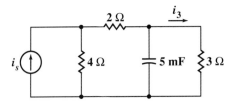

15 The current $i_s(t) = (4/a)[u(t) - u(t - a)]$ A is applied to the parallel combination of a 0.5-H inductor and a 10-Ω resistor. Find the magnitude of the inductor current at $t = 30$ ms: (*a*) if $a = 40$ ms; (*b*) if $a = 20$ ms; (*c*) if $a = 2$ ms; (*d*) in the limit as $a \to 0$.

16 Use the definition of the Fourier transform to prove the following results, where $\mathcal{F}\{f(t)\} = \mathbf{F}(j\omega)$: (*a*) $\mathcal{F}\{f(t - t_0)\} = e^{-j\omega t_0}\,\mathcal{F}\{f(t)\}$; (*b*) $\mathcal{F}\{df(t)/dt\} = j\omega\mathcal{F}\{f(t)\}$; (*c*) $\mathcal{F}\{f(kt)\} = (1/|k|)\mathbf{F}(j\omega/k)$; (*d*) $\mathcal{F}\{f(-t)\} = \mathbf{F}(-j\omega)$; (*e*) $\mathcal{F}\{tf(t)\} = j\,d[\mathbf{F}(j\omega)]/d\omega$.

17 Find $\mathcal{F}\{f(t)\}$ if $f(t)$ is given by (*a*) $4[\text{sgn }(t)]\delta(t - 1)$; (*b*) $4[\text{sgn }(t - 1)]\delta(t)$; (*c*) $4[\sin(10t - 30°)]$.

18 Find $\mathbf{F}(j\omega)$ if $f(t)$ equals (*a*) $A\cos(\omega_0 t + \phi)$; (*b*) $3\,\text{sgn }(t - 2) - 2\delta(t) - u(t - 1)$; (*c*) $(\sinh kt)u(t)$.

19 Find $f(t)$ at $t = 5$ if $\mathbf{F}(j\omega)$ equals (*a*) $3u(\omega + 3) - 3u(\omega - 1)$; (*b*) $3u(-3 - \omega) + 3u(\omega - 1)$; (*c*) $2\delta(\omega) + 3u(-3 - \omega) + 3u(\omega - 1)$.

20 Find $f(t)$ if $\mathbf{F}(j\omega)$ equals (*a*) $3/(1 + j\omega) + 3/j\omega + 3 + 3\delta(\omega - 1)$; (*b*) $(5\sin 4\omega)/\omega$; (*c*) $6(3 + j\omega)/[(3 + j\omega)^2 + 4]$.

21 Find the Fourier transform of the periodic time function shown in Fig. 18-20.

Figure 18-20

See Prob. 21.

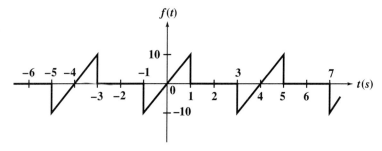

22 The periodic function $f(t)$ is defined over the period $0 < t < 4$ ms by $f_1(t) = 10u(t) - 6u(t - 0.001) - 4u(t - 0.003)$. Find $\mathbf{F}(j\omega)$.

23 If $\mathbf{F}(j\omega) = 20\sum\limits_{n=1}^{\infty}\dfrac{1}{|n|! + 1}\,\delta(\omega - 20n)$, find the value of $f(0.05)$.

24 Given an input $x(t) = 5[u(t) - u(t - 1)]$, use convolution to find the output $y(t)$ if $h(t)$ equals (*a*) $2u(t)$; (*b*) $2u(t - 1)$; (*c*) $2u(t - 2)$.

25 Let $x(t) = 5[u(t) - u(t - 2)]$ and $h(t) = 2[u(t - 1) - u(t - 2)]$. Find $y(t)$ at $t = -0.4$, $0.4, 1.4, 2.4, 3.4,$ and 4.4 by using: (*a*) the first integral of Eq. (43); (*b*) the second integral of Eq. (43).

26 The impulse response of a certain linear system is $h(t) = 3(e^{-t} - e^{-2t})$. Given the input $x(t) = u(t)$, find the output for $t > 0$.

27 The unit-impulse response and the input to a certain linear system are shown in Fig. 18-21. (*a*) Obtain an integral expression for the output that is valid in the interval $4 < t < 6$ and does not contain any singularity functions. (*b*) Evaluate the output at $t = 5$.

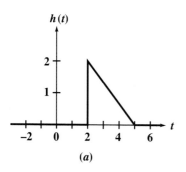

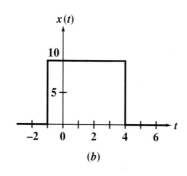

Figure 18-21

See Prob. 27.

28 Given an input signal $x(t) = 5e^{-(t-2)}u(t - 2)$ and the impulse response $h(t) = (4t - 16)[u(t - 4) - u(t - 7)]$, find the value of the output signal at (a) $t = 5$; (b) $t = 8$; (c) $t = 10$.

29 When an input $\delta(t)$ is applied to a linear system, the output is $\sin t$ for $0 < t < \pi$, and zero elsewhere. Now, if the input $e^{-t}u(t)$ is applied, specify the numerical value of the output at t equal to (a) 1; (b) 2.5; (c) 4.

30 Let $x(t) = 0.8(t - 1)[u(t - 1) - u(t - 3)]$ and $h(t) = 0.2(t - 2)[u(t - 2) - u(t - 3)]$. Evaluate $y(t)$ for (a) $t = 3.8$; (b) $t = 4.8$.

31 A signal $x(t) = 10e^{-2t}u(t)$ is applied to a linear system for which the impulse response is $h(t) = 10e^{-2t}u(t)$. Find the output $y(t)$.

32 An impulse is applied to a linear system, generating the output $h(t) = 5e^{-4t}u(t)$ V. What percentage of the 1-Ω energy in this response: (a) occurs during the time interval $0.1 < t < 0.8$ s? (b) lies in the frequency band $-2 < \omega < 2$ rad/s?

33 If $\mathbf{F}(j\omega) = 2/[(1 + j\omega)(2 + j\omega)]$, find (a) the total 1-Ω energy present in the signal, and (b) the maximum value of $f(t)$.

34 Find $\mathbf{F}^{-1}(j\omega)$ if $\mathbf{F}(j\omega)$ equals (a) $1/[(j\omega)(2 + j\omega)(3 + j\omega)]$; (b) $(1 + j\omega)/[(j\omega)(2 + j\omega)(3 + j\omega)]$; (c) $(1 + j\omega)^2/[(j\omega)(2 + j\omega)(3 + j\omega)]$; (d) $(1 + j\omega)^3/[(j\omega)(2 + j\omega)(3 + j\omega)]$.

35 In Sec. 18-7 the impulse response $h(t) = 2e^{-t}u(t)$, was hypothesized. Let us develop one network which has such a response. (a) Determine $\mathbf{H}(j\omega) = \mathbf{V}_o(j\omega)/\mathbf{V}_i(j\omega)$. (b) By inspecting either $h(t)$ or $\mathbf{H}(j\omega)$, note that the network has a single energy-storage element. Arbitrarily selecting an RC circuit with $R = 1$ Ω, $C = 1$ F, to provide the necessary time constant, determine the form of the circuit to give $\frac{1}{2}h(t)$ or $\frac{1}{2}\mathbf{H}(j\omega)$. (c) Place an ideal voltage amplifier in cascade with the network to provide the proper multiplicative constant. What is the gain of the amplifier?

36 Let $h(t) = 10e^{-20(t-5)}u(t - 5)$ for a certain linear system. (a) Use the statements of Prob. 16 (if necessary) to help find $\mathbf{H}(j\omega)$. (b) Find the system output $y(t)$ if the input is $x(t) = \text{sgn}\,(t)$.

37 A certain linear system has the system function $\mathbf{H}(j\omega) = 20/(j\omega + 8)$. Find the system output $v_o(t)$ if the input is $v_i(t)$ equal to (a) $\delta(t)$ V; (b) $u(t)$ V; (c) $100 \cos 6t$ V.

38 Find $v_o(t)$ for the circuit of Fig. 18-22.

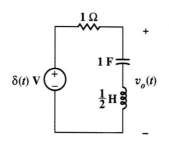

Figure 18-22

See Prob. 38.

39 Find $v_C(t)$ for the circuit illustrated in Fig. 18-23.

Figure 18-23

See Prob. 39.

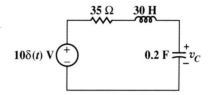

40 Let $f(t) = 5e^{-2t}u(t)$ and $g(t) = 4e^{-3t}u(t)$. (a) Find $f(t) * g(t)$ by working in the time domain. (b) Find $f(t) * g(t)$ by using convolution in the frequency domain.

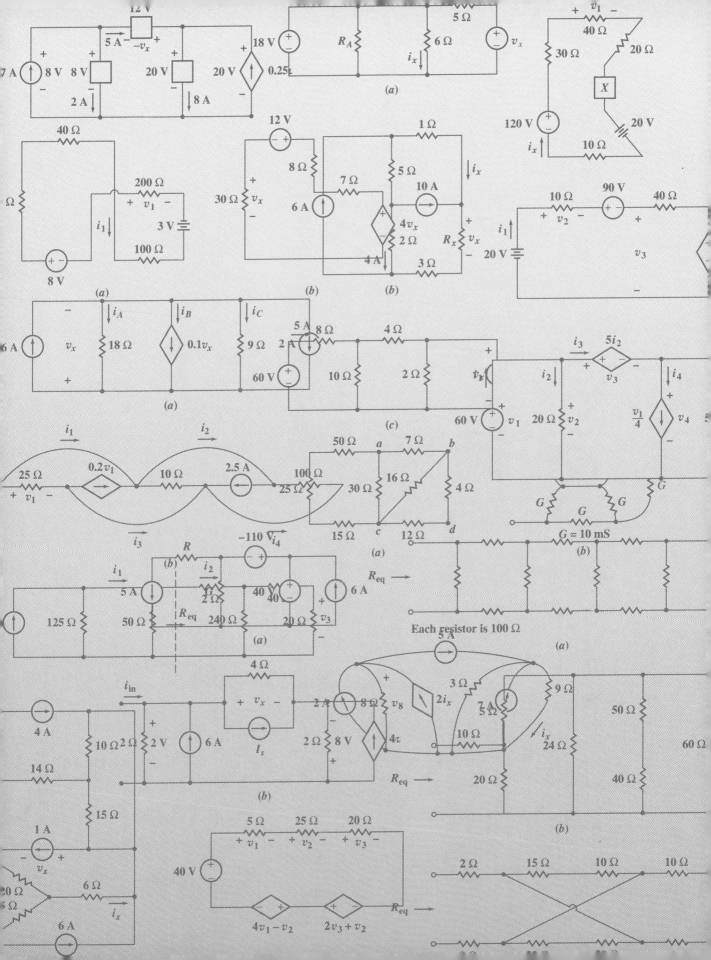

Laplace Transform Techniques

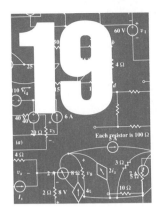

As we begin this last chapter on the Laplace transform, it may be beneficial to pause long enough to review our progress to date, first because we have covered a considerable amount of ground since page 1 and some tying together is advisable, and second because we need to see how the Laplace transform fits into our hierarchy of analytical methods.

Our constant goal has been one of analysis: given some forcing function at one point in a linear circuit, determine the response at some other point. For the first several chapters, we played only with dc forcing functions and responses of the form $V_0 e^0$. However, after the introduction of inductance and capacitance, the sudden dc excitation of simple RL and RC circuits produced responses varying exponentially with time: $V_0 e^{\sigma t}$. When we considered the RLC circuit, the responses took on the form of the exponentially varying sinusoid, $V_0 e^{\sigma t} \cos(\omega t + \theta)$. All of this work was accomplished in the time domain, and the dc forcing function was the only one we considered.

As we advanced to the use of the sinusoidal forcing function, the tedium and complexity of solving the integrodifferential equations caused us to begin casting about for an easier way to work problems. The phasor transform was the result, and we might remember that we were led to it through consideration of a complex forcing function of the form $V_0 e^{j\theta} e^{j\omega t}$. As soon as we concluded that we did not need the factor containing t, we were left with the phasor $V_0 e^{j\theta}$; we had arrived at the frequency domain.

After gaining some facility with sinusoidal steady-state analysis, a modicum of pleasant cerebration led us to apply a forcing function of the form $V_0 e^{j\theta} e^{(\sigma + j\omega)t}$, and we thereby invented the complex frequency $\mathbf{s}$, relegating all our previous functional forms to special cases: dc ($\mathbf{s} = 0$), exponential ($\mathbf{s} = \sigma$), sinusoidal ($\mathbf{s} = j\omega$), and exponential sinusoid ($\mathbf{s} = \sigma + j\omega$). Such was our status at the end of Chap. 15.

We then returned to the time domain and spent some time learning how to cast our equations into the standard form demanded by state-variable methods. For circuits of first-order complexity, we were able to solve the resultant equations for a variety of forcing functions, including forms we earlier had to avoid. Although higher-order systems were considered only briefly, we did see that a digital computer might be programmed to use numerical methods on the matrix expressions.

Continuing to strive for mastery over an ever greater number of different types of forcing functions, we turned next to nonsinusoidal periodic functions. Here the infinite series developed by Fourier, $a_0 + \Sigma(a_n \cos n\omega_0 t + b_n \sin n\omega_0 t)$, was found to be capable of representing almost any periodic function in which we might be interested. Thanks to linearity and superposition, the response could then be found as the sum of the responses to the individual sinusoidal terms.

Remembering our success with the complex exponential function, we next developed the complex form of the Fourier series, $\Sigma \mathbf{c}_n e^{jn\omega_0 t}$, suitable again only for periodic functions. However, by letting the period of a periodic sequence of pulses increase without limit, we arrived at a form applicable to a single pulse, the inverse Fourier transform:

$$v(t) = \frac{1}{2\pi} \int_{-\infty}^{\infty} e^{j\omega t} \mathbf{V}(j\omega)\, d\omega$$

This then is our current status. What more could we possibly need? It turns out that there are several things. First, there are a few time functions for which a Fourier transform does not exist, such as the increasing exponential, many random signals, and other time functions that are not absolutely integrable. Second, we have not yet permitted initial energy storage in the networks whose transient or complete response we desired to compute by Fourier transform methods. Both of these objections are overcome with the Laplace transform, and in addition we shall note a simpler nomenclature and ease of manipulation.

19-2

Definition of the Laplace transform

We shall present the Laplace transform as a development or evolution of the Fourier transform, but it is also possible to define it directly. For those who prefer to look ahead, Eqs. (5) and (6) are direct definitions of the two-sided Laplace transform, while Eqs. (7) and (8) define the one-sided Laplace transform that is used in the remainder of this chapter.

We begin our development of the Laplace transform by using the inverse Fourier transform to interpret $v(t)$ as the sum (integral) of an infinite number of terms, each having the form

$$\left[\frac{\mathbf{V}(j\omega)\, d\omega}{2\pi} \right] e^{j\omega t}$$

Upon comparing this with the form of the complex forcing function,

$$[V_0 e^{j\theta}]\, e^{j\omega t}$$

which led us to the phasor $V_0 e^{j\theta}$, we should note that both bracketed terms are complex quantities in general; therefore, $(1/2\pi)\mathbf{V}(j\omega)\, d\omega$ can also be interpreted as some kind of phasor. Of course, the frequency differential leads to a vanishingly small amplitude, but we add an infinite number of such terms together when we perform the integration.

Our final step, therefore, parallels the step which we took when we introduced phasors in the complex-frequency domain; we now let the time variation have the form $e^{(\sigma + j\omega)t}$.

To do this, consider the Fourier transform of $e^{-\sigma t} v(t)$, rather than $v(t)$ itself. For $v(t)$ alone we have

$$\mathbf{V}(j\omega) = \int_{-\infty}^{\infty} e^{-j\omega t} v(t)\, dt \tag{1}$$

and
$$v(t) = \frac{1}{2\pi} \int_{-\infty}^{\infty} e^{j\omega t} \mathbf{V}(j\omega)\, d\omega \tag{2}$$

Letting
$$g(t) = e^{-\sigma t} v(t) \tag{3}$$

we see that
$$\mathbf{G}(j\omega) = \int_{-\infty}^{\infty} e^{-j\omega t} e^{-\sigma t} v(t)\, dt$$

$$= \int_{-\infty}^{\infty} e^{-(\sigma + j\omega)t} v(t)\, dt$$

or
$$\mathbf{G}(j\omega) = \mathbf{V}(\sigma + j\omega) = \int_{-\infty}^{\infty} e^{-(\sigma + j\omega)t} v(t)\, dt \tag{4}$$

through comparison with Eq. (1). Taking the inverse Fourier transform, we obtain

$$g(t) = \frac{1}{2\pi} \int_{-\infty}^{\infty} e^{j\omega t} \mathbf{G}(j\omega)\, d\omega$$

$$= \frac{1}{2\pi} \int_{-\infty}^{\infty} e^{j\omega t} \mathbf{V}(\sigma + j\omega)\, d\omega$$

With the help of Eq. (3), we have

$$e^{-\sigma t} v(t) = \frac{1}{2\pi} \int_{-\infty}^{\infty} e^{j\omega t} \mathbf{V}(\sigma + j\omega)\, d\omega$$

or, upon transferring $e^{-\sigma t}$ inside the integral,

$$v(t) = \frac{1}{2\pi} \int_{-\infty}^{\infty} e^{(\sigma + j\omega)t} \mathbf{V}(\sigma + j\omega)\, d\omega$$

We now replace $\sigma + j\omega$ by the single complex variable $\mathbf{s}$, and, since σ is a constant and $d\mathbf{s} = j\, d\omega$, we have

$$v(t) = \frac{1}{2\pi j} \int_{\sigma_0 - j\infty}^{\sigma_0 + j\infty} e^{\mathbf{s}t} \mathbf{V}(\mathbf{s})\, d\mathbf{s} \tag{5}$$

where the real constant σ_0 is included in the limits to ensure convergence of this improper integral. In terms of $\mathbf{s}$, Eq. (4) may be written

$$\mathbf{V}(\mathbf{s}) = \int_{-\infty}^{\infty} e^{-\mathbf{s}t} v(t)\, dt \tag{6}$$

Equation (6) defines the *two-sided,* or *bilateral, Laplace transform* of $v(t)$. The term *two-sided* or *bilateral* is used to emphasize the fact that both positive and negative values of t are included in the range of integration. Equation (5) is the inverse Laplace transform, and the two equations constitute the two-sided Laplace transform pair.

   If we continue the process we began earlier in this section, we may note that $v(t)$ is now represented as the sum (integral) of terms of the form

$$\left[\frac{\mathbf{V}(\mathbf{s})\, d\mathbf{s}}{2\pi j} \right] e^{\mathbf{s}t} = \left[\frac{\mathbf{V}(\mathbf{s})\, d\omega}{2\pi} \right] e^{\mathbf{s}t}$$

Such terms have the same form as those we encountered when we used phasors to represent exponentially varying sinusoids,

$$[\mathbf{V}_0 e^{j\theta}]e^{st}$$

where the bracketed term was a function of **s**. Thus, the two-sided Laplace transform may be interpreted as expressing $v(t)$ as the sum (integral) of an infinite number of vanishingly small terms with complex frequency $\mathbf{s} = \sigma + j\omega$. The variable is **s** or ω, and σ should be thought of as governing the convergence factor $e^{-\sigma t}$. That is, by including this exponential term in Eq. (6), more positive values of σ ensure that the function $e^{-\sigma t}v(t)u(t)$ is absolutely integrable for almost any $v(t)$ that we might meet. The more negative values of σ are needed when $t < 0$. Thus, the two-sided Laplace transform exists for a wider class of functions $v(t)$ than does the Fourier transform. The exact conditions required for the existence of the (one-sided) Laplace transform are given in the following section.

In many of our circuit analysis problems, the forcing and response functions do not exist forever in time, but rather they are initiated at some specific instant that we usually select as $t = 0$. Thus, for time functions that do not exist for $t < 0$, or for those time functions whose behavior for $t < 0$ is of no interest, the time-domain description can be thought of as $v(t)u(t)$. The defining integral for the Laplace transform is taken with the lower limit at $t = 0^-$ in order to include the effect of any discontinuity at $t = 0$, such as an impulse or a higher-order singularity. The corresponding Laplace transform is then

$$\mathbf{V(s)} = \int_{-\infty}^{\infty} e^{-st}v(t)u(t)\,dt = \int_{0^-}^{\infty} e^{-st}v(t)\,dt$$

This defines the *one-sided Laplace transform* of $v(t)$, or simply the *Laplace transform* of $v(t)$, one-sided being understood. The inverse transform expression remains unchanged, but when evaluated, it is understood to be valid only for $t > 0$. Here then is the definition of the Laplace transform pair that we shall use from now on:

$$\mathbf{V(s)} = \int_{0^-}^{\infty} e^{-st}v(t)\,dt \tag{7}$$

$$v(t) = \frac{1}{2\pi j}\int_{\sigma_0 - j\infty}^{\sigma_0 + j\infty} e^{st}\mathbf{V(s)}\,d\mathbf{s} \tag{8}$$

$$v(t) \Leftrightarrow \mathbf{V(s)}$$

These are memorable expressions.

The script $\mathscr{L}$ may also be used to indicate the direct or inverse Laplace transform operation:

$$\mathbf{V(s)} = \mathscr{L}\{v(t)\} \qquad \text{and} \qquad v(t) = \mathscr{L}^{-1}\{\mathbf{V(s)}\}$$

Drill Problems

19-1. Let $f(t) = -6e^{-2t}[u(t + 3) - u(t - 2)]$. Find (a) $\mathbf{F}(j\omega)$; (b) two-sided $\mathbf{F(s)}$; (c) one-sided $\mathbf{F(s)}$.

$$Ans: \frac{6}{2 + j\omega}[e^{-4 - j2\omega} - e^{6 + j3\omega}]; \frac{6}{2 + \mathbf{s}}[e^{-4 - 2\mathbf{s}} - e^{6 + 3\mathbf{s}}]; \frac{6}{2 + \mathbf{s}}[e^{-4 - 2\mathbf{s}} - 1]$$

19-2. Find the Laplace transform of each of the following functions, giving also the minimum value of σ for which the transform exists: (a) $2u(t - 3)$; (b) $5u(t + 3)$; (c) $4e^{-3t}$; (d) $6e^{-3t}u(t)$; (e) $7e^{-3t}u(t - 3)$.

$$Ans: \frac{2}{\mathbf{s}}e^{-3\mathbf{s}}, \sigma > 0; \frac{5}{\mathbf{s}}, \sigma > 0; \frac{4}{\mathbf{s}+3}, \sigma > -3; \frac{6}{\mathbf{s}+3}, \sigma > -3; \frac{7}{\mathbf{s}+3}e^{-3\mathbf{s}-9}, \sigma > -3$$

In this section we shall begin to build up a catalog of Laplace transforms for those time functions most frequently encountered in circuit analysis, just as we did for Fourier transforms. This will be done, at least initially, by utilizing the definition,

$$\mathbf{V}(\mathbf{s}) = \int_{0^-}^{\infty} e^{-\mathbf{s}t}v(t)\,dt = \mathcal{L}\{v(t)\}$$

which, along with the expression for the inverse transform,

$$v(t) = \frac{1}{2\pi j}\int_{\sigma_0-j\infty}^{\sigma_0+j\infty} e^{\mathbf{s}t}\mathbf{V}(\mathbf{s})\,d\mathbf{s} = \mathcal{L}^{-1}\{\mathbf{V}(\mathbf{s})\}$$

establishes a one-to-one correspondence between $v(t)$ and $\mathbf{V}(\mathbf{s})$. That is, for every $v(t)$ for which $\mathbf{V}(\mathbf{s})$ exists, there is a unique $\mathbf{V}(\mathbf{s})$. At this point, we may be looking with some trepidation at the rather ominous form given for the inverse transform. Fear not! As we shall see shortly, an introductory study of Laplace transform theory does not require actual evaluation of this integral. By going from the time domain to the frequency domain and taking advantage of the uniqueness just mentioned, we shall be able to generate a catalog of transform pairs that will already contain the corresponding time function for nearly every transform that we wish to invert.

Before we start evaluating some transforms, however, we must pause to consider whether there is any chance that the transform may not even exist for some $v(t)$ that concerns us. In our study of Fourier transforms, we were forced to take a kind of back-door approach in finding several transforms. This was true with time functions that were not absolutely integrable, such as the unit-step function. In the case of the Laplace transform, the use of realizable linear circuits and real-world forcing functions almost never leads to troublesome time functions. Technically, a set of conditions sufficient to ensure the absolute convergence of the Laplace integral for Re $(\mathbf{s}) > \sigma_0$ is

1 The function $v(t)$ is integrable in every finite interval $t_1 < t < t_2$, where $0 \le t_1 < t_2 < \infty$.
2 $\lim_{t\to\infty} e^{-\sigma_0 t}|v(t)|$ exists for some value of σ_0.

Time functions that do not satisfy these conditions are seldom encountered by the circuit analyst.[1]

19-3

Laplace transforms of some simple time functions

[1] Examples of such functions are e^{t^2} and e^{e^t}, but not t^n or n^t. For a somewhat more detailed discussion of the Laplace transform and its applications, refer to Clare D. McGillem and George R. Cooper, *Continuous and Discrete Signal and System Analysis,* 2d ed., Holt, Rinehart and Winston, New York, 1984, chap. 5.

Now let us look at some specific transforms. In a somewhat vengeful mood, we first examine the Laplace transform of the unit-step function $u(t)$, which caused us some earlier trouble. From the defining equation, we may write

$$\mathscr{L}\{u(t)\} = \int_{0^-}^{\infty} e^{-st} u(t)\, dt = \int_{0}^{\infty} e^{-st}\, dt$$

$$= -\frac{1}{\mathbf{s}} e^{-st} \Big|_{0}^{\infty} = \frac{1}{\mathbf{s}}$$

since Re $(\mathbf{s}) > 0$, to satisfy condition 2. Thus,

$$u(t) \Leftrightarrow \frac{1}{\mathbf{s}} \tag{9}$$

and our first Laplace transform pair has been established with great ease.

Another singularity function whose transform is of considerable interest is the unit-impulse function $\delta(t - t_0)$, for which $t_0 > 0^-$:

$$\mathscr{L}\{\delta(t - t_0)\} = \int_{0^-}^{\infty} e^{-st} \delta(t - t_0)\, dt = e^{-st_0}$$

$$\delta(t - t_0) \Leftrightarrow e^{-st_0} \tag{10}$$

In particular, note that we obtain

$$\delta(t) \Leftrightarrow 1 \tag{11}$$

for $t_0 = 0$.

Recalling our past interest in the exponential function, we examine its transform,

$$\mathscr{L}\{e^{-\alpha t} u(t)\} = \int_{0^-}^{\infty} e^{-\alpha t} e^{-st}\, dt$$

$$= -\frac{1}{\mathbf{s} + \alpha} e^{-(\mathbf{s}+\alpha)t} \Big|_{0}^{\infty} = \frac{1}{\mathbf{s} + \alpha}$$

and therefore,

$$e^{-\alpha t} u(t) \Leftrightarrow \frac{1}{\mathbf{s} + \alpha} \tag{12}$$

It is understood that Re $(\mathbf{s}) > -\alpha$.

As a final example, for the moment, let us consider the ramp function $tu(t)$. We obtain

$$\mathscr{L}\{tu(t)\} = \int_{0^-}^{\infty} t e^{-st}\, dt = \frac{1}{\mathbf{s}^2}$$

$$tu(t) \Leftrightarrow \frac{1}{\mathbf{s}^2} \tag{13}$$

either by a straightforward integration by parts or by use of a table of definite integrals.

In order to accelerate the process of deriving more Laplace transform pairs, we shall pause to develop several useful theorems in the following section.

19-3. Determine $\mathbf{F(s)}$ if $f(t)$ equals (a) $4\delta(t) - 3u(t)$; (b) $4\delta(t - 2) - 3tu(t)$; (c) $[u(t)][u(t - 2)]$.   *Ans: $(4\mathbf{s} - 3)/\mathbf{s}$; $4e^{-2\mathbf{s}} - (3/\mathbf{s}^2)$; $e^{-2\mathbf{s}}/\mathbf{s}$*

19-4. Determine $f(t)$ if $\mathbf{F(s)}$ equals (a) 10; (b) $10/\mathbf{s}$; (c) $10/\mathbf{s}^2$; (d) $10/[\mathbf{s}(\mathbf{s} + 10)]$; (e) $10\mathbf{s}/(\mathbf{s} + 10)$.   *Ans: $10\delta(t)$; $10u(t)$; $10tu(t)$; $u(t) - e^{-10t}u(t)$; $10\delta(t) - 100e^{-10t}u(t)$*

19-4

Several basic theorems for the Laplace transform

The further evaluation of Laplace transforms is facilitated by applying several basic theorems. One of the simplest and most obvious is the linearity theorem: The Laplace transform of the sum of two or more time functions is equal to the sum of the transforms of the individual time functions. For two time functions we have

$$\mathcal{L}\{f_1(t) + f_2(t)\} = \int_{0^-}^{\infty} e^{-st}[f_1(t) + f_2(t)]\,dt$$

$$= \int_{0^-}^{\infty} e^{-st}f_1(t)\,dt + \int_{0^-}^{\infty} e^{-st}f_2(t)\,dt$$

$$= \mathbf{F}_1(\mathbf{s}) + \mathbf{F}_2(\mathbf{s})$$

As an example of the use of this theorem, suppose that we have a Laplace transform $\mathbf{V(s)}$ and want to know the corresponding time function $v(t)$. It will often be possible to decompose $\mathbf{V(s)}$ into the sum of two or more functions, say, $\mathbf{V}_1(\mathbf{s})$ and $\mathbf{V}_2(\mathbf{s})$, whose inverse transforms, $v_1(t)$ and $v_2(t)$, are already tabulated. It then becomes a simple matter to apply the linearity theorem and write

$$v(t) = \mathcal{L}^{-1}\{\mathbf{V(s)}\} = \mathcal{L}^{-1}\{\mathbf{V}_1(\mathbf{s}) + \mathbf{V}_2(\mathbf{s})\}$$

$$= \mathcal{L}^{-1}\{\mathbf{V}_1(\mathbf{s})\} + \mathcal{L}^{-1}\{\mathbf{V}_2(\mathbf{s})\} = v_1(t) + v_2(t)$$

As a specific example, let us determine the inverse Laplace transform of

$$\mathbf{V(s)} = \frac{1}{(\mathbf{s} + \alpha)(\mathbf{s} + \beta)}$$

Although it is possible to substitute this expression in the defining equation for the inverse transform, it is much easier to utilize the linearity theorem. Using the partial-fraction expansion, we can split the given transform into the sum of two simpler transforms,

$$\mathbf{V(s)} = \frac{A}{(\mathbf{s} + \alpha)} + \frac{B}{(\mathbf{s} + \beta)}$$

where A and B may be found by any of several methods. Perhaps the quickest solution is obtained by recognizing that

$$A = \lim_{\mathbf{s} \to -\alpha}\left[(\mathbf{s} + \alpha)\mathbf{V(s)} - \frac{(\mathbf{s} + \alpha)}{(\mathbf{s} + \beta)}B\right]$$

$$= \lim_{\mathbf{s} \to -\alpha}\left[\frac{1}{(\mathbf{s} + \beta)} - 0\right] = \frac{1}{\beta - \alpha}$$

Similarly,

$$B = \frac{1}{\alpha - \beta}$$

and therefore,

$$\mathbf{V(s)} = \frac{1/(\beta - \alpha)}{(\mathbf{s} + \alpha)} + \frac{1/(\alpha - \beta)}{(\mathbf{s} + \beta)}$$

We have already evaluated inverse transforms of the form shown on the right, and thus

$$v(t) = \frac{1}{\beta - \alpha} e^{-\alpha t} u(t) + \frac{1}{\alpha - \beta} e^{-\beta t} u(t)$$

$$= \frac{1}{\beta - \alpha} (e^{-\alpha t} - e^{-\beta t}) u(t)$$

If we wished, we could now include this as a new entry in our catalog of Laplace pairs,

$$\frac{1}{\beta - \alpha} (e^{-\alpha t} - e^{-\beta t}) u(t) \Leftrightarrow \frac{1}{(\mathbf{s} + \alpha)(\mathbf{s} + \beta)} \tag{14}$$

It is noteworthy that a transcendental function of t in the time domain transforms into a simpler rational function of $\mathbf{s}$ in the frequency domain. Simplifications of this kind are of paramount importance in transform theory. In this example also note that we made use of the fact that

$$kv(t) \Leftrightarrow k\mathbf{V(s)} \tag{15}$$

where k is a constant of proportionality. This result is obviously a direct consequence of the definition of the Laplace transform.

We are now able to consider two theorems that might be considered collectively the *raison d'être* for Laplace transforms in circuit analysis—the time differentiation and integration theorems. These will help us transform the derivatives and integrals appearing in the time-domain circuit equations.

Let us look at time differentiation first by considering a time function $v(t)$ whose Laplace transform $\mathbf{V(s)}$ is known to exist. We want the transform of the first derivative of $v(t)$,

$$\mathscr{L}\left\{\frac{dv}{dt}\right\} = \int_{0^-}^{\infty} e^{-\mathbf{s}t} \frac{dv}{dt} dt$$

This can be integrated by parts:

$$U = e^{-\mathbf{s}t} \qquad dV = \frac{dv}{dt} dt$$

with the result

$$\mathscr{L}\left\{\frac{dv}{dt}\right\} = v(t)e^{-\mathbf{s}t} \Big|_{0^-}^{\infty} + \mathbf{s} \int_{0^-}^{\infty} e^{-\mathbf{s}t} v(t) dt$$

The first term on the right must approach zero as t increases without limit; otherwise $\mathbf{V(s)}$ would not exist. Hence,

$$\mathscr{L}\left\{\frac{dv}{dt}\right\} = 0 - v(0^-) + \mathbf{sV(s)}$$

and

$$\frac{dv}{dt} \Leftrightarrow \mathbf{sV(s)} - v(0^-) \tag{16}$$

Similar relationships may be developed for higher-order derivatives:

$$\frac{d^2v}{dt^2} \Leftrightarrow \mathbf{s}^2\mathbf{V}(\mathbf{s}) - \mathbf{s}v(0^-) - v'(0^-) \tag{17}$$

$$\frac{d^3v}{dt^3} \Leftrightarrow \mathbf{s}^3\mathbf{V}(\mathbf{s}) - \mathbf{s}^2v(0^-) - \mathbf{s}v'(0^-) - v''(0^-) \tag{18}$$

where $v'(0^-)$ is the value of the first derivative of $v(t)$ evaluated at $t = 0^-$, $v''(0^-)$ is the initial value of the second derivative of $v(t)$, and so forth. When all initial conditions are zero, we see that differentiating once with respect to t in the time domain corresponds to one multiplication by $\mathbf{s}$ in the frequency domain; differentiating twice in the time domain corresponds to multiplication by $\mathbf{s}^2$ in the frequency domain, and so on. Thus, differentiation in the time domain is equivalent to multiplication in the frequency domain. This is a substantial simplification! We should also begin to see that, when the initial conditions are not zero, their presence is still accounted for. A simple example will serve to demonstrate this.

Example 19-1 Suppose that we have the series RL circuit shown in Fig. 19-1. We wish to find $i(t)$.

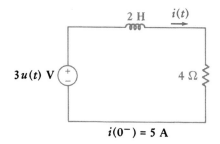

Figure 19-1

A circuit which is analyzed by transforming the differential equation $2\ di/dt + 4i = 3u(t)$ into $2[\mathbf{sI}(\mathbf{s}) - i(0^-)] + 4\mathbf{I}(\mathbf{s}) = 3/\mathbf{s}$.

Solution: The network is driven by a unit-step voltage, and we assume an initial value of the current (at $t = 0^-$) of 5 A.[2] Using KVL to write the single loop equation in the time domain, we have

$$2\frac{di}{dt} + 4i = 3u(t)$$

Instead of solving this differential equation as we have done previously, we first transform to the frequency domain by taking the Laplace transform of each term:

$$2[\mathbf{sI}(\mathbf{s}) - i(0^-)] + 4\mathbf{I}(\mathbf{s}) = \frac{3}{\mathbf{s}}$$

We next solve for $\mathbf{I}(\mathbf{s})$, substituting $i(0^-) = 5$:

$$(2\mathbf{s} + 4)\mathbf{I}(\mathbf{s}) = \frac{3}{\mathbf{s}} + 10$$

and

$$\mathbf{I}(\mathbf{s}) = \frac{1.5}{\mathbf{s}(\mathbf{s} + 2)} + \frac{5}{\mathbf{s} + 2}$$

[2] We could have established this current by letting the source be $20u(-t) + 3u(t)$ V, or $20 - 17u(t)$ V, or other expressions of this nature.

Now,

$$\lim_{s \to 0} [sI(s)] = 0.75$$

and

$$\lim_{s \to -2} [(s + 2)I(s)] = -0.75 + 5$$

and thus

$$I(s) = \frac{0.75}{s} + \frac{4.25}{s + 2}$$

We then use our known transform pairs to invert:

$$i(t) = 0.75u(t) + 4.25e^{-2t}u(t)$$

$$= (0.75 + 4.25e^{-2t})u(t)$$

Our solution for $i(t)$ is complete. Both the forced response $0.75u(t)$ and the natural response $4.25e^{-2t}u(t)$ are present, and the initial condition was automatically incorporated into the solution. The method illustrates a very painless way of obtaining the complete solution of many differential equations. ∎

The same kind of simplification can be accomplished when we meet the operation of integration with respect to time in our circuit equations. Let us determine the Laplace transform of the time function described by $\int_{0^-}^{t} v(x)\,dx$,

$$\mathscr{L}\left\{\int_{0^-}^{t} v(x)\,dx\right\} = \int_{0^-}^{\infty} e^{-st}\left[\int_{0^-}^{t} v(x)\,dx\right]dt$$

Integrating by parts, we let

$$u = \int_{0^-}^{t} v(x)\,dx \qquad dw = e^{-st}\,dt$$

$$du = v(t)\,dt \qquad w = -\frac{1}{s7}e^{-st}$$

Then

$$\mathscr{L}\left\{\int_{0^-}^{t} v(x)\,dx\right\} = \left\{\left[\int_{0^-}^{t} v(x)\,dx\right]\left[-\frac{1}{s}e^{-st}\right]\right\}_{t=0^-}^{t=\infty} - \int_{0^-}^{\infty} -\frac{1}{s}e^{-st}v(t)\,dt$$

$$= \left[-\frac{1}{s}e^{-st}\int_{0^-}^{t} v(x)\,dx\right]_{0^-}^{\infty} + \frac{1}{s}V(s)$$

But, since $e^{-st} \to 0$ as $t \to \infty$, the first term on the right vanishes at the upper limit, and when $t \to 0^-$, the integral in this term likewise vanishes. This leaves only the $V(s)/s$ term, so that

$$\int_{0^-}^{t} v(x)\,dx \Leftrightarrow \frac{V(s)}{s} \tag{19}$$

and thus integration in the time domain corresponds to division by s in the frequency domain. Once more, a relatively complicated calculus operation in the time domain simplifies to an algebraic operation in the frequency domain.

Example 19-2   As an example of how this helps us in circuit analysis, we shall determine $v(t)$ for $t > 0$ in the series RC circuit shown in Fig. 19-2. It is assumed

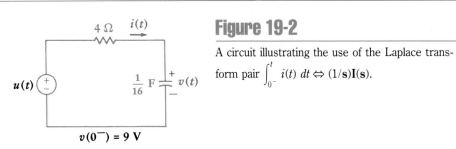

Figure 19-2

A circuit illustrating the use of the Laplace transform pair $\int_{0^-}^{t} i(t)\,dt \Leftrightarrow (1/s)\mathbf{I}(s)$.

$v(0^-) = 9\text{ V}$

that there was some initial energy stored in the capacitor prior to $t = 0^-$, so that $v(0^-) = 9$ V.

Solution: We first write the single loop equation,

$$u(t) = 4i(t) + 16\int_{-\infty}^{t} i(t)\,dt$$

In order to apply the time-integration theorem, we must arrange for the lower limit of integration to be 0^-. Thus, we set

$$16\int_{-\infty}^{t} i(t)\,dt = 16\int_{-\infty}^{0^-} i(t)\,dt + 16\int_{0^-}^{t} i(t)\,dt$$

$$= v(0^-) + 16\int_{0^-}^{t} i(t)\,dt$$

Therefore,

$$u(t) = 4i(t) + v(0^-) + 16\int_{0^-}^{t} i(t)\,dt$$

We next take the Laplace transform of both sides of this equation. Since we are utilizing the one-sided transform, $\mathcal{L}\{v(0^-)\}$ is simply $\mathcal{L}\{v(0^-)u(t)\}$, and thus

$$\frac{1}{s} = 4\mathbf{I}(s) + \frac{9}{s} + \frac{16}{s}\mathbf{I}(s)$$

and solving for $\mathbf{I}(s)$,

$$\mathbf{I}(s) = \frac{-2}{s+4}$$

the desired result is immediately obtained,

$$i(t) = -2e^{-4t}u(t)$$ ∎

Example 19-3 Find $v(t)$ for this same circuit.

Solution: This time we simply write a single nodal equation,

$$\frac{v(t) - u(t)}{4} + \frac{1}{16}\frac{dv}{dt} = 0$$

Taking the Laplace transform, we obtain

$$\frac{\mathbf{V}(s)}{4} - \frac{1}{4s} + \frac{1}{16}s\mathbf{V}(s) - \frac{v(0^-)}{16} = 0$$

or
$$V(s)\left(1 + \frac{s}{4}\right) = \frac{1}{s} + \frac{9}{4}$$

Thus, $$V(s) = \frac{4}{s(s + 4)} + \frac{9}{s + 4} = \frac{1}{s} - \frac{1}{s + 4} + \frac{9}{s + 4} = \frac{1}{s} + \frac{8}{s + 4}$$

and taking the inverse transform,

$$v(t) = (1 + 8e^{-4t})u(t)$$

we quickly obtain the desired capacitor voltage without recourse to the usual differential equation solution.

To check this result, we note that $(\frac{1}{16})\, dv\,/\,dt$ should yield the previous expression for $i(t)$. For $t > 0$,

$$\frac{1}{16}\frac{dv}{dt} = \frac{1}{16}(-32)e^{-4t} = -2e^{-4t}$$

which is correct. ∎

To illustrate the use of both the linearity theorem and the time-differentiation theorem, not to mention the addition of a most important pair to our forthcoming Laplace transform table, let us establish the Laplace transform of $\sin \omega t\, u(t)$. We could use the defining integral expression with integration by parts, but this is needlessly difficult. Instead, we use the relationship

$$\sin \omega t = \frac{1}{2j}(e^{j\omega t} - e^{-j\omega t})$$

The transform of the sum of these two terms is just the sum of the transforms, and each term is an exponential function for which we already have the transform. We may immediately write

$$\mathcal{L}\{\sin \omega t\, u(t)\} = \frac{1}{2j}\left(\frac{1}{s - j\omega} - \frac{1}{s + j\omega}\right) = \frac{\omega}{s^2 + \omega^2}$$

$$\sin \omega t\, u(t) \Leftrightarrow \frac{\omega}{s^2 + \omega^2} \qquad (20)$$

We next use the time-differentiation theorem to determine the transform of $\cos \omega t\, u(t)$, which is proportional to the derivative of $\sin \omega t$. That is,

$$\mathcal{L}\{\cos \omega t\, u(t)\} = \mathcal{L}\left\{\frac{1}{\omega}\frac{d}{dt}[\sin \omega t\, u(t)]\right\} = \frac{1}{\omega}s\frac{\omega}{s^2 + \omega^2}$$

and $$\cos \omega t\, u(t) \Leftrightarrow \frac{s}{s^2 + \omega^2} \qquad (21)$$

Drill Problems

19-5. Find $f(t)$ if $F(s)$ equals (a) $1/[s(s + 1)(s + 2)]$; (b) $s/[(s + 1)(s + 2)]$; (c) $(s + 1)/[s(s + 2)]$.

Ans: $(0.5 - e^{-t} + 0.5e^{-2t})u(t)$; $(-e^{-t} + 2e^{-2t})u(t)$; $0.5(1 + e^{-2t})u(t)$

19-6. Use Laplace transform methods to find $i(t)$ if (a) $2(di/dt) + 8i = 6e^{-2t}u(t)$, $i(0^-) = 1$ A; (b) $d^2i/dt^2 + 3(di\,/\,dt) + 2i = 4u(t)$, $i'(0^-) = 5$ A/s, $i(0^-) = 0$; (c) $i(t)$ is the current indicated in Fig. 19-3a.

Ans: $(1.5e^{-2t} - 0.5e^{-4t})u(t)$; $(2 + e^{-t} - 3e^{-2t})u(t)$; $(0.25 + 4.75e^{-20t})u(t)$ (all A)

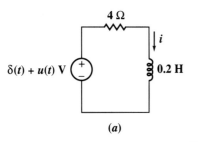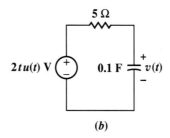

Figure 19-3

(a) See Drill Prob. 19-6. (b) See Drill Prob. 19-7.

19-7. Find $v(t)$ at $t = 0.8$ s if (a) $1.5 \int_{0^-}^{t} v(t)\, dt + 6(dv/dt) = u(t)$ and $v(0^-) = 0.5$ V; (b) $10 \int_{-\infty}^{t} v(t)\, dt + 4v(t) = 8 + 2\delta(t)$, $\int_{-\infty}^{0^-} v(t)\, dt = 0.8$; (c) $v(t)$ is the voltage defined in Fig. 19-3b.   *Ans:* 0.590; −0.1692; 0.802 (all V)

In our study of the Fourier transform, we discovered that the Fourier transform of $f_1(t) * f_2(t)$, the convolution of two time functions, was simply the product of the transforms of the individual functions. This then led to the very useful concept of the system function, defined as the ratio of the transform of the system output to the transform of the system input. Exactly the same fortunate circumstances exist when we work with the Laplace transform, as we shall now see.

Convolution was first defined by Eq. (40) of Chap. 18, and we therefore may write

$$f_1(t) * f_2(t) = \int_{-\infty}^{\infty} f_1(\lambda) f_2(t - \lambda)\, d\lambda$$

Now we let $\mathbf{F}_1(\mathbf{s})$ and $\mathbf{F}_2(\mathbf{s})$ be the Laplace transforms of $f_1(t)$ and $f_2(t)$, respectively, and consider the Laplace transform of $f_1(t) * f_2(t)$,

$$\mathscr{L}\{f_1(t) * f_2(t)\} = \mathscr{L}\left\{\int_{-\infty}^{\infty} f_1(\lambda) f_2(t - \lambda)\, d\lambda\right\}$$

As we discovered in our earlier look at convolution, one of these time functions will often be the forcing function that is applied at the input terminals of a linear circuit, and the other will be the unit-impulse response of the circuit. That is, the response of a linear circuit is just the convolution of the input and the impulse response.

Since we are now dealing with time functions that do not exist prior to $t = 0^-$ (the definition of the Laplace transform forces us to assume this), the lower limit of integration can be changed to 0^-. Then, using the definition of the Laplace transform, we get

$$\mathscr{L}\{f_1(t) * f_2(t)\} = \int_{0^-}^{\infty} e^{-st} \left[\int_{0^-}^{\infty} f_1(\lambda) f_2(t - \lambda)\, d\lambda\right] dt$$

Since e^{-st} does not depend upon λ, we can move this factor inside the inner integral. If we do this and also reverse the order of integration, the result is

$$\mathscr{L}\{f_1(t) * f_2(t)\} = \int_{0^-}^{\infty} \left[\int_{0^-}^{\infty} e^{-st} f_1(\lambda) f_2(t - \lambda)\, dt\right] d\lambda$$

19-5

Convolution again

Continuing with the same type of trickery, we note that $f_1(\lambda)$ does not depend upon t, and so it can be moved outside the inner integral:

$$\mathscr{L}\{f_1(t) * f_2(t)\} = \int_{0^-}^{\infty} f_1(\lambda) \left[\int_{0^-}^{\infty} e^{-st} f_2(t - \lambda) \, dt \right] d\lambda$$

We then make the substitution $x = t - \lambda$, in the bracketed integral (where we may treat λ as a constant) and, while we are at it, remove the factor e^{-st}:

$$\mathscr{L}\{f_1(t) * f_2(t)\} = \int_{0^-}^{\infty} e^{-st} f_1(\lambda) \left[\int_{0^-}^{\infty} e^{-st} f_2(x) \, dx \right] d\lambda$$

The bracketed term is $\mathbf{F}_2(\mathbf{s})$; it is not a function of t or λ. It can therefore be removed from the integrand having λ as its variable. What remains is the Laplace transform $\mathbf{F}_1(\mathbf{s})$, and we have

$$f_1(t) * f_2(t) \Leftrightarrow \mathbf{F}_1(\mathbf{s}) \mathbf{F}_2(\mathbf{s}) \tag{22}$$

which is the desired result. Stated slightly differently, we may conclude that the inverse transform of the product of two transforms is the convolution of the individual inverse transforms, a result that is sometimes useful in obtaining inverse transforms.

Let us apply the convolution theorem to the first example of Sec. 19-4, in which we were given the transform

$$\mathbf{V}(\mathbf{s}) = \frac{1}{(\mathbf{s} + \alpha)(\mathbf{s} + \beta)}$$

and obtained the inverse transform by a partial-fraction expansion. We now identify $\mathbf{V}(\mathbf{s})$ as the product of two transforms,

$$\mathbf{V}_1(\mathbf{s}) = \frac{1}{\mathbf{s} + \alpha}$$

$$\mathbf{V}_2(\mathbf{s}) = \frac{1}{\mathbf{s} + \beta}$$

where

$$v_1(t) = e^{-\alpha t} u(t)$$

and

$$v_2(t) = e^{-\beta t} u(t)$$

The desired $v(t)$ can be immediately expressed as

$$v(t) = \mathscr{L}^{-1}\{\mathbf{V}_1(\mathbf{s})\mathbf{V}_2(\mathbf{s})\} = v_1(t) * v_2(t) = \int_{0^-}^{\infty} v_1(\lambda) v_2(t - \lambda) \, d\lambda$$

$$= \int_{0^-}^{\infty} e^{-\alpha \lambda} u(\lambda) e^{-\beta(t-\lambda)} u(t - \lambda) \, d\lambda = \int_{0^-}^{t} e^{-\alpha \lambda} e^{-\beta t} e^{\beta \lambda} \, d\lambda$$

$$= e^{-\beta t} \int_{0^-}^{t} e^{(\beta - \alpha)\lambda} \, d\lambda = e^{-\beta t} \frac{e^{(\beta-\alpha)t} - 1}{\beta - \alpha} u(t)$$

and finally,

$$v(t) = \frac{1}{\beta - \alpha} (e^{-\alpha t} - e^{-\beta t}) u(t)$$

which is the same result that we obtained before, using the partial-fraction expansion. Note that it is necessary to insert the unit step $u(t)$ in the result because all (one-sided) Laplace transforms are valid only for nonnegative time.

Was the result easier to obtain by this method? Not unless one is in love with convolution integrals. The partial-fraction-expansion method is usually simpler, assuming that the expansion itself is not too cumbersome.

As we have noted several times before, the output $v_o(t)$ at some point in a linear circuit can be obtained by convolving the input $v_i(t)$ with the unit-impulse response $h(t)$. However, we must remember that the impulse response results from the application of a unit impulse at $t = 0$ *with all initial conditions zero*. Under these conditions, the Laplace transform of the output is

$$\mathscr{L}\{v_o(t)\} = \mathbf{V}_o(\mathbf{s}) = \mathscr{L}\{v_i(t) * h(t)\} = \mathbf{V}_i(\mathbf{s})[\mathscr{L}\{h(t)\}]$$

Thus, the ratio $\mathbf{V}_o(\mathbf{s})/\mathbf{V}_i(\mathbf{s})$ is equal to the transform of the impulse response, which we shall denote by $\mathbf{H}(\mathbf{s})$,

$$\mathscr{L}\{h(t)\} = \mathbf{H}(\mathbf{s}) = \frac{\mathbf{V}_o(\mathbf{s})}{\mathbf{V}_i(\mathbf{s})} \tag{23}$$

The expression $\mathbf{H}(\mathbf{s})$ was first introduced in Sec. 12-7 as the ratio of the forced response to the forcing function that caused it, and we termed it a *transfer function.* We now use the same symbol $\mathbf{H}(\mathbf{s})$ and the same name to represent the ratio of the Laplace transform of the output (or response) to the Laplace transform of the input (or forcing function) when all initial conditions are zero. The equivalence of these two descriptions of the transfer function $\mathbf{H}(\mathbf{s})$ is demonstrated in the next-to-last section of this chapter.

From Eq. (23) we see that the impulse response and the transfer function make up a Laplace transform pair,

$$h(t) \Leftrightarrow \mathbf{H}(\mathbf{s}) \tag{24}$$

This is an important fact that we shall utilize later to analyze the behavior of some circuits that would previously have baffled us.

It is evident that there is considerable similarity between the transfer function of Laplace transform theory,

$$\mathbf{H}(\mathbf{s}) = \frac{\mathbf{V}_o(\mathbf{s})}{\mathbf{V}_i(\mathbf{s})} = \frac{\mathscr{L}\{v_o(t)\}}{\mathscr{L}\{v_i(t)\}}$$

and the system function defined earlier for Fourier transforms,

$$\mathbf{H}(j\omega) = \frac{\mathbf{V}_o(j\omega)}{\mathbf{V}_i(j\omega)} = \frac{\mathscr{F}\{v_o(t)\}}{\mathscr{F}\{v_i(t)\}}$$

More advanced treatments of this subject[3] show that $\mathbf{H}(j\omega) = \mathbf{H}(\mathbf{s})\,|_{\mathbf{s}=j\omega}$ if all the poles of $\mathbf{H}(\mathbf{s})$ lie in the left half of the $\mathbf{s}$ plane. If any poles of $\mathbf{H}(\mathbf{s})$ are located on the $j\omega$ axis, then $\mathbf{H}(j\omega)$ also contains two delta functions for each of these poles. And if $\mathbf{H}(\mathbf{s})$ has any poles in the RHP, then $\mathbf{H}(j\omega)$ does not exist.

---

[3] McGillem and Cooper, op. cit., pp. 262–265.

Drill Problems

19-8. Make use of convolution in the time domain to find the inverse Laplace transform of $\mathbf{F}_1(\mathbf{s})\mathbf{F}_2(\mathbf{s})$ equal to (a) $[5/\mathbf{s}][2/(5\mathbf{s} + 1)]$; (b) $[5/\mathbf{s}^2][2/(5\mathbf{s} + 1)]$. (c) Let $f_1(t) = \cos 8t\, u(t)$, $f_2(t) = 5u(t)$, and find $f_1(t) * f_2(t)$.

          Ans: $10(1 - e^{-0.2t})u(t)$; $50(0.2t - 1 + e^{-0.2t})u(t)$; $\frac{5}{8}\sin 8t\, u(t)$

19-9. Find $\mathbf{H}(\mathbf{s})$ and $h(t)$ for the circuit of Fig. 19-4 if the output is taken as (a) $v_L(t)$; (b) $i_L(t)$; (c) $i_s(t)$.

          Ans: $0.8\mathbf{s}/(\mathbf{s} + 2)$, $0.8\delta(t) - 1.6e^{-2t}u(t)$ V/V; $0.2/(\mathbf{s} + 2)$, $0.2e^{-2t}u(t)$ A/V;

          $0.02(\mathbf{s} + 10)/(\mathbf{s} + 2)$, $0.02\delta(t) + 0.16e^{-2t}u(t)$ A/V

Figure 19-4

See Drill Prob. 19-9.

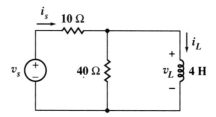

19-6

Time-shift and periodic functions

By now we have obtained a number of entries for the catalog of Laplace transform pairs that we agreed to construct earlier. Included are the transforms of the impulse function, the step function, the exponential function, the ramp function, the sine and cosine functions, and the sum of two exponentials. In addition, we have noted the consequences in the **s** domain of the time-domain operations of differentiation, integration, and convolution; and we have used the Laplace transform to define what we mean by a transfer function. These results, plus several others, are collected together for quick reference in Tables 19-1 and 19-2. Some of the tabulated relationships are unfamiliar, however, and we shall now develop a few additional useful theorems to help us obtain them.

As we have seen in some of our earlier transient problems, not all forcing functions begin at $t = 0$. What happens to the transform of a time function if that function is simply shifted in time by some known amount? In particular, if the transform of $f(t)u(t)$ is the known function $\mathbf{F}(\mathbf{s})$, then what is the transform of $f(t - a)u(t - a)$, the original time function delayed by a seconds (and not existing for $t < a$)? Working directly from the definition of the Laplace transform, we get

$$\mathscr{L}\{f(t - a)u(t - a)\} = \int_{0^-}^{\infty} e^{-\mathbf{s}t}f(t - a)u(t - a)\, dt = \int_{a^-}^{\infty} e^{-\mathbf{s}t}f(t - a)\, dt$$

for $t \geq a^-$. Choosing a new variable of integration, $\tau = t - a$, we obtain

$$\mathscr{L}\{f(t - a)u(t - a)\} = \int_{0^-}^{\infty} e^{-\mathbf{s}(\tau+a)}f(\tau)\, d\tau = e^{-a\mathbf{s}}\mathbf{F}(\mathbf{s})$$

Therefore,

$$f(t - a)u(t - a) \Leftrightarrow e^{-a\mathbf{s}}\mathbf{F}(\mathbf{s}) \qquad (a \geq 0) \qquad (25)$$

This result is known as the *time-shift theorem*, and it simply states that if a time function is delayed by a time a in the time domain, the result in the frequency domain is a multiplication by $e^{-a\mathbf{s}}$.

Table 19-1

Laplace transform pairs

$f(t) = \mathcal{L}^{-1}\{\mathbf{F(s)}\}$	$\mathbf{F(s)} = \mathcal{L}\{f(t)\}$
$\delta(t)$	1
$u(t)$	$\dfrac{1}{\mathbf{s}}$
$tu(t)$	$\dfrac{1}{\mathbf{s}^2}$
$\dfrac{t^{n-1}}{(n-1)!}u(t),\, n = 1, 2, \ldots$	$\dfrac{1}{\mathbf{s}^n}$
$e^{-\alpha t}u(t)$	$\dfrac{1}{\mathbf{s} + \alpha}$
$te^{-\alpha t}u(t)$	$\dfrac{1}{(\mathbf{s} + \alpha)^2}$
$\dfrac{t^{n-1}}{(n-1)!}e^{-\alpha t}u(t),\, n = 1, 2, \ldots$	$\dfrac{1}{(\mathbf{s} + \alpha)^n}$
$\dfrac{1}{\beta - \alpha}(e^{-\alpha t} - e^{-\beta t})u(t)$	$\dfrac{1}{(\mathbf{s} + \alpha)(\mathbf{s} + \beta)}$
$\sin \omega t\, u(t)$	$\dfrac{\omega}{\mathbf{s}^2 + \omega^2}$
$\cos \omega t\, u(t)$	$\dfrac{\mathbf{s}}{\mathbf{s}^2 + \omega^2}$
$\sin(\omega t + \theta)u(t)$	$\dfrac{\mathbf{s} \sin \theta + \omega \cos \theta}{\mathbf{s}^2 + \omega^2}$
$\cos(\omega t + \theta)u(t)$	$\dfrac{\mathbf{s} \cos \theta - \omega \sin \theta}{\mathbf{s}^2 + \omega^2}$
$e^{-\alpha t} \sin \omega t\, u(t)$	$\dfrac{\omega}{(\mathbf{s} + \alpha)^2 + \omega^2}$
$e^{-\alpha t} \cos \omega t\, u(t)$	$\dfrac{\mathbf{s} + \alpha}{(\mathbf{s} + \alpha)^2 + \omega^2}$

Example 19-4   As an example of the application of this theorem, let us determine the transform of the rectangular pulse described by $v(t) = u(t - 2) - u(t - 5)$.

Solution: This pulse has unit value for the time interval $2 < t < 5$, and has zero value elsewhere. We know that the transform of $u(t)$ is just $1/\mathbf{s}$, and since $u(t - 2)$ is simply $u(t)$ delayed by 2 s, the transform of this delayed function is $e^{-2\mathbf{s}}/\mathbf{s}$. Similarly, the transform of $u(t - 5)$ is $e^{-5\mathbf{s}}/\mathbf{s}$. It follows, then, that the desired transform is

$$\mathbf{V(s)} = \frac{e^{-2\mathbf{s}} - e^{-5\mathbf{s}}}{\mathbf{s}}$$

It was not necessary to revert to the definition of the Laplace transform in order to determine $\mathbf{V(s)}$. ■

The time-shift theorem is also useful in evaluating the transform of periodic time functions. Suppose that $f(t)$ is periodic with a period T for positive values

Table 19-2

Laplace transform operations

Operation	$f(t)$	$F(s)$
Addition	$f_1(t) \pm f_2(t)$	$\mathbf{F}_1(\mathbf{s}) \pm \mathbf{F}_2(\mathbf{s})$
Scalar multiplication	$kf(t)$	$k\mathbf{F}(\mathbf{s})$
Time differentiation	$\dfrac{df}{dt}$	$\mathbf{s}\mathbf{F}(\mathbf{s}) - f(0^-)$
	$\dfrac{d^2f}{dt^2}$	$\mathbf{s}^2\mathbf{F}(\mathbf{s}) - \mathbf{s}f(0^-) - f'(0^-)$
	$\dfrac{d^3f}{dt^3}$	$\mathbf{s}^3\mathbf{F}(\mathbf{s}) - \mathbf{s}^2f(0^-) - \mathbf{s}f'(0^-) - f''(0^-)$
Time integration	$\displaystyle\int_{0^-}^{t} f(t)\,dt$	$\dfrac{1}{\mathbf{s}}\mathbf{F}(\mathbf{s})$
	$\displaystyle\int_{-\infty}^{t} f(t)\,dt$	$\dfrac{1}{\mathbf{s}}\mathbf{F}(\mathbf{s}) + \dfrac{1}{\mathbf{s}}\displaystyle\int_{-\infty}^{0^-} f(t)\,dt$
Convolution	$f_1(t) * f_2(t)$	$\mathbf{F}_1(\mathbf{s})\mathbf{F}_2(\mathbf{s})$
Time shift	$f(t - a)u(t - a),$ $a \geq 0$	$e^{-a\mathbf{s}}\mathbf{F}(\mathbf{s})$
Frequency shift	$f(t)e^{-at}$	$\mathbf{F}(\mathbf{s} + a)$
Frequency differentiation	$-tf(t)$	$\dfrac{d\mathbf{F}(\mathbf{s})}{d\mathbf{s}}$
Frequency integration	$\dfrac{f(t)}{t}$	$\displaystyle\int_{\mathbf{s}}^{\infty} \mathbf{F}(\mathbf{s})\,d\mathbf{s}$
Scaling	$f(at), a \geq 0$	$\dfrac{1}{a}\mathbf{F}\left(\dfrac{\mathbf{s}}{a}\right)$
Initial value	$f(0^+)$	$\displaystyle\lim_{\mathbf{s}\to\infty} \mathbf{s}\mathbf{F}(\mathbf{s})$
Final value	$f(\infty)$	$\displaystyle\lim_{\mathbf{s}\to 0} \mathbf{s}\mathbf{F}(\mathbf{s})$, all poles of $\mathbf{s}\mathbf{F}(\mathbf{s})$ in LHP
Time periodicity	$f(t) = f(t + nT),$ $n = 1, 2, \ldots$	$\dfrac{1}{1 - e^{-T\mathbf{s}}}\mathbf{F}_1(\mathbf{s}),$ where $\mathbf{F}_1(\mathbf{s}) = \displaystyle\int_{0^-}^{T} f(t)e^{-\mathbf{s}t}\,dt$

of t. The behavior of $f(t)$ for $t < 0$ has no effect on the (one-sided) Laplace transform, as we know. Thus, $f(t)$ can be written as

$$f(t) = f(t - nT) \qquad n = 0, 1, 2, \ldots$$

If we now define a new time function which is nonzero only in the first period of $f(t)$,

$$f_1(t) = [u(t) - u(t - T)]\,f(t)$$

then the original $f(t)$ can be represented as the sum of an infinite number of such functions, delayed by integral multiples of T. That is,

$$f(t) = [u(t) - u(t - T)]f(t) + [u(t - T) - u(t - 2T)]f(t)$$

$$+ [u(t - 2T) - u(t - 3T)]f(t) + \cdots$$

$$= f_1(t) + f_1(t - T) + f_1(t - 2T) + \cdots$$

or
$$f(t) = \sum_{n=0}^{\infty} f_1(t - nT)$$

The Laplace transform of this sum is just the sum of the transforms,

$$\mathbf{F}(\mathbf{s}) = \sum_{n=0}^{\infty} \mathcal{L}\{f_1(t - nT)\}$$

so that the time-shift theorem leads to

$$\mathbf{F}(\mathbf{s}) = \sum_{n=0}^{\infty} e^{-nT\mathbf{s}}\mathbf{F}_1(\mathbf{s})$$

where

$$\mathbf{F}_1(\mathbf{s}) = \mathcal{L}\{f_1(t)\} = \int_{0^-}^{T} e^{-\mathbf{s}t}\, f(t)\, dt$$

Since $\mathbf{F}_1(\mathbf{s})$ is not a function of n, it can be removed from the summation, and $\mathbf{F}(\mathbf{s})$ becomes

$$\mathbf{F}(\mathbf{s}) = \mathbf{F}_1(\mathbf{s}) \,[1 + e^{-T\mathbf{s}} + e^{-2T\mathbf{s}} + \cdots]$$

When we apply the binomial theorem to the bracketed expression, it simplifies to $1/(1 - e^{-T\mathbf{s}})$. Thus, we conclude that the periodic function $f(t)$, with period T, has a Laplace transform expressed by

$$\mathbf{F}(\mathbf{s}) = \frac{\mathbf{F}_1(\mathbf{s})}{1 - e^{-T\mathbf{s}}} \qquad (26)$$

where

$$\mathbf{F}_1(\mathbf{s}) = \mathcal{L}\{[u(t) - u(t - T)]f(t)\} \qquad (27)$$

is the transform of the first period of the time function.

To illustrate the use of this transform theorem for periodic functions, let us apply it to the familiar rectangular pulse train, Fig. 19-5. We may describe this periodic function analytically:

$$v(t) = \sum_{n=0}^{\infty} V_0[u(t - nT) - u(t - nT - \tau)] \qquad t > 0$$

The function $\mathbf{V}_1(\mathbf{s})$ is simple to calculate:

$$\mathbf{V}_1(\mathbf{s}) = V_0 \int_{0^-}^{\tau} e^{-\mathbf{s}t}dt = \frac{V_0}{\mathbf{s}}(1 - e^{-\mathbf{s}\tau})$$

Now, to obtain the desired transform, we just divide by $(1 - e^{-\mathbf{s}T})$:

$$\mathbf{V}(\mathbf{s}) = \frac{V_0}{\mathbf{s}}\frac{(1 - e^{-\mathbf{s}\tau})}{(1 - e^{-\mathbf{s}T})} \qquad (28)$$

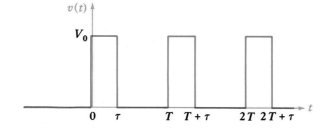

Figure 19-5

A periodic train of rectangular pulses for which $\mathbf{F}(\mathbf{s}) = (V_0/\mathbf{s})(1 - e^{-\mathbf{s}\tau})/(1 - e^{-\mathbf{s}T})$.

We should note how the two theorems described in this section show up in the transform in Eq. (28). The $(1 - e^{-sT})$ factor in the denominator accounts for the periodicity of the function, the $e^{-s\tau}$ term in the numerator arises from the time delay of the negative square wave that turns off the pulse, and the $V_0/\mathbf{s}$ factor is, of course, the transform of the step functions involved in $v(t)$.

Drill Problems

19-10. Find the Laplace transform of the time function shown in (a) Fig. 19-6a; (b) Fig. 19-6b; (c) Fig. 19-6c.

$Ans: (5/\mathbf{s})(2e^{-2\mathbf{s}} - e^{-4\mathbf{s}} - e^{-5\mathbf{s}}); \; 4e^{-\pi\mathbf{s}} + 2[e^{-\pi\mathbf{s}} + e^{-2\pi\mathbf{s}}]/[\mathbf{s}^2 + 1]; \; (2.5/\mathbf{s}^2)(e^{-2\mathbf{s}} - e^{-4\mathbf{s}}) - (5/\mathbf{s})e^{-4\mathbf{s}}$

Figure 19-6

See Drill Prob. 19-10.

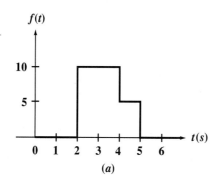

(a)

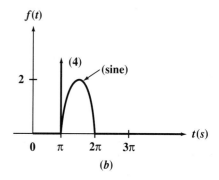

(b)

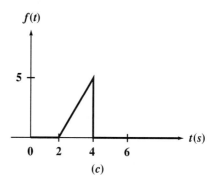

(c)

19-11. Determine the Laplace transform of the periodic function shown in (a) Fig. 19-7a; (b) Fig. 19-7b; (c) Fig. 19-7c.

$$Ans: \frac{2e^{-s}}{1 + e^{-2s}}; \; \frac{3(1 + e^{-s} - 2e^{-2s})}{\mathbf{s}(1 - e^{-3s})}; \; \frac{8}{\mathbf{s}^2 + \pi^2/4}; \; \frac{\mathbf{s} + (\pi/2)e^{-s} + (\pi/2)e^{-3s} - \mathbf{s}e^{-4s}}{1 - e^{-4s}}$$

Figure 19-7

See Drill Prob. 19-11.

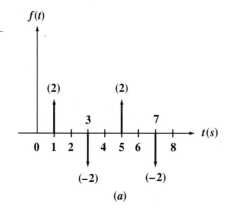

(a)

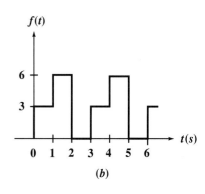

(b)

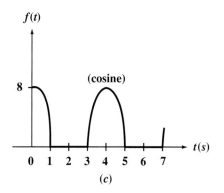

Figure 19-7

(*continued*)

Other theorems appearing in Table 19-2 specify the results in the time domain of simple operations on $\mathbf{F(s)}$ in the frequency domain. We shall obtain several easily in this section and then see how they may be applied to derive additional transform pairs.

The first new theorem establishes a relationship between $\mathbf{F(s)} = \mathscr{L}\{f(t)\}$ and $\mathbf{F(s} + a)$. We consider the Laplace transform of $e^{-at}f(t)$,

$$\mathscr{L}\{e^{-at}f(t)\} = \int_{0^-}^{\infty} e^{-\mathbf{s}t} e^{-at} f(t)\, dt = \int_{0^-}^{\infty} e^{-(\mathbf{s}+a)t} f(t)\, dt$$

Looking carefully at this result, we note that the integral on the right is identical to that defining $\mathbf{F(s)}$ with one exception: $(\mathbf{s} + a)$ appears in place of $\mathbf{s}$. Thus,

$$e^{-at}f(t) \Leftrightarrow \mathbf{F(s} + a) \tag{29}$$

We conclude that replacing $\mathbf{s}$ by $(\mathbf{s} + a)$ in the frequency domain corresponds to multiplication by e^{-at} in the time domain. This is known as the *frequency-shift* theorem. It can be put to immediate use in evaluating the transform of the exponentially damped cosine function that we used extensively in previous work. Beginning with the known transform of the cosine function,

$$\mathscr{L}\{\cos \omega_0 t\} = \mathbf{F(s)} = \frac{\mathbf{s}}{\mathbf{s}^2 + \omega_0^2}$$

then the transform of $e^{-at} \cos \omega_0 t$ must be $\mathbf{F(s} + a)$:

$$\mathscr{L}\{e^{-at} \cos \omega_0 t\} = \mathbf{F(s} + a) = \frac{\mathbf{s} + a}{(\mathbf{s} + a)^2 + \omega_0^2} \tag{30}$$

Next let us examine the consequences of differentiating $\mathbf{F(s)}$ with respect to $\mathbf{s}$. The result is

$$\frac{d}{d\mathbf{s}} \mathbf{F(s)} = \frac{d}{d\mathbf{s}} \int_{0^-}^{\infty} e^{-\mathbf{s}t} f(t)\, dt = \int_{0^-}^{\infty} -t e^{-\mathbf{s}t} f(t)\, dt = \int_{0^-}^{\infty} e^{-\mathbf{s}t} [-t f(t)]\, dt$$

which is clearly the Laplace transform of $[-tf(t)]$. We therefore conclude that differentiation with respect to $\mathbf{s}$ in the frequency domain results in multiplication by $-t$ in the time domain, or

$$-tf(t) \Leftrightarrow \frac{d}{d\mathbf{s}} \mathbf{F(s)} \tag{31}$$

19-7

Shifting, differentiation, integration, and scaling in the frequency domain

Suppose now that $f(t)$ is the unit-ramp function $tu(t)$, whose transform we know is $1/\mathbf{s}^2$. We can use our newly acquired frequency-differentiation theorem to determine the inverse transform of $1/\mathbf{s}^3$ as follows:

$$\frac{d}{d\mathbf{s}}\left(\frac{1}{\mathbf{s}^2}\right) = -\frac{2}{\mathbf{s}^3} \Leftrightarrow -t\,\mathscr{L}^{-1}\left\{\frac{1}{\mathbf{s}^2}\right\} = -t^2 u(t)$$

and

$$\frac{t^2 u(t)}{2} \Leftrightarrow \frac{1}{\mathbf{s}^3} \tag{32}$$

Continuing with the same procedure, we find

$$\frac{t^3}{3!}u(t) \Leftrightarrow \frac{1}{\mathbf{s}^4} \tag{33}$$

and in general

$$\frac{t^{(n-1)}}{(n-1)!}u(t) \Leftrightarrow \frac{1}{\mathbf{s}^n} \tag{34}$$

The effect on $f(t)$ of integrating $\mathbf{F}(\mathbf{s})$ with respect to $\mathbf{s}$ may be shown by beginning with the definition once more,

$$\mathbf{F}(\mathbf{s}) = \int_{0^-}^{\infty} e^{-\mathbf{s}t} f(t)\, dt$$

performing the frequency integration from $\mathbf{s}$ to ∞,

$$\int_{\mathbf{s}}^{\infty} \mathbf{F}(\mathbf{s})\, d\mathbf{s} = \int_{\mathbf{s}}^{\infty}\left[\int_{0^-}^{\infty} e^{-\mathbf{s}t} f(t)\, dt\right] d\mathbf{s}$$

interchanging the order of integration,

$$\int_{\mathbf{s}}^{\infty} \mathbf{F}(\mathbf{s})\, d\mathbf{s} = \int_{0^-}^{\infty}\left[\int_{\mathbf{s}}^{\infty} e^{-\mathbf{s}t}\, d\mathbf{s}\right] f(t)\, dt$$

and performing the inner integration,

$$\int_{\mathbf{s}}^{\infty} \mathbf{F}(\mathbf{s})\, d\mathbf{s} = \int_{0^-}^{\infty}\left[-\frac{1}{t}e^{-\mathbf{s}t}\right]_{\mathbf{s}}^{\infty} f(t)\, dt = \int_{0^-}^{\infty} \frac{f(t)}{t}e^{-\mathbf{s}t}\, dt$$

Thus,

$$\frac{f(t)}{t} \Leftrightarrow \int_{\mathbf{s}}^{\infty} \mathbf{F}(\mathbf{s})\, d\mathbf{s} \tag{35}$$

For example, we have already established the transform pair

$$\sin \omega_0 t\, u(t) \Leftrightarrow \frac{\omega_0}{\mathbf{s}^2 + \omega_0^2}$$

Therefore,

$$\mathscr{L}\left\{\frac{\sin \omega_0 t\, u(t)}{t}\right\} = \int_{\mathbf{s}}^{\infty} \frac{\omega_0\, d\mathbf{s}}{\mathbf{s}^2 + \omega_0^2} = \tan^{-1}\frac{\mathbf{s}}{\omega_0}\Bigg|_{\mathbf{s}}^{\infty}$$

and we have

$$\frac{\sin \omega_0 t\, u(t)}{t} \Leftrightarrow \frac{\pi}{2} - \tan^{-1}\frac{\mathbf{s}}{\omega_0} \tag{36}$$

We next develop the time-scaling theorem of Laplace transform theory by evaluating the transform of $f(at)$, assuming that $\mathscr{L}\{f(t)\}$ is known. The procedure is very simple:

$$\mathscr{L}\{f(at)\} = \int_{0^-}^{\infty} e^{-st} f(at)\, dt = \frac{1}{a} \int_{0^-}^{\infty} e^{-(s/a)\lambda}\, d\lambda$$

where the change of variable $at = \lambda$ has been employed. The last integral is recognizable as $1/a$ times the Laplace transform of $f(t)$, except that $\mathbf{s}$ is replaced by $\mathbf{s}/a$ in the transform. It follows that

$$f(at) \Leftrightarrow \frac{1}{a}\mathbf{F}\left(\frac{\mathbf{s}}{a}\right) \tag{37}$$

As an elementary example of the use of this time-scaling theorem, consider the determination of the transform of a 1-kHz cosine wave. Assuming we know the transform of a 1-rad/s cosine wave,

$$\cos t\, u(t) \Leftrightarrow \frac{\mathbf{s}}{\mathbf{s}^2 + 1}$$

the result is

$$\mathscr{L}\{\cos 2000\pi t\, u(t)\} = \frac{1}{2000\pi}\, \frac{\mathbf{s}/2000\pi}{(\mathbf{s}/2000\pi)^2 + 1} = \frac{\mathbf{s}}{\mathbf{s}^2 + (2000\pi)^2}$$

which is correct.

The time-scaling theorem offers us some computational advantages, for it enables us to work initially in a slowed-down world, where time functions may extend for several seconds and periodic functions may possess periods of the order of magnitude of a second. Practical engineering work, however, usually involves time functions that vary much more rapidly with time than our examples might have indicated. The time slowdown is used only to simplify the computational aspects of our problems. These results may then be translated easily to the real world through the use of the time-scaling theorem.

Time scaling is also useful, even essential, in dealing with the analog computer. Here we must work with slowed-down time functions because of the frequency-response limitations of the electronic circuitry in the analog computer.

Drill Problems

19-12. Find (a) $\mathscr{L}\{e^{-2t}\sin\,(5t\,+\,0.2\pi)\,u(t)\}$; (b) $\mathscr{L}\{t\,\sin\,(5t\,+\,0.2\pi)\,u(t)\}$; (c) $\mathscr{L}\{\sin^2 5t\,u(t)\}$; (d) $\mathscr{L}\{\sin^2 5t\,u(t)/t\}$.

$$Ans: \frac{0.588\mathbf{s} + 4.05}{\mathbf{s}^2 + 4\mathbf{s} + 29};\ \frac{0.588\mathbf{s}^2 + 8.09\mathbf{s} - 14.69}{(\mathbf{s}^2 + 25)^2};\ \frac{50}{\mathbf{s}(\mathbf{s}^2 + 100)};\ \frac{1}{4}\ln\frac{\mathbf{s}^2 + 100}{\mathbf{s}^2}$$

19-13. Find $f(t)$ if $\mathbf{F}(\mathbf{s})$ equals (a) $\ln\dfrac{\mathbf{s} + 2}{\mathbf{s} + 3}$; (b) $\dfrac{9}{\mathbf{s}^2 + 10\mathbf{s} + 50}$; (c) $\dfrac{2\mathbf{s}}{(\mathbf{s}^2 + 4)^2)}$.

$$Ans:\ (1/t)\,(e^{-2t} - e^{-3t})u(t);\ 1.8e^{-5t}\sin 5t\, u(t);\ \tfrac{1}{2} t\sin 2t\, u(t)$$

19-14. If $\mathscr{L}\{f(t)\} = 1/(\mathbf{s}^2 + 2\mathbf{s} + 2)$, find (a) $f(2t)$; (b) $\mathscr{L}\{f(2t)\}$. (c) Given $\mathbf{F}(\mathbf{s}) = 24/(\mathbf{s} + 3)^4$, find $f(t)$.

$$Ans:\ e^{-2t}\sin 2t\, u(t);\ 2/(\mathbf{s}^2 + 4\mathbf{s} + 8);\ 4t^3 e^{-3t}u(t)$$

19-8

The initial-value and final-value theorems

The last two fundamental theorems that we shall discuss are known as the initial-value and final-value theorems. They will enable us to evaluate $f(0^+)$ and $f(\infty)$ by examining the limiting values of $\mathbf{s}\mathbf{F}(\mathbf{s})$.

To derive the initial-value theorem, we consider the Laplace transform of the derivative once again,

$$\mathcal{L}\left\{\frac{df}{dt}\right\} = \mathbf{s}\mathbf{F}(\mathbf{s}) - f(0^-) = \int_{0^-}^{\infty} e^{-\mathbf{s}t}\frac{df}{dt}\,dt$$

We now let $\mathbf{s}$ approach infinity. By breaking the integral into two parts,

$$\lim_{\mathbf{s}\to\infty}[\mathbf{s}\mathbf{F}(\mathbf{s}) - f(0^-)] = \lim_{\mathbf{s}\to\infty}\left(\int_{0^-}^{0^+} e^{0}\frac{df}{dt}\,dt + \int_{0^+}^{\infty} e^{-\mathbf{s}t}\frac{df}{dt}\,dt\right)$$

we see that the second integral must approach zero in the limit, since the integrand itself approaches zero. Also, $f(0^-)$ is not a function of $\mathbf{s}$, and it may be removed from the left limit:

$$-f(0^-) + \lim_{\mathbf{s}\to\infty}[\mathbf{s}\mathbf{F}(\mathbf{s})] = \lim_{\mathbf{s}\to\infty}\int_{0^-}^{0^+} df = \lim_{\mathbf{s}\to\infty}[f(0^+) - f(0^-)]$$

$$= f(0^+) - f(0^-)$$

and finally,

$$f(0^+) = \lim_{\mathbf{s}\to\infty}[\mathbf{s}\mathbf{F}(\mathbf{s})]$$

or

$$\lim_{t\to 0^+} f(t) = \lim_{\mathbf{s}\to\infty}[\mathbf{s}\mathbf{F}(\mathbf{s})] \tag{38}$$

This is the mathematical statement of the *initial-value theorem*. It states that the initial value of the time function $f(t)$ can be obtained from its Laplace transform $\mathbf{F}(\mathbf{s})$ by first multiplying the transform by $\mathbf{s}$ and then letting $\mathbf{s}$ approach infinity. Note that the initial value of $f(t)$ that is obtained is the limit from the right.

The initial-value theorem, along with the final-value theorem that we shall consider in a moment, is useful in checking the results of a transformation or an inverse transformation. For example, when we first calculated the transform of $\cos \omega_0 t\, u(t)$, we obtained $\mathbf{s}/(\mathbf{s}^2 + \omega_0^2)$. After noting that $f(0^+) = 1$, we can make a partial check on the validity of this result by applying the initial-value theorem:

$$\lim_{\mathbf{s}\to\infty}\left(\mathbf{s}\,\frac{\mathbf{s}}{\mathbf{s}^2 + \omega_0^2}\right) = 1$$

and the check is accomplished.

The final-value theorem is not quite as useful as the initial-value theorem, for it can be used only with a certain class of transforms, those whose poles lie entirely within the left half of the $\mathbf{s}$ plane, except for a simple pole at $\mathbf{s} = 0$. We again consider the Laplace transform of df/dt,

$$\int_{0^-}^{\infty} e^{-\mathbf{s}t}\frac{df}{dt}\,dt = \mathbf{s}\mathbf{F}(\mathbf{s}) - f(0^-)$$

this time in the limit as $\mathbf{s}$ approaches zero,

$$\lim_{\mathbf{s}\to 0}\int_{0^-}^{\infty} e^{-\mathbf{s}t}\frac{df}{dt}\,dt = \lim_{\mathbf{s}\to 0}[\mathbf{s}\mathbf{F}(\mathbf{s}) - f(0^-)] = \int_{0^-}^{\infty}\frac{df}{dt}\,dt$$

We assume that both $f(t)$ and its first derivative are transformable. Now, the last term of this equation is readily expressed as a limit,

$$\int_{0^-}^{\infty} \frac{df}{dt}\,dt = \lim_{t\to\infty} \int_{0^-}^{t} \frac{df}{dt}\,dt = \lim_{t\to\infty} [f(t) - f(0^-)]$$

By recognizing that $f(0^-)$ is a constant, a comparison of the last two equations shows us that

$$\lim_{t\to\infty} f(t) = \lim_{s\to 0} [\mathbf{s}\mathbf{F}(\mathbf{s})] \qquad (39)$$

which is the *final-value theorem*. In applying this theorem, it is necessary to know that $f(\infty)$, the limit of $f(t)$ as t becomes infinite, exists, or—what amounts to the same thing—that the poles of $\mathbf{F}(\mathbf{s})$ all lie *within* the left half of the $\mathbf{s}$ plane except for a simple pole at the origin. The product $\mathbf{s}\mathbf{F}(\mathbf{s})$ thus has *all* of its poles lying *within* the left half plane.

Example 19-5 As a straightforward example of the application of this theorem, let us consider the function $f(t) = (1 - e^{-at})u(t)$, where $a > 0$.

Solution: We see immediately that $f(\infty) = 1$. The transform of $f(t)$ is

$$\mathbf{F}(\mathbf{s}) = \frac{1}{\mathbf{s}} - \frac{1}{\mathbf{s} + a} = \frac{a}{\mathbf{s}(\mathbf{s} + a)}$$

Multiplying by $\mathbf{s}$ and letting $\mathbf{s}$ approach zero, we obtain

$$\lim_{s\to 0} [\mathbf{s}\mathbf{F}(\mathbf{s})] = \lim_{s\to 0} \frac{a}{\mathbf{s} + a} = 1$$

which agrees with $f(\infty)$. ∎

If $f(t)$ is a sinusoid, however, so that $\mathbf{F}(\mathbf{s})$ has poles on the $j\omega$ axis, then a blind use of the final-value theorem might lead us to conclude that the final value is zero. We know, however, that the final value of either $\sin \omega_0 t$ or $\cos \omega_0 t$ is indeterminate. So, beware of $j\omega$-axis poles!

We now have all the tools to apply the Laplace transform to the solution of problems that we either could not solve previously, or that we solved with considerable stress and strain.

19-15. Without finding $f(t)$ first, determine $f(0^+)$ and $f(\infty)$ for each of the following transforms: (*a*) $4e^{-2s}(\mathbf{s} + 50)/\mathbf{s}$; (*b*) $(\mathbf{s}^2 + 6)/(\mathbf{s}^2 + 7)$; (*c*) $(5\mathbf{s}^2 + 10)/[2\mathbf{s}(\mathbf{s}^2 + 3\mathbf{s} + 5)]$.

Drill Problem

Ans: 0, 200; ∞, 0; 2.5, 1

Earlier in this chapter we defined the transfer function $\mathbf{H}(\mathbf{s})$ as the Laplace transform of the impulse response $h(t)$, initial energy being zero throughout the circuit. Before we can make the most effective use of the transfer function, we need to show that it may be obtained very simply for any linear circuit by frequency-domain analysis. We do this by an argument similar to that presented for the Fourier transform and $\mathbf{H}(j\omega)$ in Sec. 18-9. But we shall be brief.

We work in the complex-frequency domain and apply an input $v_i(t) = Ae^{\sigma_x t} \cos(\omega_x t + \theta)$, which is $\mathbf{V}_i(\mathbf{s}_x) = Ae^{j\theta}$ in phasor form. Both A and θ are

19-9

The transfer function H(s)

functions of $\mathbf{s}_x = \sigma_x + j\omega_x$, the complex frequency of the excitation. The response is $v_o(t) = Be^{\sigma_x t} \cos(\omega_x t + \phi)$, $\mathbf{V}_o(\mathbf{s}_x) = Be^{j\phi}$. Thus,

$$\frac{\mathbf{V}_o(\mathbf{s}_x)}{\mathbf{V}_i(\mathbf{s}_x)} = \mathbf{G}(\mathbf{s}_x) = \frac{B}{A}e^{j(\phi-\theta)} \tag{40}$$

To determine $\mathbf{H}(\mathbf{s})$, we need to find the ratio of the transform of the output to the transform of the input,

$$\mathbf{H}(\mathbf{s}) = \frac{\mathcal{L}\{Be^{\sigma_x t} \cos(\omega_x t + \phi)\}}{\mathcal{L}\{Ae^{\sigma_x t} \cos(\omega_x t + \theta)\}}$$

The required transform is obtained from Table 19-1 by replacing $\mathbf{s}$ by $\mathbf{s} - \sigma_x$ in the transform for $\cos(\omega t + \theta)$. We have

$$\mathbf{H}(\mathbf{s}) = \frac{B\left\{\dfrac{(\mathbf{s} - \sigma_x)\cos\phi - \omega_x \sin\phi}{(\mathbf{s} - \sigma_x)^2 + \omega_x^2}\right\}}{A\left\{\dfrac{(\mathbf{s} - \sigma_x)\cos\theta - \omega_x \sin\theta}{(\mathbf{s} - \sigma_x)^2 + \omega_x^2}\right\}}$$

At $\mathbf{s} = \mathbf{s}_x$, this simplifies to

$$\mathbf{H}(\mathbf{s}_x) = \frac{B}{A}\frac{j\omega_x \cos\phi - \omega_x \sin\phi}{j\omega_x \cos\theta - \omega_x \sin\theta}$$

$$= \frac{B}{A}\frac{\cos\phi + j\sin\phi}{\cos\theta + j\sin\theta} = \frac{B}{A}e^{j(\phi-\theta)}$$

which is identical to Eq. (40). Since there is no special significance to the x subscript, it follows that

$$\mathbf{H}(\mathbf{s}) = \mathbf{G}(\mathbf{s}) \tag{41}$$

Thus, we may find $\mathbf{H}(\mathbf{s})$ by using normal frequency-domain methods with all elements expressed in terms of their impedances at a complex frequency $\mathbf{s}$.

Let us see how these results help us.

Example 19-6 We shall find both $h(t)$ and an output voltage for the circuit shown in Fig. 19-8a. At this time we assume that there is no initial energy storage in the network.

Solution: We first construct the frequency-domain circuit, shown in Fig. 19-8b. The ratio of $\mathbf{V}_o(\mathbf{s})/\mathbf{V}_i(\mathbf{s})$ may be found by determining $\mathbf{Z}_i(\mathbf{s})$, the impedance of the three parallel branches at the right,

$$\mathbf{Z}_i(\mathbf{s}) = \frac{1}{\mathbf{s}/24 + \frac{1}{30} + 1/(24 + 48/\mathbf{s})} = \frac{120(\mathbf{s} + 2)}{5\mathbf{s}^2 + 19\mathbf{s} + 8}$$

and then using voltage division:

$$\frac{\mathbf{V}_o(\mathbf{s})}{\mathbf{V}_i(\mathbf{s})} = \frac{\mathbf{Z}_i(\mathbf{s})}{20 + \mathbf{Z}_i(\mathbf{s})} = \frac{6(\mathbf{s} + 2)}{5\mathbf{s}^2 + 25\mathbf{s} + 20}$$

Thus,

$$\mathbf{H}(\mathbf{s}) = \frac{1.2(\mathbf{s} + 2)}{(\mathbf{s} + 1)(\mathbf{s} + 4)}$$

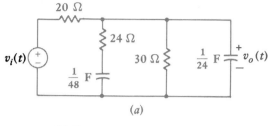

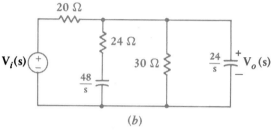

Figure 19-8

(a) An example in which the transfer function $\mathbf{H(s)} = \mathbf{V}_o(\mathbf{s})/\mathbf{V}_i(\mathbf{s})$ is to be obtained by frequency-domain analysis. Initial conditions are all zero. (b) The frequency-domain circuit.

To find $h(t)$, we need $\mathcal{L}^{-1}\{\mathbf{H(s)}\}$:

$$\mathcal{L}^{-1}\{\mathbf{H(s)}\} = \mathcal{L}^{-1}\left\{\frac{0.4}{\mathbf{s}+1} + \frac{0.8}{\mathbf{s}+4}\right\}$$

and

$$h(t) = (0.4e^{-t} + 0.8e^{-4t})u(t)$$

Thus, if $v_i(t) = \delta(t)$, then

$$v_o(t) = h(t) = (0.4e^{-t} + 0.8e^{-4t})u(t).$$ ∎

Example 19-7 Using the same circuit, next let us find the output if the input is $v_i(t) = 50 \cos 2t\, u(t)$ V.

Solution: Now we may make use of the transfer function concept:

$$\mathbf{V}_o(\mathbf{s}) = \mathbf{H(s)}\, \mathbf{V}_i(\mathbf{s})$$

where

$$\mathbf{V}_i(\mathbf{s}) = \mathcal{L}\{50 \cos 2t\, u(t)\} = \frac{50\mathbf{s}}{\mathbf{s}^2 + 4}$$

and

$$\mathbf{V}_o(\mathbf{s}) = \frac{1.2(\mathbf{s}+2)}{(\mathbf{s}+1)(\mathbf{s}+4)}\,\frac{50\mathbf{s}}{\mathbf{s}^2 + 4}$$

Expanding in partial fractions,

$$\mathbf{V}_o(\mathbf{s}) = \frac{-4}{\mathbf{s}+1} + \frac{-8}{\mathbf{s}+4} + \frac{6+j6}{\mathbf{s}+j2} + \frac{6-j6}{\mathbf{s}-j2}$$

$$= \frac{-4}{\mathbf{s}+1} + \frac{-8}{\mathbf{s}+4} + \frac{12\mathbf{s}+24}{\mathbf{s}^2+4}.$$

and

$$v_o(t) = [-4e^{-t} - 8e^{-4t} + 12 \cos 2t + 12 \sin 2t]u(t)$$ ∎

The solution is straightforward, and we should feel confident of being able to find the response for any input that is Laplace-transformable. We consider the presence of initial energy storage in the next (and final) section.

Drill Problem

19-16. For the circuit of Fig. 19-9, determine $h(t)$ if the output is (a) i_s; (b) v_L.

Ans: $0.125\delta(t) + 2(1 - t)e^{-4t}u(t)$ A; $0.75\delta(t) + 4(t - 1)e^{-4t}u(t)$ V

Figure 19-9

See Drill Prob. 19-16.

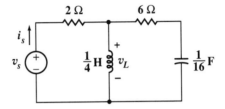

19-10

The complete response

When initial energy is present in a circuit, the Laplace transform method may be used to obtain the complete response by any of several different methods. We shall consider two of them.

The first is the more fundamental, for it involves writing the differential equations for the network and then taking the Laplace transform of those equations. The initial conditions appear when a derivative or an integral is transformed. The second technique requires each initial capacitor voltage or inductor current to be replaced by an equivalent dc source, often called an *initial-condition generator*. The elements themselves then carry no initial energy, and the transfer function procedure of the preceding section can be followed.

Let us illustrate the differential equation approach by considering the same circuit we just analyzed, but with nonzero initial conditions this time.

Example 19-8 The circuit is shown in Fig. 19-10. We let $v_1(0^-) = 10$ V and $v_2(0^-) = 25$ V, and we seek $v_2(t)$.

Solution: The differential equations for this circuit may be obtained by writing nodal equations in terms of v_1 and v_2. At the v_1 node,

$$\frac{v_1 - v_2}{24} + \frac{1}{48}v_1' = 0$$

or

$$2v_2 = 2v_1 + v_1' \qquad (42)$$

while at the v_2 node,

$$\frac{v_2 - 50\cos 2t\, u(t)}{20} + \frac{v_2 - v_1}{24} + \frac{v_2}{30} + \frac{v_2'}{24} = 0$$

or

$$v_1 = v_2' + 3v_2 - 60\cos 2t\, u(t) \qquad (43)$$

Figure 19-10

The response $v_2(t)$ is obtained for this network with the initial conditions $v_1(0^-) = 10$ V, $v_2(0^-) = 25$ V.

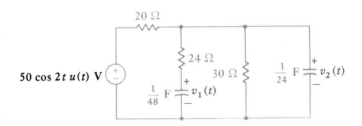

Identifying v_2 as the desired response, we eliminate v_1 and v_1' by taking the derivative of Eq. (43), remembering that $du(t)/dt = \delta(t)$:

$$v_1' = v_2'' + 3v_2' + 120 \sin 2t\, u(t) - 60\delta(t) \qquad (44)$$

and substituting Eqs. (43) and (44) in Eq. (42):

$$2v_2 = 2[v_2' + 3v_2 - 60 \cos 2t\, u(t)]$$

$$+ [v_2'' + 3v_2' + 120 \sin 2t\, u(t) - 60\delta(t)]$$

or $\qquad v_2'' + 5v_2' + 4v_2 = (120 \cos 2t - 120 \sin 2t)u(t) + 60\delta(t)$

We now take the Laplace transform,

$$\mathbf{s}^2\mathbf{V}_2(\mathbf{s}) - \mathbf{s}v_2(0^-) - v_2'(0^-) + 5\mathbf{s}\mathbf{V}_2(\mathbf{s}) - 5v_2(0^-) + 4\mathbf{V}_2(\mathbf{s}) = \frac{120\mathbf{s} - 240}{\mathbf{s}^2 + 4} + 60$$

and collect terms:

$$(\mathbf{s}^2 + 5\mathbf{s} + 4)\mathbf{V}_2(\mathbf{s}) = \mathbf{s}v_2(0^-) + v_2'(0^-) + 5v_2(0^-) + \frac{120\mathbf{s} - 240}{\mathbf{s}^2 + 4} + 60$$

We have $v_2(0^-) = 25$, so that

$$(\mathbf{s}^2 + 5\mathbf{s} + 4)\mathbf{V}_2(\mathbf{s}) = 25\mathbf{s} + 125 + v_2'(0^-) + \frac{120\mathbf{s} - 240}{\mathbf{s}^2 + 4} + 60$$

and we need a value for $v_2'(0^-)$. This we may obtain from the two circuit equations (42) and (43) by evaluating each term at $t = 0^-$. Actually, we need use only Eq. (43) in this problem:

$$v_1(0^-) = v_2'(0^-) + 3v_2(0^-) - 0$$

and $\qquad\qquad v_2'(0^-) = -65$

Thus,

$$\mathbf{V}_2(\mathbf{s}) = \frac{25\mathbf{s} + 120 + 120[(\mathbf{s} - 2)/(\mathbf{s}^2 + 4)]}{(\mathbf{s} + 1)(\mathbf{s} + 4)}$$

$$= \frac{25\mathbf{s}^3 + 120\mathbf{s}^2 + 220\mathbf{s} + 240}{(\mathbf{s} + 1)(\mathbf{s} + 4)(\mathbf{s}^2 + 4)} \qquad (45)$$

$$= \frac{\frac{23}{3}}{\mathbf{s} + 1} + \frac{\frac{16}{3}}{\mathbf{s} + 4} + \frac{12\mathbf{s} + 24}{\mathbf{s}^2 + 4}$$

from which the time-domain response is obtained:

$$v_2(t) = (\tfrac{23}{3}e^{-t} + \tfrac{16}{3}e^{-4t} + 12 \cos 2t + 12 \sin 2t)u(t) \qquad \blacksquare$$

Before considering the use of initial-condition generators, let us determine the frequency-domain equivalent of an inductor L with initial current $i(0^-)$. The time-domain network of Fig. 19-11a is described by

$$v(t) = Li'$$

Figure 19-11

(*a*) An inductor L with initial current $i(0^-)$ is shown in the time domain. (*b, c*) Frequency-domain networks that are equivalent to part *a* for Laplace transform analysis.

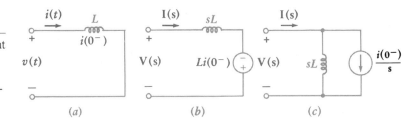

(*a*) (*b*) (*c*)

and therefore,

$$\mathbf{V(s)} = \mathbf{s}L\mathbf{I(s)} - Li(0^-) \tag{46}$$

or

$$\mathbf{I(s)} = \frac{\mathbf{V(s)}}{\mathbf{s}L} + \frac{i(0^-)}{\mathbf{s}} \tag{47}$$

The frequency-domain equivalents can be read directly from Eqs. (46) and (47), and they are shown in Figs. 19-11*b* and *c*, respectively. It may be helpful to note that the voltage source in Fig. 19-11*b* is the transform of an impulse, while the current source in Fig. 19-11*c* is the transform of a step.

Equivalent networks for an initially charged capacitor are obtained by a similar procedure; the results are shown in Fig. 19-12.

Figure 19-12

(*a*) A capacitor C with initial voltage $v(0^-)$ is shown in the time domain. (*b, c*) Frequency-domain networks that are equivalent to part *a* for Laplace transform analysis.

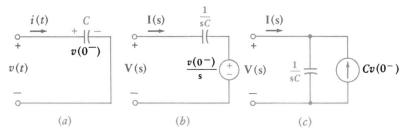

(*a*) (*b*) (*c*)

We may now use these results to construct a frequency-domain equivalent of the circuit shown in Fig. 19-10, including the effect of the initial conditions. The result is shown in Fig. 19-13. Current sources are used for the initial conditions to expedite the writing of nodal equations.

Figure 19-13

The frequency-domain equivalent of the circuit of Fig. 19-10. The current sources $\frac{10}{48}$ and $\frac{25}{24}$ provide initial time-domain voltages of 10 and 25 V across the $\frac{1}{48}$-F and $\frac{1}{24}$-F capacitors, respectively.

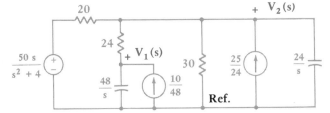

Example 19-9 Let us again find $v_2(t)$, this time making use of the circuit of Fig. 19-13.

Solution: We must now use both the superposition principle and the concept of the transfer function to see that $\mathbf{V_2(s)}$ is composed of the sum of three terms, one due to each source acting alone. Moreover, each of these sources has a transfer function to $\mathbf{V_2(s)}$. Each transfer function could be obtained by applying the standard frequency-domain analysis methods. We need not solve three little problems, however, for we can accomplish the frequency-domain analysis with all three sources operating. We do this by writing two nodal equations:

$$\frac{\mathbf{V}_1(\mathbf{s}) - \mathbf{V}_2(\mathbf{s})}{24} + \frac{\mathbf{s}\mathbf{V}_1(\mathbf{s})}{48} - \frac{10}{48} = 0$$

or $$(\mathbf{s} + 2)\mathbf{V}_1 - 2\mathbf{V}_2 = 10 \qquad (48)$$

and

$$\frac{\mathbf{V}_2(\mathbf{s}) - 50\mathbf{s}/(\mathbf{s}^2 + 4)}{20} + \frac{\mathbf{V}_2(\mathbf{s}) - \mathbf{V}_1(\mathbf{s})}{24} + \frac{\mathbf{V}_2(\mathbf{s})}{30} - \frac{25}{24} + \frac{\mathbf{s}\mathbf{V}_2(\mathbf{s})}{24} = 0$$

or $$(\mathbf{s} + 3)\mathbf{V}_2 - \mathbf{V}_1 = \frac{60\mathbf{s}}{\mathbf{s}^2 + 4} + 25 \qquad (49)$$

Using Eq. (48) to eliminate $\mathbf{V}_1$ in Eq. (49), we have

$$\mathbf{V}_1 = \frac{10 + 2\mathbf{V}_2}{\mathbf{s} + 2}$$

and then

$$\mathbf{V}_2 = \frac{25\mathbf{s}^3 + 120\mathbf{s}^2 + 220\mathbf{s} + 240}{(\mathbf{s} + 1)(\mathbf{s} + 4)(\mathbf{s}^2 + 4)}$$

This agrees with Eq. (45), and we need not repeat the inverse transform operation. The two methods check. ∎

In comparing their use, it is probably true that the transfer function method with initial-condition generators is a little faster. This becomes a safer statement to make as the complexity of the network increases. However, we must also bear in mind that the closer we stay to fundamentals, the less apt we are to become confused with special techniques and procedures. Certainly the differential-equation approach is the more basic. The use of the Laplace transform may then be thought of simply as a handy method of solving linear differential equations.

In finishing our discussion of the Laplace transform, it is helpful to compare its uses with those of the Fourier transform. The latter, of course, does not exist for as wide a variety of time functions; there are also more j's scattered about in the transform expressions, and the Fourier transform is more difficult to apply to circuits containing initial energy storage. Thus, transient problems are more easily handled with the Laplace transform. However, when spectral information about a signal is desired, such as the distribution of energy across the frequency band, the Fourier transform is the more convenient.

And now we approach the last few paragraphs of the last section of the last chapter. Looking back at some six hundred pages of linear circuit analysis, we should ask ourselves what we have accomplished. Are we prepared to tackle a really practical problem, or have we just been tilting at quixotic windmills? Can we analyze a multistage active filter, a complicated telemetry receiver, or a large interconnected power grid? Before we confess that we cannot, or at least admit that there are other things we would rather do, let us consider how far we could go in such a complex problem.

We certainly have developed substantial skills in writing accurate sets of equations to describe the behavior of increasingly complicated linear circuits. This in itself is no mean accomplishment, for we have done so in both the frequency and time domains. Our only real limitation lies in the difficulty of obtaining numerical results as the complexity of the circuit increases. For that reason we concentrated on those examples which suited computational abilities that were only human. However, we did develop an increased understanding of the most important analysis techniques in the process.

In the last several chapters we have been developing those techniques which facilitate the process of obtaining specific numerical answers. We could now write a set of time-domain or frequency-domain equations for any large-scale circuit we choose, but the numerical solution would strain our accuracy and stretch our patience. If our discussion were to proceed any further, we would have to resort to inhuman means to solve these equations. This is the point at which we turn to the computer. Its patience is virtually unlimited, its accuracy is astonishing, and its speed is almost beyond comprehension. The detailed procedure by which the computer arrives at its numerical results, however, is generated by us, and if we make an error in the program, the computer will proceed to generate erroneous results, also with speed, accuracy, and patience.

In summary, then, we can now produce sets of accurate descriptive equations that characterize almost any given linear circuit configuration, and we have some modest abilities for solving the simpler sets of these equations. Some of us have moved a bit further and have started utilizing simple computer techniques like SPICE, described in App. 5, to analyze linear circuits. Our next step, if we were to continue, would be to study those large-scale computer programs which can turn the state-variable equations or a set of transform equations into useful data. But we must stop sometime, so let it be now.

Drill Problems

19-17. Let $v_C(0^-) = 2$ V in the circuit shown in Fig. 19-14. Write a suitable differential equation, take the Laplace transform of both sides, and find $v_C(t)$ if $v_s(t)$ equals (a) $9u(t)$ V; (b) $9 \cos t\, u(t)$ V; (c) $9e^{-t}u(t)$ V.

Ans: $(6 - 4e^{-5t})u(t)$; $\frac{1}{13}[-49e^{-5t} + 75 \cos t + 15 \sin t]u(t)$; $(7.5e^{-t} - 5.5e^{-5t})u(t)$ (all V)

Figure 19-14

See Drill Probs. 19-17 and 19-18.

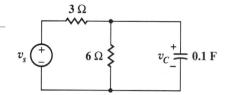

19-18. Rework Drill Prob. 19-17 by first installing a suitable initial-condition generator in the circuit and then working in the frequency domain.

Ans: (Same as for Drill Prob. 19-17)

Problems

1 Specify the range of σ over which the Laplace transform exists if $f(t)$ equals (a) $t + 1$; (b) $(t + 1)u(t)$; (c) $e^{50t}u(t)$; (d) $e^{50t}u(t - 5)$; (e) $e^{-50t}u(t - 5)$.

2 For each of the following functions, determine the Fourier transform, the two-sided Laplace transform, and the one-sided Laplace transform: (a) $8e^{-2t}[u(t + 3) - u(t - 3)]$; (b) $8e^{2t}[u(t + 3) - u(t - 3)]$; (c) $8e^{-2|t|}[u(t + 3) - u(t - 3)]$.

3 Using the one-sided Laplace transform, find $\mathbf{F(s)}$ if $f(t)$ equals (a) $[2u(t - 1)][u(3 - t)]u(t^3)$; (b) $2u(t - 4)$; (c) $3e^{-2t}u(t - 4)$; (d) $3\delta(t - 5)$; (e) $4\delta(t - 1)[\cos \pi t - \sin \pi t]$.

4 Use the definition of the (one-sided) Laplace transform to find $\mathbf{F(s)}$ if $f(t)$ equals (a) $[u(5 - t)][u(t - 2)]u(t)$; (b) $4u(t - 2)$; (c) $4e^{-3t}u(t - 2)$; (d) $4\delta(t - 2)$; (e) $5\delta(t) \sin (10t + 0.2\pi)$.

5 Determine $f(t)$ if $\mathbf{F(s)}$ equals (a) $[(\mathbf{s} + 1)/\mathbf{s}] + [2/(\mathbf{s} + 1)]$; (b) $(e^{-\mathbf{s}} + 1)^2$; (c) $2e^{-(\mathbf{s}+1)}$; (d) $2e^{-3\mathbf{s}} \cosh 2\mathbf{s}$.

6 Use the definition of the Laplace transform to calculate the value of $\mathbf{F}(1 + j2)$ if $f(t)$ equals (a) $2u(t - 2)$; (b) $2\delta(t - 2)$; (c) $e^{-t}u(t - 2)$.

7 Given the following expressions for $\mathbf{F(s)}$, find $f(t)$: (a) $5/(\mathbf{s} + 1)$; (b) $5/(\mathbf{s} + 1) - 2/(\mathbf{s} + 4)$; (c) $18/[(\mathbf{s} + 1)(\mathbf{s} + 4)]$; (d) $18\mathbf{s}/[(\mathbf{s} + 1)(\mathbf{s} + 4)]$; (e) $18\mathbf{s}^2/[(\mathbf{s} + 1)(\mathbf{s} + 4)]$.

8 If $f(0^-) = -3$ and $15u(t) - 4\delta(t) = 8f(t) + 6f'(t)$, find $f(t)$ by taking the Laplace transform of the differential equation, solving for $\mathbf{F(s)}$, and inverting to find $f(t)$.

9 (a) Find $v_C(0^-)$ and $v_C(0^+)$ for the circuit shown in Fig. 19-15. (b) Obtain an equation for $v_C(t)$ that holds for $t > 0$. (c) Use Laplace transform techniques to solve for $\mathbf{V}_C(\mathbf{s})$ and then find $v_C(t)$.

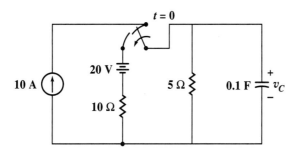

Figure 19-15

See Prob. 9.

10 Find $f(t)$ if $\mathbf{F(s)}$ equals (a) $\dfrac{2}{\mathbf{s}} - \dfrac{3}{\mathbf{s} + 1}$; (b) $\dfrac{2\mathbf{s} + 10}{\mathbf{s} + 3}$; (c) $3e^{-0.8\mathbf{s}}$; (d) $\dfrac{12}{(\mathbf{s} + 2)(\mathbf{s} + 6)}$; (e) $\dfrac{12}{(\mathbf{s} + 2)^2(\mathbf{s} + 6)}$.

11 Given the differential equation $12u(t) = 20f_2'(t) + 3f_2(t)$, where $f_2(0^-) = 2$, take its Laplace transform, solve for $\mathbf{F}_2(\mathbf{s})$, and then find $f_2(t)$.

12 (a) Determine $i_C(0^-)$ and $i_C(0^+)$ for the circuit of Fig. 19-16. (b) Write an equation for $i_C(t)$ in the time domain that is valid for $t > 0$. (c) Use Laplace transform methods to solve for $\mathbf{I}_C(\mathbf{s})$ and then find the inverse transform.

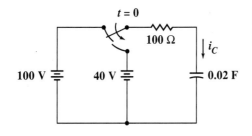

Figure 19-16

See Prob. 12.

13 A resistor R, a capacitor C, an inductor L, and an ideal current source $i_s = 100e^{-5t}u(t)$ A are in parallel. Let the voltage v be across the source with the positive reference at the terminal at which $i_s(t)$ leaves the source. Then $i_s = v' + 4v + 3\int_{0^-}^{t} v\, dt$. (a) Find R, L, and C. (b) Use Laplace transform techniques to find $v(t)$.

14 Given the two differential equations $x' + y = 2u(t)$ and $y' - 2x + 3y = 8u(t)$, where $x(0^-) = 5$ and $y(0^-) = 8$, find $x(t)$ and $y(t)$.

15 Obtain a single integrodifferential equation in terms of i_C for the circuit of Fig. 19-17, take the Laplace transform, solve for $\mathbf{I}_C(\mathbf{s})$, and then find $i_C(t)$ by making use of the inverse transform.

Figure 19-17

See Prob. 15.

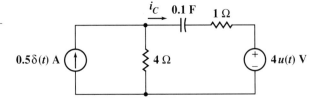

16 The impulse response of a certain linear system is $h(t) = 5 \sin \pi t \, [u(t) - u(t-1)]$. An input signal $x(t) = 2[u(t) - u(t-2)]$ is applied. Find and sketch the output $y(t)$.

17 Let $f_1(t) = e^{-5t}u(t)$ and $f_2(t) = (1 - e^{-2t})u(t)$. Find $y(t) = f_1(t) * f_2(t)$ by (a) convolution in the time domain; (b) $\mathscr{L}^{-1}\{\mathbf{F}_1(\mathbf{s})\mathbf{F}_2(\mathbf{s})\}$.

18 When an impulse $\delta(t)$ V is applied to a certain two-port network, the output voltage is $v_o(t) = 4u(t) - 4u(t-2)$ V. Find and sketch $v_o(t)$ if the input voltage is $2u(t-1)$ V.

19 Let $h(t) = 2e^{-3t}u(t)$ and $x(t) = u(t) - \delta(t)$. Find $y(t) = h(t) * x(t)$ by (a) using convolution in the time domain; (b) finding $\mathbf{H}(\mathbf{s})$ and $\mathbf{X}(\mathbf{s})$ and then obtaining $\mathscr{L}^{-1}\{\mathbf{H}(\mathbf{s})\mathbf{X}(\mathbf{s})\}$.

20 Use the time-shift theorem to find the transform of $f(t) = 5(t-1)[u(t-1) - u(t-2)]$.

21 A periodic function with $T = 5$ has the values: $f(t) = 0$, $0 < t < 1$ and $3 < t < 5$, and $f(t) = 3 + \delta(t-3)$, $1 < t < 3^+$. Find its Laplace transform.

22 Find $\mathbf{F}(\mathbf{s})$ for the pulse shown in Fig. 19-18 by using: (a) the time-shift theorem; (b) the definition of the Laplace transform.

Figure 19-18

See Prob. 22.

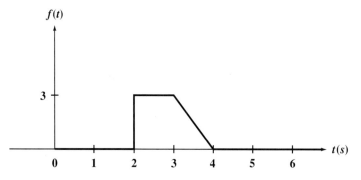

23 Find $\mathbf{F}(\mathbf{s})$ for the periodic function sketched in Fig. 19-19.

Figure 19-19

See Prob. 23.

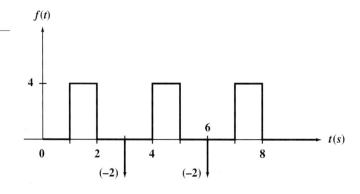

24 Let $T = 4$ for the periodic waveform shown in Fig. 19-20. (a) Find $\mathbf{F(s)}$. (b) Find $\mathbf{F}(0.3)$.

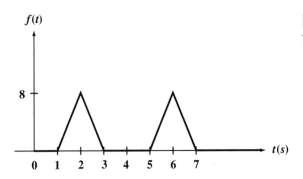

Figure 19-20

See Prob. 24.

25 A periodic function with a period $T = 6$ has a waveform that may be defined by its first period, $f_1(t) = 4[\sin 0.4\pi t][u(t) - u(t - 2.5)]$. Find $\mathbf{F(s)}$.

26 Determine $f(t)$ if $\mathbf{F(s)}$ equals (a) $\dfrac{8}{s^2 + 4}$; (b) $\dfrac{8}{s^2 + 4}\, e^{-0.5\pi s}$; (c) $\dfrac{8}{s^2 + 4}\,(1 + e^{-0.5\pi s})$;

(d) $\dfrac{8}{s^2 + 4}\,(1 + e^{-0.5\pi s})/(1 - e^{-1.25\pi s})$.

27 Find the Laplace transform of (a) $(t - 2)u(t - 1)$; (b) $20 \sin (5t - 1)\, u(t)$;

(c) $\displaystyle\sum_{n=0}^{\infty} 8(t - 5n)^2[u(t - 5n) - u(t - 1 - 5n)]$.

28 Find the Laplace transform of $f(t)$ equal to (a) $5t \cos (3t + \pi/3)\, u(t)$; (b) $(4/t) \sin^2 3t\, u(t)$.

29 Find $\mathbf{F(s)}$ if $f(t)$ equals (a) $t \sin \omega t\, u(t)$; (b) $(1/t) \sin \omega t\, u(t)$. (c) Find $f(t)$ if $\mathbf{F(s)} = s^2 e^{-2s}/[(s + 5)^2 + 100]$.

30 Find $f(t)$ if $\mathbf{F(s)}$ equals (a) $4 \ln (s + 3) - 4 \ln (s + 2)$; (b) $(2s + 50)/(s^2 + 10s + 50)$. (c) Find $f(3t)$ if $\mathbf{F(s)} = 4/(s + 2)^4$.

31 If $f_1(t) = (t - 1)u(t)$ and $f_2(t) = (t - 1)u(t - 1)$, use any method of your choice to find: (a) $\mathbf{F}_1(\mathbf{s})$; (b) $\mathbf{F}_2(\mathbf{s})$; (c) $f_1(t) * f_2(t)$.

32 Given the differential equation $v' + 6v + 9 \displaystyle\int_{0^-}^{t} v(z)\, dz = 24(t - 2)u(t - 2)$, let $v(0^-) = 0$ and find $v(t)$.

33 Find $f(0^+)$ and $f(\infty)$ for a time function whose Laplace transform is (a) $5(s^2 + 1)/(s^3 + 1)$; (b) $5(s^2 + 1)/(s^4 + 16)$; (c) $(s + 1)(1 + e^{-4s})/(s^2 + 2)$.

34 Find $f(\infty)$ and $f(0^+)$ for a time function whose Laplace transform is (a) $5(s^2 + 1)/(s + 1)^3$; (b) $5(s^2 + 1)/[s(s + 1)^3]$; (c) $(1 - e^{-3s})/s^2$.

35 Find both the initial and final values (or show that they do not exist) of the time functions corresponding to (a) $\dfrac{8s - 2}{s^2 + 6s + 10}$; (b) $\dfrac{2s^3 - s^2 - 3s - 5}{s^3 + 6s^2 + 10s}$; (c) $\dfrac{8s - 2}{s^2 - 6s + 10}$;

(d) $\dfrac{8s^2 - 2}{(s + 2)^2(s + 1)(s^2 + 6s + 10)}$.

36 Let $f(t) = (1/t) (e^{-at} - e^{-bt})u(t)$. (a) Find $\mathbf{F(s)}$. (b) Evaluate both sides of the equation $\displaystyle\lim_{t \to 0^+} f(t) = \lim_{s \to \infty} [\mathbf{sF(s)}]$.

37 Find $\mathbf{H(s)} = \mathbf{V}_o(\mathbf{s})/\mathbf{V}_s(\mathbf{s})$ and $h(t)$ for the circuit shown in Fig. 19-21. Assume that all initial conditions are zero.

Figure 19-21

See Prob. 37.

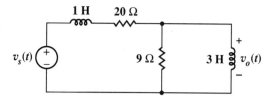

38 Network N in Fig. 19-22 contains no energy storage at $t = 0^-$. If $\mathbf{H}(\mathbf{s}) = \mathbf{I}_o(\mathbf{s})/\mathbf{V}_s(\mathbf{s}) = 0.3/(\mathbf{s} + 1)(\mathbf{s} + 2)$, find $i_o(t)$ when $v_s(t) = 2 \cos t \, u(t)$ V.

Figure 19-22

See Prob. 38.

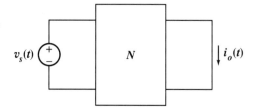

39 Refer to the circuit of Fig. 19-23 and find (a) $\mathbf{H}(\mathbf{s}) = \mathbf{I}_2(\mathbf{s})/\mathbf{V}_s(\mathbf{s})$; (b) $h(t)$; (c) $i_2(t)$ if $v_s(t) = 132u(t)$ V.

Figure 19-23

See Prob. 39.

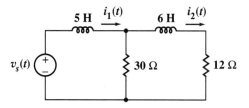

40 Find $i(t)$ in the circuit of Fig. 19-24.

Figure 19-24

See Prob. 40.

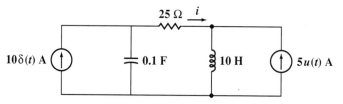

41 Two linear two-ports are connected in cascade. The input to the first stage is $x(t)$, its output $y(t)$ is the input to the second stage, and the second stage output is $z(t)$. Assume no initial energy storage. If $\mathbf{H}_1(\mathbf{s}) = \mathbf{Y}(\mathbf{s})/\mathbf{X}(\mathbf{s}) = 10/[\mathbf{s}(\mathbf{s} + 4)]$, while $\mathbf{H}_2(\mathbf{s}) = \mathbf{Z}(\mathbf{s})/\mathbf{Y}(\mathbf{s}) = 20(\mathbf{s} + 8)/[\mathbf{s}(\mathbf{s} + 10)]$, find $z(t)$ when $x(t) = \delta(t)$.

42 Determine $v_C(t)$ for the circuit of Fig. 19-25 if $v_s(t) = 4te^{-t}u(t)$ V.

Figure 19-25

See Prob. 42.

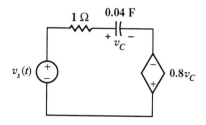

43 Let $v_s(t) = 8tu(t)$ V and $v_C(0^-) = 5$ V in the circuit shown in Fig. 19-26. (a) Draw a frequency-domain equivalent for this circuit which includes a voltage source to supply the initial capacitor voltage. (b) Find $\mathbf{V}_C(\mathbf{s})$ and $v_C(t)$.

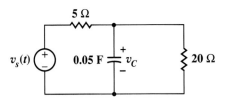

Figure 19-26

See Prob. 43.

44 Let $i_{L2}(0^-) = 2$ A and $i_{L3}(0^-) = 0$ in the circuit of Fig. 19-27. Find $v_o(t)$.

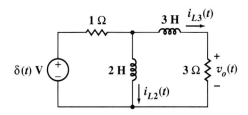

Figure 19-27

See Prob. 44.

45 Let $i_s(t) = 3tu(t)$ A and $v_C(0^-) = 8$ V in the circuit of Fig. 19-28. (a) Construct a frequency-domain equivalent circuit that includes a current source to supply $v_C(0^-)$. (b) Find $\mathbf{V}_C(\mathbf{s})$ and $v_C(t)$.

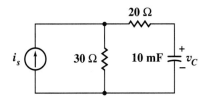

Figure 19-28

See Prob. 45.

46 Let $v_s = -5 + 12u(t) + 3\delta(t)$ V for the circuit shown in Fig. 19-29, and use Laplace transform techniques to find $i_L(t)$ for $t > 0$.

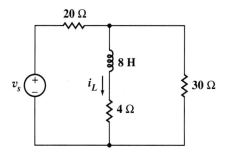

Figure 19-29

See Prob. 46.

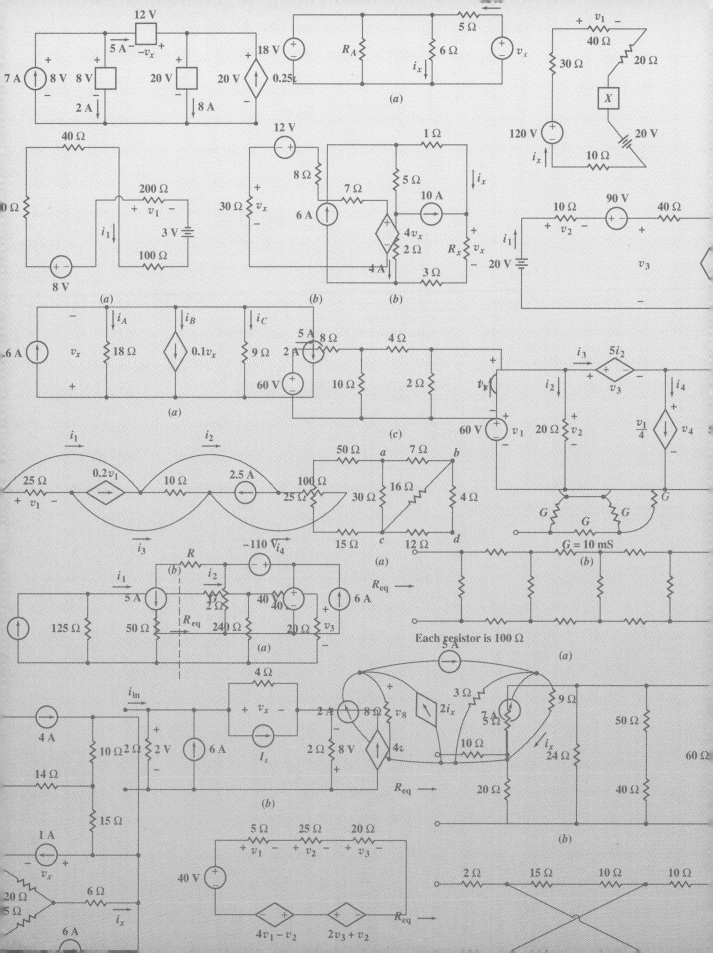

Part Seven:
Appendixes

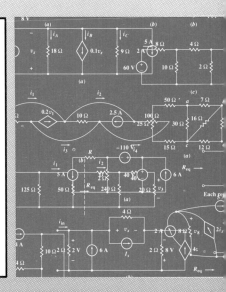

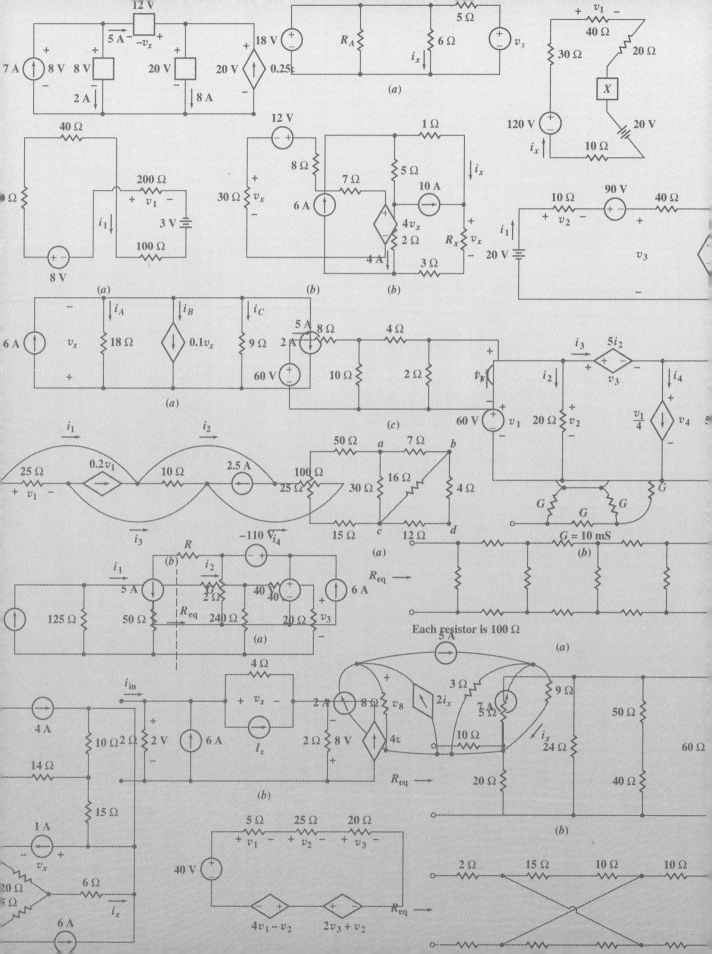

Determinants

In Sec. 2-2 of Chap. 2 we obtained a system of three equations for the four-node circuit of Fig. 2-2:

$$7v_1 - 3v_2 - 4v_3 = -11 \qquad (2\text{-}3)$$

$$-3v_1 + 6v_2 - 2v_3 = 3 \qquad (2\text{-}4)$$

$$-4v_1 - 2v_2 + 11v_3 = 25 \qquad (2\text{-}5)$$

This set of equations could have been solved by a systematic elimination of the variables. This procedure is lengthy, however, and may never yield answers if done unsystematically for a greater number of simultaneous equations. A much more orderly method involves using determinants and Cramer's rule, as discussed in most courses in college algebra. The use of determinants has the additional advantages that it leads naturally into the expression of the circuit elements in terms of matrices, and it establishes a method of analyzing a general circuit which will be helpful in proving general theorems. It should be pointed out that the number of arithmetic steps required to solve a large set of simultaneous equations by determinants is excessive; a digital computer would be programmed to use another method. This appendix consists of a brief review of the determinant method and nomenclature.

Consider Eqs. (2-3), (2-4), and (2-5). The array of the constant coefficients of the equations,

$$[\mathbf{G}] = \begin{bmatrix} 7 & -3 & -4 \\ -3 & 6 & -2 \\ -4 & -2 & 11 \end{bmatrix}$$

is called a *matrix*; the symbol $\mathbf{G}$ has been selected since each element of the matrix is a conductance value. A matrix has no "value"; it is merely an ordered array of elements. We use a letter in boldface type to represent a matrix, and we enclose the array itself by square brackets. Basic algebraic operations using matrices are discussed in Appendix 2.

The *determinant* of a square matrix *does* have a value, however. To be precise, we should say that the determinant of a matrix *is* a value, but common usage enables us to speak of both the array itself and its value as the determinant. We shall symbolize a determinant by Δ, and employ a suitable subscript to denote the matrix to which the determinant refers. Thus,

$$\Delta_G = \begin{vmatrix} 7 & -3 & -4 \\ -3 & 6 & -2 \\ -4 & -2 & 11 \end{vmatrix}$$

Note that simple vertical lines are used to enclose the determinant.

The value of any determinant is obtained by expanding it in terms of its minors. To do this, we select any row j or any column k, multiply each element in that row or column by its minor and by $(-1)^{j+k}$, and then add the products. The *minor* of the element appearing in both row j and column k is the determinant which is obtained when row j and column k are removed; it is indicated by Δ_{jk}.

As an example, let us expand the determinant Δ_G along column 3. We first multiply the (-4) at the top of this column by $(-1)^{1+3} = 1$ and then by its minor:

$$(-4)(-1)^{1+3} \begin{vmatrix} -3 & 6 \\ -4 & -2 \end{vmatrix}$$

and then repeat for the other two elements in column 3, adding the results:

$$\Delta_G = (-4) \begin{vmatrix} -3 & 6 \\ -4 & -2 \end{vmatrix} - (-2) \begin{vmatrix} 7 & -3 \\ -4 & -2 \end{vmatrix} + 11 \begin{vmatrix} 7 & -3 \\ -3 & 6 \end{vmatrix}$$

The minors now contain only two rows and two columns. They are of *order* 2, and their values are easily determined by expanding in terms of minors again, here a trivial operation. Thus, for the first determinant, we expand along the first column by multiplying (-3) by $(-1)^{1+1}$ and its minor, which is merely the element (-2), and then multiplying (-4) by $(-1)^{2+1}$ and by 6. Thus,

$$\begin{vmatrix} -3 & 6 \\ -4 & -2 \end{vmatrix} = (-3)(-2) - 6(-4) = 30$$

It is usually easier to remember the result for a second-order determinant as "upper left times lower right minus upper right times lower left." Finally,

$$\Delta_G = -4[(-3)(-2) - 6(-4)] + 2[7(-2) - (-3)(-4)] + 11[7(6) - (-3)(-3)]$$

$$= -4(30) + 2(-26) + 11(33)$$

$$= 191$$

For practice, let us expand this same determinant along the first row:

$$\Delta_G = 7 \begin{vmatrix} 6 & -2 \\ -2 & 11 \end{vmatrix} - (-3) \begin{vmatrix} -3 & -2 \\ -4 & 11 \end{vmatrix} + (-4) \begin{vmatrix} -3 & 6 \\ -4 & -2 \end{vmatrix}$$

$$= 7(62) + 3(-41) - 4(30)$$

$$= 191$$

The expansion by minors is valid for a determinant of any order.

Repeating these rules for evaluating a determinant in more general terms, we would say, given a matrix [**a**],

$$[\mathbf{a}] = \begin{bmatrix} a_{11} & a_{12} & \cdots & a_{1N} \\ a_{21} & a_{22} & \cdots & a_{2N} \\ \cdots & \cdots & \cdots & \cdots \\ a_{N1} & a_{N2} & \cdots & a_{NN} \end{bmatrix}$$

that Δ_a may be obtained by expansion in terms of minors along any row j:

$$\Delta_a = a_{j1}(-1)^{j+1}\Delta_{j1} + a_{j2}(-1)^{j+2}\Delta_{j2} + \cdots + a_{jN}(-1)^{j+N}\Delta_{jN}$$

$$= \sum_{n=1}^{N} a_{jn}(-1)^{j+n}\Delta_{jn}$$

or along any column k:

$$\Delta_a = a_{1k}(-1)^{1+k}\Delta_{1k} + a_{2k}(-1)^{2+k}\Delta_{2k} + \cdots + a_{Nk}(-1)^{N+k}\Delta_{Nk}$$

$$= \sum_{n=1}^{N} a_{nk}(-1)^{n+k}\Delta_{nk}$$

The *cofactor* C_{jk} of the element appearing in both row j and column k is simply $(-1)^{j+k}$ times the minor Δ_{jk}. Thus, $C_{11} = \Delta_{11}$, but $C_{12} = -\Delta_{12}$. We may now write

$$\Delta_a = \sum_{n=1}^{N} a_{jn}C_{jn} = \sum_{n=1}^{N} a_{nk}C_{nk}$$

As an example, let us consider this fourth-order determinant:

$$\Delta = \begin{vmatrix} 2 & -1 & -2 & 0 \\ -1 & 4 & 2 & -3 \\ -2 & -1 & 5 & -1 \\ 0 & -3 & 3 & 2 \end{vmatrix}$$

We find

$$\Delta_{11} = \begin{vmatrix} 4 & 2 & -3 \\ -1 & 5 & -1 \\ -3 & 3 & 2 \end{vmatrix} = 4(10 + 3) + 1(4 + 9) - 3(-2 + 15) = 26$$

$$\Delta_{12} = \begin{vmatrix} -1 & 2 & -3 \\ -2 & 5 & -1 \\ 0 & 3 & 2 \end{vmatrix} = -1(10 + 3) + 2(4 + 9) + 0 = 13$$

and $C_{11} = 26$, whereas $C_{12} = -13$. Finding the value of Δ for practice, we have

$$\Delta = 2C_{11} + (-1)C_{12} + (-2)C_{13} + 0$$

$$= 2(26) + (-1)(-13) + (-2)(3) + 0 = 59$$

We next consider Cramer's rule, which enables us to find the values of the unknown variables. Let us again consider Eqs. (2-3), (2-4), and (2-5); we define the determinant Δ_1 as that determinant which is obtained when the first column of Δ_G is replaced by the three constants on the right-hand sides of the three equations. Thus,

$$\Delta_1 = \begin{vmatrix} -11 & -3 & -4 \\ 3 & 6 & -2 \\ 25 & -2 & 11 \end{vmatrix}$$

We expand along the first column:

$$\Delta_1 = -11\begin{vmatrix} 6 & -2 \\ -2 & 11 \end{vmatrix} - 3\begin{vmatrix} -3 & -4 \\ -2 & 11 \end{vmatrix} + 25\begin{vmatrix} -3 & -4 \\ 6 & -2 \end{vmatrix}$$

$$= -682 + 123 + 750 = 191$$

Cramer's rule then states that

$$v_1 = \frac{\Delta_1}{\Delta_G} = \frac{191}{191} = 1 \text{ V}$$

and

$$v_2 = \frac{\Delta_2}{\Delta_G} = \frac{\begin{vmatrix} 7 & -11 & -4 \\ -3 & 3 & -2 \\ -4 & 25 & 11 \end{vmatrix}}{191} = \frac{581 - 63 - 136}{191} = 2 \text{ V}$$

and finally,

$$v_3 = \frac{\Delta_3}{\Delta_G} = \frac{\begin{vmatrix} 7 & -3 & -11 \\ -3 & 6 & 3 \\ -4 & -2 & 25 \end{vmatrix}}{191} = \frac{1092 - 291 - 228}{191} = 3 \text{ V}$$

Cramer's rule is applicable to a system of N simultaneous linear equations in N unknowns; for the ith variable v_i:

$$v_i = \frac{\Delta_i}{\Delta_G}$$

Drill Problem

A1-1. Evaluate: (a) $\begin{vmatrix} 2 & -3 \\ -2 & 5 \end{vmatrix}$; (b) $\begin{vmatrix} 1 & -1 & 0 \\ 4 & 2 & -3 \\ 3 & -2 & 5 \end{vmatrix}$; (c) $\begin{vmatrix} 2 & -3 & 1 & 5 \\ -3 & 1 & -1 & 0 \\ 0 & 4 & 2 & -3 \\ 6 & 3 & -2 & 5 \end{vmatrix}$.

(d) Find i_2 if $5i_1 - 2i_2 - i_3 = 100$, $-2i_1 + 6i_2 - 3i_3 - i_4 = 0$, $-i_1 - 3i_2 + 4i_3 - i_4 = 0$, and $-i_2 - i_3 = 0$.   *Ans:* 4; 33; −411; 1.266

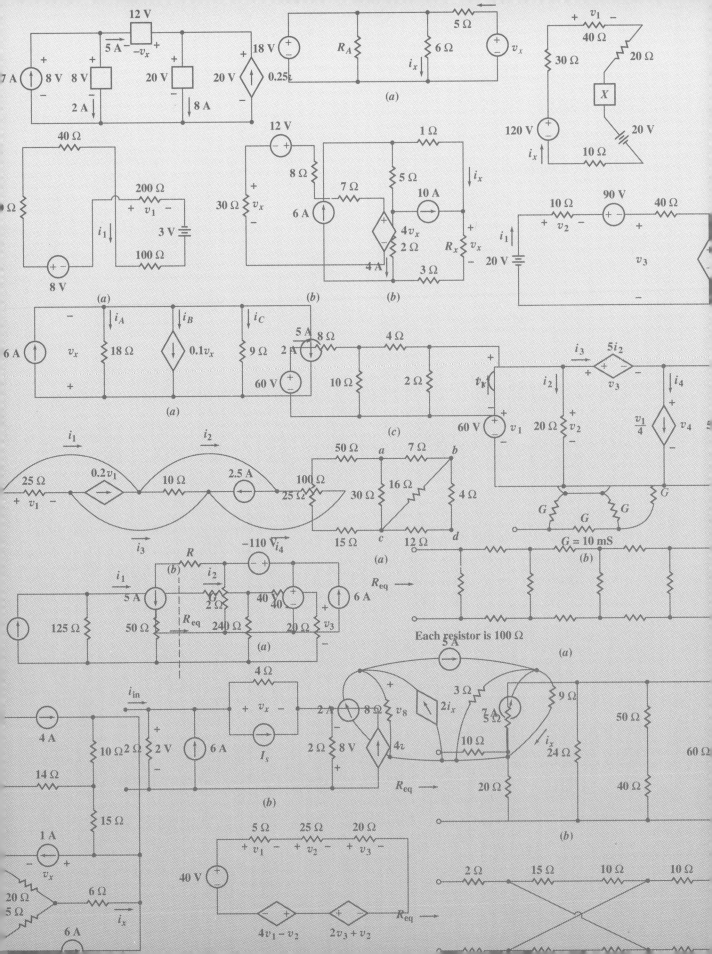

Matrices

We defined a matrix in Appendix 1 as an ordered array of elements, such as the conductance matrix $[\mathbf{G}]$,

$$[\mathbf{G}] = \begin{bmatrix} 7 & -3 & -4 \\ -3 & 6 & -2 \\ -4 & -2 & 11 \end{bmatrix}$$

We use a letter in boldface type to represent a matrix and enclose the symbol within square brackets, as we do the array itself.[1]

A matrix having m rows and n columns is called an $(m \times n)$ (pronounced "m by n") matrix. Thus,

$$[\mathbf{A}] = \begin{bmatrix} 2 & 0 & 5 \\ -1 & 6 & 3 \end{bmatrix}$$

is a (2×3) matrix, and the $[\mathbf{G}]$ matrix of our example is a (3×3) matrix. An $(n \times n)$ matrix is a *square matrix* of order n.

An $(m \times 1)$ matrix is called a *column matrix,* or a *vector.* Thus,

$$[\mathbf{V}] = \begin{bmatrix} \mathbf{V}_1 \\ \mathbf{V}_2 \end{bmatrix}$$

is a (2×1) column matrix of phasor voltages, and

$$[\mathbf{I}] = \begin{bmatrix} \mathbf{I}_1 \\ \mathbf{I}_2 \end{bmatrix}$$

is a (2×1) phasor-current vector.

A $(1 \times n)$ matrix is known as a *row vector.*

Two $(m \times n)$ matrices are equal if their corresponding elements are equal. Thus, if a_{jk} is that element of $[\mathbf{A}]$ located in row j and column k and b_{jk} is the element at row j and column k in matrix $[\mathbf{B}]$, then $[\mathbf{A}] = [\mathbf{B}]$ if and only if $a_{jk} = b_{jk}$ for all $1 \le j \le m$ and $1 \le k \le n$. Thus, if

$$\begin{bmatrix} \mathbf{V}_1 \\ \mathbf{V}_2 \end{bmatrix} = \begin{bmatrix} \mathbf{z}_{11}\mathbf{I}_1 + \mathbf{z}_{12}\mathbf{I}_2 \\ \mathbf{z}_{21}\mathbf{I}_1 + \mathbf{z}_{22}\mathbf{I}_2 \end{bmatrix}$$

then $\mathbf{V}_1 = \mathbf{z}_{11}\mathbf{I}_1 + \mathbf{z}_{12}\mathbf{I}_2$ and $\mathbf{V}_2 = \mathbf{z}_{21}\mathbf{I}_1 + \mathbf{z}_{22}\mathbf{I}_2$.

[1] Chapter 16 uses a simpler notation, as described in Sec. 16-4.

Two $(m \times n)$ matrices may be added by adding corresponding elements. Thus,

$$\begin{bmatrix} 2 & 0 & 5 \\ -1 & 6 & 3 \end{bmatrix} + \begin{bmatrix} 1 & 2 & 3 \\ -3 & -2 & -1 \end{bmatrix} = \begin{bmatrix} 3 & 2 & 8 \\ -4 & 4 & 2 \end{bmatrix}$$

An $(n \times n)$ square matrix has a determinant whose value is determined by the rules given in Appendix 1 for expanding determinants. The determinant of $[\mathbf{G}]$ is symbolized by Δ_G or det $[\mathbf{G}]$. Therefore, if

$$[\mathbf{G}] = \begin{bmatrix} 7 & -3 & -4 \\ -3 & 6 & -2 \\ -4 & -2 & 11 \end{bmatrix}$$

then

$$\Delta_G = \det [\mathbf{G}] = \begin{vmatrix} 7 & -3 & -4 \\ -3 & 6 & -2 \\ -4 & -2 & 11 \end{vmatrix}$$

Expanding on the first column, we find

$$\det [\mathbf{G}] = 7(66 - 4) + 3(-33 - 8) - 4(6 + 24) = 191$$

There is only one other matrix operation that is needed in our work: matrix multiplication. Let us consider the matrix product $[\mathbf{A}][\mathbf{B}]$, where $[\mathbf{A}]$ is an $(m \times n)$ matrix and $[\mathbf{B}]$ is a $(p \times q)$ matrix. If $n = p$, the matrices are said to be *conformal,* and their product exists. That is, matrix multiplication is defined only for the case where the number of columns of the first matrix in the product is equal to the number of rows in the second matrix.

The formal definition of matrix multiplication states that the product of the $(m \times n)$ matrix $[\mathbf{A}]$ and the $(n \times q)$ matrix $[\mathbf{B}]$ is an $(m \times q)$ matrix having elements c_{jk}, $1 \le j \le m$ and $1 \le k \le q$, where

$$c_{jk} = a_{j1}b_{1k} + a_{j2}b_{2k} + \cdots + a_{jn}b_{nk}$$

That is, to find the element in the second row and third column of the product, we multiply each of the elements in the second row of $[\mathbf{A}]$ by the corresponding element in the third column of $[\mathbf{B}]$ and then add the n results. For example, given the (2×3) matrix $[\mathbf{A}]$ and the (3×2) matrix $[\mathbf{B}]$,

$$\begin{bmatrix} a_{11} & a_{12} & a_{13} \\ a_{21} & a_{22} & a_{23} \end{bmatrix} \begin{bmatrix} b_{11} & b_{12} \\ b_{21} & b_{22} \\ b_{31} & b_{32} \end{bmatrix}$$

$$= \begin{bmatrix} (a_{11}b_{11} + a_{12}b_{21} + a_{13}b_{31}) & (a_{11}b_{12} + a_{12}b_{22} + a_{13}b_{32}) \\ (a_{21}b_{11} + a_{22}b_{21} + a_{23}b_{31}) & (a_{21}b_{12} + a_{22}b_{22} + a_{23}b_{32}) \end{bmatrix}$$

The result is a (2×2) matrix.

As a numerical example of matrix multiplication, we take

$$\begin{bmatrix} 3 & 2 & 1 \\ -2 & -2 & 4 \end{bmatrix} \begin{bmatrix} 2 & 3 \\ -2 & -1 \\ 4 & -3 \end{bmatrix} = \begin{bmatrix} 6 & 4 \\ 16 & -16 \end{bmatrix}$$

where $6 = (3)(2) + (2)(-2) + (1)(4)$, $4 = (3)(3) + (2)(-1) + (1)(-3)$, and so forth.

Matrix multiplication is not commutative. For example, given the (3×2)

matrix $[\mathbf{C}]$ and the (2×1) matrix $[\mathbf{D}]$, it is evident that the product $[\mathbf{C}][\mathbf{D}]$ may be calculated, but the product $[\mathbf{D}][\mathbf{C}]$ is not even defined.

As a final example, let

$$[\mathbf{t}_A] = \begin{bmatrix} 2 & 3 \\ -1 & 4 \end{bmatrix}$$

and

$$[\mathbf{t}_B] = \begin{bmatrix} 3 & 1 \\ 5 & 0 \end{bmatrix}$$

so that both $[\mathbf{t}_A][\mathbf{t}_B]$ and $[\mathbf{t}_B][\mathbf{t}_A]$ are defined. However,

$$[\mathbf{t}_A][\mathbf{t}_B] = \begin{bmatrix} 21 & 2 \\ 17 & -1 \end{bmatrix}$$

while

$$[\mathbf{t}_B][\mathbf{t}_A] = \begin{bmatrix} 5 & 13 \\ 10 & 15 \end{bmatrix}$$

Drill Problem

A2-1. Given $[\mathbf{A}] = \begin{bmatrix} 1 & -3 \\ 3 & 5 \end{bmatrix}$, $[\mathbf{B}] = \begin{bmatrix} 4 & -1 \\ -2 & 3 \end{bmatrix}$, $[\mathbf{C}] = \begin{bmatrix} 50 \\ 30 \end{bmatrix}$, and $[\mathbf{V}] = \begin{bmatrix} \mathbf{V}_1 \\ \mathbf{V}_2 \end{bmatrix}$, find (a) $[\mathbf{A}] + [\mathbf{B}]$; (b) $[\mathbf{A}][\mathbf{B}]$; (c) $[\mathbf{B}][\mathbf{A}]$; (d) $[\mathbf{A}][\mathbf{V}] + [\mathbf{B}][\mathbf{C}]$; (e) $[\mathbf{A}]^2 = [\mathbf{A}][\mathbf{A}]$; (f) det $[\mathbf{A}]$.

$$Ans: \begin{bmatrix} 5 & -4 \\ 1 & 8 \end{bmatrix}; \begin{bmatrix} 10 & -10 \\ 2 & 12 \end{bmatrix}; \begin{bmatrix} 1 & -17 \\ 7 & 21 \end{bmatrix}; \begin{bmatrix} \mathbf{V}_1 - 3\mathbf{V}_2 + 170 \\ 3\mathbf{V}_1 + 5\mathbf{V}_2 - 10 \end{bmatrix};$$

$$\begin{bmatrix} -8 & -18 \\ 18 & 16 \end{bmatrix}; 14$$

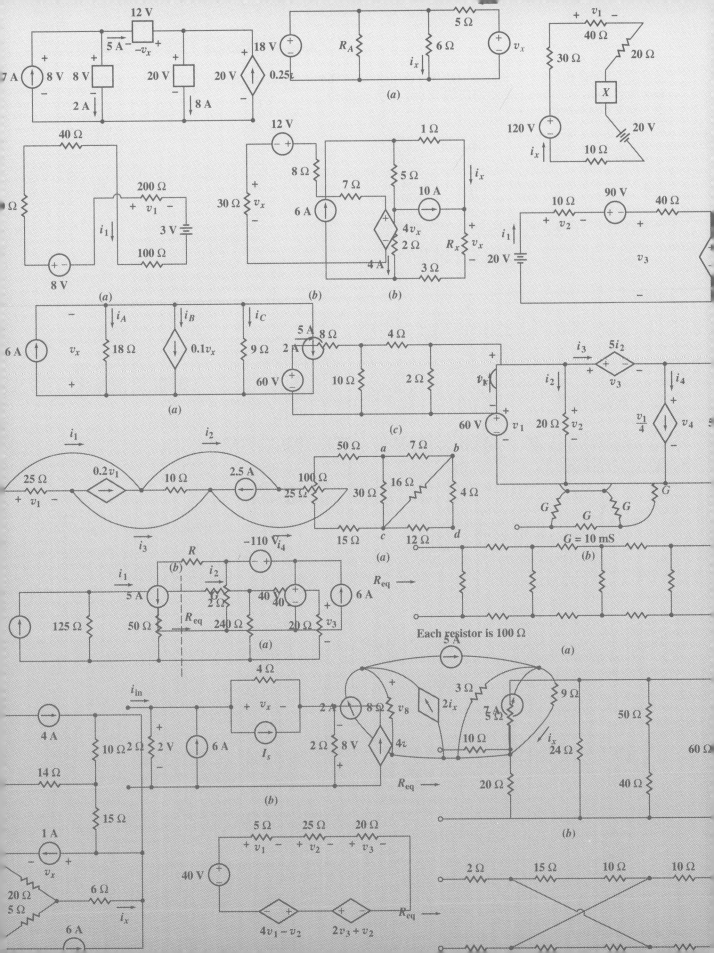

A Proof of Thévenin's Theorem

We shall prove Thévenin's theorem in the same form in which it is stated in Sec. 2-6 of Chap. 2, repeated here for reference:

> Given any linear circuit, rearrange it in the form of two networks A and B that are connected together by two resistanceless conductors. If either network contains a dependent source, its control variable must be in that same network. Define a voltage v_{oc} as the open-circuit voltage which would appear across the terminals of A if B were disconnected so that no current is drawn from A. Then all the currents and voltages in B will remain unchanged if A is killed (all independent voltage sources and independent current sources in A replaced by short circuits and open circuits, respectively) and an independent voltage source v_{oc} is connected, with proper polarity, in series with the dead (inactive) A network.

We shall effect our proof by showing that the original A network and the Thévenin equivalent of the A network both cause the same current to flow into the terminals of the B network. If the currents are the same, then the voltages must be the same; in other words, if we apply a certain current, which we might think of as a current source, to the B network, then the current source and the B network constitute a circuit which has a specific input voltage as a response. Thus, the current determines the voltage. Alternatively we could, if we wished, show that the terminal voltage at B is unchanged, because the voltage also determines the current uniquely. If the input voltage and current to the B network are unchanged, then it follows that the currents and voltages *throughout* the B network are also unchanged.

Let us first prove the theorem for the case where the B network is inactive (no independent sources). After this step has been accomplished, we may then use the superposition principle to extend the theorem to include B networks which contain independent sources. Each network may contain dependent sources, provided that their control variables are in the same network.

The current i, flowing in the upper conductor from the A network to the B network in Fig. A3-1a, is therefore caused entirely by the independent sources present in the A network. Suppose now that we add an additional voltage source v_x, which we shall call the Thévenin source, in the conductor in which i is

Figure A3-1

(a) A general linear network A and a network B that contains no independent sources. Controls for dependent sources must appear in the same part of the network. (b) The Thévenin source is inserted in the circuit and adjusted until $i = 0$. No voltage appears across network B and thus $v_x = v_{oc}$. The Thévenin source thus produces a current $-i$ while network A provides i. (c) The Thévenin source is reversed and network A is killed. The current is therefore i.

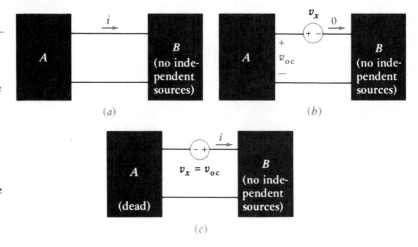

measured, as shown in Fig. A3-1b, and then adjust the magnitude and time variation of v_x until the current is reduced to zero. By our definition of v_{oc}, then, the voltage across the terminals of A must be v_{oc}, since $i = 0$. Network B contains no independent sources, and no current is entering its terminals; therefore, there is no voltage across the terminals of the B network, and by Kirchhoff's voltage law the voltage of the Thévenin source is v_{oc} volts, $v_x = v_{oc}$. Moreover, since the Thévenin source and the A network jointly deliver no current to B, and since the A network by itself delivers a current i, superposition requires that the Thévenin source acting by itself deliver a current of $-i$ to B. The source acting alone in a reversed direction, as shown in Fig. A3-1c, therefore produces a current i in the upper lead. This situation, however, is the same as the conclusion reached by Thévenin's theorem: the Thévenin source v_{oc} acting in series with the inactive A network is equivalent to the given network.

Now let us consider the case where the B network may be an active network. We now think of the current i, flowing from the A network to the B network in the upper conductor, as being composed of two parts, i_A and i_B, where i_A is the current produced by A acting alone and the current i_B is due to B acting alone.

Figure A3-2

Superposition enables the current i to be considered as the sum of two partial responses.

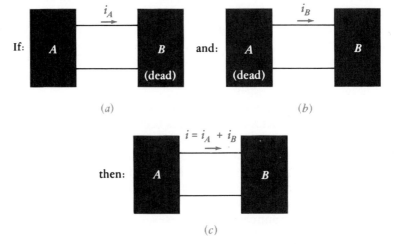

Our ability to divide the current into these two components is a direct consequence of the applicability of the superposition principle to these two *linear* networks; the complete response and the two partial responses are indicated by the diagrams of Fig. A3-2.

The partial response i_A has already been considered; if network B is inactive, we know that network A may be replaced by the Thévenin source and the inactive A network. In other words, of the three sources which we must keep in mind—those in A, those in B, and the Thévenin source—the partial response i_A occurs when A and B are dead and the Thévenin source is active. Preparing for the use of superposition, we now let A remain inactive, but turn on B and turn off the Thévenin source; by definition, the partial response i_B is obtained. Superimposing the results, the response when A is dead and both the Thévenin source and B are active is $i_A + i_B$. This sum is the original current i, and the situation wherein the Thévenin source and B are active but A is dead is the desired Thévenin equivalent circuit. Thus the active network A may be replaced by its Thévenin source, the open-circuit voltage, in series with the inactive A network, regardless of the status of the B network; it may be either active or inactive.

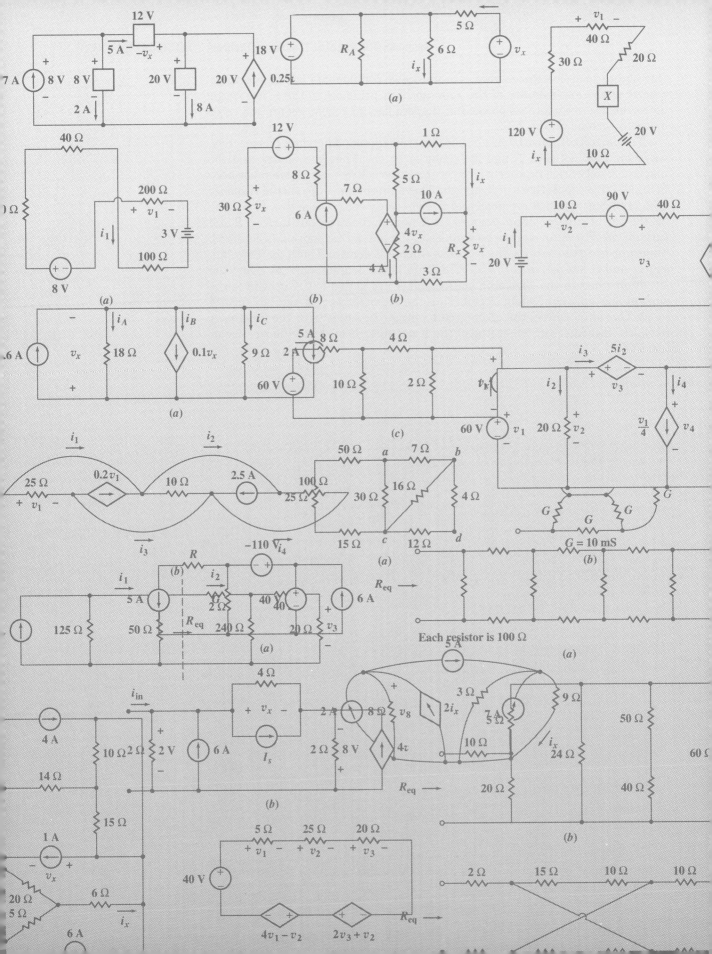

Complex Numbers

This appendix includes sections covering the definition of a complex number, the basic arithmetic operations for complex numbers, Euler's identity, and the exponential and polar forms of the complex number. We first introduce the concept of a complex number.

The complex number

Our early training in mathematics dealt exclusively with real numbers, such as 4, $-\frac{2}{7}$, and π. Soon, however, we began to encounter algebraic equations, such as $x^2 = -3$, which could not be satisfied by any real number. Such an equation can be solved only through the introduction of the *imaginary unit,* or the *imaginary operator,* which we shall designate[1] by the symbol j. By definition, $j^2 = -1$, and thus $j = \sqrt{-1}$, $j^3 = -j$, $j^4 = 1$, and so forth. The product of a real number and the imaginary operator is called an *imaginary number,* and the sum of a real number and an imaginary number is called a *complex number.* Thus, a number having the form $a + jb$, where a and b are real numbers, is a complex number.[2]

We shall designate a complex number by means of a special single symbol; thus, $\mathbf{A} = a + jb$. The complex nature of the number is indicated by the use of boldface type; in handwritten material, a bar over the letter is customary. The complex number $\mathbf{A}$ just shown is described as having a *real component* or *real part a* and an *imaginary component* or *imaginary part b*. This is also expressed as

$$\text{Re}\,[\mathbf{A}] = a \qquad \text{Im}\,[\mathbf{A}] = b$$

The imaginary component of $\mathbf{A}$ is *not jb*. By definition, the imaginary component is a real number.

It should be noted that all real numbers may be regarded as complex numbers having imaginary parts equal to zero. The real numbers are therefore

---

[1] Mathematicians designate the imaginary operator by the symbol i, but it is customary to use j in electrical engineering in order to avoid confusion with the symbol for current.

[2] The choice of the words *imaginary* and *complex* is unfortunate. They are used here and in the mathematical literature as technical terms to designate a class of numbers. To interpret imaginary as "not pertaining to the physical world" or complex as "complicated" is neither justified nor intended.

included in the system of complex numbers, and we may now consider them as a special case. When we define the fundamental arithmetic operations for complex numbers, we should therefore expect them to reduce to the corresponding definitions for real numbers if the imaginary part of every complex number is set equal to zero.

Since any complex number is completely characterized by a pair of real numbers, such as a and b in the previous example, we can obtain some visual assistance by representing a complex number graphically on a rectangular, or cartesian, coordinate system. By providing ourselves with a real axis and an imaginary axis, as shown in Fig. A4-1, we form a *complex plane,* or *Argand diagram,* on which any complex number can be represented as a single point. The complex numbers $\mathbf{M} = 3 + j1$ and $\mathbf{N} = 2 - j2$ are indicated. It is important to understand that this complex plane is only a visual aid; it is not at all essential to the mathematical statements which follow.

Figure A4-1

The complex numbers $\mathbf{M} = 3 + j1$ and $\mathbf{N} = 2 - j2$ are shown on the complex plane.

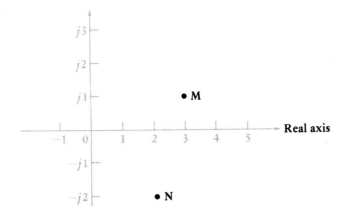

We shall define two complex numbers as being equal if, and only if, their real parts are equal and their imaginary parts are equal. Graphically, then, to each point in the complex plane there corresponds only one complex number, and conversely, to each complex number there corresponds only one point in the complex plane. Thus, suppose we are given the two complex numbers:

$$\mathbf{A} = a + jb \qquad \text{and} \qquad \mathbf{B} = c + jd$$

Then, if

$$\mathbf{A} = \mathbf{B}$$

it is necessary that

$$a = c \qquad \text{and} \qquad b = d$$

A complex number expressed as the sum of a real number and an imaginary number, such as $\mathbf{A} = a + jb$, is said to be in *rectangular* or *cartesian* form. Other forms for a complex number will appear shortly.

Let us now define the fundamental operations of addition, subtraction, multiplication, and division for complex numbers. The sum of two complex numbers is defined as the complex number whose real part is the sum of the real parts of the two complex numbers and whose imaginary part is the sum of the imaginary parts of the two complex numbers. Thus,

$$(a + jb) + (c + jd) = (a + c) + j(b + d)$$

For example,

$$(3 + j4) + (4 - j2) = 7 + j2$$

The difference of two complex numbers is taken in a similar manner; for example,

$$(3 + j4) - (4 - j2) = -1 + j6$$

Addition and subtraction of complex numbers may also be accomplished graphically on the complex plane. Each complex number is represented as a vector, or directed line segment, and the sum is obtained by completing the parallelogram, illustrated by Fig. A4-2a, or by connecting the vectors in a head-to-tail manner, as shown in Fig. A4-2b. A graphical sketch is often useful as a check for a more exact numerical solution.

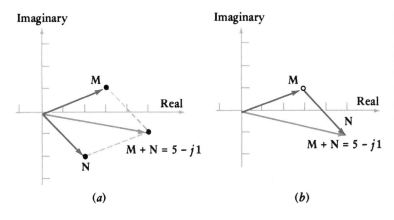

(a) (b)

Figure A4-2

(a) The sum of the complex numbers $\mathbf{M} = 3 + j1$ and $\mathbf{N} = 2 - j2$ is obtained by constructing a parallelogram. (b) The sum of the same two complex numbers is found by a head-to-tail combination.

The product of two complex numbers is defined by

$$(a + jb)(c + jd) = (ac - bd) + j(bc + ad)$$

This result may be easily obtained by a direct multiplication of the two binomial terms, using the rules of the algebra of real numbers, and then simplifying the result by letting $j^2 = -1$. For example,

$$(3 + j4)(4 - j2) = 12 - j6 + j16 - 8j^2$$
$$= 12 + j10 + 8$$
$$= 20 + j10$$

It is easier to multiply the complex numbers by this method, particularly if we immediately replace j^2 by -1, than it is to substitute in the general formula which defines the multiplication.

Before defining the operation of division for complex numbers, we should define the conjugate of a complex number. The *conjugate* of the complex number $\mathbf{A} = a + jb$ is $a - jb$ and is represented as $\mathbf{A}^*$. The conjugate of any complex number is therefore easily obtained by merely changing the sign of the imaginary part of the complex number. Thus, if

$$\mathbf{A} = 5 + j3$$

then

$$\mathbf{A}^* = 5 - j3$$

It is evident that the conjugate of any complicated complex expression may be found by replacing every complex term in the expression by its conjugate, which may be obtained by replacing every j in the expression by $-j$.

The definitions of addition, subtraction, and multiplication show that the following statements are true: the sum of a complex number and its conjugate is a real number; the difference of a complex number and its conjugate is an imaginary number; and the product of a complex number and its conjugate is a real number. It is also evident that if $\mathbf{A}^*$ is the conjugate of $\mathbf{A}$, then $\mathbf{A}$ is the conjugate of $\mathbf{A}^*$; in other words, $\mathbf{A} = (\mathbf{A}^*)^*$. A complex number and its conjugate are said to form a *conjugate complex pair* of numbers.

We now define the quotient of two complex numbers:

$$\frac{\mathbf{A}}{\mathbf{B}} = \frac{(\mathbf{A})(\mathbf{B}^*)}{(\mathbf{B})(\mathbf{B}^*)}$$

and thus

$$\frac{a + jb}{c + jd} = \frac{(ac + bd) + j(bc - ad)}{c^2 + d^2}$$

We multiply numerator and denominator by the conjugate of the denominator in order to obtain a denominator which is real; this process is called *rationalizing the denominator*. As a numerical example,

$$\frac{3 + j4}{4 - j2} = \frac{(3 + j4)(4 + j2)}{(4 - j2)(4 + j2)}$$

$$= \frac{4 + j22}{16 + 4} = 0.2 + j1.1$$

The addition or subtraction of two complex numbers which are each expressed in rectangular form is a relatively simple operation; multiplication or division of two complex numbers in rectangular form, however, is a rather unwieldy process. These latter two operations will be found to be much simpler when the complex numbers are given in either exponential or polar form. These forms will be introduced in Secs. A4-3 and A4-4.

Drill Problems

A4-1. Let $\mathbf{A} = -4 + j5$, $\mathbf{B} = 3 - j2$, and $\mathbf{C} = -6 - j5$, and find (a) $\mathbf{C} - \mathbf{B}$; (b) $2\mathbf{A} - 3\mathbf{B} + 5\mathbf{C}$; (c) $j^5\mathbf{C}^2(\mathbf{A} + \mathbf{B})$; (d) $\mathbf{B}$ Re $[\mathbf{A}] + \mathbf{A}$ Re $[\mathbf{B}]$.

Ans: $-9 - j3$; $-47 - j9$; $27 - j191$; $-24 + j23$

A4-2. Using the same values for $\mathbf{A}$, $\mathbf{B}$, and $\mathbf{C}$ as in the previous problem, find (a) $[(\mathbf{A} - \mathbf{A}^*)(\mathbf{B} + \mathbf{B}^*)^*]^*$; (b) $(1/\mathbf{C}) - (1/\mathbf{B})^*$; (c) $(\mathbf{B} + \mathbf{C})/(2\mathbf{BC})$.

Ans: $-j60$; $-0.329 + j0.236$; $0.0662 + j0.1179$

A4-2
Euler's identity

In Chap. 8 we encounter functions of time which contain complex numbers, and we are concerned with the differentiation and integration of these functions with respect to the real variable t. We differentiate and integrate such functions with respect to t by exactly the same procedures we use for real functions of time. That is, the complex constants are treated just as though they were real constants when performing the operations of differentiation or integration. If

$\mathbf{f}(t)$ is a complex function of time, such as

$$\mathbf{f}(t) = a \cos ct + jb \sin ct$$

then

$$\frac{d\mathbf{f}(t)}{dt} = -ac \sin ct + jbc \cos ct$$

and

$$\int \mathbf{f}(t)\,dt = \frac{a}{c} \sin ct - j\frac{b}{c} \cos ct + \mathbf{C}$$

where the constant of integration $\mathbf{C}$ is a complex number in general.

In Chap. 19 it is necessary to differentiate or integrate a function of a complex variable with respect to that complex variable. In general, the successful accomplishment of either of these operations requires that the function which is to be differentiated or integrated satisfy certain conditions. All our functions do meet these conditions, and integration or differentiation with respect to a complex variable is achieved by using methods identical to those used for real variables.

At this time we must make use of a very important fundamental relationship known as Euler's identity (pronounced "oilers"). We shall prove this identity, for it is extremely useful in representing a complex number in a form other than rectangular form.

The proof is based on the power series expansions of $\cos\theta$, $\sin\theta$, and e^z, given toward the back of your favorite college calculus text:

$$\cos\theta = 1 - \frac{\theta^2}{2!} + \frac{\theta^4}{4!} - \frac{\theta^6}{6!} + \cdots$$

$$\sin\theta = \theta - \frac{\theta^3}{3!} + \frac{\theta^5}{5!} - \frac{\theta^7}{7!} + \cdots$$

or

$$\cos\theta + j\sin\theta = 1 + j\theta - \frac{\theta^2}{2!} - j\frac{\theta^3}{3!} + \frac{\theta^4}{4!} + j\frac{\theta^5}{5!} - \cdots$$

and

$$e^z = 1 + z + \frac{z^2}{2!} + \frac{z^3}{3!} + \frac{z^4}{4!} + \frac{z^5}{5!} + \cdots$$

so that

$$e^{j\theta} = 1 + j\theta - \frac{\theta^2}{2!} - j\frac{\theta^3}{3!} + \frac{\theta^4}{4!} + \cdots$$

We conclude that

$$e^{j\theta} = \cos\theta + j\sin\theta \tag{1}$$

or, if we let $z = -j\theta$, we find that

$$e^{-j\theta} = \cos\theta - j\sin\theta \tag{2}$$

By adding and subtracting Eqs. (1) and (2), we obtain the two expressions which we used without proof in our study of the underdamped natural response of the parallel and series RLC circuits,

$$\cos\theta = \tfrac{1}{2}(e^{j\theta} + e^{-j\theta}) \tag{3}$$

$$\sin\theta = -j\tfrac{1}{2}(e^{j\theta} - e^{-j\theta}) \tag{4}$$

Drill Problems

A4-3. Use Eqs. (1) through (4) to evaluate: (a) e^{-j1}; (b) e^{1-j1}; (c) $\cos(-j1)$; (d) $\sin(-j1)$.　　　　　Ans: $0.540 - j0.841$; $1.469 - j2.29$; 1.543; $-j1.175$

A4-4. Evaluate at $t = 0.5$: (a) $(d/dt)(3\cos 2t - j2\sin 2t)$; (b) $\int_0^t (3\cos 2t - j2\sin 2t)\,dt$. Evaluate at $\mathbf{s} = 1 + j2$: (c) $\int_{\mathbf{s}}^{\infty} \mathbf{s}^{-3}\,d\mathbf{s}$; (d) $(d/d\mathbf{s})[3/(\mathbf{s}+2)]$.

Ans: $-5.05 - j2.16$; $1.262 - j0.460$; $-0.06 - j0.08$; $-0.0888 + j0.213$

A4-3

The exponential form

Let us now take Euler's identity

$$e^{j\theta} = \cos\theta + j\sin\theta$$

and multiply each side by the real positive number C:

$$Ce^{j\theta} = C\cos\theta + jC\sin\theta \tag{5}$$

The right-hand side of Eq. (5) consists of the sum of a real number and an imaginary number and thus represents a complex number in rectangular form; let us call this complex number $\mathbf{A}$, where $\mathbf{A} = a + jb$. By equating the real parts

$$a = C\cos\theta \tag{6}$$

and the imaginary parts

$$b = C\sin\theta \tag{7}$$

then squaring and adding Eqs. (6) and (7),

$$a^2 + b^2 = C^2$$

or

$$C = +\sqrt{a^2 + b^2} \tag{8}$$

and dividing Eq. (7) by Eq. (6):

$$\frac{b}{a} = \tan\theta$$

or

$$\theta = \tan^{-1}\frac{b}{a} \tag{9}$$

we obtain the relationships of Eqs. (8) and (9), which enable us to determine C and θ from a knowledge of a and b. For example, if $\mathbf{A} = 4 + j2$, then we identify a as 4 and b as 2 and find C and θ:

$$C = \sqrt{4^2 + 2^2} = 4.47$$

$$\theta = \tan^{-1}\tfrac{2}{4} = 26.6°$$

We could use this new information to write $\mathbf{A}$ in the form

$$\mathbf{A} = 4.47\cos 26.6° + j4.47\sin 26.6°$$

but it is the form of the left-hand side of Eq. (5) which will prove to be the more useful:

$$\mathbf{A} = Ce^{j\theta} = 4.47e^{j26.6°}$$

A complex number expressed in this manner is said to be in *exponential form*. The real positive multiplying factor C is known as the *amplitude* or *magnitude*, and the real quantity θ appearing in the exponent is called the *argument* or *angle*. A mathematician would always express θ in radians and would write

$$\mathbf{A} = 4.47e^{j0.464}$$

but engineers customarily work in terms of degrees. The use of the degree symbol (°) in the exponent should make confusion impossible.

To recapitulate, if we have a complex number which is given in rectangular form,

$$\mathbf{A} = a + jb$$

and wish to express it in exponential form,

$$\mathbf{A} = Ce^{j\theta}$$

we may find C and θ by Eqs. (8) and (9). If we are given the complex number in exponential form, then we may find a and b by Eqs. (6) and (7).

When $\mathbf{A}$ is expressed in terms of numerical values, the transformation between exponential (or polar) and rectangular forms is available as a built-in operation on most hand-held scientific calculators.

One question will be found to arise in the determination of the angle θ by using the arctangent relationship of Eq. (9). This function is multivalued, and an appropriate angle must be selected from various possibilities. One method by which the choice may be made is to select an angle for which the sine and cosine have the proper signs to produce the required values of a and b from Eqs. (6) and (7). For example, let us convert

$$\mathbf{V} = 4 - j3$$

to exponential form. The amplitude is

$$C = \sqrt{4^2 + (-3)^2} = 5$$

and the angle is

$$\theta = \tan^{-1}\frac{-3}{4} \tag{10}$$

A value of θ has to be selected which leads to a positive value for $\cos\theta$, since $4 = 5\cos\theta$, and a negative value for $\sin\theta$, since $-3 = 5\sin\theta$. We therefore obtain $\theta = -36.9°$, $323.1°$, $-396.9°$, and so forth. Any of these angles is correct, and we usually select that one which is the simplest, here, $-36.9°$. We should note that the solution of Eq. (10), $\theta = 143.1°$, is not correct, because $\cos\theta$ is negative and $\sin\theta$ is positive.

A simpler method of selecting the correct angle is available if we represent the complex number graphically in the complex plane. Let us first select a complex number, given in rectangular form, $\mathbf{A} = a + jb$, which lies in the first quadrant of the complex plane, as illustrated in Fig. A4-3. If we draw a line from the origin to the point which represents the complex number, we shall have constructed a right triangle whose hypotenuse is evidently the amplitude of the exponential representation of the complex number. In other words, $C = \sqrt{a^2 + b^2}$. Moreover, the counterclockwise angle which the line makes with the positive real axis is seen to be the angle θ of the exponential representation, because $a = C\cos\theta$ and $b = C\sin\theta$. Now if we are given the rectangular form of a complex number which lies in another quadrant, such as $\mathbf{V} = 4 - j3$, which

Figure A4-3

A complex number may be represented by a point in the complex plane through choosing the correct real and imaginary parts from the rectangular form, or by selecting the magnitude and angle from the exponential form.

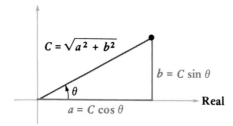

is depicted in Fig. A4-4, the correct angle is graphically evident, either $-36.9°$ or $323.1°$ for this example. The sketch may often be visualized and need not be drawn.

Figure A4-4

The complex number $\mathbf{V} = 4 - j3 = 5e^{-j36.9°}$ is represented in the complex plane.

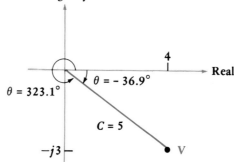

If the rectangular form of the complex number has a negative real part, it is often easier to work with the negative of the complex number, thus avoiding angles greater than 90° in magnitude. For example, given

$$\mathbf{I} = -5 + j2$$

we write

$$\mathbf{I} = -(5 - j2)$$

and then transform $(5 - j2)$ to exponential form:

$$\mathbf{I} = -Ce^{j\theta}$$

where

$$C = \sqrt{29} = 5.39 \quad \text{and} \quad \theta = \tan^{-1}\frac{-2}{5} = -21.8°$$

We therefore have

$$\mathbf{I} = -5.39e^{-j21.8°}$$

The negative sign may be removed from the complex number by increasing or decreasing the angle by 180°, as shown by reference to a sketch in the complex plane. Thus, the result may be expressed in exponential form as

$$\mathbf{I} = 5.39e^{j158.2°} \quad \text{or} \quad \mathbf{I} = 5.39e^{-j201.8°}$$

Note that use of an electronic calculator in the inverse tangent mode always yields angles having magnitudes less than 90°. Thus, both $\tan^{-1}[(-3)/4]$ and

$\tan^{-1}[3/(-4)]$ come out as $-36.9°$. Calculators that provide rectangular-to-polar conversion, however, give the correct angle in all cases.

One last remark about the exponential representation of a complex number should be made. Two complex numbers, both written in exponential form, are equal if, and only if, their amplitudes are equal and their angles are equivalent. Equivalent angles are those which differ by multiples of $360°$. For example, if $\mathbf{A} = Ce^{j\theta}$ and $\mathbf{B} = De^{j\phi}$, then if $\mathbf{A} = \mathbf{B}$, it is necessary that $C = D$ and $\theta = \phi \pm (360°)n$, where $n = 0, 1, 2, 3, \ldots$

Drill Problems

A4-5. Express each of the following complex numbers in exponential form, using an angle lying in the range $-180° < \theta \le 180°$: (a) $-18.5 - j26.1$; (b) $17.9 - j12.2$; (c) $-21.6 + j31.2$.   *Ans:* $32.0e^{-j125.3°}$; $21.7e^{-j34.3°}$; $37.9e^{j124.7°}$

A4-6. Express each of these complex numbers in rectangular form: (a) $61.2e^{-j111.1°}$; (b) $-36.2e^{j108°}$; (c) $5e^{-j2.5}$.
    Ans: $-22.0 - j57.1$; $11.19 - j34.4$; $-4.01 - j2.99$

A4-4

The polar form

The third (and last) form in which we may represent a complex number is essentially the same as the exponential form, except for a slight difference in symbolism. We use an angle sign ($\underline{/\ \ }$) to replace the combination e^j. Thus, the exponential representation of a complex number $\mathbf{A}$,

$$\mathbf{A} = Ce^{j\theta}$$

may be written somewhat more concisely as

$$\mathbf{A} = C\underline{/\theta}$$

The complex number is now said to be expressed in *polar* form, a name which suggests the representation of a point in a (complex) plane through the use of polar coordinates.

It is apparent that transformation from rectangular to polar form or from polar form to rectangular form is basically the same as transformation between rectangular and exponential form. The same relationships exist between C, θ, a, and b.

The complex number

$$\mathbf{A} = -2 + j5$$

is thus written in exponential form as

$$\mathbf{A} = 5.39e^{j111.8°}$$

and in polar form as

$$\mathbf{A} = 5.39\underline{/111.8°}$$

In order to appreciate the utility of the exponential and polar forms, let us consider the multiplication and division of two complex numbers represented in exponential or polar form. If we are given

$$\mathbf{A} = 5\underline{/53.1°} \quad \text{and} \quad \mathbf{B} = 15\underline{/-36.9°}$$

then the expression of these two complex numbers in exponential form

$$\mathbf{A} = 5e^{j53.1°} \quad \text{and} \quad \mathbf{B} = 15e^{-j36.9°}$$

enables us to write the product as a complex number in exponential form whose amplitude is the product of the amplitudes and whose angle is the algebraic sum of the angles, in accordance with the normal rules for multiplying two exponential quantities:

$$(\mathbf{A})(\mathbf{B}) = (5)(15)e^{j(53.1° - 36.9°)}$$

or

$$\mathbf{AB} = 75e^{j16.2°} = 75\underline{/16.2°}$$

From the definition of the polar form, it is evident that

$$\frac{\mathbf{A}}{\mathbf{B}} = 0.333\underline{/90°}$$

Addition and subtraction of complex numbers are accomplished most easily by operating on complex numbers in rectangular form, and the addition or subtraction of two complex numbers given in exponential or polar form should begin with the conversion of the two complex numbers to rectangular form. The reverse situation applies to multiplication and division; two numbers given in rectangular form should be transformed to polar form, unless the numbers happen to be small integers. For example, if we wish to multiply $(1 - j3)$ by $(2 + j1)$, it is easier to multiply them directly as they stand and obtain $(5 - j5)$. If the numbers can be multiplied mentally, then time is wasted in transforming them to polar form.

We should now endeavor to become familiar with the three different forms in which complex numbers may be expressed and with the rapid conversion from one form to another. The relationships among the three forms seem almost endless, and the following lengthy equation summarizes the various interrelationships

$$\mathbf{A} = a + jb = \mathrm{Re}\,[\mathbf{A}] + j\,\mathrm{Im}\,[\mathbf{A}] = Ce^{j\theta} = \sqrt{a^2 + b^2}\,e^{j\tan^{-1}(b/a)}$$
$$= \sqrt{a^2 + b^2}\,\underline{/\tan^{-1}(b/a)}$$

Most of the conversions from one form to another can be done quickly with the help of a calculator.

We shall find that complex numbers are a convenient mathematical artifice which facilitates the analysis of real physical situations.

Inevitably in a physical problem a complex number is somehow accompanied by its conjugate.

Drill Problems

A4-7. Express the result of each of these complex-number manipulations in polar form, using six significant figures just for the pure joy of calculating: (a) $[2 - (1\underline{/-41°})]/(0.3\underline{/41°})$; (b) $50/(2.87\underline{/83.6°} + 5.16\underline{/63.2°})$; (c) $4\underline{/18°} - 6\underline{/-75°} + 5\underline{/28°}$.

Ans: $4.691\,79\underline{/-13.2183°}$; $6.318\,33\underline{/-70.4626°}$; $11.5066\underline{/54.5969°}$

A4-8. Find $\mathbf{Z}$ in rectangular form if (a) $\mathbf{Z} + j2 = 3/\mathbf{Z}$; (b) $\mathbf{Z} = 2\ln(2 - j3)$; (c) $\sin \mathbf{Z} = 3$.

Ans: $\pm 1.414 - j1$; $2.56 - j1.966$; $1.571 \pm j1.763$

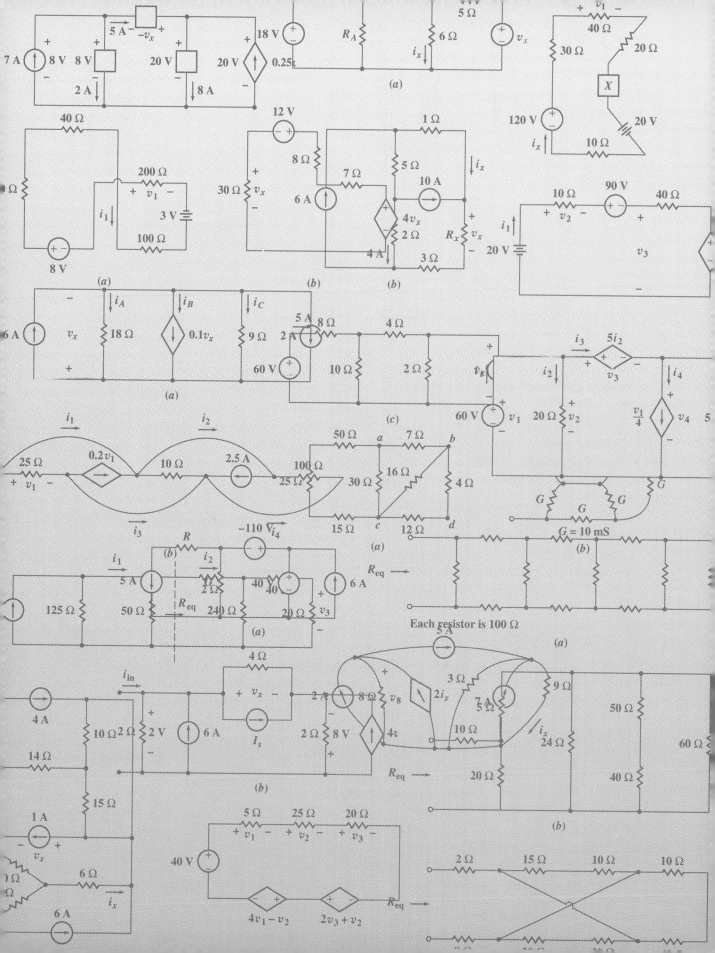

A SPICE Tutorial

Introduction

SPICE is an acronym for *Simulation Program with Integrated Circuit Empha-sis*. It is a very powerful program that will carry out many different procedures, although we will use it only for the basic types of dc, ac, and transient analysis. This appendix merely introduces the basic SPICE analyses, but much more information is available in the references listed at the end. For example, these references describe how to determine how sensitive an output value is to a change in a component value; how to obtain plots of the output versus a source value; how to find an ac output as a function of the source frequency; how to make noise and distortion analyses on circuits; how to make use of accurate nonlinear models of electronic devices; and how to show temperature effects on an electronic circuit.

In order to use SPICE, a digital computer must be available. Most electrical engineering departments of universities have SPICE installed on the system provided for student usage. Versions of SPICE are also available for personal computers, such as PSpice[1] for IBM-compatible PCs and SPICE/STUDENT VERSION[2] for the PC,[3] as well as PSpice for some Macintosh computers.

We first consider the dc analysis of a linear circuit. The methods we describe in Secs. A5-2 and A5-3 are sufficient to use SPICE to analyze any of the circuits in the examples and problems appearing in Chaps. 1 and 2 of the text.

A5-2

The dc analysis: circuit elements

Only seven circuit elements are used in the basic dc circuit for SPICE. These are the resistor, the two independent sources, and the four dependent sources: voltage source with voltage as a control parameter, voltage source with current as a control parameter, current source with voltage as a control parameter, and current source with current as a control parameter.

The location of an element in the circuit is specified by listing the two nodes to which it is connected. Each node is identified by an integer. The integers may or may not be consecutive; however, 0 is reserved for the ground or refer-ence node. The order in which the two nodes at the terminals of an element are listed is important. The first node is that one at which the + reference for

---

[1] Available from the Microsim Corporation, 2515 S. Western Ave., Suite 203, San Pedro, CA 90732.
[2] Contact the Intusoft Corporation, P.O. Box 6607, San Pedro, CA 90734.
[3] PC is a commonly used abbreviation for "personal computer."

the voltage across the circuit element is located. Thus, an element connected between nodes 3 (first-named) and 5 (last-named) has the voltage v_{35} with the + reference at node 3. The element current i_{35} flows from node 3 (first-named) through the element to node 5 (last-named).

A resistor is identified by a capital[4] R followed by no more than seven additional letters or integers. Thus, R1, ROUTPUT, and R5PRIME are satisfactory names for resistors. R2-6, R5/0, and R5 0 are not, since each name contains a symbol which is neither a letter nor an integer. The specification for each resistor appears as a single line of data. If the specification is too long to fit on one line, then a continuation sign (+) must be used at the beginning of the next line. After the resistor name there must be one or more blanks or spaces, followed by the integer number identifying the first node, followed by one or more blanks and then the number of the second node. At least one other blank precedes the resistor value. The value may be given as an integer, such as 5 or 1000, as a decimal number, such as 87.25, or as a floating-point number, such as 5E1 or 1E3. The value is given in ohms unless it is followed immediately by a capital letter specifying a scale factor. These letters and scale factors are

F	femto-	10^{-15}
P	pico-	10^{-12}
N	nano-	10^{-9}
U	micro-	10^{-6}
M	milli-	10^{-3}
K	kilo-	10^{3}
MEG	mega-	10^{6}
G	giga-	10^{9}
T	tera-	10^{12}

Note that M indicates milli- or 10^{-3}; mega- or 10^{6} is shown by MEG. The line of data

`ROUT 6 2 10K`

thus specifies a resistance called ROUT (R_{out}), connected between nodes 6 and 2, with a value of 10 kΩ. This value may be expressed in equivalent forms as 10 000, 1E4, 0.01E6, .01MEG, and so forth. Additional letters after scale factors do not affect the value. Thus, 10KOHM and 1E4OHMS also refer to the same value.

Some errors may be minimized by giving each resistor a name that reflects its location in the circuit rather than its use or value. Thus, if the resistor we called ROUT were named R62 instead, then incorrect node specification would be quite unlikely.

There are two independent sources, the independent voltage source and the independent current source. The independent voltage source is identified by a name beginning with V and followed by any combination of no more than seven additional letters or numbers. The name is followed by a blank, the node to which the positive reference of the source is connected, another blank, and then the node at which the negative terminal is located. Another blank precedes the letters DC, which are followed by a blank and the numerical value of the

---

[4] Most of the larger SPICE and PSpice installations will accept either upper- or lowercase letters. Systems intended for personal computers tend to require capital letters. Since it is easier to type in lowercase, it is wise to check the requirements of the system you will be using.

source voltage in volts. The nine scale factors listed in the previous table may also be used. Thus, all the following lines of data represent dc voltage sources:

```
VIN 6 0 DC 1.5
V2 1 2 DC 10M
VCC 4 3 DC 9
```

The SPICE program uses node voltages as its variables, and these voltages are the natural outputs of the calculation. However, we often desire the value of a current in a specific branch in the circuit. The only current that the SPICE program can calculate is the current through an independent voltage source. The reference direction for this current is from the first-named node, *through the source*, to the last-named node. Thus, if we want to know the value of the current in some branch of the circuit that does not contain an independent voltage source, we insert a 0-volt source there. We then ask for the value of this source current when we specify the outputs we desire. Such a source might be specified as

```
VI3 2 24
```

Thus, the independent voltage source VI3 has its positive reference at node 2 and its negative reference at node 24. This latter node is one that we have introduced into the circuit in order that we might determine the value of i_3. The term DC and the value 0 may be omitted; they are selected by default.

Independent current sources have names beginning with I. The current flows from the first-named node, *through the source,* to the last-named node. Examples of current sources might be

```
IIN 1 0 DC 1M
ISOURCE 2 5 DC .01
```

As an example, let us consider the circuit shown in Fig. A5-1a. The three nodes have been numbered, and the current i_{30} is desired. We shall therefore insert a 0-volt voltage source called v_{30} in the branch carrying i_{30}, as shown in Fig. A5-1b. The data list for this circuit therefore appears as

```
VS 1 0 DC 80
IS 0 2 DC 5
R20 1 2 20
R30 2 4 30
V30 4 0
```

Now let us turn to the dependent sources. Each depends on a current or voltage at some other point in the circuit. There are therefore four different

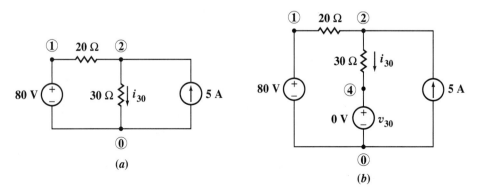

(*a*)   (*b*)

Figure A5-1

(*a*) An example circuit in which the current i_{30} is desired. (*b*) A 0-volt independent voltage source is inserted in series with the 30-Ω resistor, and a new node 4 is created.

types of dependent sources. We first consider the two sources that are voltage-controlled.

A voltage-controlled voltage source has a name beginning with E, followed by no more than seven letters and numerals. Next is the node at which the + reference is located, and then the second node. The two nodes that define the control voltage come next, + reference first. The last item is the numerical factor by which the control voltage must be multiplied to obtain the source voltage. In brief, we have

$$\text{Ename } +\text{node } - \text{ node } +\text{controlNode } -\text{controlNode gainFactor}$$

A voltage-controlled current source has a name beginning with G. Its location is specified by the two nodes defining the direction in which the current passes through the source. The nodes defining the voltage control come next, and the last value is the conductance (in siemens) by which the control voltage must be multiplied to obtain the source current. In concise terms,

$$\text{Gname IintoSource IexitSource } +\text{controlNode } -\text{controlNode Gfactor}$$

Figure A5-2 shows a circuit containing both types of voltage-controlled dependent sources. If we give the sources names that reflect their location in the circuit, the two lines of data for the dependent sources are

```
E30 3 0 2 0 0.8
G12 1 2 1 0 1M
```

Note that the Gfactor is 10^{-3} S, not 10^6 S.

Figure A5-2

A circuit used as an example illustrating the presence of voltage-controlled dependent sources.

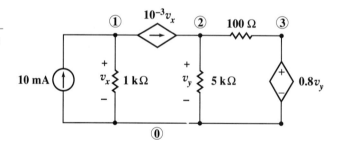

The order in which the sources (both types) and resistors are listed is immaterial. However, the SPICE program always interprets the *first* line of data as the title of the analysis. Hence, if we forget to put a title first, then the first line of data is used as a title, and the data in that line will be lost for the analysis.

The two current-controlled dependent sources both require the installation of a 0-volt independent voltage source in the branch where the control current is located. These 0-volt sources must be given names having a first letter V. Of course, it is not necessary to add the 0-volt source if there already is an independent voltage source in the controlling branch.

The current-controlled voltage source has a name beginning with an H, followed by the two nodes specifying its location. These are followed by the name of the 0-volt source (or other independent voltage source) whose current is the controlling current, and the value by which the controlling current must be multiplied to obtain the dependent source voltage. This has the dimensions of a resistance. Thus, we have

Hname +node −node VcontrolCurrent Rfactor

Finally, the current-controlled current source has a name beginning with F, followed by the two nodes defining the direction of current flow through the dependent source, the name of the 0-volt source (or other independent voltage source) in the control branch, and the numerical value by which the controlling current must be multiplied to obtain the dependent-source current. In short,

Fname IintoSource IexitSource Vcontrol Current gainFactor

To illustrate these points, let us change our last circuit to the circuit shown in Fig. A5-3a, in which both sources are current-controlled. Since there are two different controlling currents, it is necessary to install 0-volt independent voltage sources in those branches. We name these VX and VY and show them in Fig. A5-3b. The two new nodes are numbered 10 and 11, for no particular reason.

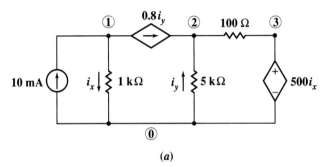

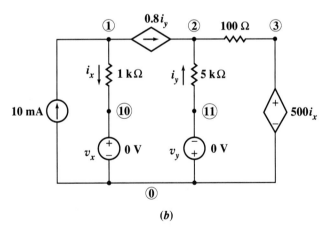

Figure A5-3

(a) A circuit containing two current-controlled dependent sources. (b) The required 0-volt independent sources are inserted in the controlling branches.

If we entitle our circuit "Current-Control Example," then our data appear as follows:

```
CURRENT-CONTROL EXAMPLE
ISOURCE 0 1 DC 10M
R1 1 10 1K
R5 2 11 5K
R100 2 3 100
VX 10 0
VY 0 11
F12 1 2 VY 0.8
H30 3 0 VX 500
```

As an aid to another person reading your program, it is helpful to use "comments" whenever a little explanation might help. Any line beginning with an asterisk (*) will be printed or displayed with the program, but will otherwise be ignored by the computer. For example, immediately after the title, lines such as the following might help:

```
*THIS CIRCUIT CONTAINS BOTH TYPES OF CURRENT-
*CONTROLLED SOURCES.
```

As an aid in remembering the requisite first letter in the name of each kind of source, the following chart may be useful.

Kind of source	Name begins with letter
Independent voltage	V
Independent current	I
Dependent voltage	
(Voltage-controlled)	E
(Current-controlled)	H
Dependent current	
(Voltage-controlled)	G
(Current-controlled)	F

A5-3

The dc analysis: control and operating statements

In addition to incorporating all the circuit data in the computer program, it is necessary to specify the operations that are to be performed. This is done by control statements. At this time, we shall look at the simpler control statements required for dc analysis. Note that each CONTROL statement begins with a period (.).

The listing

```
.OP
```

instructs the computer to calculate the dc voltage between each node and the reference node. The letters OP suggest that an operating point is being determined, perhaps for an electronic circuit. In a three- or four-node circuit, this is not an unreasonable amount of data, but as the number of nodes increases, we find that we are getting a lot of output data that we really do not need. Besides increasing the size of the output file, a lot of unnecessary paper is required when a printed copy of the output is requested. Instead of the .OP statement, we may request specific outputs with the .PRINT statement, which we now describe.

The print control statement consists of .PRINT followed by a space and DC, another space, and the desired node voltage or node voltages (separated by at least one space), where each voltage is given in the form V(1) or V(10), for example. The voltage between two nodes may also be named. That is, V(1,3) would provide the node-to-node voltage $v_{1,3}$. In addition, current values may be requested, provided that there is an independent voltage source (0-volt or otherwise) in the branch specified. The appropriate form is I(VX), where VX is the name of the appropriate voltage source. Thus, to obtain values for the voltage at node 3, the voltage between nodes 1 and 3, and the current i_x in the circuit of Fig. A5-3, the control statement would be

```
.PRINT DC V(3) V(1,3) I(VX)
```

Note that the .PRINT command does not result in printing of anything on paper. It is merely available in the computer's memory. If there is a printer connected to the computer, then the production of a printed output requires commands that are not part of the SPICE program.

Perhaps the most important control statement is

```
.END
```

which *must* be the last line in every SPICE program.

Thus, the first line must be a suitable title, and the last line must be the .END statement. The remainder of the program may be listed in any order.

If we collect this material, a suitable program for the circuit shown in Fig. A5-3 could be

```
CURRENT-CONTROL EXAMPLE
*THIS CIRCUIT CONTAINS BOTH TYPES OF CURRENT-
*CONTROLLED SOURCES
ISOURCE 0 1 DC 10M
R1 1 10 1K
R5 2 11 5K
R100 2 3 100
VX 10 0
VY 0 11
F12 1 2 VY 0.8
H30 3 0 VX 500
.PRINT DC V(3) V(1,3) I(VX)
.END
```

In case you wish to try this example yourself, the values are V(3) = v_3 = 5.418 V, V(1,3) = $v_1 - v_3$ = 5.418 V, and I(VX) = i_x = 10.837 mA.

A5-1. Write a SPICE program for the circuit shown in Fig. A5-4 and determine the values of (a) i_A; (b) i_B; (c) the voltage across the dependent current source, (+) reference at the top.   *Ans:* 3 A; 2 A; −4.8 V

Drill Problem

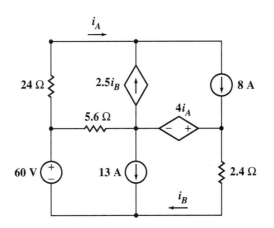

Figure A5-4

See Drill Prob. A5-1.

A5-4

The transient analysis: two more circuit elements

As of this time we have seven circuit elements: the resistor, two independent sources, and four dependent sources. We will now augment this collection with the inductor and the capacitor; later, we will provide some time-varying capabilities for the two independent sources. The following sections on transient analysis in this appendix present material that can be used to analyze any example or problem in Chaps. 4 through 6.

The specification for an inductor parallels that for a resistor, except that a name beginning with L must be chosen. Thus, a 6-H inductor connected between nodes 5 and 9 might appear as

```
LBIG 5 9 6
```

where the value is given in henrys. In addition to this information, it may be desirable to specify the initial value of the inductor current, $i_L(0^+)$. Thus, with an initial current of 10 mA flowing through the inductor from node 5 to node 9 at $t = 0^+$, the previous line of data would appear as

```
LBIG 5 9 6 IC=10M
```

where IC stands for "initial condition." If an initial condition is specified for any inductor or capacitor in the data list for a circuit, then initial values must be given for every inductor and capacitor. If no initial conditions are given, then the transient analysis program first performs a DC analysis to determine them. Also, if initial conditions are provided along with the element values, then it is necessary to include the specific term UIC ("use initial conditions") when we order the transient analysis, as we shall see in Sec. A5-6.

A 12-nF capacitor connected between nodes 2 and 0, carrying an initial voltage of 10 V, $v_C(0^+) = 10$ V, requires a name beginning with C, and might be listed as

```
COUT 2 0 12N IC=10
```

if we plan to use the UIC command, or

```
COUT 2 0 12N
```

if we intend for the SPICE program to determine the initial conditions automatically by running a preliminary dc analysis.

A5-5

The transient analysis: time-varying sources

The independent voltage and the independent current source may each be given in any of five different time-varying formats. These include pulse, exponential, sine, frequency-modulation, and piecewise linear specifications. In each of these cases, the values of the source voltage or current are given only for $t > 0$. Values for $t < 0$ are not specified and are not part of the transient analysis.

All the transient analyses considered in Chaps. 4, 5, and 6 require only the piecewise linear formulation, and this is accomplished by the following type of statement for an independent voltage source, say, VIN:

```
VIN 3 1 PWL(T1 V1 T2 V2.......TN VN)
```

where the source has its positive reference at node 3 and its negative reference at node 1. The expression PWL stands for "piecewise linear." The value of VIN at time T1 is V1, at time T2 it is V2, and so forth. Source values may be positive, negative, or zero; successive values of time must form a series of increasing positive values. As a simple example, consider the expression

`VIN 3 1 PWL(0 0 1U 100 1M 100)`

This piecewise linear function is 0 at $t = 0$, rises linearly to 100 V at $t = 1$ μs, and remains constant at 100 V until $t = 1$ ms, as shown in Fig. A5-5.

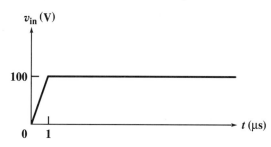

Figure A5-5

The piecewise linear expression PWL (0 0 1U 100 1M 100) is a reasonable approximation for a 100-V step function occurring at $t = 0$ in a circuit for which 1 μs is a relatively short time.

This voltage might be used as a reasonable approximation to a 100-V step function occurring at $t = 0$. The 1-μs rise time should be negligible in comparison with the circuit time constant. Also, the function is defined only out to 1 ms. We assume that this time is much greater than the time constant.

Transient analysis requires a .PRINT command similar to that for dc analysis, except that the term DC is replaced by TRAN. From one to eight output variables may be requested. These may be node voltages, node-to-node voltages, or currents through independent voltage sources.

Instead of getting a single value, however, we are now dealing with transient values and we shall obtain a listing of values as a function of time. The parameters controlling the analysis and the output are specified by the .TRAN command, often represented symbolically as

`.TRAN TSTEP TSTOP TSTART TMAX UIC`

The time interval employed in the listing is TSTEP, the first parameter specified. The second parameter, TSTOP, is the maximum value of time to be used. The third parameter, TSTART, is optional; it specifies the time at which the analysis begins. If it is omitted, $t = 0$ is the default value. TMAX is also an optional parameter. The numerical technique used in the SPICE analysis employs a variable time interval which is larger when the output is relatively constant, and smaller when it is changing more rapidly. The maximum value of this analysis interval may be specified as TMAX. If this value is omitted, the default value is (TSTOP − TSTART)/50 or TSTEP, whichever is smaller. Finally, we shall include the statement UIC, since we will be supplying the initial values. Thus, for a circuit having a time constant τ, we might select TSTEP = 0.1τ and a maximum time of 10τ. If the value of the time constant were 5 ms, then the command would be

`.TRAN 0.5M 50M UIC`

where the time interval or time step is 0.5 ms and the maximum value of time considered is 50 ms. Since TSTART is not specified, the analysis begins at $t = 0$; since TMAX is not specified, the largest time interval used in the numerical

A5-6

The transient analysis: operation and command statements

analysis will be TSTEP = 0.5 ms, this value being less than $(50 - 0)/50 =$ 1 ms.

As a first example of transient analysis, let us determine the voltage at node 1 in the circuit shown in Fig. A5-6. The time constant is L/R_{eq}, where $R_{eq} = 150 + 50 = 200\ \Omega$. Thus, we have a time constant of $0.005/200 = 25\ \mu s$. Let us select 2 μs as our time increment and 100 μs as the final time. We shall also specify a rise time for the current step of 1 ns, and define the source value out to our maximum time of 100 μs. Since the initial inductor current is zero, the SPICE program is

```
RL TRANSIENT EXAMPLE
R150 1 0 150
R50 1 2 50
L5 2 0 5M IC=0
ISTEP 0 1 PWL(0 0 1N 2M 100U 2M)
.TRAN 2U 100U UIC
.PRINT TRAN V(1)
.END
```

Figure A5-6

A circuit used as an example of SPICE transient analysis.

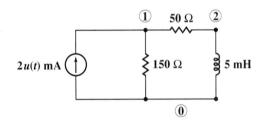

This program yields values for v_1 at 2-μs intervals from $t = 0$ to $t = 100\ \mu s$. To indicate the degree of accuracy of the SPICE program, we may compare several output values with the true values calculated from the expression $v_1 = (0.075 + 0.225e^{-t/25\times10^{-6}})u(t)$ V. At $t = 0$, each value is 0.3000 V; at $t = 10\ \mu s$, SPICE gives 0.2261 V, compared with the true value of 0.2258 V; at $t = 40\ \mu s$, both values are 0.1204 V; and at $t = 100\ \mu s$, SPICE yields 0.079 10 V, as contrasted with 0.079 12 V.

Another command that is available in transient analysis is one that accomplishes plotting:

```
.PLOT TRAN V(1)
```

Up to eight output variables may be listed, separated by spaces. Each one given in the .PLOT command must also appear in the .PRINT command. There is also an optional statement in the .PLOT command that specifies maximum and minimum values; this is most useful when several variables appear on the same plot, and we shall ignore it.

The resultant plot is not one with very high quality, particularly on the larger computers. Some of the routines for personal computers are good.

We select a source-free circuit as a second transient example. The circuit is shown in Fig. A5-7. The initial value of i_L is chosen as 80 mA. The downward current in the 50-Ω resistor is $1.6i_L$, and the voltage across that element is therefore $80i_L$. The equivalent resistance offered to the inductor is $80i_L/i_L = 80$

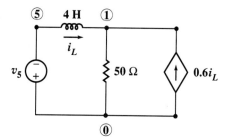

Figure A5-7

A source-free RL circuit with $i_L(0^+) = 80$ mA.

Ω. The exact solution is therefore $i_L = 80e^{-20t}$ mA for $t > 0$. The nodes are numbered in Fig. A5-7, and we have the following SPICE program:

```
SOURCE-FREE RL CIRCUIT
R50 1 0 50
L4 5 1 4 IC=80M
V5 0 5
F1 0 1 V5 0.6
.TRAN 0.01 0.2 UIC
.PRINT TRAN I(V5)
.END
```

Note that the controlled source depends on the inductor current, and it is necessary to include the 0-V source V5 in that branch.

A few values of i_L are compared in the following table. The small differences are negligible.

t (s)	$i_{L,\text{exact}}$ (mA)	$i_{L,\text{SPICE}}$ (mA)
0.01	65.5	65.5
0.02	53.6	53.65
0.05	29.4	29.4
0.1	10.83	10.82
0.15	3.98	3.98
0.2	1.465	1.462

A5-2. Write a SPICE program for the circuit of Fig. A5-8 to determine values of v_C at t equal to (a) 5 ms; (b) 10 ms; (c) 50 ms.   *Ans:* 150.2 V; 255 V; 486 V

Drill Problem

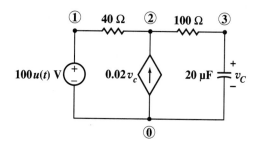

Figure A5-8

See Drill Prob. A5-2. An RC transient example.

A5-7

The ac analysis: circuit elements

Each independent voltage and current source in ac analysis is characterized as a phasor having an amplitude and a phase angle. The frequency is specified in a CONTROL statement, as we shall see in the next section. The frequency for every source must be the same. The amplitude and phase of each source is given with the source statement:

```
VIN 6 0 AC 120 30
```

Thus, the source named VIN has its positive voltage reference at node 6 and its negative reference at node 0. The term AC signifies that it is a phasor source with an amplitude of 120 V and a phase angle of 30°, or $\mathbf{V}_{in} = 120\underline{/30°}$ V. If the phase angle is omitted, it is assumed to be zero. If the amplitude and phase angle are both omitted, the default value is $1\underline{/0°}$ V. A 0-V source inserted for current measurement should be labeled AC 0 0.

Coupled coils may also be specified.[5] Thus, if inductors L5 and L8 have already been defined, then it remains only to name a coefficient of coupling, such as K58, and give its value, say, 0.4:

```
K58 L5 L8 0.4
```

The coefficient of coupling must be greater than zero and equal to or less than unity. The dot convention is used by defining the nodes for each inductor with the dotted terminal listed first.

A5-8

The ac analysis: operation and command statements, single-frequency operation

The ac control statement begins with the specification .AC and continues with four additional terms which describe the range of frequencies to be used, and the manner in which that range is to be covered. In this section we shall look at the simplest case, in which a single frequency is selected. The required specification is

```
.AC LIN 1 F F
```

The term LIN actually requests a linear variation of frequency (in Hz), here utilizing only one point and thus covering the range from F to F. For example, to obtain a phasor analysis of a circuit at $\omega = 100$ rad/s, we must specify

```
.AC LIN 1 15.9155 15.9155
```

since $100/2\pi = 15.9155$. It is pretty obvious that the SPICE program was written more for practicing engineers than for electrical engineering students (and professors).

One other command that changes slightly is the .PRINT command. The phrase AC replaces DC or TRAN, and we may elect to obtain the magnitude, the phase angle, the real part, the imaginary part, or the magnitude expressed in dB (20 times the logarithm to the base 10 of the magnitude). This is accomplished by appending the letters M, P, R, I, or DB, respectively, to the V or I. Thus,

```
.PRINT AC VM(3) VP(3) VM(2,4)
```

would yield the magnitude and phase of the voltage at node 3 and the magnitude of the voltage between nodes 2 and 4, while

```
.PRINT AC IDB(VIN)
```

would give the current magnitude through VIN in dB.

---

[5] Coupled coils may also be included in a transient analysis.

This section of Appendix 5 may well be postponed until frequency response is met in Chap. 13 and in some of the following chapters.

Three choices are offered to us: we may utilize frequencies that are spaced linearly over a given range, or uniformly spaced in a logarithmic manner throughout a decade or throughout an octave (two-to-one frequency range). We also specify the number of points to be used over the linear range, or we specify the number of intervals into which a decade or an octave is to be divided. For the linear analysis, we give a starting frequency value f_A and an ending value f_B. In either of the logarithmic analyses, we give a starting frequency value f_A and a limiting maximum value f_B.

Let us consider a few examples. The statement

`.AC LIN 6 2000 3000`

requests the use of 6 frequencies uniformly (linearly) spaced from 2000 to 3000 Hz. Again note that all frequencies in SPICE are given in hertz, not radians per second. This analysis would therefore provide results at the starting and stopping frequencies and four intermediate frequencies that are uniformly spaced. The six frequencies used would be 2000, 2200, 2400, 2600, 2800, and 3000 Hz. In general, if f_A and f_B are the lower and upper ends of the frequency range, and if N frequencies are desired, then any frequency f_i is given by

$$f_i = f_A + (f_B - f_A)\frac{i-1}{N-1} \qquad i = 1, 2, 3, \ldots, N$$

We see that $f_1 = f_A$ and $f_N = f_B$. Note that the number of frequencies selected always includes the endpoints, f_A and f_B.

If we select decade analysis, then we let N_D be the number of equal intervals into which a decade is divided. The frequencies applied would include the beginning frequency f_A and also the beginning frequencies of all the remaining intervals. The maximum frequency f_B would also be applied if it is the endpoint of the last interval. As an example, consider

`.AC DEC 10 2 20`

where each decade on a logarithmic scale is to be divided into 10 equal intervals. Since the specified range from $f_A = 2$ Hz to $f_B = 20$ Hz is a decade, the 10 equal intervals extend from 2 Hz to 20 Hz. In general, the frequencies that define the endpoints of the intervals would be

$$f_i = (f_A)10^{(i-1)/N_D} \qquad i = 1, 2, 3, \ldots, N_D$$

so that $f_1 = f_A$. For this example, we have $f_i = (2)10^{(i-1)/10}$, so that $f_1 = 2$, $f_2 = 2.5179$, $f_3 = 3.1698$, and so forth, up to $f_{10} = 15.8866$, and $f_{11} = 20$ Hz. Since f_A and f_B are separated by exactly one decade, the 11 frequencies from 2 to 20 Hz would be used. If f_B had been equal to 30 Hz, then the frequency $f_{12} = 25.179$ would also have been applied. Since $f_{13} = 31.698$, it is greater than the new specified maximum, $f_B = 30$ Hz, and would not be used.

Using the octave formulation, we let N_O be the number of intervals into which an octave is divided. Thus,

`.AC OCT 10 5 15`

would use 10 frequencies per octave beginning at 5 Hz and continuing up to the highest frequency that is not greater than 15 Hz. The individual frequencies would be given by

$$f_i = (f_A)\, 2^{(i-1)/N_O} \qquad i = 1, 2, 3, \ldots, N_O$$

Frequency values for the preceding command would be $f_1 = 5$, $f_2 = 5.3589$, $f_3 = 5.744\,35$, $\ldots$, $f_{11} = 10$, $f_{12} = 10.7177$, and so forth, up to $f_{15} = 14.1421$. Since $f_{16} = 15.1572$ Hz, it would not be included. Note that 15 different frequencies will be applied, 11 of which appear in the octave from 5 to 10 Hz.

In each of the three general cases, the first frequency used must be f_A, but the last is not necessarily f_B. Instead, it is the largest f_i that is not greater than f_B. With the linear scale, both f_A and f_B are included.

Now let us try our hand at an ac analysis in which results are desirable over a range of frequencies. Figure A5-9 shows a series RLC circuit that is resonant at $\omega = 50$ rad/s or $f = 7.9577$ Hz. It has a Q of 10 and a bandwidth

Figure A5-9

A series resonant circuit that is used as an example to illustrate frequency analysis over various frequency ranges.

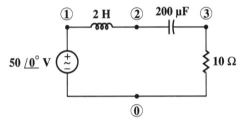

of 5 rad/s, or about 0.8 Hz. We might wish to obtain data over a linear range of about three bandwidths centered at 8 Hz, approximately equal to the resonant frequency. A suitable SPICE program would be

```
RLC EXAMPLE
VIN 1 0 AC 50
L1 1 2 2
C1 2 3 200U
R1 3 0 10
.AC LIN 11 7 9
.PRINT AC VM(3) VP(3)
.END
```

The 11 applied frequencies would be 7, 7.2, 7.4, $\ldots$, 8.8, 9 Hz, and the corresponding values of $\mathbf{V}_3$ are shown in tabular form as follows:

f (Hz)	$\mathbf{V}_3$ (V)
7	$18.12\underline{/\,68.75°}$
7.2	$22.3\underline{/\,63.5°}$
7.4	$28.3\underline{/\,55.5°}$
7.6	$36.8\underline{/\,42.6°}$
7.8	$46.4\underline{/\,21.8°}$
8	$49.7\underline{/\,-6.05°}$
8.2	$42.9\underline{/\,-31.0°}$
8.4	$33.9\underline{/\,-47.3°}$
8.6	$27.1\underline{/\,-57.2°}$
8.8	$22.2\underline{/\,-63.6°}$
9	$18.78\underline{/\,-67.9°}$

To cover a greater frequency range, it would be advisable to use one of the logarithmic sequences. For example, the command

```
.AC DEC 10 2 20
```

would divide the decade from 2 to 20 Hz into 10 equal intervals, the first beginning at 2 Hz and the last ending at 20 Hz. At 2 Hz, $\mathbf{V}_3 = 1.341\underline{/88.5°}$; at 7.962 Hz (very close to resonance), we have $50.0\underline{/-0.06°}$ V; and at 20 Hz, we obtain an output voltage of $2.361\underline{/-87.3°}$ V.

A5-3. Find the phasor voltages $\mathbf{V}_{out}$ and $\mathbf{V}_x$ and the phasor current $\mathbf{I}_y$ at $\omega = 4$ krad/s in the circuit shown in Fig. A5-10.

Drill Problems

Ans: $1.231\underline{/-28.0°}$ V; $0.627\underline{/45.5°}$ V; $10.39\underline{/-22.6°}$ mA

A5-4. Find $|\mathbf{V}_{out}|$ over a logarithmic range of frequencies extending from 100 Hz to 10 kHz for the circuit of Fig. A5-10.
Ans: 100 Hz: 1.007 V; 251 Hz: 1.045 V; 501 Hz: 1.163 V; 1995 Hz: 0.7865 V; 5010 Hz: 0.1029 V; 10 kHz: 0.008 98 V

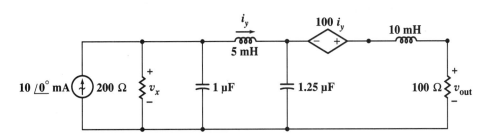

Figure A5-10

See Drill Probs. A5-3 and A5-4.

Conant, Roger C.: *Engineering Circuit Analysis with PSPICE and PROBE,* McGraw-Hill Book, 1993. (This paperback text covers the use of PSPICE and its graphical companion PROBE in linear circuit analysis. All features of PSPICE and PROBE which are applicable to linear circuits are covered, with an emphasis on their use in deepening students' understanding of circuit behavior. Text includes a $5\frac{1}{4}$-in. disk of PSPICE files.)

Meares, L. G., and C. E. Hymowitz: "Simulating with SPICE," Intusoft Corp., 1988. (This book accompanies IS-SPICE/Student Version, available from the Intusoft Corp., 2515 S. Western Ave., Suite 203, San Pedro, CA 90732.)

Nagel, L. W.: "SPICE2: A Computer Program to Simulate Semiconductor Circuits," Memorandum ERL-M520, Electronics Research Laboratory, College of Engineering, University of California, Berkeley, CA 94720, 1975. (This is the original publication for SPICE2.)

Thorpe, Thomas W.: *Computerized Circuit Analysis with SPICE,* Wiley, 1992. (This paperback describes SPICE 2G.6, the version of SPICE in use in 1992. It includes numerous examples.)

Tuinenga, Paul W.: *SPICE: A Guide to Circuit Simulation and Analysis using PSpice,* Prentice-Hall, Englewood Cliffs, N.J., 1992. (This popular paperback is a very readable introduction and manual for PSpice. It includes circuit examples.)

Vladimirescu, A., K. Zhang, A. R. Newton, D. O. Pederson, and A. Sangiovanni-Vincentelli: "SPICE Version 2G User's Guide," Dept. of E. E. and C. S., University of California, Berkeley, CA 94720, 1981. (This is the original publication for SPICE 2G.)

References

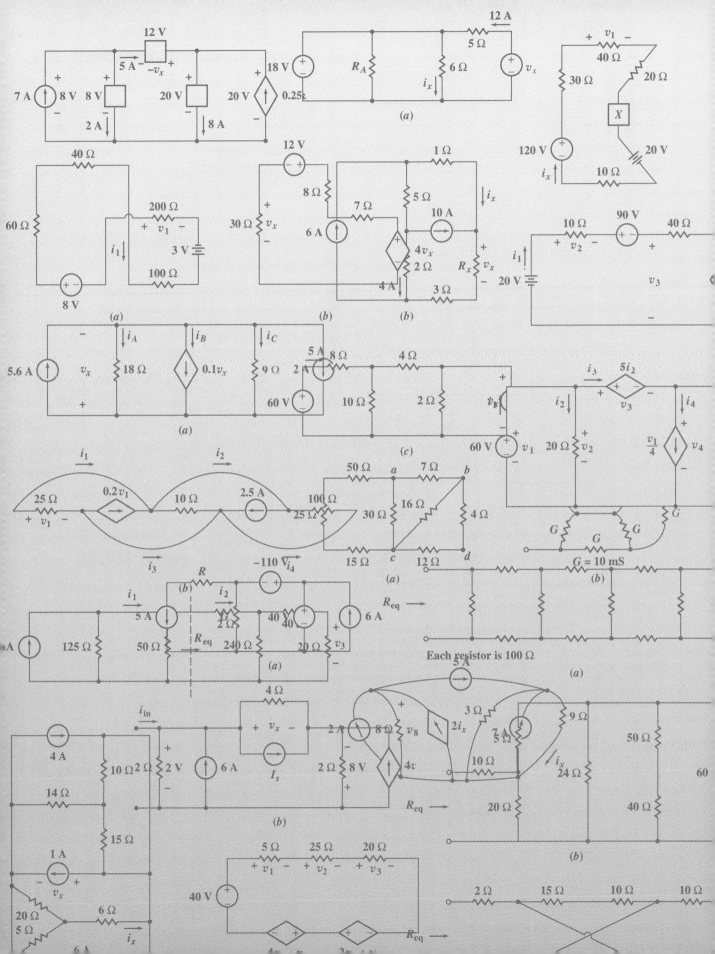

Answers to Odd-Numbered Problems

1 (a) 1317 km/h; (b) 92.1 kJ; (c) 13.33 days; (d) planet Krypton
3 (a) 40 C; (b) 40.5 C, 2.12 s; (c) 24.7 A; (d) see Fig. P1-3

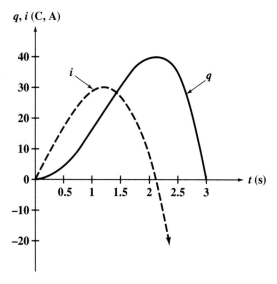

Chapter 1

Figure P1-3

t (s)	q (C)	i (A)
0	0	0
0.25	1.117	8.875
0.5	4.375	17.00
0.75	9.49	23.6
1	16.00	28.0
1.25	23.2	29.4
1.5	30.4	27.0
1.75	36.4	20.1
2	40.0	8.00
2.25	40.0	−10.125
2.5	34.4	−35.0
2.75	21.7	−67.4
3	0	−108.0

5 (a) 0.8 A; (b) 0; (c) see Fig. P1-5

Figure P1-5

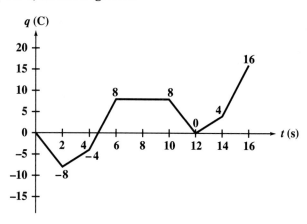

7 (a) 16.55 nW; (b) 0

9 (a) 72.7 W; (b) -36.3 W; (c) 27.6 W

11 (a) 10 V; (b) 0; (c) 0; (d) 2 W; (e) 2 W

13 34 Ω and 0.09 S

15 (a) $v_1 = 60$ V, $v_2 = 60$ V, $i_2 = 3$ A, $v_3 = 15$ V, $v_4 = 45$ V, $v_5 = 45$ V, $i_5 = 9$ A, $i_4 = 15$ A, $i_3 = 24$ A, $i_1 = 27$ A; (b) $p_1 = -1620$ W, $p_2 = 180$ W, $p_3 = 360$ W, $p_4 = 675$ W, $p_5 = 405$ W; $\Sigma = 0$

17 (a) 8 V, -4 V, -12 V; (b) 14 V, 2 V, -6 V; (c) 2 V, -10 V, -18 V

19 (a) -1 A; (b) 1 A; (c) -2 A

21 (a) 0.57 Ω; (b) 1.003 Ω; (c) 0.12 Ω

23 $p_{40\,\text{v}} = 80$ W, $p_{5\,\Omega} = 20$ W, $p_{25\,\Omega} = 100$ W, $p_{20\,\Omega} = 80$ W, $p_{2v_3+v_2} = -260$ W, $p_{4v_1-v_2} = -20$ W

25 0.571 mA

27 (a) 0.4 W; (b) 0.6 W; (c) 0.556 W; (d) -0.6 W

29 (a) 3 A; (b) 24 V; (c) 15 W

31 (a) 60 Ω; (b) 213 Ω; (c) 51.8 Ω

33 $p_{2.5} = 250$ W, $p_{30} = 187.5$ W, $p_6 = 337.5$ W, $p_5 = 180$ W, $p_{20} = 45$ W

35 (a) 0.850 mS; (b) 135.9 mS

37 (a) $v_sR_2(R_3 + R_4)/[R_1(R_2 + R_3 + R_4) + R_2(R_3 + R_4)]$; (b) $v_sR_1(R_2 + R_3 + R_4)/[\text{den.}]$; (c) $v_sR_2/[\text{den.}]$

39 (a) 42 A; (b) 11.90 V; (c) 0.238 for both

41 $V_sR_3R_5/[R_2(R_3 + R_4 + R_5) + R_3(R_4 + R_5)]$

43 (a) From left to right: 10, -30, 16, 16, -21, 36, -27 W; (b) -2 V

45 (a) $1.000\,01V_s$; (b) $-0.000\,001V_s$; (c) $1.000\,01V_s$; (d) V_s

Chapter 2

1 (a) -8.39 V; (b) 32

3 (a) 19.57, 18.71, -11.29 V; (b) 16

5 63.1 V

7 148.1 V, 178.3 W

9 2.79 A

11 -384 W

13 2 mA: 5 mW; 4 V: -6 mW; $1000i_3$: 4.5 mW; 6 V: 9 mW; $0.5i_2$: -5.62 mW

15 (a) 3 A; (b) 17 A; (c) -8 A; (d) -27 A

17 -0.75 A

19 (a) 1.3 A; (b) 1 A: 60 W; 200 Ω: 18 W; 100 V: -130 W; 50 Ω: 32 W; 0.5 A: 20 W

21 (a) 200 V; (b) 125 W; (c) 80 Ω

23 (a) 1.6 cos 400t A in parallel with, or 40 cos 400t V in series with, 25 Ω; (b) 11.25 mA in parallel with, or 90 V in series with, 8 kΩ

25 (a) 75 V in series with 12.5 Ω; (b) 72 W; (c) 112.5 W

27 (a) 69.3 V in series with 7.32 Ω; (b) 59.5 V in series with 16.59 Ω

29 0 A in parallel with 10.64 Ω

31 192.3 Ω

33 (a) 16 Ω; (b) 5 V

35 (a) 1-4-3, 2-1-4, 6-1-4, 3-4-5, 4-5-2, 4-5-6, 4-3-6, and 4-3-2; (b) 3-4-5, 3-5-6, and 3-4-6

37 $5i_2$, $0.2v_3$, 100 V, and 50 Ω; voltages are $5i_2$, $0.2v_3$, 100 V, and v_3; $(-v_3 + 0.2v_3 - 5i_2)/45 - v_3/50 - 0.02v_1 + (-v_3 + 0.2v_3 - 100)/30 = 0$, $v_3 = 50i_2$, $v_1 = 0.2v_3 - 100$; -0.377 A

39 240 V, 30 Ω, 6 Ω, 60 V; $30i_1 - 42i_2 = 360$, $39i_1 - 30i_2 = 600$; 19.51 A

41 (a) 3 Ω, 7 Ω, 6 Ω, 2 Ω; (b) 1.352 A

43 -2.5 V, 25 kΩ
45 3.3333 A

Chapter 3

1 (a) See Fig. P3-1; (b) 40^- ms; (c) 20^+ and 40^+ ms; (d) 2.5 J

Figure P3-1

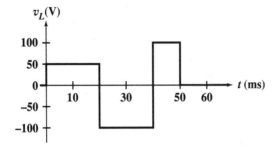

3 (a) $4t^2 + 4t$ V; (b) $4t + 4t^2 + 5$ A
5 (a) 2 A; (b) 5.63 J; (c) 1 A
7 (a) $-0.12 \sin 400t$ mA; (b) 6.4 μJ; (c) $400(1 - e^{-100t})$ V; (d) $500 - 400e^{-100t}$ V
9 (a) 2 kΩ; (b) proof
11 (a) $10.005 \sin 10t + 0.0005(1 - \cos 10t)$ V; (b) $10 \sin 10t$ V
13 (a) 11.38 Ω; (b) 11.38 H; (c) 8.79 F
15 (a) (4 Cs in series), in $\parallel$ with C, in $\parallel$ with C; (b) 1 C in series with (3 Cs in $\parallel$); (c) [(2 Cs in series), in $\parallel$ with 4 Cs], in series with 2 Cs
17 (a) 20.5 mJ; (b) 91.6 mJ; (c) 3.28 J
19 (a) $-6.4e^{-80t}$ mA; (b) $80e^{-80t} - 60$ V; (c) $20e^{-80t} + 60$ V
21 (a) v_{20}, i_L, i_C, and v_C; $i_L(0) = 12$ A and $v_C(0) = 2$ V; (b) $20v_{20} + [1/(5 \times 10^{-6})]$ $\int_0^t (v_{20} - v_C)dt + 12 = i_s$, $[1/(5 \times 10^{-6})] \int_0^t (v_C - v_{20})dt - 12 + 10v_C +$ $8 \times 10^{-3} \, dv_C/dt = 0$; (c) $(i_L - i_s)/20 + 5 \times 10^{-6} \, di_L/dt + (i_L - i_C)/10 = 0$, $(i_C - i_L)/10 + [1/(8 \times 10^{-3})] \int_0^t i_C \, dt + 2 = 0$

23 See circuit diagram, Fig. 3-23.

Figure P3-23

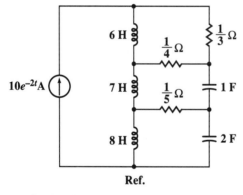

Ref.

25 Tree voltages are, beginning at top center: $0.2v_x$, v_x, v_C, and $-40e^{-20t}$.
(a) $(1/0.05) \int_0^t (0.2v_x + v_x + v_C - 40e^{-20t})dt + (v_x/50) - 0.02e^{-20t} = 0$,
$(-v_x/50) + 10^{-6}dv_C/dt + (v_C - 40e^{-20t})/100 = 0$; (b) $0.05di_L/dt + 0.2v_x +$
$50 \, (i_L + 0.02e^{-20t}) + 10^6 \int_0^t (i_L + i_{100} + 0.02e^{-20t})dt = 40e^{-20t}$, $100i_{100} +$
$10^6 \int_0^t (i_L + i_{100} + 0.02e^{-20t})dt = 40e^{-20t}$
27 30.0000 V, 0.6000 A; (a) 0.900 J; (b) 0.009 J

Chapter 4

1 (*a*) $2e^{-400t}$ A; (*b*) 36.6 mA; (*c*) 1.733 ms
3 (*a*) 1.289 mA; (*b*) 7.71 mA
5 (*a*) 2.68 A; (*b*) 1.889 A
7 (*a*) 2.30, 4.61, and 6.91; (*b*) $t/\tau = 2$
9 $R_1 = 7.82\ \Omega, R_2 = 13.86\ \Omega$
11 $10e^{-50t}$ A
13 (*a*) 0.893 A; (*b*) 0.661 A
15 (*a*) $192e^{-125t}$ V; (*b*) 18.42 ms
17 (*a*) 69.3 μs; (*b*) 34.7 μs
19 (*a*) 0.29 A; (*b*) 0.2 A; (*c*) 0.05 A; (*d*) 0.277 A; (*e*) 0.0335 A
21 (*a*) -6 mA; (*b*) $12e^{-100t}$ mA
23 $20e^{-250,000t}$ V
25 (*a*) 87.6 V; (*b*) $87.6e^{-2540t}$ V
27 (*a*) 100, 0, and 0 V; (*b*) 100, 0, and 100 V; (*c*) 0.08 s; (*d*) $100e^{-12.5t}$ V; (*e*) $5e^{-12.5t}$ mA; (*f*) $(20e^{-12.5t} + 80)$ and $(-80e^{-12.5t} + 80)$ V; $(64 + 16) + 20 = 100$ mJ
29 (*a*) 20 mA; (*b*) $20e^{-10,000t} - 2e^{-5000t}$ mA

Chapter 5

1 1, 0.6, -0.4, 0.6 A
3 (*a*) 1; (*b*) 12; (*c*) 1.472
5 2.5, 3, 2.5, 2, -2 A
7 (*a*) $(2 - 2e^{-200,000t})u(t)$ mA; (*b*) $6e^{-200,000t}\ u(t)$ V
9 (*a*) 10 A; (*b*) $8 + 2e^{-10t}$ A
11 (*a*) 80 mA; (*b*) $80(1 - e^{-25t})$ mA, $t > 0$; (*c*) $80(2 - e^{-25t})$ mA, $t > 0$; (*d*) $16(\cos 50t + 2 \sin 50t - e^{-25t})$ mA, $t > 0$

Figure P5-23

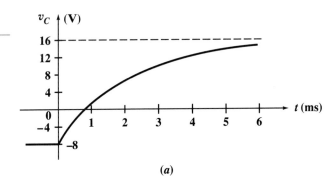

(*a*)

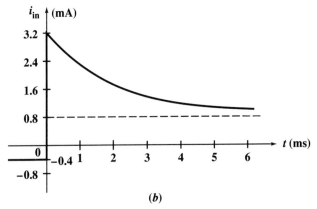

(*b*)

13 (a) 0, 0; (b) 0, 200 V; (c) 1 A, 100 V; (d) 0.551 A, 144.9 V

15 $0.1 + (0.1 - 0.1e^{-9000t})u(t)$ A

17 (a) 3 A; (b) 2.4 A; (c) 2.63 A

19 (a) $20(1 - e^{-40t})u(t)$ A; (b) $(10 - 8e^{-40t})u(t)$ A

21 $2.5u(-t) + (10 + 7.5e^{-100,000t})u(t)$ mA

23 (a) $-8u(-t) + (16 - 24e^{-500t})u(t)$ V, see Fig. P5-23; (b) $-0.4u(-t) + (0.8 + 2.4e^{-500t})$ mA, see Fig. P5-23

25 6.32 and 15.662 V

27 (a) 80 V; (b) $80 + 160e^{-100,000t}$ V; (c) 80 V; (d) $80 - 32e^{-20,000t}$ V

29 0.693 ms

31 $e^{-0.1t}\,u(t)$ V

33 $5(e^{-t} - 1)u(t)$ V

35 (a) -7.20 A; (b) 7.813 A

Chapter 6

1 4.95 Ω, 1.443 H, 14.43 mF

3 (a) $-120e^{-2t} + 160e^{-8t}$ V; (b) $-16e^{-2t} + 3e^{-8t}$ A

5 $2u(-t) + (2.25e^{-2000t} - 0.25e^{-6000t})u(t)$ A

7 (a) $20.25e^{-4t} - 2.25e^{-36t}$ V; (b) $0.50625e^{-4t} - 0.00625e^{-36t}$ A; (c) 1.181 s

9 (a) 8 mH; (b) 0.931 A; (c) 24.0 ms

11 $e^{-4000t}(-2 \cos 2000t + 4 \sin 2000t)$ A

13 (a) $e^{-5000t}(200 \cos 10^4 t + 100 \sin 10^4 t)$ V;
(b) $10 - e^{-5000t}(10 \cos 10^4 t - 7.5 \sin 10^4 t)$ mA

15 $0.6e^{-100t} \sin 1000t$ mA

17 $e^{-4t}(10 \cos 2t + 20 \sin 2t)$ A

19 $2u(-t) + (2.25e^{-2000t} - 0.25e^{-6000t})u(t)$ V

21 (a) $0.5e^{-10t}$ A; (b) $100e^{-10t}$ V

23 $[10 - e^{-4t}(20 \sin 2t + 10 \cos 2t)]u(t)$ A

25 $e^{-4000t}(2 \cos 2000t - 4 \sin 2000t)$ A

27 $12 - e^{-t}(t + 2)$ V

29 (a) $2.5e^{-500t} - 22.5e^{-1500t}$ mA; (b) $-2.5e^{-500t} + 22.5e^{-1500t}$ mA

31 10.378 Ω, 2.145 s

33 See diagram, Fig. 6-33.

35 (a) 46.55 V, 11.03 V; (b) 46.77 V, 11.368 V

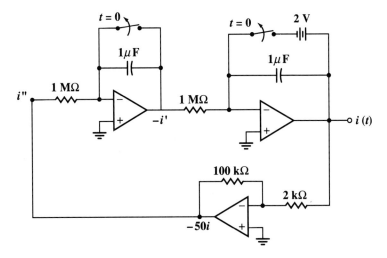

Figure P6-33

Chapter 7

1 (a) 8.5 sin (291t + 325°); (b) 8.5 cos (291t − 125°);
   (c) −4.88 cos 291t + 6.96 sin 291t
3 (a) amplitude $f(t)$ = 58.3, amplitude $g(t)$ = 57.0; (b) 133.8°
5 0.671 cos (500t − 26.6°) A
7 (a) 25.8 μs; (b) 10.12 and 25.8 μs; (c) 15.71 and 25.8 μs
9 5.88 cos (500t − 61.9°) mA
11 1.414 cos (400t − 45°) + 1.342 cos (200t − 26.6°) A
13 (a) $V_m \cos \omega t = Ri + \frac{1}{C}\int i\, dt$, $-\omega V_m \sin \omega t = R\, di/dt + i/C$;
   (b) $(\omega C V_m / \sqrt{1 + \omega^2 C^2 R^2})\cos\left[\omega t + \tan^{-1}(1/\omega CR)\right]$

Chapter 8

1 (a) $-1.710 - j4.70$; (b) $-5.64 + j2.05$; (c) $-5.34 + j12.31$; (d) $107.7\underline{/-158.2°}$;
   (e) $1.087\underline{/-101.4°}$
3 $34.9e^{j(40t+53.6°)}$ V
5 (a) 2.5 cos (500t + 42°) A; (b) 2.5 cos (500t − 48°) A; (c) $2.5e^{j(500t+42°)}$ A;
   (d) $3.37e^{j(500t+53.8°)}$ A
7 (a) −4.29 A; (b) 3.75 A; (c) $50\underline{/-130°}$ V; (d) $36.1\underline{/56.3°}$ V; (e) $72.3\underline{/-63.9°}$ V
9 35.5 cos (500t + 58.9°) V
11 0.457 and 2.19 rad/s
13 $75.1\underline{/-4.11°}$ V
15 (a) $22 - j6$ Ω; (b) $9.6 + j2.8$ Ω
17 (a) $15.00\underline{/33.1°}$ V; (b) $15.00\underline{/-73.1°}$ V; (c) $9\underline{/-20°}$ V; (d) $20.1\underline{/43.4°}$ V
19 (a) 2260 rad/s; (b) 3220 rad/s; (c) 3350 rad/s; (d) 3930 and 573 rad/s
21 (a) 1.437 μF; (b) 8.96 μF
23 $0.5 - j0.5$ Ω, 2 Ω ‖ 2 H
25 (a) 100 krad/s; (b) 100 krad/s; (c) 102.1 krad/s; (d) 52.2 and 133.0 krad/s
27 (a) 8 Ω and 250 μF; (b) 5 Ω and 100 μF
29 (a) $28.57\underline{/0°}$ V; (b) $90.24\underline{/25.52°}$ V; (c) $64.71\underline{/-49.67°}$ V

Chapter 9

1 $34.4\underline{/23.6°}$ V
3 70.7 cos ($10^3 t$ − 45°) V
5 1.213 cos (100t − 76.0°) A
7 (a) 15.72 cos ($10^3 t$ + 122.0°) A, see Fig. P9-7; (b) 15.72 cos ($10^3 t$ + 122.0°) V,
   see Fig. P9-7.
9 $57.3\underline{/-55.0°}$ V in series with $4.70 - j6.71$ Ω
11 (a) 5 cos ($10^3 t$ + 90°) V; (b) 11.79 cos ($10^3 t$ + 135°) V
13 1.414 cos (200t + 45°) + 0.5 cos (100t + 90°) V
15 (a) $57.3\underline{/-76.8°}$ A, $25.6\underline{/-140.2°}$ A, $51.2\underline{/-50.2°}$ A, $143.1\underline{/13.24°}$ V,
   $51.2\underline{/-140.2°}$ V, $51.2\underline{/-140.2°}$ V; (b) see Fig. P9-15.
17 $I_1 = 5\underline{/-40.5°}$ A, $I_2 = 7\underline{/27.7°}$ A, see Fig. P9-17.
19 Sketch ($|\mathbf{V}_{out}|$ = 9.71 V at ω = 1050 rad/s). See Fig. P9-19.
21 Sketch ($\mathbf{Z}_{in} = 35.6 - j19.2 = 40.4\underline{/-28.3°}$ Ω at ω = 7500 rad/s). See Fig.
   P9-21.
23 Zeros: 0, −1000, +1000 rad/s; poles: −500, +500, ∞ rad/s; see Fig. P9-23.
25 $\mathbf{Z}_{in} = 2 + j0 = 2\underline{/0°}$ Ω, all ω

Figure P9-7

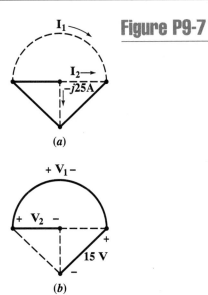

(a)

(b)

Figure P9-15

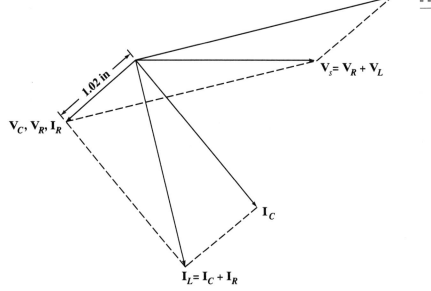

Figure P9-17

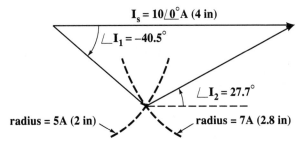

Figure P9-19

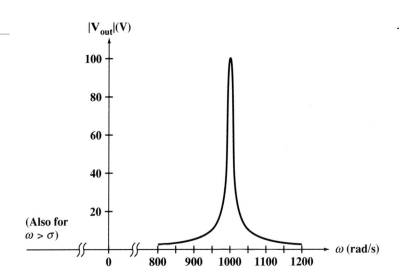

ω	$|V_{out}|$
800	2.78
850	3.60
900	5.26
950	10.21
975	19.87
980	24.5
990	45.0
1000	100
1010	44.5
1020	24.0
1025	19.36
1050	9.71
1100	4.76
1150	3.10
1200	2.27

Figure P9-21

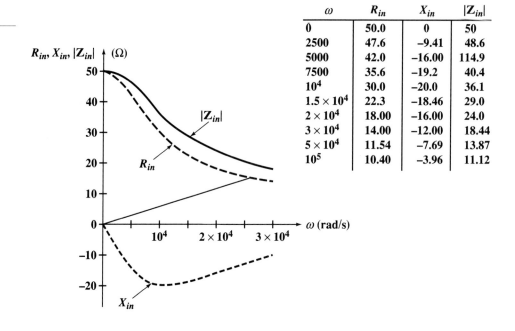

ω	R_{in}	X_{in}	$|Z_{in}|$
0	50.0	0	50
2500	47.6	−9.41	48.6
5000	42.0	−16.00	114.9
7500	35.6	−19.2	40.4
10^4	30.0	−20.0	36.1
1.5×10^4	22.3	−18.46	29.0
2×10^4	18.00	−16.00	24.0
3×10^4	14.00	−12.00	18.44
5×10^4	11.54	−7.69	13.87
10^5	10.40	−3.96	11.12

Figure P9-23

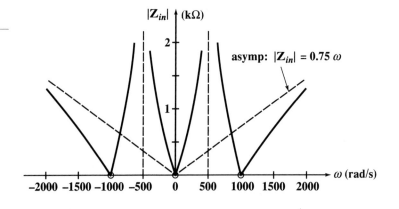

Chapter 10

1 $p_s = 116.9$ W, $p_R = 136.6$ W, $p_C = -19.69$ W

3 (a) -8 W; (b) -0.554 W; (c) 0.422 W

5 $-483, 297, 0, 185.9, 0$ W

7 (a) 10.875 W; (b) 20.75 W

9 26.2 W

11 (a) $8 + j14$ Ω; (b) 180 W

13 96 W

15 8.94 Ω, 38.6 W

17 (a) 12.59; (b) 12.25; (c) 10 V

19 (a) 8.5; (b) 12.42

21 (a) 42.7; (b) 25.0; (c) 7.32; (d) 55.2; (e) 80.2 W

23 (a) 655 W; (b) 320 W; (c) 335 W; (d) 800 VA; (e) 320 VA: (f) 568 VA; (g) 0.590 lag

25 $AP_A = 1229$ VA, $AP_B = 773$ VA, $AP_C = 865$ VA, $AP_D = 865$ VA, $AP_s = 3022$ VA

27 (a) 0.872 lag; (b) 692 μF

29 $-824 + j294, 0 - j765, 588 + j0, 0 + j471, 235 + j0$ VA

31 (a) $375 - j331$ VA; (b) $500 - j441$ VA; (c) $567 - j500$ VA

33 (a) 15.62 A rms; (b) 0.919 lag; (c) $3.30 + j1.417$ kVA

Chapter 11

1 (a) $3.04\underline{/171.2°}$ A; (b) $6.33\underline{/-152.1°}$ A

3 $39.8, 35.9, 21.9$ A

5 (a) $22.8, 0$ A; (b) $34.4, 22.8, 12$ A

7 (a) 91.5 μF; (b) 6.68 kVA

9 (a) $15.13\underline{/-16.70°}$ A; (b) $2270\underline{/0.219°}$ V; (c) $143.6 + j43.7$ Ω; (d) 98.6%

11 (a) $2.97\underline{/16.99°}$ A; (b) 52.8 W; (c) 1991 W; (d) 0.956 lead

13 (a) 0.894 lag; (b) 22.2 μF; (c) $+541$ VAR

15 (a) $233\underline{/20.7°}$ V; (b) $17.21 + j9.00$ kVA

17 (a) $242\underline{/30°}$ V; (b) $24.0\underline{/-0.964°}$ A; (c) $41.6\underline{/-31.0°}$ A

19 (a) $33.9\underline{/45.2°}$ A; (b) $53.0\underline{/-157.0°}$ A; (c) $25.2\underline{/-7.64°}$ A; (d) $6.10 + j3.34$ kVA

21 (a) 81.06 V; (b) 245.6 V; (c) 165.3 V

Chapter 12

1 (a) $8.06e^{-3t}\cos(15t - 60.3°)$ A; (b) $8.06e^{-3t}\cos(15t - 60.3°)$ A; (c) -4.13 A; (d) -4.13 A

3 (a) $(40 - 40e^{-12.5t} + 20e^{-5t})u(t)$ V; (b) $0, -5, -12.5$ s^{-1}

5 $-125 \pm j11,180$ s^{-1}

7 (a) $(16s^2 + 50s + 4000)/(s^2 + 80s)$ Ω; (b) $0.1584 - j4.67$ Ω; (c) $6.85\underline{/-114.3°}$ Ω; (d) 0.909 Ω; (e) 1 Ω

9 (a) $185.1\underline{/-47.6°}$ V; (b) $185.1e^{-3t}\cos(4t - 47.6°)$ V

11 (a) $(10^6s + 25 \times 10^9)/(s^2 + 25,000s + 5 \times 10^7)$ Ω; (b) $-32,071$ and $-17,929$ Np/s

13 (a) $1.6(\sigma^2 + 32.5\sigma + 50)/(\sigma + 4)$ Ω; (b) zeros: $-1.619, -30.9$; poles: $-4, \pm\infty$ s^{-1}; (c) and (d) see Fig. P12-13.

15 (a) $5(s + 1)(s + 4)/6(s + 1.5)$ Ω; (b) poles: $-1.5, \pm\infty$; zeros: $-1, -4$ s^{-1}; (c) see Fig. 12-15; (d) $2(s + 1)(s + 4)/(s^2 + 7.4s + 7.6)$, poles at -1.232 and -6.17 s^{-1}, zeros at -1 and -4 s^{-1}, see Fig. P12-15.

17 (a) zeros at $s = -2.5$ and -3, poles at $s = \pm j4$; (b) $4.69, 10$; (c) 15.15 cm; (d) see Fig. P12-17.

19 $(s + 0.5)^2/[(s + 0.1910)(s + 1.309)]$, double zero at -0.5, poles at -1.309 and -0.1910 s^{-1}

Figure P12-13

σ	$|\mathbf{Z}_{in}(\sigma)|$
−50	32.2
−40	15.56
−20	20
−10	46.7
0	20
10	54.3
20	73.3

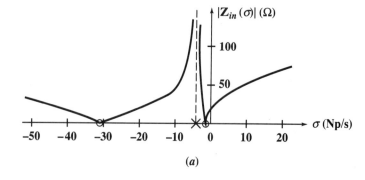

(a)

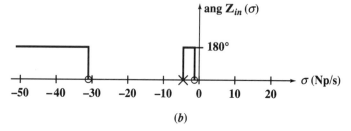

(b)

Figure P12-15

σ	$|\mathbf{Z}_{in}(\sigma)|$
0	2.22
−0.5	1.46
−1.25	2.29
−2	3.33
−3	1.11
−5	0.95
1	3.33

σ	$|\mathbf{Z}_{in}|$
−8	4.52
−7	7.50
−6	25
−5	1.82
−3.5	0.41
−3	0.71
−2	1.25
0	1.05
1	1.25
2	1.36

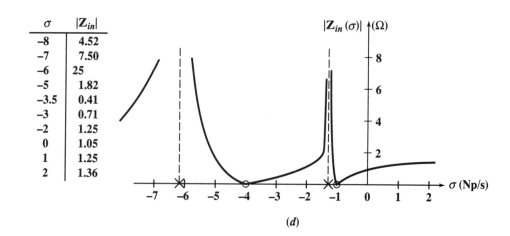

(c)

(d)

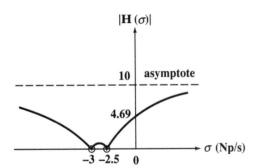
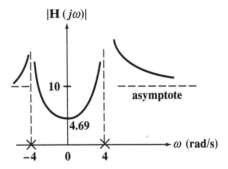

Figure P12-17

21 20 Ω and 25 mF
23 (a) $-5e^{-6t}$ A (all t); (b) $[-5e^{-6t} + e^{-2t}(5\cos 4t + 3\sin 4t)]u(t)$ A
25 (a) -1.729 and -24.1 s^{-1}; (b) $10 - 0.886e^{-1.729t} - 2.11e^{-24.1t}$ A, $t > 0$
27 (a) $2.5/(s^2 + 6.75s + 2.5)$; (b) $(1 - 1.066e^{-0.393t} + 0.0659e^{-6.36t})u(t)$ V
29 (a) 0; (b) $10/(s + 17.5)$ Ω; (c) 0; (d) $Ae^{-17.5t}$
31 (a) 400 Ω and 0; (b) 2 kΩ and 50 nF; (c) 200 kΩ and 5 nF; (d) 20 kΩ and 0, 200 Ω and 50 nF
33 (a) $R_{1A} = \infty$, $C_{1A} = 1$ nF, $R_{fA} = 10$ kΩ, $C_{fA} = 0$; $R_{1B} = 10$ kΩ, $C_{1B} = 1$ μF, $R_{fB} = 10$ kΩ, $C_{fB} = 0$; $R_{1C} = 10$ kΩ, $C_{1C} = 0$, $R_{fC} = 10$ kΩ, $C_{fC} = 0.1$ μF

Chapter 13

 1 98.5 rad/s
 3 (a) 98.5 rad/s; (b) 2.29 Ω
 5 7.52, 397 Ω, 43.9 mH, 15.75 μF
 7 (a) $(1000 - 48.4 \times 10^{-8}\,\omega^2 + j4.4 \times 10^{-4}\,\omega)/j4.4\omega$; (b) 45.5 krad/s, 10^4 Ω
 9 12.30 Ω, 15.18 mH, 5.42 mF
11 (a) 443 and 357 Hz; (b) 497 and 303 Hz
13 (a) 10^6 rad/s; (b) $15\underline{/90°}$ V; (c) $8.32\underline{/33.7°}$ V
15 (a) $1562\underline{/-38.7°}$ Ω; (b) 900 to 1100 Hz
17 (a) 50 krad/s; (b) 7.96 kHz; (c) 4; (d) 12.5 krad/s; (e) 44.1 krad/s; (f) 56.6 krad/s; (g) $65.4\underline{/-40.2°}$ Ω; (h) 4.44
19 Don't touch it! See Fig. P13-19.

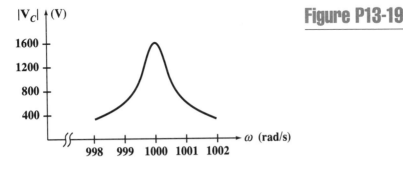

Figure P13-19

21 1.231 Ω, 30.8 μH, 4.78 μF
23 10^5 rad/s, 83.3, 1200 rad/s, 8.33 kΩ, $4.29\underline{/59.0°}$ kΩ
25 0.208 Ω at 10 krad/s
27 (a) 16.667 V; (b) 16.673 V
29 (a) $(s + 10)/[20(s + 5)]$; (b) $(s + 50)/[10(s + 25)]$; (c) I_1 in 0.2 Ω, 0.05 F, 0.4 Ω, $0.5I_1$
31 (a) I_x in 1 μF, 1250 Ω, 1.25 H, 10^3I_x; (b) $Z_{th} = -j5$ kΩ, $V_{oc} = 0$

33 (a) −13.98 dB; (b) 34.0 dB; (c) 6.45 dB; (d) 75.9; (e) 0.398; (f) 1.001

35 Amplitude: $\omega \leq 1$: 26 dB, $1 < \omega < 10$: −20 dB/dec, $10 \leq \omega \leq 100$: 6 dB, $\omega > 100$: −20 dB/dec; phase: $\omega \leq 0.1$: 0°, $0.1 < \omega < 1$: −45°/dec, $1 \leq \omega \leq 100$: −45°, $100 < \omega < 1000$: −45°/dec, $\omega \geq 1000$: −90°

37 (a) $\omega \leq 2$: 90°, $2 < \omega < 10$: −45°/dec, $10 \leq \omega \leq 100$: 58.5°, $100 < \omega < 200$: −135°/dec, $200 < \omega < 1000$: −90°/dec, $1000 < \omega < 10^4$: −135°/dec, $\omega \geq 10^4$: −180°; (b) (2 rad/s, 90°), (10, 58.5°), (100, 58.5°), (200, 17.9°), (10^3, −45°), (10^4, −180°); (c) 2: 85.1°, 10: 67.4°, 100: 39.2°, 200: 35.2°, 10^3: −49.6°, 10^4, −163.3°

39 (a) $25\mathbf{s}/(10\mathbf{s}^2 + 25\mathbf{s} + 1000)$; (b) ampl: $\omega = 1$: −32 dB, $\omega < 10$: 20 dB/dec, $\omega > 10$: −20 dB/dec, $\omega = 10$: 0 dB with a little rounding; phase: see Fig. P13-39; (c) −15.68 dB, −80.5°

Figure P13-39

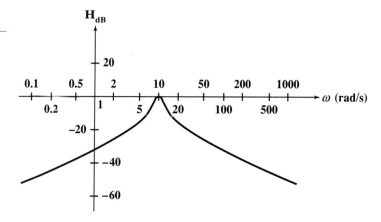

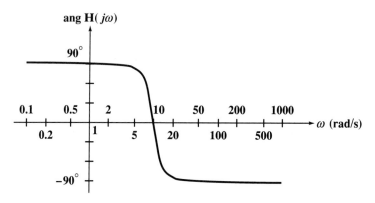

41 (a) $-0.1\mathbf{s}/(1 + 0.1\mathbf{s})^2$; (b) $\omega < 10$: +20 dB/dec, $\omega = 10$: 0 dB, $\omega > 10$: −20 dB/dec; (c) $\omega < 1$: 270°, $1 \leq \omega \leq 100$: −90°/dec, $\omega > 100$: 90°

43 6.570, 8.517, 11.30, 15.75, 24.96, 49.90, 9.484, 5.074, 12.63, 18.71, and 24.35 V

Chapter 14

1 (a) 1 and 4, 2 and 3; (b) 3 and 1, 2 and 4; (c) 1 and 3, 2 and 4

3 (a) −10.40 W; (b) $P_{50} = 5.63$, $P_{2000} = 4.77$ W; (c) 0 each; (d) 0

5 (a) $\mathbf{V}_{oc} = 145.5\underline{/-166.0°}$ V, $\mathbf{Z}_{th} = 105.9 + j76.5$ Ω; (b) 25.0 W

7 $30t/(t^2 + 0.01)^2$ μA

9 (a) $100 = (6 + 5s)\mathbf{I}_1 - 2s\mathbf{I}_2 - 6\mathbf{I}_3$, $-2s\mathbf{I}_1 + (4 + 5s)\mathbf{I}_2 - 4s\mathbf{I}_3 = 0$,
   $-6\mathbf{I}_1 - 4s\mathbf{I}_2 + (11 + 6s)\mathbf{I}_3 = 0$; (b) -5 A

11 $1.260/\underline{-60.2°}$ A

13 $2.16k^2/(k^4 - 1.82k^2 + 1.188)$ W

15 (a) 0.842 W; (b) 0.262 W; (c) 1.104 W

17 (a) $1.661/\underline{41.6°}$; (b) $0.392/\underline{-79.7°}$; (c) $2.22/\underline{0.051°}$

19 (a) 0.447; (b) 3.16 A

21 (a) $2s/(11s^2 + 145s + 300)$ S; (b) $2.26(e^{-2.57t} - e^{-10.61t})$ A

23 $3s/(17s + 15)$

25 (a) $\mathbf{V}_{oc} = 0$, $\mathbf{Z}_{th} = 82.5 + j0.312$ Ω; (b) $\mathbf{V}_{oc} = 40.0/\underline{0.0917°}$ V,
   $\mathbf{Z}_{th} = 3.20 + j0.00512$ Ω

27 (a) 250 W; (b) 68.0 W

29 $\mathbf{V}_{oc} = 0$, $\mathbf{Z}_{th} = 4 + j0$ Ω

31 (a) $1.099/\underline{0°}$ A; (b) $3.30/\underline{0°}$ A; (c) $4.40/\underline{180°}$ A; (d) 30.2 W; (e) 21.7 W;
   (f) 58.0 W

33 $P_1 = 4$, $P_4 = 4$, $P_{48} = 3$, $P_{400} = 9$ W

35 (a) $P_{AB} = 111.1$, $P_{CD} = 62.5$ W; (b) $P_{AB} = P_{CD} = 1.736$ W

1 (a) 851 W; (b) 873 W; (c) 701 W

3 2.21 Ω

5 (a) 6.71 Ω; (b) 6.71 Ω

7 $-R_x$

9 141.8 and -76.6 mS

11 $\begin{bmatrix} 0.04 & -0.04 \\ 0.04 & -0.03 \end{bmatrix}$ (S)

13 (a) 32, -320, and 50 Ω; (b) 60 Ω

15 Exp. 3: 4 A, -8 A; Exp. 4: -8.33 V, -22.2 V; Exp. 5: -58.3 V, -55.6 V;
   $\begin{bmatrix} 0.2 & -0.3 \\ -0.4 & 0.15 \end{bmatrix}$ (S)

17 9.90 Ω

19 (a) 55.6; (b) -9.62; (c) 534; (d) 3.46 Ω; (e) 34.6 Ω

21 (a) See Fig. P15-21; (b) see Fig. P15-21, $[\mathbf{y}]_{new} = \begin{bmatrix} 3 & -2 \\ 8 & 6 \end{bmatrix}$ (mS)

Chapter 15

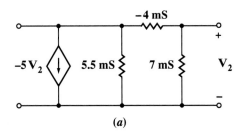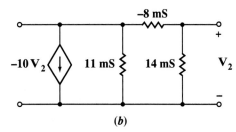

Figure P15-21

23 $\begin{bmatrix} 7.55 & 1.132 \\ -4.53 & 11.32 \end{bmatrix}$ (Ω)

25 (a) -2; (b) 4; (c) 8; (d) 1 Ω; (e) 1.333 Ω

27 $\begin{bmatrix} 133.1/\underline{-47.6°} & 94.2/\underline{-2.64°} \\ 9420/\underline{86.8°} & 565/\underline{-3.60°} \end{bmatrix}$ (Ω)

29 (a) $\begin{bmatrix} 10\,\Omega & -2 \\ 20 & 0.2\,\text{S} \end{bmatrix}$; (b) $\begin{bmatrix} 42.3\,\Omega & -1.667 \\ 16.67 & 0.1667\,\text{S} \end{bmatrix}$

31 (a) 1.2; (b) 9.6 Ω; (c) -0.24 S

33 (a) $\begin{bmatrix} 1000\,\Omega & 0.01 \\ 10 & 200\,\mu\text{S} \end{bmatrix}$; (b) 8.57 k$\Omega$

35 (a) $\begin{bmatrix} 6 & -4 \\ 8 & 38 \end{bmatrix}$; (b) $\begin{bmatrix} 22 & 16 \\ 14 & 22 \end{bmatrix}$; (c) $\begin{bmatrix} 0 & 26 & 46 & -4 \\ -13 & 13 & 21 & 1 \end{bmatrix}$;

(d) $\begin{bmatrix} -3 & -2 & 9 \\ -3 & -19 & 22 \end{bmatrix}$; (e) $\begin{bmatrix} -6 & 64 & -34 \\ -138 & -738 & 908 \end{bmatrix}$

37 $\begin{bmatrix} 2.12 & 3.85\,\Omega \\ 0.350\,\text{S} & 1 \end{bmatrix}$

39 (a) $\begin{bmatrix} 1 & 2\,\Omega \\ 0 & 1 \end{bmatrix}$; (b) proof, $\begin{bmatrix} 1 & 10\,\Omega \\ 0 & 1 \end{bmatrix}$

41 (a) $\begin{bmatrix} 3.33 & 133.3\,\Omega \\ 0.1667\,\text{S} & 9.17 \end{bmatrix}$; (b) $\begin{bmatrix} 10 & 133.3\,\Omega \\ 0.625\,\text{S} & 9.17 \end{bmatrix}$

Chapter 16

1 $i_1' = -26.4i_1 + 260i_2 - 3.6i_3 + 50 + 20\cos 10t$; $i_2' = 6.7i_1 - 85i_2 + 0.8i_3 - 12.5 - 5\cos 10t$; $i_3' = 9.3i_1 - 87i_2 + 1.2i_3 - 17.5 - 7\cos 10t$

3 $x' = -2.5z + 1.5$, $y' = -\frac{1}{2}x - \frac{1}{3}y - \frac{5}{12}z + \frac{1}{4} - \frac{t}{6}$, $z' = -\frac{2}{3}y - \frac{5}{6}z + \frac{1}{2} - \frac{t}{3}$

5 $i_L' = -300i_L - 5v_C + 500\cos 120\pi t$, $v_C' = 6.8 \times 10^5 i_L$

7 (a) $i_L' = -7500i_L + 470v_C$, $v_C' = -5000i_L - 100v_C + 5000i_s$; (b) $v_1' = -3v_1 + v_2 + 2v_s$, $v_2' = 0.5v_1 - 0.7v_2 + 0.2v_s$, $v_3' = 0.08v_2 - 0.08v_3$

9 (a) $v_1' = 10i_1 - 10i_2$, $v_2' = -5i_1$, $i_1' = -2v_1 + 2v_2$, $i_2' = 20v_1 - 40i_2 + 40i_s + 20v_s$; (b) $v_1' = 10i_1 - 10i_2$, $v_2' = -5i_1$, $i_1' = -2v_1 + 2v_2$, $i_2' = 20v_1 - 24i_2 + 12v_s$

11 $v' = -0.4v + 2i + 8i_s$, $i' = -v - 20i + 20i_s$

13 (a) $i_{L1}' = -i_{L1} - 2i_{L2} - 3v_C + 2t$, $i_{L2}' = 4i_{L1} - 5i_{L2} + 6v_C + 3t^2$, $v_C' = 7i_{L1} - 8i_{L2} - 9v_C + 1 + t$; (b) see Fig. P16-13.

Figure P16-13

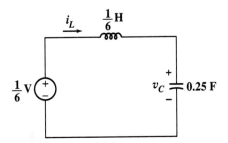

15 $v_{o1}' = v_{C1} + v_{C2} + 10$, $v_{o2}' = -v_{C1} - 4v_{C2} + 2i_L$, $i_{R1}' = 9v_{C2} - 4i_L$, $i_{R2}' = 3v_{C1} + 10$

17 $\mathbf{a} = \begin{bmatrix} -\frac{1}{27} & -\frac{1}{6} \\ \frac{1}{3} & -3 \end{bmatrix}$, $\mathbf{f} = \begin{bmatrix} -\frac{1}{108}v_s + \frac{1}{18}i_s \\ \frac{1}{3}v_s - 2i_s \end{bmatrix}$

19 $8 \times 10^{-3}t - 16 \times 10^{-6}(1 - e^{-5000t})$ A

21 $0 \le t \le 0.5$: $31.0\cos\pi t + 39.0\sin\pi t - 31.0e^{-2.5t}$ V, $t \ge 0.5$: $105.0e^{-2.5t}$ V

23 $t < 0$: 0; $0 < t < 0.5$: $4(e^{-t} - e^{-2t})$; $t > 0.5$ s: $2.59e^{-2t}$ A

25 (a) $x' = x + 2y - z + u(t)$, $y' = -y + 3z + \cos t$, $z' = -2x - 3y - z - u(t)$; (b) $\begin{bmatrix} 1.6 \\ -2.3 \\ 1.3 \end{bmatrix}$; (c) $\begin{bmatrix} 1.6175 \\ -2.2951 \\ 1.26 \end{bmatrix}$

$$27 \begin{bmatrix} \frac{2}{3}e^{-2t} + \frac{1}{3}e^{-5t} & \frac{2}{3}e^{-2t} - \frac{2}{3}e^{-5t} \\ \frac{1}{3}e^{-2t} - \frac{1}{3}e^{-5t} & \frac{1}{3}e^{-2t} + \frac{2}{3}e^{-5t} \end{bmatrix}$$

Chapter 17

1 (a) 3.00 V; (b) 4.96 V; (c) 0.02 s; (d) −2.46 V

3 (a) 1.200; (b) 1.932; (c) −0.0458

5 0, 1.061, 1.061

7 (a) $\frac{1}{8}$ s; (b) 0.0796

9 $\dfrac{2V_m}{\pi} + \dfrac{4V_m}{3\pi} \cos 10\pi t - \dfrac{4V_m}{15\pi} \cos 20\pi t + \dfrac{4V_m}{35\pi} \cos 30\pi t - \dfrac{4V_m}{63\pi} \cos 40\pi t + \dots.$

11 (a) 0.2 sin 1000πt + 0.6 sin 2000πt + 0.4 sin 3000πt; (b) 0.529; (c) 1.069

13 (a) 5.09; (b) −0.679, −2.72; (c) −4 < t < 0 ms: 8 sin 125$\pi|t|$; (d) −3.40, 0

15 $b_{\text{even}} = 0$, $b_1 = 0.246$, $b_3 = 0.427$, $b_5 = 0.1342$

17 (a) $1.25 + \displaystyle\sum_{(n=1,\text{odd})}^{\infty} \dfrac{0.255}{n^2 + 0.16} \left(\dfrac{1}{n} \sin 5nt - 2.5 \cos 5nt \right);$

(b) $-0.554e^{-2t} + 1.25 + \displaystyle\sum_{(n=1,\text{odd})}^{\infty} \dfrac{0.255}{n^2 + 0.16} \left(\dfrac{1}{n} \sin 5nt - 2.5 \cos 5nt \right)$

19 (a) $5 + \dfrac{20}{\pi} \displaystyle\sum_{(n=1,\text{odd})}^{\infty} \dfrac{1}{1 + 400n^2} \left(\dfrac{1}{n} \sin 5nt - 20 \cos 5nt \right);$

(b) $Ae^{-t/4}$; (c) $-4.61e^{-t/4} + 5 + \dfrac{20}{\pi} \displaystyle\sum_{(n=1,\text{odd})}^{\infty} \dfrac{1}{1 + 400n^2} \left(\dfrac{1}{n} \sin 5nt - 20 \cos 5nt \right)$

21 $\mathbf{c}_n = 2 \times 10^4 \left\{ \dfrac{1}{160n^2\pi^2} [e^{-j0.4n\pi} (1 + j0.4n\pi) - 1] + \dfrac{j}{400n\pi} (e^{-j0.8n\pi} - e^{-j0.4n\pi}) \right\},$

$\mathbf{c}_0 = 30$, $\mathbf{c}_{\pm 1} = 24.9 \underline{/ \mp 88.6°}$, $\mathbf{c}_{\pm 2} = 13.31 \underline{/ \pm 177.4°}$

23 (a) 1 + 0.4 cos $\omega_0 t$ + cos 2$\omega_0 t$ − 2 cos 3$\omega_0 t$ + 0.4 sin $\omega_0 t$ − 0.5 sin 2$\omega_0 t$ + 4 sin 3$\omega_0 t$, ω_0 = 400π rad/s; (b) −332 mV

25 (a) −j4.24 V; (b) 15.75 W

Chapter 18

1 (a) See Fig. P18-1; (b) $\dfrac{10}{\omega}$ (sin 3ω + sin 2ω)

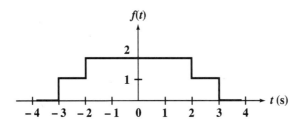

$f(t)$

3 $\dfrac{5}{\omega^2}$ (1 − cos 4ω) or 10(sin 2$\omega/\omega)^2$

5 (a) 16; (b) 13.73; (c) −0.291

7 (a) 0.664; (b) 0.0849; (c) −0.389; (d) 0.398; (e) −77.7°

9 (a) 0.391 J; (b) 10/(ω^2 + 16); (c) 0.391 and 0.0977 J/Hz

11 (a) 1.273; (b) 8.80; (c) 1.557

13 (a) 6$\delta(t)$ V, 3(1 − $e^{-5t/6}$)$u(t)$ A, 15$e^{-5t/6}u(t)$ V, and 6$\delta(t)$ + 15$e^{-5t/6}u(t)$ V; (b) proof

15 (a) 45.1; (b) 54.0; (c) 44.8; (d) 43.9 A

17 (a) 4$e^{-j\omega}$; (b) −4; (c) −$j4\pi[e^{-j\pi/6} \delta(\omega - 10) - e^{j\pi/6} \delta(\omega + 10)]$

19 (a) 0.1039$\underline{/-106.5°}$; (b) −0.1039$\underline{/-106.5°}$; (c) 0.362$\underline{/15.99°}$

21 $2\pi \sum\limits_{-\infty}^{\infty} \left[\dfrac{j10}{n\pi} \cos \dfrac{n\pi}{2} - \dfrac{j20}{n^2\pi^2} \sin \dfrac{n\pi}{2} \right] \delta\left(\omega - \dfrac{\pi n}{2} \right)$

23 1.386

25 (a) 0, 0, 4, 10, 6, 0; (b) 0, 0, 4, 10, 6, 0

27 (a) $\dfrac{20}{3} \int_2^5 (5 - z)dz$; (b) 30

29 (a) 0.335; (b) 0.741; (c) 0.221

31 $100te^{-2t}\,u(t)$

33 (a) $\frac{1}{3}$ J; (b) 0.5

35 (a) $\dfrac{2}{1 + j\omega}$; (b) see Fig. P18-35; (c) 2

Figure P18-35

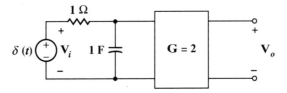

37 (a) $20e^{-8t}\,u(t)$ V; (b) $(2.5 - 2.5e^{-8t})u(t)$ V; (c) $200 \cos (6t - 36.9°)$ V

39 $2(e^{-t/6} - e^{-t})u(t)$ V

Chapter 19

1 (a) $\sigma > 0$; (b) $\sigma > 0$; (c) $\sigma > 50$; (d) $\sigma > 50$; (e) $\sigma > 0$

3 (a) $\dfrac{1}{\mathbf{s}} (e^{-\mathbf{s}} - e^{-3\mathbf{s}})$; (b) $\dfrac{2}{\mathbf{s}} e^{-4\mathbf{s}}$; (c) $\dfrac{3}{\mathbf{s}+2} e^{-4\mathbf{s}-8}$; (d) $3e^{-5\mathbf{s}}$; (e) $-4e^{-\mathbf{s}}$

5 (a) $\delta(t) + u(t) + 2e^{-t}u(t)$; (b) $\delta(t-2) + 2\delta(t-1) + \delta(t)$; (c) $2e^{-1}\delta(t-1)$; (d) $\delta(t-1) + \delta(t-5)$

7 (a) $5e^{-t}\,u(t)$; (b) $(5e^{-t} - 2e^{-4t})u(t)$; (c) $6(e^{-t} - e^{-4t})u(t)$; (d) $6(4e^{-4t} - e^{-t})u(t)$; (e) $18\delta(t) + 6(e^{-t} - 16e^{-4t})u(t)$

9 (a) 50 V and 50 V; (b) $0.1v_C' + 0.3v_C = 2$; (c) $\dfrac{50\mathbf{s} + 20}{\mathbf{s}(\mathbf{s} + 3)}$, $\dfrac{1}{3}(20 + 130e^{-3t})$ $u(t)$ V

11 $\dfrac{12 + 40\mathbf{s}}{\mathbf{s}(20\mathbf{s} + 3)}$, $(4 - 2e^{-0.15t})u(t)$

13 (a) $\frac{1}{4}\ \Omega$, 1 F, $\frac{1}{3}$ H; (b) $(75e^{-3t} - 12.5e^{-t} - 62.5e^{-5t})u(t)$ V

15 $4u(t) + i_C + 10 \int_{0^-}^{\infty} i_C\,dt + 4[i_C - 0.5\delta(t)] = 0$, $\dfrac{2\mathbf{s} - 4}{5\mathbf{s} + 10}$, $0.4\delta(t) - 1.6e^{-2t}\,u(t)$

17 (a) and (b) $(\frac{1}{5} - \frac{1}{3}\,e^{-2t} + \frac{2}{15}\,e^{-5t})u(t)$

19 (a) and (b) $(\frac{2}{3} - \frac{8}{3}\,e^{-3t})u(t)$

21 $[3e^{-\mathbf{s}} - 3e^{-3\mathbf{s}} + \mathbf{s}e^{-3\mathbf{s}}]/[\mathbf{s}(1 - e^{-5\mathbf{s}})]$

23 $\dfrac{4e^{-\mathbf{s}} - 4e^{-2\mathbf{s}} - 2\mathbf{s}e^{-3\mathbf{s}}}{\mathbf{s}(1 - e^{-3\mathbf{s}})}$

25 $1.6\pi\ (1 + e^{-2.5\mathbf{s}})/[(\mathbf{s}^2 + 0.16\pi^2)(1 - e^{-6\mathbf{s}})]$

27 (a) $e^{-\mathbf{s}}\left(\dfrac{1}{\mathbf{s}^2} - \dfrac{1}{\mathbf{s}} \right)$; (b) $\dfrac{54.0 - 16.83\mathbf{s}}{\mathbf{s}^2 + 25}$; (c) $\dfrac{8[2 - e^{-\mathbf{s}}(2 + 2\mathbf{s} + \mathbf{s}^2)]}{\mathbf{s}^3(1 - e^{-5\mathbf{s}})}$

29 (a) $2\omega\mathbf{s}/(\mathbf{s}^2 + \omega^2)^2$; (b) $\tan^{-1} \dfrac{\omega}{\mathbf{s}}$; (c) $\delta(t-2) - 10e^{-5(t-2)}\,[\cos 10(t-2)]$ $u(t-2) - 7.5e^{-5(t-2)} \sin [10(t-2)]\,u(t-2)$

31 (a) $\dfrac{1}{\mathbf{s}^2} - \dfrac{1}{\mathbf{s}}$; (b) $e^{-\mathbf{s}}\dfrac{1}{\mathbf{s}^2}$; (c) $\left[\dfrac{(t-1)^3}{6} - \dfrac{(t-1)^2}{2} \right]u(t-1)$

33 (a) 5, indeterminate; (b) 0, indeterminate; (c) 1, indeterminate

35 (a) $f(0^+) = 8, f(\infty) = 0$; (b) $\infty, -0.5$; (c) 8, indeterminate, (d) 0, 0

37 $\dfrac{9\mathbf{s}}{\mathbf{s}^2 + 32\mathbf{s} + 60}$ and $\dfrac{9}{14}(15e^{-30t} - e^{-2t})u(t)$

39 (a) $\dfrac{1}{\mathbf{s}^2 + 13\mathbf{s} + 12}$; (b) $\dfrac{1}{11}(e^{-t} - e^{-12t})u(t)$ A/V; (c) $(11 - 12e^{-t} + e^{-12t})u(t)$ A

41 $(-9 + 40t + 8.33e^{-4t} + 0.667e^{-10t})u(t)$

43 (a) See Fig. P19-43; (b) $\dfrac{5(\mathbf{s}^2 + 6.4)}{\mathbf{s}^2(\mathbf{s} + 5)}$ and $(-1.28 + 6.4t + 6.28e^{-5t})u(t)$ V

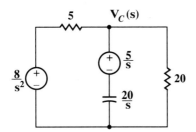

Figure P19-43

45 (a) See Fig. P19-45; (b) $\dfrac{8\mathbf{s}^2 + 180}{\mathbf{s}^2(\mathbf{s} + 2)}$ and $(-45 + 90t + 53e^{-2t})u(t)$ V

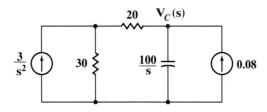

Figure P19-45

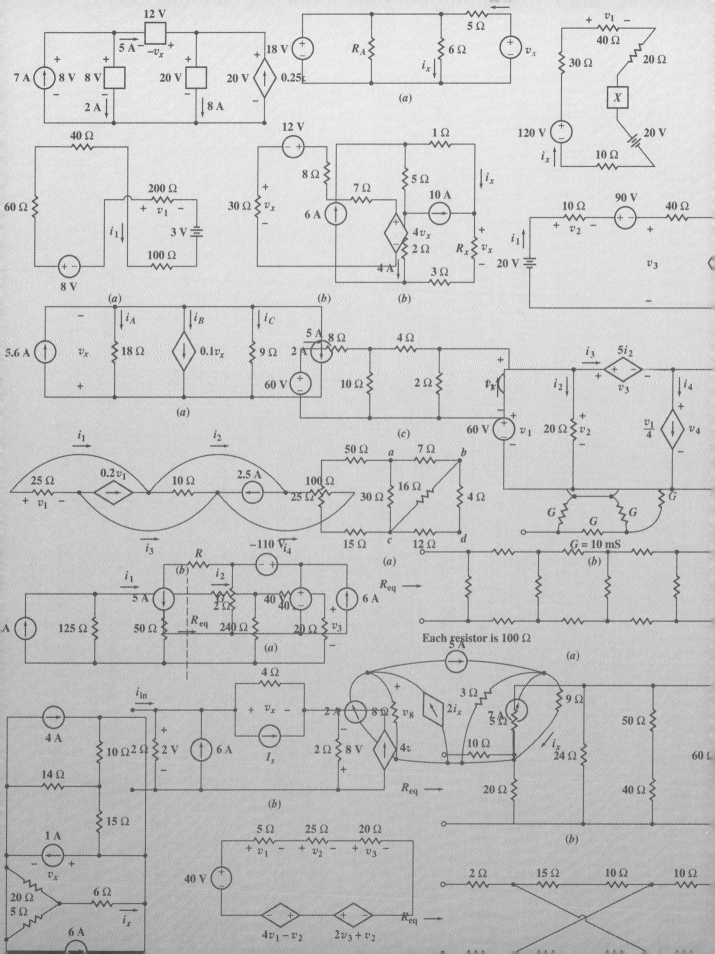

Index

Standard abbreviations

alternating current	ac	meter-kilogram-second	mks
ampere	A	mho (siemen)	℧
coulomb	C	minute	min
cycle per second	cps (avoid)	neper	Np
decibel	dB	newton	N
degree Celsius	°C	newton-meter	N · m
direct current	dc	ohm	Ω
electronvolt	eV	pound-force	lbf
farad	F	power factor	PF
foot	ft	radian	rad
gram	g	resistance-inductance-	
henry	H	capacitance	*RLC*
hertz	Hz	revolutions per second	rps
hour	h	root-mean-square	rms
inch	in	second	s
joule	J	volt	V
kelvin	K	voltampere	VA
kilogram	kg	watt	W
meter	m	watthour	Wh

[Note: Standard prefixes of the decimal system are tabulated in Sec. 1-2.]